P9-DEM-892

GREEN BUILDING

Principles and Practices in Residential Construction

GREEN BUILDING

Principles and Practices in Residential Construction

ABE KRUGER
CEM, LEED AP Homes + ND
President, Kruger Sustainability Group

CARL SEVILLE
LEED AP Homes
President, Seville Consulting, LLC
Contributing Editor, Green Building Advisor
Green Building Curmudgeon

Australia • Brazil • Japan • Korea • Mexico • Singapore • Spain • United Kingdom • United States

Green Building: Principles and Practices in Residential Construction

Abe Kruger and Carl Seville

Vice President, Editorial: Dave Garza
Director of Learning Solutions: Sandy Clark
Senior Acquisitions Editor: Jim DeVoe
Managing Editor: Larry Main
Product Manager: Ohlinger Publishing Services
Editorial Assistant: Cris Savino
Vice President, Marketing: Jennifer Baker
Marketing Director: Debbie Yarnell
Marketing Manager: Erin Brennan
Production Director: Wendy Troeger
Production Manager: Mark Bernard
Art Director: Casey Kirchmayer
Technology Project Manager: Joe Pliss

Library of Congress Control Number: 2011939713

ISBN-13: 9781111135959
ISBN-10: 1111135959

Delmar
5 Maxwell Drive
Clifton Park, NY 12065-2919
USA

Cengage Learning is a leading provider of customized learning solutions with office locations around the globe, including Singapore, the United Kingdom, Australia, Mexico, Brazil, and Japan. Locate your local office at: **international.cengage.com/region**

Cengage Learning products are represented in Canada by Nelson Education, Ltd.

To learn more about Delmar, visit **www.cengage.com/delmar**

Purchase any of our products at your local college store or at our preferred online store **www.cengagebrain.com**

Notice to the Reader

Publisher does not warrant or guarantee any of the products described herein or perform any independent analysis in connection with any of the product information contained herein. Publisher does not assume, and expressly disclaims, any obligation to obtain and include information other than that provided to it by the manufacturer. The reader is expressly warned to consider and adopt all safety precautions that might be indicated by the activities described herein and to avoid all potential hazards. By following the instructions contained herein, the reader willingly assumes all risks in connection with such instructions. The publisher makes no representations or warranties of any kind, including but not limited to, the warranties of fitness for particular purpose or merchantability, nor are any such representations implied with respect to the material set forth herein, and the publisher takes no responsibility with respect to such material. The publisher shall not be liable for any special, consequential, or exemplary damages resulting, in whole or part, from the readers' use of, or reliance upon, this material.

Printed in the United States of America
1 2 3 4 5 6 7 15 14 13 12 11

Contents

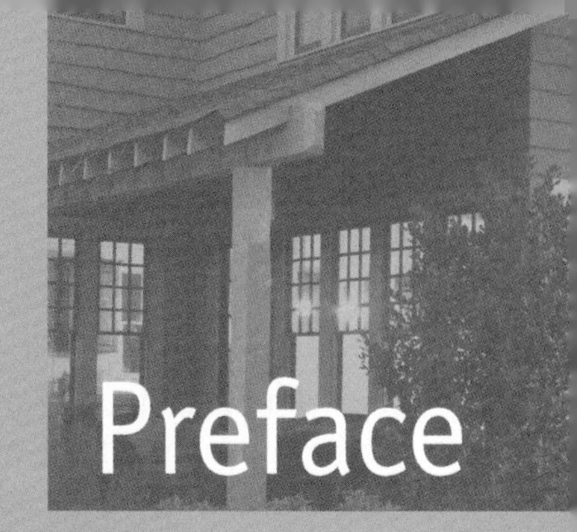

Preface

Introduction

Residential green building has been evolving for several decades, reaching its first level of maturity with the development of building certification programs. As these programs move into their second and third generations of development, we believe the time has come for a comprehensive text that specifically addresses green building principles as they apply to single-family homes. In the marketplace, green building has experienced steady growth, with wider acceptance in some markets than others. In 2005, green building was a small, burgeoning market, comprising approximately 2% of commercial and residential construction.[1] This percentage represented a total value of $10 billion ($3 billion for residential and $7 billion for commercial). By 2013, McGraw-Hill Construction estimates the overall green building market may reach between $96 and $140 billion for residential and commercial buildings. The recent financial crisis has significantly reduced the pace of new construction, but green building continues to increase its share of the market.

The current state of residential green building as a distinct discipline is primarily limited to training and designation offered by professional organizations and to certification of individual buildings. Limited college and post-graduate level training is currently available in sustainable residential construction. We hope that *Green Building: Principles and Practices in Residential Construction* will provide a foundation for future programs on the subject.

---

[1] McGraw-Hill Construction (2009). 2009 Green Outlook: Trends Driving Change Report.

About the Cover

The cover photo is from the Glenwood Park community in Atlanta, Georgia. Glenwood Park is an entire community built with green building principles. Each residence adheres to the EarthCraft House construction standard developed by Southface Energy Institute and the Greater Atlanta Home Builders Association. Additional images of the community appear in Chapter 3 (Figures 3.8a and 3.8b). The photo was taken by Abby Smith (http://www.abbysmithphotography.com).

Approach

We have used our varied experience in building, remodeling, building science, and green home evaluation to create this comprehensive introductory text on green homes. Our approach in the book is to provide an overview of the concepts for green building, followed by detailed methods for incorporating materials and methods into specific projects as well as real-world examples of implementation. Residential green building as a discipline has developed in the field with little college-level training available. As students see career opportunities in residential green building, educational programs are needed to prepare them for the industry. Most existing books on the subject are either focused on commercial construction or target the consumer. *Green Building: Principles and Practices in Residential Construction* is designed to serve students seeking careers in the residential construction industry as well as industry professionals.

This book is structured to provide anyone with an interest in home construction and renovation with a guide to understanding both the principles and implementation of green building. It gives both beginning and advanced students, as well as experienced professionals, useful information that they can incorporate into their studies and practices.

ORGANIZATION

Green Building: Principles and Practices in Residential Construction is divided into five sections, beginning with an introduction to the concepts, followed by sections that roughly track the sequence of a construction project:

- Section One—What Is Green Building and Why Does It Make Sense?
- Section Two—Structural Systems
- Section Three—Exterior Finishes
- Section Four—Interior Systems
- Section Five—Mechanical Systems

Section One includes four chapters, beginning with a Green Building: An Overview, followed by The House as a System, then Planning Green From the Start, and closing with Insulation and Air Sealing. **Section Two** has chapters that cover Foundations, Floors and Exterior Walls, and Roofs and Attics. **Section Three** begins with Fenestration, followed by Exterior Wall Finishes, Outdoor Living Spaces, and finally Landscaping. **Section Four** is a single chapter, Interior Finishes. **Section Five** begins with a chapter on Heating, Ventilation, and Air Conditioning, followed by chapters on Electrical, Plumbing, and Renewable Energy. We close with a short Epilogue that provides both a recap of the text and a look into the future of the industry. The content may be used as presented, or the chapters can be rearranged to accommodate alternate formats for traditional or individualized instruction.

Each chapter begins with an outline of the green elements covered in the text, then moves on to the subject's effect on the whole-house system and an overview of materials and methods; the chapter ends with a section on remodeling considerations. Because a core tenant of green building is designing the structure for its particular climate, we emphasize throughout the chapters any regional issues that should be considered. Certain chapters vary from this structure, particularly those that cover mechanical systems in which materials and methods may be more intertwined than other areas.

KEY FEATURES

This book includes many features to assist students as they progress through the chapters:

1

Green Building: An Overview

This chapter explores definitions of green building, its importance from an environmental perspective, and its context within the design and construction industry. We provide a brief history of green building and the organizations that have helped create the guidelines and standards for the industry. The current versions of these specific green building programs are presented, along with likely national trends.

LEARNING OBJECTIVES

Upon completion of this chapter, the student should be able to:

- Define green building
- Explain the environmental impacts of the residential construction industry as a whole
- Describe the benefits of green building for builders, homeowners, and contractors
- Describe the relationship between green building programs and building codes
- Define the ENERGY STAR for Homes program
- Describe the Home Energy Rating System (HERS) and its relationship to green building programs
- Explain blower door and duct leakage testing
- Calculate simple payback and cash flow analysis for green building improvements

Green Building Principles

- Energy Efficiency
- Resource Efficiency
- Durability
- Water Efficiency
- Indoor Environmental Quality
- Reduced Community Impact
- Homeowner Education and Maintenance
- Sustainable Site Development

Defining Green Building

We define green building as a set of design, construction, and maintenance techniques and practices that minimize a building's total environmental impact. Decisions made while planning, building, renovating, and maintaining homes have long-term direct effects on many different aspects of our environment—air quality, health, natural resources, land use, water quality, and energy use. These decisions may also produce indirect effects on other aspects of our environment, such as factors that contribute to global warming.

The materials used to construct, remodel, and maintain a house all have an impact on the environment, as does the energy used to heat, cool, light, and run equipment, and the amount of water used during the home's lifetime (Figure 1.1). Neighborhood design affects how much land is consumed, how far people drive, and the amount of water pollution caused by runoff from roofs, lawns, and roads. Green building strives to reduce these negative impacts.

Energy Generation and Use

Residential buildings consume approximately 22% of the energy produced in the United States for heating, cooling,

green building an enviro... sustainable building, designed, constructed, and opera... impacts.

Learning Objectives: A clear set of learning objectives provides an overview of the chapter material and can be used by students to check whether they have understood and retained important points.

"Green Building Principles" Icons: A unique feature of this book are the "8 Principles" icons that help describe our core principles of green building. Located at the beginning of each chapter, these icons serve as a reminder of what the principles are and present an efficient way to note the specific green practices covered in a particular chapter.

FROM EXPERIENCE

What Is a Not So Big House?

Sarah Susanka, FAIA. I first coined the term "not so big" in my 1998 book, *The Not So Big House*, in an attempt to help describe an alternative to our ever-increasing house size. I wanted to make people aware that size has almost nothing to do with the qualities of home that most homeowners are seeking when they build or remodel.

What I knew as a residential architect was that many of my clients wanted a better house than their existing one and assumed that better must automatically mean bigger. It's just not so. In fact, in the vast majority of cases, bigger just means bigger, and the new homeowners end up being disappointed that their new house doesn't really feel like the dream home they'd thought they were building.

But Not So Big® doesn't mean small. In fact, it's not about mandating any specific size of house at all. Household needs differ, so the assessment about how much space is needed can only be made by the people who will eventually live there. Instead, it's about focusing on quality rather than quantity and about tailoring the house for the way we actually live, rather than designing for a more formal way of life that no longer reflects our current needs.

I tell people that a good rule of thumb in right-sizing a home to make it Not So Big is to aim for about a third less space than they think they need but to budget about the same amount of money as they would have for their larger vision of home, reapportioning dollars out of square footage and into the quality and character of the interior space and building envelope.

By eliminating rooms that get used only a few times a year, such as the formal living room and dining room, and by designing the house so that every space is in use every day, there's a natural reduction in the home's size without any sense of something being lost. If we're not using those spaces anyway, why build them?

Photo by Cheryl Mohr

Susanka is a member of the College of Fellows of the American Institute of Architects (FAIA) and a senior fellow of the Design Futures Council. Susanka is the author of nine best-selling books, including The Not So Big House *(Taunton, 1998),* The Not So Big Life *(Random House, 2007),* Not So Big Remodeling *(Taunton, 2009), and most recently,* More Not So Big Solutions for Your Home *(Taunton, 2010).*

In addition, the walls, windows, roof, and foundation of the house are designed to be highly energy-efficient and are built using sustainable materials and building practices. The house should also be designed to maintain an excellent indoor air quality that can provide a healthy and comfortable platform for everyday life.

I point out to my readers that a smaller but better designed house actually lives larger than one that's significantly bigger because the spaces work together as an integrated whole, perfectly supporting the lives of the inhabitants. It's a strategy that will appeal to not only the original homeowners but also to future generations, providing a delightful as well as comfortable environment for all their lives.

Lastly, a Not So Big house is a house that is beautiful and that inspires those who live within its walls. Beauty really does matter in terms of sustainability because people tend to take good care of places they find beautiful and delightful, so making a home Not So Big should really be one of the first steps in sustainable design and construction.

Some of the key features of a Not So Big house or remodel are as follows:

- Designed for comfort and livability—for the way we *really* live
- Designed to be as energy efficient and sustainable as possible
- Designed for our human scale (rather than for giants)
- Designed to last for centuries rather than for decades
- Designed in all three dimensions, with plenty of ceiling height variety to define and articulate activity areas
- Designed to be just the right size to accommodate the homeowners' needs—not too big, and not too small either
- Designed to be beautiful as well as functional and to inspire its occupants every day

Note: *Not So Big® is a registered trademark of Susanka Studios.*

"From Experience" Features: These boxes highlight industry leaders discussing a variety of important issues, and also practicing professionals sharing their knowledge of the industry and their success in employing specific techniques in their projects.

"Did You Know?" Features: Located throughout the text, these boxed features highlight unique or critical issues that deserve special attention, as well as tables that provide comparisons between different materials and technologies for quick reference.

Did You Know?

Radiant Barriers

Radiant barriers reduce the amount of heat entering or leaving the home by way of radiation. Attic radiant barriers are most effective in warm and hot climates with ductwork in the attic. During the summer, radiant barriers can significantly reduce attic temperatures and, in some cases, may reduce cooling loads in the house. Radiant barriers usually consist of a thin sheet or coating of a highly reflective material (usually aluminum) applied to one or both sides of a substrate material. Any of several types of substrates may be used, including kraft paper, plastic film, cardboard, plywood, or oriented strand board (OSB) sheathing, and rigid insulation materials. Although many radiant barrier products provide an "effective" or "equivalent" R-value, these products provide very little reduction in conductive heat flow and therefore often have R-values of less than R-1. The radiant barrier may be laminated directly to the roof decking (Figure 2.13a) or be fastened to the underside of the roof rafters (Figure 2.13b). See Chapter 7 for more information.

Figure 2.13a Radiant barriers may be applied in the factory to the roof decking for easy installation. LP's TechShield is OSB with a perforated radiant barrier that reduces radiant heat transfer and allows moisture drying.

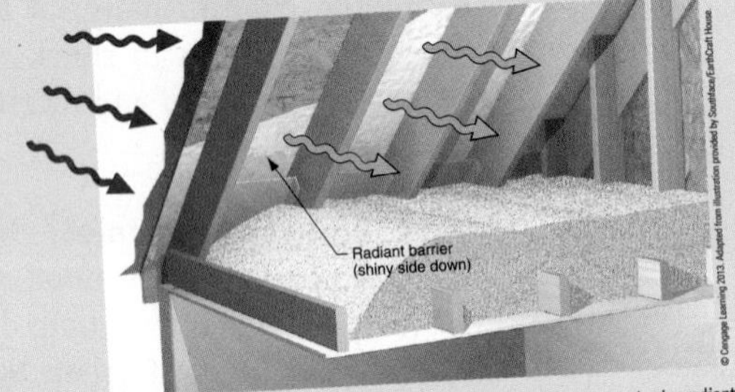

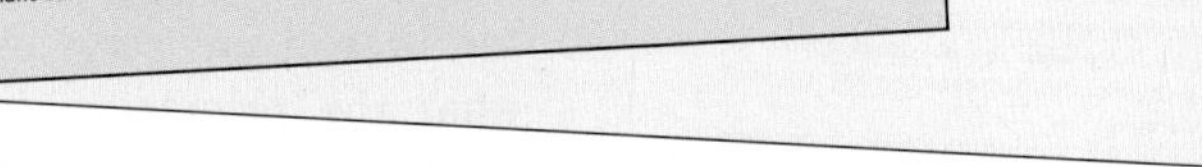

Figure 2.13b Radiant barriers may be applied to the underside of roofing rafters to retard solar radiant heat transfer.

In-Text Glossary: Definitions of key concepts are provided on the page where the concept is first mentioned and defined. A paginated list of the key concepts also appears at the end of each chapter. The glossary at the end of the book contains a complete list of the key concepts together with their definitions.

98 Section 1 What is Green Building and Why Does It Make Sense?

Insulation baffles
Air circulation from soffit vent to ridge vent
Insulation
Air barrier
Ceiling insulation batts or blown-in
All wall penetrations to be sealed with caulk
Block ceiling cavity with rigid insulation and seal edges

Figure 4.18 Attic knee walls must be fully air sealed with a backside air barrier. Blocking below knee walls prevents attic air from communicating with floor systems. Baffles in ventilated cathedral ceilings direct air around insulation.

Fiberglass Insulation

Fiberglass batt insulation is available in two forms: as *unfaced batts* or as *faced batts*, with foil or kraft paper facing. Faced batts are installed in walls by stapling the facing to the sides, or preferably, to the face of the studs. Faced batts are available in widths designed for friction fit between standard wall stud dimensions. Unfaced batts fit in wall cavities by friction (Figure 4.19a). Floor installations use metal clips, or tiger teeth, to hold the batts in place (Figure 4.19b). On ceilings, batts are stapled in place before the drywall is installed, or loose-laid between joists afterwards.

While readily available outside the United States, rigid fiberglass board insulation is not a common product; however, interest and availability is increasing. Unlike rigid foam boards, it does not require the use of added fire retardants.

Blown-in fiberglass can be installed in walls and above ceilings. Ceiling insulation is sprayed on loosely to a specified depth to obtain the correct R-value. Wall insulation can be applied with or without an acrylic binder and in varying densities, which provide different R-values. Insulation sprayed with a binder adheres to stud cavities and is scraped even with studs after installation. When no binder is used, fabric mesh is installed on the walls, and then the insulation is sprayed into the cavity through the holes of the mesh (see Figure 4.20).

Historically, fiberglass batt insulation was manufactured with phenol formaldehyde (PF) binders to hold the fibers together, but some products now use acrylic or bio-based binders without any PF. Although most of the PF dissipates during manufacture, some continues to release into the wall cavity, which could be a problem for extremely chemically sensitive people. Off-gassing is the process by which many chemicals volatilize, or let off molecules in a gas form into the air. Many manufacturers now offer product lines that use non-PF binders or no binders at all.

Loose-fill fiberglass insulation uses no binders in manufacture, but this feature is not without its drawbacks. Fiberglass fibers can become airborne during installation, and inhalation of these particles poses the risk for potential lung problems. Fiberglass insulation is frequently manufactured with a minimum of 20%, and often more, recycled content, although removed insulation and installation scraps are not normally recycled.

Mineral Wool Insulation

Mineral wool insulation refers to either *slag wool* or *rock wool*. Slag wool is made from an iron ore blast furnace waste product. Rock wool is produced from natural basalt rock. One leading manufacturer uses a 50/50 mix of these two sources. The majority of the

faced batts batt insulation that contain a foil or kraft paper vapor retarder covering.
unfaced batts cotton or fiberglass batt insulation that does not contain a vapor retarder covering.

phenol formaldehyde is a potentially harmful chemical binder commonly used in fiberglass insulation and engineered wood products.
off-gassing the process by which many chemicals volatilize, or let off molecules in a gas form into the air; see also *volatile organic compounds*.
mineral wool insulation a manufactured wool-like material consisting of fine inorganic fibers made from slag and used as loose fill or formed into blanket, batt, block, board, or slab shapes for thermal and acoustical insulation; also known as *rock wool* or *slag wool*.
slag wool another name for *mineral wool*.
rock wool another name for *mineral wool*.

End-of-Chapter Features: Included at the end of each chapter are the following components: *Summary, Review Questions, Critical Thinking Questions, Key Terms* with corresponding page numbers, and a list of *Additional Resources* that pertain to the chapter topics.

An Extensive Art Program: A comprehensive collection of vivid illustrations and photos helps bring key concepts to life, enabling the reader to understand complex concepts more easily.

Supplements

Spend Less Time Planning and More Time Teaching with Delmar, Cengage Learning's Instructor Resources to accompany *Green Building: Principles and Practices in Residential Construction*, preparing for class and evaluating students has never been easier!

This invaluable instructor CD-ROM is intended to assist you, as the instructor, in classroom preparation and management. Included within this electronic resource are tools that help reinforce the important building techniques introduced in the book as well as provide the necessary materials for evaluation of student comprehension of critical concepts:

- An Instructor's Manual including Lecture Outlines with corresponding PowerPoint slides, Chapter Summaries, answers to the end-of-chapter Review Questions and Critical Thinking Questions, and Additional Resources help prepare you for class.
- PowerPoint Presentations highlight critical concepts in each chapter to enhance classroom lectures. PowerPoint presentations also correlate to the Lesson Outlines in the Instructor's Manual, allowing for a seamless presentation of the content of the book.
- A Testbank in ExamView format includes hundreds of questions and enables you to edit, delete, or add your own questions as well as create your own tests using the questions provided. This flexible format makes this feature a handy tool for evaluating your students on the concepts presented in each chapter.
- An Image Library containing illustrations from the book enables you to supplement and enhance your classroom presentations.
- Link to delmarlearning.com and click on building trades to review other Delmar Learning titles available in the construction fields.

The use of these tools, along with *Green Building: Principles and Practices in Residential Contraction* will assist you as you guide your students down the path to success!

Order #: 9781111135959

CourseMate

A CourseMate is available to accompany *Green Building: Principles and Practices in Residential Construction*. Visit www.login.cengage.com and enter your single sign-on (SSO) login information to access CourseMate to accompany *Green Building: Principles and Practices in Residential Construction.*

Course Mate includes **Engagement Tracker**, a first-of-its-kind tool that monitors student engagement in the course.

Instructors also have access to the student resources on CourseMate, including:

- an interactive eBook
- interactive teaching and learning tools including:
  - quizzes
  - flashcards
  - videos
  - and more

For the Student

CourseMate

A CourseMate is available to accompany *Green Building: Principles and Practices in Residential Construction*. To access additional course materials including CourseMate, please visit www.cengagebrain.com. At the CengageBrain.com home page, search for the ISBN of your title (from the back cover of your book) using the search box at the top of the page. This will take you to the product page where these resources can be found.

The CourseMate to accompany *Green Building: Principles and Practices in Residential Construction* includes:

- an interactive eBook
- interactive teaching and learning tools including:
  - quizzes
  - flashcards
  - videos
  - and more

About the Authors

Abe Kruger

Abe Kruger is a certified Building Performance Institute (BPI) Building Analyst, Certified Energy Manager (CEM), and Home Energy Rating System (HERS) trainer and Rater as well as an active member of the Residential Energy Services Network (RESNET) National Technical Committee. Mr. Kruger is a designated Leadership in Energy and Environmental Design (LEED) Accredited Professional (AP) Homes and LEED AP Neighborhood Development (ND). He has conducted energy efficiency and conservation training for builders, renovators, home inspectors, contractors, and homeowners around the country. Mr. Kruger provides green building training and presentations at regional and national conferences, including RESNET, Affordable Comfort, Inc. (ACI), Green Prints, and Greenbuild, and he is accredited to perform EarthCraft House, LEED for Homes, and ENERGY STAR certifications. Mr. Kruger also assists with designing, implementing, and evaluating utility-run energy efficiency programs. In 2009, Mr. Kruger founded Kruger Sustainability Group to provide green building training, consulting, and curriculum development for colleges, companies, utilities, and nonprofit organizations.

http://www.KrugerSustainabilityGroup.com
abe@KrugerSustainabilityGroup.com

Carl Seville

Carl Seville has honed his expertise in sustainable construction through over 30 years as a contractor, educator, and consultant in the residential construction industry. He trains construction industry and allied professionals throughout the country on sustainable construction practices and certifies single and multifamily buildings under the LEED for Homes, EarthCraft House, and National Green Building Programs. His groundbreaking work has been recognized with numerous awards, including the Energy Value Housing Award (2009), two Green Advocate of the Year Awards (2005, 2007), two National Green Building Awards (2004, 2006), and the EarthCraft House Leadership award (2006). Mr. Seville is a Building Performance Institute (BPI) building analyst and a Home Energy Rating System (HERS) rater, and he holds the Leadership in Energy and Environmental Design (LEED) AP Homes and Green Rater designations.

http://www.greencurmudgeon.com
http://www.sevilleconsulting.com

Acknowledgments

We would like to thank and acknowledge many professionals who reviewed and/or contributed to the manuscript of our *Green Building: Principles and Practices in Residential Construction* text:

Michael Anschel—CEO Verified Green, Inc.
Lee Ball—Appalachian State University
Richard Bruce—Missouri State University
Christina Corley—Southface Energy Institute
Joe Dusek—Triton College
George Ford—West Carolina University
Tim Gibson—John A. Logan College
Eric A. Holt—Perdue University
Gary Klein—Affiliated International Management
Carlos Martin
Stephen McCormick—Santa Fe Community College
Luke Morton
Ed Moore—York Technical College
Amy Musser—Vandemusser Design, PLLC
Norma Nusz Chandler—South Dakota State University
Cindy Ojczyk—Verified Green, Inc.
Ashley B. Richards, Jr.—Richards & Company, Inc.
Lingguang Song, Ph.D—University of Houston
Alex Wilson—Building Green, Inc.
Robert A. Wozniak—Pennsylvania College of Technology
Peter Yost—Building Green, Inc.

Special Contributors

Green Building Advisors: Martin Holliday, Dan Morrison
Building Green: Alex Wilson, Peter Yost
Southface Energy Institute
Building Better Homes for videos in the accompanying CourseMate

Dedications

In addition to the reviewers and contributors, we would like to provide a special thank you to Southface Energy Institute and our amazing production team. In very different ways, we discovered the green gospel through Southface. Abe started his professional career at Southface and used it as a launching pad for future endeavors. Carl both taught and attended classes there over many years and helped build Southface's Resource Center. He was also intimately involved in the development of the EarthCraft House Renovation program.

At Cengage, we would like to thank James Devoe and Cristopher Savino. At Ohlinger Publishing Services we're grateful to Erin Curtis, Monica Ohlinger, and Brooke Wilson. Throughout the whole process, Erin has been an incredible resource, and we cannot thank her enough for her patience and constant optimism. Additional thanks go to the production and art teams.

Abe Kruger

This book would not be possible without the support and assistance from friends, family, and numerous industry professionals. I am incredibly thankful and constantly humbled by the passionate people who make up the green building industry.

I am forever grateful for the encouragement and amazing patience of Anne Rogers. How she managed to live through 3 years of near non-stop discussions of "the book" I will never know!

Ed Moore at York Technical College in Rock Hill, South Carolina initially brought this project to my attention. I am thankful for this and the support of his colleague, Rodney H. Trump.

Carl Seville

In addition to the help and support I have received from countless friends and industry associates over the years, I wish to send special thanks to my children, Paula Seville and Alex Cullen, for the many hours they spent listening to me talk about green building. I am grateful for their feedback from their years spent living in some of my green remodeling experiments.

Introduction to Green Building: Principles and Practices in Residential Construction

Green Building Overview

Green building is a set of design, construction, and building operation practices that minimize a building's total environmental impact. Decisions made while planning, building, renovating, and maintaining homes have long-term direct impacts on many different aspects of our environment—air quality, health, natural resources, land use, water quality, and energy use. At the same time, our building decisions have major economic implications, from the cost of land and materials to the labor and financing required to build.

Buildings are a primary point of consumption of energy, water, and raw materials. Residential buildings account for approximately 21% of all primary energy use in the United States, while commercial buildings represent another 19%.[1] Internationally, residential buildings use approximately 15% of primary energy.[2] Buildings are also responsible for a significant portion of air and water pollution.

The Eight Principles of Green Building

Although there is no universal definition of green building, we have identified eight green building principles that should always be considered when designing, building, or maintaining a home. These principles are similar to the approach defined by the United States Green Building Council's Leadership in Energy and Environmental Design (LEED) rating system and other green home rating programs.

- **Energy Efficiency:** Reducing the energy required to live in a house by designing it from the beginning with reduced consumption and increased efficiency in mind through appropriate equipment selection and high-quality construction methods
- **Resource Efficiency:** Reducing the total quantity of materials required to build or remodel a house, including selecting materials that are extracted, processed, and delivered to the job site with the least environmental impact and energy use; reusing previously used materials; and recycling of construction waste
- **Durability:** Using materials and methods that require less maintenance and increase the life of the structure; by reducing the frequency of repair and replacement, less waste is generated, and fewer materials are needed through the life of a house
- **Water Efficiency:** Reducing the amount of water used inside and outside the house through increased efficiency and minimizing opportunities for more consumption
- **Indoor Environmental Quality:** Improving occupant health by controlling moisture, toxic materials, and pollutants inside the house
- **Reduced Community Impact:** Limiting negative economic effects on the local community through responsible development and construction practices; considering how the selection of materials has an impact on the health and economic conditions of the global community—workers and local residents where products are extracted and manufactured for use in homes
- **Homeowner Education and Maintenance:** Educating homeowners and occupants to operate and maintain their homes to remain efficient, healthy, and durable throughout their lifetime
- **Sustainable Site Development:** Avoiding development of environmentally sensitive areas, orienting lots and homes to take advantage of the sun, promoting building near public transit and amenities to reduce driving, carefully managing the site

[1] U.S. Department of Energy, 2008 Buildings Energy Data Book, Section 1.1.1, 2008, http://buildingsdatabook.eren.doe.gov/

[2] http://www.eia.doe.gov/oiaf/ieo/world.html

during construction to reduce silt runoff and maintain native vegetation, and providing for permanent storm water management to reduce contaminant runoff from site to public waterways

Because all of the concepts will not be emphasized in every chapter, we have developed a series of icons to represent the different principles. The beginning of each chapter will list the icons that correspond to the principles covered in that chapter. These icons serve as a reminder of what the principles are and present an efficient way to note the green practices covered in a particular chapter.

Approach to Green Building in This Book

In this book, we cover green building in a "best practices" approach to low-rise residential construction. We examine the considerations that must be taken into account during the design, site development, and construction phases of the project, and we present the options available in material use—all with an eye toward building a truly green house.

Many green building techniques are simply those that consciously include and enforce good quality. Throughout the text, we describe how to simply build better homes that provide comfortable, safe, durable, and efficient living environments. Readers may be surprised to see that the term *green* is used sparingly throughout the text.

The book is broken into five sections that roughly follow the construction schedule. Section One defines green building, establishing why green building is desirable, and explaining the science behind green. At the foundation of green building is applied building science. For a home to operate efficiently and effectively, moisture, heat and air flow must be controlled (see Chapter 2). Section Two covers the structural systems of a home. Here we explore foundations, floors, walls, ceilings, and roofs. Section Three delves into exterior finishes, including windows and doors, cladding, outdoor spaces, and landscaping. Section Four examines interior finishes, and Section Five explores mechanical systems, including heating, ventilation, and air conditioning as well as electrical, plumbing, and renewable energy.

A Final Note

Regardless of your professional path, we hope you take away from this book an appreciation for the complexity of the issues and a firmer understanding of what it means to be truly green—recognizing that it is not just a buzzword. Now is an exciting time to be in the construction industry because of the rapidly changing products and techniques that are being developed to build better, greener homes.

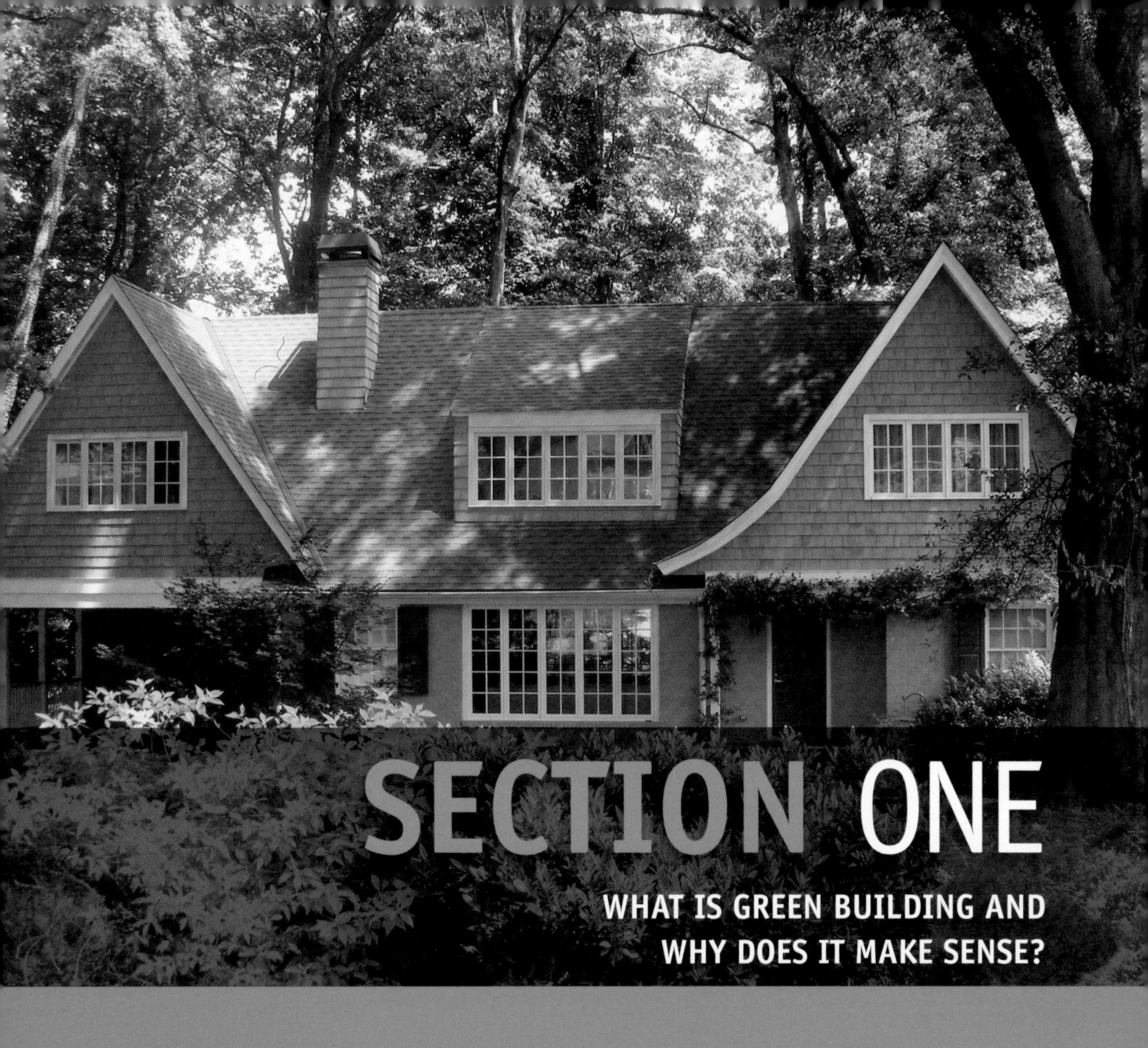

SECTION ONE

WHAT IS GREEN BUILDING AND WHY DOES IT MAKE SENSE?

1

Green Building: An Overview

This chapter explores definitions of green building, its importance from an environmental perspective, and its context within the design and construction industry. We provide a brief history of green building and the organizations that have helped create the guidelines and standards for the industry. The current versions of these specific green building programs are presented, along with likely national trends.

LEARNING OBJECTIVES

Upon completion of this chapter, the student should be able to:

- Define green building
- Explain the environmental impacts of the residential construction industry as a whole
- Describe the benefits of green building for builders, homeowners, and contractors
- Describe the relationship between green building programs and building codes
- Define the ENERGY STAR for Homes program
- Describe the Home Energy Rating System (HERS) and its relationship to green building programs
- Explain blower door and duct leakage testing
- Calculate simple payback and cash flow analysis for green building improvements

Green Building Principles

Energy Efficiency

Resource Efficiency

Durability

Water Efficiency

Indoor Environmental Quality

Reduced Community Impact

Homeowner Education and Maintenance

Sustainable Site Development

Defining Green Building

We define green building as a set of design, construction, and maintenance techniques and practices that minimize a building's total environmental impact. Decisions made while planning, building, renovating, and maintaining homes have long-term direct effects on many different aspects of our environment—air quality, health, natural resources, land use, water quality, and energy use. These decisions may also produce indirect effects on other aspects of our environment, such as factors that contribute to global warming.

The materials used to construct, remodel, and maintain a house all have an impact on the environment, as does the energy used to heat, cool, light, and run equipment, and the amount of water used during the home's lifetime (Figure 1.1). Neighborhood design affects how much land is consumed, how far people drive, and the amount of water pollution caused by runoff from roofs, lawns, and roads. Green building strives to reduce these negative impacts.

Energy Generation and Use

Residential buildings consume approximately 22% of the energy produced in the United States for heating, cooling,

---

green building an environmentally sustainable building, designed, constructed, and operated to minimize the total environmental impacts.

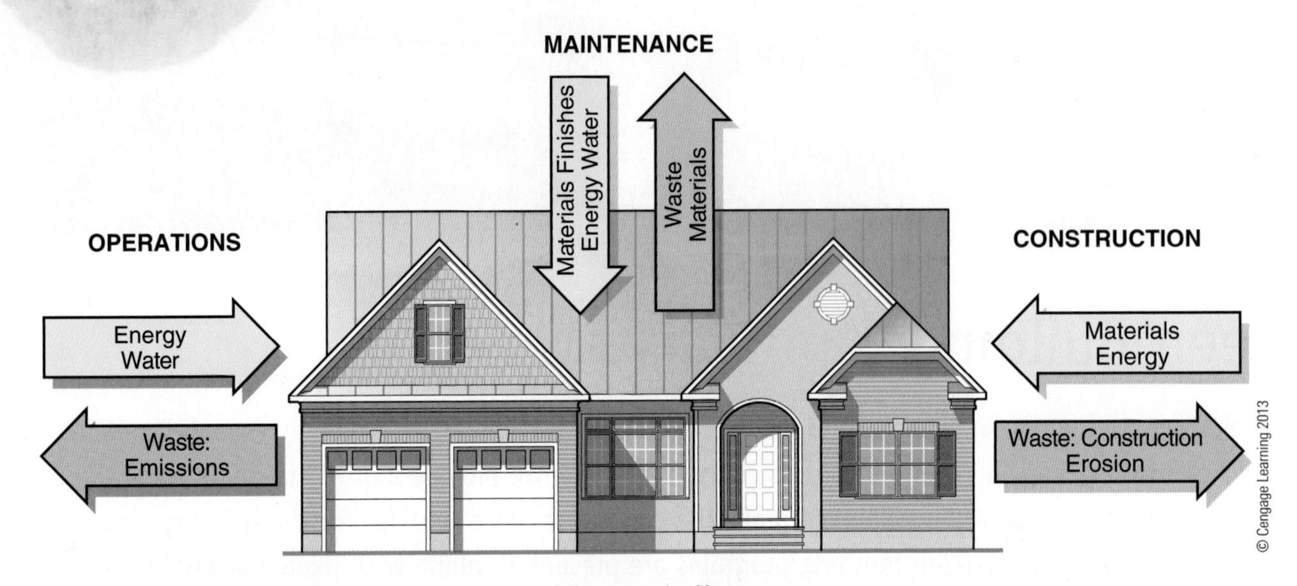

Figure 1.1 Home construction has numerous environmental effects.

cooking, water heating, and operating electrical devices.[1] This volume of energy demand and supply presents several problems. Electrical power generation carries significant environmental consequences, including air pollution and greenhouse gas emissions, not to mention the consumption of natural resources in the construction of new power generating facilities. Direct air pollutants include volatile organic compounds (VOC) and small dust particles (aerosols) that may contribute immediately to health and environmental degradation. Greenhouse gases (GHG) are atmospheric gases, such as carbon dioxide (CO_2), sulfur oxides (SO_x), and nitrous oxides (NO_x), that are also emitted directly but lead to increasing global temperatures—that is, the "greenhouse effect." Coal plants add pollutants to the atmosphere, mining operations can cause permanent damage to the land, and fly ash, a by-product of power generation, can foul waterways if not properly stored and managed. The operation of hydroelectric plants does not pollute the air, but it reduces the availability and quality of water for downstream communities and river habitats. Nuclear power, while not emitting pollutants, is produced in plants that are costly to build and operate, and the issue of waste disposal is not yet resolved. Both coal and nuclear plants use significant amounts of water to operate, although hydroelectric plants lose water through evaporation from reservoirs, further depleting supplies available for drinking and irrigation. According to the U.S. Department of Energy (DOE), the national weighted average for thermoelectric and hydroelectric water use is 2.0 gallons (7.6 L) of evaporated water per kilowatt-hour (kWh) of electricity.[2] A kilowatt-hour is a unit of electric energy equal to 1000 watts operating for one hour; therefore, 1 kWh would operate a 100-watt light bulb for 10 hours.

In addition to electricity, many homes use oil and natural gas for space and water heating (see Chapter 13 for a discussion on fuel usage). While natural gas is an efficient energy source that produces fewer pollutants than oil or coal, all three are nonrenewable resources that are subject to scarcity, increasing costs to extract, price fluctuation, and, potentially, the ultimate exhaustion of all supplies.

Renewable energy sources, such as solar, geothermal, and wind, provide alternatives with less environmental impact as compared to nonrenewable options. Renewable

volatile organic compounds (VOC) chemical compounds that have a high vapor pressure and low water solubility; many VOCs are man-made chemicals that are used and produced in the manufacture of paints, pharmaceuticals, refrigerants, and building materials; VOCs are common indoor air pollutants and ground water contaminants.

greenhouse gas (GHG) any of the atmospheric gases, such as carbon dioxide (CO_2), sulfur oxides (SO_x), and nitrous oxides (NO_x), that contribute to the greenhouse effect.

Department of Energy (DOE) the federal department responsible for maintaining the national energy policy of the United States.

renewable energy electricity generated from resources that are unlimited, rapidly replenished, or naturally renewable (e.g., wind, water, sun, geothermal [ground heat], wave, and refuse) and not from the combustion of fossil fuels.

[1] U.S. Energy Information Administration, Annual Energy Review 2009.

[2] P. Torcellini, N. Long, and R. Judkoff, Consumptive Water Use for U.S. Power Production. NREL/TP-550-33905. December 2003.

Did You Know?

Electrical generation and distribution in the United States are very inefficient. Typical coal-fired power plants are approximately 30% to 35% efficient, and distribution losses are approximately 7% to 10%. Consequently, for every 10 units of energy that go into a coal plant, only 3 to 4 units are actually delivered to a home. Even more energy is required to mine and transport the coal to power plants. Saving energy at the point of use significantly magnifies the impact in terms of both efficiency and pollution reduction (see Figure 1.2). In this example, you can see the overall efficiency of converting coal (chemical energy) into light energy in the home. The total efficiency of the system is the percent efficiency of each component multiplied together.

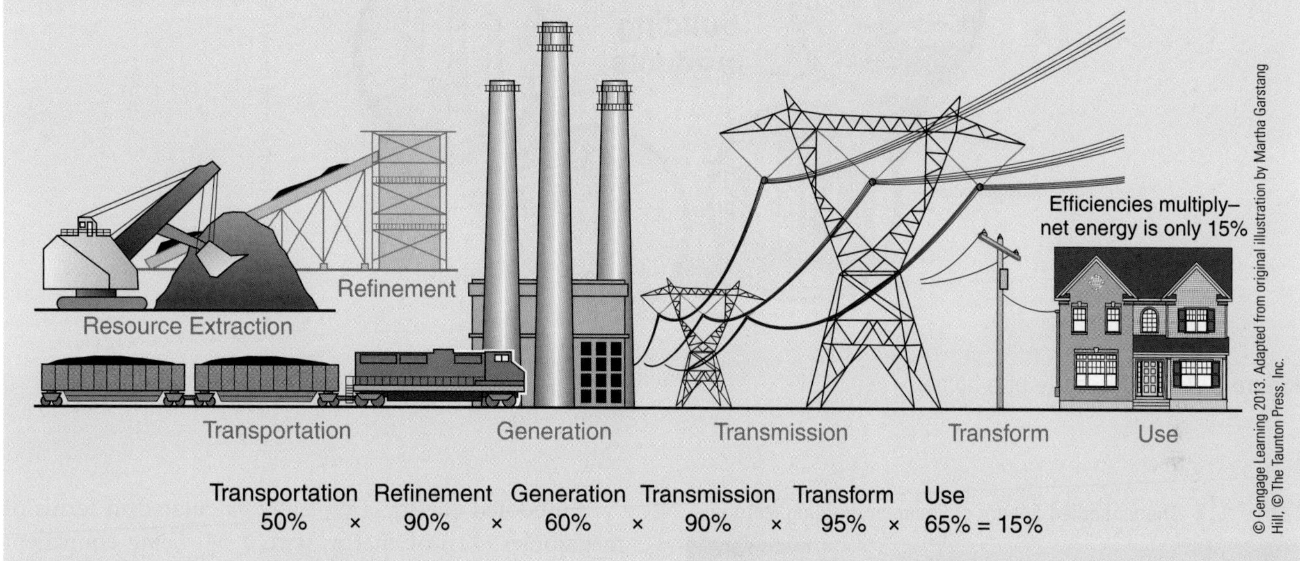

Figure 1.2 Current electrical generation and distribution systems are quite inefficient.

energy production does not create air pollution and uses fuel sources (i.e., the sun, the earth, and wind) that never diminish. While traditional energy sources still dominate the marketplace, renewable energy is an increasingly common option. Renewable energy sources include solar panels and wind turbines on individual homes; medium-scale installations that service a neighborhood or individual development; and large private, government, or utility projects that supplement or replace traditional power plants.

Embodied Energy in Material Production

Life cycle assessment (LCA) is a general term used to describe an analysis of all of the energy consumed to produce, sell, install, use, and dispose of any product throughout its physical existence (i.e., the *cradle-to-grave* energy sum). LCA begins by quantifying the energy used to gather raw materials from the earth to create a product and ends at the point when all materials are returned to the earth (see Figure 1.3). By including the impacts throughout the product life cycle, LCA provides a comprehensive view of the environmental aspects of the product or process and a more accurate picture of the true environmental trade-offs in product and process selection. Environmental factors typically include energy and water consumption, greenhouse gases (eg, CO_2, SO_x, NO_x), water contaminants, and raw material usage.

The total energy required to manufacture or harvest, package, and ship a material to a job site is referred to as embodied energy. Selecting products that are produced locally helps to reduce transportation energy and associated air pollution, and it may reduce costs. Specifying products that use less energy in their production is a consideration in green building (Table 1.1). Products that use large amounts of energy to produce, typically in the form of heat in kilns and ovens, include cement, gypsum board, glass, aluminum, and steel. Most of these are both very durable and recyclable, so the impact of the

life cycle assessment (LCA) process of evaluating a product or building's full environmental cost, from harvesting raw materials to final disposal.

embodied energy the energy required to manufacture or harvest, package, and ship a material to a job site; may refer to an individual material or home.

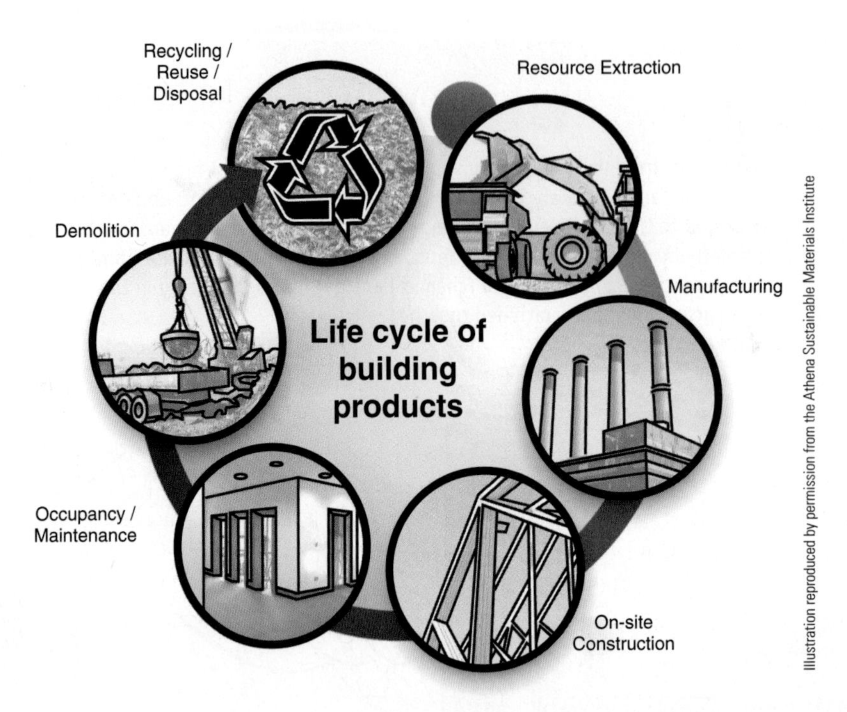

Figure 1.3 The life cycle of a home.

Table 1.1 The Embodied Energy of Common Building Materials

Material	Unit	Energy Coefficient Mj per Unit
Timber, rough	m^3	848
Timber, air-dry, treated	m^3	1,200
Timber Glulam	m^3	4,500
Timber, kiln-dry, treated	m^3	4,692
Timber, form work	m^3	283
Plywood	m^3	9,440
Building paper	m^2	75
Gypsum board	m^3	5,000
Glass	kg	31.5
Structural steel	kg	59
Aluminum	kg	145
Fiberglass batts	kg	150
Asphalt, strip shingle	m^2	280

Source: *Buchanan, Andrew H. and Honey, Brian G. "Energy and Carbon Dioxide Implications of Building Construction."* Energy and Buildings. *1994; 20:205–217; reprinted with permission from Elsevier.*

embodied energy is spread out over many years. Lumber and other such products may have much lower embodied energy but, depending on how they are used, may need more frequent replacement—spreading the energy use over fewer years. While plastics are long-lasting, they are typically energy-intensive. Compared with other materials, a low percentage of plastics are recycled; large volumes of plastics end up in landfills or waterways where they do not decompose and can endanger wildlife.

Embodied energy is typically calculated in terms of megajoules (Mj) of energy, with 1 Mj being equivalent to about 948 British thermal units (Btu). Part of the challenge of assessing and making decisions based on embodied energy is a lack of current data. As buildings become more energy-efficient, their embodied energy becomes a more significant percentage of the total energy they consume.

Every building is a complex combination of many processed materials, each of which contributes to the building's total embodied energy. When specifying building materials, architects should consider the embodied energy of not only individual materials but also their required assemblies. For example, polyurethane spray foam insulation should be compared with fiberglass batt insulation plus the additional air sealing materials that are required to achieve similar home tightness. Similarly, exterior sheathing with an integrated weather resistant barrier should be compared with oriented-strand board (OSB) plus house wrap.

Resource Use

Home construction and maintenance consumes large quantities of materials. The National Association of Home Builders (NAHB) and the U.S. Department of Energy (DOE) estimate that construction of the average home uses tons of materials and produces 8,000 lbs of construction waste (see Table 1.2 and Table 1.3).

Table 1.2 Materials Used in the Construction of a 2272-ft^2 Single-Family Home

13,837 board-feet of lumber	12 interior doors
13,118 square feet of sheathing	6 closet doors
19 tons of concrete	2 garage doors
3,206 square feet of exterior siding material	1 fireplace
3,103 square feet of roofing material	1 heating and cooling system
3,061 square feet of insulation	3 toilets, 3 bathroom sinks 2 bathtubs, 1 shower stall
6,050 square feet of interior wall material	1 washer, 1 dryer
2,335 square feet of interior ceiling material	15 kitchen cabinets, 5 other cabinets
226 linear feet of ducting	1 kitchen sink
19 windows	
4 exterior doors (3 hinged, 1 sliding)	1 range, 1 refrigerator, 1 dishwasher, 1 garbage disposal, 1 range hood
2,269 square feet of flooring material	

Sources: *NAHB.* 2004 Housing Facts, Figures and Trends. *Feb. 2004, p. 7. Courtesy of NAHB Research Center's Annual Builder Practices Survey. Data for appliances and heating, ventilation, and air conditioning courtesy of D&R International.*

Table 1.3 Typical Construction Waste Estimated for a 2000-ft^2 Home

Material	Weight, lbs	Volume, yd^3*	Percentage of Total Waste
Wood, solid sawn	1,600	6	20%
Wood, engineered	1,400	5	18%
Drywall	2,000	6	25%
Cardboard, old corrugated	600	20	8%
Metals	150	1	2%
Vinyl (polyvinyl chloride)†	150	1	2%
Masonry‡	1,000	1	13%
Hazardous materials	50	–	1%
Other	1,050	11	13%
Total	**8,000**	**50**	**100%**

*Volumes are highly variable due to compressibility and captured air space in waste materials. Due to rounding, sum does not add up to total.

†Assuming three sides of exterior clad in vinyl siding.

‡Assuming a brick veneer on home's front facade.

Source: *NAHB,* Residential Construction Waste: From Disposal to Management, *Oct. 1996,* http://www.nahb.org. *Courtesy of NAHB Research Center, Annual Builder Practices Survey.*

Impact of Buildings and Material Use on Air Quality

Not only does the selection of wood harvesting, hard-surface paving, and roofing materials influence energy production, these materials also affect our air quality. As trees help to cool the environment and provide wildlife habitats, large-scale removal through clear-cutting of forests contributes to higher air temperatures and decreased local rainfall. Dark colored paving and roofs absorb heat, contributing to the heat island effect. The **heat island effect** is the phenomenon of urban areas being warmer than rural areas. This effect is due primarily to the increased use of materials that absorb heat during the day and release it after sunset, raising the air temperatures on warm days and increasing energy demands because of greater air conditioning usage.

Water Use

The United Nations Development Program estimates that the average American uses 152 gallons of water per day. By comparison, average daily consumption is 102 gallons in Italy and 39 gallons in the United Kingdom (see Figure 1.4). Population growth, increased use, and changes in climate patterns have contributed to water shortages in many regions. Yard irrigation, toilets, showers, electrical generation, and plumbing leaks combine to make the United States the world's primary water consumer. Techniques to combat these issues include selecting drought-tolerant plants and water- conserving fixtures, designing efficient piping systems, incorporating water reclamation methods, and even conserving electricity. All these strategies help to save water and increase the available supply for everyone.

Sustainable Development

Most new residential construction since World War II has been concentrated in suburban areas on previously undeveloped land referred to as **greenfield developments**. Compared with denser urban areas, suburban development requires more roads, parking lots, and garages to accommodate an automobile-based lifestyle. With zoning regulations often mandating long distances between residential and commercial areas, residents typically rely on their cars to cover this ground. As a result, the associated air pollution increases while healthy activities,

---

heat island effect the phenomenon of urban areas being warmer than more rural areas. primarily due to the increased use of materials that effectively retain heat.

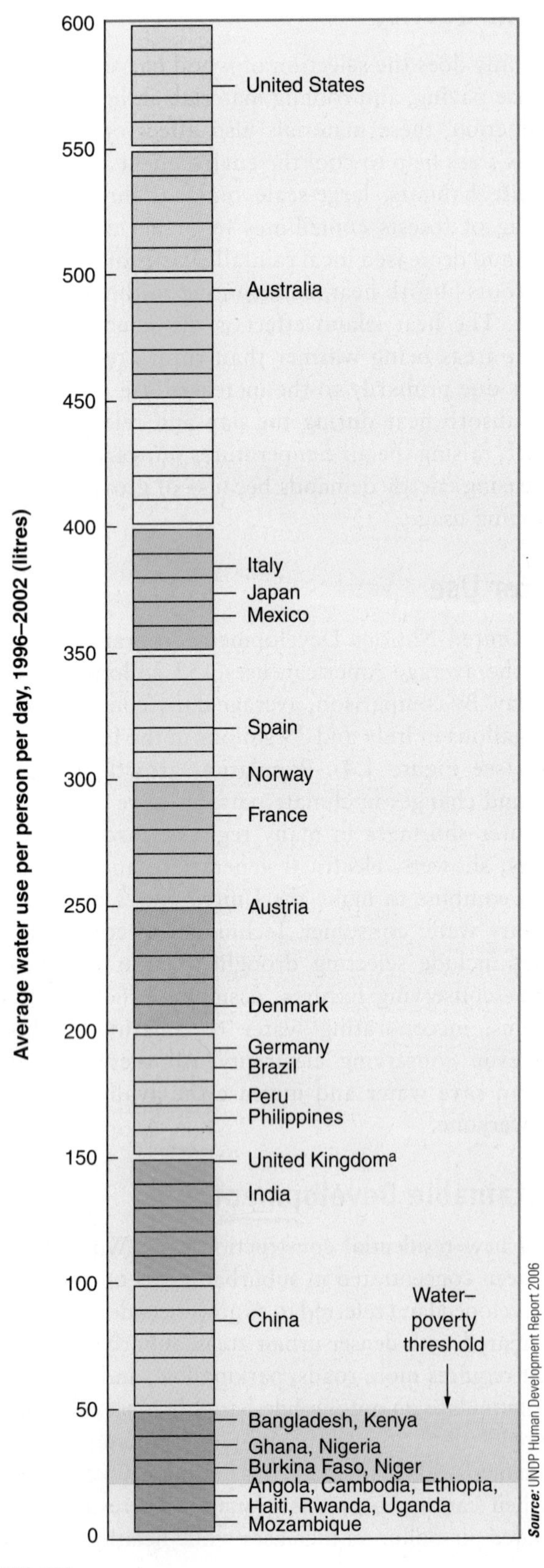

[a]OFWAT 2001.

Figure 1.4 Water usage around the world.

such as biking and walking, are not always feasible means of commuting. Suburban single family homes are usually larger than homes or apartments in urban areas, consuming more materials to build and more energy to operate per person. Many also have large yards that use substantial amounts of water for irrigation. Sustainable development practices, such as traditional neighborhood development (TND), also known as new urbanism, work to combine residential, retail, and office uses in close proximity to reduce driving miles and encourage walking. TNDs feature smaller homes and yards than those in the suburbs, resulting in material, energy, and water savings. Transit-oriented development (TOD) is the design of communities that encourage new development within walking distance of mass transit to further reduce passenger car use and promote healthier lifestyles. Figure 1.5 and Figure 1.6 illustrate the key differences between TND and urban sprawl.

Courtesy Scarlett Photography

Figure 1.5 Traditional neighborhood developments promote walking through compact, mixed-use designs.

---

sustainable a pattern of resource use that aims to meet human needs while preserving the environment so that these needs can be met, not only in the present but also for future generations.

traditional neighborhood development (TND) the design of a complete neighborhood or town using traditional town planning principles that emphasize walkability, public space, and mixed-use development; see also *new urbanism*.

new urbanism an urban design strategy that promotes walkable neighborhoods that contain a range of housing and job types; highly influenced by traditional neighborhood development and transit-oriented development.

transit-oriented development (TOD) the design of communities to be within walking distance of public transit, mixing residential, retail, office, open space, and public uses in a way that makes it convenient to travel on foot or by public transportation instead of by car.

© Used under license from Shutterstock, 2010/Shawn Kashou

Figure 1.6 Urban sprawl encourages driving that negatively impacts humans and the environment.

A Brief History of Green Building

Green building has its roots in the early solar builders, most of whom began working during the first oil crisis in the late 1970s. Although the scale of the solar movement was limited through the 1980s and early 1990s, many industry leaders worked on research projects involving high-performance buildings that helped to establish the core concepts of building science, a key component of green building.

One of the first organized green home programs to appear was in Austin, Texas, where Austin Energy, the city-owned electric utility, recognized the need to reduce electricity demand to avoid the construction of another power plant. Austin Energy's residential energy efficiency program was founded in 1985 and, in 1991, evolved into Austin Energy Green Building®. Through the 1990s,

FROM EXPERIENCE

Green Building: An Old Concept

Ron Jones, Co-Founder and President of Green Builder® Media. For as long as people have been building—most of modern human history—the relationship between the built environment and the natural environment has generally been viewed from the direction of what effect nature has on our buildings. Since buildings are designed and constructed to serve man and his purposes, the impact of earthquakes, wildfires, floods, weather events (especially storms), wind patterns, general climate conditions...even sunlight...on buildings, have understandably been of concern to us.

Courtesy Ron Jones, Green Builder®

Ron Jones is founder and editorial director of Green Builder® Media. He is an award-winning builder, architectural designer, policy advocate, and internationally recognized green building expert.

In the natural evolution of things, it is not surprising that we have more recently begun to explore that relationship from both directions, and that we are starting to at least partially understand the effects that the built environment has on the planet—not only the natural resources from which we derive our energy and building materials, but the natural systems that surround us.

We have now begun to acknowledge that the actual activities of the development and construction process—easily the most conspicuously consumptive endeavors of mankind—are not the only factors creating long-term effects on the natural world. In addition, we must factor in the operations, maintenance and eventual de-construction of our buildings to fairly and accurately determine the total cost of a building—the cost not only to ourselves but to the entire global system.

What we now commonly refer to as "green building" is by no means a new concept. For as long as there have been designers and builders, contingents of the industry have considered the larger implications of the act of building and the results that their activities would have beyond the perimeter of their structures. These individuals do not ignore the implications for the air, the water, the land, and all the other species that depend on these elements for life, just as we do. Their decisions to think, plan, and act responsibly have all been expressions of sustainability.

In the end, it is about balance, about respecting the resources that we have available to us, and about using those resources in such a way that ensures their continued existence for future generations of our species and all others as well. This is not something we do not already know; we simply fail to remember sometimes. There is a story from the first Americans about the relationship of man to his resources (including his own time and energy) that says:

A foolish man builds a large fire and then stands away from the heat, while a wise man builds a small fire and sits close to its warmth.

Today's green building industry owes a great debt to these pioneers:

Advanced Energy Corporation, Raleigh, North Carolina

http://www.advancedenergy.org

Building Science Corporation, Westford, Massachusetts

http://www.buildingscience.com

Canada Mortgage and Housing Corporation, Ottawa, Ontario

http://www.cmhc-schl.gc.ca/

Florida Solar Energy Center, Cocoa, Florida

http://www.fsec.ucf.edu/en/

Rocky Mountain Institute, Snowmass, Colorado

http://www.rmi.org

Solar Living Institute, Hopland, California

http://www.solarliving.org

Southface Energy Institute, Atlanta, Georgia

http://www.southface.org

several more local programs appeared, including Built Green® Colorado, Built Green® Washington, and EarthCraft House™.

Courtesy of Austin Energy Green Building

Courtesy of EarthCraft House

Courtesy of Built Green® Washington

In 1995, the U.S. Environmental Protection Agency (EPA) introduced the ENERGY STAR for Homes program. This certification provides owners the assurance that their homes are more efficient than standard construction. ENERGY STAR strictly evaluates the energy efficiency of a home and was not designed as a green building program. The EPA originally developed the ENERGY STAR program to certify energy-efficient electronics and appliances. Today, over 50 different types of products may achieve the ENERGY STAR certification, and the logo is one of the most recognized brands in America (Table 1.4).

U.S. Environmental Protection Agency, ENERGY STAR program

In the late 1990s and early 2000s, local green home certification programs began appearing across the United States. Approximately 100 different local and regional programs are currently available throughout the country (Table 1.5). Filling a void in the marketplace, these green building programs essentially became the definition and identity of green building nationwide.

The U.S. Green Building Council (USGBC) was established in 1993 and released its Leadership in Energy and Environmental Design (LEED) program in 1998. This national green certification program launched a rating system that takes into account the eight principles of green building (as previously discussed, but with different labels and category groupings) within the context of all types of commercial construction. Today, the USGBC comprises 78 local affiliates, more than 20,000 member companies and organizations, and more than 100,000 LEED Accredited Professionals. USGBC introduced the pilot version of LEED for Homes in 2004, and the program was officially released nationally in 2008. LEED for Homes is designed to be among the most stringent green home programs, aiming this program at the top 25% of builders in the United States.

ENERGY STAR a joint program of the U.S. Environmental Protection Agency and the U.S. Department of Energy that sets standards for energy-efficient products and buildings.

U.S. Green Building Council (USGBC) a Washington, D.C.–based nonprofit that promotes green building and developed the LEED rating systems; see also *LEED*.

Leadership in Energy and Environmental Design (LEED) a system to categorize and certify the level of environmentally sustainable construction in sustainable buildings.

Courtesy of the U.S. Green Building Council (USGBC)

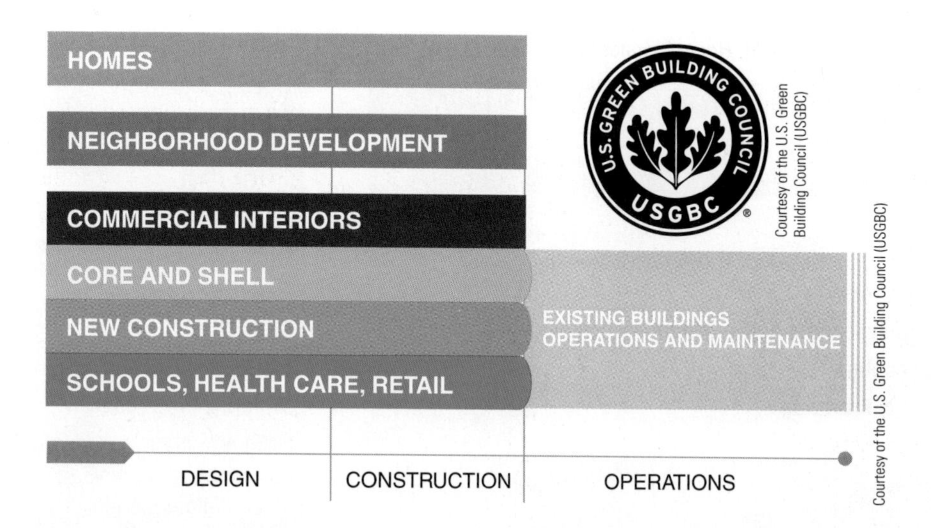

Courtesy of the U.S. Green Building Council (USGBC)

Courtesy of the U.S. Green Building Council (USGBC)

FROM EXPERIENCE

Green Begins With Blue: ENERGY STAR Blue

Sam Rashkin, R.A., National Director, ENERGY STAR for Homes. *Choosing a sustainable new home can be daunting, with approximately 100 green building programs employing myriad rating systems, point categories, minimum requirements, and "shades" (e.g., tiers or rigor). Here's some advice to get you on the right track: energy efficiency is the most critical first step. Why is that?*

Courtesy Sam Rashkin, R.A.

Sam Rashkin was the manager of ENERGY STAR for Homes since its start in 1996 until 2011. He is now the Chief Architect, U.S. DOE Building Technologies Program. During his 20-plus years as a licensed architect, he specialized in energy-efficient design and completed over 100 residential projects.

First, the underlying air, thermal, and moisture control measures associated with energy efficiency provide compelling performance advantages. Airtight construction, sealed ductwork, and comprehensive air barriers all help to minimize drafts, control surface temperatures, reduce external noise, and block humidity, dust, pollen, and pests from entering the home. High-performance windows reduce noise, unwanted solar heat gain and losses, and exposure to damaging ultraviolet (UV) sunlight. Similarly, improved insulation installation practices with minimal thermal bridging prevent unwanted heat loss. Properly sized and installed heating and cooling systems distribute space conditioning effectively to all rooms, operate at maximum efficiency, and enhance moisture control. As a result, energy-efficient construction delivers improved room-by-room comfort; quieter living conditions; reduced exposure to moisture problems; superior indoor air quality; and less fading of finishes and furnishings. Substantially reduced heating and cooling equipment, ductwork, and framing can also contribute to resource efficiency.

Second, all of these performance advantages can be achieved for lower ownership cost since the small increase in the monthly mortgage attributed to energy-efficiency measures is easily offset by monthly utility savings. These savings are likely to increase over time given projections for rapidly escalating utility costs. Better performance for lower cost is good, which is why energy efficiency should be the first consideration for any green home.

The easy way to take this first step is with the U.S. EPA's ENERGY STAR program. EPA introduced ENERGY STAR as the symbol for energy efficiency in 1992 for electronics and appliances, and it's now available on over 60 product categories. Since ENERGY STAR Qualified Homes became available in 1996, over 1 million homes have been certified. EPA estimates home buyers who purchased these homes have saved over $1 billion on their energy bills while reducing greenhouse gas emissions by 22 billion pounds. As ENERGY STAR Qualified Homes continue to get more rigorous with the release of third-generation specifications, homebuyers can look for the widely recognized blue logo to be assured their homes include a comprehensive package of building science measures along with third-party verification. In addition, they can look for an increasing number of green home programs that use ENERGY STAR Qualified Homes as a minimum requirement (e.g., LEED® for Homes, EarthCraft). So, you could say, "Green" begins with "Blue"—the blue ENERGY STAR logo.

Table 1.4 ENERGY STAR Home Product Categories

Appliances
Clothes washers
Dehumidifiers
Dishwashers
Freezers
Refrigerators
Room air cleaners and purifiers
Water coolers
Building Products
Roof products
Windows
Doors
Skylights
Computers and Electronics
Audio/visual equipment
Battery chargers
Combination units (TV/DVD or VCR/DVD)
Computers
Cordless phones
Digital-to analog converter boxes
Displays
External power adapters
Imaging equipment
Set-top boxes and cable boxes
Televisions
Heating and Cooling
Air conditioning, central
Air conditioning, room
Boilers
Fans, ventilating
Furnaces
Heat pumps, air source
Heat pumps, geothermal
Lighting
Decorative light strings
Fans, ceiling
Light bulbs (compact fluorescent lamps [CFLs])
Light fixtures
Residential light-emitting diode (LED) lighting
Plumbing
Water heater, gas condensing
Water heater, heat pump
Water heater, high-efficiency gas storage
Water heater, solar
Water heater, whole-home gas, tankless

Source: http://www.energystar.gov

Table 1.5 Green Building Programs Across the United States

Alabama
EarthCraft House™
energy right®

Arizona
Scottsdale Green Building Program
TEP Guarantee Home

California
California Green Builder
Earth Advantage®
GreenPoint Rated Homes/ Green Building in Alameda County
Santa Monica Green Building Program
San Jose Green Building Program (in progress)

Colorado
Built Green Colorado®
City of Aspen/Pitkin County Efficient Building Program
City of Boulder Green Points Program

Florida
EarthCraft House™
Florida Green Building Coalition
Good Cents

Georgia
EarthCraft House™
energy right®
Good Cents
Right Choice

Hawaii
Hawaii BuiltGreen™

Kansas
Build Green Program of Kansas City

Kentucky
energy right®

Louisiana
Power Miser Homes

Michigan
Green Built™ Michigan

Minnesota
Minnesota GreenStar

Mississippi
energy right®

Missouri
Build Green Program of Kansas City

New Jersey
New Jersey Affordable Green
New Jersey ENERGY STAR Homes

North Carolina
energy right®
NC HealthyBuilt Homes

Oregon
Earth Advantage®

South Carolina
EarthCraft House™

Tennessee
EarthCraft House™
EcoBUILD
energy right®

Texas
Austin Green Building Program
Build San Antonio Green®
Frisco Green Building Program
Good Cents
San Bernard Electric Coop New Home Program

Vermont
Burlington Electric Department's Residential New Construction Program
Vermont Builds Greener

Virginia
Arlington County Green Home Choice Program
EarthCraft House™
Energy Saver Home

Washington
Built Green® Washington
BUILT SMART
Earth Advantage®

Wisconsin
Green Built™ Home
Wisconsin ENERGY STAR Homes Program

In 2005, the National Association of Home Builders (NAHB) released *NAHB Model Green Home Building Guidelines*, a book designed to provide builders with guidance on making their projects green. In 2007, the NAHB and the International Code Council (ICC) partnered to develop the National Green Building Standard. In 2008 the NAHB Research Center began certifying homes under the guidelines through the NAHB Green program. The American National Standards Institute (ANSI)–approved ICC 700-2008 National Green Building Standard defines green building for single and multifamily homes, residential remodeling projects, and site development projects. The NAHB Research Center began certifying single and multifamily homes and renovations under the new standard in 2009. Certification under the original 2005 Green Home Guidelines was discontinued in 2010.

The Future

As energy codes become more stringent, higher performance homes will become standard. Some states, particularly California, have rigorous energy codes and emissions limits for building materials that are designed to improve indoor environmental quality. As people learn more about the benefits of green homes, the demand for green building will continue to increase. Many industry leaders believe that basic green building will ultimately become the minimum standard, and "green" as a differentiator will begin to fade away. This transition, however, will likely evolve slowly over many years.

Green Home Certifications

Certifications are most easily separated into three categories: the *buildings* themselves; the *people* who evaluate, build, and remodel those buildings; and the *products* used in the process. In this section, we will discuss the process of certification as well as the various certification programs that exist today.

Building Certification

Most green building programs require third-party verification by independent professionals who are trained and authorized to inspect and certify homes. Site inspections by independent verifiers help ensure high-quality construction practices, leading to more efficient, healthier, and more durable buildings. Inspections may include verification of correct insulation installation, complete and correct air sealing measures, proper flashing and weather barriers, installation of heating, ventilation, and air conditioning (HVAC) equipment, and resource-efficient construction methods. The most stringent programs require field testing of the house in addition to visual inspections. These inspections and testing provide the builder and homeowner assurance that the home will be as efficient and as healthy as planned. These tests can help identify defects that, if not corrected, would significantly diminish a home's performance.

Certification Levels

Some programs have a single level of certification, although others have multiple levels. LEED for Homes starts at the Certified level and offers Silver, Gold, and Platinum awards to projects that meet increasingly more stringent requirements. The National Green Building Standard levels are Bronze, Silver, Gold, and Emerald. ENERGY STAR does not have different levels; instead, each home must meet the minimum certification requirements to obtain the label.

What Can Be Certified?

All green building programs certify new single-family homes. Many certify multifamily buildings, and some also certify renovations and additions. LEED for Homes only certifies existing homes that are completely gutted from the interior or the exterior so that the insulation and air sealing can be visually verified. The National Green Building Standard allows certification of renovations and additions. Several local programs, including EarthCraft House, Minnesota GreenStar, and Build It Green in California (among others), certify renovation projects. For existing homes that are not undergoing major renovations, the Home Performance with ENERGY STAR program is available to guide owners and contractors to make repairs that improve the performance, durability, and indoor environmental quality.

Certification Programs

Most certification programs use checklists or worksheets that have minimum requirements or prerequisites, awarding points for individual criteria met or products used on the project. The certification level is determined by the numeric score after all the items and the building performance are certified. Some local programs allow self-certification by the builder, but most require a third party to confirm the work is complete and correct.

National Association of Home Builders (NAHB) national trade association representing home builders.

Most certification programs require performance testing of every house in a new development or, at the very least, a sampling of a group of homes to achieve certification. Performance testing typically includes blower door and duct leakage testing (see Chapter 4 for more information on blower door testing and Chapter 13 for additional information about duct leakage testing). The blower door is a large fan installed in a door that depressurizes or pressurizes the entire house, providing a measurement of the amount of air that leaks into or out of the house (Figure 1.7). The duct leakage test uses a smaller fan that pressurizes or depressurizes the ducts and similarly determines the amount of leakage (Figure 1.8). The amount of whole-house and duct leakage is a major factor in determining the efficiency

Figure 1.8 The duct blaster test calculates duct leakage and helps identify duct leaks.

Figure 1.7 The blower door test calculates envelope leakage and helps identify air leakage pathways.

of a house. The leakier the house, the less efficient it is. Both tests are valuable diagnostic tools that help identify potential problems in both new and existing homes.

ENERGY STAR

ENERGY STAR has been revised several times since its initial release. The newest version, ENERGY STAR Version 3, was released in 2010 and is being phased in gradually through 2011; this version will replace the earlier versions in all market-rate homes permitted after January 1, 2012 and in affordable housing beginning January 1, 2013. An interim standard, ENERGY STAR version 2.5, was used in 2011 to phase in the more significant program updates. The U.S. EPA has also created additional certifications beyond energy efficiency, including WaterSense®, Indoor airPLUS, and Advanced Lighting Package. New homes are ENERGY STAR–certified through the Home Energy Rating System (HERS). Individuals known as HERS raters inspect, test, and certify homes as meeting ENERGY STAR standards.

---

blower door a diagnostic tool designed to measure the air tightness of buildings and to identify air leakage locations.

duct leakage test a diagnostic tool designed to measure the air tightness of heating and air conditioning duct systems and to identify air leakage locations.

Home Energy Rating System (HERS) a nationally recognized measurement of a home's energy efficiency.

HERS rater a nationally accredited individual who performs HERS Ratings and evaluates the energy efficiency of homes; sometimes referred to as Home Energy Rater.

ENERGY STAR Certification Process ENERGY STAR homes can be certified through two different routes: the *Performance Path* or the *Prescriptive Path*. The Performance Path uses energy modeling software to calculate the efficiency of the home and to estimate the amount of energy it will require to operate. Similar to a construction estimate, the HERS rating is calculated after all the details of the house are entered into the application. These variables should include walls, floors, windows, equipment, and site orientation, as well as the results of the blower door and duct leakage tests.

Unlike the Performance Path, the Prescriptive Path is not derived from a software model of the home. For certification through this route, the home must contain specific items from a list of construction specifications that include insulation levels, HVAC equipment efficiencies, and window types. The Prescriptive Path also requires that the house meet specific minimum performance through blower door and duct leakage testing. Both paths require an inspection before drywall is installed, with findings recorded on a document called the Thermal Enclosure System Rater Checklist. This inspection is mandated to confirm that the insulation is installed correctly and that air leaks and duct systems are properly sealed (Figure 1.9 and Figure 1.10, page 18). A thermal bypass occurs when heat is allowed to move around or through ("bypass") insulation and is frequently due to missing air barriers or gaps between the air barriers and the insulation.

In addition to the Thermal Enclosure System Rater Checklist, the rater must fill out a separate HVAC inspection checklist, and the HVAC installer must complete a similar inspection checklist. The builder must also complete a Water Management System Checklist. At completion of the project, another inspection that includes performance testing will complete the process, providing the HERS rater with the information they need to certify the home as compliant with ENERGY STAR standards. The Prescriptive Path offers builders a less expensive route to certification; however, it offers less flexibility in choosing which energy efficiency methods to include in a home.

ENERGY STAR certification has served as either the foundation or a model for the core of many green building programs, which may require this certification or an equivalent performance level in their programs. Both the National Green Building Standard and the LEED for Homes program allow builders to choose between their own Prescriptive or Performance Paths to certification. Both programs accept a HERS rating for the Performance Path and provide extra points for exceeding the base level of ENERGY STAR. They also have their own Prescriptive Path requirements with associated minimum requirements and points available for exceeding them. LEED for Homes projects are managed by Green Raters who inspect, collect data, and deliver the information to the USGBC through organizations known as LEED for Homes Providers. When certification is complete, the USGBC delivers the certification documents to the project team. National Green Building Standard projects are verified by individuals who are authorized by the NAHB Research Center to confirm that projects comply with their requirements. Similar to LEED for Homes' Green Raters and Providers, these verifiers inspect, collect data, and report to the NAHB Research Center, which delivers certification documents to the builder. Green Raters and NAHB verifiers may offer performance testing, or they may use independent HERS raters for this work. With the introduction of ENERGY STAR Version 3, most local and national green building programs are reviewing their requirements to determine how to incorporate some or all of the new ENERGY STAR requirements into their programs.

Home Performance with ENERGY STAR Developed by the U.S. EPA specifically for existing homes, Home Performance with ENERGY STAR (HPwES) is best described as a house "tune-up" program. Although HPwES was designed as a national program, each local market must have a program sponsor (Figure 1.11, page 19). Electric and gas utilities are the most common program sponsors, although state energy offices and regional nonprofits may be sponsors. Under HPwES, local energy auditors and contractors are trained to

U.S. Environmental Protection Agency, ENERGY STAR program

Thermal Enclosure System Rater Checklist an inspection of building details for thermal bypasses; for a home to be qualified as ENERGY STAR, a TBC must be completed by a certified HERS Rater.

thermal bypass the movement of heat around or through insulation, frequently due to missing air barriers or gaps between the air barriers and the insulation.

Home Performance with ENERGY STAR (HPwES) a joint program of the U.S. Environmental Protection Agency and the U.S. Department of Energy that sets standards evaluating and improving the energy efficiency of existing homes.

ENERGY STAR Qualified Homes, Version 3 (Rev. 03)
Thermal Enclosure System Rater Checklist

Home Address: ______________________ City: ______________________ State: ________

Inspection Guidelines	Must Correct	Builder Verified[1]	Rater Verified	N/A
1. High-Performance Fenestration				
1.1 *Prescriptive Path:* Fenestration shall meet or exceed ENERGY STAR requirements[2]	☐	☐	☐	☐
1.2 *Performance Path:* Fenestration shall meet or exceed 2009 IECC requirements[2]	☐	☐	☐	☐
2. Quality-Installed Insulation				
2.1 Ceiling, wall, floor, and slab insulation levels shall meet or exceed 2009 IECC levels[3,4,5]	☐	☐	☐	☐
2.2 All ceiling, wall, floor, and slab insulation shall achieve RESNET-defined Grade I installation or, alternatively, Grade II for surfaces with insulated sheathing (see checklist item 4.4.1 for required insulation levels)	☐	☐	☐	☐
3. Fully-Aligned Air Barriers[6]				
At each insulated location noted below, a complete air barrier shall be provided that is fully aligned with the insulation as follows: At interior surface of ceilings in all Climate Zones; also, at interior edge of attic eave in all Climate Zones using a wind baffle that extends to the full height of the insulation. Include a baffle in every bay or a tabbed baffle in each bay with a soffit vent that will also prevent wind washing of insulation in adjacent bays. At exterior surface of walls in all Climate Zones; and also at interior surface of walls for Climate Zones 4-8[7, 8]. At interior surface of floors in all Climate Zones, including supports to ensure permanent contact and blocking at exposed edges[9,10]				
3.1 Walls				
3.1.1 Walls behind showers and tubs	☐	☐	☐	☐
3.1.2 Walls behind fireplaces	☐	☐	☐	☐
3.1.3 Attic knee walls / Sloped attics[11]	☐	☐	☐	☐
3.1.4 Skylight shaft walls	☐	☐	☐	☐
3.1.5 Wall adjoining porch roof	☐	☐	☐	☐
3.1.6 Staircase walls	☐	☐	☐	☐
3.1.7 Double walls	☐	☐	☐	☐
3.1.8 Garage rim / band joist adjoining conditioned space	☐	☐	☐	☐
3.1.9 All other exterior walls	☐	☐	☐	☐
3.2 Floors				
3.2.1 Floor above garage	☐	☐	☐	☐
3.2.2 Cantilevered floor	☐	☐	☐	☐
3.2.3 Floor above unconditioned basement or vented crawlspace	☐	☐	☐	☐
3.3 Ceilings				
3.3.1 Dropped ceiling/soffit below unconditioned attic	☐	☐	☐	☐
3.3.2 Sloped ceilings[11]	☐	☐	☐	☐
3.3.3 All other ceilings	☐	☐	☐	☐
4. Reduced Thermal Bridging				
4.1 For insulated ceilings with attic space above (i.e., non-cathedralized ceilings), uncompressed insulation extends to the inside face of the exterior wall below at the following levels: CZ 1 to 5: ≥ R-21; CZ 6 to 8: ≥ R-30[12]	☐	☐	☐	☐
4.2 For slabs on grade in CZ 4 and higher, 100% of slab edge insulated to ≥ R-5 at the depth specified by the 2009 IECC and aligned with thermal boundary of the walls[4,5]	☐	☐	☐	☐
4.3 Insulation beneath attic platforms (e.g., HVAC platforms, walkways) ≥ R-21 in CZ 1 to 5; ≥ R-30 in CZ 6 to 8	☐	☐	☐	☐
4.4 Reduced thermal bridging at walls (rim / band joists are exempted) using one of the following options:				
4.4.1 Continuous rigid insulation, insulated siding, or combination of the two; ≥ R-3 in Climate Zones 1 to 4, ≥ R-5 in Climate Zones 5 to 8[13,14], **OR**;	☐	☐	☐	☐
4.4.2 Structural Insulated Panels (SIPs), **OR**;	☐	☐	☐	☐
4.4.3 Insulated Concrete Forms (ICFs), **OR**;	☐	☐	☐	☐
4.4.4 Double-wall framing[15], **OR**;	☐	☐	☐	☐
4.4.5 Advanced framing, including all of the items below:				
4.4.5a All corners insulated ≥ R-6 to edge[16], **AND**;	☐	☐	☐	☐
4.4.5b All headers above windows & doors insulated[17], **AND**;	☐	☐	☐	☐
4.4.5c Framing limited at all windows & doors[18], **AND**;	☐	☐	☐	☐
4.4.5d All interior / exterior wall intersections insulated to the same R-value as the rest of the exterior wall[19], **AND**;	☐	☐	☐	☐
4.4.5e Minimum stud spacing of 16" o.c. for 2 x 4 framing in all Climate Zones and, in Climate Zones 5 through 8, 24" o.c. for 2 x 6 framing unless construction documents specify other spacing is structurally required[20]	☐	☐	☐	☐

U.S. Environmental Protection Agency, ENERGY STAR program

Figure 1.9 A HERS Rater must review all items on the Thermal Enclosure System Rater Checklist. Any items not in compliance with the checklist must be corrected prior to achieving the ENERGY STAR.

ENERGY STAR Qualified Homes, Version 3 (Rev. 03) Thermal Enclosure System Rater Checklist

Inspection Guidelines	Must Correct	Builder Verified[1]	Rater Verified	N/A
5. Air Sealing				
5.1 Penetrations to unconditioned space fully sealed with solid blocking or flashing as needed and gaps sealed with caulk or foam				
5.1.1 Duct / flue shaft	☐	☐	☐	☐
5.1.2 Plumbing / piping	☐	☐	☐	☐
5.1.3 Electrical wiring	☐	☐	☐	☐
5.1.4 Bathroom and kitchen exhaust fans	☐	☐	☐	☐
5.1.5 Recessed lighting fixtures adjacent to unconditioned space ICAT labeled and fully gasketed. Also, if in insulated ceiling without attic above, exterior surface of fixture insulated to $\geq$ R-10 in CZ 4 and higher to minimize condensation potential.	☐	☐	☐	☐
5.1.6 Light tubes adjacent to unconditioned space include lens separating unconditioned and conditioned space and are fully gasketed [21]	☐	☐	☐	☐
5.2 Cracks in the building envelope fully sealed				
5.2.1 All sill plates adjacent to conditioned space sealed to foundation or sub-floor with caulk. Foam gasket also placed beneath sill plate if resting atop concrete or masonry and adjacent to conditioned space.	☐	☐	☐	☐
5.2.2 At top of walls adjoining unconditioned spaces, continuous top plates or sealed blocking using caulk, foam, or equivalent material	☐	☐	☐	☐
5.2.3 Sheetrock sealed to top plate at all attic/wall interfaces using caulk, foam, or equivalent material. Either apply sealant directly between sheetrock and top plate or to the seam between the two from the attic above. Construction adhesive shall not be used	☐	☐	☐	☐
5.2.4 Rough opening around windows & exterior doors sealed with caulk or foam[22]	☐	☐	☐	☐
5.2.5 Marriage joints between modular home modules at all exterior boundary conditions fully sealed with gasket and foam	☐	☐	☐	☐
5.2.6 All seams between Structural Insulated Panels (SIPs) foamed and/or taped per manufacturer's instructions	☐	☐	☐	☐
5.2.7 In multi-family buildings, the gap between the drywall shaft wall (i.e. common wall) and the structural framing between units fully sealed at all exterior boundary conditions	☐	☐	☐	☐
5.3 Other Openings				
5.3.1 Doors adjacent to unconditioned space (e.g., attics, garages, basements) or ambient conditions gasketed or made substantially air-tight	☐	☐	☐	☐
5.3.2 Attic access panels and drop-down stairs equipped with a durable $\geq$R-10 insulated cover that is gasketed (i.e., not caulked) to produce continuous air seal when occupant is not accessing the attic [23]	☐	☐	☐	☐
5.3.3 Whole-house fans equipped with a durable $\geq$R-10 insulated cover that is gasketed and either installed on the house side or mechanically operated [23]	☐	☐	☐	☐

Rater Name: ____________ Rater Pre-Drywall Inspection Date: ________ Rater Initials: ______

Rater Name: ____________ Rater Final Inspection Date: ________ Rater Initials: ______

Builder Employee: ____________ Builder Inspection Date: ________ Builder Initials: ______

U.S. Environmental Protection Agency, ENERGY STAR program

Figure 1.9 (Concluded) A HERS Rater must review all items on the Thermal Enclosure System Rater Checklist. Any items not in compliance with the checklist must be corrected prior to achieving the ENERGY STAR.

evaluate and improve existing homes in terms of energy efficiency, durability, moisture control, HVAC systems, and indoor environmental quality. This program includes third-party testing and provides the homeowner with a description of the completed improvements to assure them that their house will perform better. This program is currently available in about 25 different areas and continues to expand into new markets.

Local and regional programs, most of which pre-date LEED and NAHB, have similar requirements for certification; however, not all programs require performance testing, and some allow builders to self-certify projects. The most rigorous programs require third-party verification as well as performance testing.

Green home certification programs will continue to change as the national programs gain market share and new energy codes are introduced. We may see the disappearance of some local and regional certification programs as codes and national programs evolve.

Individual Designations

Individual designations, distinct from building certifications, are available to individuals who are certified as having completed training and testing by several organizations. These programs vary from basic introductions to green building to extensive training and field testing in building performance concepts.

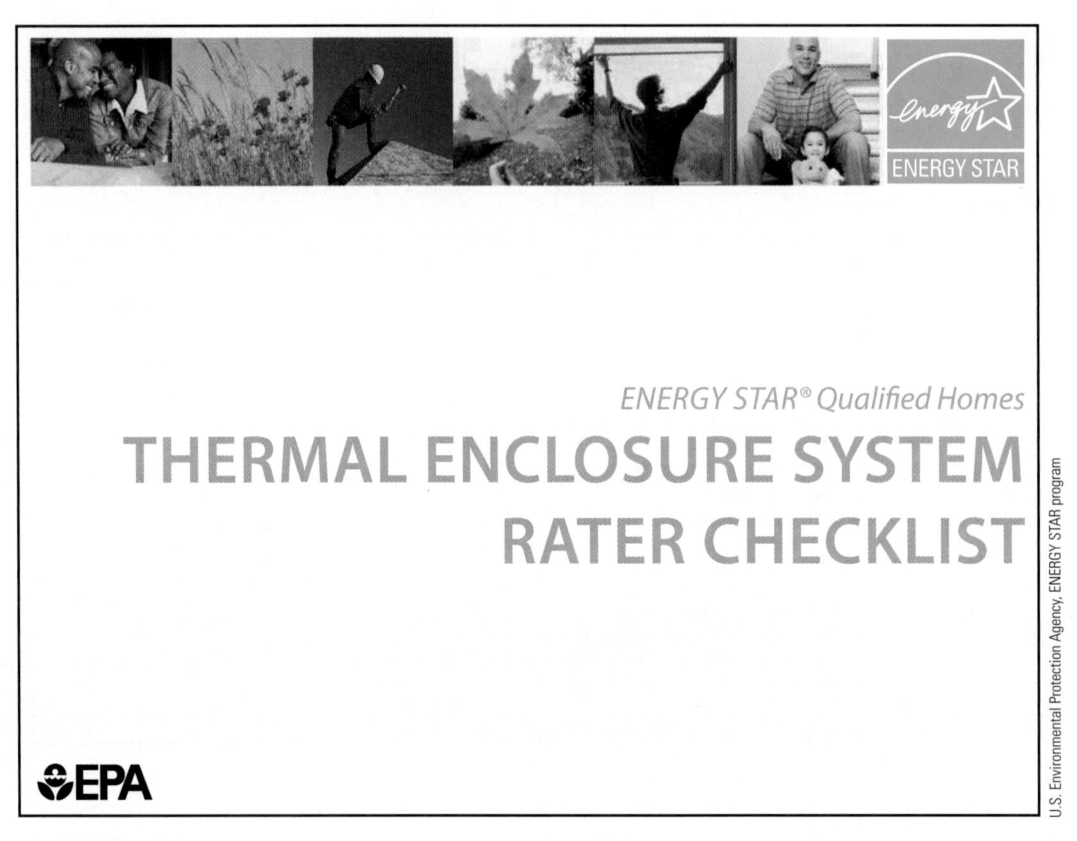

U.S. Environmental Protection Agency, ENERGY STAR program

Figure 1.10 The ENERGY STAR Thermal Enclosure System Rater Checklist must be verified by a HERS Rater. In addition to the checklist, the EPA produces a guide that includes descriptions and references for performing field verifications.

Did You Know?

Utility and Comfort Guarantee Programs

In addition to green building and energy efficiency programs, some builders opt to provide utility and comfort guarantees for their homes. Builders may develop their own programs or work with national programs like Comfort Home and Environments for Living (EFL). North Carolina has a statewide program exclusively for low-income homes called SystemVision. In general, these programs require that a home exceed ENERGY STAR specifications in exchange for guaranteeing certain utility bills and degree of comfort within the home.

ENVIRONMENTS FOR Living®

SystemVision™ by Advanced Energy

Courtesy of Advanced Energy

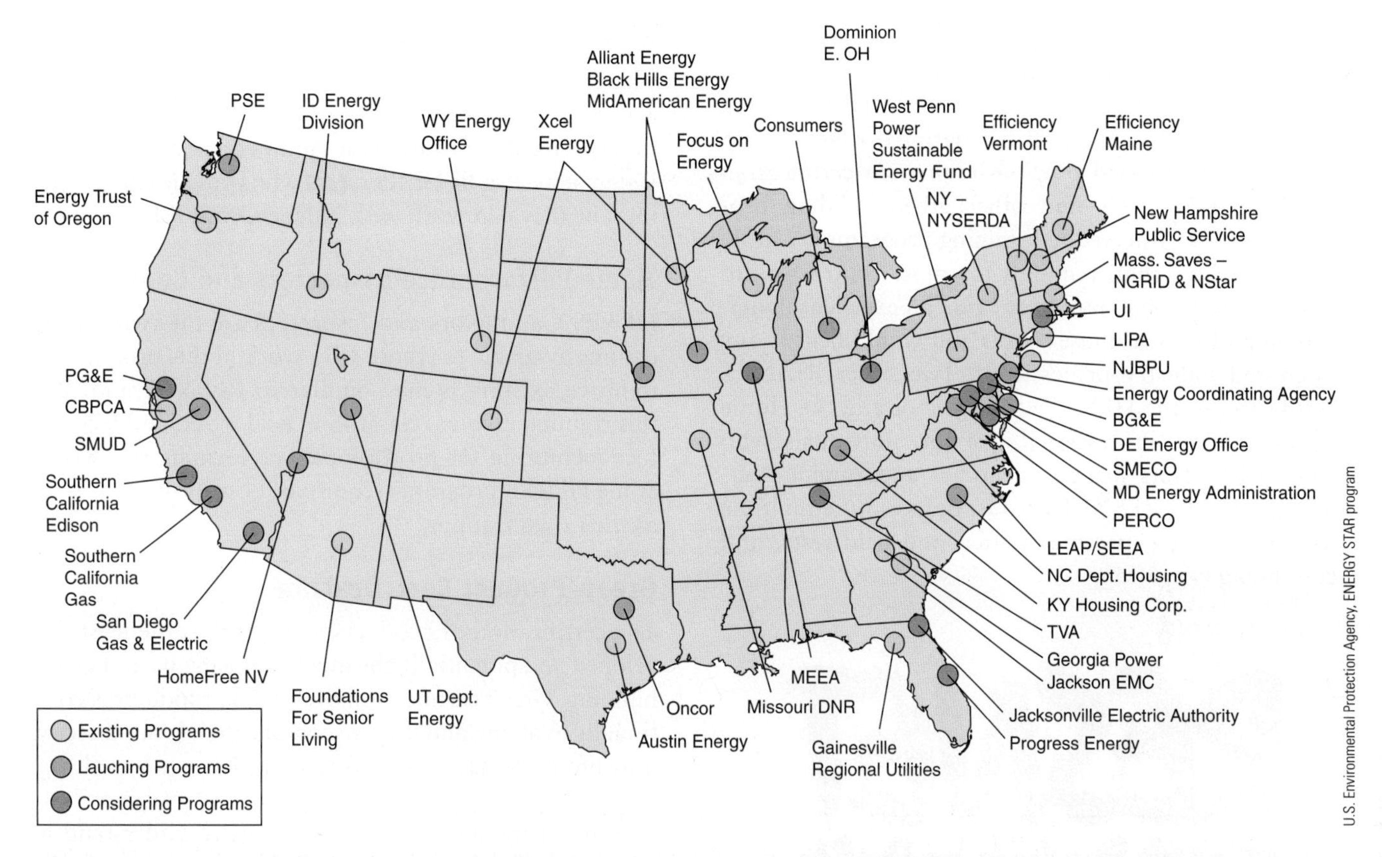

Figure 1.11 Home Performance with ENERGY STAR programs as of October 2010.

Available designations for the residential green market include USGBC's LEED Accredited Professional (AP) Homes, the NAHB Certified Green Professional (CGP) and Master Certified Green Professional (Master CGP), the National Association of the Remodeling Industry (NARI) Green Certified Professional (GCP), and Green Advantage Certification.

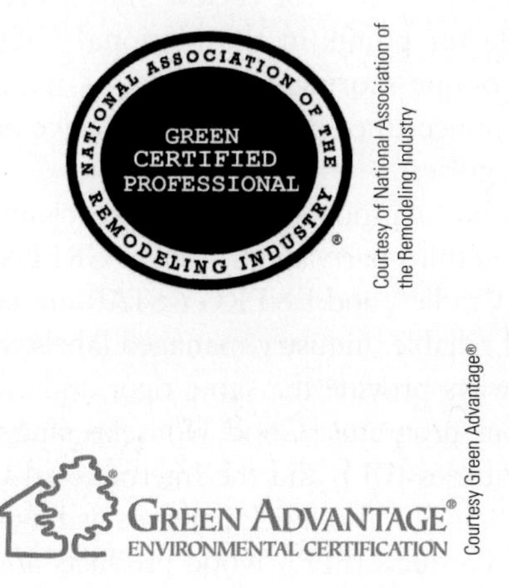

Many local green programs offer their own training and designation programs for builders, remodelers, and other industry professionals. Build It Green in California offers Certified Green Building Professional (CGBP) training to anyone involved in the industry, including contractors, architects, designers, manufacturers, and suppliers. These designations provide an introduction to green building rather than proving that the holder is an accomplished professional. They represent the beginning of an individual's education, and one of their primary benefits is that they provide differentiation in the marketplace.

Training and certifications for advanced contractors and affiliated professionals involved in testing and certifying homes include those available from The Residential Energy Services Network (commonly known as RESNET), The Building Performance Institute (known as BPI), and HPwES.

RESNET

In April 1995, the National Association of State Energy Officials and Energy Rated Homes of America founded the Residential Energy Services Network (RESNET)

LEED Accredited Professional (AP) Homes an individual who has passed a national test and displayed the knowledge necessary to participate in the LEED design and certification process.

Residential Energy Services Network (RESNET) a nonprofit member-based organization that strives to ensure the success of the building energy performance certification industry, set the standards of quality, and increase the opportunity for ownership of high-performance buildings.

to develop a national market for home energy rating systems and energy-efficient mortgages. RESNET developed and continues to maintain the HERS Rating standards and the HERS Rater national exam. HERS Raters, as mentioned previously, provide certification for ENERGY STAR and other green building programs. RESNET accredited training programs are more comprehensive than required for most contractors to build a green home, but do provide an excellent understanding of how a house works and would benefit any interested industry professional. Two other RESNET certifications, the HERS Field Inspector and the Home Energy Survey Professional, are for individuals who work with raters inspecting homes and running performance tests. RESNET certifications require multiple days of training, passing written and field tests, and continuing education.

Courtesy RESNET

Building Performance Institute

The Building Performance Institute (BPI) provides certification that is similar to RESNET, including classroom and field training and testing. They train and certify individuals as Building Analysts or as Envelope, Manufactured Housing, Heating, and Air Conditioning or Heat Pump Professionals.

Courtesy of Building Performance Institute, Inc

Green Raters

The USGBC offers the Green Rater designation. Green raters provide field inspections and collect documentation specifically for LEED for Homes certification. Green Raters may also be HERS raters who provide energy ratings, or they can work with independent HERS raters.

Home Performance Inspectors and Contractors

HPwES Contractors and Inspectors are the two certifications available for those who work in this program to improve existing homes. Similar to RESNET and BPI, this training involves classroom, field work, and face-to-face mentoring for professionals performing this work. Some HPwES programs require BPI certification as well as their own training.

Green Product Certification

The certification of products as "green" is the least developed and potentially the most confusing part of green building. Green certification of building products is continually evolving and is quite complicated. It ties directly into life cycle assessment (LCA), which is an evolving field of study.

One of the most important things to understand is which organizations certify products as green and which don't. *The USGBC and LEED do not certify products.* If you see a manufacturer claiming that their product is "LEED certified," they are misrepresenting themselves. Some products, such as linoleum, low-VOC finishes, or wood certified by the Forest Stewardship Council (FSC), are eligible for points in the LEED program, but individual products are never LEED certified.

NAHB *does* certify products. Manufacturers can submit their products for review and, if accepted, they are eligible for points in the National Green Building Standard online scoring tool. Also, as with the LEED program, noncertified products can receive points in different categories.

Thousands of products on the market claim to be green. Independent third-party labels, such as GREENGUARD℠, Cradle to Cradle®, and ENERGY STAR are generally rigorous and reliable. Industry-managed labels, while useful, do not always provide the same rigor and critical eye as independent programs. Good Housekeeping, Underwriters Laboratories (UL), and the International Code Council (ICC ES SAFE) also label materials as Environmentally Preferable Products (EPP). Wood products are certified as being sustainably harvested by such organizations as the FSC and the Sustainable Forestry Institute (SFI).

Building Performance Institute (BPI) provides nationally recognized training, certification, accreditation, and quality-assurance programs for contractors and business.

environmentally preferable products (EPP) products that have a reduced effect on human health and the environment when compared with traditional products or services that serve the same purpose.

Web resources for green building products include:

BuildingGreen

http://www.buildinggreen.com/

Green Building Advisor

http://www.greenbuildingadvisor.com/

NAHBGreen

http://nahbgreen.org/

LEED Reference Guide

http://www.usgbc.org

ICC SAFE

http://www.iccsafe.org/

GREENGUARD

http://greenguard.org/

Cradle to Cradle

http://www.c2ccertified.com/

ENERGY STAR

http://www.energystar.gov/

Most green building programs have specifications for EPPs, including allowable levels of VOCs in paints and finishes, the amount of added urea formaldehyde in composite wood products, and the percentage of pre- and post-consumer recycled content. Local extraction and manufacturing criteria and other specific guidelines can also help determine appropriate products to use in a green project.

The Case for Green Homes

Green building is no longer a fringe movement, and it makes for good business. There is considerable consumer demand for healthy, high-performance homes. Building professionals who don't pay attention to this change may lose market share to those capable of meeting the demand. Occupants of green homes are typically more satisfied than owners of standard homes. They report lower energy bills; fewer allergy problems from reduced levels of dust, mold, and other irritants; increased comfort; and lower maintenance costs. In addition to these tangible benefits, many gain personal satisfaction from knowing that their house has less impact on the environment.

Builders and remodelers can benefit from going green. Satisfied customers can lead to more business through referrals and a reputation of high-quality construction. Green homes generally have fewer callbacks for HVAC and comfort problems, drywall cracks, paint peeling, and other issues that can be improved through green building techniques. Reduced callbacks increase builder and subcontractor profits. Tax incentives, energy-efficient mortgages, and energy-improvement mortgages can help offset the costs of upgraded construction and certification. Energy modeling software can show expected utility savings of energy upgrades and help explain to clients how upfront investments in efficiency can provide immediate positive cash flow through energy savings.

Energy Mortgages

An energy mortgage is a mortgage that credits a home's energy efficiency savings with respect to the monthly payment amount. This allows the buyer of an energy-efficient new home to purchase a higher-quality home because of the lower operating costs. Energy mortgages recognize that energy-efficient homes cost homeowners less to operate on a monthly basis than standard homes because they use less energy. Home buyers who choose energy-efficient homes can afford a slightly higher mortgage payment because they will have lower energy costs. For existing homes in which energy efficiency can be improved, this concept allows the money saved in monthly utility bills to finance energy improvements. Although some energy mortgages provide lower interest rates and no additional down payment requirements, this is not a universal benefit of all energy mortgage products.

To qualify for one of these loans, a HERS Rating must be performed to document energy efficiency and energy savings. Fannie Mae, Freddie Mac, the U.S. Department of Housing and Urban Affairs' Federal Housing Administration (FHA), and the Veteran's Administration all have energy mortgage programs. There are two types of energy mortgages:

- **Energy improvement mortgages (EIMs)** finance the energy upgrades of an existing home in the mortgage loan by using monthly energy savings.
- **Energy-efficient mortgages (EEMs)** use the energy savings from a new energy-efficient home to increase the home buying power of consumers, capitalizing the energy savings in the appraisal.

---

energy improvement mortgage (EIM) finances the energy upgrades of an existing home in the mortgage loan by using monthly energy savings.

energy-efficient mortgage (EEM) uses the energy savings from a new energy-efficient home to increase the home buying power of consumers, capitalizing the energy savings in the appraisal.

Table 1.6 The Cost Effectiveness of Energy-Efficient Homes

Standard Home		Energy-Efficient Home	
Purchase price	$250,000	Purchase price	$253,000
Monthly PITI	$1,331	Monthly PITI	$1,351
Down payment	$50,000	Down payment	$50,000
Energy improvements	-0-	Energy improvements	$3,000
Loan amount	$200,000	Loan amount	$203,000
Interest	7%	Interest	7%
Monthly energy savings	–	Monthly energy savings	$60
Average energy bill	$120	Average energy bill	$60
Total Expenses	**$1,451**	**Total Expenses**	**$1,411**

The total cost of ownership is used to calculate the cost-effectiveness of energy-efficient upgrades. Here, a $250,000 home is used as the base case, but the same assessment can be performed on homes at any price point.

Simple Payback vs. Cash Flow Analysis

Table 1.6 illustrates the cost-effectiveness of energy efficiency upgrades. The table compares a home built to standard code with one that is energy-efficient. Although the cost of making a home energy-efficient doesn't need to be high, there often is a small premium. In this example, we're assuming the energy-efficient home costs approximately 1.5% more. The two homes are new construction, but only one incorporates energy-conserving details throughout the construction process. For an added $3,000, a builder can provide air sealing, properly installed insulation, tightly sealed and properly designed ductwork, additional attic insulation, and more efficient windows. Since the home is more expensive, the monthly loan amount is higher than that for the standard home; however, the home is more efficient and therefore the actual cost of ownership (monthly principal, taxes, and insurance [PITI] plus utility bills) has decreased.

The traditional measure of housing expenses, PITI, is really just the beginning. Utilities, maintenance, and repairs can cost additional thousands of dollars every year. Green buildings reduce these operating and ownership expenses. In this example, the energy-efficient home actually costs $40 less per month to own and operate. For this home, what is the payback on the $3,000 energy efficiency upgrades?

Payback

The payback of energy efficiency upgrades is traditionally calculated through simple payback. Simple payback is the cost of upgrade divided by energy savings. For example, Table 1.6 lists the upgrades cost as $3,000 and the energy savings as $60 a month. Therefore, divide the upgrade costs by the monthly energy savings, as shown, to determine the simple payback.

Upgrade costs: $3,000

Monthly energy savings: $60

Payback = $3,000/60 = 50 months or 4.16 years

Simple payback works well for cash purchases that are not financed over time. Simple payback is often used for considering smaller energy efficiency upgrades. These upgrades may include installing compact fluorescent lamps (CFLs) or ENERGY STAR appliances, or by adding insulation and air sealing. As most homeowners do not pay cash for their homes, simple payback is a misleading metric. If the down payment remains unchanged (which it will with many EEMs), the cost-effectiveness of energy-efficient homes should be evaluated on the basis of cash flow analysis.

Cash Flow Analysis

Cash flow analysis compares the reduced cost of ownership from utility savings to the increase in PITI. In the previous example, the standard home's monthly cost of ownership is $1,451, compared with $1,411 for the energy-efficient home. Despite the slightly higher purchase price, the energy-efficient home is the cheaper home to own and operate. Performing a *cash flow analysis* illustrates this:

Standard home total monthly costs: $1,451

Energy-efficient home total monthly costs: $1,411

$1,451 – $1,411 = $40/month savings

To look at this example in another way, the homeowner spends $20 (increase in monthly PITI) to save $60 (reduced utility bills). The investment is making the homeowner money from day one.

simple payback the amount of time it takes to recover the initial investment for energy-efficiency improvements through energy savings, dividing initial installed cost by the annual energy cost savings.

Summary

Our goal in writing this book is to help you understand the benefits of green home building and remodeling for owners, contractors, and the general public as a whole. Understanding the visibly tangible, short-term benefits as well as the less tangible, longer term benefits will help you make the right decisions as an industry professional as you seek to make your work more sustainable. In the process, you can help improve your business operations and profitability, broaden your reputation for quality construction through increased customer referrals, and help reduce the negative impact our buildings have on the environment.

Review Questions

1. Residential buildings consume approximately how much of the energy produced in the United States?
   a. 15%
   b. 21%
   c. 25%
   d. 31%
2. What is the purpose of the blower door test?
   a. To calculate duct leakage
   b. To ensure structural integrity
   c. To calculate envelope leakage
   d. To measure VOC content of home
3. Which of the following is a renewable energy source?
   a. Clean coal
   b. Natural gas
   c. Garbage incineration
   d. Solar photovoltaics
4. What is ENERGY STAR?
   a. National green building program
   b. USGBC's label for energy-efficient appliances
   c. Southeast regional green building program
   d. Federal program for appliances and homes
5. A contractor must possess which credential in order to certify a home as ENERGY STAR compliant?
   a. HERS Rater
   b. NARI Green Certified Professional
   c. LEED AP HOMES
   d. NAHB Certified Green Professional
6. What is simple payback?
   a. Cost of upgrade divided by energy savings
   b. Energy savings divided by cost of upgrade
   c. When the amount of a government incentive equals or exceeds the cost of energy improvements
   d. Cost of the upgrade multiplied by the energy savings
7. Which of the following is a feature of all energy-efficient mortgages?
   a. Utility savings added to borrowers income
   b. No additional down payment
   c. Lower interest rate
   d. Additional money for larger homes
8. Which of the following is required for a home to be certified green?
   a. Straw bale walls
   b. Solar photovoltaic system
   c. Energy efficient
   d. Geothermal heat pump
9. Which of the following is a national green building program?
   a. ENERGY STAR
   b. EarthCraft House
   c. Built Green Colorado
   d. LEED for Homes
10. Which of the following building materials has the least embodied energy?
    a. Fiberglass batt
    b. Structural steel
    c. Gypsum board
    d. Kiln-dried, treated timber

Critical Thinking Questions

1. What is a green home?
2. Why are green buildings important?
3. Do green buildings cost more?
4. What are the characteristics of a green building material?
5. What is the qualification process for an ENERGY STAR home?

Key Terms

blower door, 14
Building Performance Institute (BPI), 20
Department of Energy (DOE), 4
duct leakage test, 14
embodied energy, 5
energy efficient mortgage (EEM), 21
energy improvement mortgage (EIM), 21
ENERGY STAR, 10
environmentally preferable products (EPP), 20
green building, 3
greenhouse gas (GHG), 4
heat island effect, 7
Home Energy Rating System (HERS), 14
HERS rater, 14
Home Performance with ENERGY STAR (HPwES), 15
Leadership in Energy and Environmental Design (LEED), 10
LEED Accredited Professional (AP) Homes, 19
life cycle assessment (LCA), 5
National Association of Home Builders (NAHB), 13
new urbanism, 8
renewable energy, 4
Residential Energy Services Network (RESNET), 19
simple payback, 22
sustainable, 8
thermal bypass, 15
Thermal Enclosure System Rater Checklist, 15
traditional neighborhood development (TND), 8
transit-orientated development (TOD), 8
U.S. Green Building Council (USGBC), 10
volatile organic compounds (VOC), 4

Additional Resources

Built Green Colorado: http://www.builtgreen.org/

Built Green Washington: http://www.builtgreenwashington.org/

EarthCraft House: http://www.earthcrafthouse.com/

ENERGY STAR: http://www.energystar.com

NAHB Green Building Program: http://www.nahbgreen.org/

Rashkin, Sam, Retooling the U.S. Housing Industry: How it Got Here, Why It's Broken, and How to Fix It. Delmar Cengage Learning, 2010.

RESNET: http://www.resnet.us/

USGBC: http://www.usgbc.org/

2

The House as a System

Homes are complex systems with numerous interconnected components. This chapter presents the importance of "systems thinking" and the science behind how buildings operate. This area of study, building science, focuses on heat and moisture transfer and their effects on energy efficiency, comfort and durability. Ventilation and building material selection are also covered. Later chapters provide specific strategies and construction techniques for moisture, heat, and ventilation control. The core of green building is accurately applied building science and well-executed management of the house as an integrated system. Green building professionals pay close attention to building science principles in their design and specifications, assuring that these buildings are efficient, healthy, and durable.

LEARNING OBJECTIVES

Upon completion of this chapter the student should be able to:

- Describe the core components of building science
- Define the building envelope
- Explain the difference between R-value and U-value
- Calculate the average R-value for a building assembly
- Describe the methods of heat transfer
- Calculate the conductive heat transfer through a building assembly
- Describe the methods of moisture transfer

Green Building Principles

Energy Efficiency

Resource Efficiency

Durability

Indoor Environmental Quality

Building Science

Building science is the study of the interaction of building systems and components, occupants, and the surrounding environment, focusing on the flows of heat, air, and moisture. Building science is a broad area of study that also includes structural systems, life safety, lighting, and acoustics. The application of building science is one of the key components of green building. Of particular importance are the ways in which the management of heat, air, and moisture affect the efficiency and durability of the structure and the comfort and health of the occupants. To create a green home, you must first understand how heat, air, and moisture interact within the interior and exterior of a building, its subsystems, and its occupants. In this chapter, we will focus on thermodynamics (heat flow) and hydrodynamics (moisture flow).

building science the study of the interaction of building systems and components, occupants, and the surrounding environment; focuses on the flows of heat, air, and moisture.

FROM EXPERIENCE

House as a System Approach to Construction and Retrofit

Laura Capps, Director of Residential Green Building Services, Southface Energy Institute, Atlanta, Georgia. Ever wonder why one room in your house is always too hot or cold? Or why that ceiling water spot keeps showing up every summer, even though you cannot find a leak? Maybe you have moldy shoes in the closet? Or, perhaps your two-year-old has a persistent cough, but your doctor cannot find anything wrong with her. These are all common symptoms of a failing home, but fortunately, all can be prevented through the *House as a System* approach to construction and retrofit.

Courtesy of Southface/EarthCraft House

As the director of Residential Green Building Services for Southface Energy Institute, Laura works on the development and delivery of several regional and national green building programs. With a focus on education, Laura collaborates with builders, contractors, and verifiers on residential market transformation.

Similar to a fishbowl, your home works as a system. Each of its components is connected to and impacted by all of its other components. When you fill the fishbowl with water, you must also check its pH, temperature, salt content, filtration system, and variety of other factors to ensure the longevity and health of the fish you have chosen. If any of those variables are overlooked, you are likely to end up with sick fish, algae growth, and other problems. The same is true for your home. As you design and build it, you are wise to take into consideration the home's location, layout, materials, construction plans, and, most importantly, the well-being of its occupants.

The *House as a System* approach to design and construction provides contractors and future homeowners with a comprehensive, building science–based methodology for creating healthy, comfortable, and durable homes. While building scientists strive to optimize a home's design and performance, their methodologies also provide the necessary foundation for building "green." In fact, a home can be constructed from all the green products in the world, but if those products are incorrectly installed or work at cross-purposes, or if the home leaks conditioned air or wastes water, its occupants will pay the price in uncomfortable surroundings, high utility bills, and possibly even sickness due to poor indoor air quality.

Let us take a building scientist's approach to our home with the room that is too hot or too cold. A scientist would diagnose one of a few possible causes: The home could be missing insulation, its insulation may have been incorrectly installed, or possibly too much framing material was used during construction. Alternative causes may be that the room suffers from poorly distributed or undersized heating and cooling equipment.

A summertime water spot may be coming from condensation on the cold water pipes that run through your attic. Ventilated attics harbor warm, humid summer air that can cause condensation on cold water pipes—condensation that then drips onto the drywall below. Moldy spots in closets may be due to poor air circulation or to over-framed corners that may have been left uninsulated and now allow warm moist air to condense. Chronic coughs can arise from poor indoor air quality and may be attributed to a wide range of problems, including poorly vented combustion appliances, high indoor humidity levels that are hospitable to dust mites and mold, or even the off-gassing of construction materials that can irritate occupants who have chemical sensitivities.

So while green homes are often thought to be built around expensive and elaborate new technologies, such as solar panels and green roofs, a green home really begins with the design you choose! An energy-, water- and resource-efficient design also takes into account the building site, the home's orientation, and the durability of its materials and construction. Indoor air quality should also be an important consideration. Regardless of the design, the home's future occupants should be educated on how to get maximum benefit from its construction. Building science underpins each of these decisions, sets the stage for green building, and provides a solid *House as a System* approach to construction and retrofit.

Principles of Energy

Buildings use energy for heating air and water, cooling and dehumidifying air, lighting, and operating appliances. Much of this energy is lost or wasted through inefficiencies. Energy is a measurable quantity of heat, work, or light. *Potential energy* is stored energy, like a gallon of gasoline or a ton of coal. *Kinetic energy* is transitional energy, like a flame.

Energy is measured in many different units, which can be easily converted. Common units of measurements are calories, joules, Btu, and kWh. A *calorie* is the amount of energy it takes to raise the temperature of 1 gram of water by 1°C (1.8°F). The calorie has largely been replaced by the *joule*, which is an *International System of Units* (*SI* or *metric*) unit. One calorie is equal to 4.184 joules. Although nearly the entire world uses SI units, the United States still uses the English customary system (also called the American system or, sometimes, "English units").

Heat can be measured in British thermal units (Btu), which is the amount of heat required to raise a pound of water 1°F. For example, when 1 pound of water (approximately 1 pint) is heated from 68°F to 69°F, 1 Btu of heat energy is absorbed into the water (see Figure 2.1). This is roughly equivalent to the amount of heat provided by one kitchen match. In the United States, the most common form of measurement for heat energy is the Btu and for electricity is kWh.

Laws of Thermodynamics

The *first law* of thermodynamics states that energy is neither created nor destroyed. Energy simply moves from place to place and changes form. The *second law* of thermodynamics states that heat moves from high temperature regions to low temperature regions.

Temperature

The term *temperature* describes how hot or cold something is. Outdoor temperatures are constantly changing with the weather and seasons. Geographic regions can be characterized by the amount of heating and cooling required for a home. Estimates of how much heating or cooling is needed for any region can be determined by the number of heating degree-days or cooling degree-days.

Heating degree days (HDD) are defined relative to a base temperature—the outside temperature above which a building needs no heating. HDD can be seen as the average difference between inside and outside over the course of a day. This temperature is also known as the *heating balance point* and generally assumed to be 65°F. For example, if the average outdoor temperature is 15°F for January 1, then the HDD is 65°F − 15°F = 50 degree days for that day.

Cooling degree days (CDD) are calculated similarly by determining the difference on days that are warmer than the *cooling balance point* of 65°F, and therefore requiring more cooling within buildings. HDD and CDD provide a simple way to determine how much heating a home in a particular location needs over a specific time period (e.g., day, month, or year). By using the HDD or CDD in conjunction with average insulation values, an estimate of the annual amount of energy required to heat and cool a building can be calculated (see page 31 for example).

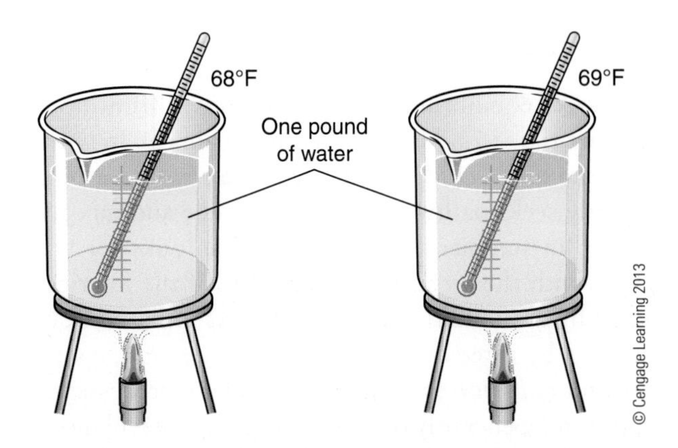

Figure 2.1 One British thermal unit (Btu) of heat energy is required to raise the temperature of 1 pound of water from 68°F to 69°F.

> Sources for local HDD and CDD figures can be found on several websites, including:
>
> Degree Days.net
>
> http://www.degreedays.net/
>
> National Weather Service Climate Prediction Center
>
> http://www.cpc.noaa.gov/products/analysis_monitoring/cdus/degree_days/

---

energy a measurable quantity of heat, work, or light.

British thermal units (Btu) the amount of heat required to raise a pound of water 1°F.

heating degree-day (HDD) a measure of how cold a location is over a period of time relative to a base temperature, most commonly specified as 65°F.

cooling degree-day (CDD) a measure of how warm a location is over a period of time relative to a base temperature, most commonly specified as 65°F.

Heat Flow

Heat seeks equilibrium and always moves from hot to cold (per the second law of thermodynamics). In winter, when the air is cold outside, warm air inside a house naturally seeks pathways to move outside and mix with cold air. In summer, the hot air outside looks for ways to mix with the air inside, particularly if the interior is cooled by air conditioning. Heat moves into and out of buildings by three different routes: *conduction*, *convection*, and *radiation* (see Figure 2.2). These processes can happen independently or simultaneously. We will discuss each of these routes in the coming sections.

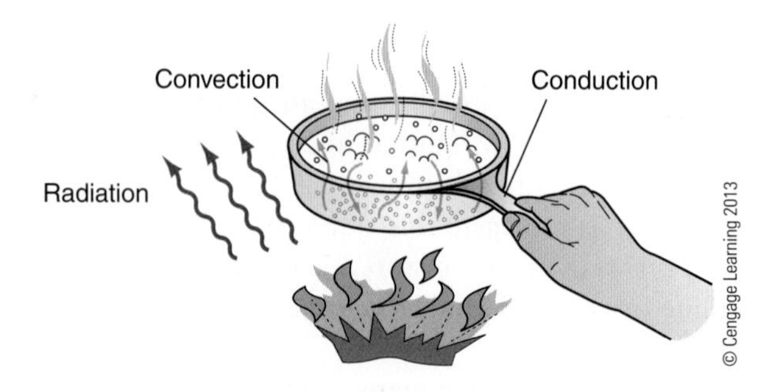

Figure 2.2 The different types of heat transfer can occur independently or simultaneously. Here all three forms of heat transfer are occurring.

Conduction

Conduction is the movement of heat through a solid. Materials conduct heat at different rates. For instance, metal is a very good conductor of heat. Glass is also good conductor, wood and plastics are poor conductors, and insulation is a very poor conductor. If you pick up a hot pan with a metal handle, it can burn your hand. A glass of boiling water is hot to the touch. A metal ladle in a pot of hot soup will get hot, but a wood or plastic handle will stay cool enough to serve the soup. On the other hand, a thin polystyrene foam cup of steaming coffee stays cool to the touch because the foam cup is a very poor conductor of heat (see Figure 2.3).

One measurement of conduction is the measure of a building material's resistance to heat flow, usually called its **R-value.** Perhaps the most common product to be labeled with an R-value is insulation, which you expect to be rated higher than other more conductive materials. Practically everything, however, has an R-value, as shown in Table 2.1. The R-value states how much heat transmits through 1 square foot of a surface (e.g., a wall, floor, or ceiling) in one hour with a 1°F difference between opposite surfaces. R-value is measured in Btu per hour per degree Fahrenheit (°F) per square foot. You may see this written as ft^2°F hr/Btu or Btu/hr ft^2 °F.

R-values are calculated for a given thickness of material. For example, a fiberglass batt has an R-value of 3.5 per inch. Thus, doubling the insulation thickness will double its R-value. The R-value of materials in series or stacked against each other in the direction of heat flow can be added together. For example, an attic that has an R-13 fiberglass batt covered with R-19 of blown cellulose has a nominal R-value of R-32 (19 + 13 = 32) (Figure 2.4). This calculation does not take into account ceiling joists and other obstructions that displace insulation, or the quality of the installation.

When materials are installed in a series within a building assembly, the R-values are added together to determine the total R-value of a wall, floor, or ceiling component. Figure 2.5 shows a 2 × 4 wood-framed wall that is insulated with R-13 fiberglass batts and has 4" brick cladding. Using Table 2.1, we can calculate the R-value for the wall cavity and the stud. The cavity is R-0.44 + R-0.7 + R-13 + R-0.45 = R-14.59. The total R-value for

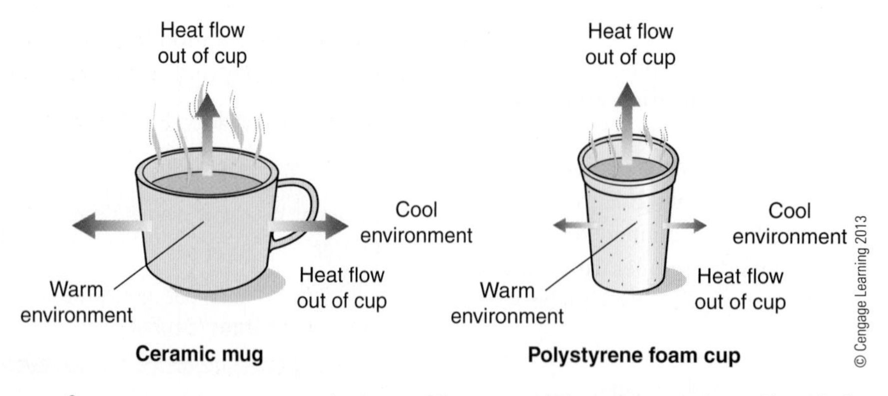

Figure 2.3 Polystyrene foam cups are poor conductors of heat, enabling them to keep liquids hot or cold.

---

conduction the transfer of heat from one substance to another by direct contact.

R-value quantitative measure of resistance to heat flow or conductivity, the reciprocal of U-value.

Table 2.1 R-Values of Common Building Materials

Material	R/Inch	R/Thickness
Insulation Materials		
Fiberglass, batt	3.14–4.30	
Fiberglass, blown (attic)	2.20–4.30	
Cellulose, blown (attic)	3.13	
Cellulose, blown (wall)	3.70	
Concrete, autoclaved aerated	1.05	
Polystyrene, expanded (beadboard)	4.00	
Polystyrene, extruded	5.00	
Polyurethane, open cell (foamed-in-place)	3.4–3.8	
Polyurethane, closed cell (foamed-in-place)	6.25	
Polyisocyanurate (foil-faced)	7.20	
Construction materials		
Concrete block, 4"		0.80
Concrete block, 12"		1.28
Brick, common 4"		0.80
Concrete, poured	0.08	
Lumber, soft wood	1.25	
2 × 4 (3 1/2")		4.38
Sheathing materials		
Plywood	1.25	
1/4"		0.31
5/8"		0.77
Fiberboard	2.64	
1/2"		1.32
25/32"		2.06
Oriented strand board (OSB)	1.6	
7/16"		0.70
Siding materials		
Hardboard, 1/2"		0.34
Brick, 4"		0.44
Interior Finish Materials		
Gypsum Board (drywall 1/2")		0.45
Windows		
Single-pane glass		0.91
Double-pane insulating glass (3/16" air space)		1.61
Double-pane insulating glass (with suspended film and low-emissivity [low-E])		4.05
Doors		
Wood, hollow-core flush (1 3/4")		2.17

Note: *See Appendix page 483 for a more extensive list.*

the stud is R-0.44 + R-0.7 + R-4.4 + R-0.45 = R-5.99. Later in this chapter, we discuss how to calculate the overall R-value for the full surface of this wall.

Unlike most building materials, windows are labeled with a U-value instead of an R-value. The **U-value** is the thermal transmission or conductance rather than the resistance of a material. Mathematically, the U-value is the inverse of the R-value. U-values are typically used for building assemblies, such as window or door units and wall assemblies. The U-values of windows, for example, take into account the whole window assembly (i.e., the frame, glazing, air space, and any films or coatings).

For example, an R-13 fiberglass batt has a U-value of 1/13 or 0.077.

$$\text{U-value} = 1/\text{R-value}$$
$$\text{R-value} = 1/\text{U-value}$$

Unlike R-values, U-values cannot be added together. To calculate the overall U-value of an assembly, you must first calculate the R-values of each component and then convert to U-values. Once the components' U-values are calculated, you can calculate the average U-value of the assembly. The average U-value is written as $U_{average}$.

The $U_{average}$ is calculated by adding the UA (U-value multiplied by area) of each component and dividing by the total area:

$$U_{average} = [(U_1A_1) + (U_2A_2)]/A_{total}$$
$$U_{average} = \text{area-weighted average U-value}$$
$$U = 1/\text{R-value}$$
$$A = \text{area (ft}^2\text{)}$$

This equation is represented in Table 2.2a, Table 2.2b, and Table 2.2c.

To calculate the $U_{average}$ for this wall section, we must first break it up into two areas (Figure 2.6). For the purposes of computation, it does not matter which component is entered first or second into the equation. Table 2.3a through Table 2.3c shows an example of how to calculate the R-value and U-value averages.

Calculating Conductive Heat Flow Heat flow in terms of conduction only (that is, through a solid object) is calculated by using the following equation:

$$Q = U \times A \times \Delta T$$

for which

Q = heat flow (Btu/hr)
U = 1/R-value
A = area (square feet)
ΔT = temperature difference across component (°F)

U-value thermal transmittance or thermal conductance of a material; the reciprocal of R-value.

Figure 2.4 Combinations of insulation types are commonly found in homes, especially in older attics. Here, the attic contains an R-13 fiberglass batt covered with R-19 blown cellulose. The total nominal R-value of R-32.

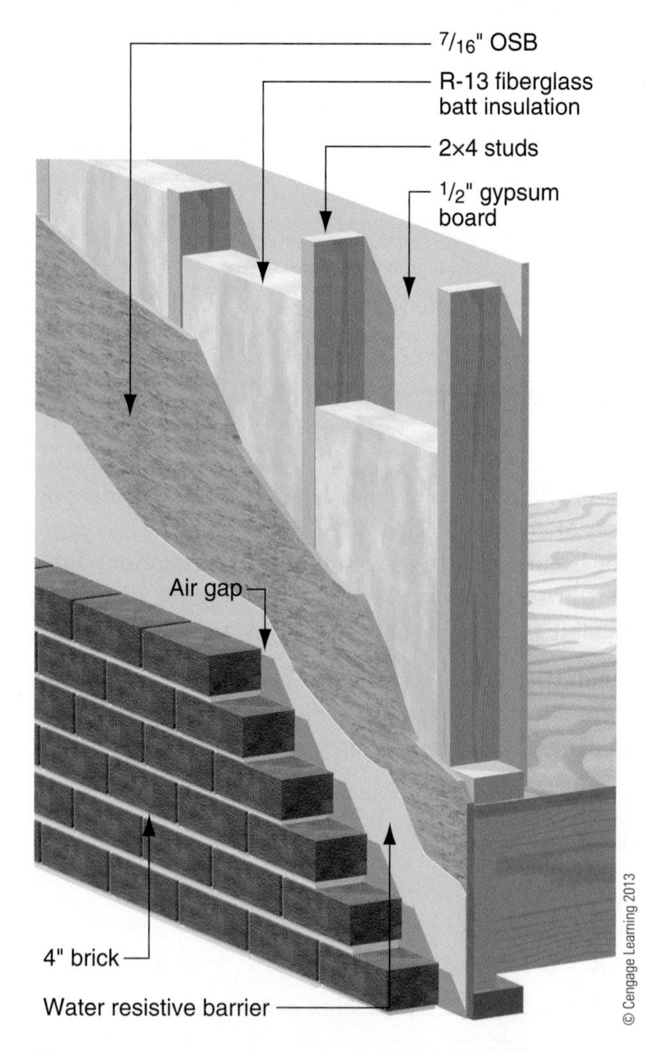

Figure 2.5 The R-value of materials in series can be added together. Here, you see the R-value of individual materials and the total wall.

Table 2.2a R-Value/U-Value Average Worksheet

Component Description	R-Value	U-Value (1 ÷ R-Value)	Area (ft²)	UA (U-Value × Area)
			TOTAL AREA =	TOTAL UA =

Table 2.2b Calculating Weighted-Average R-Value

Total Area	÷	Total UA	=	Weighted-Average R-Value

Table 2.2c Calculating Weighted-Average U-Value

Total UA	÷	Total Area	=	Weighted-Average U-Value

Table 2.3a R-Value/U-Value Average Worksheet Example

Component Description	R-Value	U-Value (1 ÷ R-Value)	Area (ft²)	UA (U-Value × Area)
Wall	13	0.077	80.45	6.19
Window	1.79	0.56	15.55	8.71
			TOTAL AREA = 96	TOTAL UA = 14.90

Table 2.3b Calculating Weighted-Average R-Value Example

96		14.90		6.44
Total Area	÷	Total UA	=	Weighted-Average R-Value

Table 2.3c Calculating Weighted-Average U-Value Example

14.90		96		0.155
Total UA	÷	Total Area	=	Weighted-Average U-Value

Now, let us try an example. Calculate the conductive heat flow through the following walls.

Wall #1

Review Figure 2.7. The wall measures 8′ × 12′ and is insulated with R-13. For now, treat the wall as having continuous insulation (no studs). The temperature inside the home is 72°F, and the outside air temperature is 89°F.

$Q = U \times A \times \Delta T$

$U = 1/13 = 0.077$

$A = (8' \times 12') = 96\ ft^2$

$\Delta T = 89 - 72 = 17$

$Q = 0.077 \times 96\ ft^2 \times 17 = 125.66\ Btu/hr$

Wall #2

Take a look at Figure 2.8. The wall is identical to the previous one, except it now contains a window. For now, treat the wall as having continuous insulation (no studs). Break the wall into a wall section and a window section. Calculate the heat flow through both and add for the total.

$\mathbf{Q_{Wall} = 0.077 \times 80.4\ ft^2 \times 17}$

$\mathbf{= 105.24\ Btu/hr}$

$U = 1/13 = 0.077$

$A = (8' \times 12') - (3.33' \times 4.67') = 80.4\ ft^2$

$\Delta T = 89 - 72 = 17$

$\mathbf{Q_{Window} = 0.56 \times 15.6\ ft^2 \times 17}$

$\mathbf{= 148.5\ Btu/hr}$

$U = 0.56$

$A = (3.33' \times 4.67') = 15.6\ ft^2$

$\Delta T = 89 - 72 = 17$

$Q_{Wall} = 0.077 \times 80.4\ ft^2 \times 17$

$= 105.24\ Btu/hr$

$Q_{Window} = 0.56 \times 15.6\ ft^2 \times 17$

$= 148.5\ Btu/hr$

$Q_{Total} = 105.24\ Btu/hr + 148.5\ Btu/hr$

$= 254\ Btu/hr$

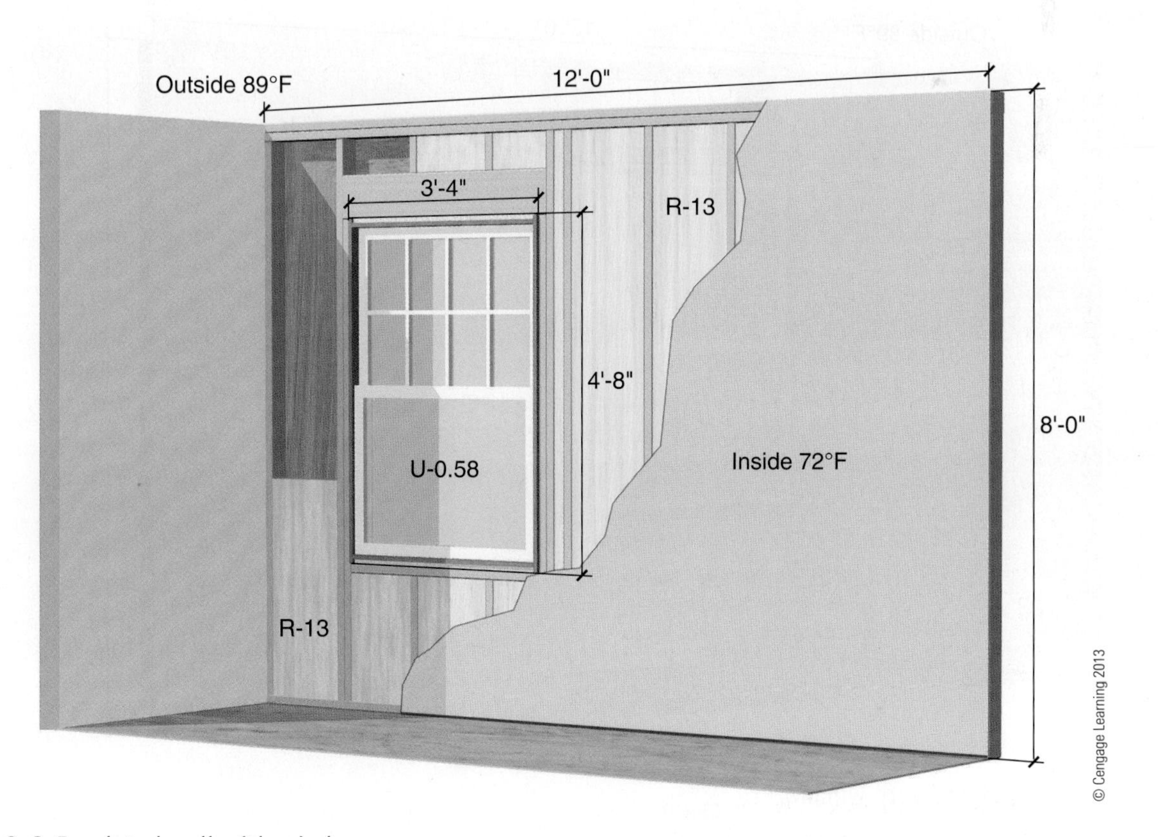

Figure 2.6 Insulated wall with window.

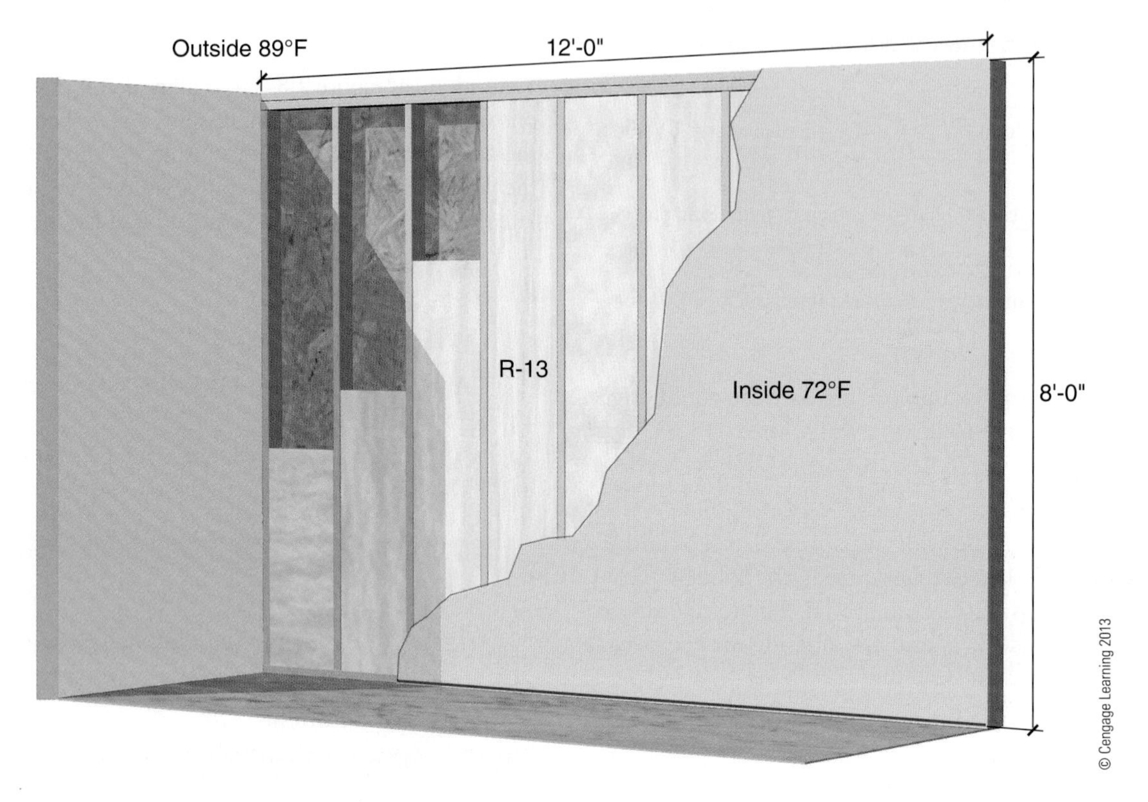

Figure 2.7 Insulated wall.

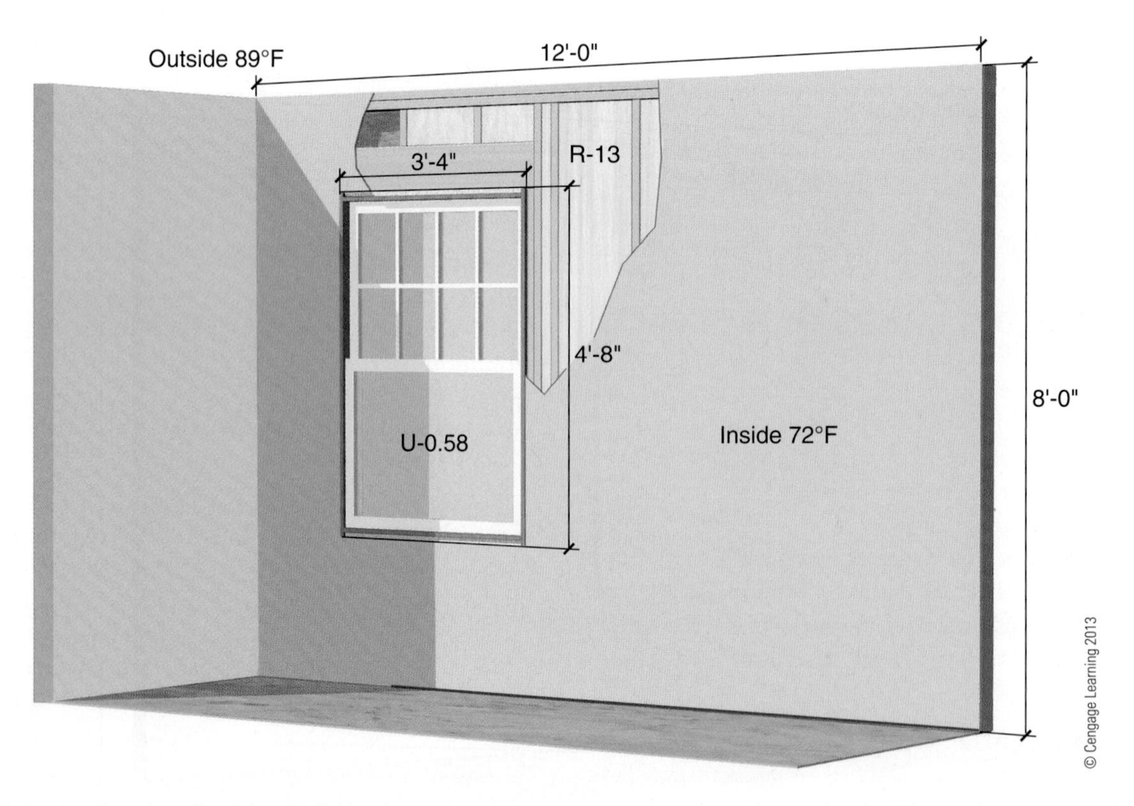

Figure 2.8 Insulated wall with window.

Calculating Seasonal Heat Flow As discussed previously on page 27, a home's seasonal heating losses can be calculated using the HDD and the $U_{average}$.

$$\text{Heat flow} = \text{U-value} \times \text{Area} \times \text{HDD} \times \text{Change days to hours}$$

$$Q = U \times A \times HDD \times 24$$

Q = heat flow (Btu/hr)

U = 1/R-value

A = area (square feet)

HDD = heating degree days for the specific location

Using Wall #1 from the previous example, what is the seasonal heat flow? The wall is in a location with 4,350 HDDs.

$$Q = U \times A \times HDD \times 24$$

$$U = 1/13 = 0.077$$

$$A = (8' \times 12') = 96 \text{ ft}^2$$

$$HDD = 4350$$

$$Q = 0.077 \times 96 \times 4350 \times 24$$

$$= 771{,}724.80 \text{ Btu/h}$$

Convection

Convection is the transfer of heat through a fluid (liquid or gas). In buildings, convection usually refers to heat moving through the air. Just like in conduction, heat in the air moves from hot to cold seeking equilibrium. To illustrate, a forced-air gas furnace is an example of convection. House air is drawn into the return air of the furnace by the fan. This air is then forced out the fan outlet and over the furnace heat exchanger, which exchanges heat to the air from a gas flame. The air is then forced into the ductwork and distributed throughout the home (see Figure 2.9).

Air leaks in a house allow heat to move in or out (depending on the interior and exterior temperatures), increasing the amount of energy needed to heat or cool the home. Air leaking in and out of homes is referred to as infiltration and exfiltration, respectively (Figure 2.10). Sealing these air leaks reduces heat movement through convection, helping to improve a building's energy efficiency. Convection and air movement can occur in relatively small spaces. Even within walls, uninsulated air spaces can contain convective loops (Figure 2.11a). Such loops can also occur in air-permeable attic insulation (Figure 2.11b). Convective loops occur when air (or another medium) continuously circulates around in an enclosed space as it is heated and cooled.

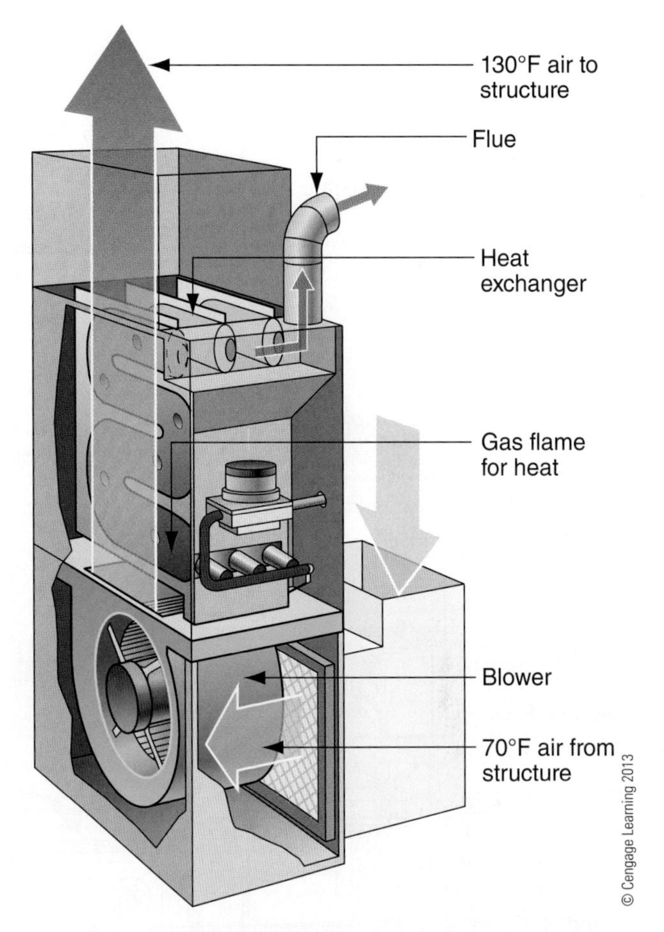

Figure 2.9 Air from the room enters the fan at 70°F. The fan forces the air across the hot heat exchanger and out into the structure at 130°F.

Radiation

Radiation is the transfer of heat from one surface to another via electromagnetic waves. The most common example of radiant heat is direct sunlight. Sun shining on our skin warms us as electromagnetic radiation from the sun strikes our skin, causing the molecules in the skin to accelerate. Solar energy (light from the sun) hitting a roof will heat up a structure, and sunlight shining through windows heats up the air and materials inside our homes. All the solar energy that hits an object is either *reflected*, *absorbed*, or *transmitted*, depending on its particular characteristics (Figure 2.12). Most materials absorb solar energy,

---

convection the transfer of heat through a fluid (liquid or gas).

infiltration the uncontrolled process by which air or water flows through the building envelope into the home.

exfiltration air flow outward through a wall, building envelope, window, or other material.

convective loop the continuous circulation of air (or another liquid) in an enclosed space as it is heated and cooled.

radiation heat energy that is transferred through air.

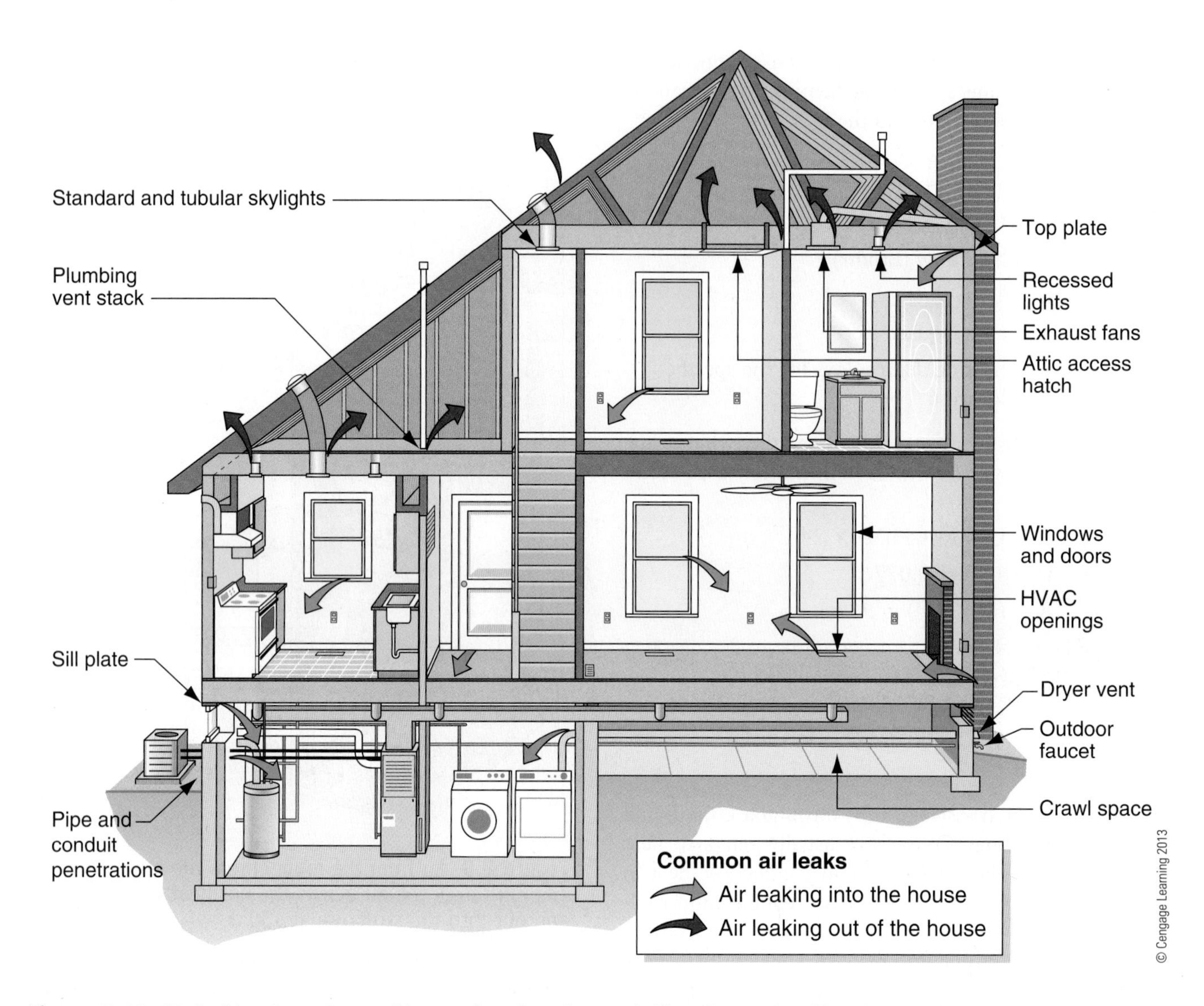

Figure 2.10 Air leaking in and out of homes is referred to as *infiltration* and *exfiltration*.

but metals reflect it very efficiently. Darker materials absorb more energy than lighter ones. During the heating season, buildings can take advantage of radiation, allowing the sun to help heat living spaces. In the summer, shading buildings from excess radiation helps to keep them cooler. Exterior shading can be created through planting trees or by incorporating awnings, roof overhangs, and other elements into the design. Recent advancements in building materials, such as radiant barrier roof sheathing, help reduce unwanted heat from radiation on surfaces that cannot be effectively shaded. Another relatively new product, low-emissivity (low-E) window coating, restricts the level of radiant heat that transfers through the glass without interfering with the passage of light.

Fuel Types

Homes use a range of fuels for space and water heating (Table 2.4). The fuel selection is often driven by local availability, but the most common fuels are natural gas,

Table 2.4 Comparison of the Energy Content of Fuels

Fuel Type	Fuel Unit	Approximate Energy Content per Unit
#2 Fuel oil	Gallon	139,000 Btu
Natural gas	Therm	100,000 Btu
Electricity	kWh	3,412 Btu
Propane	Gallon	92,000 Btu

low emissivity (low-E) a surface that radiates, or emits, low levels of radiant energy.

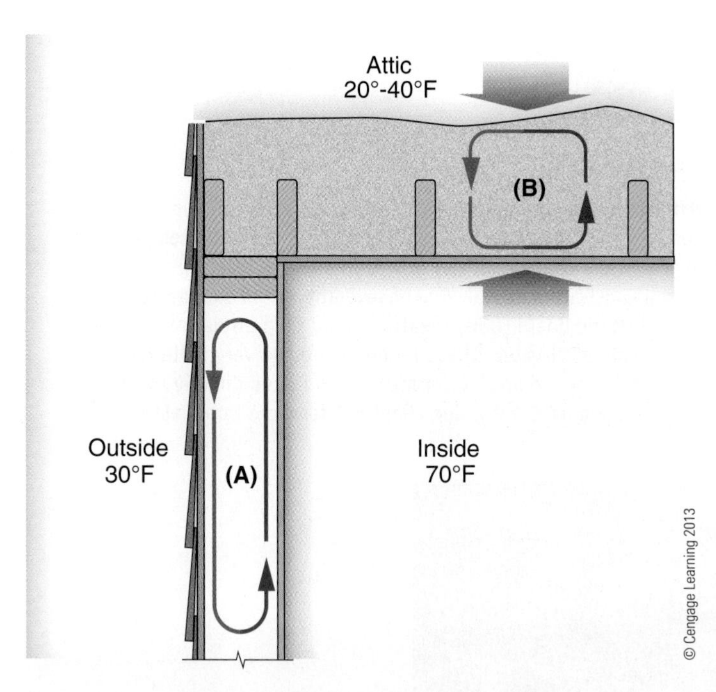

Figure 2.11 Uninsulated air spaces within walls can contain convective loops (A). Figure (B) shows that convective loops can also occur within the insulation of ceilings.

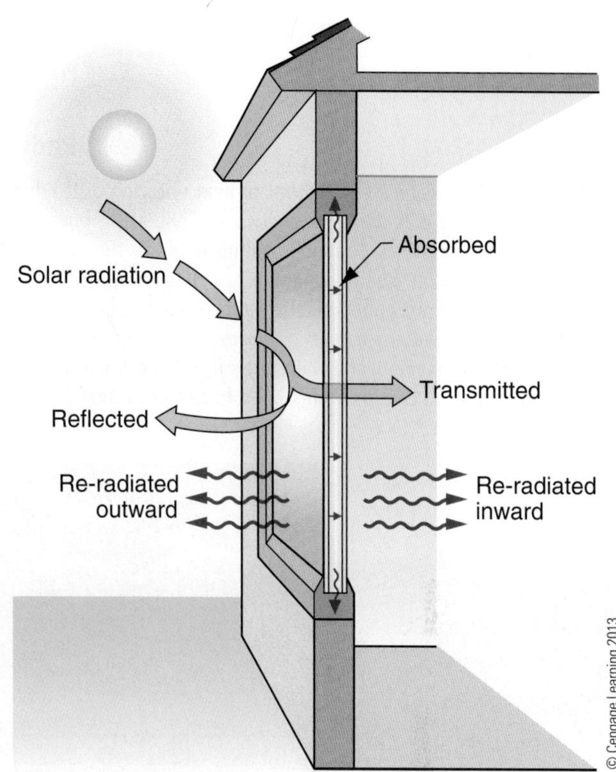

Figure 2.12 Materials absorb, reflect, or transmit radiation, depending on the type of receiving surface and the incoming radiation angle.

fuel oil, electricity, and propane. The fuels are supplied in a range of units, but all fuels can be compared by analyzing their Btu content.

Air Flow

In addition to managing conduction, convection, and radiation, building professionals need to understand and regulate air movement and heat flow within and around a home.

Air flow is one of the most important forces that can affect building performance. Just as heat always seeks to disperse into colder areas, air always moves from areas of high pressure to ones of low pressure. Uncontrolled air flow can increase energy consumption and draw moisture and other pollutants into the home. In order for uncontrolled air flow in a structure to occur, there must be holes that allow movement and a pressure difference as the driving force to move the air. Driving forces can be either natural (such as wind) or manmade (Figure 2.14a and Figure 2.14b). Manmade forces can be heating and air conditioning (HVAC) systems, vent fans, and even fireplaces. Negative pressures draw in the same outside air as vent fans, although positive pressure can force conditioned air outside, wasting energy in the process. Vent fans (i.e., fans that suck air out of the house), particularly large ones like those installed in attics or kitchens, can create significant negative pressures. This pressure draws in outside air—along with its various pollutants, including mold, pollen, dust, and carbon monoxide from attached garages. HVAC duct systems may create additional pressures in the house, both positive and negative.

A common cause of air flow in homes is called the stack effect. This convection effect results because cold air is denser than hot air, so hot air rises. Homes that have holes at the top to let the hot air exit, and holes at the bottom that let in colder air, create natural convection (Figure 2.15a). In summer, the stack effect can actually reverse as hot attic air enters the home and forces relatively cool, air conditioned air out at the bottom (Figure 2.15b). Every building has a point at which the inside and outside pressure are equal. Known as the neutral pressure plane, this phenomenon commonly occurs around the midpoint of a home's height.

This air movement reduces efficiency by allowing uncontrolled movement of hot and cold air in and out of the house. When intentionally managed, this air flow can provide natural ventilation that increases a home's efficiency by reducing the need for air conditioning.

---

stack effect the draft established in a building from air infiltrating low and exfiltrating high.

Did You Know?

Radiant Barriers

Radiant barriers reduce the amount of heat entering or leaving the home by way of radiation. Attic radiant barriers are most effective in warm and hot climates with ductwork in the attic. During the summer, radiant barriers can significantly reduce attic temperatures and, in some cases, may reduce cooling loads in the house. Radiant barriers usually consist of a thin sheet or coating of a highly reflective material (usually aluminum) applied to one or both sides of a substrate material. Any of several types of substrates may be used, including kraft paper, plastic film, cardboard, plywood, or oriented strand board (OSB) sheathing, and rigid insulation materials. Although many radiant barrier products provide an "effective" or "equivalent" R-value, these products provide very little reduction in conductive heat flow and therefore often have R-values of less than R-1. The radiant barrier may be laminated directly to the roof decking (Figure 2.13a) or be fastened to the underside of the roof rafters (Figure 2.13b). See Chapter 7 for more information.

Courtesy of Miller Custom Homes

Figure 2.13a Radiant barriers may be applied in the factory to the roof decking for easy installation. LP's TechShield is OSB with a perforated radiant barrier that reduces radiant heat transfer and allows moisture drying.

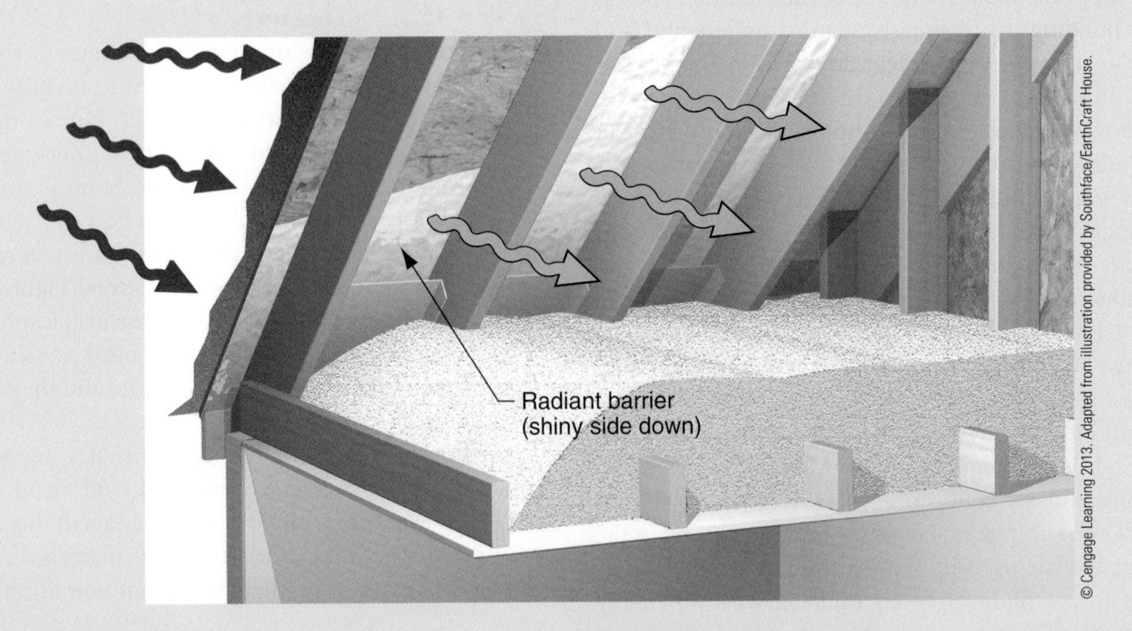

© Cengage Learning 2013. Adapted from illustration provided by Southface/EarthCraft House.

Figure 2.13b Radiant barriers may be applied to the underside of roofing rafters to retard solar radiant heat transfer.

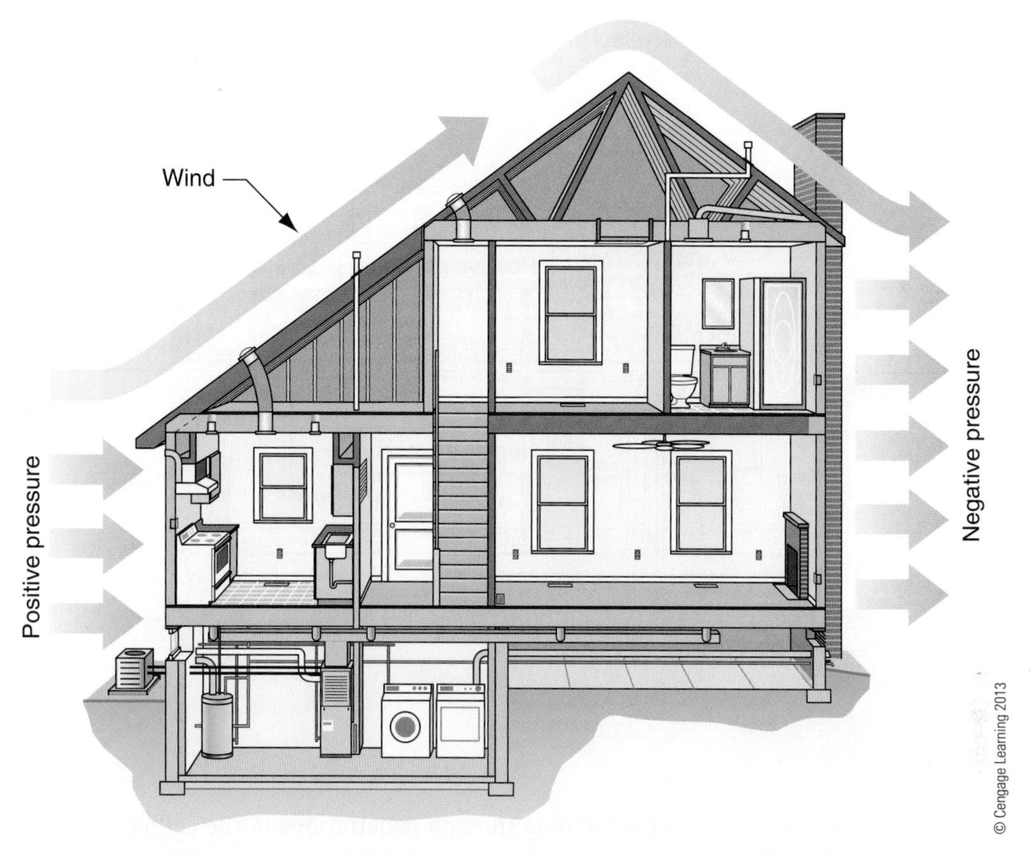

Figure 2.14a Air is pushed into the home on the windward side, creating a positive pressure. On the leeward side, negative pressure is drawing air out of the home.

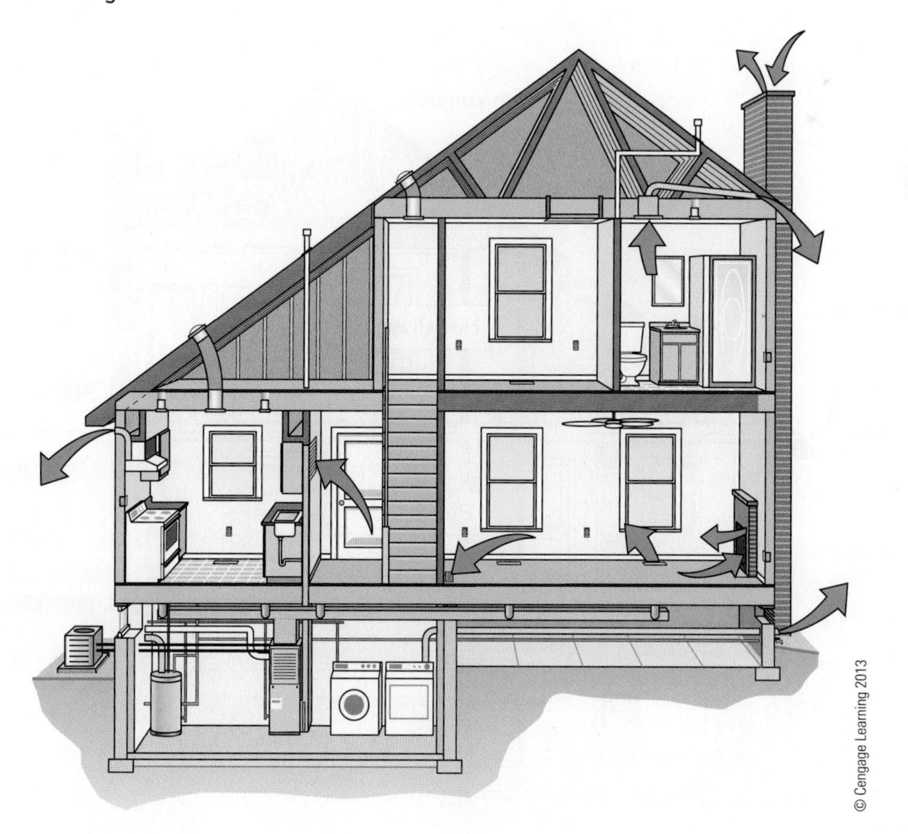

Figure 2.14b House pressures are affected by the operation of mechanical systems. The HVAC system, clothes dryer, exhaust fans, and even fireplace can cause the homes pressure to go positive or negative.

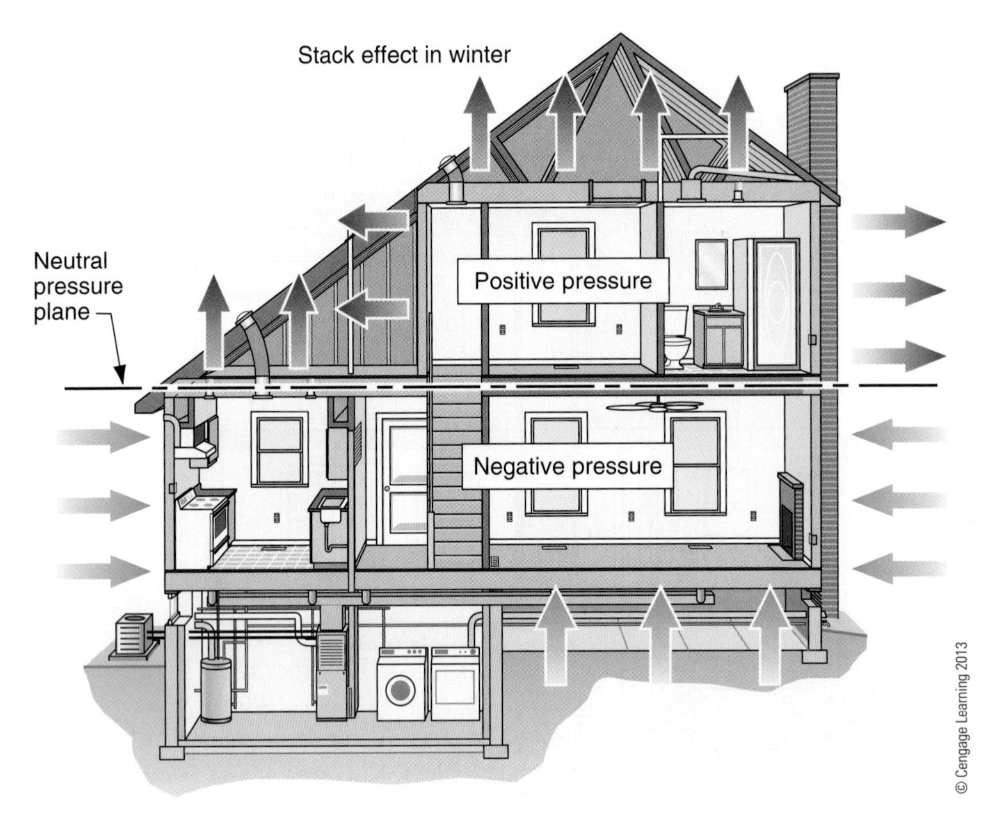

Figure 2.15a In the winter, warm air leaves the building through penetrations in the ceiling, while cold outdoor air enters through cracks in the lower level.

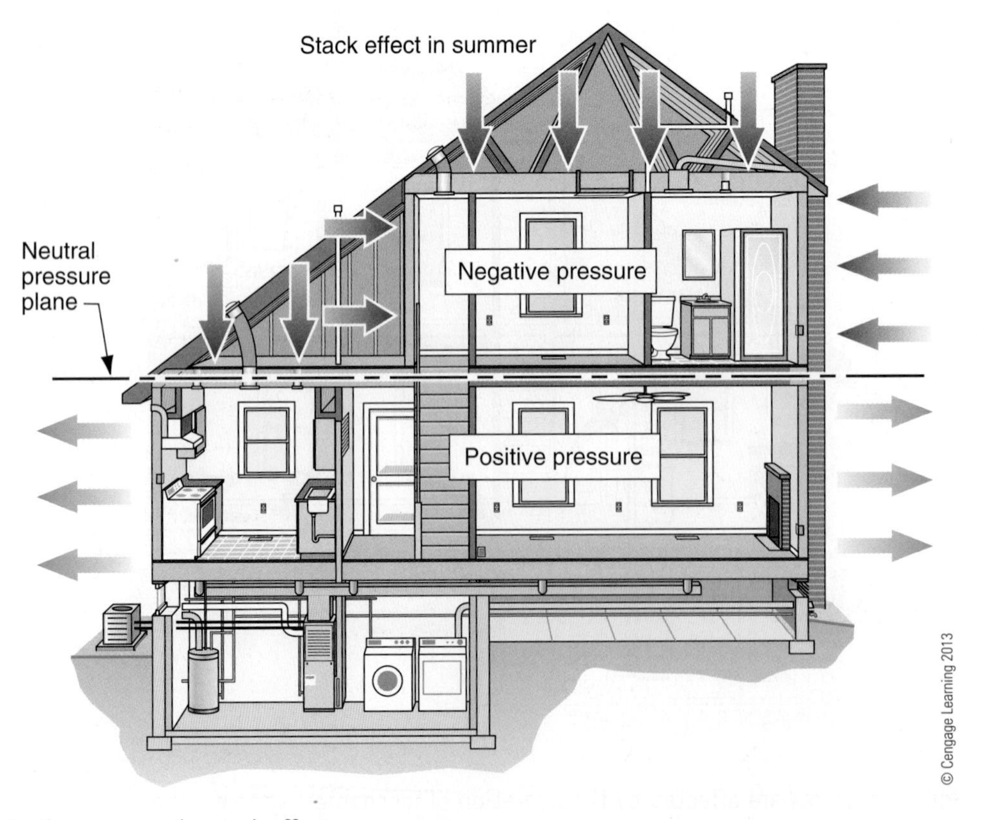

Figure 2.15b In the summer, the stack effect can reverse.

Moisture Flow

Moisture is a critical element in green building, and understanding how to manage its movement into and out of homes is a key piece of the green building puzzle. There are two types of moisture: *bulk moisture* (water) and *vapor*. Both bulk moisture and vapor move from areas of high concentration to low, from wet to dry (Figure 2.16).

Bulk Moisture

When allowed into buildings, bulk moisture causes structural damage and mold growth that can lead to health and comfort problems for occupants. Sources of bulk moisture include rain and groundwater entering through basement walls and floors, clogged gutters or storm drains, leaky pipes, roof leaks, improper

FROM EXPERIENCE

Unintended Consequences in Energy-Efficient Homes

Bruce Harley, Technical Director, Conservation Services Group, Westboro, Massachusetts. A common subject in the construction world is energy efficiency and its unintended consequences: detrimental effects on building durability and air quality. It is easy for builders and popular media to portray homes that are "built too tight" or have "too much insulation" as causing mold or structural damage, but the relationships among energy, moisture, air quality and building durability are often misunderstood or even ignored. One way to clear up some of the misconceptions about energy efficiency and moisture damage is to compare the moisture performance of new, energy-efficient homes with that of old, inefficient ones.

Courtesy of Bruce Harley

Bruce Harley is a nationally recognized authority on energy-efficient residential construction and retrofit. With over two decades of experience in efficient building construction, research, technical and program policy, training, and building modeling, he has written two hands-on books: Insulate and Weatherize *(2002), and* Cut Your Energy Bills Now *(2008). In 2007, he received a Legacy Award from the Energy and Environmental Building Association for his pioneering work in the field of building science.*

Most homes built before the 1930s had little or no insulation and leaked badly. These homes "breathed": heat, air, and water vapor could move easily through walls and roofs, and areas that got wet could dry out easily. If a little water leaked in around a bad window or chimney flashing, heat loss into the wall cavity or attic would dry it out fast. This is really what people mean when they say a house "has to breathe"—we want a home to have a large "drying potential."

Construction practices have changed rapidly since World War II. Exterior sheathing materials like plywood and OSB slow air movement and act as a condensing surface for water vapor in cold weather. These materials are also easily damaged when they get wet (especially OSB); unlike the solid wood boards we formerly used, they delaminate, lose shear and fastener strength, and grow mold.

Next, consider the moisture implications of insulation in a cold climate. Exterior sheathings in an uninsulated house are warmer, allowing them to dry faster. As soon as a moderate amount of insulation is added to the cavity, the sheathing stays much closer to outdoor temperature, so insulated building cavities are much slower to dry out when they do get wet.

Energy efficiency does not cause moisture problems, but air sealing and insulation do change the moisture dynamics and tend to make a building less forgiving of moisture that is already being generated inside or leaking from outside the house. Deciding not to make a house "too efficient" in the hope that it will "breathe" is uninformed and doomed to failure. Deliberate control of moisture is the best way to minimize risk: One method is to control moisture sources and indoor pollutants with mechanical ventilation. Reducing groundwater entry is also critical; at very least, leaking groundwater should be separated from the building interior with drained foundation membranes and cavities. Similarly, rainwater entry should be prevented with overhangs and proper flashing techniques. Drainage and wall cavity drying can be promoted with vented rain screen claddings, and condensing surface temperatures can be controlled with the use of spray applied foam cavity and foundation insulation or continuous rigid exterior insulation. You can also control unwanted air and moisture flows by sealing building enclosure air leaks, moving ducts and air handlers inside conditioned space, and tightly sealing any ducts that must be outside. Just as importantly, you should avoid the use of building cavities as return ducts.

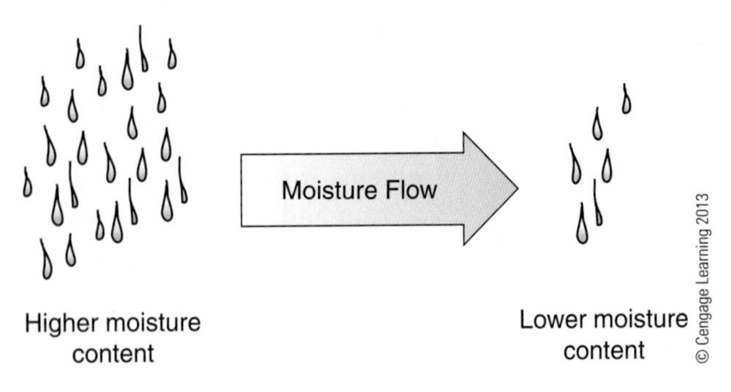

Figure 2.16 Moisture transfers from areas of relatively high concentrations to those of lower concentrations.

window flashing, ice dams, and plumbing leaks. Correct design and construction details and execution in the field are very effective at eliminating bulk moisture intrusion.

While we want to keep all bulk moisture out of our buildings, the proper level of water vapor helps keep a house comfortable. Vapor in homes comes both from internal sources (including cooking, showering, and breathing) and from external sources that include diffusion through building materials and via air transport from infiltration (Figure 2.17). Internal vapor is typically controlled through air conditioning, dehumidifiers, and ventilation in bathrooms and kitchens. Diffusion and infiltration vapor are controlled through building construction techniques.

Diffusion-Transported Vapor

Most building materials allow a certain amount of vapor to diffuse, or pass, through them, although the volume of diffused vapor is insignificant when compared with the amount transported through infiltration. The amount of water vapor that diffuses through a building assembly is affected by the following factors:

- The chemical composition of the building materials
- The thickness of the building materials
- The absolute humidity on each side of the building assembly

Building designs and specifications often include a vapor barrier, more correctly described as a vapor diffusion retarder (VDR), to reduce the amount of moisture that diffuses through building materials into the wall structure. All materials have perm ratings that describe their ability to restrict or allow vapor movement through them (Table 2.5). In general, the water vapor permeability of a material is inversely proportional to its thickness. Therefore, if you double a material's thickness, you cut its permeability in half. Although this characteristic holds true for building materials like OSB and insulation, the science is more complicated for films and coatings.

Materials can be separated into four general classes on the basis of their permeance[1]:

- Vapor impermeable: 0.1 perm or less
- Vapor semi-impermeable: greater than 0.1 perm and less than 1.0 perm
- Vapor semi-permeable: greater than 1.0 perm and less than 10.0 perms
- Vapor permeable: greater than 10.0 perm

While VDRs are often recommended on the exterior in warm humid climates and on the interior in cold climates, these barriers are not recommended in most mixed climates (Figure 2.18). When impermeable VDRs are installed in cold climates where air conditioning is used, any vapor in the wall structure can condense on the VDR within the wall. This condensation can cause structural damage and encourage mold growth as it turns into water (Figure 2.19). Condensation in walls is exacerbated when air infiltration allows moisture laden air into the wall cavity. When a VDR is installed on either the interior or exterior surface of a wall, the wall structure must be fully sealed against air leakage to reduce the amount of moisture that can condense inside the wall.

While a single sheet of drywall will only allow one third of a quart of water to pass through it during an entire season via diffusion, a single 1" hole in that same sheet of drywall will allow 30 quarts of water to pass through via air transport (Figure 2.20). Unanticipated (and therefore uncontrolled) air movement is the primary problem related to vapor entering and exiting buildings. In humid climates, this vapor transport increases relative humidity indoors, requiring air conditioning systems to work harder to keep the house comfortable. In dry climates, the vapor that occurs indoors through cooking, showering, and breathing can be transferred to the outdoors through exfiltration, making the interior uncomfortably dry. A house that is air sealed significantly reduces the amount of vapor transmission between inside and outside, reducing air conditioning costs and making the house more comfortable year round.

---

vapor diffusion retarder (VDR) a material that reduces the rate at which water vapor can move through a material.

perm rating the rate of water vapor passage through a material under fixed conditions.

---

[1]Lstiburek, J.W., Understanding Vapor Barriers, *ASHRAE Journal,* August 2004.

Figure 2.17 Moisture vapor comes from internal and external sources.

Table 2.5 Permeability Values for Common Building Materials

Material	Perm Rating
Vapor retarders	
Insulation facing, kraft	1
Plywood, 1/4" (Douglas fir, exterior glue)	0.7
Insulation facing, foil kraft laminate	0.5
Paint, vapor-retarder latex 0.0031"	0.45
Polyethylene sheet	
0.002"	0.16
0.004"	0.08
0.006"	0.06
Aluminum Foil, 0.00035"	0.05
Polyurethane foam, closed-cell spray (2.5")	0.8
Not vapor retarders	
Gypsum wall board, plain 3/8"	50
Mineral wool, unfaced 4"	30
Paint, typical latex, ~0.002"	5.5–8.6
Sheathing paper, 4.4-lb. asphalt-saturated (100 ft²)	3.3
Plywood, 1/4" (Douglas fir, interior glue)	1.9
Polyurethane foam, open-cell spray, 3"	16–17

Source: *North American Insulation Manufacturers Association (NAIMA), "*Insulation Facts #71: Use of Vapor Retarders.*" NAIMA, http://www.certainteed.com/resources/Use%20of%20Vapor%20Retarders.pdf (accessed June 7, 2010).*

Relative humidity (RH) is the percentage of the maximum moisture that the air can hold at a specific temperature (see Figure 2.21). RH ranges from 0% to 100%. At 100%, the air is saturated with moisture, and any additional moisture added at that same temperature will result in condensation on any cool surfaces. If the relative humidity is 50%, the air is holding one half the moisture it is capable of holding.

Recommended RH levels for indoor air quality and health range from 40% to 60% (Figure 2.22, page 44). Keeping homes between these levels will help avoid condensation and indoor environmental problems. When RH drops below 30%, the air can become dry, causing sinus irritations, static electricity, and drying and cracking of wood and interior finishes. In contrast, RH levels that persist over 60% for an extended time can promote mold growth, dust mites, and general discomfort as the occupants feel clammy and sticky in warm weather.

Condensation and the Dew Point

All air has a dew point, the temperature at which vapor condenses into water droplets. The dew point, which varies with the temperature and RH, can be calculated

---

relative humidity (RH) the ratio of the amount of water in the air at a given temperature to the maximum amount it could hold at that temperature; expressed as a percentage.

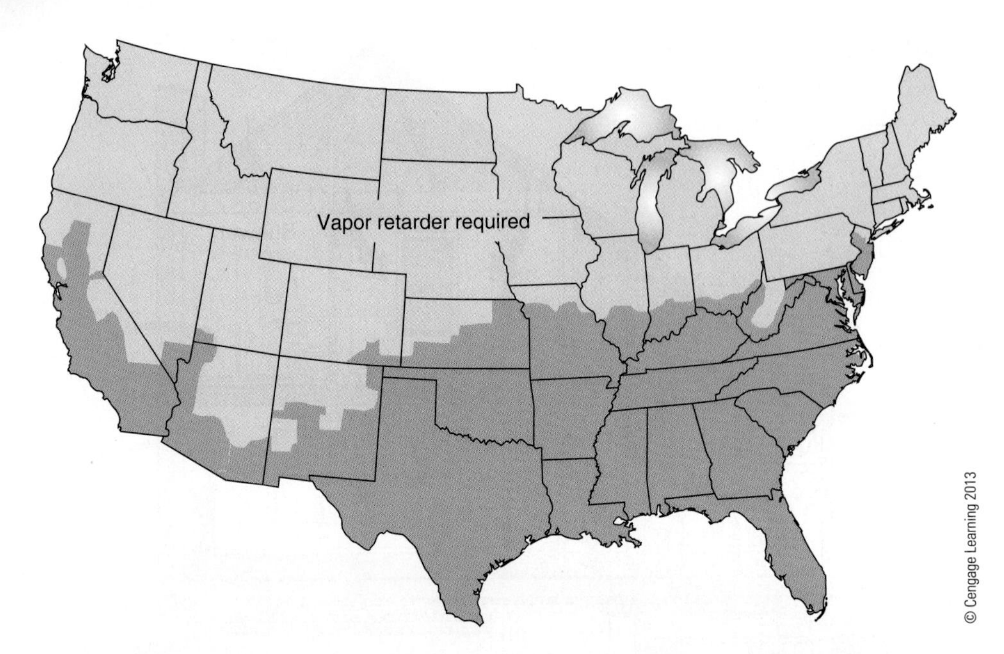

Figure 2.18 The 2006 International Residential Code and 2006 International Energy Conservation Code require vapor diffusion retarders only for climates 5 and above.

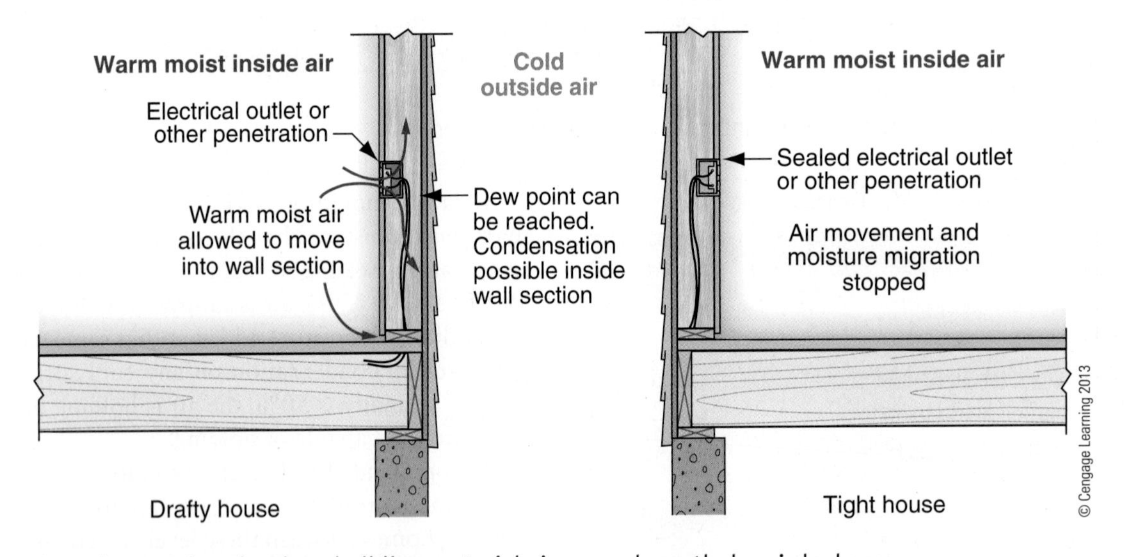

Figure 2.19 Moisture migration into building materials is caused mostly by air leakage.

using a psychrometric chart. Allowing air to reach the dew point in a building should be avoided because the resulting moisture can cause structural damage, encourage mold growth, and create health and comfort issues in the house. When the dew point is reached inside exterior walls, these risks increase significantly and create hidden moisture problems. Air leaks allow moisture-laden air into walls where it can condense on the inside of cold sheathing (in the winter) or on the inside of drywall cooled by air conditioning (in the summer). Air sealing is the process of reducing air leakage into and out of a home. Air sealing is generally accomplished by applying some combination of caulk, spray foam, weather stripping, and other air impermeable materials. Careful air sealing and limited use of vapor impermeable materials can help avoid condensation and the problems associated with it.

Psychrometric Chart

The psychrometric chart is a graph of the physical properties of moist air at a constant pressure. The psychrometric chart is particularly useful in diagnosing moisture and mold

air sealing the process of tightening the building envelope by reducing air leakage into and out of a home.

psychrometric chart a chart or graph showing the relationship between a particular sample of air's dew point temperature, dry-bulb temperature, wet-bulb temperature, humidity ratio, and relative humidity.

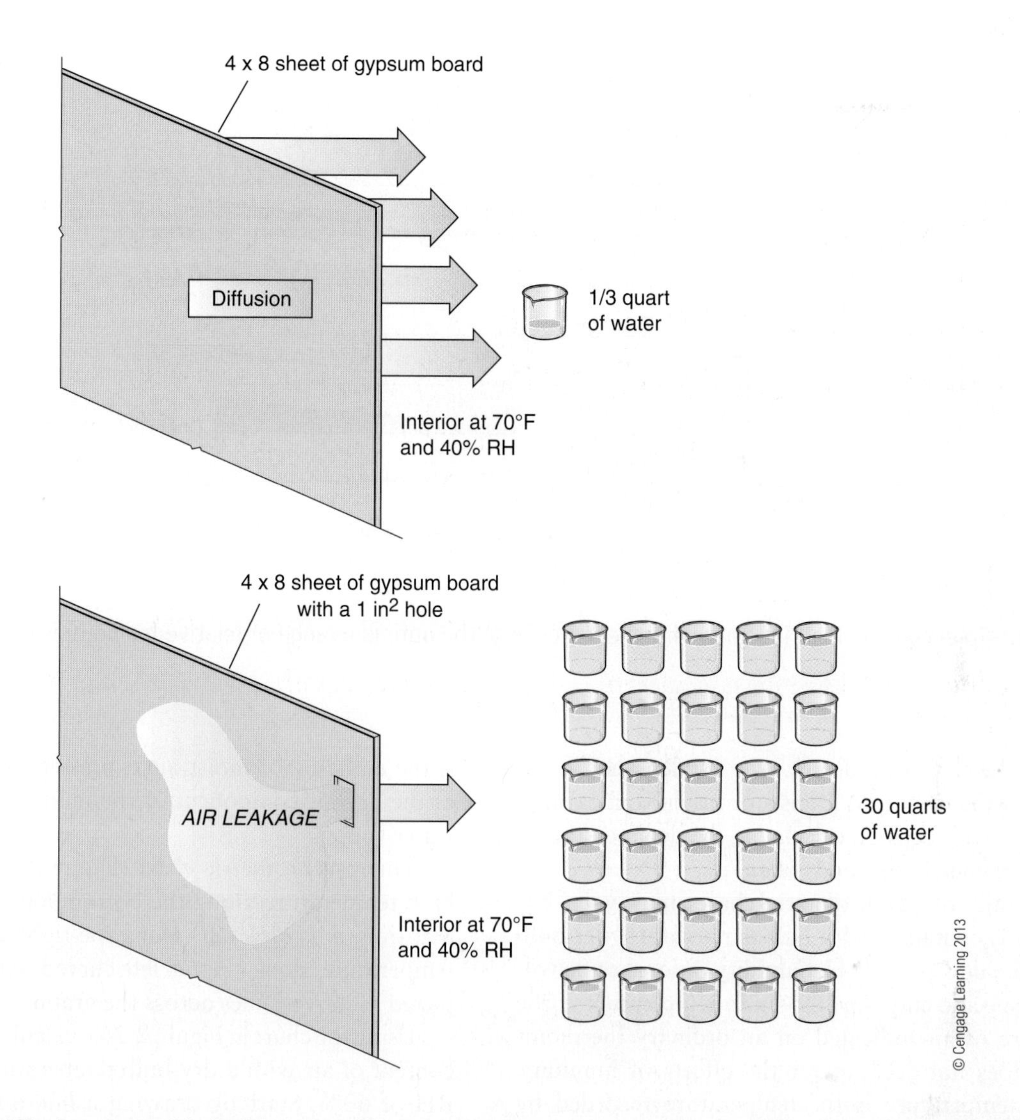

Figure 2.20 Diffusion versus air leakage. In most cold climates over an entire heating season, one third of a quart of water can be collected by diffusion through gypsum board without a vapor retarder; 30 quarts of water can be collected through air leakage.

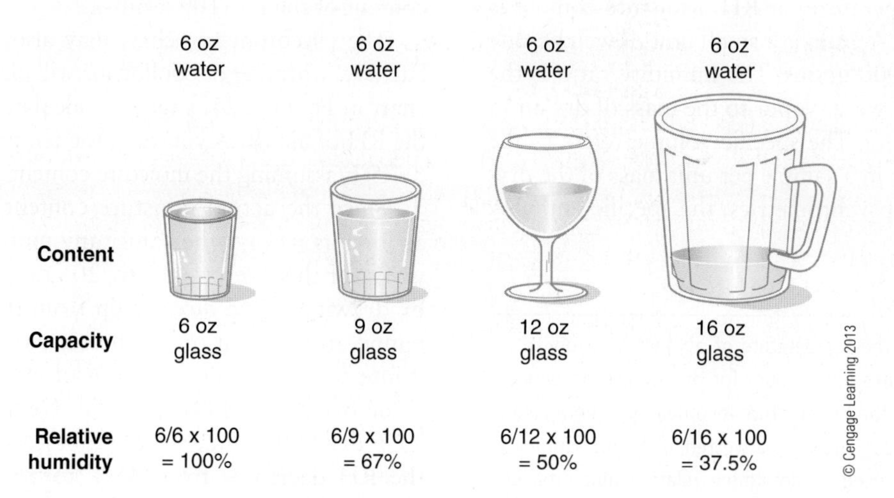

Figure 2.21 The ability of air to store moisture is directly related to temperature. Cold air is like a small glass with a relatively small storage capacity, whereas warmer air is like a large glass. When the volume of moisture remains constant, the RH rates decrease as air temperature increases.

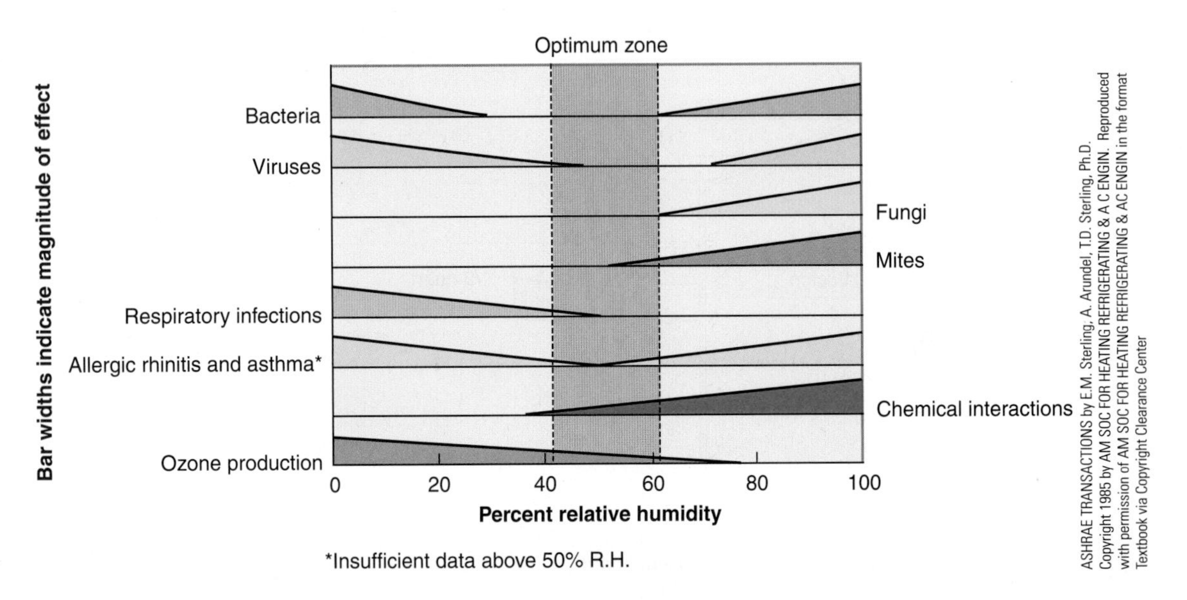

Figure 2.22 To encourage high indoor environmental quality, the optimum range of relative humidity is 40% to 60%.

problems. The chart can identify why condensation is occurring on windows or floor joists or inside wall cavities. For now, we will focus on calculating the dew point, RH, and moisture content of a body of air.

The details of psychrometric charts may vary, but they typically contain the dry-bulb temperature, wet-bulb temperature, dew point, RH, humidity ratio, specific volume, and specific enthalpy. **Dry-bulb temperature** is the temperature of air indicated on an ordinary thermometer and does not account for the effects of humidity. **Wet-bulb temperature** is the temperature recorded by either a sling thermometer or an aspirating psychrometer. Both instruments use a thermometer covered with a wetted sleeve. The dew point can be calculated with any two of the following three variables: dry-bulb temperature, wet-bulb temperature, or RH. Moisture content is measured in grains. A **grain** is a small unit of weight; one pound contains 7,000 grains. The humidity ratio is the ratio of the mass of water vapor to the mass of dry air in a volume of moist air. The specific volume refers to the volume of moist air in a sample per unit mass of the dry-air component. In psychrometrics, the specific enthalpy is the enthalpy of moist air expressed as per unit mass of the dry-air component of the mixture of dry air and water vapor.

This psychrometric chart (Figure 2.23) has the dry-bulb temperature along the bottom axis, moisture content (in grains of water) along the right axis, dew point temperature along the far left curved axis, and RH depicted as curved lines across the graph.

Using the chart in Figure 2.24a, calculate the moisture content of air with a dry-bulb temperature of 85°F and RH of 60%. Start by drawing a line directly up from the 85°F (dry-bulb) point on the bottom axis, and stop when it intersects with the 60% RH line. To the immediate left (Figure 2.24c) of this point is the dew point (70°), and to the right (Figure 2.24b) is the moisture content of the air (109 grains).

The psychrometric chart may also be used to calculate how warming or cooling air will affect RH. Using the chart in Figure 2.25, page 49, calculate what happens to the RH of air when you raise the temperature from 45°F to 75°F, assuming the moisture content stays the same.

Since the actual moisture content does not matter as long as it stays the same, any moisture content will work for this exercise. Here, 20 grains is selected. Start by drawing a line directly up from the 45°F (dry-bulb temperature) point on the bottom axis, and stop when it intersects with the 20 grains of moisture. Record the value of the curved line (45%). Next, do the same for 75°F dry-bulb temperature (15%). As the air is heated, the RH decreases from 45% to 15% because warm air can hold more moisture and therefore becomes less saturated by 20 grains. The reverse is true when air is cooled.

---

dry-bulb temperature the temperature of air indicated on an ordinary thermometer; does not account for the effects of humidity.

wet-bulb temperature the temperature recorded by a thermometer whose bulb has been covered with a wetted wick and whirled on a sling psychrometer; used to determine relative humidity, dew point, and enthalpy.

grain a unit of measurement for moisture content; 1 pound contains 7,000 grains.

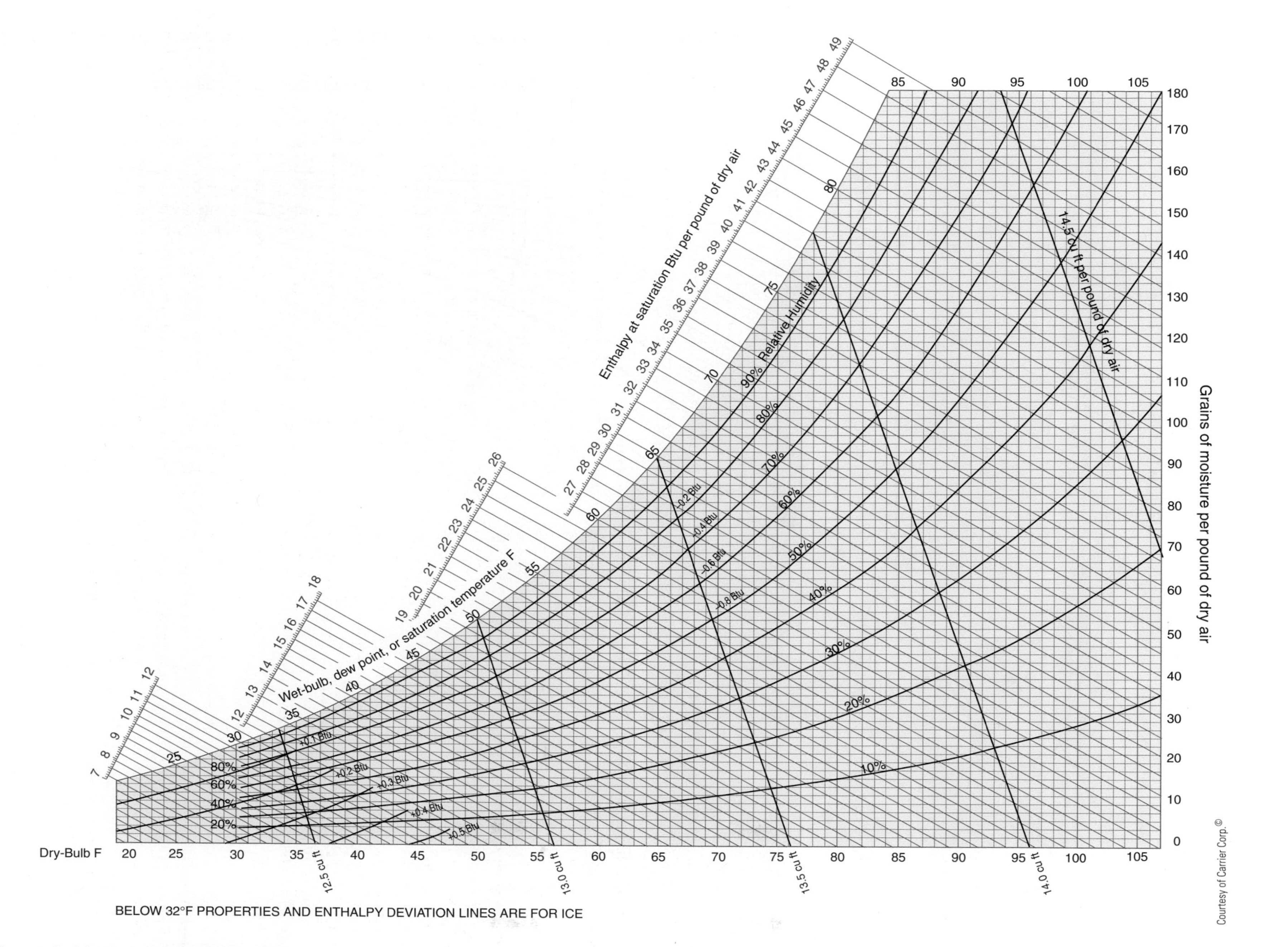

Figure 2.23 The psychrometric chart.

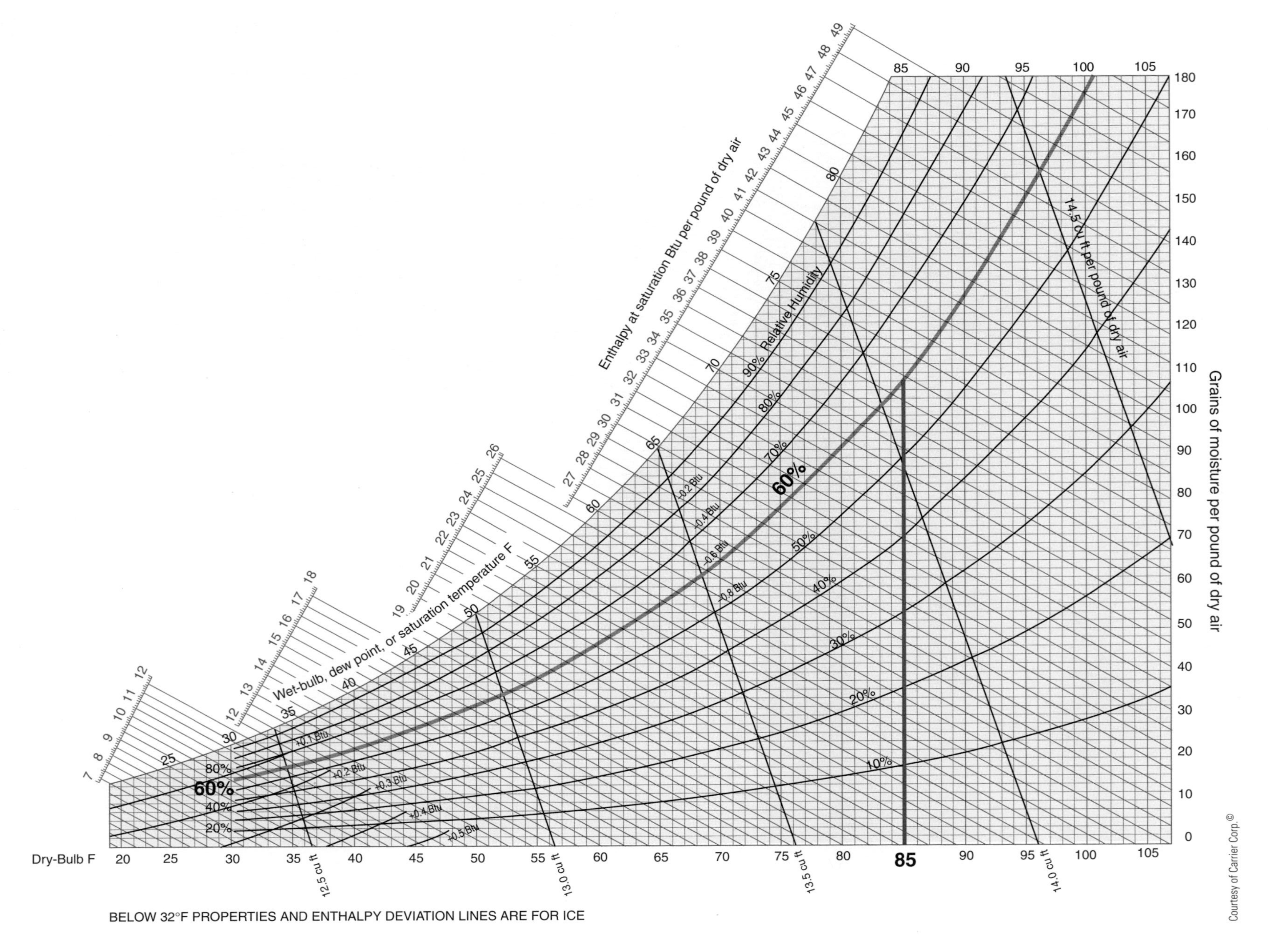

Figure 2.24a First, draw a line from 85°F up to where it intersects with the 60% RH curve.

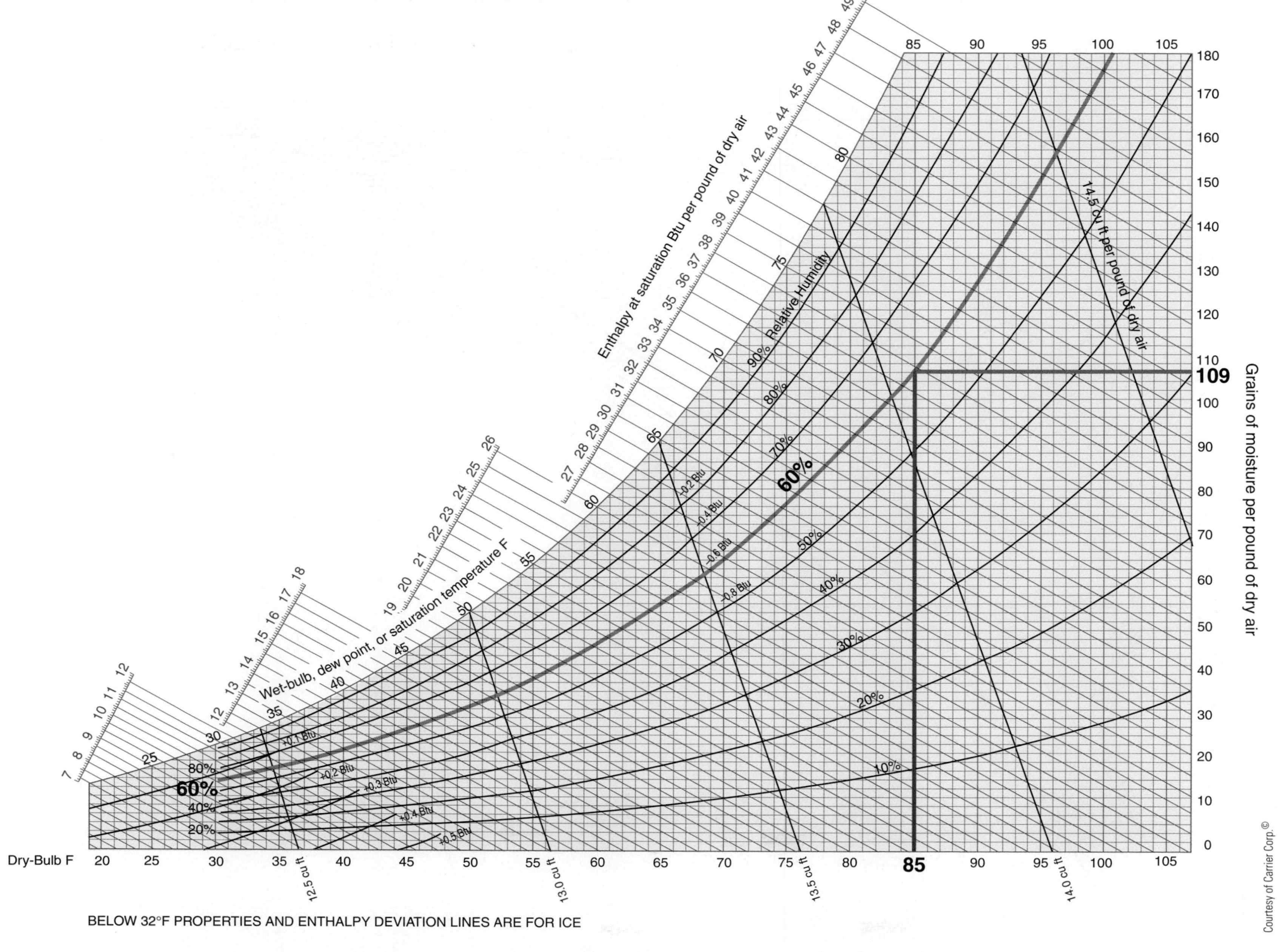

Figure 2.24b Second, draw a line to the right. The point at which the line intersects the right vertical axis is the moisture content of the air.

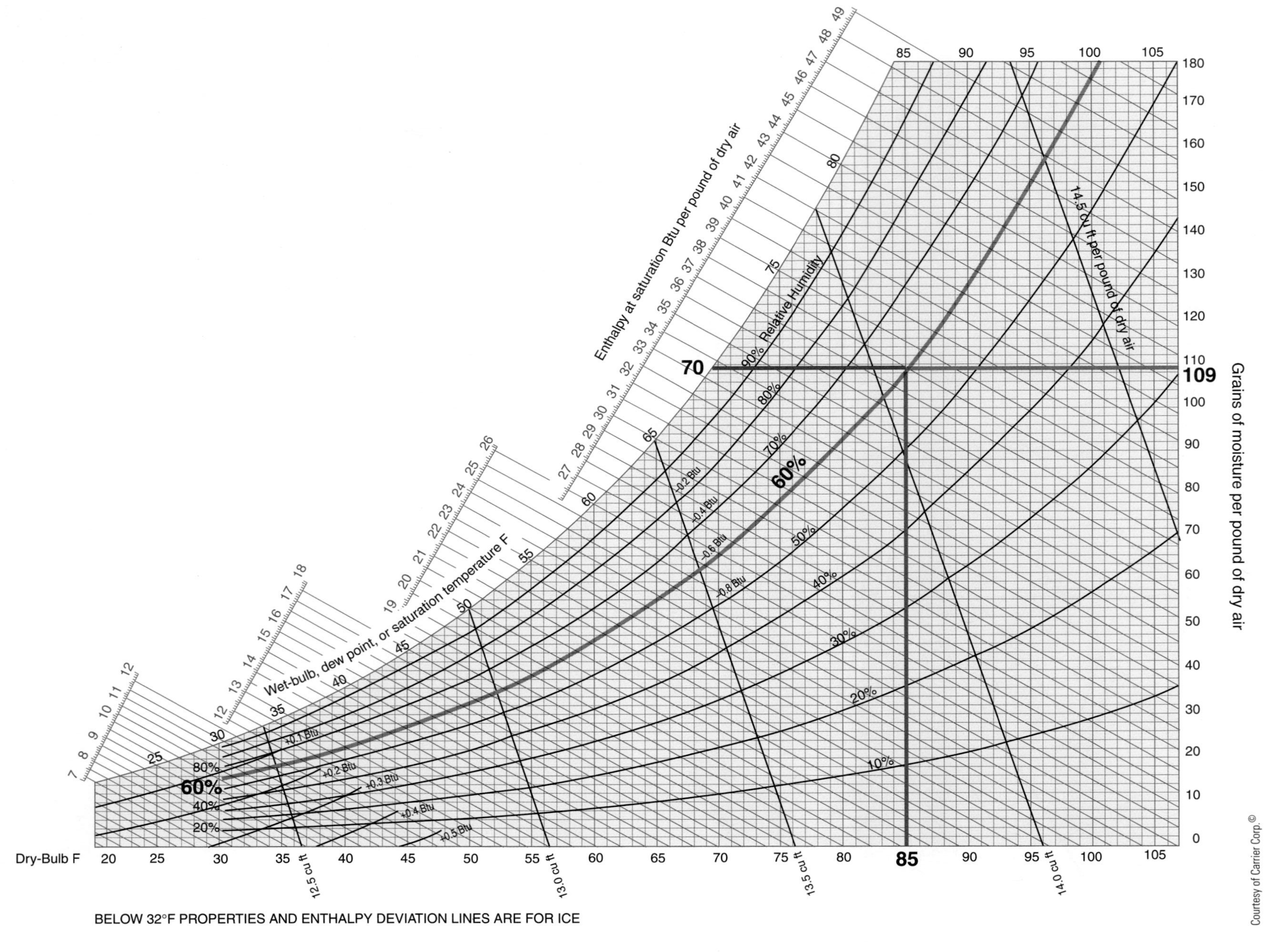

Figure 2.24c Third, draw a line to the left. The point at which the line intersects the left curved vertical axis is the dew point for the air.

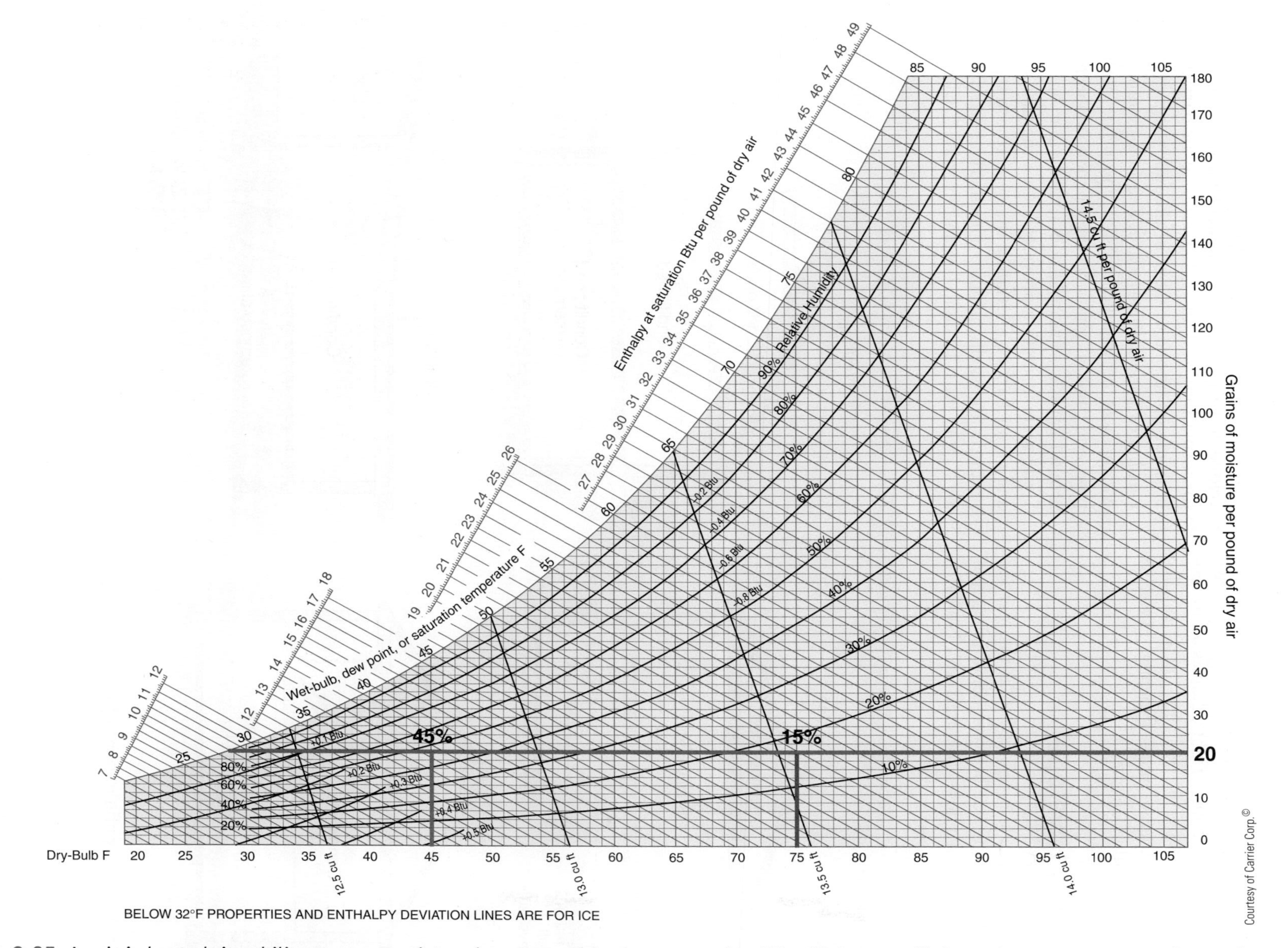

Figure 2.25 As air is heated, its ability to store moisture increases. This also means that RH will decrease if the moisture content remains unchanged. To illustrate this point, select a moisture content for the air. The amount does not matter as long as it remains constant in the exercise. Here, we select 20 grains. First, draw a line from 45°F up to where it intersects the 20-grains line. Next, draw a line at 75°F up to where it insects the 20-grains line. As the air is heated, the RH decreased from 45% to 15%.

Relative Humidity and Air Conditioning

Air conditioning can both cool and dehumidify a house. Houses in mixed and humid climates that have high infiltration rates, and consequently higher RH in the summer, require more dehumidification and larger AC systems than tight homes. Tighter homes allow less moisture to infiltrate to the interior, allowing for smaller cooling systems that operate more efficiently. (HVAC is covered in more detail in Chapter 13).

The Building Envelope

The building envelope, also referred to as the *thermal envelope*, is the dividing line between conditioned and unconditioned space in a house (Figure 2.26). Typically, the building envelope includes the exterior walls, ceilings, and floors; however, many homes are built with conditioned or semi-conditioned attics, basements, and crawl spaces where the roof and foundation walls and floor serve as the building envelope. The building envelope consists of two components: a thermal barrier (or insulation), and an air barrier. The air barrier may be made up of one or more of the following: drywall, hardboard panels, foam board, and housewrap, all of which must be installed using appropriate sealants, tape, and caulk at joints between panels or sheets and around gaps at windows, doors, and other openings. Spray foam insulation provides both a thermal barrier and an air barrier, alleviating the need to use separate products for each purpose. The thermal barrier and air barrier must be complete and in contact with each

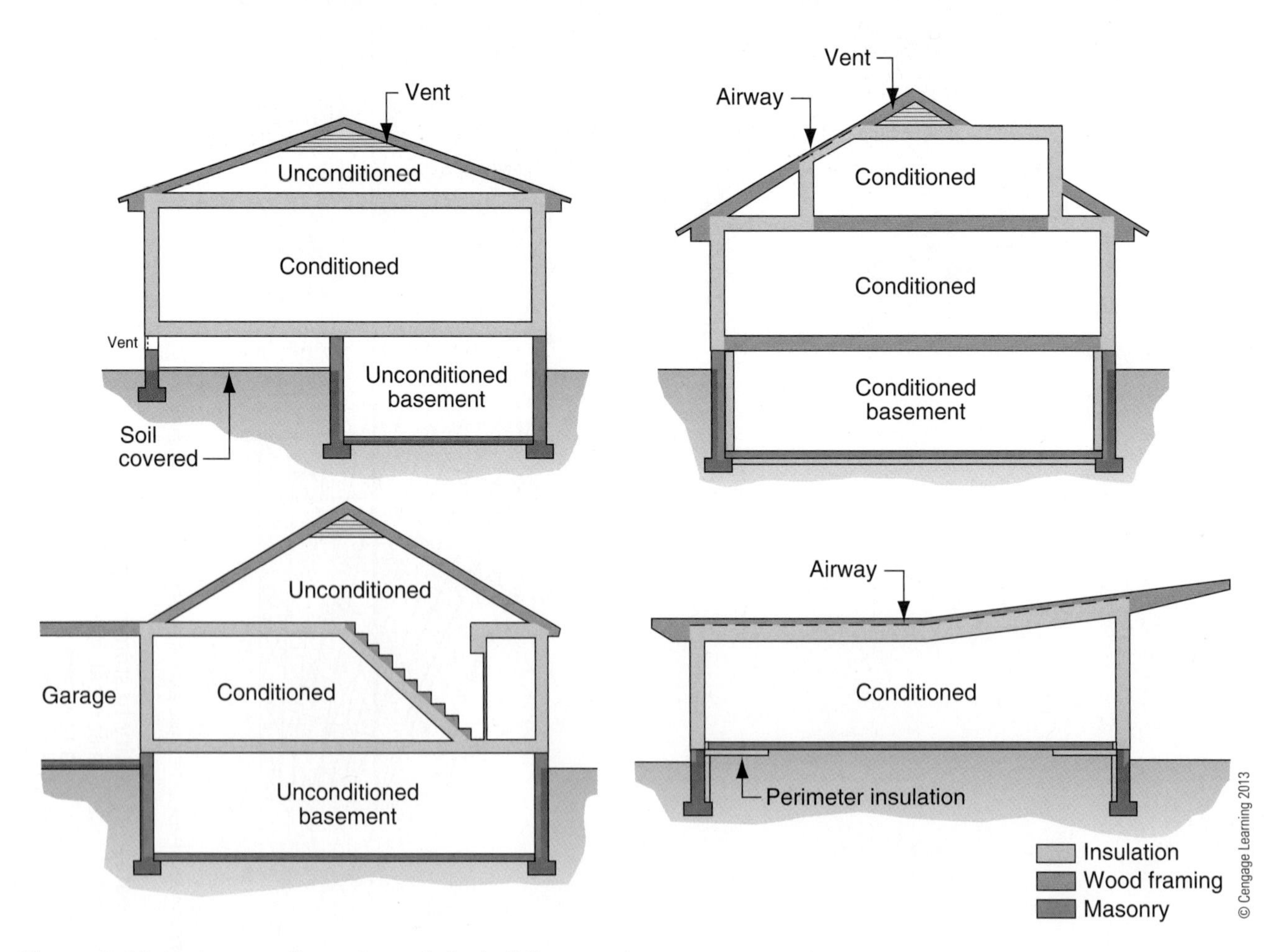

Figure 2.26 Various configurations of the building envelope.

building envelope the separation between the interior and the exterior environments of a building, consisting of an air barrier and thermal barrier that are continuous and in contact.

thermal barrier a boundary to heat flow (i.e., insulation).

air barrier a protective, air-resistant material that controls air leakage into and out of the building envelope.

other on all sides. Air-permeable insulation that does not have an air barrier on all sides does not perform at its rated R-value, reducing the efficiency of the installation (Figure 2.27).

To be effective in keeping the house comfortable and efficient, the building envelope must be continuous and complete (Figure 2.28a and Figure 2.28b). The building envelope on most existing homes (and many new ones) is neither. Creating a complete building envelope requires careful attention to selection and installation of insulation and air sealing materials.

A complete building envelope would be easy to create in a house with no windows, doors, or complicated roofs and ceilings, but it would not be very practical. Large expanses of walls, ceilings, and floors are simple to insulate and air seal properly. The challenge arises where there are transitions to different materials, such as windows, doors, vents, lights, and wires; areas behind bathtubs, above dropped ceilings, and around other architectural details are also more difficult to seal. (See chapter 4 for detailed discussions of insulation and air sealing.)

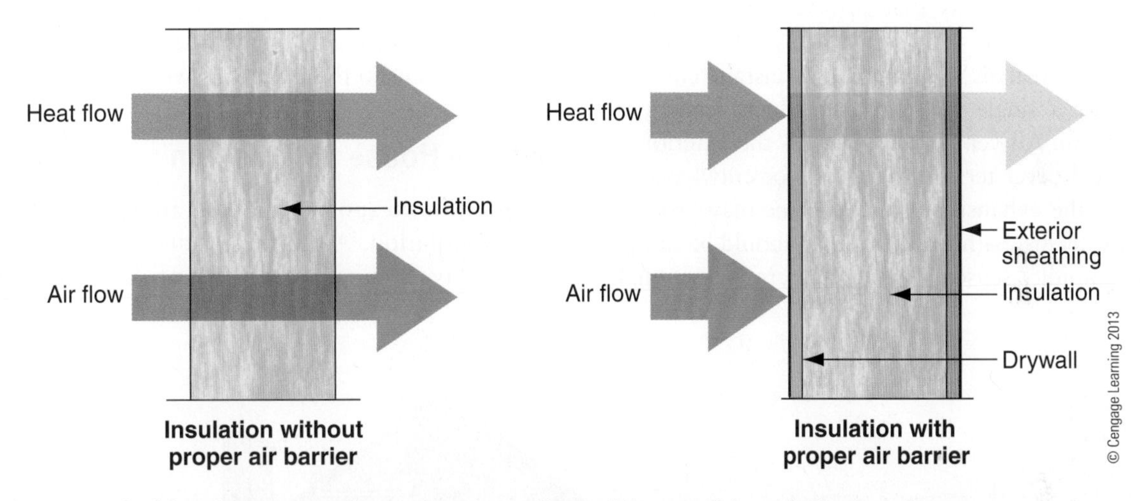

Figure 2.27 Vertical insulation must be enclosed on all six sides to achieve desired performance.

Figure 2.28a Poorly installed insulation with multiple gaps, voids, and misalignment with the air barrier. In many places, the insulation is no longer in contact with the air barrier (subfloor), which creates a thermal bypass.

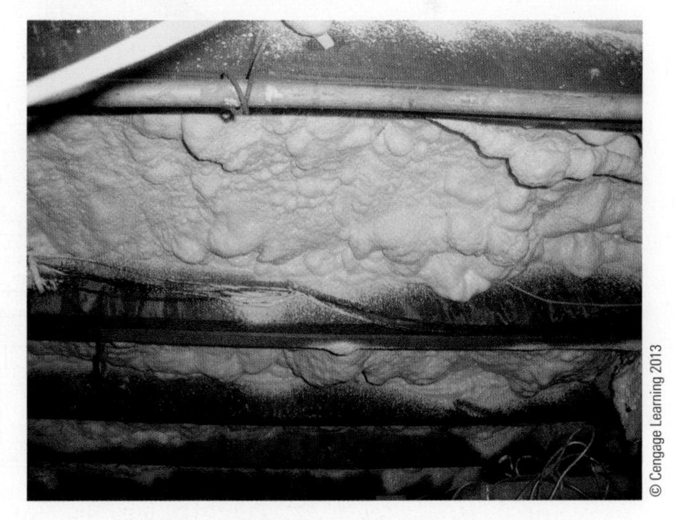

Figure 2.28b Spray polyurethane foam applied to the underside of a frame floor over a crawl space. The insulation is in permanent contact with the subfloor and provides both thermal and air barriers.

House Ventilation

All homes require ventilation to remove excess moisture, filter pollutants (e.g., dust, dirt, and chemicals), and replace stale air with fresh air from outdoors. As homes become more efficient through improved air sealing, the need for ventilation becomes more critical. When designing high-performance homes, an important maxim to remember is "build it tight and ventilate it right."

Types of Ventilation

Spot ventilation, such as that provided by bathroom and kitchen fans, works to remove moisture and odors where they are created. Vent fans can be installed in each bathroom, or a single fan may have ducts leading to each bathroom. All vents should run to the exterior of the house and never terminate in attics or crawl spaces. Terminating the exhausts inside the home may produce moisture problems. Bathroom vent fans should be run on timers or humidity sensors so they operate long enough after a bath or shower to completely remove the extra moisture.

Kitchen vent fans should be as small as possible while still removing moisture and odors while cooking. Exhaust fans must be ducted to the exterior because recirculating fans do not remove moisture or most pollutants. Large fans work best with multiple or variable speed controls that allow them to operate at low speeds. In very tight homes, large fans may require separate makeup air systems. Otherwise, a window should be opened while the fan is operating to avoid depressurizing the house, which can cause backdrafting of fireplaces and open-combustion appliances (Figure 2.29).

Homes with workshops, heavily used laundry rooms, or other moisture- and pollutant-producing areas should have vent fans in these areas as well.

Whole-House Ventilation

In addition to spot ventilation, all homes require general ventilation. Although a house can be ventilated by merely opening windows and doors, this method

Figure 2.29 Although seemingly harmless, large commercial-grade range hoods can depressurize the home and cause backdrafting. Large range hoods should be designed with integrated makeup air.

is subject to significant variations with temperature, wind direction, and wind speed. Consequently, the technique is not a reliable or consistent way to provide a measurable amount of air exchanges. Mechanical ventilation is the generally preferred method to provide fresh air in a home. Ventilation systems can utilize an air handler and duct system, a separate fan and duct system, or a combination of the two. In extreme climates, air-to-air heat exchangers known as energy recovery ventilators (ERV) or heat recovery ventilators (HRV) can save energy by pretreating cold, hot, and humid air before it enters the house (Figure 2.30a and Figure 2.30b). (Ventilation systems are covered in greater depth in Chapter 13.)

Indoor Comfort

While comfort can be very subjective and differs from person to person in the same house, it is a function of air temperature, radiant temperature, air movement, and the home's RH. These variables are all subject to being controlled through the home's design and operation; proper management of these factors can lead to maximum comfort for the building's occupants.

There are generally accepted ranges of temperature and RH that are considered to be comfortable by most individuals. As noted in the comfort charts (Figure 2.31), winter temperatures between 65°F and 77°F with RH levels ranging between 22% and 70% are generally considered to be comfortable for most people. Summer ranges of between 72°F and 80°F and 20% to 70% RH are considered comfortable. When the temperature is higher, lower RH levels generally keep people more comfortable. As shown in Figure 2.22, however, RH levels below 40% and above 60% may produce indoor environmental problems.

Another factor in comfort is air movement. Wind or fans can reduce the perceived temperature by providing convective cooling directly on one's skin, increasing comfort in warm weather and decreasing it in colder weather.

Finally, radiant heat affects our comfort level. Sitting in the shade on a warm day will help you feel more comfortable than if you were to sit in the sun; conversely, sitting in the sun on a cold day is preferable to sitting in the shade. By properly maintaining temperature and RH in a house, using fans appropriately for cooling, and managing radiant heat, we can keep homes more comfortable and use less energy in the process.

Figure 2.30a Heat-recovery ventilator in winter.

Figure 2.30b Energy-recovery ventilator in winter.

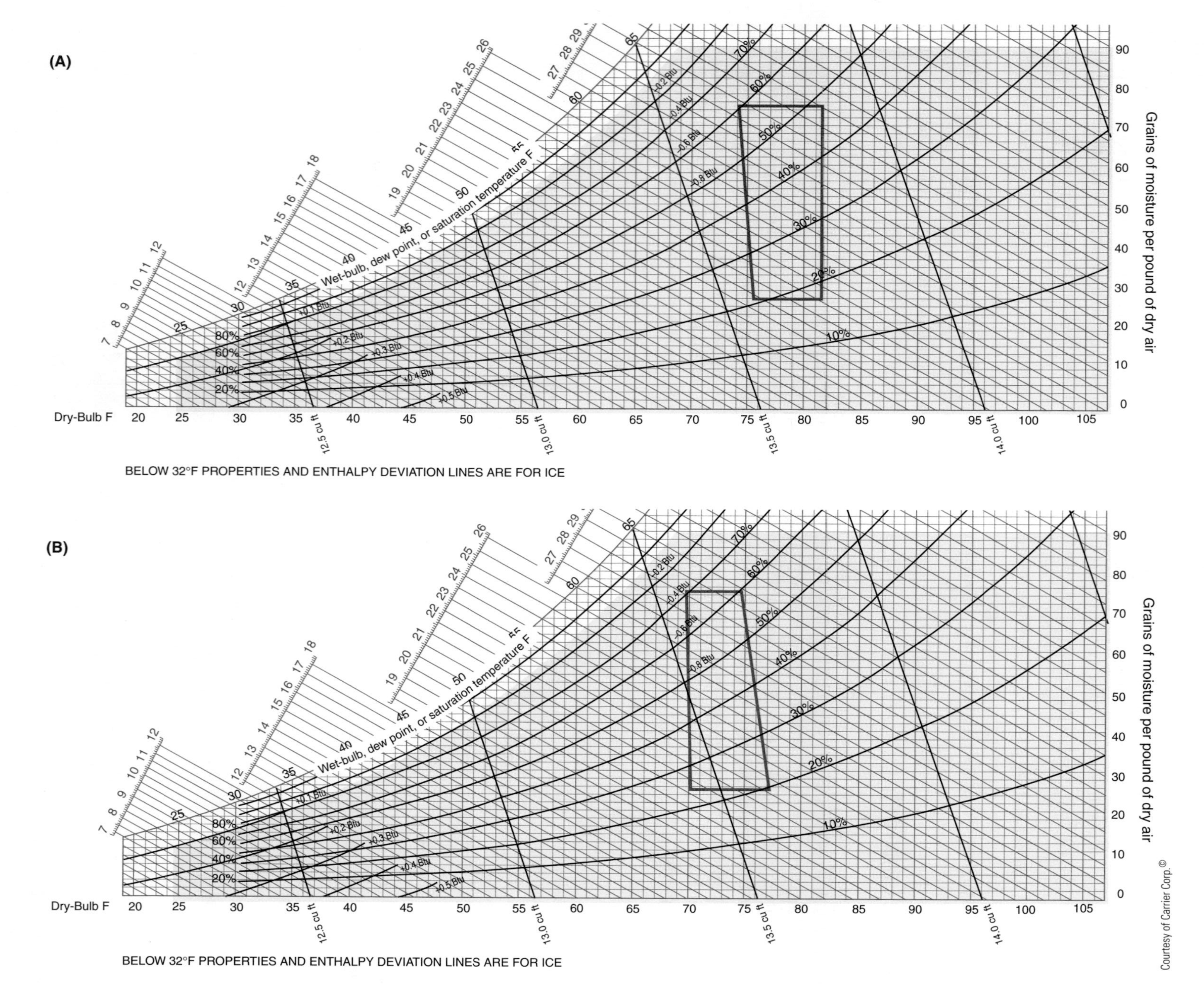

Figure 2.31 Comfort charts for (A) summer and (B) winter.

Summary

Building science is at the core of green building. With the exception of very few regions where there is little need for any heating, cooling, or humidity control, every home must be designed to properly manage heat, air, and moisture. Understanding how these key components of building science make a home energy efficient, durable, and healthy is the key to green building.

Review Questions

1. In which direction does heat flow?
   a. Up
   b. From a cold surface to a cold surface
   c. Down
   d. From a hot surface to a cold surface
2. Where will air leakage occur if the stack effect is the only driving force in a house?
   a. Above the neutral plane
   b. At the neutral plane
   c. Below the neutral plane
   d. A and C
3. What is the primary mode of heat transfer through an exterior wall assembly in a thoroughly airsealed home?
   a. Convection
   b. Evaporation
   c. Radiation
   d. Conduction
4. What is the primary mode of heat transfer through an east-facing window on a summer day?
   a. Convection
   b. Evaporation
   c. Radiation
   d. Conduction
5. A 100-ft^2 wall with an R-value of 10 has a 10-ft^2 window with a U-value of 1.0. What is the approximate UA for this wall?
   a. 9 c. 18
   b. 10 d. 19
6. What is the area weighted average R-value of this same wall?
   a. 0.19 c. 0.20
   b. 5.26 d. 5.0
7. What is the cooling load, expressed in Btu/hr, through this wall if the window is shaded, the indoor temperature is 75°F, and the outdoor temperature is 105°F?
   a. 7,500 c. 2,200
   b. 570 d. 750
8. If this wall was built in a climate of 3,250 CDD, how many Btu of cooling load would occur from conduction over the course of a summer?
   a. 780,000
   b. 426,800
   c. 1,482,000
   d. 61,750
9. What is the R-value of a window with a U-value of 0.37?
   a. 2.7 c. 0.19
   b. 0.37 d. 3.7
10. What is the dew point temperature for a body of air with an air temperature of 80°F and 50% RH?
    a. 60°F c. 70°F
    b. 40°F d. 50°F

Critical Thinking Questions

1. Why might condensation form on the interior of a single-pane window during the winter? What are some possible solutions?
2. What factors affect occupant comfort?
3. What forms of heat transfer are occurring when an attic heats up on a summer day?
4. Define the building envelope and list four common "holes."

Key Terms

air barrier, 50
air sealing, 42
building envelope, 50
building science, 25
British thermal units (Btu), 27
cooling degree-day (CDD), 27
conduction, 28
convection, 33
convective loop, 33
dry-bulb temperature, 44
energy, 27
exfiltration, 33
grain, 44
heating degree-day (HDD), 27
infiltration, 33
low emissivity (low-E), 34
perm rating, 40
psychrometric chart, 42
radiation, 33
relative humidity (RH), 41
R-value, 28
stack effect, 35
thermal barrier, 50
U-value, 29
vapor diffusion retarder (VDR), 40
wet-bulb temperature, 44

Additional Resources

Bruce Harley. *Insulate and Weatherize Your Home: Expert Advice from Start to Finish*. Taunton Press, 2002.

Joeseph Lstiburek. *Builder's Guide to Cold Climates: Details for Design and Construction*. Building Science Press, 2006.

Joeseph Lstiburek. *Builder's Guide to Hot-Dry & Mixed-Dry Climates,* 6th ed. Energy & Environmental Building Association, 2004.

Joeseph Lstiburek. *Builder's Guide to Hot-Humid Climates*. Building Science Press, 2005.

Joeseph Lstiburek. *Builder's Guide to Mixed-Humid Climates*. Building Science Press, 2005.

R. Christopher Mathis, *Insulating Guide*, Building Science Press, 2007.

Energy & Environmental Building Alliance (EEBA), http://www.eeba.org/

Home Energy Magazine, http://www.homeenergy.org/

3

Planning for Green From the Start

Green building is most effective when applied at the earliest stages of a design. This overarching philosophy of creating a good building must be carried from design through construction. All too often, a homeowner or builder decides to make their home "green" after construction has begun. While there are many opportunities for improvements throughout construction, the greatest benefits can only be reaped during the design phase.

For many builders and contractors, green building is a new and challenging concept. This chapter covers the important design decisions and early construction choices that lead to a green home. Green building is more than simply the structure—site selection and development are important factors in the home's overall environmental impact.

LEARNING OBJECTIVES

Upon completion of this chapter the student should be able to:

- Describe the integrated design process
- Define different types of construction sites
- Define proper solar orientation
- Describe methods of conserving existing natural environment
- Describe construction and demolition waste management strategies
- Identify flexible design strategies

Green Building Principles

Energy Efficiency

Resource Efficiency

Durability

Water Efficiency

Indoor Environmental Quality

Reduced Community Impact

Homeowner Education and Maintenance

Sustainable Site Development

Integrated Design

Green building requires professionals to change the way they think about construction, starting with the integrated design process. Much of green building today is limited to applying green materials and methods to standard home plans that are no different from those that have been built for many years. Although these homes can be considered "greener" than standard homes, they could have been much better at little or no extra cost had different decisions been made at the very beginning. **Integrated design** relies on collaboration between all project stakeholders (the designer, builder, homeowner, and trade contractors) in the initial planning stages of a project.

---

integrated design a collaborative method for designing buildings that emphasizes the development of a holistic design.

Traditional Design vs. Integrated Design

To understand integrated design, we need to compare it with the standard design and construction process. Most buildings go through an established set of steps, starting with design, then pricing, and followed by value engineering, contract negotiations, and construction. Figure 3.1a illustrates the traditional design process. This system relies on a designer to pass information on to a builder, who must then relay this information to trade contractors and ultimately to the homeowner. Since many projects using this process end up over budget and do not meet performance expectations, we must assume that this top-down approach is flawed. Contractors always complain about decisions and details that architects and designers include in their drawings presented for pricing or construction. Designers complain that projects are not built as intended. Mechanical contractors are frustrated because no one left them space to fit equipment or run ductwork. Every one of these points are valid, but more importantly, these problems can be avoided by having everyone involved in the project from the very beginning. Figure 3.1b, on the other hand, shows the communication benefits of an integrated design approach.

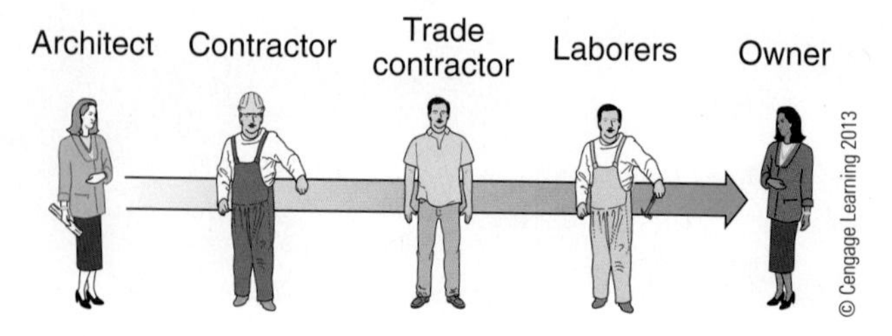

Figure 3.1a The traditional design process does not encourage collaboration among trades and design team members.

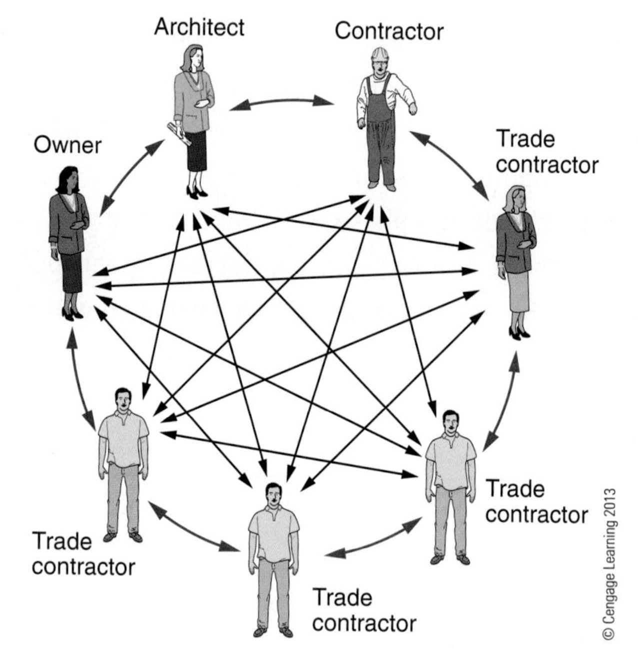

Figure 3.1b The preferable integrated design process encourages collaboration among all involved parties.

Systems Thinking and the Design Phase

Each decision made during design affects the construction outcome, and those made without input from the entire team will often have a negative effect on the job. Consider how you plan a road trip. If your route includes 10 turns and you make the first one wrong, you will not get to your destination regardless of how well you navigate the rest of the trip. Successful projects require making good decisions from the very beginning, which is something done best when working together with the entire team. As we saw in Chapter 2, the home is a complex system of interrelated components. How the framer constructs the home has a direct effect on where the heating, ventilating, and air conditioning (HVAC) contractor is able to run the ductwork and how well the envelope can be sealed and insulated.

Savings From Good Design

One example of the effect that early decisions can have on the cost and performance of a new house is the site's solar orientation. In almost any climate, placing a house with a large wall of unshaded windows facing west will significantly increase the cooling load, requiring a larger HVAC system than if the house had been designed to reduce or shade those west-facing windows. An identical house facing a different direction could save up to 30% in HVAC installation costs as well as benefit from lower power bills for the life of the house (see Figure 3.2). A decision like this can save money and energy with no up-front extra costs.

Charrette

Integrated design is typically accomplished through a charrette, an intensive meeting very early in the design process that includes representatives of every discipline involved in designing and building the project (see Figures 3.3 and 3.4). Examples of skills that should be represented include architects, interior designers, landscape designers and installers, the construction manager, the green rater, the mechanical contractor or engineer, carpenters,

---

solar orientation the cardinal direction in which the home and its glazing faces.

charrette a design meeting consisting of all project stakeholders.

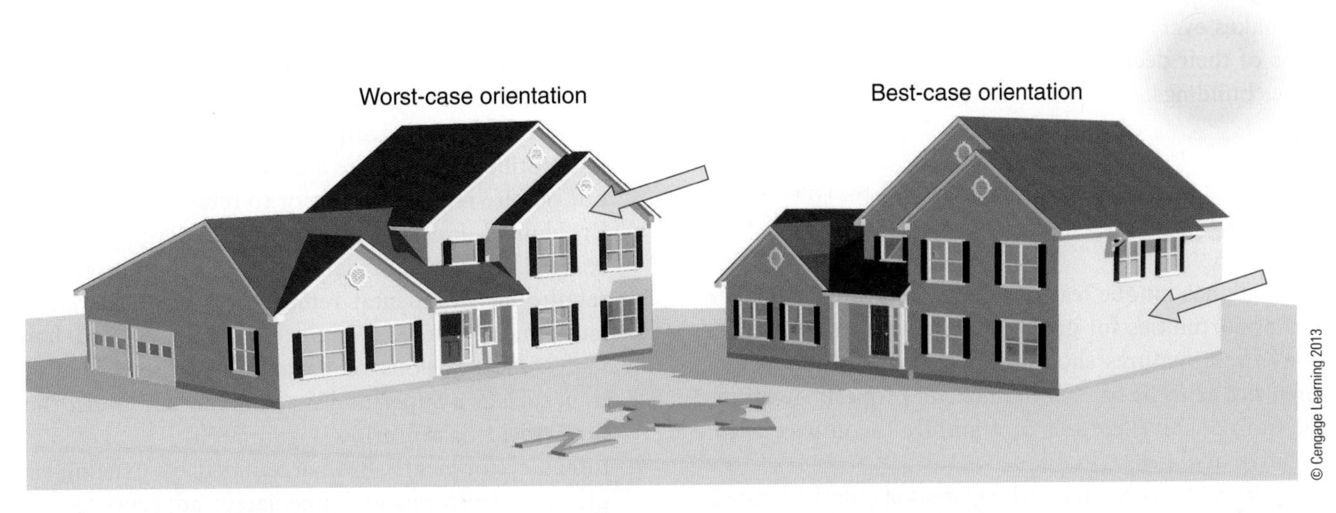

Figure 3.2 A HERS Rating was performed on a standard 2,200-ft^2 new home built in the southeastern United States. The house was modeled facing the worst-case orientation (on the basis of energy consumption) and then rotated to the most efficient orientation. The selection of a site appropriate house orientation allows for cost savings by reducing the air conditioning system from 3.5 to 3 tons and saves an estimated $50 per year on utility bills. Overall comfort is also improved by limiting the amount of late afternoon sun shining in unshaded west-facing windows.

Figure 3.3 People involved in a charrette.

charrette (shar-ette')*n.* **1.** A small cart. **2.** A collection of ideas. During the 19th century, students of l'Ecole des Beaux Arts in Paris would ride in the cart sent to retrieve their final art and architecture projects. While en route to the school in the cart, students frantically worked together to complete or improve these projects. The meaning of the word has evolved to imply a collection of ideas or a session of intense brainstorming. **3.** An intensely focused activity intended to build consensus among participants, develop specific design goals and solutions for a project, and motivate participants and stakeholders to be committed to reaching those goals. Participants represent all those who can influence the project design decisions. [Fr. *charrette*]

Figure 3.4 Definition of a charrette.

plumbers, electricians, and irrigation contractors. Each of these people will have important information and ideas to contribute, and those contributions will make it a better project. One side benefit of involving the entire team from the beginning is developing a sense of ownership in the project. When trade contractors are asked for their input, they become invested in the project's success and will work harder to make sure that everything is done right.

Integrated design challenges the status quo. Such comments as "We never do it this way," "This is a waste of time," and "I can't take the time out of my schedule for a full-day meeting" are common. Frequently, both the homeowner and the contractors express a sense of urgency to start a project, either to "get it done" or to "get some income flowing in." The value of advanced planning is not fully appreciated. Starting a project too soon often results in extended construction schedules as hastily made decisions must be reevaluated and changed during the process. These late, often incorrect, decisions also lead to poorly performing buildings. That poor performance remains for the life of the building with damaging impact to the occupant and the environment. By making informed decisions before construction starts, the team can accurately design for all needs and compress the schedule while giving greater assurance to the owners that their building will perform as designed for years to come.

The long-term benefits of an efficient, healthy, and durable building that is completed on time and at a reasonable price are often sacrificed in the name of expedience. Incorporating integrated design into a project makes everyone think about the long-term implications of their decisions, leading to a better process and better buildings.

Site Selection

Each site has unique features that will create challenges and opportunities for green home design and construction. These features should be identified through thorough site analysis so the building can properly integrate the site's positive features or minimize the impact of its negative qualities.

Buildings, roads, and parking areas should be located to minimize negative environmental impacts. Avoid developing land that is ecologically sensitive (including wetlands or rare habitats), prime farmland, culturally or archeologically significant, or vulnerable to wildfire or floods. Sites with available renewable energy sources (e.g., solar, wind, geothermal, or biomass) are preferable. Sites should allow for orienting the home for proper solar gain and shading.

Environmentally sensitive lands include the following:

- Land within the 100-year floodplain as defined by Federal Emergency Management Agency (FEMA) (Figure 3.5)
- Land that is specifically identified as habitat for any species on federal or state threatened or endangered lists
- Land within 100 feet of any body of water
- Areas identified as of special concern by state or local jurisdiction
- Land that was public parkland
- Land identified by state Natural Resources Conservation Services soil surveys to contain "prime soils," "unique soils," or "soils of state significance" (Figure 3.6)

Renovating existing buildings should always be considered before looking for new building sites. If no suitable existing building can be found, previously developed or infill sites should be evaluated next. These strategies all preserve farmland and ecologically valuable natural areas while limiting urban sprawl. These options also tend to have lower infrastructure costs because transportation access and utilities, such as sewage, electricity, and gas, are usually already in place. Finally, locating close to planned or existing schools, businesses, entertainment, and retail is convenient and can potentially reduce automobile use by residents.

The following list describes the different types of sites:

- **Greenfield development**: sites that are previously undeveloped, such as forest or farm land; should be avoided whenever possible to preserve natural resources (Figure 3.7a).
- **Grayfield development**: previously developed and underutilized real estate assets or land; typically do not require remediation prior to reuse.
- **Brownfield development**: previously developed sites contaminated by hazardous waste or pollution that require environmental remediation prior to reuse; land that is more severely contaminated and has high concentrations of hazardous waste or pollution, such as a Superfund site, does not fall under the brownfield classification (Figure 3.7b).
- **Infill development**: sites surrounded by existing or planned development immediately adjacent to the property boundaries; they can be greenfield, grayfield, or brownfield, and they either already contain infrastructure on site or can easily make these connections (Figure 3.8a and Figure 3.8b on page 64).
- **Edge development**: sites with 25% or more of the property boundary adjacent to existing development.

Density

One way to decrease construction's environmental impact is to decrease the footprint of all the homes in any new development. This can be done by increasing the number of dwelling units on the site. This density is achieved by placing homes on smaller lots or by building *multifamily* projects. Multifamily units are typically more energy-efficient than single-family detached homes. Individual units have smaller heating and cooling loads because walls, floors, and ceilings are typically against other conditioned units. *Mixed-use* projects are a sustainable option when building multifamily housing. Mixed-use buildings contain a commercial component (office or retail) that is typically on the ground level. By providing residential and commercial services nearby, residents are more likely to walk instead of drive.

---

greenfield development a previously undeveloped land, in a city or rural area that is currently used for agriculture, landscape design, or wilderness area.

grayfield development previously developed and underutilized real estate assets or land.

brownfield development an abandoned or underused industrial or commercial facility available for re-use.

infill development the insertion of additional housing units into an already approved subdivision or neighborhood.

edge development a site with 25% or less of the property boundary adjacent to existing development.

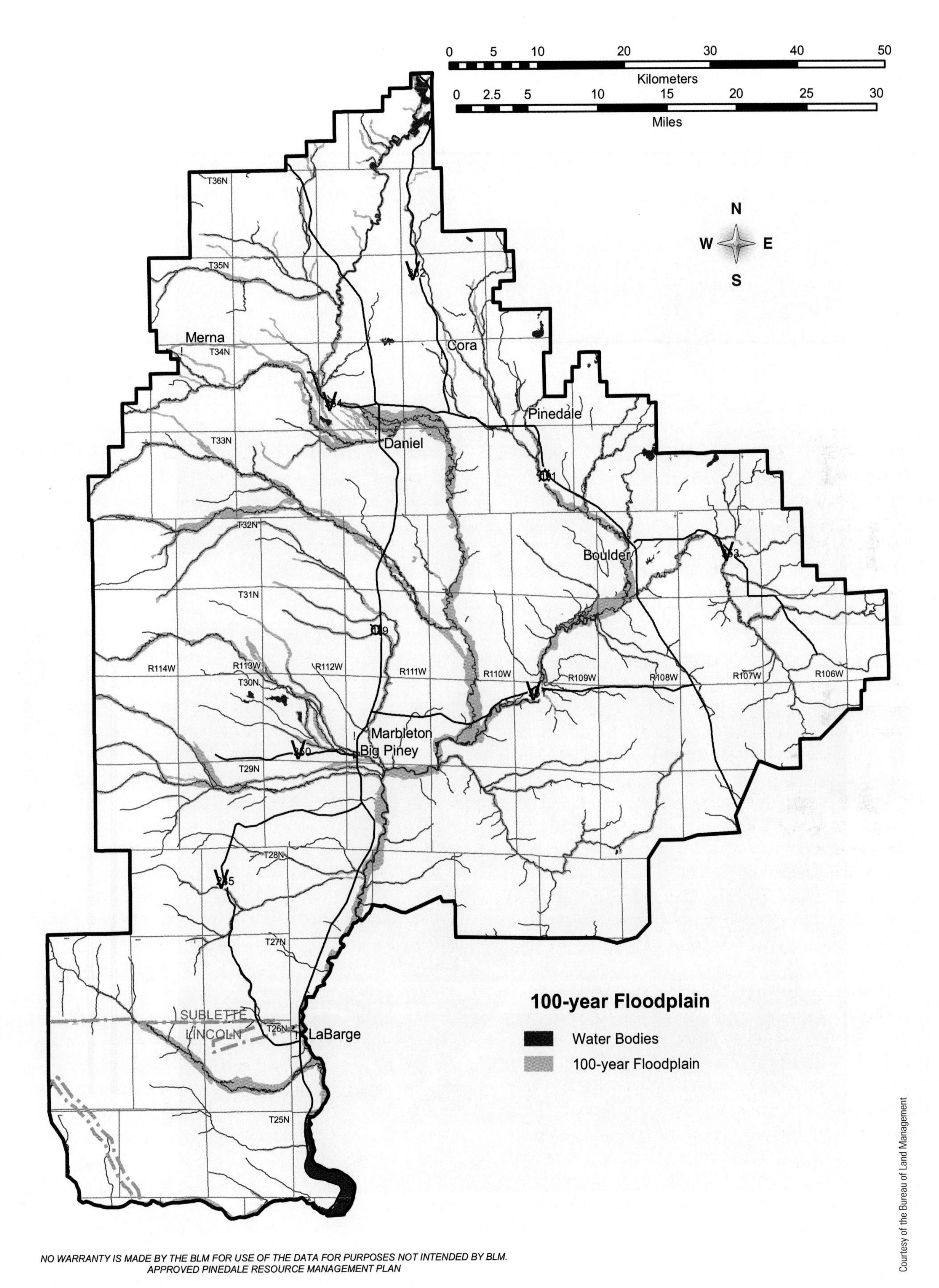

Figure 3.5 A 100-year flood map showing FEMA flood zones.

Figure 3.6 Soil map for a community in Georgia.

Map Unit Legend

Appling and Jeff Davis Counties, Georgia (GA601)			
Map Unit Symbol	**Map Unit Name**	**Acres in AOI**	**Percent of AOI**
Bf	Bayboro loam	14.5	7.2%
LL	Leefield soils	83.7	41.7%
Ls	Leefield loamy sand	0.9	0.5%
Mn	Mascotte sand	6.9	3.4%
Oa	Olustee sand	10.0	5.0%
Pl	Pelham loamy sand	84.9	42.2%
Totals for Area of Interest		**200.9**	**100.0%**

Figure 3.6 (Concluded) Soil map for a community in Georgia.

Figure 3.7a Greenfield sites, such as farms, should be preserved whenever possible.

Figure 3.7b This abandoned warehouse is a preferred site because the site has already been disturbed and is bordered by other developments.

Site Development

Site clearing and earth moving can contribute to runoff and erosion. Home construction is a major source of water pollution because of the erosion caused by soil changes, the chemicals used during various construction phases, and litter left during and after construction. Once an adequate site is selected, an erosion control plan should be designed and implemented to protect natural ecosystems. The plan should be developed for the entire site to address erosion control, no disturbance zones, and staging of construction material.

Did You Know?

Costs of Going Green

Construction costs on infill or previously developed sites are often greater than those on greenfields. Such sites are typically more expensive because of their proximity to amenities or community centers. Furthermore, the pricetag on infill sites may be larger because of additional regulatory requirements and costs for demolition, soil remediation, and additional permitting or legal requirements. Although the sites may be more costly, the location-based amenities often increase the sales price of the home, and state or regional subsidies may be available to help reduce construction costs.

Figure 3.8a Glenwood Park is an infill community in Atlanta, Georgia. This is the former industrial site prior to redevelopment.

Figure 3.8b Glenwood Park as of October 2007.

FROM EXPERIENCE

Sprawl and Health

Howard Frumkin, M.D., Dr.P.H., Dean, School of Public Health, University of Washington. In recent years, health professionals have increasingly recognized the health consequences of urban sprawl. As metropolitan areas spread outward over great distances, several common attributes emerge. Land is developed with low density. Different land uses—residential, commercial, recreational, and so on—are separated from each other. Travel demand is high, and trip distances are long. Connectivity—the ease of getting from one point to another—is low when winding roads in disjointed subdivisions replace traditional gridlike arrangements. Roads are built without sidewalks, and automobile travel eclipses walking, bicycling, and mass transit.

Howard Frumkin, M.D., Dr.P.H., M.P.H., is the dean of the School of Public Health at the University of Washington. He was formerly the director of the National Center for Environmental Health and Agency for Toxic Substances and Disease Registry at the U.S. Centers for Disease Control and Prevention. He is an internist, an environmental and occupational medicine specialist, and an epidemiologist.

This pattern has several direct and indirect health effects. First, with the loss of "active transportation," routine physical activity is designed out of sprawling communities. Sedentary lifestyles raise the risk of many ailments, from cardiovascular disease and cancers to osteoporosis and depression. Second, large amounts of driving contribute to local and regional air pollution, which aggravates heart and lung diseases. Third, more time in automobiles increases the probability of being in a car crash—and car crashes are the leading killer of young people in the United States. Sprawl may also threaten health in other ways: by undermining "social capital" (the social connections that glue communities together and that incidentally have strong health benefits), by compromising mental health (think of road rage), and by contributing to greenhouse gas emissions.

For these reasons, developers should seek out design strategies that increase density and encourage mixed land use, connectivity, pedestrian infrastructure, mass transit, and parks and greenspaces. These strategies not only create more environmentally friendly and liveable communities but also promote better health.

For further information, see Frumkin, H., Frank, L., Jackson, R.J. *Urban Sprawl and Public Health: Designing, Planning, and Building for Healthier Communities*. Washington, D.C.: Island Press, 2004.

Controlling Erosion

For larger sites, use phased grading to limit the extent and exposure of soils to erosion. Erosion is the removal of solids (i.e., sediment, soil, rock, and other particles) by wind, water, or ice in the natural environment. Once the site is disturbed, many techniques can prevent affected soils from entering sewers and waterways. Silt fencing keeps sediment and runoff onsite (Figure 3.9). A silt fence is a temporary sediment barrier made of woven, synthetic filtration fabric and supported by either wood or steel posts. Most jurisdictions require silt fencing as a minimum erosion control measure and provide best management practices for additional methods of erosion control. Additionally, straw bales, silt fencing, silt sacks, and rock filters can protect storm sewer inlets (Figure 3.10). Disturbed topsoil should be stockpiled and covered to prevent erosion and allow future reuse. Implementing terracing, retaining walls, and restabilization techniques prevent long-term erosion.

Stormwater

Stormwater is the flow of water that results from precipitation following rainfall or as a result of snowmelt. Impervious surfaces like building rooftops, driveways, sidewalks, and paved streets prevent stormwater from naturally percolating into the ground. As the runoff flows over the land or impervious surfaces, it accumulates debris, chemicals, sediment, automotive fluids, or other pollutants

Figure 3.9 A silt fence installed along the edge of a property for erosion control.

From CAUFIELD. *Going Green with the International Residential Code, 1E.* © 2011 Delmar Learning, a part of Cengage Learning, Inc. Reproduced by permission. www.cengage.com/permissions

Courtesy of Filtrexx

Figure 3.10 A filter sock preventing sediment from entering the storm water system.

before it is discharged into nearby streams and rivers. Identifying sources of storm water pollution and keeping this pollution away from storm drains and ditches is the best and most economical way to keep storm water clean.

Stormwater pollution control is regulated by the Clean Water Act amendments of 1987. The amendments authorized the U.S. Environmental Protection Agency (EPA) to expand the National Pollutant Discharge Eliminated System (NPDES) program to cover storm water discharges. Under the NPDES, operators of certain construction projects may be required to obtain authorization to discharge stormwater, depending on the amount of land that will be disturbed and the potential threat to the water quality standard. The NPDES is a federal program that is implemented on a state level. The U.S. EPA has granted authorized state environmental agencies to administer the program in all states except Massachusetts, New Mexico, Alaska, Idaho, and New Hampshire; the U.S. EPA oversees the program in these five states, at certain federal facilities, and on Native American lands. The primary method to control stormwater discharges is the use of best management practices (BMPs).

Stormwater Management

Whenever possible, permanent storm water systems should be designed and implemented to capture and reuse stormwater. Captured water may be used for site irrigation or building use. This strategy is explored more in Chapter 15 (plumbing).

Before releasing stormwater from the site, low-impact methods of treating the quality of the water and slowing the rate of release should also be examined.

---

erosion the removal of solids (e.g., sediment, soil, rock, and other particles) by wind, water, or ice in the natural environment.

stormwater the flow of water that results from precipitation following rainfall or as a result of snowmelt.

best management practices (BMPs) strategies for keeping soil and other pollutants out of streams and lakes; BMPs are designed to protect water quality and prevent new pollution.

Did You Know?

Pollution Effects

Excess sediment caused from erosion can cloud water and prevent sunlight from reaching aquatic plants. Plants without adequate sunlight are not able to grow. Excess nutrients from fertilizer runoff can cause **algal blooms** in which the algae population of an aquatic system grows exponentially within a short time period. As dead algae decompose, oxygen—necessary for other aquatic organisms to live—is removed from the water. Algae can also block sunlight and release harmful toxins. Additional effects include the following:

- Bacteria and pathogens can enter recreational waters and create human health risks.
- Garbage and debris (e.g., cigarette butts, six-pack rings, plastic bags) can harm or kill aquatic organisms and wildlife.
- Household hazardous wastes (insecticides, pesticides, paint, solvents, used motor oil, etc.) can poison aquatic life.
- Land animals and people can become sick or die from eating diseased seafood.
- Polluted stormwater threatens fresh drinking water sources, threatening human health and increasing drinking water treatment costs.

These precautions will prevent the quick release of pollutants from impervious surfaces into nearby water bodies. Chapter 11 examines landscaping techniques that can be used to reduce impervious surfaces and to filter pollution from stormwater runoff.

Prevent Site Disturbance

Site disturbance during construction should be minimized through tree or plant preservation areas and selective excavation. Construction drawings should contain clearly labeled construction and disturbance boundaries. The site should be clearly marked to prevent unnecessary disturbance of other soils. Construction materials and vehicles can compact soils and must be kept out of undisturbed areas.

Furthermore, the site's natural vegetation and topography should be used whenever possible. When situating the house, driveways can be aligned with natural topography to minimize grading and reduce cut and fill. If properly protected during construction, existing trees can help shade the home, thus reducing energy needs in the summer.

Preserving existing vegetation may not be appropriate for previously developed sites, especially brownfields. In these instances, the focus should be on rehabilitating the site and restoring natural areas by using local plant species.

Minimize Slope Disturbance

Slope disturbance is the process of destabilizing a slope through grading or site development. Typically defined as having a grade of 25% or greater, steep slopes present serious erosion challenges and should be left undeveloped whenever possible. Hydrological soil stability studies are helpful tools in guiding site development. If slopes are disturbed during construction, tiers, erosion blankets, compost blankets, filter socks, berms, or other measures should be used to keep soil stabilized (Figure 3.11a and Figure 3.11b).

Courtesy of Filtrexx

Figure 3.11a Freshly installed filter socks provide slope stabilization.

Courtesy of Filtrexx

Figure 3.11b The same slope stabilization project, a couple of months later.

algal bloom the rapid, excessive growth of the algae population in an aquatic system within a short time period.

slope disturbance a process of destabilizing a slope through grading and site development.

Staging of Construction Material

The erosion control plan should indicate the area on which construction equipment and materials will be stored. The staging area should be selected carefully to prevent soil disturbance and compaction of sensitive areas.

House Design

Design decisions, such as home orientation, window location and shading, and structural complexity, all contribute to energy consumption and the overall environmental impact of a building. Because the size of the home directly impacts the amount of materials used and the waste created during construction, square footage should be no larger than is needed to achieve occupant satisfaction.

Home Orientation

A home's solar orientation significantly impacts energy use and occupant comfort. Human dwellings originally utilized the sun for heat and illumination. The Native Americans of Mesa Verde in southwestern Colorado designed their cliff dwellings to catch the warm rays of the sun in the winter and to take advantage of the hill's cool shadows during the summer (Figure 3.12). The practice of designing a home to utilize the sun's energy for heating and cooling is called passive solar design.

Passive solar design uses a building's components to collect, store, and distribute the sun's energy to reduce the demand for heating. Passive solar uses natural methods of heat flow (radiation, convection, and conduction) and does not employ mechanical equipment. These systems do not have a high initial cost or long-term payback period. Ideally, passive solar heating should be incorporated into the building's initial plan. The home design can be simple, but knowledge of solar geometry, window technology, and local climate is needed to achieve the greatest gain. Chapter 16 covers passive solar strategies in more detail.

Figure 3.12 Passive solar design at Mesa Verde's Cliff Palace.

passive solar design the practice of designing a home to utilize the sun's energy for heating and cooling.

Window Orientation

Select, orient, and size glazing to optimize winter heat gain and minimize summer heat gain for the specific climate (Figure 3.13). Glazing is the transparent part of a wall or door assembly that is usually made of glass, or occasionally plastic. In the northern hemisphere, the sun moves across the southern sky at a lower angle during the winter because of the seasonal shift of the tilt in the earth's axis. During this time, the sun's rays also strike the earth at a shallower angle. South-facing windows allow the winter sun's energy to penetrate further into the home, which provides heating and illumination benefits. In the southern hemisphere, the seasons are reversed, so the majority of glazing should be arranged on the north side of the home.

The building's southern exposure must be clear of large obstacles (e.g., tall buildings or tall trees) that block the sunlight. Although a true southern exposure is desired to maximize solar contribution, it is neither mandatory nor always feasible. If the building faces within 30° of due south, the south-facing glazing will receive about 90% of the optimal winter solar heat gain.

To accommodate the added glazing on the south side of the home, passive solar homes typically employ rectangular floor plans that are elongated on an east–west axis. This floor plan also encourages little glazing on the east–west sides of the home (Figure 3.14).

If designed improperly, glazing can be an energy penalty on the home. Glass naturally allows sunlight through and traps long-wave heat radiation. The greenhouse effect is the buildup of heat in an interior space caused by energy input through a transparent membrane (such as glass) and is why homes heat up on sunny

glazing a transparent part of a wall or door assembly that is usually made of glass or plastic.

greenhouse effect the buildup of heat in an interior space caused by energy input through a transparent membrane such as glass; also refers to process by which planets maintain temperature through the presence of an atmosphere containing gas that absorbs and emits infrared radiation.

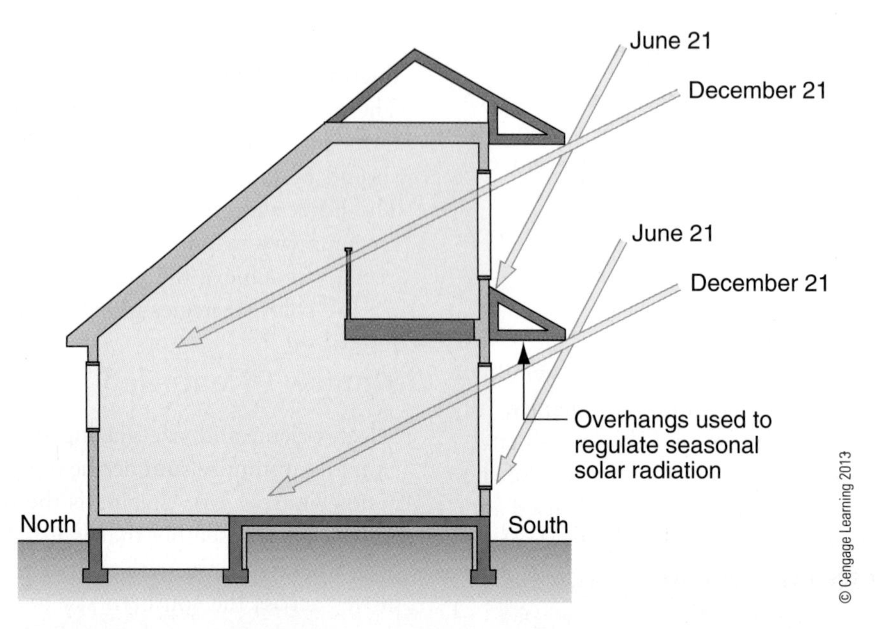

Figure 3.13 Passive solar design utilizes south-facing windows with proper overhangs to control heat gain in the heating season and to avoid excessive heat gain in the cooling season.

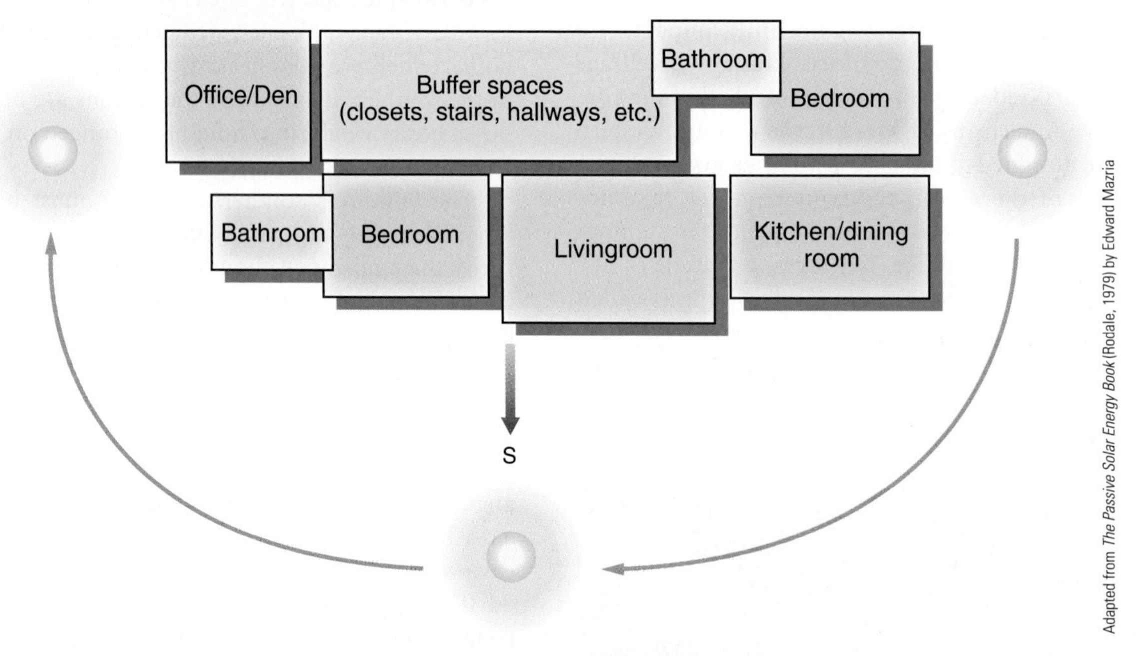

Figure 3.14 Place rooms within a home according to their heating and lighting requirements. Rooms with high heating and lighting requirements should be orientated southeast, south, and southwest. Along the north face of the building, locate spaces with minimal heating and lighting requirements, such as corridors, closets, and garages.

days (Figure 3.15). Greenhouse effect also refers to the process by which planets maintain temperature through the presence of an atmosphere containing gas that absorbs and emits infrared radiation. Windows are not perfect insulators, however, and allow heat to escape. The challenge is to properly size the south-facing glazing to balance heat gain and heat loss properties without overheating. Increasing the glazing area can increase building energy loss and lead to excessive heat gain. See Chapter 8 for more information about windows.

Sun-Tempered Design

Sun-tempered design is a strategy in which the house and most of the glazing are oriented toward the south. The total window area should be limited to approximately 8% of south facing floor area unless rooms contain such elements as ceramic tile or masonry floors or

---

sun-tempered design a passive solar design strategy where the majority of glazing is oriented on the north-south axis.

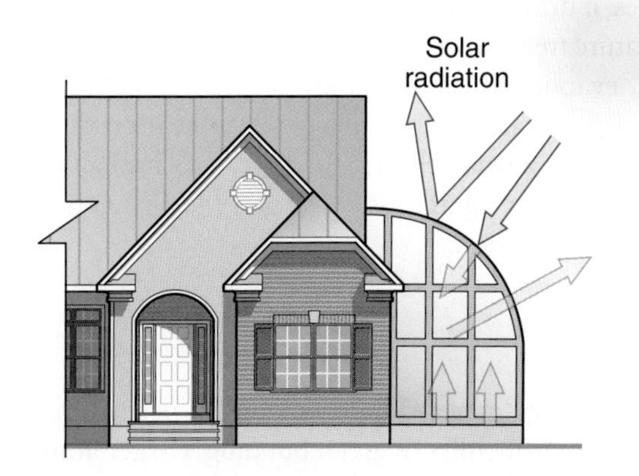

Figure 3.15 The greenhouse effect in the Earth's atmosphere is similar to a glass greenhouse attached to a home or sunroom with lots of glazing. The solar radiation from the sun passes through the glass and heats up the ground. Infrared heat energy from the ground is partially reflected by the glass and trapped inside the greenhouse.

Figure 3.16 Window overhangs come in different forms. This multiple family project uses slotted overhangs on the south-facing windows. The shadow from the early afternoon sun is clearly visible.

walls that absorb heat. If more of these heat-retaining materials are installed, additional south-facing glazing may be included (see Chapter 16 for more details). This simple shift in window location is a great strategy for cold climates and costs nothing beyond good planning.

Window Overhangs

Shading provided by landscaping, overhangs, shutters, and solar window screens helps lower heat gain on windows that receive full sun. An overhang is an architectural feature, such as a soffit, awning, or porch, that protects glazing from the sun (Figure 3.16). South-facing overhangs should be sized to shade windows in summer and allow solar gain in winter. To size window overhangs, you must take into account the home's geographical location (Figure 3.17). Interior shading may be used but is not as effective as exterior shading at blocking solar radiation. Although interior shading does reduce the heating somewhat, it still allows significant amounts of heat into the house. Window overhangs also help direct water away from the home and improve the home's durability.

Figure 3.17 Evapotranspiration is the sum of ground surface evaporation and plant transpiration.

overhang an architectural feature, such as a soffit, awning, or porch, that protects glazing from the sun and walls and doors from rain.

In the northern hemisphere, the sun is highest in the sky around June 21 and the lowest around December 22. These days are referred to as the summer and winter solstices, respectively.

Landscape for Shading

Landscaping can also provide shade for south-, east-, and west-facing windows from summer heat gain. Deciduous trees are ideal because they provide the most shading exactly when the home needs it in the summer. Mature deciduous trees permit most winter sunlight (60% or more) to pass while providing dappled shade throughout summer.

The tree canopy shades and cools the surface temperature below. These cooler surfaces transmit less heat into buildings and the atmosphere, thereby saving energy and reducing the heat island effect. One study examined this effect in two similar buildings and found that shade trees helped to reduce wall and roof surface temperatures by 20°F to 45°F (11°C to 25°C).[1] Landscaping is covered in more detail in Chapter 11.

Evapotranspiration

In addition to providing shade, plants also cool the air through evapotranspiration.

Evapotranspiration describes the transport of water into the atmosphere from surfaces, including soil (soil evaporation) and vegetation (transpiration). Transpiration is the evaporation of water from the aerial parts of plants, especially leaves but also stems, flowers, and roots. Evapotranspiration occurs when plants secrete or "transpire" water through pores in their leaves (Figure 3.17). As water evaporates, it draws heat and cools the air in the process. A single, mature tree with a crown of 30 feet can, with proper watering, evapotranspire up to 40 gallons of water in a day. This is equivalent to removing all the heat produced in four hours by a small electric space heater. Through this process, plants can significantly decrease the air temperature surrounding a home.

House Size

Building a house that fits the needs of the occupants is one of the core principles of green building. Larger homes require more materials to build and more energy to operate. While larger homes can be made green, the concept of a very large green house is an inherent contradiction. According to the U.S. Green Building Council, a 100% increase in the size of a home yields an increase in annual energy usage of anywhere from 15% to 50%, depending on the design, location, and occupants. That same size increase yields an increase in material usage of between 40% and 90%, depending on design and location. Larger homes also tend to be built on larger lots with more space between homes, reducing the walkability of neighborhoods and increasing the reliance on cars.

Two homes were modeled with Athena Sustainable Materials Institute's embodied energy calculator (Figure 3.18). The results in Table 3.1 illustrate the direct relationship between home size and embodied energy. The only difference between the two homes is size and the associated material usage. The construction techniques and location are the same.

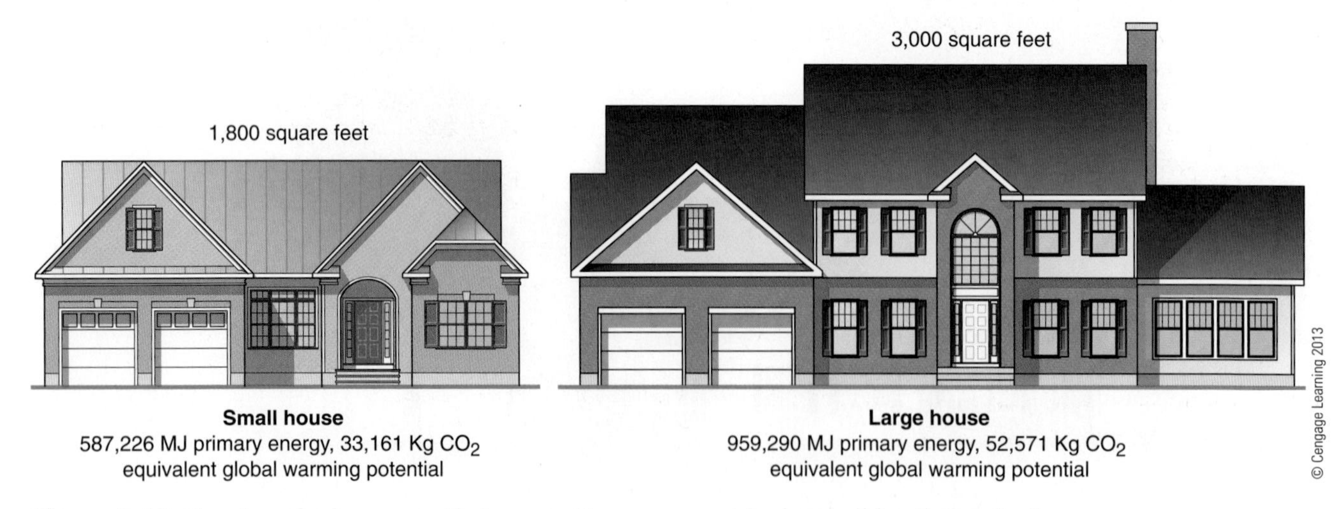

Figure 3.18 The size of a home greatly impacts the energy required to build and operate it.

[1]Akbari H., Kurn D., Bretz S., Hanford J. Peak power and cooling energy savings of shade trees. *Energy and Buildings.* 1997;25:139–148.

Table 3.1 Embodied Energy Home Comparison

Home Square Footage	Primary Energy Consumption (MJ)	Global Warming Potential (kg CO_2 equivalent)
1,800 ft^2	587,226	33,161
3,000 ft^2	959,290	52,571

Source: *Luke Morton using the ATHENA® Impact Estimator for Buildings, 2010.*

Size to Satisfy Need

Smaller homes do not require sacrifice. Through improved design, smaller homes can be more comfortable and pleasing spaces. Smaller homes may lead to smaller mortgages, which, along with smaller utility bills, can provide more financial security for homeowners.

When designing a new home or addition to an existing home, consider the basic needs of the residents and determine the minimum requirements to meet them. Avoid building unnecessary spaces that will get little if any use, and reconsider the common belief that every home must have separate living and dining rooms. In many homes, these rooms are unused much of the time. Paying for the construction, utilities, furnishing, and maintenance of rooms that are not used is a waste of financial, material, and energy resources.

Not So Big House: Sarah Susanka, Taunton Press, 1998

Figure 3.19a This large room is uninviting and feels uncomfortable.

Photo by Christian Korab, courtesy of Susanka Studios, Inc.

Figure 3.19b This home features multiple ceiling height changes and is properly scaled to accommodate the occupants.

Comfort

Another often ignored issue in house size is the comfort factor. People tend to feel most comfortable in spaces that are scaled to accommodate them appropriately. A large banquet hall can feel comfortable with the right number of people in it but empty and desolate with too few. Similarly, a large family room with a vaulted ceiling will be a pleasant place with a large group, but one or two people are more likely to spend time in a smaller room or in a cozy nook off to the side. Such techniques as altering ceiling heights, introducting natural light, constructing alcoves, and similar concepts (many of which are described in the *Not So Big House* series of books by architect Sarah Susanka) can make a small house feel larger and frequently much more comfortable to live in than a large home (Figure 3.19a and Figure 3.19b). Rooms with high ceilings may be attractive, but they waste energy by heating and cooling large volumes of unusable space in a house and can make the home less comfortable to inhabit.

Outdoor Spaces

Decks, patios, or screened or open porches are important elements to consider including in a project (Figure 3.20). Depending on the climate, these spaces can be used anywhere from 3 to 12 months out of the year, providing

Courtesy of Warren Bond Photography

Figure 3.20 This screened-in porch creates an inviting space nearly year round.

FROM EXPERIENCE

What Is a Not So Big House?

Sarah Susanka, FAIA. I first coined the term "not so big" in my 1998 book, *The Not So Big House,* in an attempt to help describe an alternative to our ever-increasing house size. I wanted to make people aware that size has almost nothing to do with the qualities of home that most homeowners are seeking when they build or remodel.

Susanka is a member of the College of Fellows of the American Institute of Architects (FAIA) and a senior fellow of the Design Futures Council. Susanka is the author of nine best-selling books, including The Not So Big House *(Taunton, 1998),* The Not So Big Life *(Random House, 2007),* Not So Big Remodeling *(Taunton, 2009), and most recently,* More Not So Big Solutions for Your Home *(Taunton, 2010).*

What I knew as a residential architect was that many of my clients wanted a better house than their existing one and assumed that better must automatically mean bigger. It's just not so. In fact, in the vast majority of cases, bigger just means bigger, and the new homeowners end up being disappointed that their new house doesn't really feel like the dream home they'd thought they were building.

But Not So Big® doesn't mean small. In fact, it's not about mandating any specific size of house at all. Household needs differ, so the assessment about how much space is needed can only be made by the people who will eventually live there. Instead, it's about focusing on quality rather than quantity and about tailoring the house for the way we actually live, rather than designing for a more formal way of life that no longer reflects our current needs.

I tell people that a good rule of thumb in right-sizing a home to make it Not So Big is to aim for about a third less space than they think they need but to budget about the same amount of money as they would have for their larger vision of home, reapportioning dollars out of square footage and into the quality and character of the interior space and building envelope.

By eliminating rooms that get used only a few times a year, such as the formal living room and dining room, and by designing the house so that every space is in use every day, there's a natural reduction in the home's size without any sense of something being lost. If we're not using those spaces anyway, why build them?

In addition, the walls, windows, roof, and foundation of the house are designed to be highly energy-efficient and are built using sustainable materials and building practices. The house should also be designed to maintain an excellent indoor air quality that can provide a healthy and comfortable platform for everyday life.

I point out to my readers that a smaller but better designed house actually lives larger than one that's significantly bigger because the spaces work together as an integrated whole, perfectly supporting the lives of the inhabitants. It's a strategy that will appeal to not only the original homeowners but also to future generations, providing a delightful as well as comfortable environment for all their lives.

Lastly, a Not So Big house is a house that is beautiful and that inspires those who live within its walls. Beauty really does matter in terms of sustainability because people tend to take good care of places they find beautiful and delightful, so making a home Not So Big should really be one of the first steps in sustainable design and construction.

Some of the key features of a Not So Big house or remodel are as follows:

- Designed for comfort and livability—for the way we *really* live
- Designed to be as energy efficient and sustainable as possible
- Designed for our human scale (rather than for giants)
- Designed to last for centuries rather than for decades
- Designed in all three dimensions, with plenty of ceiling height variety to define and articulate activity areas
- Designed to be just the right size to accommodate the homeowners' needs—not too big, and not too small either
- Designed to be beautiful as well as functional and to inspire its occupants every day

Note: *Not So Big® is a registered trademark of Susanka Studios.*

additional living space that does not require energy to heat and cool. In hot climates, porches provide an added bonus by shading windows to keep out the heat.

Designing for Mechanical Systems

Mechanical systems include heating, ventilation, and air conditioning (commonly referred to as HVAC), plumbing, and electrical systems. Their design, installation, and operations affect comfort, water and power efficiency, and durability. Understanding how each of these complex systems integrate with the entire house is critical to choosing the right system components and making the right design decisions to enhance their efficiency.

HVAC

HVAC systems provide heating, cooling, ventilation, air filtration, and humidity control; the extent of each will vary based on the local climate. Designing systems that are correctly sized ("right-sized") for the house will reduce energy use, extend the life of the equipment, and make the house more comfortable and healthy. A home should be designed to provide space for HVAC equipment to be located within the building envelope (as opposed to unconditioned attics or basements) to maximize energy efficiency. Providing central locations for HVAC equipment allows for shorter duct and pipe lengths, reducing material use and improving overall system efficiency (Figure 3.21).

Plumbing

Plumbing systems deliver hot and cold water and remove waste products from the house. Plumbing codes do an excellent job of providing these services, but they do not take into account water or energy efficiency.

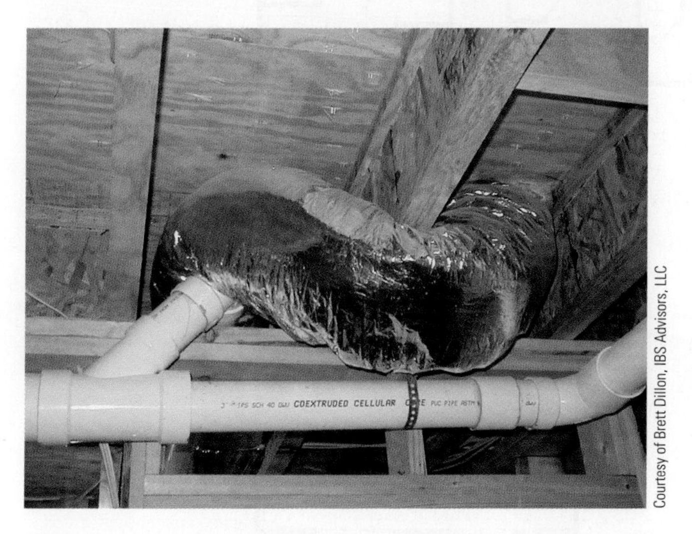

Courtesy of Brett Dillon, IBS Advisors, LLC

Figure 3.21 Pinched and twisted duct has greatly reduced airflow. This problem could have been avoided through careful design and early communication among builder, architect, and HVAC contractor.

Plumbing systems in green homes conserve water with the use of high-performance fixtures, gray water and rainwater reclamation, and efficient water heaters and piping systems. Stacked plumbing reduces the amount of pipe needed to supply fixtures and it saves energy and water by reducing hot water pipe runs (Figure 3.22). See Chapter 15 for additional information on green plumbing strategies.

Electrical

Electrical systems provide the power to run lights, appliances, and equipment. The design and installation of electrical infrastructure affects the amount of material used, its impact on the building envelope, and the occupants' ability to easily control the equipment. Similarly, equipment selection and installation influence a building's total electrical demand, the cooling load, and the impact on the building envelope.

Design for Future Needs

Family size and home usage patterns change over time. The average American family is getting smaller and spending less time in formal dining rooms. Baby boomers are also aging and facing new mobility challenges within their homes and communities. Homes should be able to adapt to these changing needs. Rather than perform a major renovation every few years, a home needs to be flexible enough to satisfy multiple homeowner requirements.

Flexible Design

Homes should be designed and built to last hundreds of years. To achieve this goal, the structure must be strong, durable, and flexible. Because future generations may have completely different needs and requirements than today's occupants, home designers are challenged to create homes that have a built-in capacity to evolve. Such flexible homes decrease the costs of homeownership and the waste generated through remodeling and demolition.

According to the National Association of Home Builders (NAHB), Americans spent $235 billion on remodeling in 2007. While some of these expenditures were for routine maintenance and upkeep, most reflected spending to accommodate changing occupant desires. Through careful planning and increased flexibility, homeowners can largely avoid costly renovations. The environmental costs of these renovations are equally steep in the form of landfill waste and resource consumption.

Flexible design is commonly referred to as *open building* or *design for disassembly* (DfD). DfD is building design that allows for easy recovery of parts, materials, and products when a home is disassembled or renovated. William McDonough, an American architect, and Michael Braungart, a German chemist, helped to

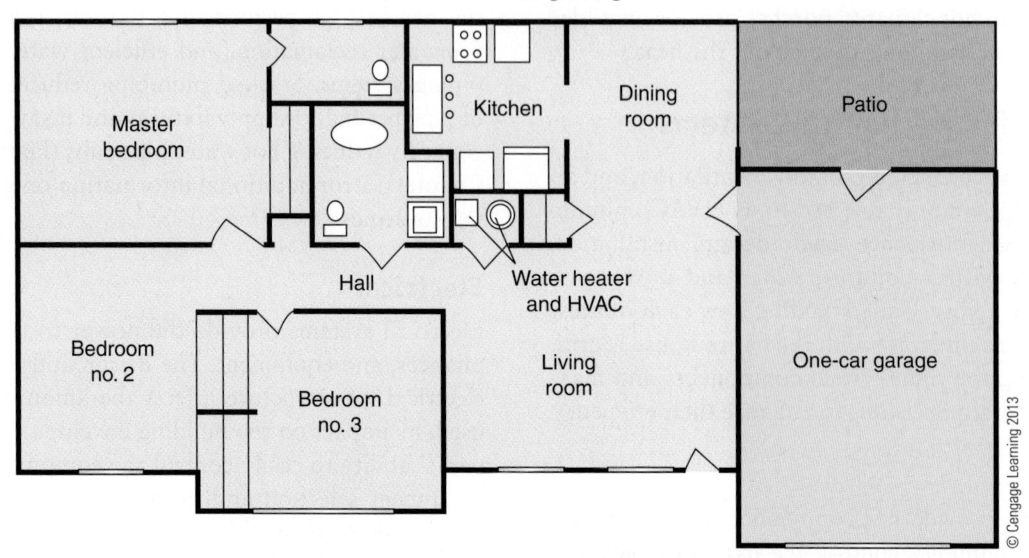

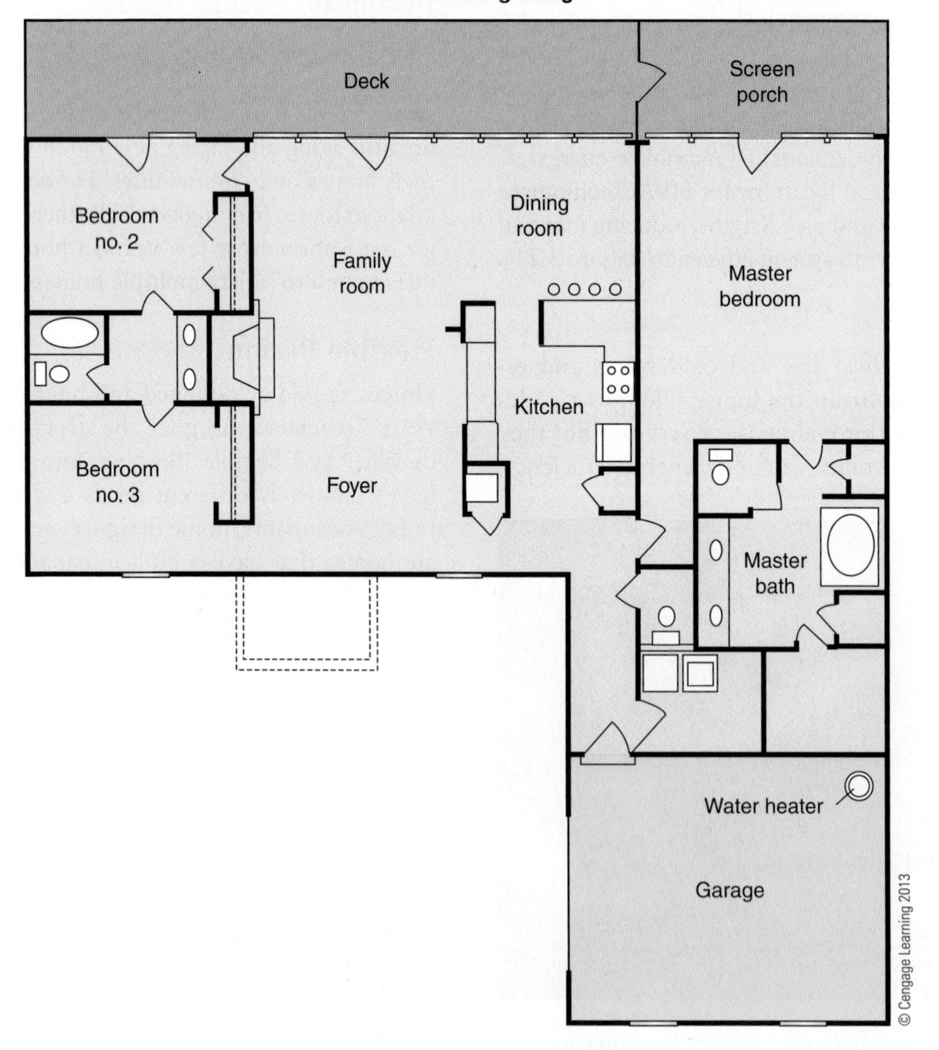

Figure 3.22 One design consideration is to incorporate a central core plumbing layout. Keeping the plumbing areas (kitchen/bathroom/laundry) near one another will save costs by reducing materials use, reducing water consumption, and improving the energy efficiency of the hot water heating system.

popularize this idea through their book *Cradle to Cradle*. Cradle to Cradle (C2C) promotes industrial design that eliminates waste by creating "closed-loop" industrial systems. Within closed-loop systems, the "waste" from one manufacturing process becomes the "fuel" or input for another. Although McDonough and Braungart do not specifically discuss C2C as it applies to home building, the theory can be readily applied to homes. Everything within a home should be easily reusable or recyclable.

Some simple techniques include designing a home so that electrical outlets can be readily added or removed, and HVAC systems can be easily accessed and replaced. More aggressive techniques include installing walls that can be relocated to alter the home's layout.

Universal Design

Universal design provides for homes designed for all people, regardless of age or abilities. Universal design grew out of the post–World War II concept called *barrier-free* living and plays an important role in flexible homes.

The United States' elderly population is growing faster than ever before. The Bureau of Labor Statistics predicts that, between 2002 and 2012, nearly 6 million people between 16 and 54 years of age will join the working population. Over the same period, the 55-and-over age bracket will swell by 18 million. Aging boomers are not quite ready for nursing facilities or assisted-living centers, and they desire homes connected to neighborhoods, cities, and hometowns. The challenge is finding homes that accommodate their aging physical needs.

Universal design promotes social and environmental sustainability. Flexible design allows homeowners to age in place, thus reducing material use and encouraging sustainable development.

The Americans with Disabilities Act (ADA) was signed into law in 1990 and prohibits, under certain circumstances, discrimination based on disability. Although most housing falls outside the purview of the law, the ADA Standards for Accessible Design is a valuable resource. Five common features in accessible design are the following (Figure 3.23):

- **No-step entry:** No one has to use stairs to enter the home or main rooms.
- **One-story living:** Places to eat, use the bathroom, and sleep are all on one level.
- **Wide doorways:** Doorways are 32 to 36 inches wide to accommodate wheelchairs and walkers.
- **Wide hallways:** Hallways are 36 to 42 inches wide for easy access between rooms.
- **Extra floor space:** Open floor plan for easy movement.

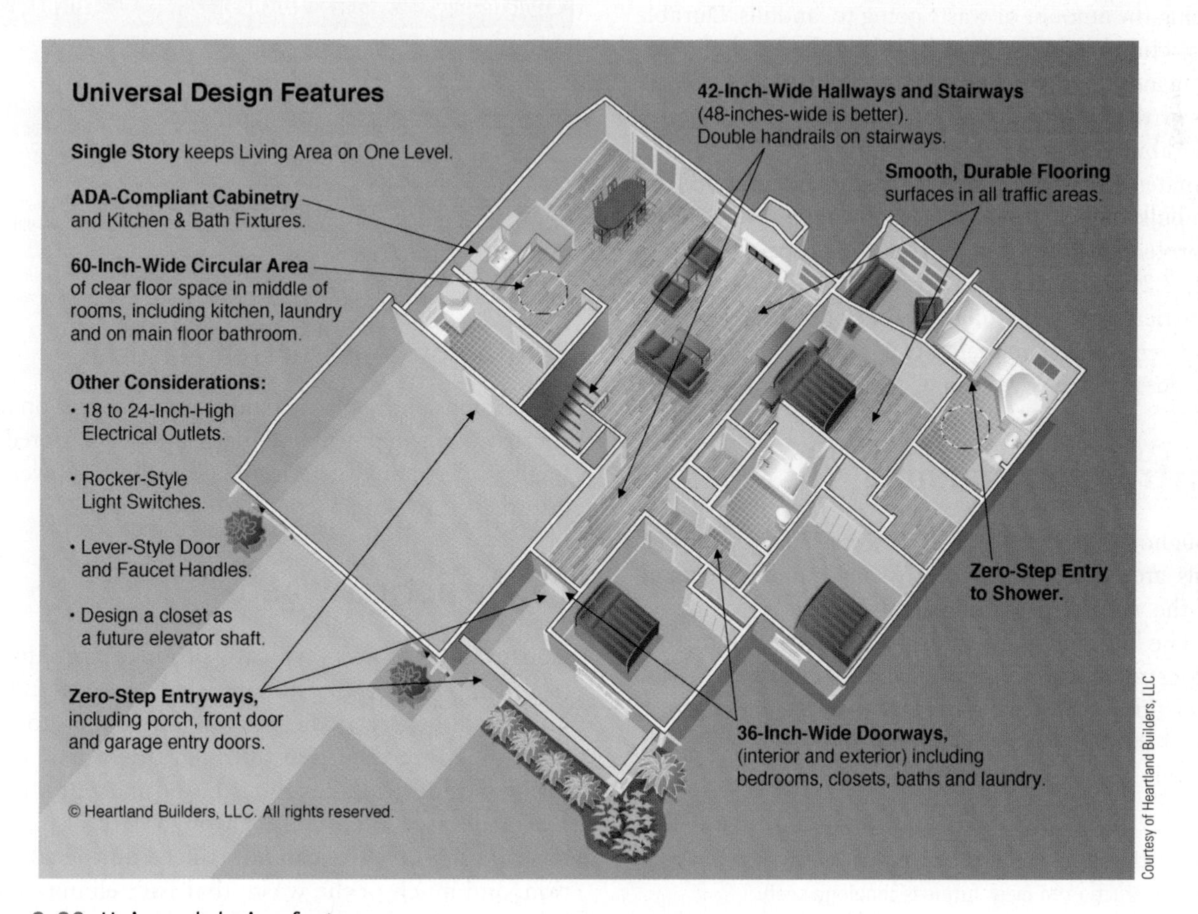

Figure 3.23 Universal design features.

The NAHB Remodelers Council, in collaboration with the AARP, NAHB Research Center, and NAHB Seniors Housing Council, developed the Certified Aging-In-Place Specialist (CAPS) designation. The NAHB reports that home modifications for the aging-in-place population represent the fastest growing segment of the residential remodeling industry. The CAPS program provides remodelers with the knowledge and skills to compete in this market.

Pre-Plumb and Pre-Wire for Solar

Pre-plumbing is the process of installing plumbing distribution systems during construction to meet future technology need and pre-wiring is the process of installing electric wiring during construction to meet future technology demands. Homes designed to accommodate solar technologies are often referred to as "solar-ready homes." By using these processes, homes can be designed so that a solar hot water or photovoltaic system can be easily added.

Durability

Durable buildings save resources by reducing labor and materials required for maintenance, repair, and replacement. Fewer materials are removed and replaced, reducing embodied energy required for their production, and reducing the amount of waste going to landfills. Durable construction methods differ based on the local climate. Rain, wind, flood risk, heat, cold, and pest control are some of the factors that affect decisions made regarding durability (Figure 3.24 and Figure 3.25). Techniques and materials that improve a structure's durability include bulk moisture management through proper flashing, protection from rain and sun by proper overhangs (Table 3.2), termite control through physical and chemical barriers, and extending the life of finishes through proper ventilation and humidity control. Details on these techniques will be covered in Chapters 5, 6, 7, 8, and 10.

Construction Planning

Throughout the construction process, building professionals are faced with numerous opportunities to minimize the negative environmental impacts of construction. The local environment is protected though managing construction processes to avoid site runoff and soil compaction. Construction waste can be significantly reduced through careful planning, and much of the unavoidable waste can be recycled.

pre-plumb the process of installing plumbing distribution systems during construction to meet future technology need.

pre-wire the process of installing electric wiring during construction to meet future technology demands.

Courtesy of Brett Dillon, IBS Advisors, LLC

Figure 3.24 The moisture damage on the window casing is from improper flashing. The flashing doesn't extend over the top of the head trim and kick out. Instead, it is about 1/4-inch short, and the water was able to wick back up under the flashing to the top and back of the trim (which hadn't been back-primed or painted). The moisture then came through and blistered the paint.

Table 3.2 Recommended Minimum Roof Overhang Widths for One- and Two-Story Wood Frame Buildings*

Climate Index	Eave Overhang (Inches)	Rake Overhang (Inches)
Less than 20	N/A	N/A
21 to 40	12	12
41 to 70	18	12
>70	≥24	≥12

*Table based on typical two-story home with vinyl or similar lap siding. Larger overhangs should be considered for taller buildings or wall systems susceptible to water penetration and rot.

Source: Durability by Design: A Guide for Residential Builders and Designers. *Prepared for U.S. Department of Housing and Urban Development, Washington, D.C.*

Staging of Construction Material

The erosion control plan should indicate the area on which construction equipment and materials will be stored. The staging area should be selected carefully to prevent soil disturbance and compaction of sensitive areas.

Construction Waste

Construction projects typically produce large amounts of waste. The NAHB estimates that a typical 2000-ft^2 house will generate four tons of waste that goes into landfills. More than two thirds of this waste, including wood, drywall, cardboard, and masonry, is fully recyclable (see Table 3.3). Through careful planning, the quantity of waste can be reduced during construction, and much of the waste that isn't eliminated can be recycled and repurposed. By reducing the total amount of debris created and recycling as much as

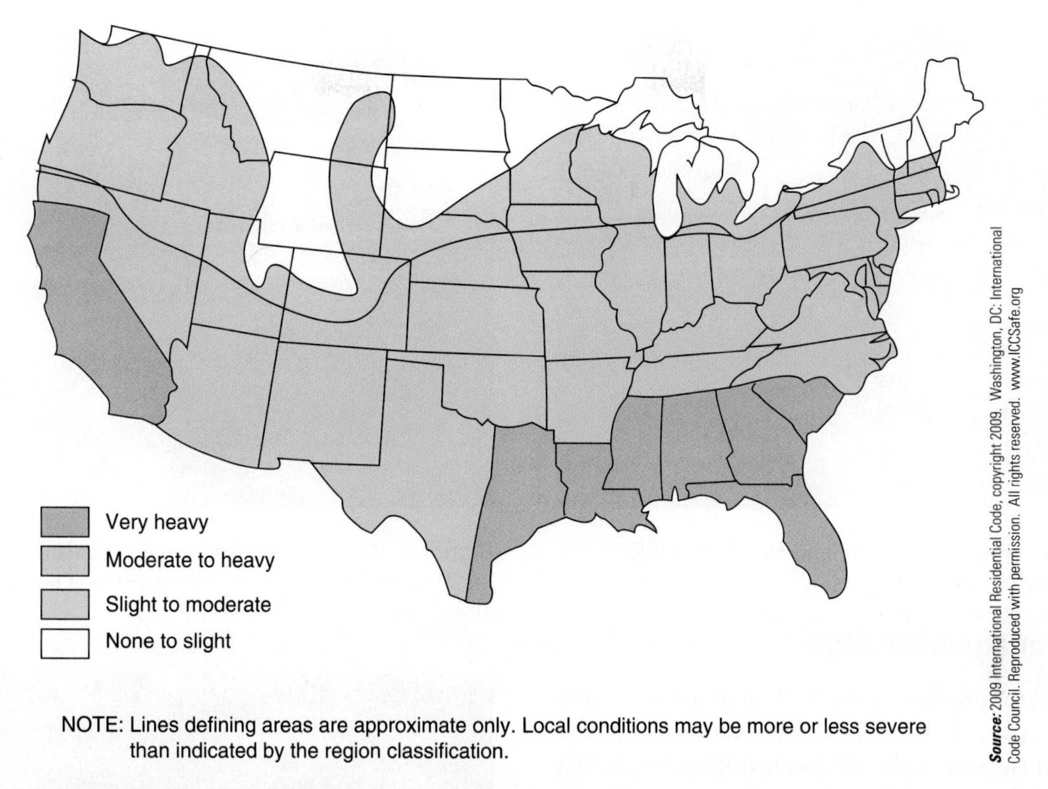

Figure 3.25 Termite infestation probability map, 2009 International Residential Code.

Table 3.3 Construction Waste Materials and Markets

Material	How Is It Recycled?	Recycling Markets
Concrete	The material is crushed, any metal reinforcement bar is removed, and the material is screened for size	Road base General fill Drainage media Pavement aggregate Slab and driveway bedding
Asphalt shingles	After removal of nails, asphalt shingles are ground and recycled into hot-mix asphalt	Asphalt binder and fine aggregate for hot-mix asphalt
Wood	Clean, untreated wood can be re-milled, chipped or ground	Feed stock for engineered particle board Boiler fuel Recovered lumber re-milled into flooring Mulch and compost Animal bedding
Drywall	Unpainted scrap drywall is typically ground or broken up	Gypsum wallboard Cement manufacture Agriculture (soil amendment)
Metal	Melted down and reformed	Metal products
Cardboard	Ground and used in new pulp stock	Paper products
Aluminum cans	Melted down and reformed	Aluminum products

Source: *Modified from* Construction Business Owner, *June 2007.* http://www.constructionbusinessowner.com/topics/environment-and-compliance/recycling-construction-materials-an-important-part-of-the-construction-process.html

possible, builders can minimize the impact on our overflowing landfills and save money by lowering dump fees (Figure 3.26).

Waste Reduction Strategies

To reduce waste, buildings can be designed to a 2-foot modular grid to reduce the amount of wood cutoffs, allowing for more accurate takeoffs and material orders. A central cut area helps to promote the use of scrap materials, and the use of pre-assembled components (such as trusses, panelized, or modular construction) also minimizes waste. Renovation and demolition projects can be deconstructed, and many of those materials recycled instead of discarded. Valuable products can be sold for a profit or donated to nonprofit organizations, providing tax deductions to the building owner.

© Cengage Learning 2013

Figure 3.26 This dumper is bound for the landfill despite being full of recyclables.

© Cengage Learning 2013

Figure 3.27a Construction waste containers do not need to be fancy. This builder uses a scrap barrel for aluminum cans and glass bottles.

Waste Management Plan

You can reduce job-site waste by writing a good waste management plan, posting it on the job-site, and enforcing it with all on-site staff and trade contractors to ensure that it is followed throughout the construction process (Figure 3.27a and Figure 3.27b).

© Cengage Learning 2013

Figure 3.27b Waste materials can be sorted in different containers or comingled, depending on the waste management company. This waste management firm sorts at a centralized facility and then processes all recyclables.

Remodeling

Remodeling is naturally a green practice. Existing homes are improved and expanded rather than being abandoned or demolished. Every square foot of living space that is reclaimed and upgraded into a more efficient and healthy living space will save thousands of dollars in energy costs, tons of waste discarded, and more tons of new material required for a replacement structure.

Large Market

According to the Joint Center for Housing Studies at Harvard University, over 800,000 room additions are constructed each year (Figure 3.28a and Figure 3.28b). Each of those projects that is remodeled green will provide the owners with decades of reduced energy bills, improved comfort, healthier air, and lower home maintenance costs. Green remodeling provides long-term benefits to homeowners, remodelers, local economies, and our environment. Every remodeling project that is not green is a missed opportunity that will last for generations and deprive homeowners of energy savings and improved indoor air quality.

Local Economy

Green remodeling is both a local and a global issue; old homes will always outnumber new homes. Older homes are typically inefficient and provide unhealthy indoor air. Inefficient homes require increased utility expenses. Utilities, being part of a global market, tend to direct money out of the local economy. Instead, if we were to invest money in making homes more efficient, that money would be spent locally on labor and materials while reducing the amount spent on utility bills.

Indoor Environmental Quality

The air inside most homes is less healthy than the air outside (Figure 3.29). Most people spend over 90% of their lives indoors. Ever-increasing allergies and asthma are a direct result of unhealthy indoor air. We are

Figure 3.28a A home in Atlanta, Georgia, prior to a major renovation.

Figure 3.28b The same home post renovation. The renovation incorporated many green elements, including a design of building up to minimize resource use.

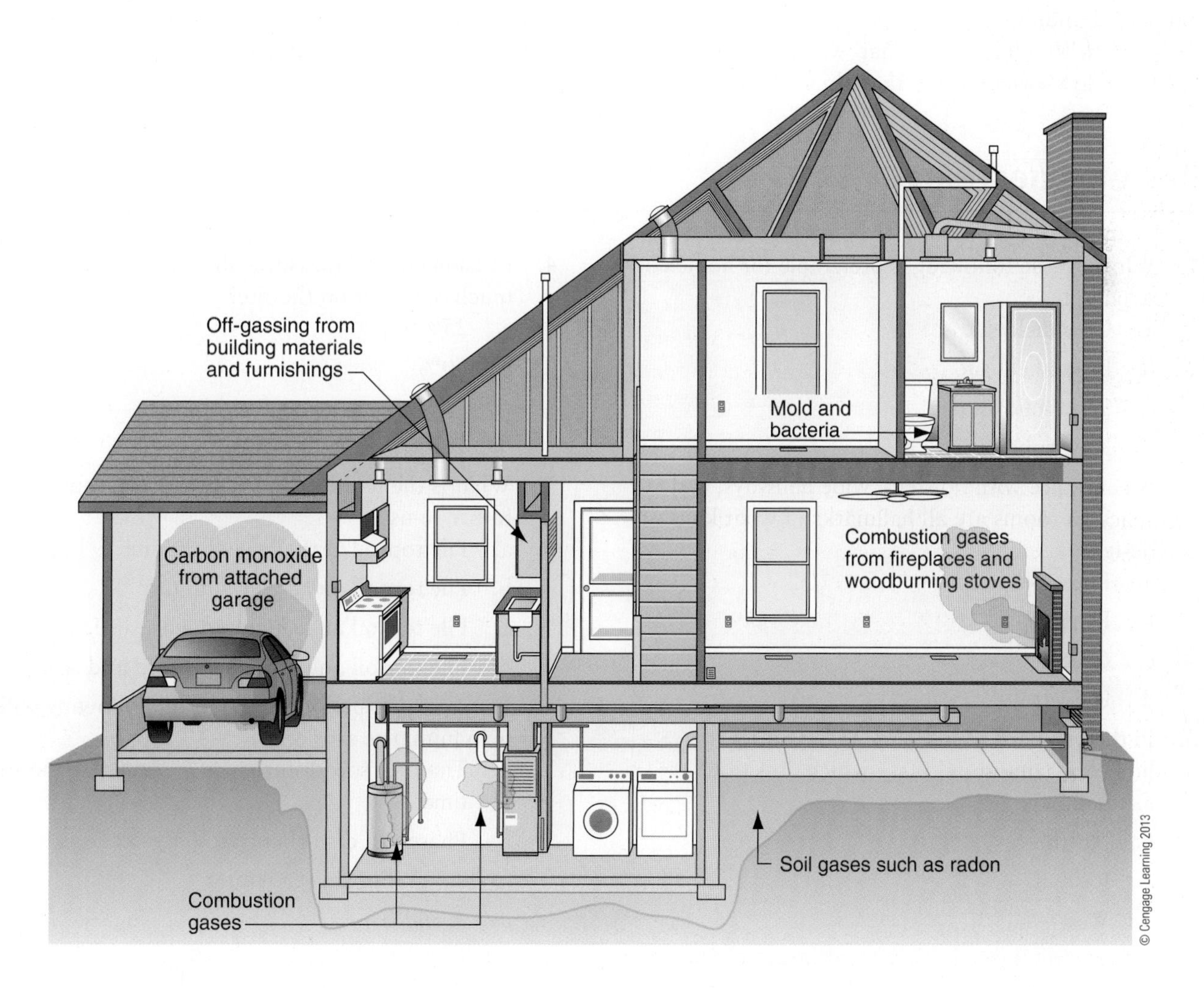

Figure 3.29 Indoor environmental pollutants come from many sources.

literally poisoning ourselves and our children by the way we build our houses. The knowledge is there to make homes healthier, and it is not rocket science. It is building science. Occupants of green homes continue to report that they are healthier and have reduced allergies and asthma. Green remodeling provides us with the ability to make almost any existing home a healthier place to live.

Renovation Planning

When planning for a green remodeling project, it is critical to evaluate and correct structural defects, to identify moisture control problems, and to recognize any hazardous materials that may be disturbed during renovations. Performance testing of the house using such tools as a blower door, duct blaster, and thermal imaging camera will provide you with information about hidden defects and provide a baseline of the house conditions against which you can measure the proposed and completed improvements. One point to keep in mind is the overall condition and value of the structure. In certain instances, deconstructing and replacing a house may be more appropriate than remodeling when the cost and effort will not deliver a project that is sufficiently healthy and efficient.

Summary

For a new home or renovation to be as green as possible, sustainable principles must be considered from the very start of the project. When early decisions are made without considering their effect on the sustainability of the project, many opportunities are lost. When this happens, buildings are often poorly located, oversized, designed with dimensions that waste materials, and have window orientations that do not maximize the sun's energy. When projects are developed using an integrated design process that includes all the stakeholders from the beginning, better decisions are made, and a project can incorporate green principles successfully. When executed correctly, an integrated design process that considers green principles from the start can ensure that a project is truly green at little or no additional cost.

Review Questions

1. Which of the following is preferable for home construction?
   a. Greenfield site
   b. Brownfield site
   c. Superfund site
   d. New subdivision
2. An entrance with no steps, wide hallways, and spacious rooms are all hallmarks of what kind of design?
   a. Energy-efficient
   b. Flexible
   c. Green
   d. Universal
3. In the northern hemisphere, the majority of a home's glazing should face which direction?
   a. South
   b. North
   c. East
   d. West
4. Erosion control measures should contain how much sediment on the site?
   a. 25%
   b. 50%
   c. 75%
   d. 100%
5. What is the best method of protecting topsoil for future re-use?
   a. Pile topsoil at the rear of the site
   b. Pile topsoil and cover with tarp
   c. Pile topsoil and surround with silt fence
   d. Pile topsoil and cover with seed and straw
6. During construction, which of the following will help minimize erosion?
   a. Time all soil disturbance to occur at the same time
   b. Phase soil disturbance to only occur as needed
   c. Do grading in the winter
   d. A and C

7. Smaller homes that meet the occupants needs will result in:
   a. Uncomfortable homes
   b. More efficient homes
   c. Cheaper mortgages
   d. Smaller carbon footprint
8. A typical 2,000-ft^2 home generates how much landfill waste?
   a. 1 ton
   b. 4 tons
   c. 8 tons
   d. 24 tons
9. Typically, how much construction waste is recyclable?
   a. None
   b. One fourth
   c. One half
   d. Two thirds
10. The average American spends how much time indoors?
   a. 25%
   b. 60%
   c. 75%
   d. 90%

Critical Thinking Questions

1. What is integrated design, and what are the benefits?
2. Why is universal home design considered green?
3. List and describe the steps a builder should take to manage erosion and stormwater on the construction site.
4. Describe three architectural decisions that affect a home's energy use.

Key Terms

algal bloom, 66
best management practices (BMPs), 65
brownfield development, 60
charrette, 58
edge development, 60
erosion, 65
glazing, 67
grayfield development, 60
greenhouse effect, 67
greenfield development, 60
infill development, 60
integrated design, 57
overhang, 69
passive solar design, 67
pre-plumb, 76
pre-wire, 76
slope disturbance, 66
solar orientation, 58
stormwater, 65
sun-tempered design, 68

Additional Resources

AARP: http://www.aarp.org/

Lifecycle Building Challenge: http://www.lifecyclebuilding.org/

Howard Frumkin, *Urban Sprawl and Public Health: Designing, Planning, and Building for Healthier Communities* (Island Press, 2004).

Edward Mazria, *The Passive Solar Energy Book* (Rodale Press, 1980).

Sarah Susanka, *The Not So Big House* (Taunton Press, 1998).

Sarah Susanka, *Not So Big Remodeling* (Taunton Press, 2009).

4

Insulation and Air Sealing

Unless a house is located in paradise, where the weather is always perfect, it will most likely need some amount of heating and cooling for portions of the year. Whenever a house is heated or cooled, insulation is essential for maintaining an energy-efficient and comfortable home. For insulation to perform properly, we must also seal the house to eliminate air movement through it. Selecting the appropriate insulation and air sealing materials and following proper installation methods is critical to creating a green home. The specific characteristics of different insulation and air sealing materials give us the information we need to make appropriate decisions for new and existing homes.

LEARNING OBJECTIVES

Upon completion of this chapter the student should be able to:

- Describe the differences between an air barrier and a thermal barrier
- Discuss the different types of insulation
- Identify common thermal bypasses in a home
- Describe methods of insulating and air sealing an existing home
- Describe available insulation and air sealing materials and how they work together to create a complete thermal envelope

Green Building Principles

 Energy Efficiency

 Resource Efficiency

 Durability

Indoor Environmental Quality

Insulation History

The *building envelope*, as defined in Chapter 2, is the dividing line between conditioned and unconditioned space consisting of a thermal barrier (insulation) and an air barrier. Few homes before the mid 20th century were purposely insulated when originally built. Following World War I, wall cavity and ceiling insulation products made of mineral wool, asbestos, and silica appeared. In the 1930s, fiberglass, cotton, and wood fiber products were introduced into the market. A combination of rising fuel costs and the introduction of national model energy codes in the 1970s increased the demand for building insulation products. At the same time that the demand for insulation began to increase, the health dangers of asbestos led to its removal from most building products. The dominant building insulation products available at that time were mineral wool, fiberglass, and cellulose.

Federal regulations introduced in the late 1970s required that insulation products provide increased fire resistance; these codes favored nonflammable glass products over paper-based cellulose, leading to the long-standing market dominance of fiberglass insulation.

---

asbestos a naturally occurring fibrous material once commonly used for fireproofing; extremely harmful when inhaled.

fiberglass insulation a blanket or rigid board insulation that is composed of glass fibers bound together with a binder.

Although cellulose manufacturers ultimately developed suitably fire-resistant products, they are only now beginning to gain significant market share. Spray polyurethane foam insulation products entered the market in the 1970s and early 1980s, but many of these contained a potentially toxic binder called urea formaldehyde (UF). Due to health and quality concerns related to formaldehyde gas emission, these products did not gain wide acceptance and were banned in several states and in Canada. Spray polyurethane foam (SPF) insulation, a foam plastic that is installed as a liquid and expands many times its original size, is now commonly available with no UF content, although some phenol-formaldehyde insulation is still available. While fiberglass insulation is the dominant product in construction, cellulose and spray foam continue to expand their shares of the market place. Each of these insulation products can be an effective component in a green home. Making the right choice for a project requires understanding how each works as part of a complete house system, selecting the right product for a project, and installing it correctly throughout the house.

Building science research has provided us with the knowledge of how to properly insulate and air seal a house, as well as an almost limitless selection of products to use for this purpose. Climate, budget, product availability, and appropriate skills available for installation help determine the products and quantities used to create a building's thermal envelope, as do the preferences of the builder, designer or architect, and, in the case of custom projects, the homeowner. A home may use only one or several different types of insulation and air sealing products in various locations to achieve the desired thermal performance.

The Future of Insulation

Aerogel insulation has recently been developed for use in skylights and wall panels. With R-values of up to 8 per inch, translucent aerogels provide for highly insulating glazing panels that allow natural light into interior spaces. The material is also being developed into thin insulation wraps and as loose-fill insulation. Other emerging technologies include insulation made from agricultural byproducts, such as straw and rice hulls. Perhaps the most interesting insulation material currently in development is Greensulate™, a biodegradable, rigid insulation board made of natural fungus that is designed to replace polystyrene products. Most of these are still new and not fully proven in field applications, but expect these products and others like them to become more readily available in the future.

Selecting Insulation

When selecting insulation, four primary factors should be taken into consideration: *performance*, *installation method and quality*, *material characteristics*, and *cost*. *Performance* includes the R-value per inch, air permeability, and vapor permeability. *Installation methods* include applied insulation materials that are applied to the structure, such as batts, sprayed or blown-in products, and continuous boards. Batt insulation is thermal or sound insulation material, made of fiberglass, mineral wool, or cotton, which comes in varying widths and thicknesses to conform to standard framing of walls, floors, and ceilings. Another method is to use structurally integrated insulation that is part of the structure itself. *Material characteristics* address recycled content, recyclability, embodied energy, extraction and production location, hazardous components, and bio-based content. Bio-based insulation is created from plants, animals, or other renewable sources. *Cost* is often one of the most difficult criteria to objectively evaluate. When weighing expenses, we must consider not only the initial costs of material but also the costs of proper installation and any additional work required for some materials to match the thermal performance of others.

Performance Characteristics

Each insulation material has its own set of performance characteristics, including R-value and resistance to air and vapor flow, all of which have an effect on how it will

---

urea formaldehyde a potentially toxic chemical commonly used as a binder or adhesive in engineered building products.

spray polyurethane foam (SPF) insulation a spray-applied insulating foam plastic that is installed as a liquid and then expands many times its original size; see also *open-cell foam* and *closed-cell foam*.

applied insulation is thermal or sound insulation material that is placed between or on top of structural members after they are installed.

batt insulation a thermal or sound insulation material, generally of fiberglass or cotton, which comes in varying widths and thickness (R-value) to conform to standard framing of walls and joists.

structurally integrated insulation an insulation system that is part of a building structure as opposed to applied to a structural component.

bio-based insulation insulation products that contain materials from renewable resources as substitutes for petroleum and other nonrenewable products.

Did You Know?

Thermal Mass: What It Is and When It Improves Comfort

Heavy or massive objects like masonry, earth, and water can hold a lot of heat. Because of this capacity to act as a heat source (warming their surroundings) or a heat sink (drawing heat from and cooling their surroundings), materials with thermal mass affect comfort both indoors and out.

Buildings in climates with large diurnal (day–night) temperature swings, like the high-elevation Southwest, offer a classic example of the time-lag effect of thermal mass. Adobe and other types of masonry walls absorb intense daytime heat, keeping temperatures comfortable inside. During the cold night, the walls pour out their accumulated heat, keeping the inside warm. By morning, the walls, if they are designed correctly, can again absorb the daytime heat.

In most of North America, under most conditions, temperatures vary over the course of 24 hours but stay either above or below the comfort level. Heating or cooling is then necessary for most buildings, so building a tight envelope with materials that insulate well, or have a high R-value, should be the top priority.

Do materials with high thermal mass also insulate well? Some manufacturers would like us to think so, wielding a metric called "effective R-value" as evidence. And indeed, the time-lag provided by thermal mass saves energy in some climate conditions, but the effect is very circumstantial. As a more general rule, the most effective thermal storage materials are fairly good conductors and thus poor insulators. A thermal mass like poured concrete insulates poorly with R-0.08 per inch, compared with R-3.7 for cellulose.

But even in climates where insulation is the priority, buildings can use thermal mass. For example, night-flush cooling and passive solar heating can be viable strategies in the same location during different seasons. Using thermal mass on the interior of a well-insulated building envelope aids both strategies because the mass can absorb solar heat during the day and release it at night.

Many uses of thermal mass can reduce energy consumption and improve comfort. In buildings that are only occupied sporadically, however, it is often more efficient to minimize the interior mass so they can warm up (or cool down) quickly when needed. Also, thermal mass can be expensive and space-intensive, so architects and builders tend to use it where it can also serve other functions: as structure, as a durable interior surface like flooring, or in a heating system like a masonry stove.

Source: "Thermal Mass: What It Is and When It Improves Comfort." November 2007; http://www.buildinggreen.com/auth/article.cfm/2007/10/30/Thermal-Mass-What-It-Is-and-When-It-Improves-Comfort/. Reprinted with permission from *Environmental Building News*.

perform in a particular installation. Understanding the characteristics of different materials is critical to selecting the right insulation for a green home.

R-Value of Insulation

The *R-value* per inch rating for different insulation materials is based on laboratory tests (refer to Table 2.1 in Chapter 2); however, the performance of a particular insulation in a building component is affected as much by the quality of the installation and the completeness of the air barrier as by the rated R-value. Insulation works by limiting heat transmission through the properties of the insulating material and its thickness. When insulation is compressed, it reduces the rated R-value as the thickness is reduced, although the R-value per inch may in fact slightly increase from reduced convection within the insulation material. This characteristic is analogous to the insulation value of a down coat—it insulates better when it is thick and fluffy than when it is compressed.

Well-installed insulation of lower R-value limits heat transfer more effectively than poorly installed, higher R-value materials do. Appendix 4A on page 121 lists the R-values of typical insulation materials and where they are commonly installed.

Air Permeability

As we learned in Chapter 2, airflow is one of the most important factors affecting building performance. **Permeability** is the measure of air or moisture flow through a material or assembly. For insulation to properly perform, air must not be allowed to move through it. Insulation is described as either air impermeable, meaning that the material itself does not allow air to pass through it, or air permeable, which allows air to pass through. The Air Barrier Association of America (AABA) has developed ASTM-approved test methods for air barrier materials and systems. They define an air barrier as a material or system that has an air permeability of no more than 0.004 cfm/ft^2 at 1.57 pascals (1.57 Pa). **Pascals** are a basic unit of measurement for pressure.

---

permeability the measure of air or moisture flow through a material or assembly.

pascal (Pa) a unit of pressure equal to one newton per square meter.

Foam insulation is the only commonly available product that is air impermeable. All air permeable insulation materials must be combined with air impermeable materials into a complete air barrier system for maximum performance.

Vapor Permeability

Vapor permeability in building materials, including insulation, is categorized in one of four ways: impermeable, semi-impermeable, semi-permeable, and permeable. Each of these categories describes how much water vapor is allowed to pass through the material. In Chapter 2, we discussed vapor diffusion retarders and where to install them in building assemblies. Although we tend to avoid vapor barriers (materials that are rated at 0.1 perm or lower, such as sheet polyethylene) in wall assemblies, closed-cell foam or other vapor-impermeable insulation may be an appropriate choice where there is a risk of moisture condensing on cold surfaces. Vapor-impermeable insulation is beneficial when applied to floors in air conditioned spaces over damp crawl spaces (Figure 4.1a and Figure 4.1b) or roofs and walls in extreme cold climates (see Figure 4.2).

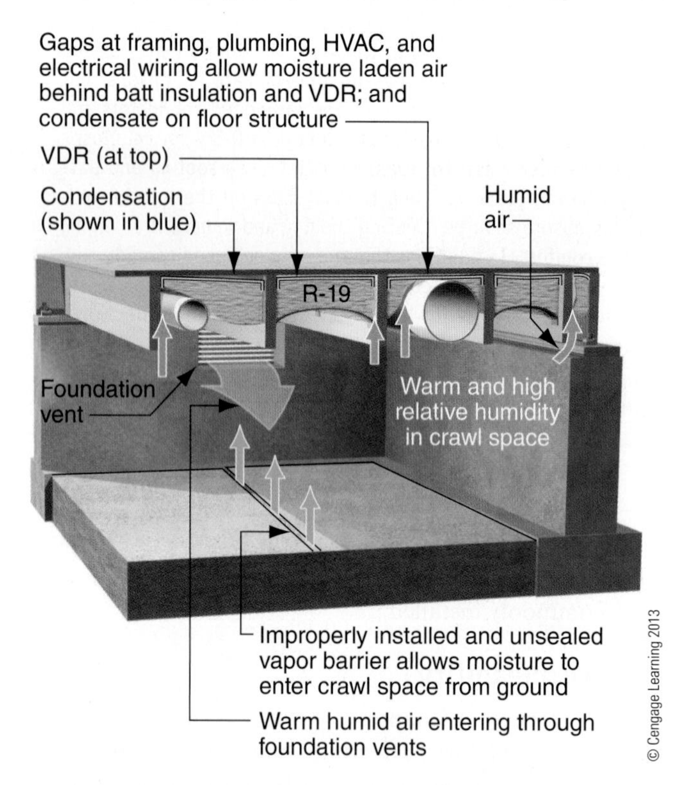

Figure 4.1a Moisture-laden air can easily pass through fiberglass batts because they are air- and vapor-permeable. Condensation may occur if this relatively warm and humid air comes in contact with a surface below the dew point. Using the psychrometric chart and knowledge of material moisture storage capacity, we can identify scenarios where condensation may occur on the underside of the subfloor, floor joists, and duct work. The risk of condensation is increased when fiberglass batts are installed with gaps, when crawl space ground covers are not completely sealed, and when crawl spaces are vented.

---

impermeable a material or assembly that does not allow air or moisture to flow through.

permeable a material or assembly that allows air or moisture to flow through.

semi-permeable a material or assembly that allows some air or moisture to flow through.

Fire Resistance

Insulation must be fire-resistant or covered with a fire-retardant barrier to remain exposed in a habitable area or when combustion appliances are nearby. Unfaced fiberglass insulation may be left exposed. Foam insulation, either spray-applied or rigid boards, must be covered with a fire-resistant covering to comply with fire codes. Coverings can be drywall, intumescent paint, or, in the case of rigid boards, foil facing. Intumescent paints are fire-retardant. Some building codes allow foam to be left exposed under certain conditions, such as in attics

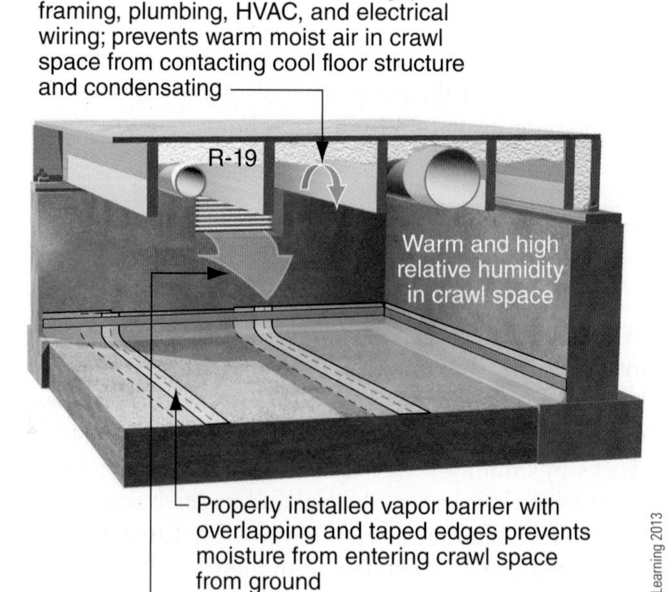

Figure 4.1b Using vapor and air impermeable insulation over crawl spaces helps to prevent condensation from forming.

---

intumescent paint a fire-retardant paint.

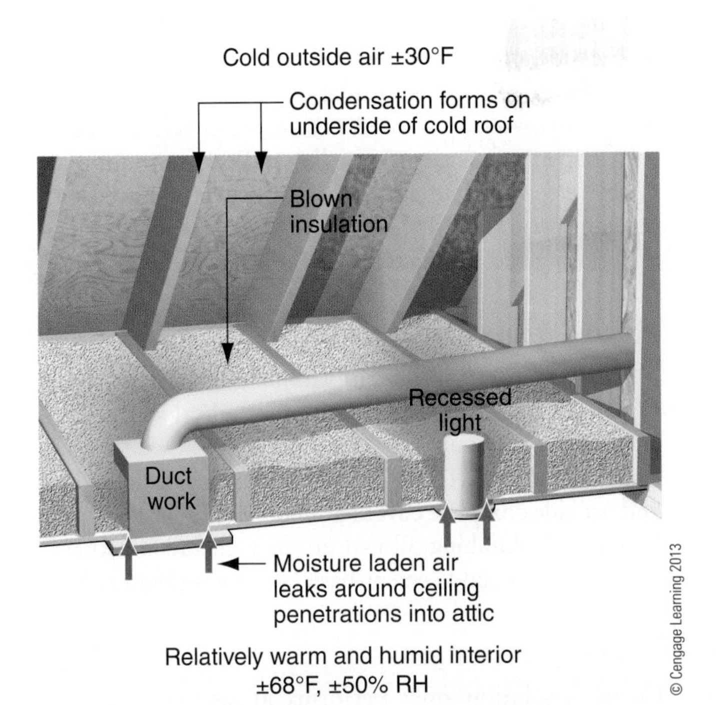

Figure 4.2 In the winter, relatively warm and humid conditioned air leaks into the attic and may condense on the underside of the roof decking.

and at rim joists. If foam insulation is designed to be left exposed in a project, you should confirm that it meets all the applicable building and fire safety codes.

Pest Management

Insulation products should be selected and installed to prevent pest entry and damage. Fiberglass insulation is not a food source for pests and poses little risk. Foam insulation is another material that is also not a food source, but it can provide pathways for termites, rodents, and other pests to enter a structure undetected. Some rigid foam products are treated with borates that deter pests. The cellular glass rigid insulation Foamglas® offers the greatest termite and rodent resistance of any insulation material. This product is widely used in Europe but only recently became available in North America as a building insulation. Because spray foam is not normally treated for pests, foundation and floor applications should leave an inspection gap to ensure that termites do not have access to the building structure.

Installation Methods and Quality

Insulation can be applied on or in the structure of a home, or it can be integrated directly into the structure itself. Applied insulation is available in batts and rigid boards, or it can be blown in place. Structurally integrated insulation can be incorporated into concrete exterior walls or load-bearing exterior wall or floor panels. By understanding the pros and cons of each installation method, the project team can make sound decisions about which products to use in a home. Correct installation is critical to achieve desired performance of applied insulation.

Batt Insulation

The most common installation method for applied insulation is fiberglass batts. Sized to fit in standard 16" and 24"framing cavities and available in thicknesses from 3½" up to 12", batt insulation must fit snugly without any compressions or gaps to work effectively. Batts must be carefully cut and fitted around all obstructions, such as wires and pipes, rather than stuffed around them. Gaps and compressions of as little as 5% of the area of a single batt can decrease its thermal performance by as much as 50% (Figure 4.3). Batts are available in fiberglass, mineral wool, cotton, and, to a lesser extent, wool (Figure 4.4).

Batt insulation is more susceptible to voids and compressions than other insulation types. Paper-faced batts are usually installed with installation tabs stapled

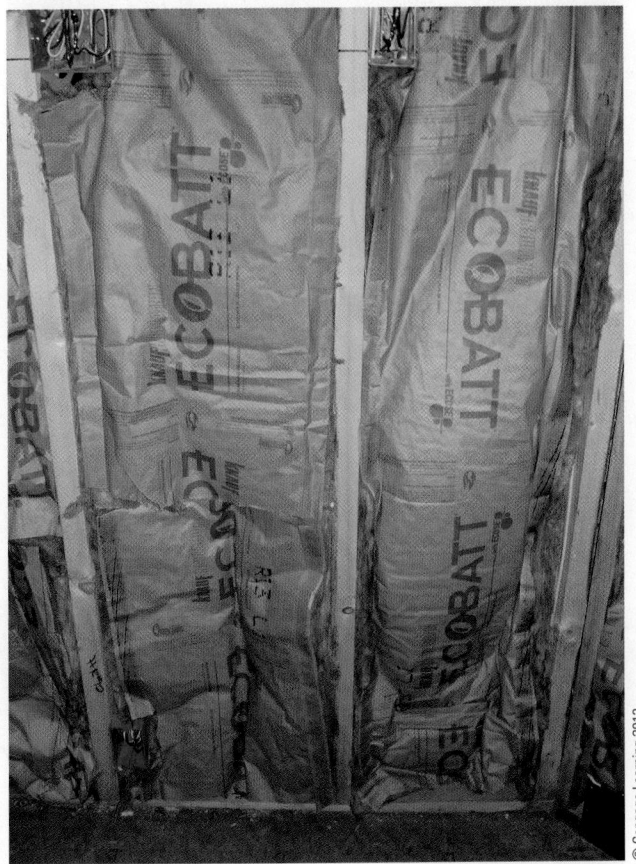

Figure 4.3 Insulation performance is greatly reduced through improper installation. These batts are compressed and will not deliver the labeled R-value.

© Cengage Learning 2013

Figure 4.4a Correctly installed unfaced fiberglass batts with steel studs.

Courtesy of Bonded Logic; UltraTouch™ Denim Insulation is a registered trademark of Bonded Logic

Figure 4.4b UltraTouch cotton batts.

Courtesy of Black Mountain USA LLC

Figure 4.4c Lambs' wool batt insulation.

to the side of studs, leaving a continuous gap at each edge. Face-stapling the insulation is more effective, but this strategy is rarely used because it covers the studs with paper, preventing drywall installers from applying adhesive to the framing. Properly installed unfaced batt insulation can avoid the compression problems associated with faced products.

Blown Insulation

Blown insulation is sprayed into framing cavities and on the top of ceilings and, when installed properly, does an excellent job of eliminating gaps and compressions common in batt installations. These products easily fill wall cavities and provide complete coverage, even behind obstacles like wiring and plumbing. Blown products include fiberglass, cellulose, foam, and cement-based products (Figures 4.5).

Proper Installation

Cavity insulation only performs at its rated R-value when it is installed to completely fill building cavities with no compressions, gaps, or voids. Insulation grading is not applied to insulation boards because they are generally not prone to the same installation problems. The Residential Energy Services Network (RESNET) has developed a grading system that provides criteria for inspectors and HERS raters to determine the quality of an insulation job. The highest quality installations are Grade 1, meaning that they are near-perfect and follow manufacturer specifications completely. Grade 2

Courtesy Johns Manville

Figure 4.5a Blown-in fiberglass wall insulation.

---

blown insulation a material composed of loose insulating fibers such as fiberglass, foam, or cellulose, that is pumped or injected into walls, roofs, and other areas.

cavity insulation insulation placed between wall studs or joists.

Figure 4.5b Blown cellulose without netting.

Figure 4.5c Open-cell SPF.

Figure 4.5d Installation of cementitious insulation. This particular product is a combination air, water, and magnesium oxychloride cement.

allows for some compression and incomplete coverage. Grade 3 is very poor quality installation with multiple gaps, voids, and areas of compression. The RESNET Standards provide clear guidance for insulation grading (Figures 4.6).

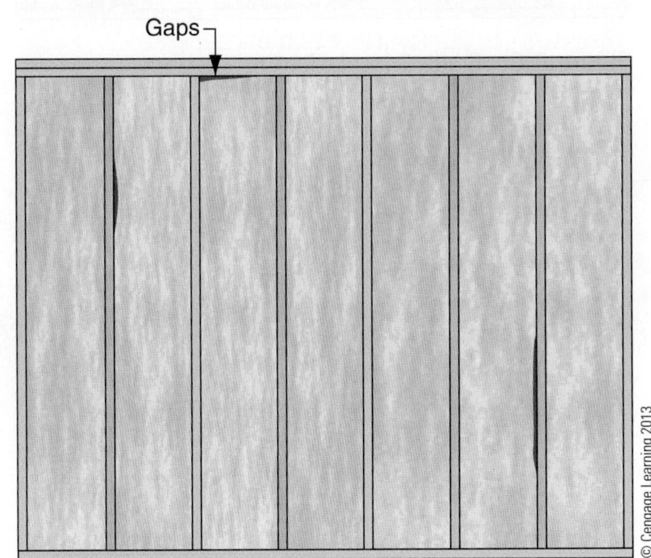

Figure 4.6a Grade 1 insulation installation.

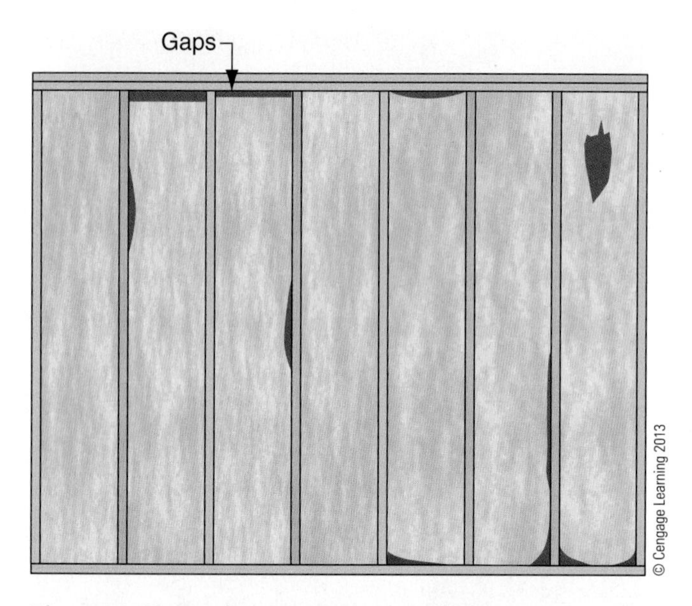

Figure 4.6b Grade 2 insulation installation.

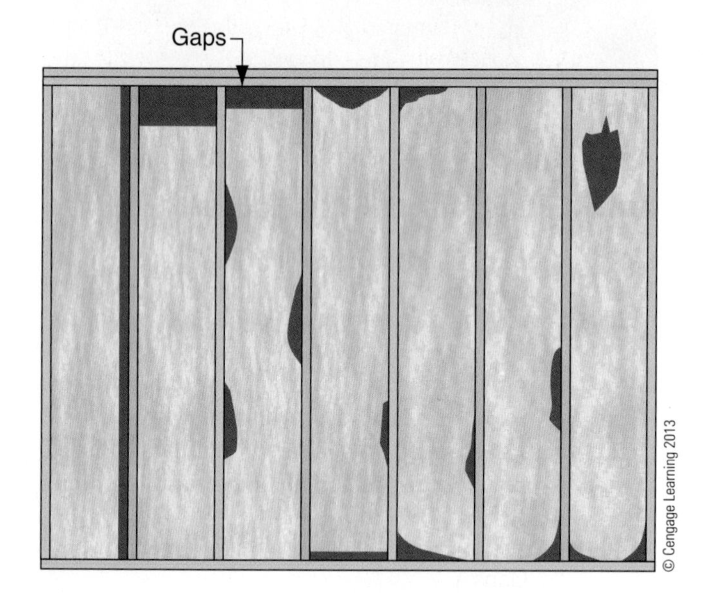

Figure 4.6c Grade 3 insulation installation.

Figure 4.7a Polyisocyanurate insulation boards are offered with a foil facing that provides an ignition barrier and improved performance.

Figure 4.7b XPS insulation boards come in a range of thicknesses.

Insulation Boards

Insulation boards are rigid insulation panels and commonly made of expanded or extruded polystyrene or polyisocyanurate foam. Expanded polystyrene (EPS) is usually white and made of expanded polystyrene beads. Extruded polystyrene (XPS) is a foam insulation board and commonly sold under the trademark STYROFOAM™ by Dow Chemical (Figures 4.7a, 4.7b, and 4.7c) or Owens Corning Foamular. EPS and XPS are typically installed on the exterior of a structure to provide a thermal break over the framing and to supplement the thermal

---

insulation board a rigid insulation product available in varying widths and thickness (R-value).

expanded polystyrene (EPS) a foam insulation board made of expanded polystyrene beads.

extruded polystyrene (XPS) a closed cell insulation foam board.

thermal break a material of low thermal conductivity placed in an assembly to reduce or prevent the flow of heat between conductive materials.

Figure 4.7c The PerformGuard® EPS insulation board is pretreated with a termiticide.

insulation installed in framing cavities. These boards can also be installed on the interior behind drywall to provide a thermal break. Thermal bridges are paths through which heat transfer occurs in areas that are not directly insulated, such as wall studs and roof rafters (Figure 4.8). In some cases, several thick layers of board are installed on the exterior in lieu of any cavity insulation. Fiberglass, mineral wool, and cellular glass boards are also available but are uncommon in the United States except in some foundation installation applications.

Structurally Integrated Insulation

Structurally integrated insulation can be on the exterior or the interior of the structure or incorporated throughout. Insulated concrete forms (ICFs) are concrete walls with foam insulation forms that remain in place after the concrete cures (see Chapter 5). ICFs have insulation on the interior and exterior (Figure 4.9). Structural insulated panels (SIP) are load-bearing wall, roof, or floor panels made of insulation foam sandwiched between two sheets of plywood or oriented-strand board (OSB)(see Chapter 5 and Figure 4.10). Straw bale construction uses baled straw from wheat, oats, barley, rye, rice, and other agricultural waste products in walls covered by stucco or earthen plaster (see Chapter 6 and Figure 4.11). Aerated autoclaved concrete (AAC) is a lightweight concrete used in precast building materials that provides insulation, structural support, and fire resistance (see Chapter 5 and Figure 4.12). Although these construction methods differ significantly from standard wood frame construction methods used in most homes, they are gaining broader acceptance, particularly in the green building industry. Structurally integrated insulation provides continuous insulation coverage with few thermal bridges.

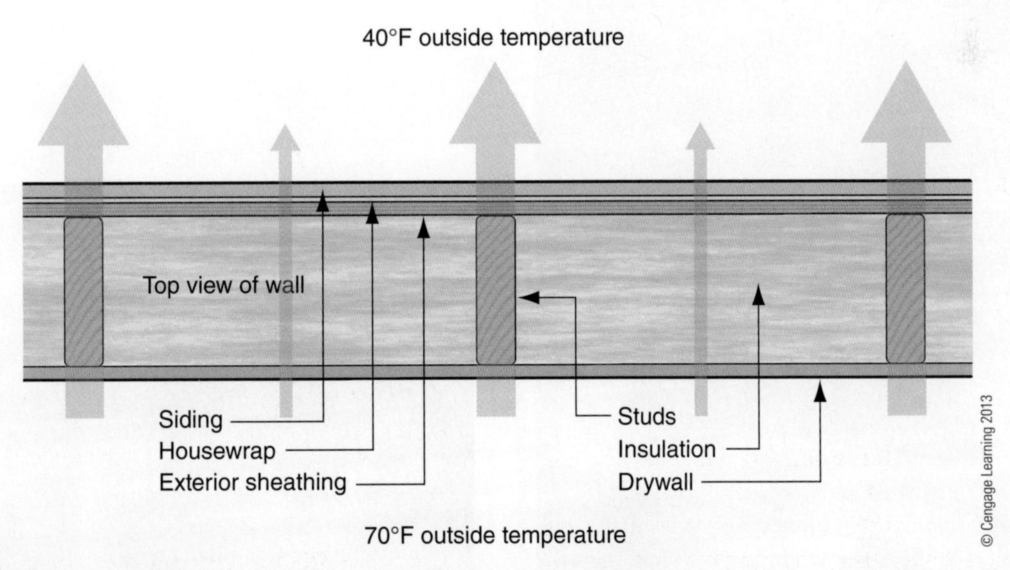

Figure 4.8 Compared with insulation, wood studs provide little resistance to heat flow. Heat flow always takes the path of least resistance, and wall studs can act as thermal bridges.

---

thermal bridge a thermally conductive material that penetrates or bypasses an insulation system, such as a metal fastener or stud.

structural insulated panels (SIP) are composite building materials made from solid foam insulation sandwiched between two sheets of oriented-strand board to create construction panels for floors, walls, and roofs.

straw bale construction using baled straw from wheat, oats, barley, rye, rice, and other agricultural waste products in walls covered by stucco or earthen plaster.

continuous insulation insulation that is not interrupted by structural members, typically placed on the outside surface of wood framing or concrete walls.

Figure 4.9 ICFs provide continuous insulation on the interior and exterior of walls with no thermal bridges.

Figure 4.10 This home is using SIPs for the exterior walls and roof.

Figure 4.11 Straw bale wall construction.

Figure 4.12 AAC blocks provide insulation and structural support. Here, AAC blocks are used for interior and exterior walls; in the United States, however, AAC exterior walls are more commonly combined with conventional stick framing on the interior walls.

Material Characteristics

Chapter 1 addressed life cycle assessment of a product or structure's full environmental cost, including the embodied energy, contaminants created in production and distribution, and the product's cumulative impact on the environment. In addition to these factors, we can consider the amount of recycled content, the recyclability of insulation, the *bio-based* content of the material, added UF, and the distance the materials traveled from initial extraction sites and production facilities. These characteristics have broad societal implications as well as direct effects on building occupants. By selecting locally produced products with lower embodied energy, builders can improve overall air quality through resource conservation and reducing pollutants from transportation fuels. Avoiding irritants and contaminants in insulation can prevent adverse health effects, both in the product installers and in the building's occupants.

Cost

The cost of insulating a house can be divided into several distinct segments: materials, installation, peripheral costs (e.g., air sealing and waste disposal), and cumulative energy costs through the lifetime of the structure. Only by taking all these factors into account can we make the best judgments on the actual cost of a product. For example, consider fiberglass batt insulation. Although this product is the most common and least expensive insulation on the market, it is the most prone to incorrect installation and associated reduced performance. Installers are typically paid by the quantity rather than by the quality of their work; with minimal quality control in the field, this factor often results in poor-quality applications. When installed correctly, batt insulation is an excellent product, but expectations for proper installation with the insulation contractor must be set in the pricing stage before work starts. The work must be inspected and any deficiencies corrected at completion, before final payment to the installer is approved. Grade 1 batt installation will cost more than the average commodity rate. When you consider the cost of high-quality insulation and the work required to install a comprehensive air barrier, the price difference between batt and blown-in or SPF insulation is smaller than the basic cost per square foot suggests. For example, R-13 fiberglass batt insulation may cost as little as $0.50/ft^2 installed, whereas spray foam may cost as much as $1.50/ft^2. Since the fiberglass batts are air permeable, every seam and crack in the exterior sheathing, wall plates, and the interior drywall must be thoroughly air sealed for batts to achieve the overall performance of spray foam. In addition, obtaining grade 1 quality with batt insulation often carries a premium price. The additional costs for installation, combined with the added sealing of sheathing and drywall, usually minimizes most (if not all) of the difference in the net costs of the insulation.

Embodied Energy

The embodied energy required to manufacture equivalent R-values of different insulation materials is another factor to consider when making product selections. The exact amount of embodied energy in a specific material can vary between manufacturers and is based on the distance covered during the transportation of both raw and finished materials from extraction to job site. Understanding the relative relationships of embodied energy between products is helpful in making decisions for a green home. Recycled and natural products, such as cellulose and wool, have the lowest embodied energy, and petroleum-based SPF and rigid foam have the highest (see Table 4.1). Keep in mind that regardless of the amount of embodied energy in a particular insulation, a properly installed product will save many times that amount of energy over the life of the building.

Table 4.1 Estimated Embodied Energy of Insulation Materials*

Insulation Material	Btu/Ft^2 R-1
Batts	
Cotton	22
Sheep's wool	25
Mineral wool	311
Fiberglass	227–345
Blown	
Cellulose, densely packed	30–49
Icynene	208
Mineral wool, loose fill	149–245
Polyurethane	899
Insulated boards	
Mineral wool	973
Polyisocyanurate	715
EPS	900–1075
XPS	1509

*Courtesy of Robert Riversong and Environmental Business News. Represents approximations of embodied energy in different insulation materials. The actual embodied energy in any product will vary with the fuel used in production, transportation distances and methods, and other factors. As the determination of embodied energy is not an exact science, these should not be used as absolutes; rather they provide a point of comparison between different materials. Regardless of the amount of embodied energy in insulation, when properly installed, it will likely save more energy over its lifetime than was used to manufacture it.

Did You Know?

2009 American Recovery and Reinvestment Act

As part of the American Recovery and Reinvestment Act of 2009 (ARRA, or "The Recovery Act") states accepting funds must adopt the 2009 International Energy Conservation Code (IECC) or better and document 90% compliance by 2017. ARRA provides additional funds for code training and enforcement. According to an analysis by the environmental consulting firm ICF International that compared the updated standards with those released in 2006, homes built to the 2009 IECC standards will save an estimated 12.2% under the simple "prescriptive" method and could save 14.7% or more under the more complicated "performance-based" method.[1] Using 2008 utility rates, this equates to a savings of approximately $163 to $437 per home (depending on where the home is located) and a national average savings of $235 per year, based on average energy costs at the time. As utility bills increase, the savings will continue to increase.

[1]An Analysis Prepared for the Energy Efficient Codes Coalition (EECC) by ICF International, January 2009.

Selecting an Air Barrier

Air sealing is equally critical to the thermal envelope as the insulation, and consideration must be given to both components before any products are chosen. When insulation is air permeable, such as fiberglass, a separate air barrier system must be designed and installed to eliminate air infiltration through the envelope. This air barrier may consist of one or a combination of the following: properly sealed housewrap, sealed exterior sheathing, or interior drywall. Housewrap, typically a plastic or spun-fiber polyethylene textile material, is primarily a water-resistive barrier (WRB) designed to act as a secondary drainage plane behind the cladding. When sealed properly, housewrap can provide resistance to airflow; however, the use of housewrap as the primary air barrier is not advised (Figure 4.13). The primary barrier must be combined with sealants at joints between framing members, at drywall edges, at mechanical penetrations, and at rim joists (see Figure 4.14 and the "Air sealing" section on pages 108–110 for more details). Conversely, if the insulation is by its nature an air barrier, such as spray foam or SIPs, then fewer areas of the building envelope need a separate air barrier. A third option, the airtight drywall approach (ADA), involves completely sealing the drywall to the framing to provide the air barrier (Figure 4.15). The drywall is sealed with caulk, glue, or gaskets to the framing and around all penetrations (primarily electrical boxes and recessed fixtures). Regardless of the air permeability of the insulation, every project must address air sealing around windows, mechanical penetrations, and at the transitions between materials.

infiltration the uncontrolled process by which air or water flows through the building envelope into the home.

housewrap a synthetic water-resistive barrier designed to shed bulk moisture and allow vapor to pass through; see also *water-resistive barriers*.

water-resistive barrier (WRB) the material behind the cladding that forms a secondary drainage plane for liquid water, commonly referred to as the weather-resistive barrier, weather-resistant barrier, and water-resistant barrier.

airtight drywall approach (ADA) an air barrier system that connects the interior finish of drywall to the homes framing to form a continuous air barrier.

Locating the Thermal Envelope

Deciding exactly where to place the *thermal envelope*, where the insulation and air sealing will be installed in contact to each other, is one of the most important decisions made when planning a house. In theory, the air barrier must be complete and continuous on all six sides (Figure 4.16) of all cavity insulation. While it is easy to accomplish six-sided air sealing in walls, the insulation is typically left exposed above the ceiling and below the floor because installing an air barrier in these locations is often impractical and not very cost-effective. Furthermore, the location of the building envelope is often flexible. For example, the building envelope may be designed to include or exclude the basement, crawl space, or attic.

Foundations

Starting from the bottom, homes with a basement or crawl space have the option of insulating the bottom of the first floor or the walls of the foundation. When the foundation can be kept dry, insulating the basement or crawl space walls provides for the best performance in most climates (Figure 4.17). If there is a risk of flooding or water infiltration through the foundation, then the insulation and air sealing should be installed on the underside of the first-floor framing, placing the basement or crawl space outside the thermal envelope. If the house is on a slab, the perimeter of the slab should be insulated in all climates; in cold climates and wherever a radiant

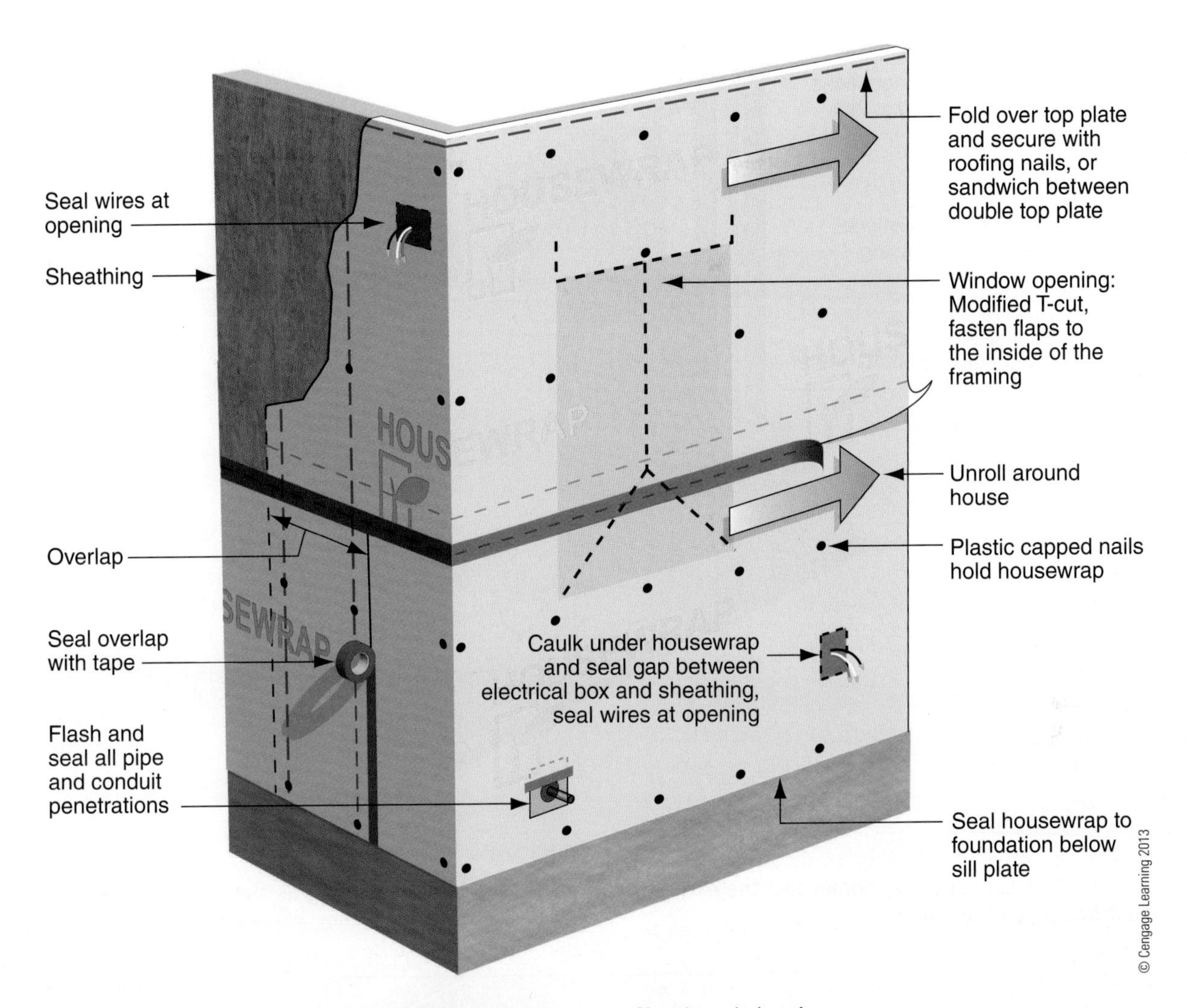

Figure 4.13 When properly installed, housewrap can be an effective air barrier.

floor heating system is installed, the entire underside of the slab should also be insulated. Foundations are covered in more detail in Chapter 5.

Walls

In all buildings, the thermal envelope is located at the exterior walls. Choosing exactly where within the wall assembly to install the air sealing and insulation has a significant effect on the building performance. Insulated sheathing on the exterior of the wall structure helps reduce thermal bridging at the framing members—a particularly effective strategy in cold climates where it can reduce or eliminate condensation within the wall structure while serving as an effective air barrier. Cavity insulation placed between the studs, combined with well-sealed drywall, provides an alternate method of creating a thermal envelope. With this strategy, however, gaskets or other sealants must be used to seal the drywall at the top and bottom plates and around all receptacles, lights, and switches. The rim joists must also be sealed with caulk or spray foam to provide a comprehensive air seal. Using spray foam installation in the stud cavities alleviates the need for most of the air sealing on either exterior sheathing or interior drywall; however, the gaps around windows and doors and between framing members must still be properly sealed.

Ceilings

The top of a house often poses the biggest challenge to completing the thermal envelope. In unconditioned attics above flat ceilings, a thick layer of properly installed batt or blown insulation in combination with the ceiling drywall can provide a good thermal barrier, provided all the details are properly addressed, including soffits, attic accesses, heat registers, light fixtures, whole-house fans, and the line where the top of the drywall meets wall-top plates. Critical areas to address include interior soffits and attic knee walls, which are the vertical walls that separate a home's

---

attic knee wall a vertical wall separating conditioned interior space from an unconditioned attic area.

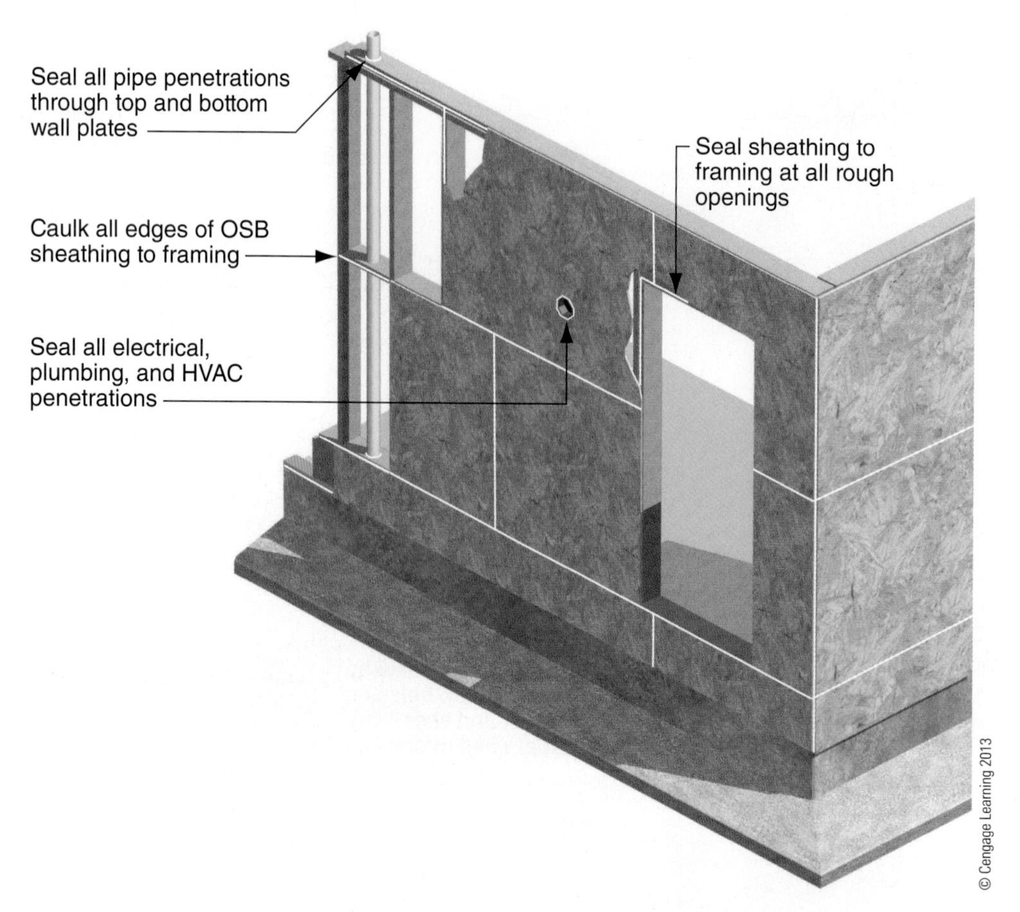

Figure 4.14 Most conventional homes use the exterior wall, floor, and ceiling framing as the air barrier. All penetrations must be completely air sealed.

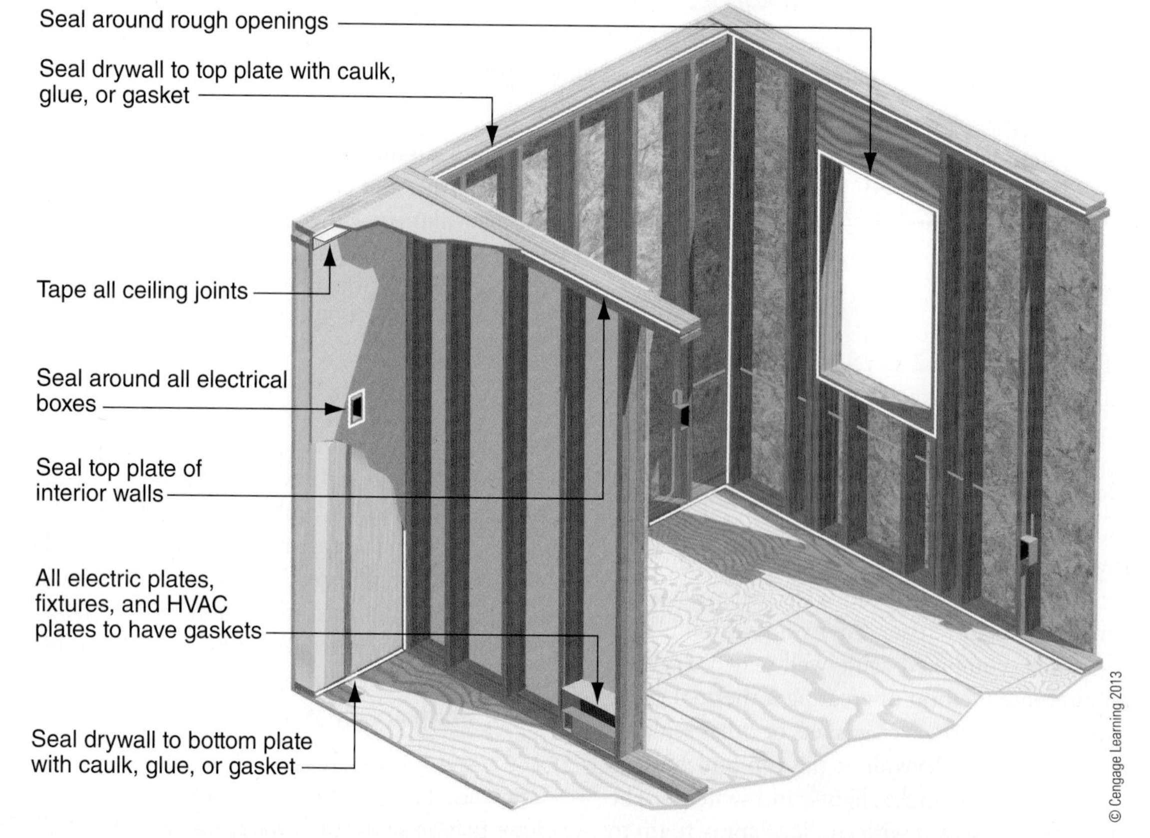

Figure 4.15 Wall and ceiling drywall can be an effective air barrier if fully air sealed.

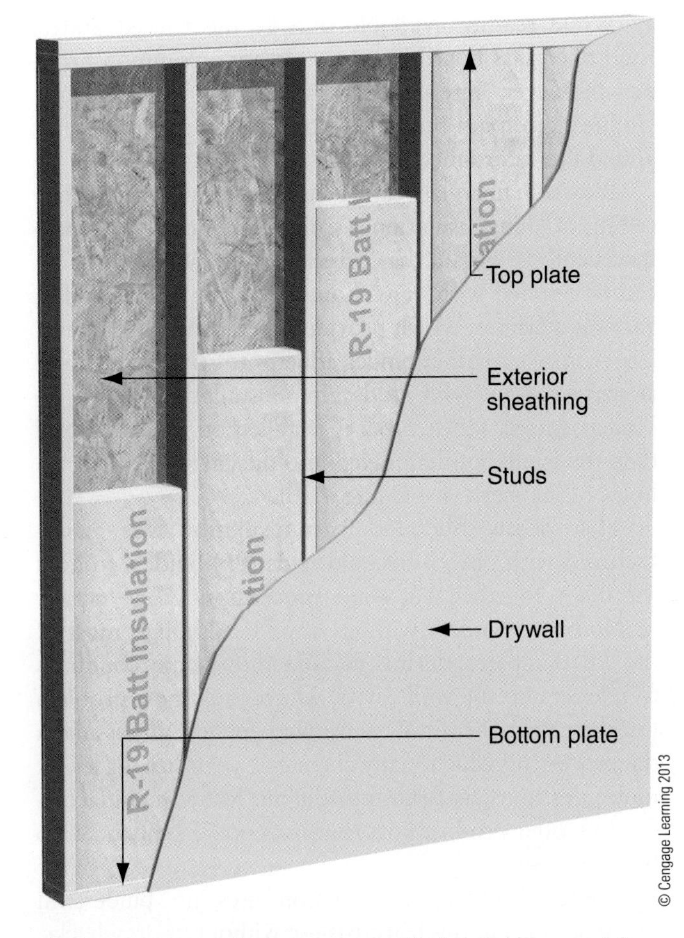

Figure 4.16 Vertical air permeable insulation must be enclosed on all six sides (top, bottom, left, right, top, and bottom) to perform as designed.

conditioned space from the unconditioned attic. Soffits must be properly air sealed to avoid creating a thermal bypass (see the section on air sealing on pages 108–111).

Knee walls that separate conditioned space attic areas must have an air seal on the unconditioned side of the wall as well as blocking below the wall (Figure 4.18). An alternative location for the top of the thermal envelope is to place it in the roof, creating a fully or semi-conditioned attic (see Figure 7.21).

Materials and Methods

As discussed earlier in this chapter, two types of insulation are available: applied or structurally integrated. A single building may use one or both of these methods; for example, structurally integrated insulation may be used for the walls in combination with applied insulation on the ceilings. Once the type of insulation is determined, then specific materials and application methods are selected. Along with the insulation, appropriate air sealing must be incorporated to complete the building envelope (see Appendix 4A, page 121 for a detailed comparison of common insulation types).

Applied Insulation Materials

Applied insulation products include fiberglass, mineral wool, cellulose, cotton, SPF, and cement-based materials. Insulation and air sealing components must be carefully specified and installed in order to perform properly in green homes.

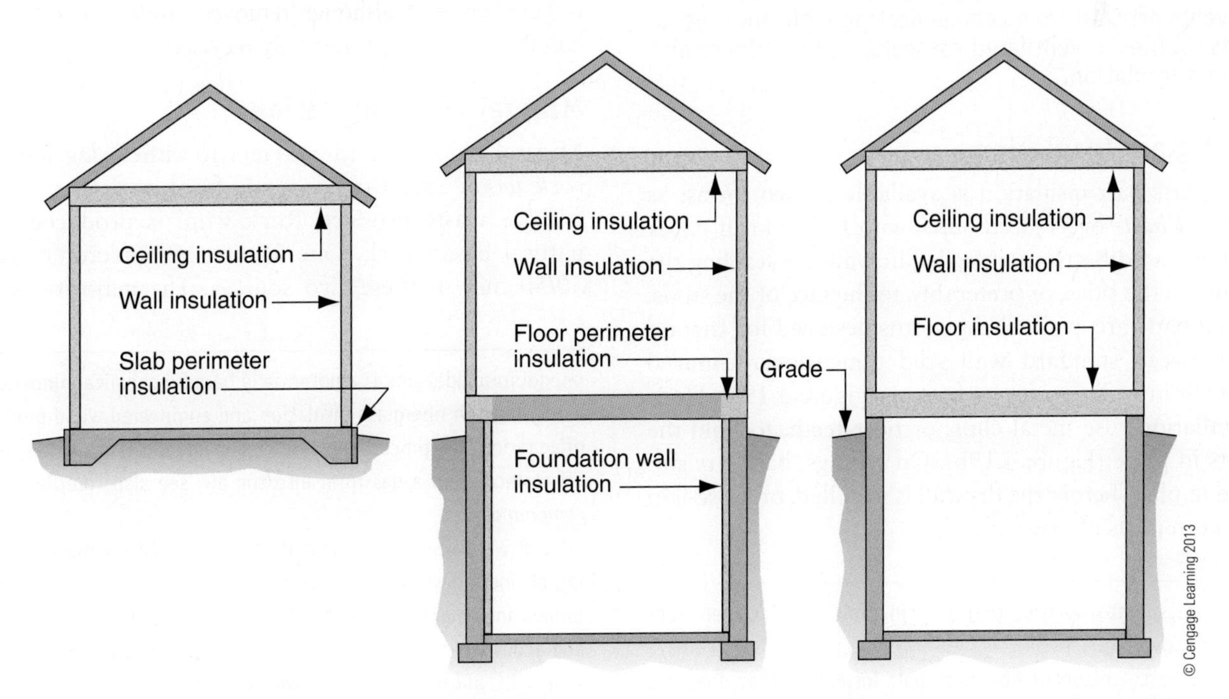

Figure 4.17 The location of the bottom of the building envelope is based on the foundation type. The building envelope may include or exclude basements and crawl spaces.

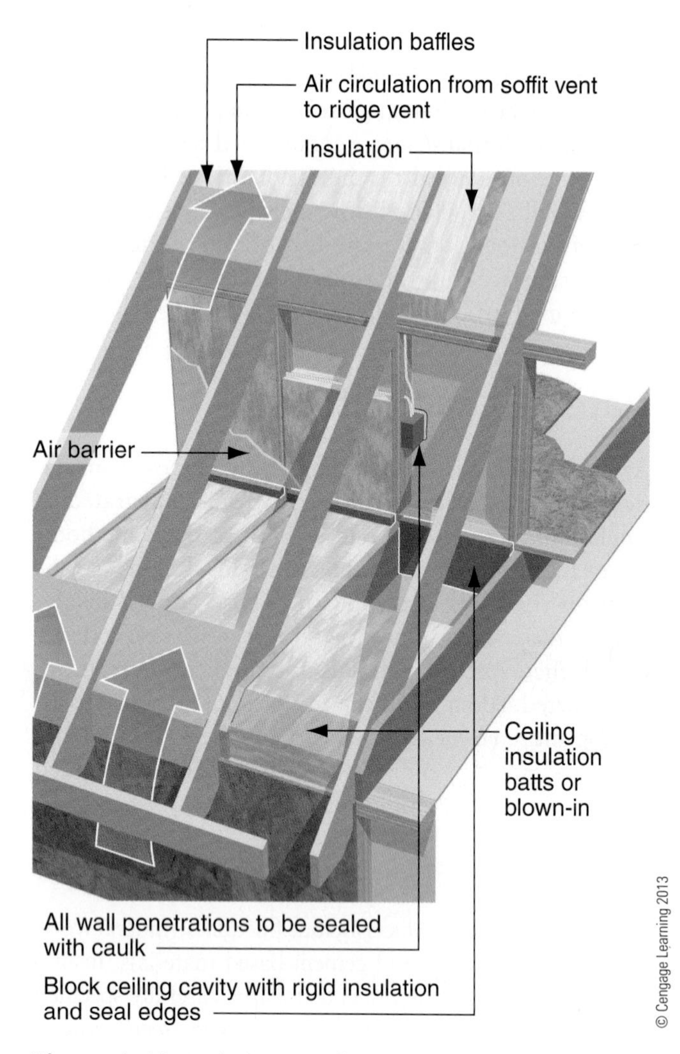

Figure 4.18 Attic knee walls must be fully air sealed with a backside air barrier. Blocking below knee walls prevents attic air from communicating with floor systems. Baffles in ventilated cathedral ceilings direct air around insulation.

Fiberglass Insulation

Fiberglass batt insulation is available in two forms: as *unfaced batts* or as *faced batts*, with foil or kraft paper facing. Faced batts are installed in walls by stapling the facing to the sides, or preferably, to the face of the studs. Faced batts are available in widths designed for friction fit between standard wall stud dimensions. Unfaced batts fit in wall cavities by friction (Figure 4.19a). Floor installations use metal clips, or tiger teeth, to hold the batts in place (Figure 4.19b). On ceilings, batts are stapled in place before the drywall is installed, or loose-laid between joists afterwards.

faced batts batt insulation that contain a foil or kraft paper vapor retarder covering.

unfaced batts cotton or fiberglass batt insulation that does not contain a vapor retarder covering.

While readily available outside the United States, rigid fiberglass board insulation is not a common product; however, interest and availability is increasing. Unlike rigid foam boards, it does not require the use of added fire retardants.

Blown-in fiberglass can be installed in walls and above ceilings. Ceiling insulation is sprayed on loosely to a specified depth to obtain the correct R-value. Wall insulation can be applied with or without an acrylic binder and in varying densities, which provide different R-values. Insulation sprayed with a binder adheres to stud cavities and is scraped even with studs after installation. When no binder is used, fabric mesh is installed on the walls, and then the insulation is sprayed into the cavity through the holes of the mesh (see Figure 4.20).

Historically, fiberglass batt insulation was manufactured with phenol formaldehyde (PF) binders to hold the fibers together, but some products now use acrylic or bio-based binders without any PF. Although most of the PF dissipates during manufacture, some continues to release into the wall cavity, which could be a problem for extremely chemically sensitive people. Off-gassing is the process by which many chemicals volatilize, or let off molecules in a gas form into the air. Many manufacturers now offer product lines that use non-PF binders or no binders at all.

Loose-fill fiberglass insulation uses no binders in manufacture, but this feature is not without its drawbacks. Fiberglass fibers can become airborne during installation, and inhalation of these particles poses the risk for potential lung problems. Fiberglass insulation is frequently manufactured with a minimum of 20%, and often more, recycled content, although removed insulation and installation scraps are not normally recycled.

Mineral Wool Insulation

Mineral wool insulation refers to either *slag wool* or *rock wool*. Slag wool is made from an iron ore blast furnace waste product. Rock wool is produced from natural basalt rock. One leading manufacturer uses a 50/50 mix of these two sources. The majority of the

phenol formaldehyde is a potentially harmful chemical binder commonly used in fiberglass insulation and engineered wood products.

off-gassing the process by which many chemicals volatilize, or let off molecules in a gas form into the air; see also *volatile organic compounds*.

mineral wool insulation a manufactured wool-like material consisting of fine inorganic fibers made from slag and used as loose fill or formed into blanket, batt, block, board, or slab shapes for thermal and acoustical insulation; also known as *rock wool* or *slag wool*.

slag wool another name for *mineral wool*.

rock wool another name for *mineral wool*.

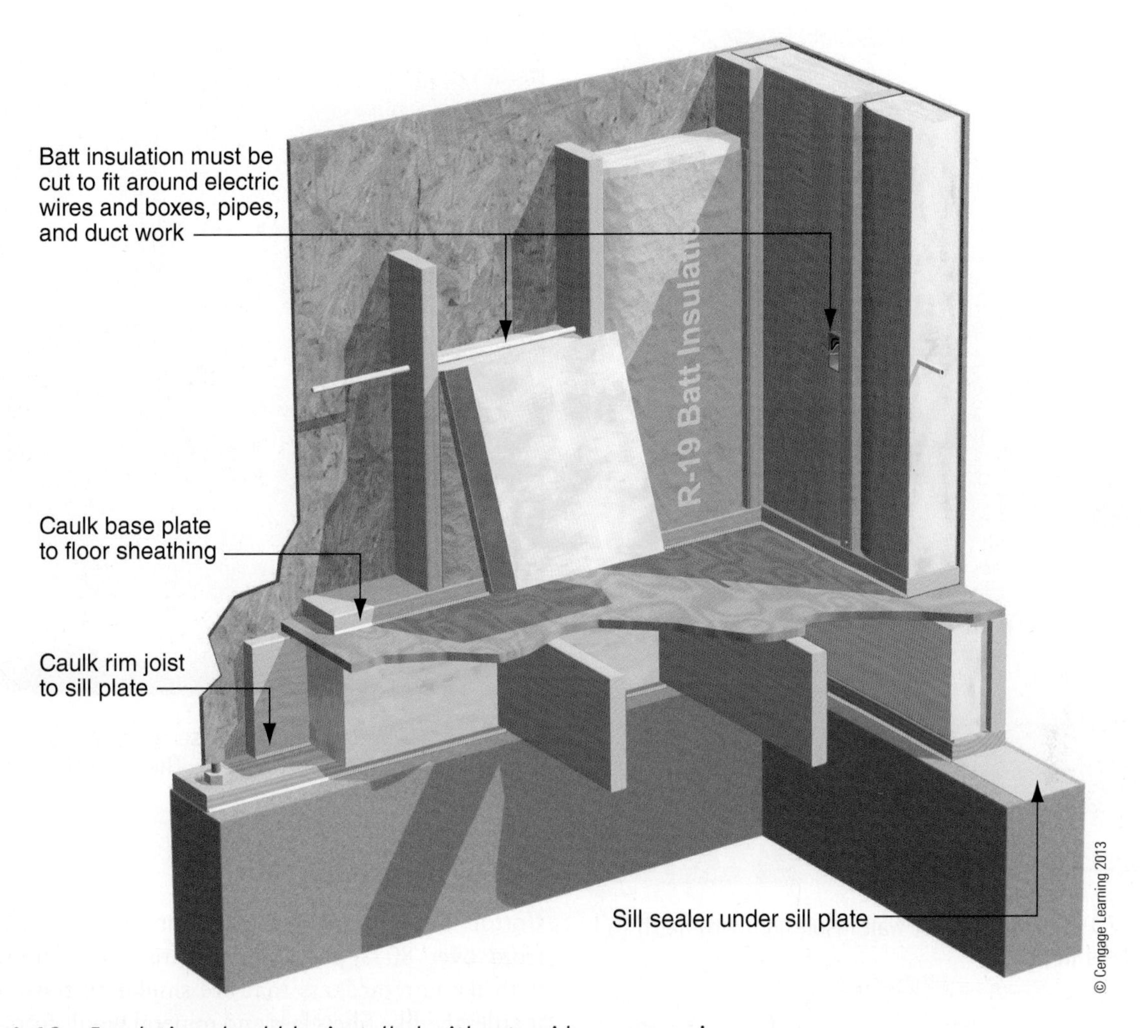

Figure 4.19a Insulation should be installed without voids, compression, or gaps.

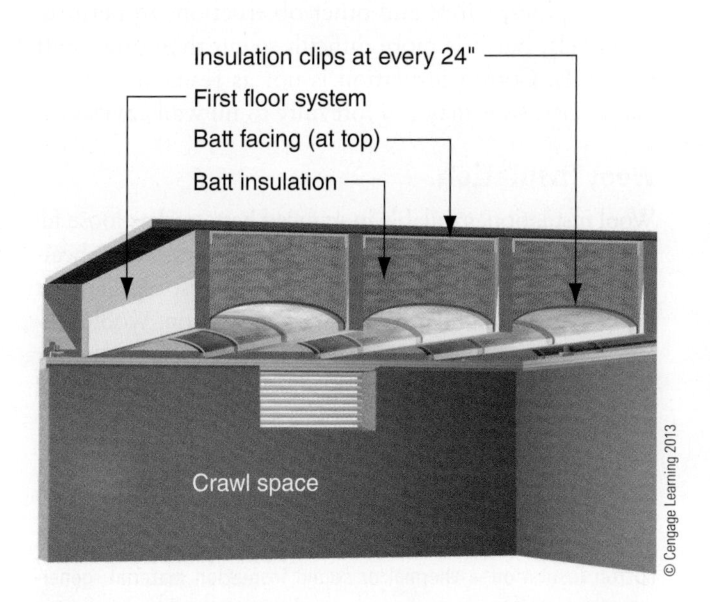

Figure 4.19b Insulation supports keep floor insulation in contact with the subfloor. Supports must be carefully installed to not overly compress the insulation.

mineral wool products available are slag products, which contain as much as 90% post-industrial recycled content. Mineral wool insulation is available in loose-fill products and in batts that are made with a starch binder. Similar to fiberglass, the fibers can become airborne and, when inhaled, are a lung irritant. Mineral wool insulation is also available in rigid boards, both unfaced and foil faced. Different densities are available, some of which may be adequate for use under concrete slabs. Like fiberglass, they are noncombustible without added flame retardants.

Cellulose Insulation

Cellulose insulation is made primarily of recycled newsprint, chemically treated to resist fire, mold, and pests. Only available as a blown-in product, cellulose can be

---

cellulose insulation made from recycled newspaper and an added fire retardant.

Courtesy of Blow-In-Blanket, LLC

Figure 4.20 Blow-in-blanket wall insulation with mesh to hold material in place.

Photo courtesy of GreenFiber

Figure 4.21 Wet-spray cellulose is installed and excess scraped off for a smooth finish. The scrapings are easily reused on site.

installed in open or closed wall cavities, above ceilings, and in rooflines. Installations include dry and wet spray. Dry cellulose is installed behind mesh fabric in walls and rooflines and above drywall ceilings in attic spaces. "Stabilized" cellulose uses a small amount of water-activated acrylic binder to prevent settling, and allows for lower-density cellulose that achieves more thickness with less weight. Wet-spray is installed in openwall cavities and scraped even with studs. Excess material is reused throughout the process (Figure 4.21). Wet-spray cellulose must be properly installed without too much water, and it should be allowed to dry for several days before drywall is installed to avoid mold formation in walls. Cellulose has a low level of embodied energy and a high quantity of recycled content, manufactured from as much as 80% postconsumer recycled newsprint. Extremely sensitive people may react to the chemicals in the ink present in cellulose, to the dust created during installation, or to the chemicals used for fire and pest treatment, although this is not generally a major concern. Flame retardants used in cellulose include borate compounds, which also serve to inhibit insects and other pests, and ammonium sulfate.

Cotton Insulation

Cotton insulation, available in unfaced batts, is made from over 80% post-industrial recycled denim waste with flame retardants that are similar to those used in textiles. Unlike fiberglass and mineral wool, cotton is not a skin or lung irritant, and it is fully recyclable. Available in limited sizes, it has the same installation challenges as other batt insulation because it must be carefully cut around pipes, wires, and other obstructions to perform effectively, but it is more difficult to cut than other batt materials. Cotton insulation is not as resilient as fiberglass batts, so it may not loft fully to fill wall cavities.

Wool Insulation

Wool insulation, available in unfaced batts and as loose fill for blown applications, is made from lamb and sheep shearings with added boron fire retardant and pest repellants. It has an R-value similar to cellulose and cotton. Wool insulation has very low embodied energy since it is minimally processed. Although not yet readily available in the United States, it is used in Europe, Canada, Australia and New Zealand. Like all insulation, wool batts must be carefully cut to fit framing cavities and trimmed around obstructions.

---

cotton insulation a thermal or sound insulation material, generally from clothing manufacturing waste, that comes in varying widths and thickness (R-value) to conform to standard framing of walls and joists.

FROM EXPERIENCE

Michael Chandler on Double Walls

Courtesy Michael Chandler, Chandler Design-Build, Inc.

Michael Chandler has been designing and building high-performance homes since 1978. He is a North Carolina licensed builder and plumber, building three to six homes per year, all green, ENERGY STAR, and under 3,000 sq. ft.

Michael Chandler, President, Chandler Design Build Increasing attention is being drawn to double-wall construction as a cost-effective, high R-value alternative to competing high R-value systems, such as ICFs, SIPs, or AAC.

Advantages of double-wall construction are simplified wiring and plumbing due to wide open access prior to insulation, which is especially important when smart home wiring requires holding low-voltage wiring well away from line voltage (Figure 4.22). After the sheetrock is installed, wires can be more easily fished through the blown-in fiber insulation used in double walls than through foam. Double-wall systems are also easier to modify for future additions.

The embodied energy and global warming impact of the double-wall system also compares favorably with that of the concrete and foam in ICFs and SIPs, and with the concrete and aluminum in AAC walls. Double walls, including exterior OSB, are framed off-site, and all scrap is diverted to an I-joist plant. The walls for a typical house go up in a day, and we're generally dried in with tar paper on the roof in two weeks.

The R-value of a 10" AAC wall is R-12, 9" ICF is R-20, and 6.5" SIP is R-23. We can expect a 12" double-wall assembly with R-46 blown-in cellulose or fiberglass insulation in a thermally broken assembly to perform better than any of these (Figure 4.23). But most significantly, the cost of double-wall construction per R-value delivered is very favorable. Our wall panel quote on a recent 2,474-ft^2 home was $12,810 total, $5/ft^2 of heated floor area for double 2 × 4 construction. The total cost for the R-46 formaldehyde-free fiberglass insulation was $1.30/ft^2, so the wall cost for our 2,500-ft^2 house was $16,000 with the 2 × 4 exterior walls. By comparison, the panel quote for a similar house with energy framed 2 × 6 walls was $3.40/ft^2, and the 5½" R-23 fiberglass insulation quote was $1.37, so the up-charge to go from 2 × 6 with R-23 to 12" double 2 × 4 walls (installed with a space between them to allow for R-46) was from $12,000 to $16,000. At $4,000, this is less than the up-charge for a solar water heater.

Courtesy of Michael Chandler, Chandler Design-Build, Inc.

Figure 4.22 A double-framed wall with a small space in between.

Courtesy of Michael Chandler, Chandler Design-Build, Inc.

Figure 4.23 A double-framed wall with blown-in fiberglass insulation.

Foam Insulation

Foam insulation can be sprayed into open-wall, roof, and floor framing cavities or applied in rigid boards on the exterior of the house structure. Spray foam is available in low-density open-cell and high-density closed-cell products (Table 4.2). Open-cell foam, which is typically installed at a rate of approximately 0.5 lb/ft^3, has a lower R-value, is vapor-permeable, and remains flexible after installation (Figure 4.24a). Closed-cell foam, which is installed at a rate of approximately 2 lb/ft^3, has a higher R-value and very low vapor-permeability, Closed-cell foam hardens after installation and can add structural integrity to a building (Figure 4.24b). Open-cell foam expands very rapidly to about 100 times its initial volume, whereas closed-cell foam does not expand significantly after application. Open-cell applications usually fill most, or all, of the framing cavity; any excess material is trimmed flush with the framing after installation. Closed-cell applications often do not fill the framing cavity and require little if any trimming.

Blowing agents are used during the application of SPF to create the air pockets that provide the insulation value. Closed-cell foam uses hydrofluorocarbon (HFC) gas that remains in the foam cells and provides a high R-value; this blowing agent does off-gas for a period of time after application. HFC is a so-called third-generation blowing agent that emerged after federal regulations mandated that manufacturers shift away from chlorofluorocarbon (CFC) and later hydrochlorofluorocarbon (HCFC) blowing agents to protect the Earth's stratospheric ozone layer. HFCs have an "ozone-depletion potential" (ODP) of zero, but they are potent greenhouse gases—hundreds of times more potent than carbon dioxide (see the box on polystyrene insulation, page 102). Unlike closed-cell foam, open-cell foam uses a water-based blowing agent that creates carbon dioxide in the foam cells. Over the course of approximately 30 days, this carbon dioxide is replaced with air as the foam cures.

Spray foam is available for consumer use in small cans and two-part containers (Figure 4.25a and Figure 4.25b) for small installations and minor air sealing; however, large applications are installed by contractors that have been licensed by manufacturers. SPF installers and other personnel on the job site should always follow the manufacturer's safety instructions, including use of appropriate protective equipment and respirators to avoid injury from chemicals that are airborne during application.

Spray foam insulation is largely made from petroleum products, although several manufacturers now offer products that substitute agricultural products, such as soy and castor oil, for a small portion of the fossil-based oils. The value of substituting soy oil for petroleum is debatable because the production of soybeans is fairly energy- and water-intensive and may not significantly reduce the overall environmental impact of the product. Products using castor oil claim to be more sustainable because they use no pesticides or irrigation in production. In all cases, the amount of soy or castor oil remains a minor percentage of the total oils in the foam.

Flash and Batt Hybrid Insulation System

A hybrid insulation system, commonly referred to as flash and batt (FAB), combines a 1" to 2" layer of closed-cell SPF that is covered with fiberglass batts to fill the framing cavity. This combination provides both an air seal and thermal insulation at a lower cost than SPF alone. As with any batt insulation, it must be installed without gaps and compression, according to RESNET grade 1 requirements to work effectively. In extremely cold climates, sufficient SPF must be applied in FAB installations to maintain the interior temperature of the wall above the dew point to avoid condensation in the wall cavity.

Table 4.2 Comparison of Open-Cell and Closed-Cell SPF

Characteristic	Open-Cell SPF	Closed-Cell SPF
Installation density	~0.5 lb/ft^3	~2.0 lb/ft^3
R-value	R-3.4/inch – R-3.8/inch	R-6.0/inch – R-7.0/inch
Vapor permeability	16 at 3" (30–35 perm at 1")	0.8 perms at 2.5"
Blowing agents	Water-based	Most use HFCs; some water-based
Indoor environmental quality	No off-gassing from cured foam	Limited off-gassing from cured foam
Source material	May contain bio-based petroleum substitutes	May contain bio-based petroleum substitutes
Additional features	–	Adds structural strength

open-cell foam a type of spray polyurethane foam installed at a rate of approximately 0.5 lb/ft^3 and sometimes referred to as "half-pound foam"; see also *spray polyurethane foam*.

closed-cell foam a type of spray polyurethane foam installed at a rate of approximately 2 lb/ft^3 and sometimes referred to as "2-pound foam"; see also *spray polyurethane foam (SPF)*.

flash and batt (FAB) a hybrid insulation system that combines a 1" to 2" layer of closed-cell SPF that is covered with fiberglass batts to fill the framing cavity.

Figure 4.24a Open-cell SPF installed in exterior walls. An additional air barrier is not required behind the tub on the exterior wall.

Figure 4.24b Closed-cell SPF installed on a poured basement wall. Open-cell SPF in the band joist.

Figure 4.25a One-part urethane spray foam.

Figure 4.25b Disposable two-part sealant.

Rigid Foam Insulation

Rigid foam boards are made from closed-cell XPS, polyisocyanurate (commonly known as polyiso), and EPS. XPS is suitable for high-moisture locations, such as foundation walls, although all three are appropriate for nonstructural wall and roof sheathing applications. Rigid foam installed on the exterior surface of a building reduces thermal bridging at framing members, improving overall energy performance of a surface. If enough rigid foam is installed on the exterior of a building, cavity insulation can be eliminated completely, although this is not a common practice in residential construction.

Although XPS boards are produced with HFC blowing agents, pentane (which is less harmful to the environment) is used to produce both EPS and polyiso. Polyiso can absorb moisture, so it should not be used below grade unless great care has been exercised to drain moisture away.

Most rigid foam boards do not provide any structural integrity when used as wall sheathing; however, Dow offers structural insulated sheathing (SIS) that consists of a thin layer of structural material laminated to a layer of polyiso, providing both a thermal break and structural integrity (Figure 4.26). Both XPS and EPS boards are manufactured with brominated flame retardants that are considered toxic chemicals, leading some green builders to consider more benign alternatives. Most polyiso is made with a chlorinated flame retardant

Figure 4.26 Dow's Structural Insulated Sheathing (SIS) combines rigid insulation with structural sheathing.

that is likely less hazardous than its brominated relatives but may still be a health hazard.

What Is Foam Insulation Made Of? Polyurethane foam insulation, both sprayed on site and formed into boards in a factory, is created by the combination of two sets of chemicals under pressure. The parts are referred to as the A side, which contains diisocyanates, and the B side, consisting of polyols, blowing agents, and other compounds. Diisocyanates are a leading cause of workplace asthma and are recognized as toxic to humans. The U.S. Occupational Safety and Health Administration (OSHA) recommends that all SPF installers and helpers wear full personal protective equipment, and all unprotected personnel should be instructed to leave the work area. The entire area should be ventilated during and after installation. The U.S. Environmental Protection Agency (EPA) has stated that, after curing, cleanup, and ventilation, SPF insulation that is enclosed and separated from the living area does not pose a long-term threat to occupant health. The health effects of SPF are, however, still under study by OSHA, EPA, and other agencies.

Polystyrene Insulation: Does It Belong in a Green Building?

Polystyrene, in both extruded and expanded forms, is very widely used as rigid insulation in North America and worldwide. In below-grade applications, owing to its good insulation value, superb moisture resistance, strength, performance, and affordability, polystyrene dominates the market.

Unfortunately, a chemical that is added to polystyrene to provide fire resistance, hexabromocyclododecane (HBCD, or HBCDD in Europe) has recently raised significant concerns. Indeed, the European Union may be on the verge of significantly restricting the use of this chemical, and both the United States and Canada are examining it very carefully. Given other environmental concerns about polystyrene, this latest development raises the question of whether this insulation material belongs in green buildings at all.

Both XPS and EPS contain HBCD at concentrations of between 0.5% and 1.2% by weight, according to the American Chemistry Council. While there is still much to learn about the health and environmental impacts of HBCD, enough information has come to light in recent years to prompt the European Chemicals Agency to classify the compound as a chemical of "very high concern" and recommend that its use be restricted. Although significant research on HBCD and its effects on human health and the environment remains to be done, some suggest that enough information is available today to support—following the precautionary principle—phasing out this chemical or seek alternatives to polystyrene insulation.

Even if the flame retardant HBCD weren't a concern, there are other life-cycle concerns with the plastic. "Polystyrene depends on some highly toxic chemicals, including known carcinogens, from the beginning to the end of its life cycle," says Tom Lent of the Healthy Building Network in Berkeley, California. Polystyrene is made by combining ethylene (made from natural gas or petroleum) and benzene (made from petroleum) to produce ethylbenzene, which is then dehydrogenated to form styrene in a process that produces by-products benzene and toluene.

In addition to HBCD and other chemical concerns, the extruded form of polystyrene, XPS, also contains the hydrofluorocarbon blowing agent HFC-134a, which is 1,400 times as potent a greenhouse gas as carbon dioxide (expanded polystyrene, EPS, does not use a blowing agent with high global warming potential). If thick layers of XPS are used to achieve very low-energy, "carbon-neutral" buildings, many decades of energy savings from that insulation may be needed to "pay back" the global warming potential that's going to be released over its life. Most closed-cell spray polyurethane foam has the same concern because it is made with HFC-245fa as the blowing agent, which also has a very high global warming potential.

While XPS and EPS are often used for exterior insulative sheathing, polyiso foam insulation works just as well—and often a little better, because it has a somewhat higher R-value per inch and may come with a reflective foil facing. Polyiso was once made with blowing agents that both depleted ozone and contributed significantly

(Continued)

to global warming, but that is no longer the case. Hydrocarbon blowing agents with very low global warming potential are used in nearly all polyiso products today.

Insulation is one of the most important components of buildings and absolutely critical in creating buildings that will minimize environmental impact. Polystyrene insulation, both XPS and EPS, has played an important role in insulating buildings, but it is the least green of common insulation materials. Polystyrene is less environmentally friendly for two reasons: the HBCD flame retardant used to impart a (moderate) level of fire resistance, and the HCFC blowing agent used in XPS.

When we can do so without sacrificing energy performance, green designers and builders may want to look to alternative materials and methods that are better from a health and environmental standpoint, such as those discussed in this article. While taking this precautionary approach, we should also continue research on these materials and others to identify assemblies that are high-performing without entailing significant environmental tradeoffs.

Adapted with permission from *Environmental Building News*. http://www.buildinggreen.com.

Cementitious Insulation

A proprietary product of Air Krete, cementitious foam insulation is a cement-based insulation that is foamed into the cavity similar to polyurethane foam products. It is an inorganic product that uses no CFC or HCFC blowing agents (it uses compressed air only) and provides a high level of pest, mold, and fire resistance. It is installed by licensed contractors in open- or closed-wall cavities and above finished ceilings. In floors and rooflines, cementitious insulation is installed behind a layer of housewrap to keep the product in place until it has set.

Selecting the Right Insulation

When comparing costs and performance of different insulation materials, it is critically important to make sure that each of them will be installed according to RESNET Grade 1 criteria and to the R-values specified. A low-cost installation that is not installed correctly reduces a home's efficiency and comfort, will result in higher energy bills, and, ultimately, cost the homeowner more through the life of the house. The different air sealing requirements of each insulation product should also be taken into account. Lower cost insulations that are more air permeable will require additional air sealing in relationship to those that are less air permeable. When you look at the total cost of properly insulating and air sealing the building envelope for each type of insulation being considered, you will have the information you need to make the best decision for your project.

How Much Insulation Should Be Installed?

The amount of insulation required is determined by the geographic location of the home and where the assembly is within the home (Figure 4.27). Building codes, such as the International Energy Conservation Code (IECC),

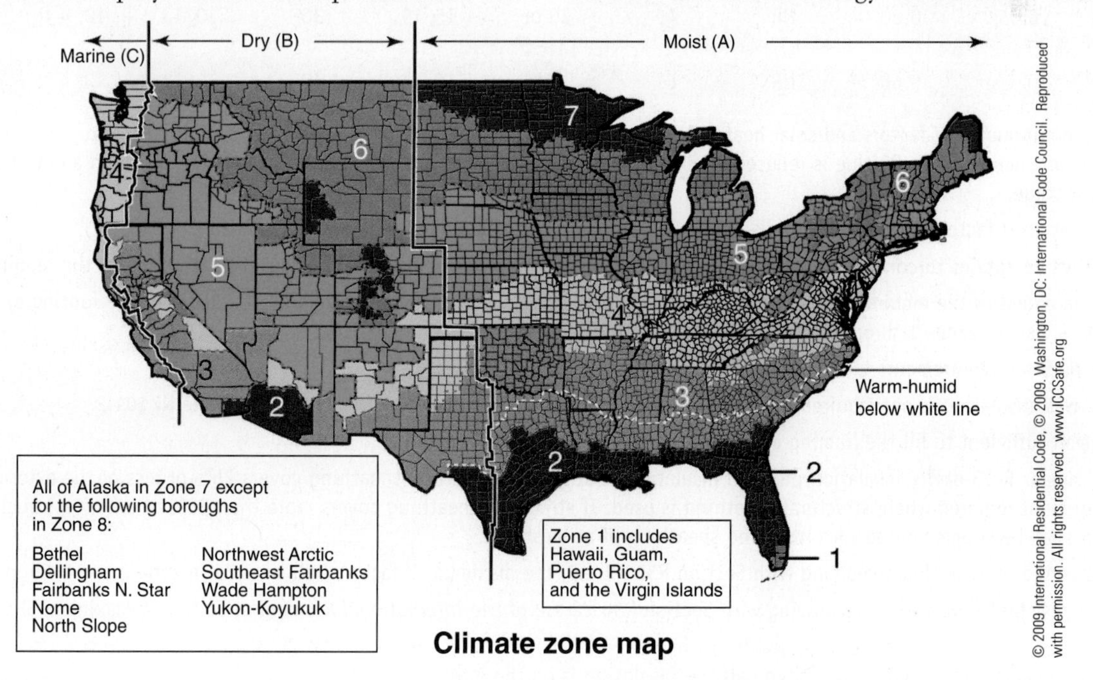

Figure 4.27 United States map showing climate zones, moisture regimes, and warm-humid designations.

provide legal minimums (Table 4.3). The Department of Energy's Oak Ridge National Laboratory (ORNL) developed the following recommended minimum R-values on the basis of an analysis of cost-effectiveness by using average local energy prices, regional average insulation costs, equipment efficiencies, climate factors, and energy savings for both the heating and cooling seasons (Table 4.4).

The R-value of wall and floor insulation is usually determined by the thickness of the available cavity. Since the R-value is reduced by voids and compression, insulation must be cut carefully around obstructions and not be thicker than the cavity it is being installed in. When insulation has gaps as small as 3% of the total area, the overall reduction in performance is as high as 30% for the entire assembly.

While installing more than the recommended minimum wall and floor insulation is always a good idea, there can be a point of diminishing returns—particularly in moderate regions where heat losses and gains are not as great as in cold climates.

Ceiling insulation usually has open attic space above, allowing greater thicknesses to be installed without compression, particularly when blown-in rather than installed as batts. No insulation project is immune to problems, however. Blown-in ceiling insulation must installed to proper depths and densities throughout the entire area to perform properly.

Table 4.3 Residential single-family homes must meet these prescriptive requirements based on climate zone or document code compliance through an acceptable alternative performance path.

Table N1102.1 Insulation and Fenestration Requirements by Component[a]

Climate Zone	Fenestration *U*-Factor	Skylight[b] *U*-Factor	Glazed Fenestration SHGC	Ceiling *R*-Value	Wood Frame Wall *R*-Value	Mass Wall *R*-Value[k]	Floor *R*-Value	Basement[c] Wall *R*-Value	Slab[d] *R*-Value And Depth	Crawl Space[c] Wall *R*-Value
1	1.2	0.75	0.35[j]	30	13	3/4	13	0	0	0
2	0.65[i]	0.75	0.35[j]	30	13	4/6	13	0	0	0
3	0.50[i]	0.65	0.35[e,j]	30	13	5/8	19	5/13[f]	0	5/13
4 except Marine	0.35	0.60	NR	38	13	5/10	19	10/13	10, 2 ft	10/13
5 and Marine 4	0.35	0.60	NR	38	20 or 13 + 5[h]	13/17	30[f]	10/13	10, 2 ft	10/13
6	0.35	0.60	NR	49	20 or 13 + 5[h]	15/19	30[g]	10/13	10, 4 ft	10/13
7 and 8	0.35	0.60	NR	49	21	19/21	30[g]	10/13	10, 4 ft	10/13

[a] *R*-values are minimums. *U*-factors and solar heat gain coefficient (SHGC) are maximums. R-19 batts compressed in to nominal 2 × 6 framing cavity such that the *R*-value is reduced by R-l or more shall be marked with the compressed batt *R*-value in addition to the full thickness *R*-value.

[b] The fenestration *U*-factor column excludes skylights. The SHGC column applies to all glazed fenestration.

[c] The first *R*-value applies to continuous insulation, the second to framing cavity insulation; either insulation meets the requirement.

[d] R-5 shall be added to the required slab edge *R*-values for heated slabs. Insulation depth shall be the depth of the footing or 2 feet, whichever is less, in zones 1 through 3 for heated slabs.

[e] There are no SHGC requirements in the Marine Zone.

[f] Basement wall insulation is not required in warm-humid locations as defined by Figure Nl 101.2 and Table Nl 101.2.

[g] Or insulation sufficient to fill the framing cavity, R-19 minimum.

[h] "13 + 5" means R-13 cavity insulation plus R-5 insulated sheathing. If structural sheathing covers 25% or less of the exterior, R-5 sheathing is not required where structural sheathing is used. If structural sheathing covers more than 25% of exterior, structural sheathing shall be supplemented with insulated sheathing of at least R-2.

[i] For impact-rated fenestration complying with Section R301.2.1.2, the maximum *U*-factor shall be 0.75 in zone 2 and 0.65 in zone 3.

[j] For impact-resistant fenestration complying with Section R301.2.1.2 of the *International Residential Code*, the maximum SHGC shall be 0.40.

[k] The second *R*-value applies when more than half the insulation is on the interior.

Source: 2009 *International Residential Code*, © 2009. Washington, DC: International Code Council. Reproduced with permission. All rights reserved. http://www.ICCSafe.org

Did You Know?

Some Pros and Cons of Loose-Fill Insulation

When properly installed, loose-fill attic insulation is a moderately priced and effective way to reduce heat loss through ceilings. However, loose-fill insulation is prone to improper installation and inadequate coverage by installers, both deliberate and unintentional. Understanding what constitutes a correct installation and inspecting the finished work are critical to high-performance loose fill insulation.

As far back as 1996, the Insulation Contractors Association of America (ICAA) recognized the problem of inadequate attic insulation work and published "A Plan to Stop Fluffing and Cheating of Loose-Fill Insulation in Attics." Fluffing, or adding extra air while blowing fiberglass insulation in ceilings, can reduce the R-value by as much as 50%. The only way to confirm that loose-fill fiberglass insulation has been installed at the proper density is to have core samples taken from random locations and the result weighed to confirm the results. Cellulose and mineral wool insulation are not affected by fluffing, but all types of blown insulation can have gaps, shallow areas, and tapering, particularly at eaves and other difficult-to-reach areas.

Cellulose insulation, due to its weight, settles as much as 20% over time, which is considered in the R-value calculations. For example, according to one manufacturer's recommendations, 11.7" of cellulose must be installed to reach a settled thickness of 10" (accounting for approximately 14% settlement) to reach an R-value of 38.[2] Cellulose manufacturers also list the area of coverage per bag of product for total R-value.

Laboratory tests have shown that fiberglass and mineral wool insulation also settle, although they lose only about 5% of their thickness; manufacturers, however, do not account for this compression in their calculations. As a general rule, blown-in glass and mineral fiber ceiling insulation should exceed the listed thickness by 5% to 10%, and cellulose by 15% to 20% to account for future settling.

When planning for loose-fill attic insulation, sections that are to have floor decking must be framed up to allow for the full depth of insulation (see Figure 4.28). Blocking must be installed at level changes to avoid tapering and must be protected from disturbance during construction as well as after occupancy.

A 1992 study by ORNL shows that as attic temperatures drop below 45 degrees, loose-fill ceiling insulation loses some of its R-value through convective currents. Cold attic air moving down through the insulation warms as it reaches the room temperature ceiling, causing it to rise back into the attic, slowly rotating air and allowing heat from the house to move into the cold attic. As the attic temperature goes below 0°F, the insulation loses as much as 50% of its R-value. Subsequent studies by insulation manufacturers have concluded that convective losses occur only in low-density, extremely air permeable materials, and less air permeable insulation made with smaller fibers has less convective losses at low temperatures.

Loose-fill cellulose insulation is less air permeable and does not have the same level of convective R-value loss as fiberglass, but due to its greater weight, it may cause sagging in drywall ceilings when framing members are wider than 16" on center (OC), or when more than R-38 is installed. In these cases, 5/8" drywall is recommended.

When combined with an effective air barrier, loose-fill insulation can be an excellent solution for many homes. But like any product, it must be specified and installed properly with all details carefully addressed, and it must be inspected after installation to perform properly.

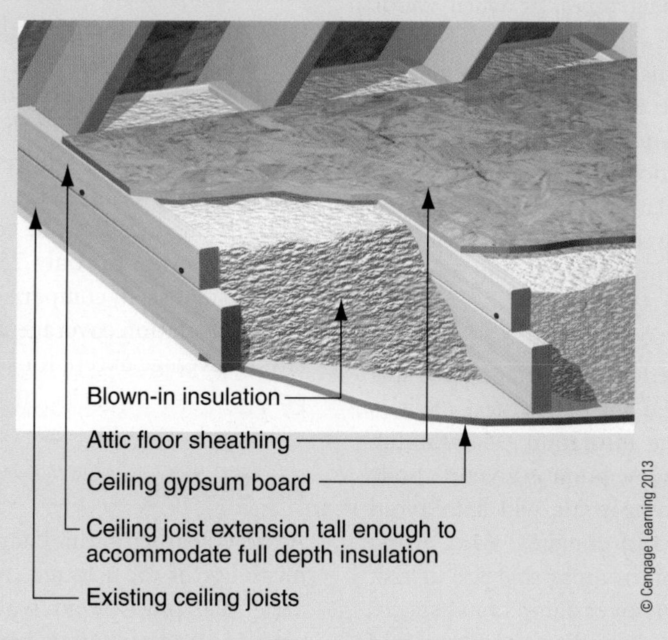

Figure 4.28 Many attic insulation products require a depth of 10" to 14" to achieve desired R-values. When installing decking over 2 × 6 ceiling joists joist extensions must be added to provide adequate vertical clearance for insulation. Here, a 2 × 4 is added to the 2 × 6 joist to provide nearly 10" of insulation depth.

[2]Nu-Wool® Engineered Cellulose Insulation Coverage Chart 6/04.

Table 4.4 The U.S. Department of Energy's recommendations for insulation levels are similar to the 2009 IECC, though more stringent and strictly voluntary.

Zone	Heating System	Attic	Cathedral Ceiling	Wall Cavity	Insulation Sheathing	Floor
1	All	R30 to R49	R22 to R38	R13 to R15	None	R13
2	Gas, oil, heat pump Electric furnace	R30 to R60	R22 to R38	R13 to R15	None	R13 R19–R25
3	Gas, oil, heat pump Electric furnace	R30 to R60	R22 to R38	R13 to R15	None R2.5 to R5	R25
4	Gas, oil, heat pump Electric furnace	R38 to R60	R30 to R38	R13 to R15	R2.5 to R6 R5 to R6	R25–R30
5	Gas, oil, heat pump Electric furnace	R38 to R60	R30 to R38 R30 to R60	R13 to R15 R13 to R21	R2.5 to R6 R5 to R6	R25–R30
6	All	R49 to R60	R30 to R60	R13 to R21	R5 to R6	R25–R30
7	All	R49 to R60	R30 to R60	R13 to R21	R5 to R6	R25–R30
8	All	R49 to R60	R30 to R60	R13 to R21	R5 to R6	R25–R30

Source: *Insulation Fact Sheet 2008* (DOE/CE-0180).

Courtesy of the Department of Energy.

Moisture Problems with Cavity Insulation

In wood frame structures, cavity insulation must be kept dry to both maintain its R-value and to protect the building from mold and rot. Cellulose and open cell SPF insulation absorb moisture, filling the insulating air spaces and promoting mold growth on wood and drywall. Fiberglass and mineral wool do not absorb water, but they can trap it in wall cavities and cause similar problems. Insulation can get wet from exterior leaks through the roof, walls, and foundation, from interior leaks such as broken plumbing pipes, or when vapor condenses in the wall cavity.

To avoid moisture problems during construction, the water-resistive barrier on the walls must be complete, the roof must be dried-in, and the structure must be completely dry before insulation is installed. Insulation that becomes wet after installation must be either dried thoroughly or replaced before covering. Exterior or interior leaks must be repaired immediately, and any wet insulation or finishes dried or replaced.

Vapor condensation is controlled by comprehensive air sealing, which reduces the amount of air-transported moisture allowed into structural cavities where it can condense in the insulation. Maintaining relative humidity so that the dew point is always above the temperature of the wall sheathing will help avoid condensation problems in cold climates. Also, using vapor-impermeable insulation in areas that are at risk of condensation (e.g., in floors over damp crawl spaces in hot climates, or in insulated rooflines in severe cold climates) will reduce the possibility of vapor condensing on cold surfaces.

Structurally Integrated Insulation

Structurally integrated insulation methods include ICFs, AAC, SIPs, and straw bale systems. ICFs are available in polyurethane foam and mineralized wood. Foam ICFs are available with some recycled content material. Mineralized wood ICFs have a lower insulation value than foam and typically employ mineral wool batts in the cavities to increase the R-value. AAC is a lightweight solid concrete block that provides insulation via air pockets within the material. SIPs are available with either polyurethane foam or agricultural fiber insulation. Finally, straw bale construction provides insulation with standard agricultural bales that are reinforced and covered with concrete.

Integrated insulation offers one major benefit: it eliminates most or all of the framing members that create thermal bridges and reduce the amount of insulation in wall, ceiling, or floor structure. A standard wood-framed wall can be composed of as much as 25% wood, allowing for only 75% of the framed area to be fully insulated. In comparison, ICF and AAC walls have 100% insulation coverage, and SIPs have complete insulation coverage except for structural members at bearing points.

Air Sealing

As noted earlier in this chapter, regardless of the air permeability of the selected insulation, every building has areas that must be air sealed to complete the thermal envelope and reduce air infiltration.

Starting again from the bottom, where the foundation walls serve as the air barrier, every hole must be fully sealed, including doors, windows, pipes, wires, and ducts. If the air barrier is the floor, then every pipe, wire, duct, and chase must be fully sealed. Current building codes typically require that these areas be sealed; however, they are often filled with mineral wool or other products that reduce the risk of fire but do not always eliminate air infiltration.

All transition points between materials in walls must be completely air sealed, including between double and triple studs, along the line where the wall plates meets floors and ceilings, around windows and doors, between masonry or concrete and wood, along rim joists, and around wires and pipes. Typical materials used for air sealing include caulk, spray foam, and gaskets.

Many of these air sealing measures can be eliminated with the installation of an exterior air barrier, such as a rigid foam that is carefully taped and sealed. Housewrap and other air infiltration barriers can reduce air infiltration to a certain extent, but neither housewrap nor rigid foam should be expected to offer a complete air seal in any building. With both products, gaps around windows and doors, floor cantilevers, and wires and pipes must still be air sealed separately.

Air sealing at the ceiling is the most challenging and among the most important areas to address (Figure 4.29). Attic accesses, such as pull-down stairs, ceiling hatches, and knee wall doors must be fully air sealed and insulated (Figure 4.30). Both commercially available products and

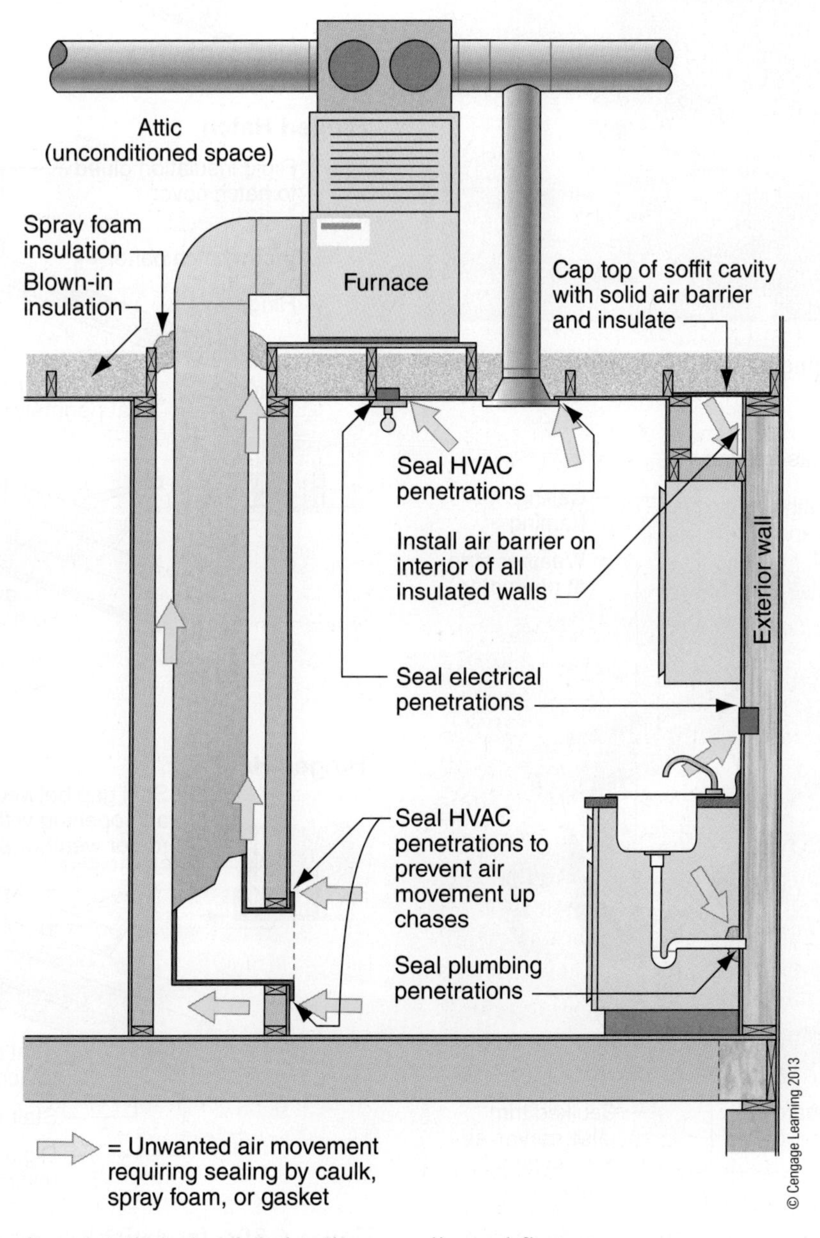

Figure 4.29 Key air sealing details in insulated ceilings, walls, and floors.

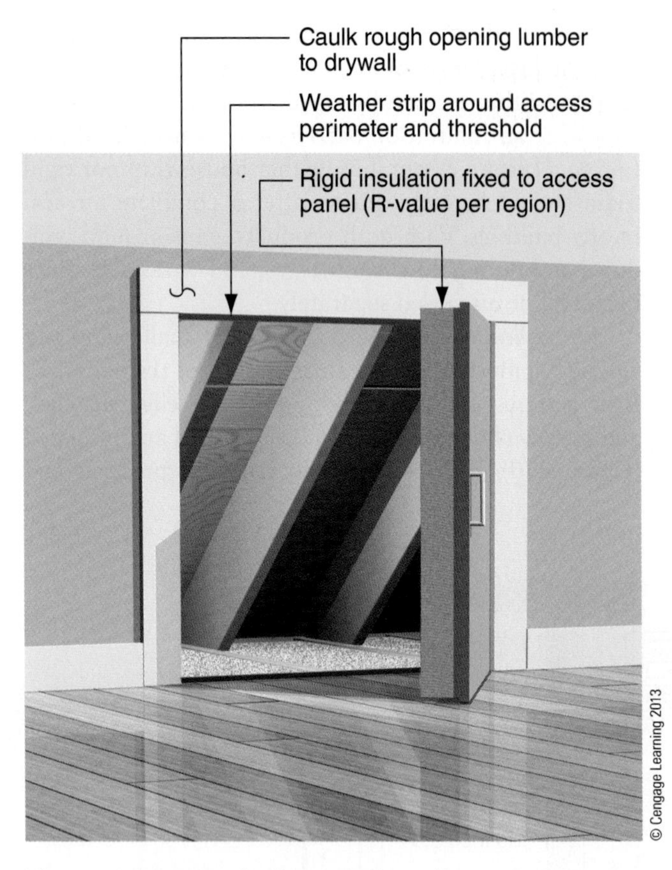

Figure 4.30a Insulating and air sealing attic knee wall doors.

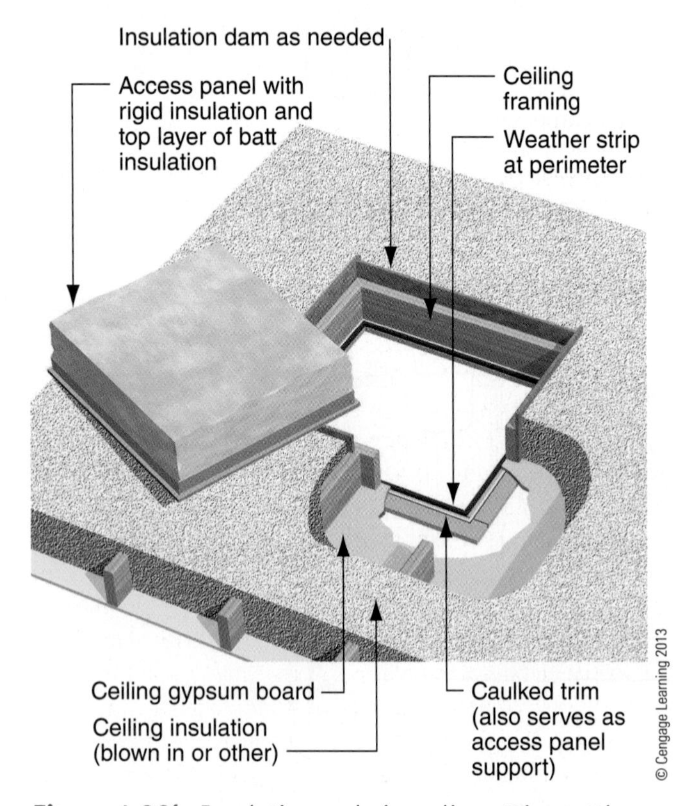

Figure 4.30b Insulating and air sealing attic scuttle holes.

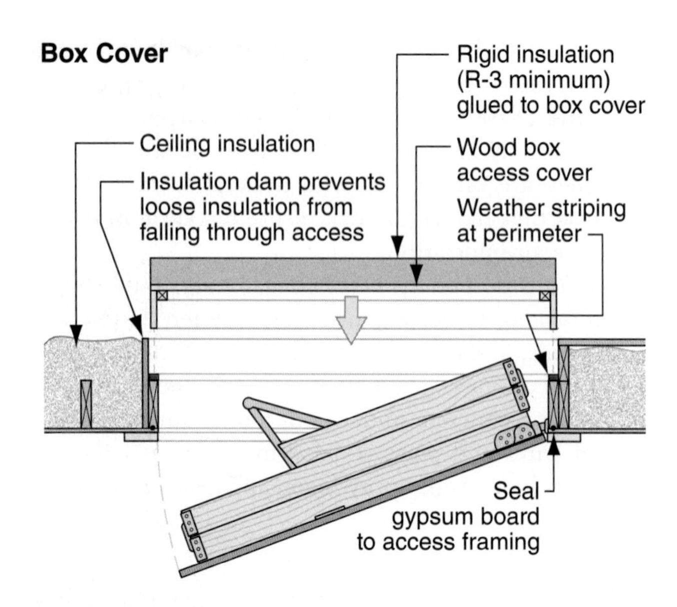

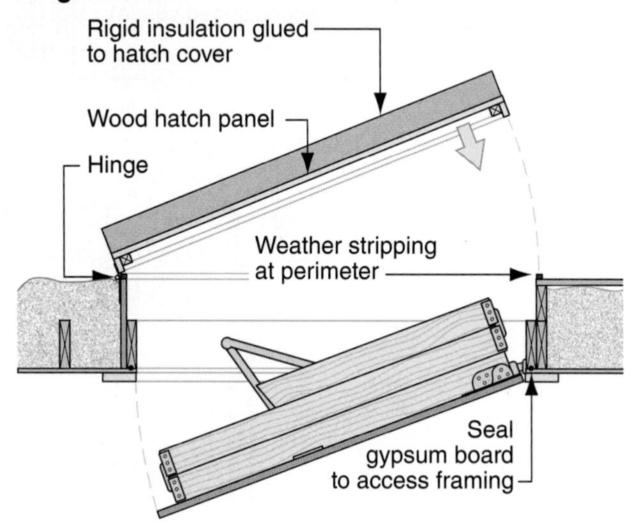

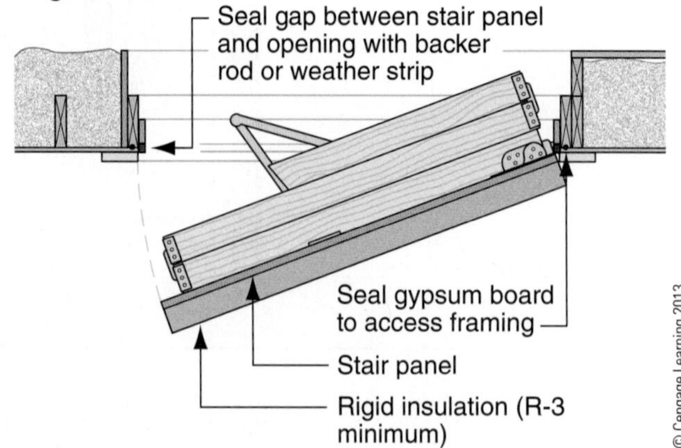

Figure 4.30c Insulating and air sealing attic pull-down stairs.

site-constructed covers are effective ways to seal these areas; however they must be maintained so that they work properly through the life of the house. Light fixtures and registers for heating, ventilating, and air conditioning (HVAC) must be sealed to the drywall with caulk or mastic to eliminate air leaks into and out of the space and to properly direct conditioned air completely into the living space. Recessed light fixtures in insulated ceilings must be airtight units that typically have gaskets that seal them to the drywall and be insulation contact rated units. Insulation contact (IC) units permit insulation to be in contact with the fixture without causing a fire hazard or gaps in the thermal envelope. Whole-house fans must have removable tight covers for out-of-season use, or insulated covers that open automatically when the fan is turned on. Drywall must be sealed at all top plates on both interior and exterior walls to eliminate air leakage from the attic into the wall cavities.

An alternative to the detailed air sealing measures required at the ceiling, the building envelope can instead be located at the roofline. This technique, particularly effective when HVAC equipment is located in the attic, eliminates most of the air sealing required between the finished space and the attic, moving it to the roofline where SPF can be sprayed on the decking, rafters, and any gable walls or SIPs. A thick layer of rigid foam can also be installed on top of the roof and on gable walls (Figures 4.31 and 4.32). Insulated rooflines, also known as cathedral ceilings, can either be in an unfinished attic or in a living space with finish material applied directly to the rafters. This subject will be addressed in detail in Chapter 7.

Figure 4.31 A sealed, conditioned attic with open-cell SPF on the roof line. By insulating the roofline, the knee walls no longer need insulation or air sealing.

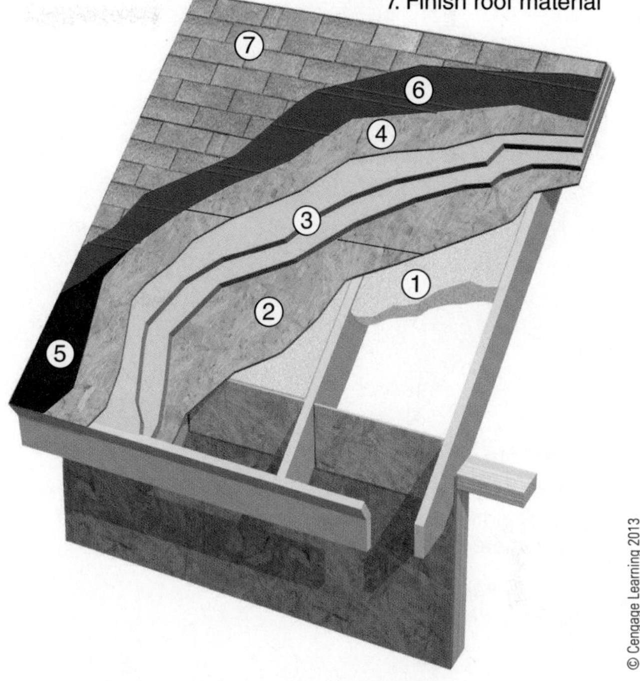

Figure 4.32 Cathedral ceilings can be insulated from above with rigid insulation, from below, or with a combination of the two. Rigid insulation eliminates any thermal bridges.

Completing the Thermal Envelope

In addition to the floors, walls, ceiling, or roof, other areas must be carefully addressed to ensure a complete and continuous thermal envelope. Paying close attention to the transitions between different materials and the changes from horizontal to vertical or sloped locations is one of the keys to creating a high-performance building.

Interior walls between conditioned and unconditioned space, such as attic knee walls and walls to garages (see Figure 4.33), must be fully sealed on all sides. Attic knee walls often have exposed fiberglass insulation on the unfinished side, which reduces the effectiveness of the insulation as well as the air seal. Exterior walls behind tubs and prefabricated fireplaces, as well as duct and plumbing chases between floors, are often left open, creating pathways for unwanted air movement (see Figure 4.34). Skylight wells pose the same problem as knee walls and must be sealed

---

insulation contact (IC) units recessed light fixtures that dissipate heat into the room, permitting insulation to be in contact with the fixture without causing a fire hazard or gaps in the thermal envelope.

cathedral ceiling an insulated vaulted ceiling with roof above; sometimes referred to as a "roof-ceiling" combination and commonly found in living rooms and attics with insulated rooflines.

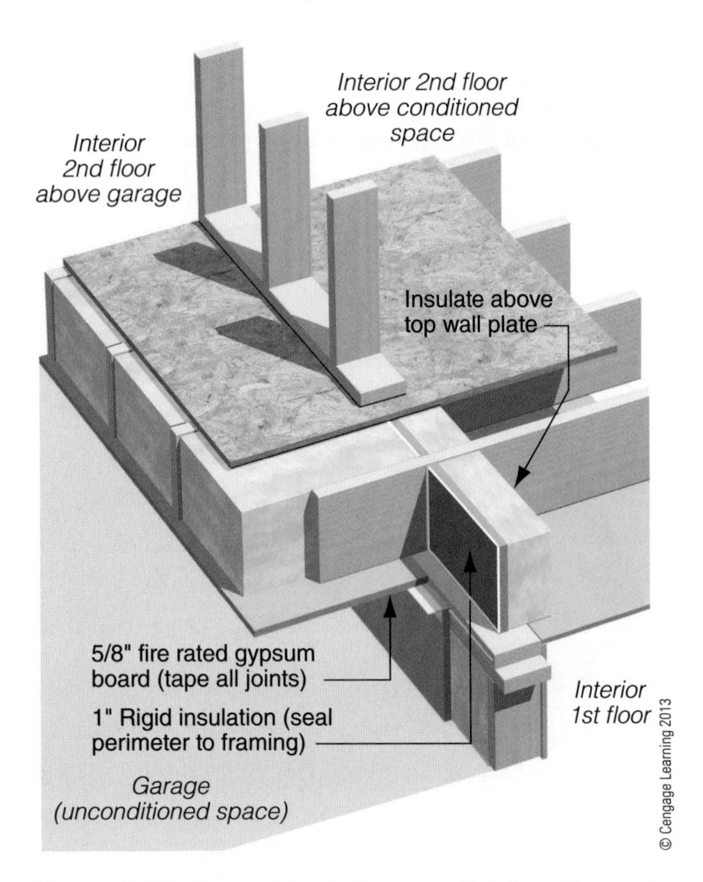

Figure 4.33 Floor joists that run parallel from the home into garages must be blocked above the garage partition wall.

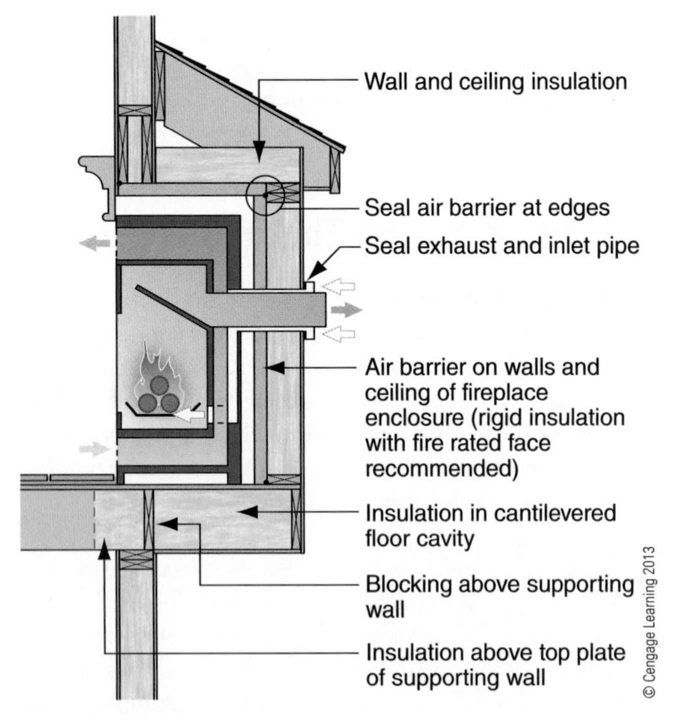

Figure 4.34 Bringing the fireplace chase within the building envelope. Insulate and air seal walls behind fireplaces and the ceiling above.

with a complete air barrier on the attic side, and tubular skylights must be insulated and air sealed to complete the building envelope (see Figure 4.35). Cantilevered floors must be fully blocked and air sealed above the exterior walls below to eliminate air movement into the floor structure (see Figure 4.36).

One of the most challenging areas to create an effective envelope is the floor of finished rooms above garages, where a poor air seal can allow carbon monoxide and other poisonous gases to enter the living space. Carbon monoxide is a colorless, odorless gas that is a by-product of combustion and is extremely dangerous. In this location, the insulation must be in full contact with the subfloor above, and the rim joists and the drywall ceiling below must be completely and carefully air sealed throughout. Finally, at the top of the house, air can leak around lights, ducts, attic access panels, whole-house fans, flues, and vent pipes if not properly sealed. ENERGY STAR has created a guide for builders that provides construction details and inspection protocols for proper insulation and air sealing of homes.

Additional critical areas for air sealing include chases between conditioned and unconditioned space, such as those constructed for flues, chimneys, ducts, and pipes (Figure 4.37). Fire-resistant materials must be used adjacent to any pipes that will get hot, such as metal furnace or fireplace flues. Combinations of wood, sheet metal, and high-temperature sealants are often used for these applications.

Finally, wherever air permeable insulation is installed in walls, an air barrier must be installed to provide the required six-sided air seal behind metal fireplaces, at framed soffits, behind tubs and showers, at knee walls, at cantilevers, and anywhere else that air could flow freely through cavity insulation. Typical materials used for this air barrier include drywall, thin sheathing, or rigid foam. In these locations, the use of an air-impermeable insulation, such as spray foam, usually eliminates the need for a separate air barrier.

Selecting Air sealing Materials

There are three types of air sealing materials: caulks and adhesives, liquid (spray) foam, and air barriers (Table 4.5). Each of the three types contains a variety of products and materials (Table 4.6), some of which are better suited than others to a particular application. For example, only fire code–approved materials and methods should be used for air sealing around a combustion appliance flue pipe. In that situation, only stove cement or fire caulk should be used in conjunction with sheet metal blocking.

---

carbon monoxide (CO) a colorless, odorless, poisonous gas that results from incomplete combustion of fuels (e.g., natural or liquefied petroleum gas, oil, wood, and coal).

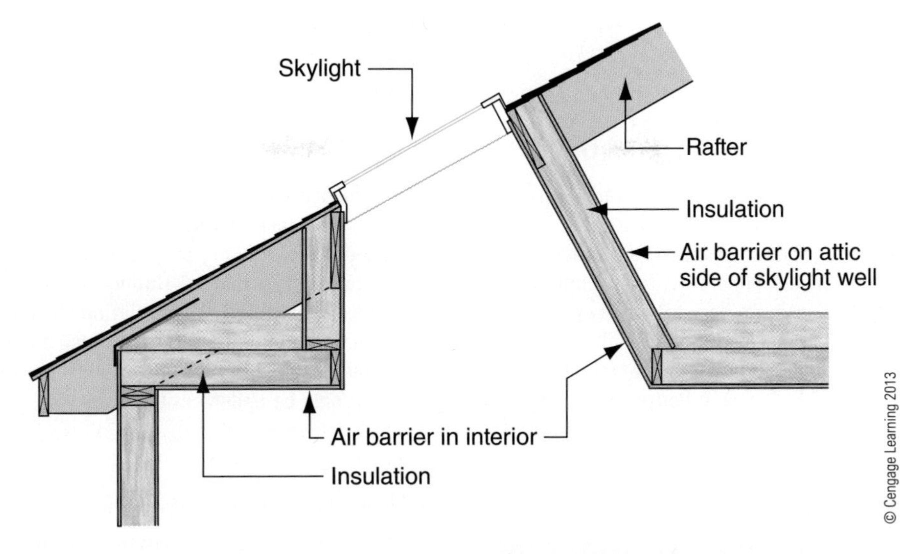

Figure 4.35 The walls of skylight shafts are insulated and sheathed with an air barrier.

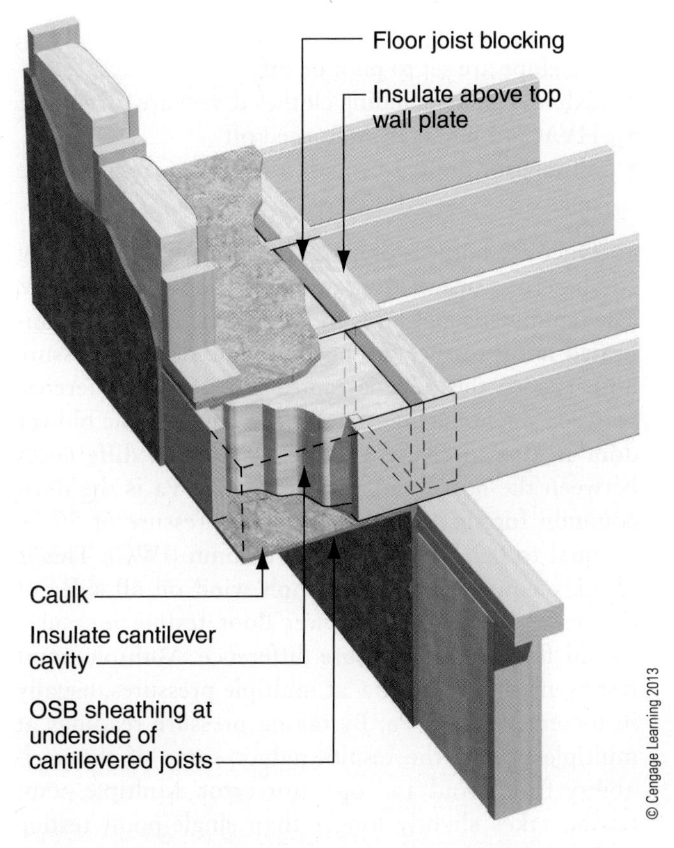

Figure 4.36 Blocking the joists between cantilevered floors prevents air infiltration.

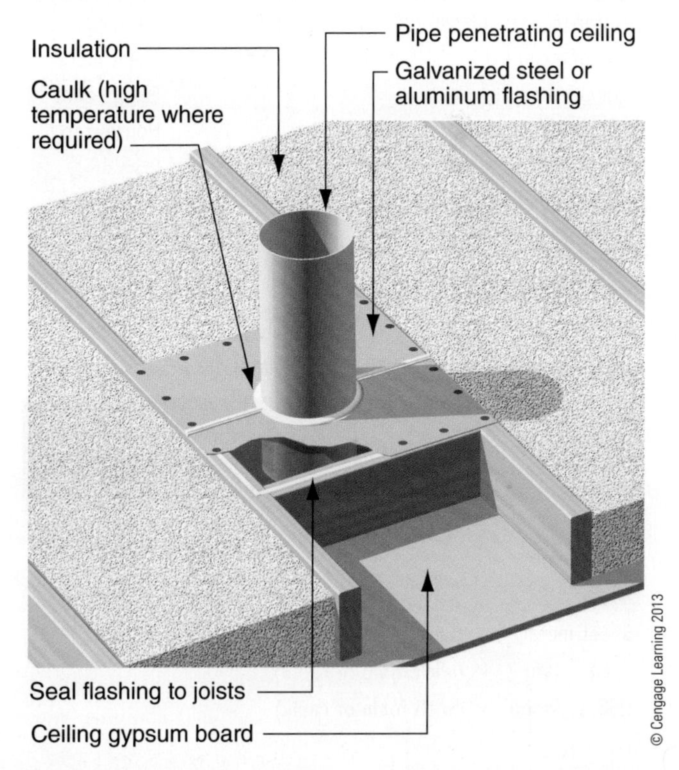

Figure 4.37 When sealing around flue pipes, always use code-approved materials that do not cause a fire risk.

Testing Envelope Leakage

The airtightness of a building is useful information to have when trying to increase energy efficiency, improve indoor environmental quality, and control building pressures and humidity. Most green building programs require envelope leakage testing to ensure building performance. Envelope leakage can be quantified and specific leaks can be identified through blower door and tracer gas testing. In existing homes, envelope leakage testing is a valuable tool for troubleshooting and identifying air leakage pathways. Blower door testing is the most common form of testing because of its relative speed, ease of operation and affordable equipment.

---

tracer gas a nontoxic gas used to measure envelope air leakage.

Did You Know?

The Myth of the "Too-Tight House"

Historically, buildings were very air permeable. Early work with high-performance buildings included increased insulation and air sealing that improved efficiency but often led to poor indoor air quality (from a lack of fresh air) and mold and mildew growth due to poor moisture management. Stories about problems such as these led many building professionals and homeowners to the conclusion that buildings should not be built tight but rather should be allowed to "breathe" through air leakage in order to be healthy. While a very tight house can have air quality problems without proper ventilation, that ventilation should never be provided through uncontrolled infiltration. Uncontrolled infiltration allows moisture and pollutants to enter the home and reduces efficiency. Controlled ventilation (discussed in detail in Chapter 13) provides fresh air when needed to keep the air inside healthy while keeping out unwanted pollutants and humidity. A house can never be too tight; it can only be underventilated.

Table 4.5 Types of Air sealing Materials

Caulking and Adhesives	Liquid Foam	Air Barrier Materials
Water-based caulks	One-part foam	Plywood
Silicone caulk	Two-part foam	Sheet metal
Polyurethane caulk		Foam board
Acoustical sealant		Housewrap
Water-soluble duct mastic		Drywall
Stove cement		
Fire-rated caulk		
Construction adhesives		

Table 4.6 Where to Use Air sealing Materials

Product	Large Holes	Small Holes	Large Linear Gaps	Small Linear Gaps
Caulk		✓		✓
Spray foam		✓	✓	✓
Rope caulk				✓
Sheet metal	✓ (with foam or caulk)			
Foam board	✓ (with foam or caulk)			
OSB/plywood	✓ (with foam or caulk)			

Blower Door Testing

As explained in Chapter 1, the *blower door* consists of a calibrated fan for measuring an airflow rate and a pressure gauge to measure the total air pressure in the home created by the air flow through the fan. The combination of air pressure and fan flow measurement is used to determine the building's air tightness. The blower door can either pressurize (push air into) or depressurize (exhaust air from) the home.

Careful preparation of the home is required to ensure accurate results and prevent harm to the home or occupants. Without proper preparation, the blower door may backdraft combustion appliances. Prior to conducting the blower door test, the following conditions must be met:

- all exterior windows and doors are closed
- fireplace dampers are closed
- combustion appliances that are within the building envelope are set to pilot or off
- exhaust fans (including clothes dryer) are turned off
- HVAC air handlers are turned off
- all interior doors to rooms with HVAC registers are open

Once the home is prepared, the blower door testing equipment is installed in an exterior door or window. The equipment measures the pressures across the calibrated fan (pressure tap inside the home and pressure hose run to the outside) and the pressure difference between the home interior and the outside. The blower door is able to create a range of pressure differences between the home and outside, but 50 Pa is the most common for single-point testing. A pressure of 50 Pa is equal to 0.2 inches of water column (IWC). This is roughly equivalent to a 20-mph wind on all sides of the home. **Single-point blower door testing** measures the air flow at one pressure difference. **Multiple-point testing** measures air flow at multiple pressures, usually in increments of 5 Pa. By taking pressure readings at multiple points, the results reduce some of the variability from wind and operator error. Multiple-point testing takes slightly longer than single-point testing and requires complicated calculations or software to analyze the results.

---

single-point blower door testing blower door testing that utilizes a single measurement of fan flow needed to create a 50-Pa change in building pressure.

multiple-point testing a blower door procedure involving testing the building over a range of pressures (typically 60 pascals to 15 pascals) and analyzing the data using a blower door test analysis computer program.

Blower door envelope tightness measurements are presented in a number of different formats, including fan flow, air changes per hour (ACH), effective leakage area (ELA), and CFM_{50}/square footage of building envelope (CFM_{50}/SFBE).

Fan Flow

One of the most basic measurements of envelope tightness is CFM_{50}, which is defined as the air flow (in cubic feet per minute) at 50 Pa pressure difference across the building envelope.

Air Changes per Hour

One of the most common ways to normalize envelope leakage is to calculate the number of times every hour the total volume of the home's air changes under the test pressure. The air changes per hour at 50 pascals (ACH_{50}) is the number of times the total volume of the home exchanges with outside air when the blower door depressurizes the home to 50 Pa. To calculate ACH_{50}, the volume of the building envelope and CFM_{50} are required. The equation for ACH_{50} is:

$$ACH_{50} = \frac{CFM_{50} \times 60}{\text{Building Volume (ft}^3\text{)}}$$

First, the CFM_{50} is multiplied by 60 to calculate the cubic feet per hour at 50 Pa. This figure is then divided by the building volume (cubic feet) to produce the ACH_{50}.

Air Changes per Hour Natural

The air changes per hour natural ($ACH_{Natural}$) roughly estimates the number of air changes per hour under normal operating conditions. $ACH_{Natural}$ utilizes an N factor developed by Lawrence Berkeley National Lab to calculate the leakage rate based on climate zone, number of stories of a building, and sheltering from wind (Figure 4.38). The equation for $ACH_{Natural}$ is as follows:

$$ACH_{Natural} = \frac{ACH_{50}}{N}$$

Zone refers to climate conditions and how well the home is protected from wind (wind-shielded, normal, or exposed). Wind significantly affects the stack effect and the amount of envelope leakage. The zone and wind shield classifications are reasonably subjective. In general, well-shielded refers to homes that are protected, such as townhomes and homes with permanent wind breaks. Exposed homes have little protection from wind. When using the table, start by locating the climate zone on the map and then match the climate zone to the table with the amount of wind shielding and the number of stories.

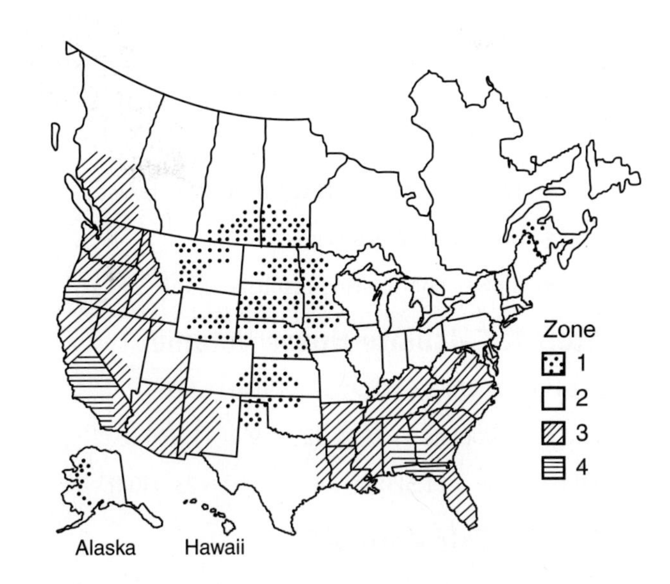

n-Factor Table

Zone ↓	# of stories →	1	1.5	2	3
1	Well-shielded	18.6	16.7	14.9	13.0
	Normal	15.5	14.0	12.4	10.9
	Exposed	14.0	12.6	11.2	9.8
2	Well-shielded	22.2	20.0	17.8	15.5
	Normal	18.5	16.7	14.8	13.0
	Exposed	16.7	15.0	13.3	11.7
3	Well-shielded	25.8	23.2	20.6	18.1
	Normal	21.5	19.4	17.2	15.1
	Exposed	19.4	17.4	15.5	13.5
4	Well-shielded	29.4	26.5	23.5	20.6
	Normal	24.5	22.1	19.6	17.2
	Exposed	22.1	19.8	17.6	15.4

Figure 4.38 $ACH_{Natural}$ map and table.

CFM_{50} and Surface Area of the Building Envelope

One of the most accurate ways to compare homes is to normalize envelope leakage rates based on the surface area of the building envelope. This procedure is known

air changes per hour at 50 pascals (ACH_{50}) the number of times that the total volume of a home is exchanged with outside air when the home is depressurized or pressurized to 50 pascals.

air changes per hour natural ($ACH_{Natural}$) the number of times the total volume of a home is exchanged with outside air under natural conditions.

STEP-by-STEP Calculating CFM_{50}/SFBE

- Two-story home located in Austin, Texas
- Normal wind exposure
- Blower door testing results = 2,468 CFM_{50}

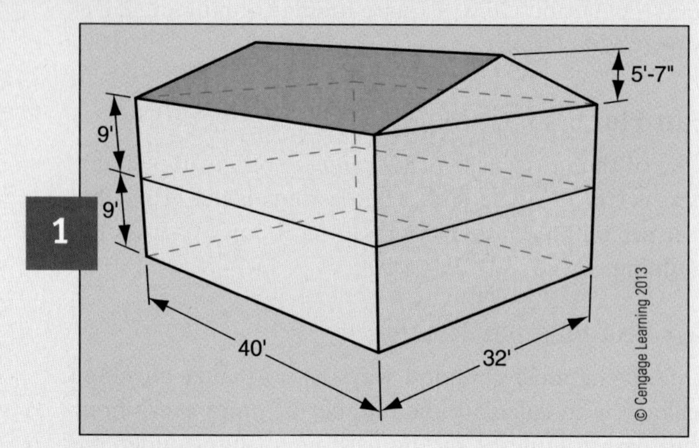

A simple two-story conventional home.

Step 1. Calculate House Volume

1st Floor Volume = 40' × 32' ×9' = 11,520 ft^3

2nd Floor Volume = 40' × 32' × 9' = 11,520 ft^3

Total Volume = 11,520 ft^3 + 11,520 ft^3 = 23,040 ft^3

Step 2. Calculate SFBE

Building Envelope Floor Area = 40' × 32' = 1,280 ft^2

Building Envelope Ceiling Area = 40' × 32' = 1,280 ft^2

Building Envelope Wall Area =

First-Floor Wall Area = (40' + 32' + 40' + 32') × 9' = 1,440 ft^2

Second-Floor Wall Area = (40' + 32' + 40' + 32') × 9' = 1,440 ft^2

Total Wall Area = 2,880 ft^2

SFBE = 1,280 ft^2 + 1,280 ft^2 + 2,880 ft^2 = 5,440 ft^2

Step 3. Calculate ACH_{50}

$$\text{If } ACH_{50} = \frac{CFM_{50} \times 60}{\text{Building Volume (ft}^3\text{)}}$$

$$\text{then } ACH_{50} = \frac{2{,}468\ CFM_{50} \times 60}{23{,}040\ (\text{ft}^3)} = 6.43$$

Step 4. Calculate $ACH_{Natural}$

$$\text{If } ACH_{Natura} = \frac{ACH_{50}}{N}$$

$$\text{then } ACH_{Natural} = 6.43 \frac{\square}{\square} 14.8$$

Step 5. Calculate CFM_{50}/SFBE

$$\frac{2{,}468\ CFM_{50}}{5{,}440\ ft^2} = 0.45$$

Chart for "N"

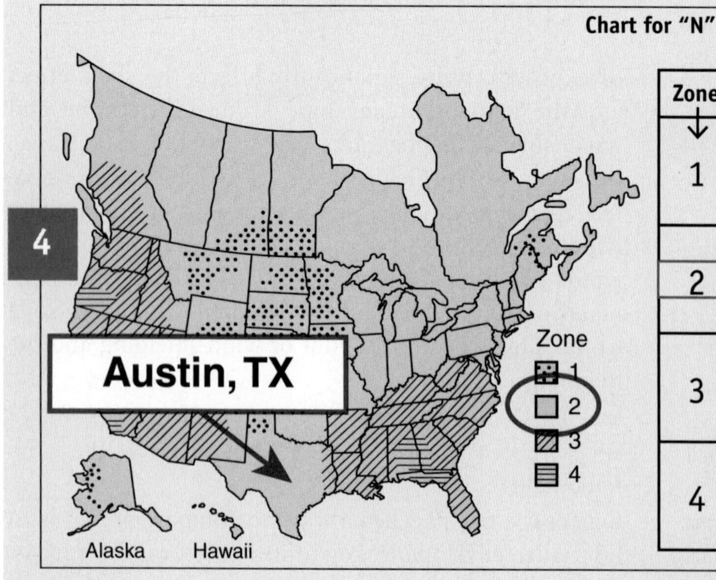

Zone ↓	# of stories→	1	1.5	2	3
1	Well-shielded	18.6	16.7	14.9	13.0
	Normal	15.5	14.0	12.4	10.9
	Exposed	14.0	12.6	11.2	9.8
2	Well-shielded	22.2	20.0	17.8	15.5
	Normal	18.5	16.7	14.8	13.0
	Exposed	16.7	15.0	13.3	11.7
3	Well-shielded	25.8	23.2	20.6	18.1
	Normal	21.5	19.4	17.2	15.1
	Exposed	19.4	17.4	15.5	13.5
4	Well-shielded	29.4	26.5	23.5	20.6
	Normal	24.5	22.1	19.6	17.2
	Exposed	22.1	19.8	17.6	15.4

as the *envelope leakage ratio (ELR)*. The surface area of building envelope (SFBE) is the total area (in ft^2) of the building envelope or building shell. The SFBE includes the walls, floors, and ceilings where heat is lost in the winter. Since air leakage and heat gain occur through the building envelope, it is a logical way to evaluate leakage rates. ACH_{50} and $ACH_{Natural}$ rely on the home's volume, which has little correlation to home heat gain or loss.

$$\text{Envelop leakage ratio} = \frac{CFM_{50}}{SFBE}$$

Effective Leakage Area

Leakage area estimates are a useful way to visualize the cumulative size of all leaks or holes in the building enclosure. These holes may be anywhere in the enclosure, including attics and crawl spaces. The formula to determine effective leakage area was developed by Lawrence Berkeley Laboratory and is used in their infiltration model. The effective leakage area (ELA) is defined as the area of a special nozzle-shaped hole (similar to the inlet of a blower door fan) that would leak the same amount of air as the building does at a pressure of 4 Pa.

How Tight Is Tight?

Envelope leakage rates vary across the country by home age, local climate, construction techniques, and code. Table 4.7 shows the approximate infiltration rates for typical wood-framed homes. To prevent unhealthy indoor air quality many green building and home performance programs require mechanical ventilation for homes tighter than 0.35 $ACH_{Natural}$.

Table 4.7 Typical Envelope Leakage Rates

Home Type	ACH_{50}	$ACH_{Natural}$	$CFM_{50}/SFBE$
High performance	1.5–4.0	0.25	0.25
Energy efficient	4.0–7.0	0.33	0.40
Standard, new	8.0–12.0	0.80	0.70
Standard, existing	11.0–21.0	1.0	1.0+
Standard, older	18.0+	1.75+	2.0+

surface area of building envelope (SFBE) the total area (in ft^2) of the building envelope or building shell.

effective leakage area (ELA) the area of a special nozzle-shaped hole (similar to the inlet of a blower door fan) that would leak the same amount of air as the building does at a pressure of 4 pascals.

Tracer Gas Testing

Unlike the blower door, tracer gas testing measures the envelope leakage rate under natural pressures. Tracer gas testing is primarily a research technique, although it is sometimes used in evaluating commercial and large multifamily buildings. Because tracer gas testing occurs under natural pressures, it provides a more accurate calculation of envelope leakage. During the test, a small quantity of a harmless gas (typically sulfur hexafluoride or carbon dioxide) is released into the building, and its concentration is measured over time. The gas can be released all at once or at intervals. Leakage rates are calculated according to the rate the tracer gas is diluted by incoming outside air. Tracer gas testing is significantly more costly and time consuming than blower door testing.

Remodeling Considerations

There are excellent opportunities to improve the insulation and air sealing in remodeling projects, but those same opportunities can create serious problems if the "house as a system" principles are not carefully considered.

Testing Existing Building Performance

All remodeling projects should begin with testing and inspection of the building envelope. A blower door test will provide information about the total air leakage and help to identify specific leaks that need to be corrected. Visual inspections, which can be augmented by infrared photography, will provide information about the locations and quantities of insulation in the walls and ceilings.

Opportunities and Challenges in Improving Existing Homes

When undertaking a green remodeling project, the existing conditions should be carefully inspected. Ideally, any changes to the building envelope that will negatively affect the structure should be avoided. One of the most important considerations to remember is that most of the energy loss in a building, particularly an older one, goes through the roof. Improving the insulation and air sealing at the ceiling plane or moving the building envelope to the roofline are options to consider in any remodeling project.

Floors above unconditioned foundations should have all gaps air sealed, and a layer of insulation should be installed tight against the subfloor (Figure 4.39a and

Figure 4.39a Duct chase before air sealing.

Figure 4.39b Duct chase after air sealing.

Figure 4.39b). Provided the basement or crawl space is dry, it could be converted into conditioned space by insulating and air sealing the foundation walls instead of the floor. Whenever an existing unconditioned attic, basement, or crawl space is changed to conditioned space, any open-combustion equipment (like furnaces and water heaters) should be separated from the breathing air. At a minimum, a combustion safety test must be performed to make sure that the equipment will not backdraft carbon monoxide into the house (see Chapters 13 and 15 for more information on this subject).

Wood-framed exterior walls pose one of the biggest challenges in existing homes. Uninsulated walls in old homes, while inefficient, are usually very durable. They were rarely weatherproofed and flashed, but the empty framing cavities and solid wood structure can withstand wetting and drying without sustaining major damage. Various techniques are available to install insulation in existing walls by blowing it through holes in the siding or interior walls. Appropriate materials include cellulose, loose-fill fiberglass, certain mixtures of SPF, and cementitious foam. However, when insulation is installed in a wall without a complete WRB, bulk moisture can work its way into the wall cavities and wet the insulation. Since the insulation itself slows the drying process, the water can remain in the walls long enough to promote mold growth and cause structural damage. Unfortunately, the siding almost always needs to be removed to be able to determine if there is a complete WRB on a wall.

Existing brick veneer walls provide the best opportunity to add insulation. While there often isn't a complete WRB, an air space between the brick and the wall sheathing will help keep rain out of the wall structure and reduce the chance of rot and mold. In moderate climates, leaving exterior walls uninsulated may not create too much of an energy penalty, provided they are carefully air sealed. In extremely cold climates, a lack of wall insulation will reduce the efficiency of a home, but this must be weighed against the long-term problems created by insulating existing walls that lack effective bulk moisture management. Unfortunately for both homeowners and remodelers, the only way to ensure long-term performance of existing walls is to remove all the siding, windows, and doors to install a complete WRB system.

The Value and Effects of Air sealing

Air sealing existing walls is not as difficult as installing insulation, nor does it pose the same challenges. The existing drywall or plaster provides an excellent air barrier except where it meets other surfaces and where holes are present. Sealing the wall to the floor and caulking around windows, doors, and electrical and plumbing openings will help reduce air leakage. Air sealing generally will not cause major moisture problems; however, after a complete air seal, blower door testing should be conducted to determine the amount of infiltration and any fresh air ventilation required for the renovated home.

One important issue to address in existing homes is the size of the HVAC systems, particularly in climates that use air conditioning. As we will discuss in Chapter 15, air-conditioning systems must not be oversized in order to work efficiently. After a significant improvement in insulation and air sealing, the existing cooling system in a house may be too large for the new efficient spaces and, if so, it can cause indoor air quality, efficiency, and durability problems.

Summary

Deciding what insulation and air sealing materials and methods to incorporate in a project must be considered with respect to the building structure, climate, skills of available installers, and project budget. Other factors to consider include the amount of recycled content, embodied energy, hazardous materials, and material's effect on the health of the installers and the final occupants.

Review Questions

1. Which of the following products is not an air barrier?
   a. EPS insulation board
   b. XPS insulation board
   c. Fiberglass batt
   d. OSB sheathing
2. Which of the following is not a component of the building envelope?
   a. Roof line over flat-ceiling insulation
   b. Insulated roofline over uninsulated ceiling
   c. Stairwell wall in uninsulated basement
   d. Insulated basement wall
3. Which of the following is not structurally integrated insulation?
   a. SIP
   b. ICF
   c. Insulated sheathing
   d. Air Krete
4. What is the typical recycled content of fiberglass insulation?
   a. 25%
   b. 50%
   c. 75%
   d. 100%
5. What is the typical recycled content of cellulose insulation?
   a. 25%
   b. 50%
   c. 75%
   d. 100%
6. According to the U.S. Department of Energy, how much insulation is recommended in a ceiling in climate zone 3?
   a. R-11
   b. R-20
   c. R-30
   d. R-38
7. Which wall construction method does not reduce thermal bridging?
   a. Double-wall framing
   b. 2 × 6 @ 24" on center with OSB sheathing
   c. 2 × 4 @ 16" on center with rigid foam sheathing
   d. Steel-stud framing
8. Which insulation has an R-value of 5 per inch?
   a. Polyisocyanurate
   b. Strawbale
   c. Extruded polystyrene (XPS)
   d. Expanded polystyrene (EPS)
9. If the blower door test of an existing house resulted in a reading of 0.30 $ACH_{Natural}$, what would be a recommended course of action?
   a. Additional air sealing
   b. Add mechanical ventilation
   c. Blower door recalibration
   d. Indoor air quality testing
10. Which is generally not a problem associated with well-sealed and insulated homes?
   a. Drafts
   b. Backdrafting of open combustion furnaces
   c. High energy bills
   d. High humidity in warm months

Critical Thinking Questions

1. Consider the different techniques involved in structurally integrated insulation versus applied insulation.
2. Discuss the difference between air permeability and vapor permeability and how they affect the choice of different insulation and air sealing materials.
3. Consider various options for creating a complete air seal between living space both above and beside an attached garage.

Key Terms

air changes per hour at 50 pascals (ACH_{50}), 115
air changes per hour natural ($ACH_{Natural}$), 115
airtight drywall approach (ADA), 94
applied insulation, 84
asbestos, 83
attic knee wall, 95
batt insulation, 84
bio-based insulation, 84
blown insulation, 88
carbon monoxide, 112
cathedral ceiling, 111
cavity insulation, 88
cellulose insulation, 99
closed-cell foam, 102
continuous insulation, 91
cotton insulation, 100
effective leakage area (ELA), 117
expanded polystyrene (EPS), 90
extruded polystyrene (XPS), 90
faced batts, 98
fiberglass insulation, 83
flash and batt (FAB), 102
housewrap, 94
impermeable, 86
infiltration, 94
insulation boards, 90
insulation contact (IC) units, 111
intumescent paint, 86
mineral wool insulation, 98
multiple-point testing, 114
off-gassing, 98
open-cell foam, 102
pascal (Pa), 85
permeable, 86
permeability, 85
phenol formaldehyde, 98
rock wool, 98
semi-permeable, 86
single-point blower door testing, 114
slag wool, 98
spray polyurethane foam (SPF) insulation, 84
straw bale, 91
structural insulated panels (SIP), 91
structurally integrated insulation, 84
surface area of building envelope (SFBE), 117
thermal break, 90
thermal bridge, 91
tracer gas, 113
unfaced batts, 98
urea formaldehyde, 84
water-resistive barrier, 94

Additional Resources

Cellulose Insulation Manufacturers Association (CIMA): http://www.cellulose.org/

North American Insulation Manufacturers Association (NAIMA): http://www.naima.org/

Oak Ridge National Lab (ORNL) Insulation Fact Sheet: http://www.ornl.gov/sci/roofs+walls/insulation/ins_08.html

Spray Polyurethane Foam Alliance (SPFA): http://www.sprayfoam.org/

Air Barrier Association of America (AABA): http://www.airbarrier.org/

RESNET: http://www.resnet.us/

APPENDIX 4A: Comparison of Common Types of Insulation

Type	Installation Methods	Application Location	R-Value per Inch	Raw Materials	Pollution from Manufacture	Indoor Environmental Quality Impacts	Additional Information
Fiberglass	Batts, loose-fill, and semi-rigid board	Floors; walls; flat, vaulted, and cathedral ceilings	3.0–4.0	Silica sand, limestone, boron, often recycled glass, PF resin, or acrylic resin	Possible formaldehyde emissions	Fibers can be irritants; possible formaldehyde emissions	Moderate embodied energy; rigid board can be a foundation drainage plane and insulator
Mineral wool	Batts, loose-fill, semi-rigid, and rigid board	Floors; walls; flat, vaulted, and cathedral ceilings	2.8–3.7	Iron ore (often blast furnace slag), natural rock, PF binder	Possible formaldehyde emissions	Fibers can be irritants	Moderate embodied energy, though often some recycled content; naturally fire resistant; rigid board can be a foundation drainage plane and insulator
Cotton	Batts	Floors; walls; flat, vaulted, and cathedral ceilings	3.0–3.7	Cotton and polyester mill scraps	Negligible	Considered safe	Few producers, so transportation pollution is higher than other insulation
Cellulose	Loose-fill, damp-spray, and dense pack	Floors, walls, flat and vaulted ceilings	3.6–4.0	Recycled newspapers, telephone directories, borates, ammonium sulfate	Negligible	Fibers and chemicals can be irritants	High recycled content and low embodied energy
Open-cell SPF	Spray	Floors; walls; flat, vaulted, and cathedral ceilings	3.6–4.3	Petroleum and soybeans; water as blowing agent; non-brominated flame-retardant	Negligible	Toxic during installation (respirators or supplied air required)	High embodied energy
Closed-cell SPF	Spray	Floors; walls; flat, vaulted, and cathedral ceilings	5.8–6.8	Petroleum; HFC blowing agent; non-brominated flame retardant	Climate change potential from HFC blowing agents	Toxic during installation (respirators or supplied air required)	High embodied energy
Expanded polystyrene (EPS)	Rigid boards	Roofs, walls	3.8–4.4	Petroleum, Brominated flame retardants	Negligible	Negligible	Can provide a thermal break or an air barrier; high embodied energy

(Continued)

APPENDIX 4A (Concluded): Comparison of Common Types of Insulation

Type	Installation Methods	Application Location	R-Value per Inch	Raw Materials	Pollution from Manufacture	Indoor Environmental Quality Impacts	Additional Information
Extruded polystyrene (XPS)	Rigid boards	Roofs, walls	5	Petroleum, HFC blowing agents, brominated flame retardants	High global warming potential from HFC blowing agents	Negligible	Can provide a thermal break or an air barrier; high embodied energy
Polyisocyanurate (Polyiso)	Rigid Boards	Roofs, walls	5.6–8.0	Petroleum	Negligible	Negligible	Can provide a thermal break or an air barrier; high embodied energy
Cementitious foam insulation (Air Krete)	Spray	Walls, floors, flat and vaulted ceilings	3.9	Air, water, magnesium oxide, talc	Negligible	Negligible	Low embodied energy

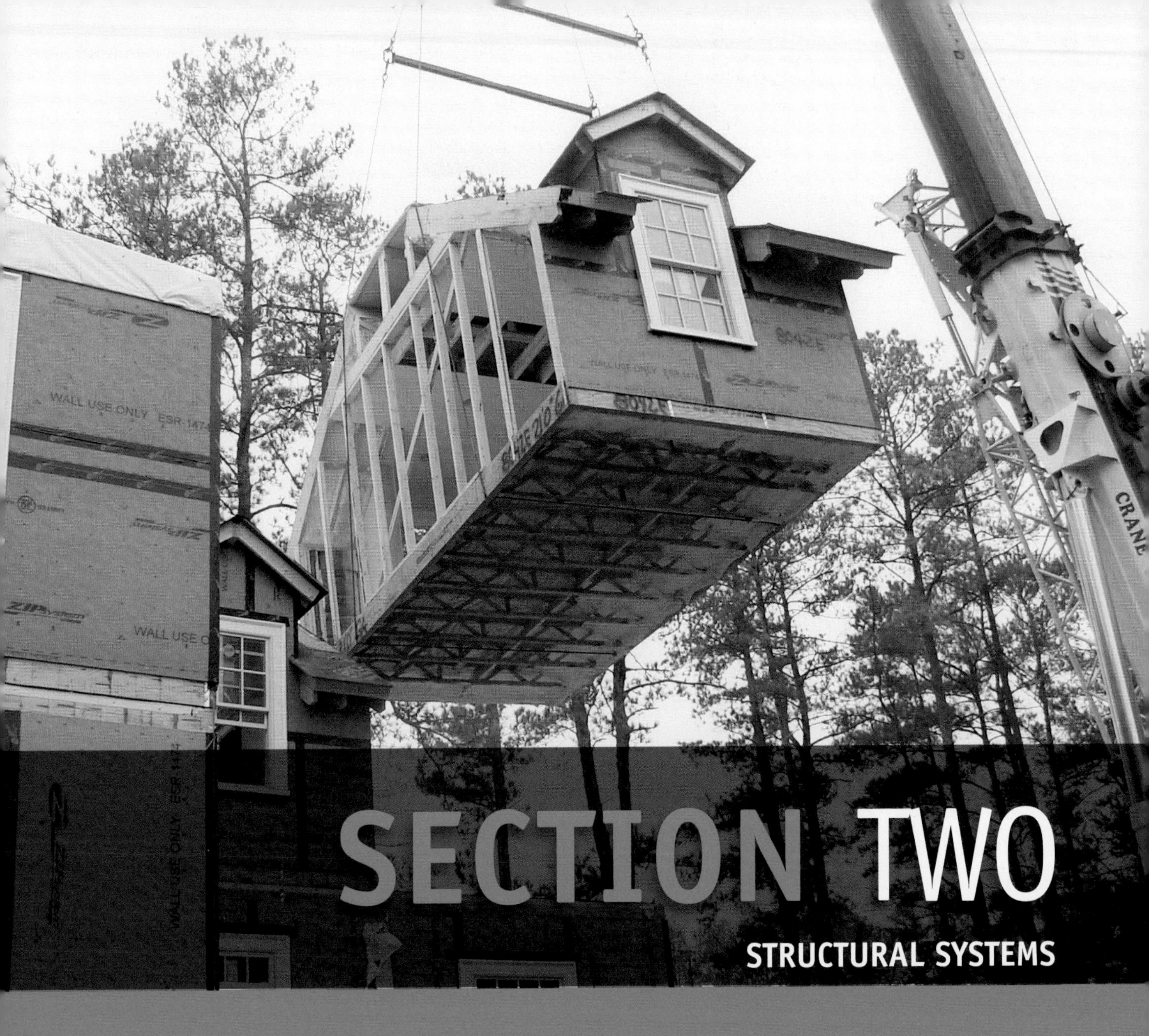

SECTION TWO

STRUCTURAL SYSTEMS

5

Foundations

With the possible exception of houseboats, every home needs a foundation attached to the ground to support it. Foundation walls, whether partially or fully below-ground, can be constructed from a wide variety of materials and methods to support the weight of the building above and enclose the basement or crawl space. The decisions made during design and construction of foundations play a key role in the energy efficiency, indoor environmental quality, resource efficiency, sustainable site development, and durability of a home. This chapter introduces the different types of foundation design and material choices and what to consider when selecting a foundation system for a green home.

LEARNING OBJECTIVES

Upon completion of this chapter the student should be able to:

- Describe the different foundation types available and their relationship to green building
- Describe the different methods to insulate foundations
- Describe the different methods of moisture management in foundation systems
- Demonstrate how to determine the resource efficiency in terms of material used in different foundation systems
- Define the difference between vented, sealed, and conditioned crawl spaces
- Describe the design of a radon vent system

Green Building Principles

Energy Efficiency

Durability

Sustainable Site Development

Indoor Environmental Quality

Resource Efficiency

Types of Foundations

There are four basic foundation designs from which to choose: basement, crawl space, slab-on-grade, and piers, each of which can be constructed using a variety of materials and methods (Figure 5.1). Basements and crawl spaces are enclosed by **foundation walls**, which may be partially or fully below-ground and support the weight of the building above. Any of these foundations can be appropriate for a green home, provided that the choice is based on sound decisions regarding the sustainability of the building. Different foundation types can be combined within a single house.

Site and design conditions that affect the foundation design include site accessibility, existing grades, water table level, existing trees to remain, flood risk, thermal mass requirements, the climate zone, and the amount of space required for the design.

---

foundation walls walls constructed partially below ground that support the weight of the building above and enclose the basement or crawl space.

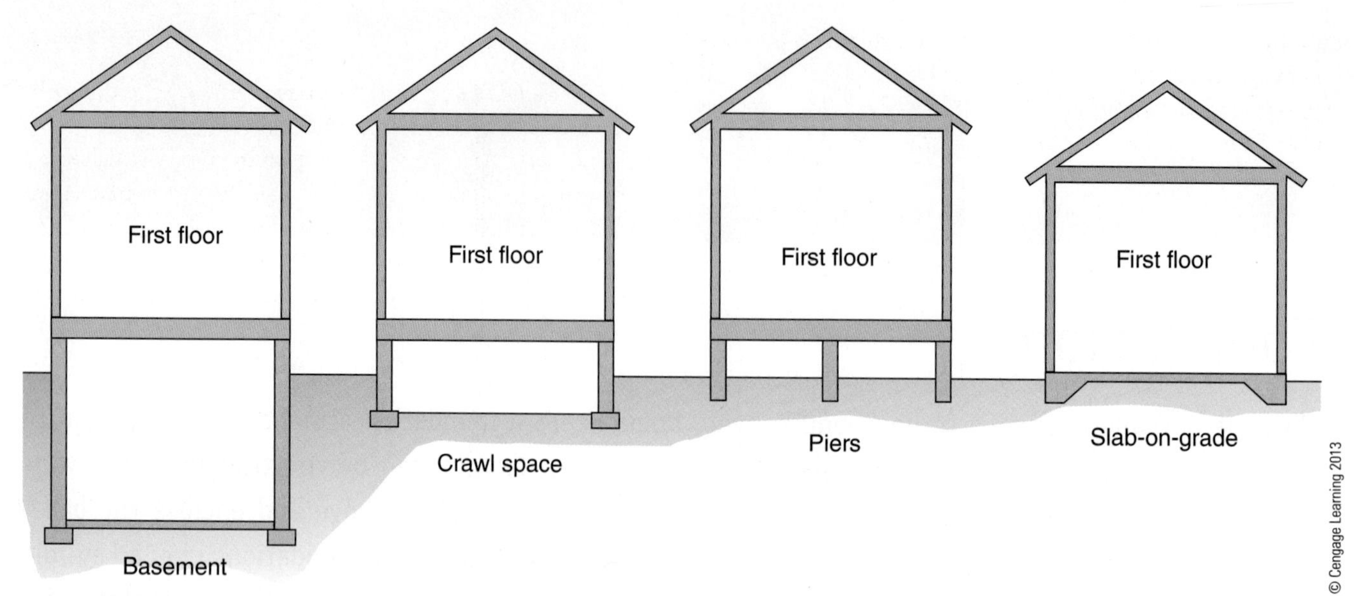

Figure 5.1 The four basic foundation designs are basement, crawl space, pier, and slab-on-grade.

Foundation Selection

Building sites that are easily accessible with large equipment and have few, if any, large trees to avoid are good candidates for basement, crawl space, or slab-on-grade foundations. Wide open, level sites are easily excavated for basements or crawl spaces; however, tight infill locations may not allow for deep excavation without undermining adjacent structures or requiring large retaining walls. Areas with high water tables should avoid basements and crawl spaces. Sites at risk for flooding should consider a pier foundation to avoid problems with water getting into the structure. Buildings close to mature trees can minimize the impact on them by using piers and grade beams instead of traditional footings around critical root zones. Homes that require thermal mass for passive solar designs are often best served by using slab-on-grade construction. Cold climates should avoid open pier foundations to minimize heat loss through the floor. Finally, the amount of space required or desired by the homeowner may suggest construction of a full basement, but keep in mind that more built space requires more materials, more labor, has a greater impact on the site, and uses more energy throughout the life of the structure. Table 5.1 breaks down the process of foundation selection.

Materials and Methods

Once the foundation design is selected, the next step is to choose which materials and methods should be employed to build it. While there are some exceptions that will be addressed later in this chapter, most foundations use poured-in-place concrete for some or all of their construction. Footings are the widened supports located at the base

Table 5.1 Selecting the Right Foundation for the Job

Foundation Type	Vegetation Density	Water Table	Flood Risk	Climate
Basement	Low	Deep	Low	Hot to cold
Crawl space	Low	Deep	Low to moderate	Hot to cold
Slab-on-Grade	Low to high	Deep to shallow	Low to moderate	Hot to cold
Piers	Low to high	Deep to shallow	Low to high	Hot to mixed

retaining wall a structure that holds back soil or rock from a building, structure, or area.

footing the widened support, usually concrete, at the base of foundation walls, columns, piers, and chimneys that distributes the weight of these elements over a larger area and prevents uneven settling.

of foundation walls, columns, piers, and chimneys that distribute the weight of these elements over a larger area and prevent uneven settling. Although footings are almost always poured-in-place concrete, walls and piers can be built with concrete or a variety of alternate materials.

Concrete is a masonry product composed of cement or *pozzolans*, sand, and gravel or other coarse aggregate (Figure 5.2). With a name derived from the Italian term for volcanic ash found on the slopes of Mount Vesuvius, pozzolan is a category of siliceous and aluminous materials that, when mixed with calcium hydroxide, give concrete its strength. Examples of pozzolans include such industrial waste products as slag and fly ash, as well as Portland cement. Portland cement, which is created by heating a mixture of materials that typically includes limestone and silica, has become the most commonly used hydraulic cement because of its strength and widespread availability.

Portland cement does require large amounts of energy to produce, so green builders may wish to lower the embodied energy in a structure by reducing the amount of concrete in a foundation as well as the amount of cement in the concrete mix. When examining alternatives, however, it is important to acknowledge that concrete is an extremely long-lasting material and can be recycled at the end of its life. These factors can be considered in determining its overall environmental impact.

Courtesy of Portland Cement Association

Figure 5.2 Concrete components: cement, water, sand, and coarse aggregate next to a cut section of hardened concrete.

concrete a masonry product composed of cement or pozzolans, sand, and gravel or other coarse aggregate.

pozzolan a material which, when combined with calcium hydroxide, exhibits cementitious properties; examples include Portland cement, coal fly ash, and ground granulated blast furnace slag.

Portland cement the most common form of cement consisting of certain minerals that form the binder in concrete and plasters; see also *pozzolan*.

Concrete

The amount of concrete and cement may be reduced by designing smaller foundation structures and replacing cement with alternative materials in the mix. A method for constructing smaller foundations can be as simple as testing the bearing capacity of the soil and designing the minimum required foundation structure for the specific site, rather than using a generic design that exceeds the requirements of the structure and uses more material. Cold climates typically require footings placed below the frost line; however, an alternate design called frost-protected shallow foundations allows for shallower footings that require less concrete. Such techniques reduce material use and help create more sustainable buildings while also lowering the cost of construction.

To reduce the amount of Portland cement in a concrete mixture, other materials must be substituted. Industrial waste products like coal fly ash from coal-fired power plants and ground granulated blast furnace slag are common materials used in concrete production.

In addition to fly ash content, other sustainable issues to consider when installing concrete foundations include substituting recycled crushed concrete for virgin stone in aggregate, using recycled-content reinforcing steel and nontoxic form-release agents, and managing ready-mix truck washout to keep residue from polluting waterways (see Figure 5.3). If concrete washout waste enters waterways, it can change the pH and harm aquatic life and the wildlife that depends on

Courtesy of RTC Supply

Figure 5.3 When not properly controlled, concrete washout waste can enter waterways and harm aquatic life and wildlife. Systems are available to control and easily dispose of washout waste.

frost-protected shallow foundations provides protection against frost damage without the need for excavating below the frost line.

coal fly ash the very fine portion or ash residue that results from the combustion of coal that may be used as a Portland cement substitute.

ground granulated blast furnace slag a by-product of iron and steel making that is used to make durable concrete in combination with ordinary Portland cement or other pozzolanic materials.

Did You Know?

Using Fly Ash in Concrete

Long before the invention of Portland cement, the Romans created impressive concrete structures by using lime and a volcanic ash (with properties that were first discovered in Pozzuoli, Italy) that reacted with the lime and hardened the concrete. Coal fly ash, which is the particulate matter collected by pollution-control equipment from the smokestacks of coal-burning power plants, has a similar pozzolanic effect because of its silica and aluminum content. Other widely used pozzolans are blast-furnace slag and silica fume.

Portland and similar cements react with water (hydrate) to create a gel that hardens by absorbing carbon dioxide. As it hardens, the cement binds aggregates (typically sand and crushed stone) together, creating concrete. When Portland cement cures, it leaves behind some hydrated lime. Adding fly ash or another pozzolan allows that lime to cure as well (as in the Roman walls), making the concrete stronger and less porous.

Fly ash and other pozzolans increase the durability of concrete and can also be used to shrink its environmental footprint by reducing the amount of Portland cement in the mix. Although nearly a ton of carbon dioxide is emitted to produce each ton of Portland cement, fly ash is a waste byproduct of energy generation. Mixes in which up to 25% of the cement is replaced by fly ash are quite common, and some designers are specifying over 50% substitution for certain applications. High-volume fly ash mixes must be tested before each application because the chemistry of fly ash is more variable than that of Portland cement. They also have to be managed differently as they cure because they tend to cure and gain strength more slowly than mixes with more cement.

Use of fly ash in concrete in the United States is governed largely by ASTM Standard C618. This standard prohibits the use of fly ash with too much residual carbon, which indicates that the coal was not burned thoroughly. Residual carbon impedes air entrainment and reduces the concrete's freeze-thaw resistance.

Under the ASTM specifications, fly ash is classified by its chemical composition and properties as either Class F or Class C. Class F fly ash comes primarily from burning coal found in the Appalachian Mountains and southeastern United States and is purely pozzolanic in its effect. Class C fly ash tends to come from younger coals found in the western United States. This type of fly ash has cementitious (in addition to pozzolanic) properties, so using it instead of some of the cement in a mix will have little effect on early strength gain.

All fly ash consists mostly of the mineral constituents of coal that do not burn with the hydrocarbons, including a variety of potentially hazardous heavy metals such as mercury, lead, selenium, arsenic, and cadmium. There is little evidence that these metals will leach out of concrete made with fly ash, but questions remain about whether those hazardous ingredients might complicate eventual reuse or disposal of the concrete.

Source: "Using Fly Ash in Concrete." February 2009; http://www.buildinggreen.com/auth/article.cfm/2009/1/29/Using-Fly-Ash-in-Concrete/. Reprinted with permission from *Environmental Building News*.

these surface waters. Washout wastes also contain fine particles of sand and cement, which can become suspended in the waters and prevent fish from breathing.

Basements and Crawl Spaces

Basements and crawl spaces require solid foundation walls to enclose the space with enough strength to resist the horizontal pressure of the earth against them without failure. They must also be strong enough to support the structure above and be designed to keep water from entering the space.

While footings are typically poured-in-place concrete, basement and crawl space walls and piers can be made of masonry units, such as cement block or concrete masonry units (CMUs) (Figure 5.4a). Other masonry materials used for foundations include concrete or clay brick, natural stone, and autoclaved aerated concrete (AAC) (Figure 5.4b, Figure 5.4c, and Figure 5.4d). AAC is a lightweight, air-entrained block that uses less energy to produce and is more energy-efficient than standard CMUs. Traditional CMU, brick, and stone have minimal insulation value and should be insulated on the interior or exterior of walls, whereas the air pockets in AAC blocks provide built in insulation value. *NRG blocks*, a variation on the standard CMUs, are formed in two pieces with a piece of foam insulation installed between them, providing built-in insulation (Figure 5.4e). Table 5.2 compares the R-values of different foundation materials.

concrete masonry unit (CMU) a large rectangular block of concrete used in construction.

autoclaved aerated concrete (AAC) a lightweight, precast building material that provides structure, insulation, fire, and mold resistance.

Figure 5.4a CMU installation.
From SPENCE/KULTERMANN. *Construction Materials, Methods and Techniques*, 3E. © 2011 Delmar Learning, a part of Cengage Learning, Inc. Reproduced by permission. www.cengage.com/permissions

Figure 5.4b Clay masonry units are available in a wide variety of colors and textures.
From SPENCE/KULTERMANN. *Construction Materials, Methods and Techniques*, 3E. © 2011 Delmar Learning, a part of Cengage Learning, Inc. Reproduced by permission. www.cengage.com/permissions

© Cengage Learning 2013

Figure 5.4c Granite foundation laid in a random pattern, forming a textured surface.

© Cengage Learning 2013

Figure 5.4d AAC is cut and assembled on site.

Courtesy of NRG Insulated Block

Figure 5.4e NRG blocks are a variation of a CMU that have integrated foam insulation.

Table 5.2 Foundation Wall R-Values

Foundation Type	R/Inch	R/Assembly*
Autoclaved aerated concrete, 12"	1.05	12.6
Insulated concrete form	0.08	
Concrete, 6"	0.08	17
Expanded polystyrene, 4"	4.25	17
Structural insulated panel	OSB 1.24 EPS 4.25	17
Oriented-strand board	1.24	17
Expanded polystyrene, 4"	4.25	17
Concrete block 12"	0.1067	1.28
Poured concrete 12"	0.08	0.96
Permanent wood foundation (2 × 6 with R-13)		
Studs	1.25	10
Insulation	3.7	10
Oriented-strand board	1.24	10

* Based on the average R-value for the whole assembly.

Insulated Concrete Forms

One foundation type that is a hybrid of concrete and unit masonry uses insulated concrete forms (ICFs). Typically made of foam, ICFs are used to create exterior walls for both foundations (crawl spaces and basements) and above-grade locations. As ICFs are stacked in place, rebar is installed; the blocks are braced for stability, and then they are filled with concrete to create the structure (Figure 5.5 and Figure 5.6). The blocks remain in place, providing both interior and exterior insulation for the wall. Foam ICFs are a well-established building technique, provide excellent structural capacity and insulation value; however, any thermal mass value of the concrete is reduced or eliminated by its encasement in foam insulation.

One type of ICF is made with mineralized wood chips, eliminating the foam entirely. Known by their trade name Durasol®, these ICFs are made in part from recycled wood and use mineral wool insulation, providing an alternative to less sustainable foam products used in traditional ICFs. One additional benefit is their ability to absorb and release large amounts of moisture without damage or supporting mold growth.

Figure 5.5 Expanded polystyrene insulating concrete forms provide formwork and remain as insulation.

Figure 5.6 ICFs are assembled from smaller units.

Above-Grade Facade

Decorative masonry or stucco finishes are often installed on foundation walls above grade. Although stucco is typically applied directly to the foundation, traditional brick or stone, ranging between 4" to 12" thick, must be installed on a footing or brick ledge that requires additional concrete in the foundation structure (Figure 5.7). The use of thin-profile brick or cultured stone that can be applied directly to the wall without structural support reduces the amount of concrete required by eliminating the brick ledge as well as the amount of masonry (Figure 5.8). Exterior wall finishes are covered in more detail in Chapter 9.

Slab-on-Grade

For slab-on-grade construction, the perimeter footings, interior turn-downs, slab thickness, and reinforcing steel should be designed for the minimum sizes required for the structure and soil conditions. By taking project-specific structural requirements into account as part of the integrated design process, developers can save both materials and costs by constructing foundations that are no larger than necessary (see Figure 5.9).

Pier Foundations

Pier foundations support a house by providing intermediate supports between beams, eliminating the need for continuous foundation walls and reducing the materials

insulated concrete form (ICF) insulating foam or mineralized wood forms that are left in place after the concrete is poured for a foundation or wall.

pier foundation is a grid system of girders (beams), piers, and footings used in construction to elevate the superstructure above the ground plane or grade; the piers serve as columns for the superstructure.

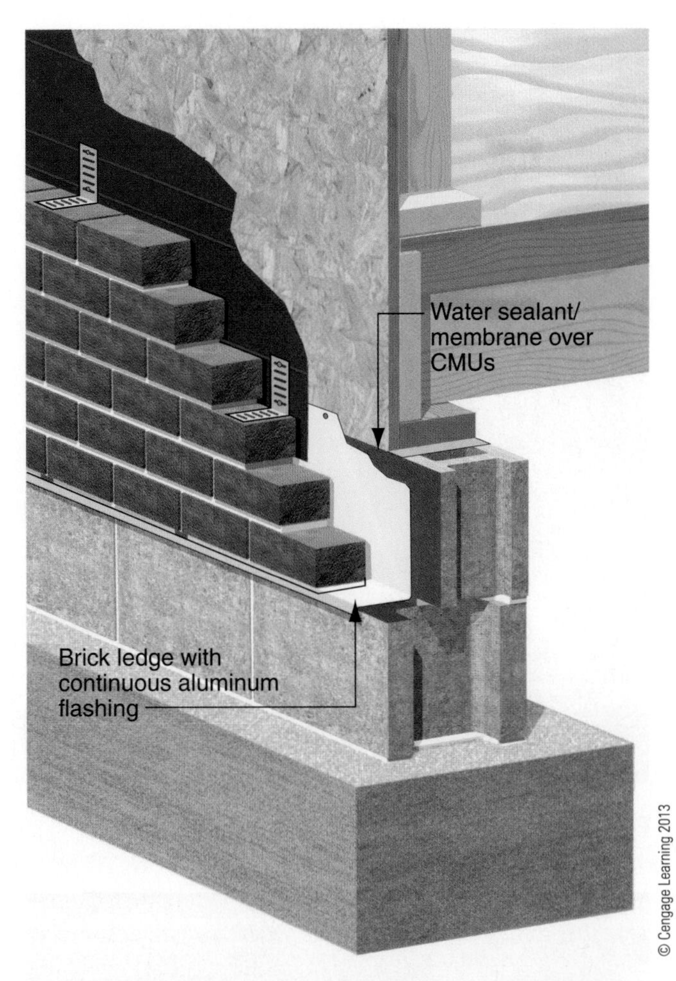

Figure 5.7a When using a CMU brick ledge, the top must be properly flashed and waterproofed to prevent water accumulation within the CMU.

Figure 5.7b Brick ledge of CMU with veneer installed on crawl space foundation.

Figure 5.8 Thin brick veneer is light enough in weight to not require additional structural support.

required to support the home. Pier foundations are the most resource-efficient foundation because they require the fewest materials to construct. An appropriate choice for moderate climates, a pier foundation is also a good solution for a structure that may be at risk for high groundwater or flooding. Properly located and installed piers can help protect root systems of large trees, minimizing the building's negative effect and prolonging their life. Piers can have concrete footings with wood, masonry, or steel posts connecting to the structure. Alternatives

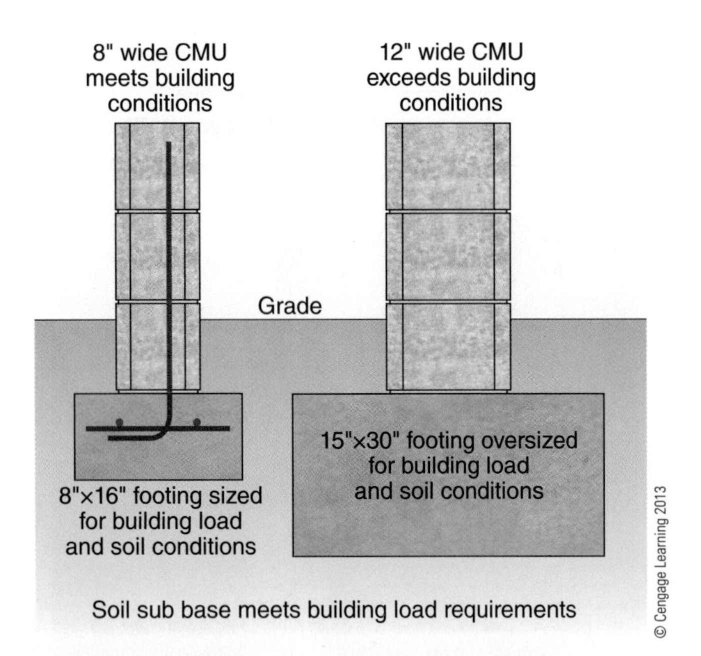

Figure 5.9 The particular application dictates the thickness of a poured concrete foundation wall and the width of its footer. For example, steel reinforcement allows thinner walls and smaller footers.

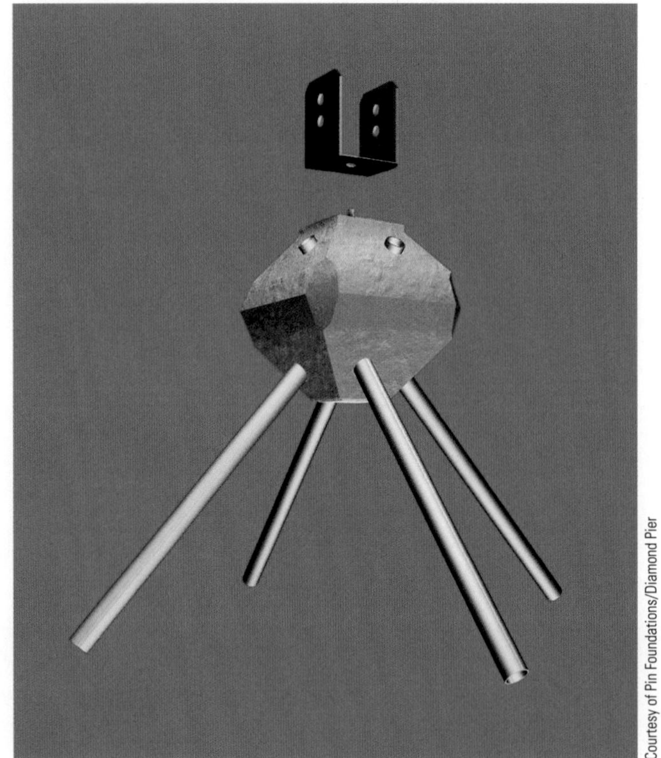

Figure 5.10a The Diamond Pier pin foundation.

include steel or wood piers that are driven or screwed into the ground to provide adequate soil-bearing capacity with minimum disruption to the site. Smaller structures, including homes, decks, and walkways, can be supported on such foundations as those made by Diamond Pier (Figure 5.10a and Figure 5.10b). These pin foundations are installed with hand or pneumatic equipment and can eliminate all excavation for the foundation.

Piers can be combined with traditional foundations by constructing grade beams over sections of soft soils, around large tree roots, and other obstructions. Grade beams are designed to span between piers or areas of solid soil as well as support the foundation and structure above them while not bearing directly on the soil below. This technique can be used with slab-on-grade, poured-in-place walls, and masonry wall construction.

Figure 5.10b Modular beam supports using the Diamond Pier pin foundation.

Alternative Foundation Wall Systems

In addition to poured-in-place concrete, ICFs, and unit masonry, other less common construction methods may be appropriate to consider for a green project. These include prefabricated foundation panels, permanent wood foundations made of chemically treated lumber, and structural insulated panels (SIPs).

Prefabricated Foundation Systems

Prefabricated foundations shift most of the work of constructing foundation walls to the factory, which saves time, reduces material usage, and eliminates most job-site waste. Conceptually, prefabricated foundations all work the same. Walls are placed on a bed of gravel rather than on a concrete footing, with the wall itself

grade beam a perimeter load-bearing support of a structure that spans between piers without relying on the ground below for support.

prefabricated foundations are foundation walls that are manufactured in a factory and assembled on site.

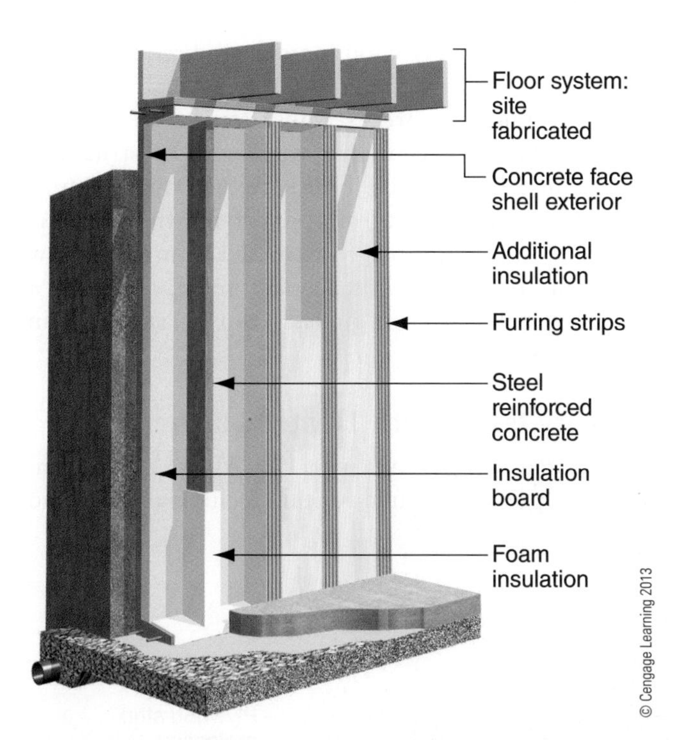

Figure 5.11a One example of a prefabricated foundation system is the Superior Wall system.

Figure 5.11b Superior Walls arriving on site.

Figure 5.11c Installation of Superior Walls.

providing the structural rigidity instead of the footing (Figure 5.11a, Figure 5.11b, and Figure 5.11c). The gravel must be placed on solid earth with adequate structural bearing capacity, then leveled accurately before the walls are placed. After the walls are installed and attached together, an interior floor is installed to anchor them together, the first floor structure is framed to provide horizontal support for the walls, and the completed foundation is waterproofed and backfilled. Prefabricated foundations can be made from precast concrete, a technique in which concrete components are cast in a factory or on the site before being lifted into their final position on a structure or of pressure-treated wood or SIPs.

Prefabricated concrete foundations are a thin skin of high-pressure concrete reinforced with concrete studs, header, and footer. Together, these elements provide all the structure needed to hold up a house while using significantly less concrete and cement. By eliminating the concrete footing as well as much of the wall concrete, these foundations may use as much as 80% less Portland cement than a standard poured-in-place wall. In addition to these savings, brick ledges are built into the walls just below grade level, eliminating the extra wall thickness required in standard poured or masonry walls.

Standard finishes include stucco, brick, or stone, which are applied directly to the walls or attached with factory-installed brick ties. These foundations typically are delivered with foam board insulation installed on the interior surface and with cutouts for windows and doors in place, ready for installation of interior drywall. During installation, wall sections are bolted together and sealed with an industrial-grade caulk. A concrete slab floor is poured after the walls are installed, and backfilling can begin as soon as the first floor framing is complete.

Permanent Wood Foundations

Foundations of wood are referred to as permanent wood foundations (PWF). They can be fabricated in a shop or on the job site by using wood that is pressure-treated to

---

precast concrete a construction technique in which concrete components are cast in a factory or on the site before being lifted into their final position on a structure.

permanent wood foundations (PWF) foundation systems consisting of pressure-treated wood walls.

withstand moisture and pest damage from ground contact. Similar in design to precast concrete walls, PWFs have top and bottom plates, studs, and plywood sheathing that provide the strength required to support the house when they are placed on compacted gravel footings (Figure 5.12). PWFs do not come insulated, but any standard wall insulation can be installed in the stud cavities. PWFs can be installed with either a concrete slab or framed wood floor at the base. They can be finished with stucco, thin brick, or traditional siding materials.

SIPs

As described in Chapter 4, SIPs are made with a thick layer of foam sandwiched between two layers of pressure-treated plywood or oriented-strand board (OSB), or a cementitious panel (Figure 5.13). Installation is similar to precast concrete or PWF foundations, with the exception that a vertical structural stud is installed at each joint between panels to provide horizontal structural support for the wall. As with PWFs, foundations constructed from SIPs allow for floors built with SIPs, wood-framed, or laid as a concrete slab. Exterior finishes can be stucco, thin brick, or siding.

Toxicity Issues with Treated Lumber

Chromated copper arsenate (CCA) is a chemical used in pressure-treated wood, including PWF and

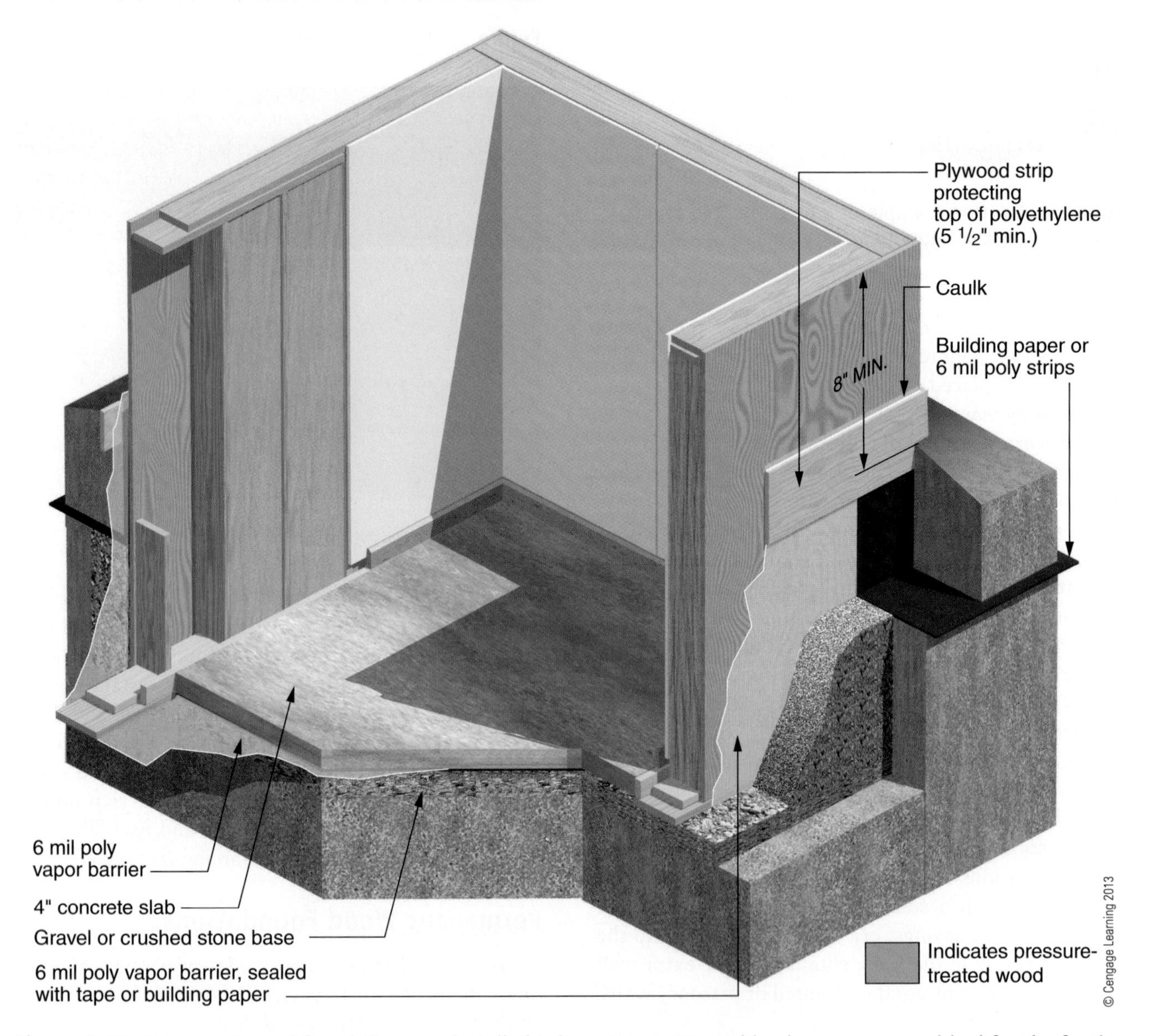

Figure 5.12 Permanent wood foundations are installed using pressure-treated lumber over a gravel bed for the footing.

---

chromated copper arsenate (CCA) a chemical wood preservative containing chromium, copper, and arsenic.

Courtesy of Insulspan

Figure 5.13 An SIP consists of insulation foam sandwiched between two sheathing materials.

SIP foundations, to protect wood from rotting due to insects and microbial agents. In 2004, CCA was banned for most residential uses due to concerns about the toxicity of the arsenic used in the process, although it is still used in foundations, docks, farm fencing, and other applications. Proper precautions should be taken in handling and disposal of CCA-treated lumber; gloves and masks should be worn, and the lumber should not be burned on the job site. Disposal of unused materials should be done in accordance with the manufacturer's recommendations and local waste management authority requirements. Furthermore, any CCA-treated lumber inside a home should be covered, painted, or sealed to avoid any occupant contact with the chemicals.

Prefabricated foundation systems can provide significant material savings, shorter construction time, integrated insulation or simple insulation installation, and an excellent air seal. When installed in moderate- to high-risk seismic design zones, prefabricated foundations are installed on poured-in-place footings instead of crushed stone to provide more resistance to earthquakes. The builder who is intending to install a prefabricated foundation may encounter resistance from building inspectors who are unfamiliar with the products. More advance planning may be needed with prefabricated foundations due to factory lead times and the need to confirm that the foundation design is accurate before fabrication starts. Factory-built panels are more difficult to change on site than poured-in-place or masonry foundations.

Moisture Management

Foundations, since they are mostly underground, are very susceptible to moisture problems: high humidity, standing water, and even structural damage to walls from excessive groundwater pressure against below-grade walls. Keeping water out of all areas of the home is critical but starts with the foundation. The process of directing groundwater away from the foundation and the home is generically known as foundation drainage (Figure 5.14).

The first line of defense is to properly grade the site, routing surface water and gutter drains away from the walls of the house and eliminating low spots that allow water to pool and seep into the ground. The ground around the house should slope away from the home at a minimum 5% grade (6" in 10') for at least the first 5'. All foundation walls below grade should have a waterproof coating applied and drainage mats installed to allow water to flow down to the footing. Perforated drain pipes should be installed beside or below all footings to collect groundwater, which must then be piped downhill to a daylight exit. Without adequate drainage, water can collect against the foundation wall, creating hydrostatic pressure that can drive water into the foundation through cracks and gaps and, in severe cases, cause structural damage.

Waterproofing systems usually consist of a membrane material to keep water out of the foundation wall and a drainage mat to keep water away from the wall. Membranes are available in both spray-on and sheet-applied materials. Some spray-applied products are available with little or no volatile organic compounds, reducing their impact on the environment. Asphalt-based damp-proofing products do not provide sufficient resistance to groundwater, nor are they durable enough to last the life of a structure. They should be avoided in favor of more durable products that will keep water out of the foundation. Drainage mats create a gap between the soil and the foundation walls (Figure 5.15). The space provided by drainage mats relieves hydrostatic pressure and provides water with a path of least resistance to drain away from the home. Drainage mats are available in several styles, some of which include a filter fabric to keep them from filling up with silt that would impair its ability to drain water rapidly. Footing drains can be flexible or rigid perforated

---

foundation drainage the process of directing groundwater away from the foundation and the home.

hydrostatic pressure the force exerted on a foundation by groundwater.

waterproofing a treatment used on concrete, masonry, or stone surfaces that prevents the passage of water under hydrostatic pressure.

damp-proofing a treatment used on concrete, masonry, or stone surfaces to repel water and reduce the absorption of water in the absence of hydrostatic pressure.

drainage mat a material that creates a gap between the soil and the foundation walls; this space relieves hydrostatic pressures and provides water with a path of least resistance to drain away from the home.

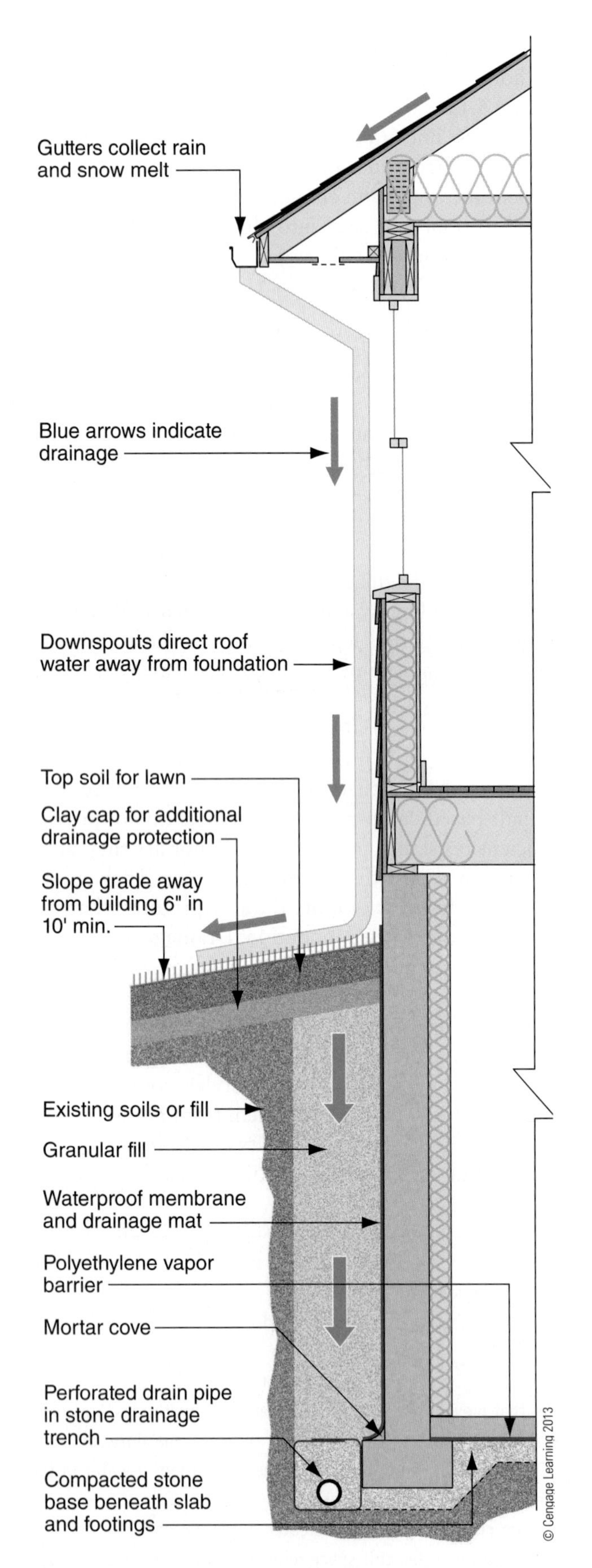

Figure 5.14 Groundwater control in a home with a basement.

Figure 5.15 A drainage mat is installed over the foundation waterproofing.

Figure 5.16 Foundation drainage systems can be integrated into the footing forms. Unlike standard wood forms, the Form-A-Drain stays in place to provide long-term foundation drainage.

pipe, with or without a filter covering or "sock." Rigid pipe is more durable and can be cleaned out with less risk of damage than flexible pipe if it becomes clogged. Drains are also available that are built into forms that remain in place after the concrete is poured (Figure 5.16). Drain pipes running from the footings to daylight should be rigid pipe whenever possible, with cleanouts provided at elbows. Gutter downspouts should be kept separate from the footing drains and directed above or below ground to keep water away from the foundation. Extra protection from water intrusion can be gained by installing perforated drain pipes below a basement slab or crawl space floor, drained out to daylight along with the footing drains.

In addition to water intrusion, foundations are susceptible to moisture wicking into the building through capillary action from footings and foundation walls or through slabs. This can be controlled through the installation of a capillary break between the top of the footing and the foundation wall or piers (Figure 5.17), and a vapor barrier below the concrete slab floor or over the dirt floor of the crawl space (Figure 5.18).

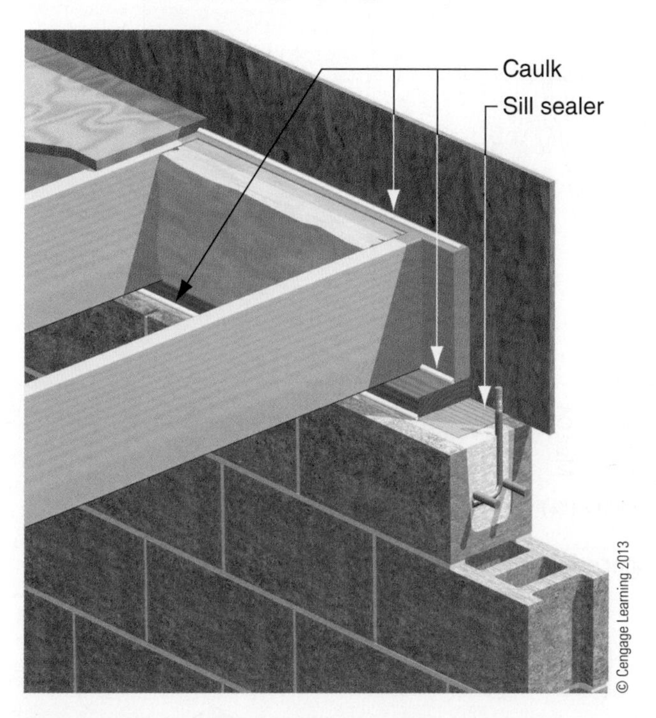

Figure 5.17 A foam gasket is placed between the foundation wall and the sill plate to form a capillary break.

Figure 5.18 DELTA®-MS is used as a below-slab vapor barrier to prevent the introduction of soil gases into the home.

capillary break an air space or material that prevents the movement of moisture between two surfaces by capillary action.

vapor barrier a Class I vapor retarder (0.1 perm or less).

Foundations and the Building Envelope

Foundations are a major factor in determining the location of the building envelope. Traditionally, crawl spaces were ventilated and basements were sealed. Crawl spaces were ventilated in the belief that the air movement would allow excess moisture to escape; however, research has shown that venting does not reduce the amount of moisture, and in fact, often makes the crawl space wetter than when not vented (Figure 5.19). The general consensus among building scientists is that all basements and crawl spaces should be sealed from the outside and from the ground below to keep out unwanted moisture and reduce energy loss through the floor of a building. Sealed crawl spaces and basements can be described as *unconditioned*, *semi-conditioned*, or fully *conditioned* (Figure 5.20). When unconditioned, the building envelope, consisting of insulation and air sealing, is the floor above the foundation. Semi-conditioned and fully conditioned spaces have insulation on the foundation walls and generally do not have insulation or air sealing in the floors and walls between them and the finished areas. Foundation areas that are cannot be made watertight should not be conditioned and must be located outside the building envelope.

A conditioned crawl space (see Figure 5.21) is a foundation without wall vents that encloses an intentionally heated or cooled space. Insulation is located at the exterior walls. Continuous exhaust, conditioned supply air only, or a dedicated dehumidifier must be installed to provide "conditioning." Although conditioned crawl spaces are code acceptable, many jurisdictions may limit their use or have additional requirements for their construction.

Foundation insulation is typically most effective when installed on the exterior of the foundation walls, although interior insulation applications are often appropriate. Exterior insulation prevents thermal bridging and provides continuous thermal coverage. Slab-on-grade floors serve as the bottom of the building envelope and should be insulated at the perimeter in all climates and completely across the bottom in cold climates, particularly when radiant heating is installed in the slab. Slabs lose heat primarily through the perimeter edge, with significant losses through the center only in heating-dominated climates (Figure 5.22, page 140). Slabs may be insulated around

conditioned crawl space is a foundation without wall vents that encloses an intentionally heated or cooled space; insulation is located at the exterior walls.

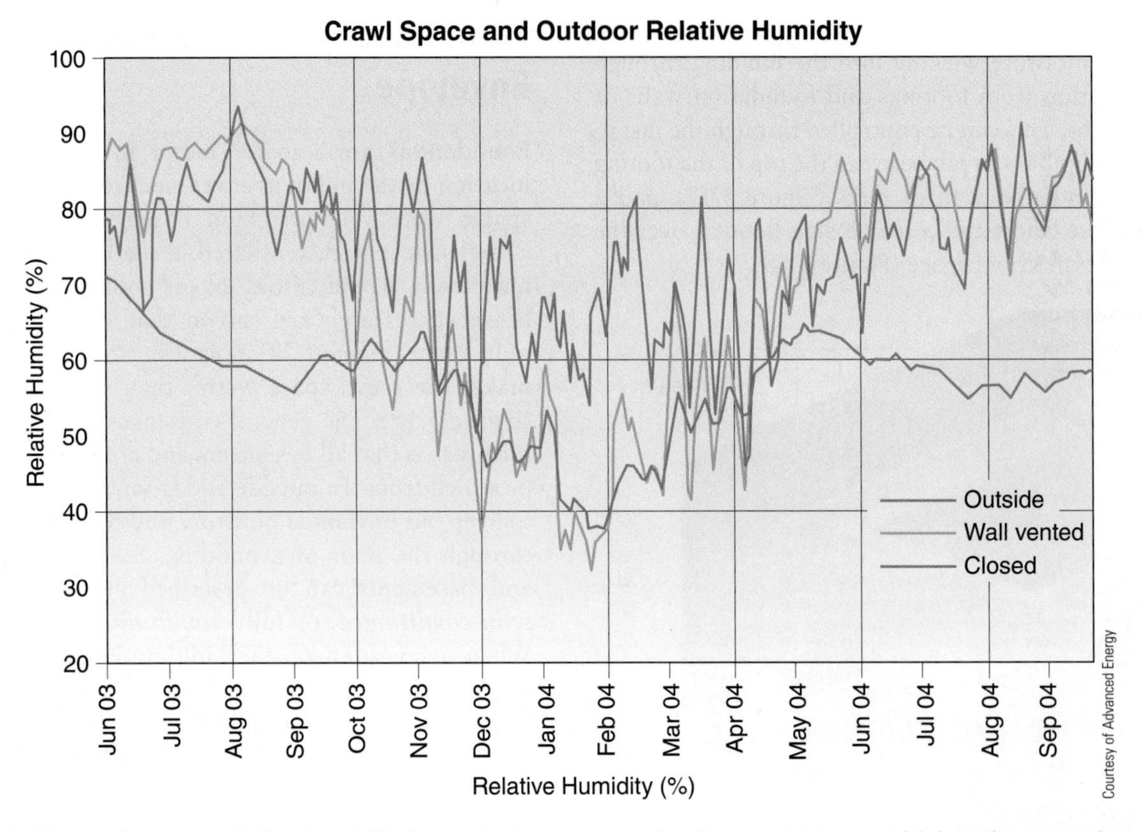

Figure 5.19 Graph shows relative humidity in vented crawl space is often as high as or higher than exterior.

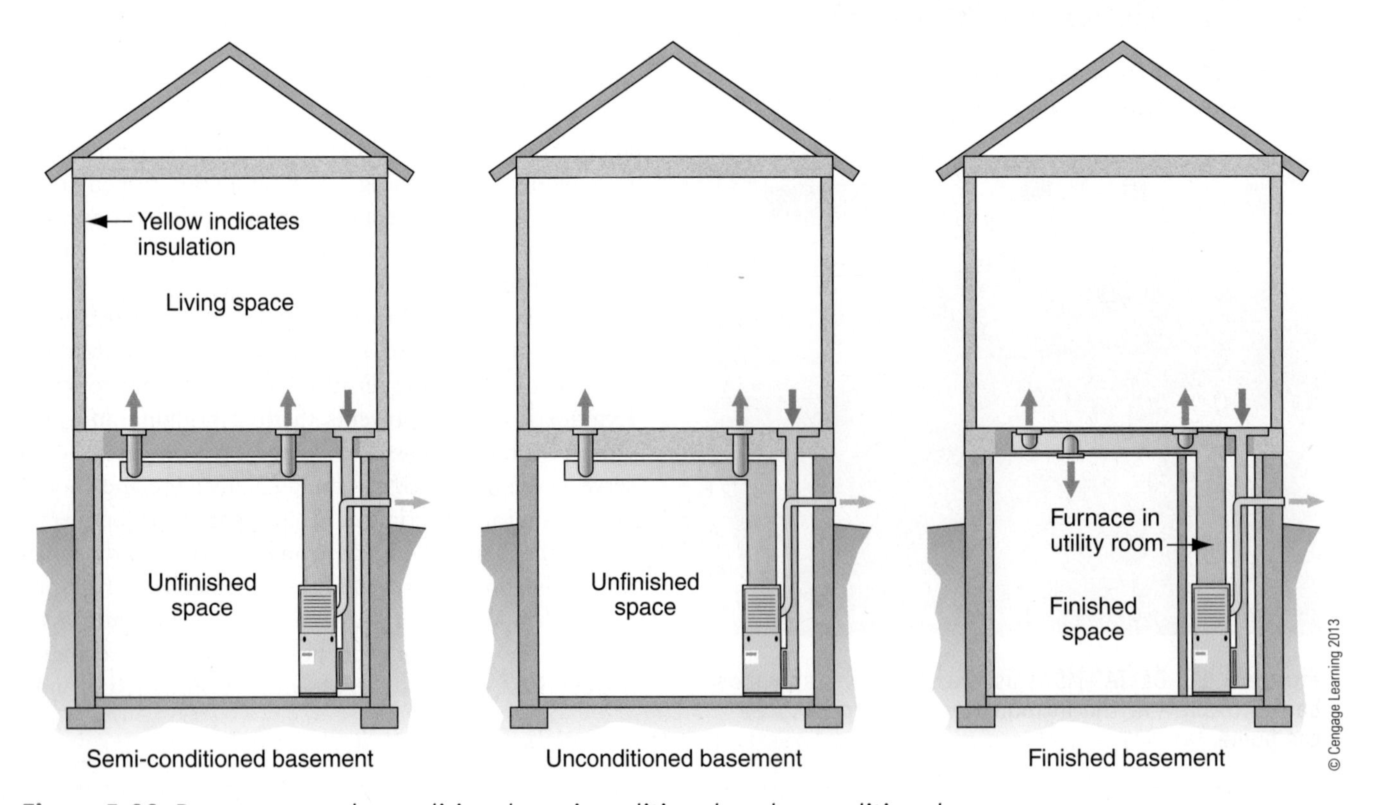

Figure 5.20 Basements may be conditioned, semi-conditioned, and unconditioned.

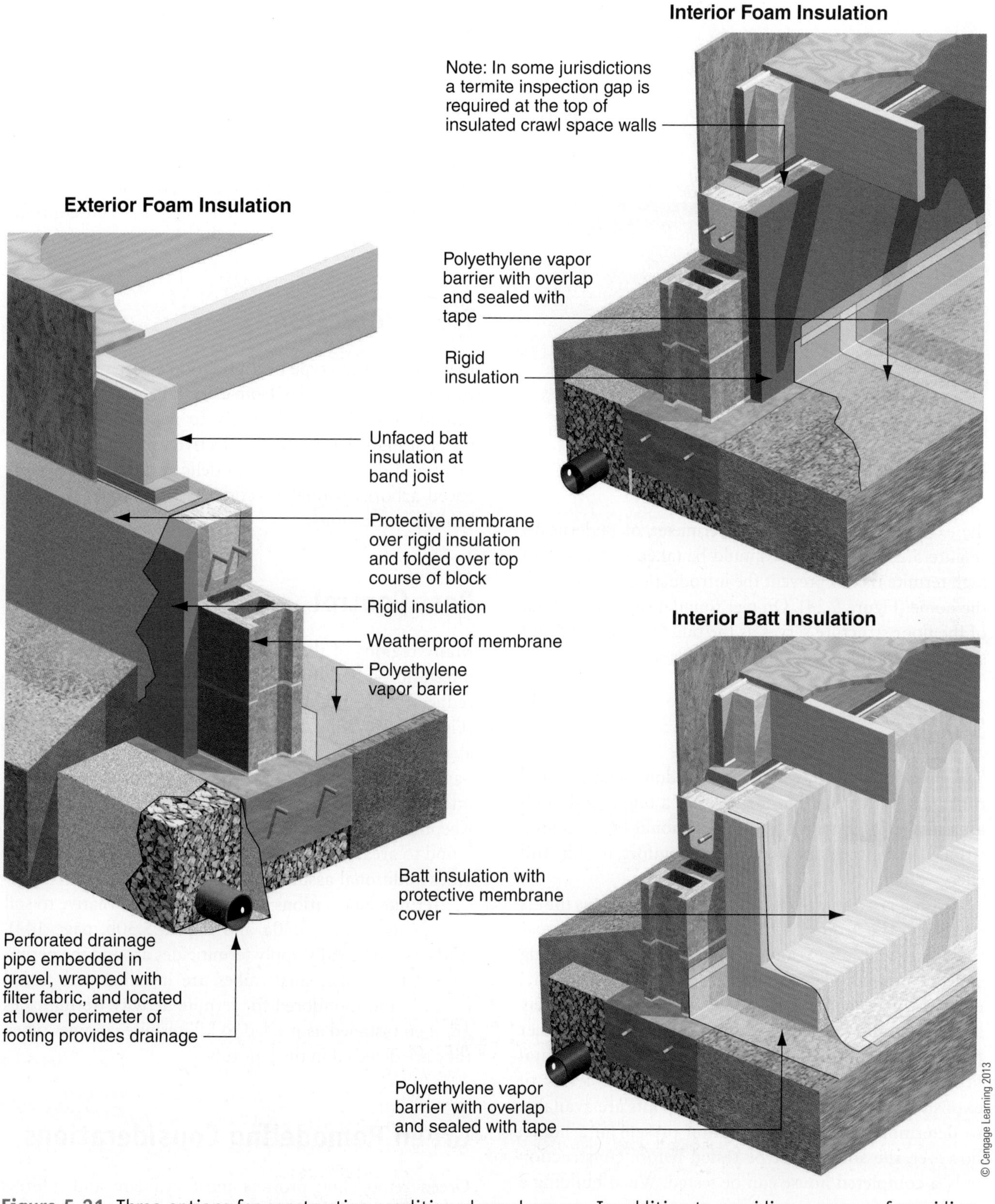

Interior Foam Insulation

Note: In some jurisdictions a termite inspection gap is required at the top of insulated crawl space walls

Polyethylene vapor barrier with overlap and sealed with tape

Rigid insulation

Exterior Foam Insulation

Unfaced batt insulation at band joist

Protective membrane over rigid insulation and folded over top course of block

Rigid insulation

Weatherproof membrane

Polyethylene vapor barrier

Perforated drainage pipe embedded in gravel, wrapped with filter fabric, and located at lower perimeter of footing provides drainage

Interior Batt Insulation

Batt insulation with protective membrane cover

Polyethylene vapor barrier with overlap and sealed with tape

Figure 5.21 Three options for constructing conditioned crawl spaces. In addition to providing a means of providing conditioning, conditioned crawl spaces should contain a drain run to daylight or a sump pump in the event of water entry. The drain should contain a backflow preventer and rodent proof screen.

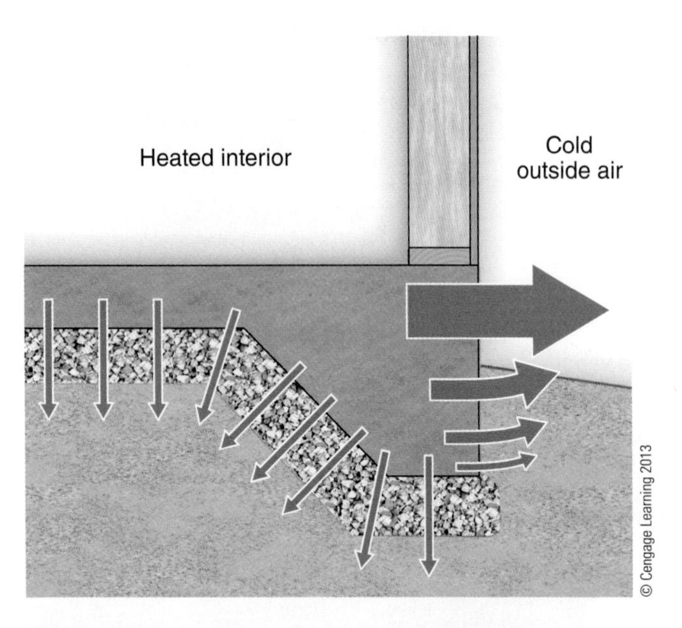

Figure 5.22 Heat loss through slab on grade.

the exterior perimeter, interior perimeter, or underneath (Figure 5.23). Precautions should be taken in areas with high termite risk to prevent the introduction of pests into the home (Figure 5.24). On pier foundations, the bottom of the floor structure serves as the building envelope and should be completely insulated and air sealed.

Soil Gas

Soil gases include air, water vapor, radon, methane, and other soil pollutants that might enter a building though air leakage. At all times, soil gases should be prevented from entering the home to improve comfort, health, and energy efficiency.

Radon is a naturally occurring radioactive gas that is present in the ground at varying concentrations across the country. Radon is one of the major causes of lung cancer, and all homes should be designed to prevent its entry into the home. The home should be completely air sealed, and soil gases should be vented to further prevent entry into the home. The U.S. Environmental Protection Agency (EPA) has set maximum levels of exposure that are considered safe. Test kits are available to determine the amount of radon present in a home; however, the soil cannot be tested before construction only a completed house can be tested. When building a new home, installation of a radon ventilation system is recommended. Such a system usually consists of gravel below the floor or basement slab (or a crawl space liner) and a vent pipe running from the gravel bed through the interior and out the roof of the house, or installed on the exterior (Figure 5.25 and Figure 5.26). With this system in place, a completed home can be tested; if it shows higher-than-recommended radon levels, a small fan can be installed in the vent pipe and run continuously to remove the excess radon from the structure.

Tree Protection

Regardless of the type of foundation selected, structures should be located away from critical root zones. The site should be excavated carefully to avoid damaging the roots of mature trees, with selective use of heavy equipment and digging by hand in delicate areas. An experienced arborist should be consulted to help maintain a healthy tree canopy.

Pest Control

Many regions of the country are susceptible to termite infestations, and foundations should be designed to reduce the possibility of termites entering the structure (Figure 5.27). Solid concrete foundations, termite shields at the top of the foundation wall, and underslab mesh barriers all help reduce the possibility of termite entry into the home (Figure 5.28 and Figure 5.29, page 144). Keeping wood structures dry and limiting the use of all wood to areas at least 12" above the exterior grade provides additional assurance against termites.

Termite bait stations are a nontoxic alternative to soil treatments (Figure 5.30a and Figure 5.30b, page 144). Rather than broadly apply termiticides around the perimeter of the home, small tubes are installed with wood ("bait") and monitored for termite activity. Termiticides are then installed as needed to kill termite colonies when they are detected in the bait tubes.

Green Remodeling Considerations

Green remodeling projects must evaluate and address deficiencies in existing foundations and determine the most appropriate changes and improvements for the home. Understanding how the following four of the eight

---

soil gas includes air, water vapor, radon, methane, and other soil pollutants that can enter a building though gaps in the foundation or crawl space floor.

radon a naturally occurring radioactive gas that is present in the ground at varying concentrations across the country.

radon ventilation systems that prevent the entry of radon and other soil gases into the home by ventilation to the outside.

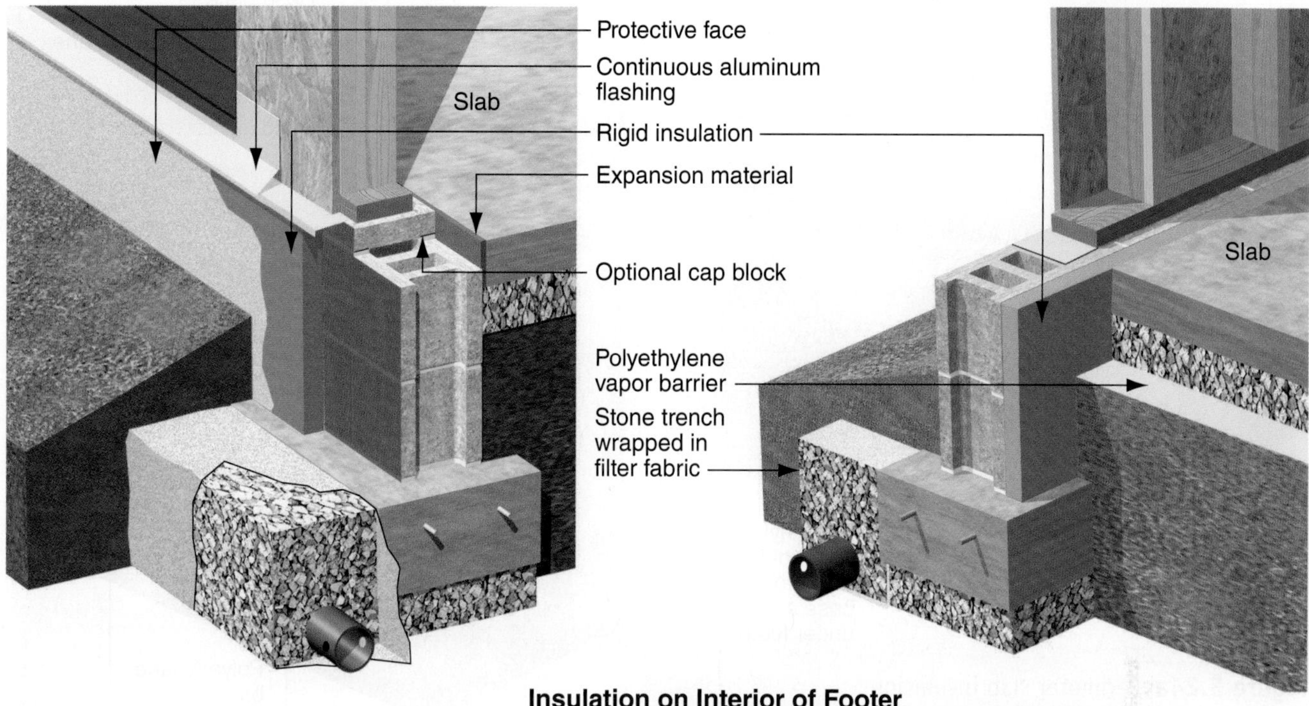

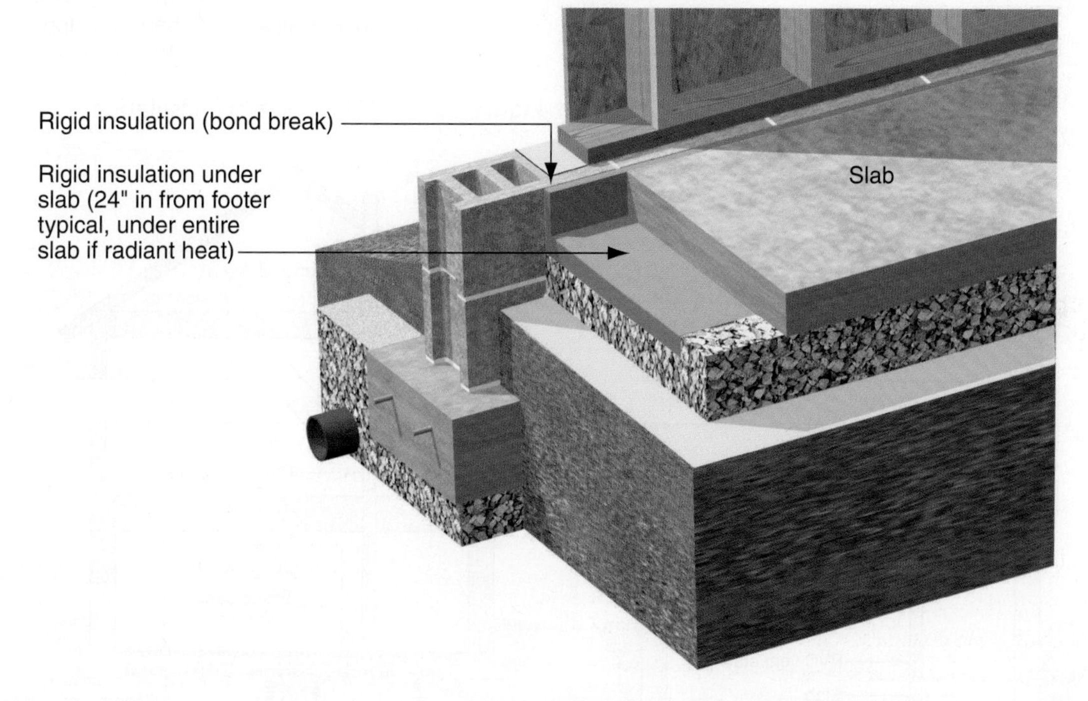

Figure 5.23 Three options for insulating slabs. Although slabs may be a major source of heat loss, local jurisdictions may not allow slab insulation due to termite risks. Details to prevent termite access must be included in all slab insulation designs.

fundamental principles apply to your project will help guide your decisions.

Energy Efficiency Many older homes do not have a complete building envelope between the house and the foundation or at the perimeter of the foundation. The key question for basements and crawl spaces is, can you create a complete thermal barrier at the foundation walls? If not, can you do this at the floor line, or through a combination of the floor and some of the foundation walls? In cold climates, a study by nonprofit consulting firm Advanced Energy suggests that you may be better

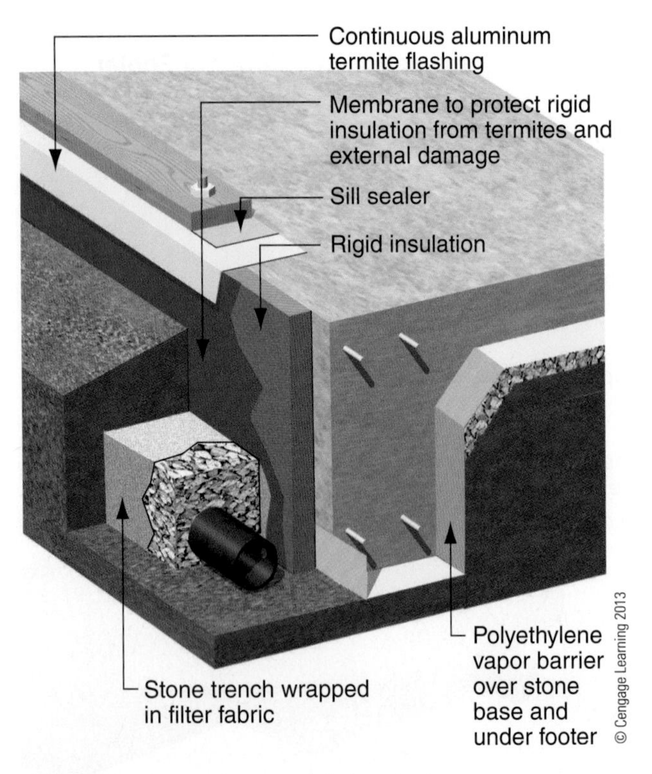

Figure 5.24a Perimeter slab insulation.

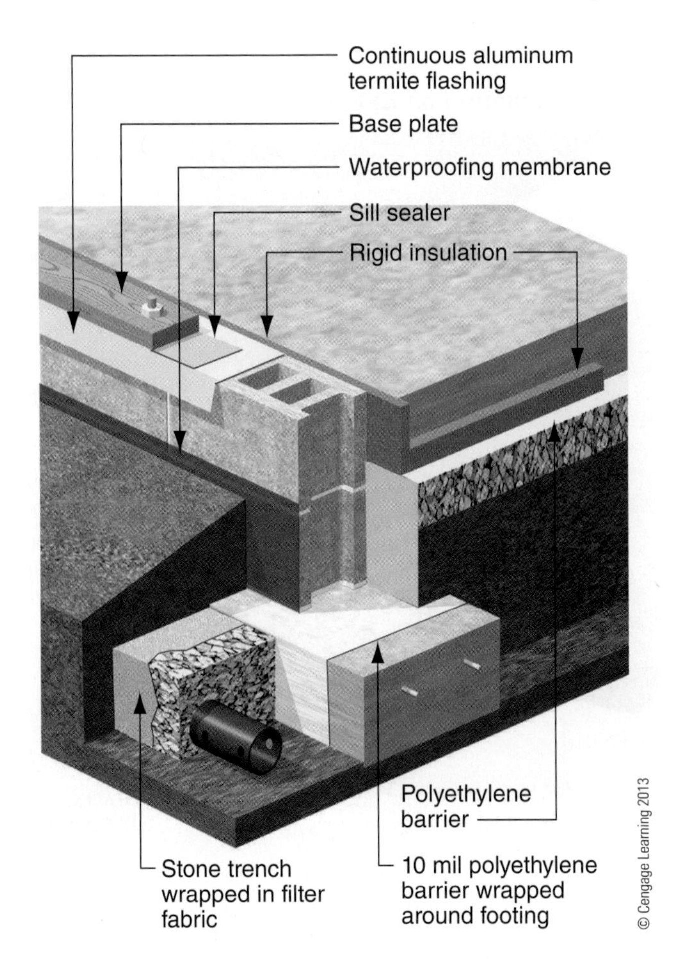

Figure 5.24b "Floating slab" insulation

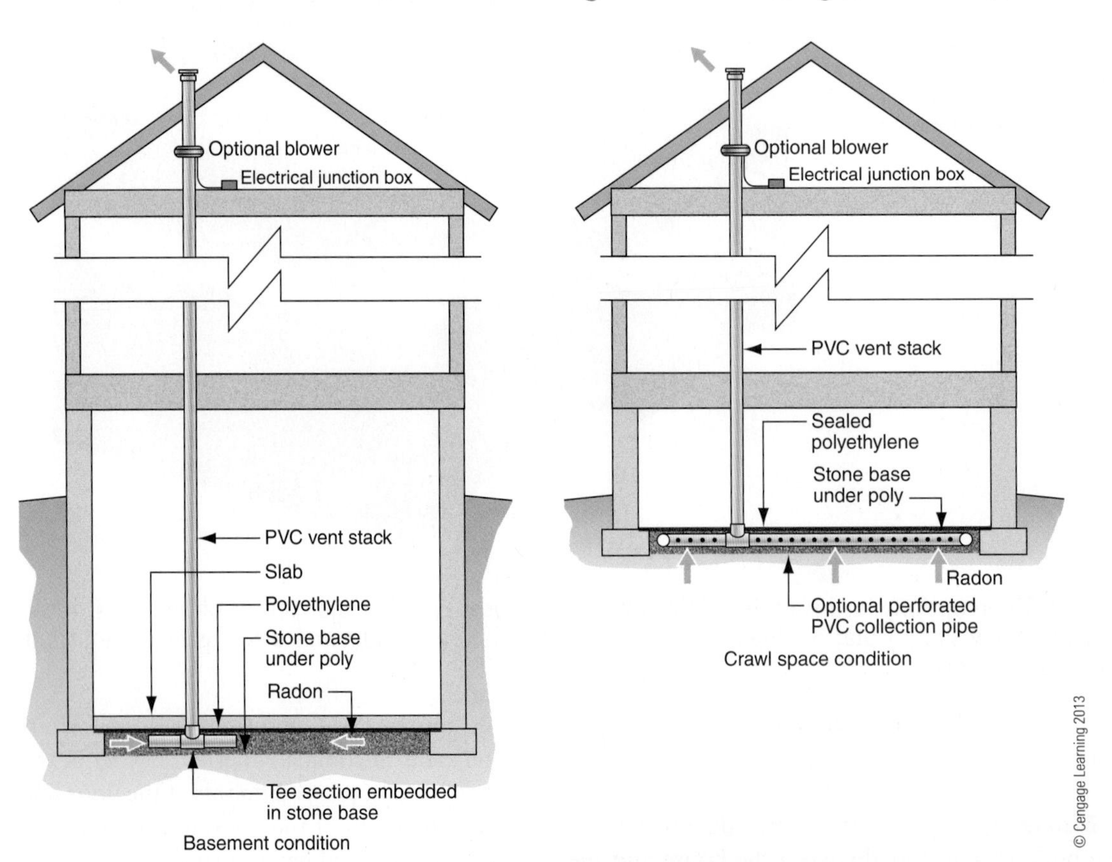

Figure 5.25 Radon ventilation systems are typically run vertically up through an interior wall near a plumbing stack. Depending on the radon risk, multiple below-slab intakes may be required.

Figure 5.26a Radon ventilation fan in an attic with exhaust pipe run through interior walls to the crawl space.

Figure 5.26b A radon ventilation system run along the exterior of the home.

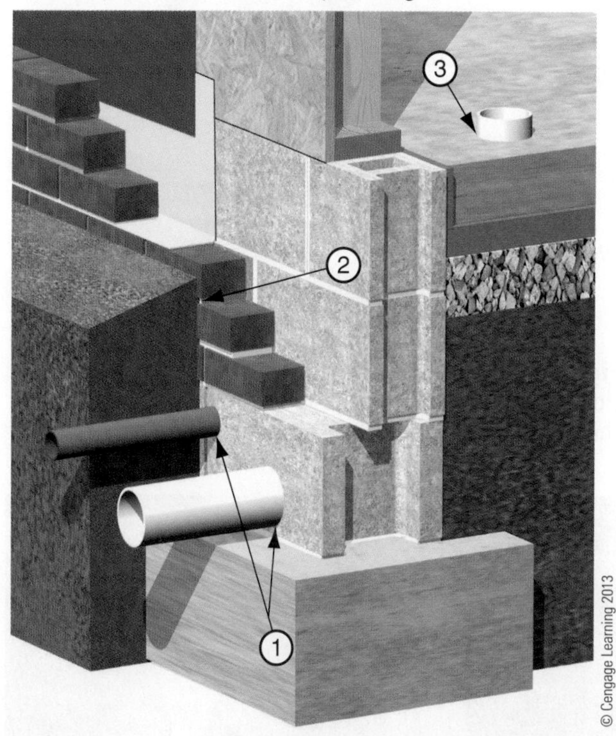

Figure 5.27 Termites may gain entry to homes behind brick facades and through plumbing and other slab penetrations.

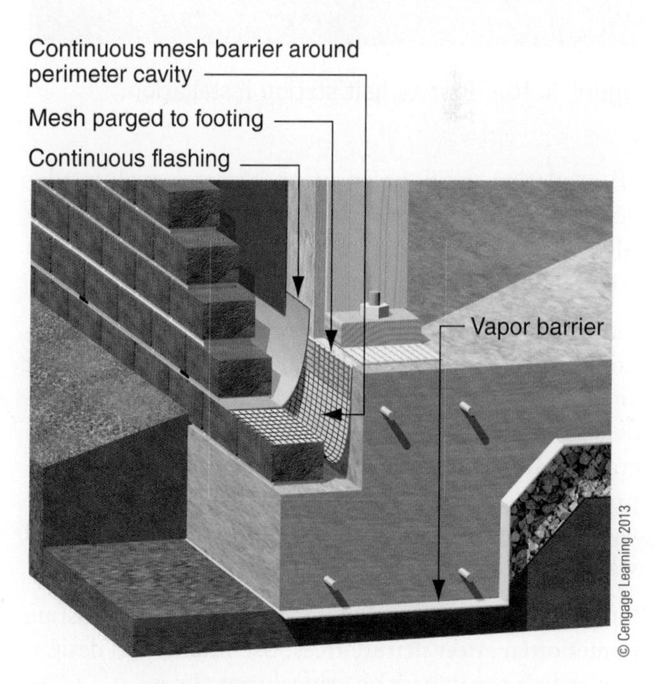

Figure 5.28 Stainless steel mesh installed behind brick veneer to prevent termite entry.

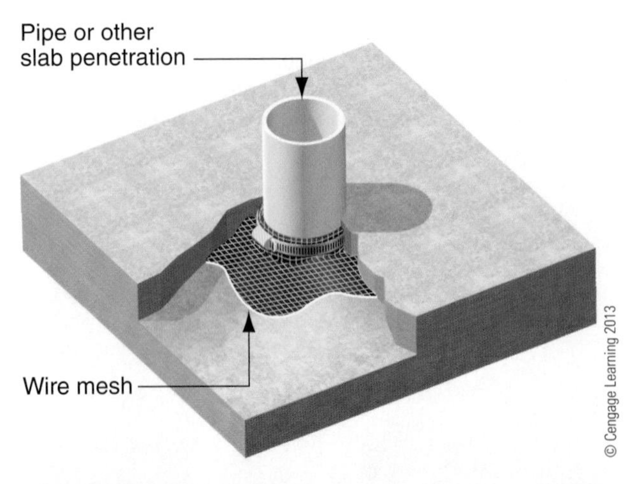

Figure 5.29 Metal mesh is installed around penetrations to eliminate the gaps in the slab created during the "thermal process" of the concrete curing.

Figure 5.30a Termite bait station installation.

Figure 5.30b The top portion is a bait cartridge that contains both a termite bait matrix and termiticide. The bottom portion is the monitoring base that allows easy checking for termite activity.

off insulating the floor; in warm climates, wall insulation may be more appropriate. Adding insulation to the edge of slab floors helps reduce heat loss and keep the house more comfortable in cold weather.

Durability Water and excess moisture can lead to structural and maintenance problems. Identify and correct any foundation leaks, and consider installing dehumidifiers in humid climates to eliminate excess moisture in crawl spaces and basements. Interior or exterior waterproofing systems can keep the foundation dry and provide long-term assurance through an installer's extended warranty.

Sustainable Site Development Additions to existing homes often affect mature trees. Use foundation designs that reduce the impact on critical root zones.

Indoor Environmental Quality Test homes for radon, and install mitigation systems if the levels are too high. Controlling water and moisture reduces the chance of mold growth, and a complete building envelope helps keep the air clean and free from contaminants. If there are open combustion furnaces or water heaters in an existing basement or crawl space, you should not enclose the space within the building envelope unless the equipment is replaced with sealed combustion systems. Alternatively, the equipment can be enclosed in a combustion closet that isolates it from the home air and provides exterior fresh air to the closet.

Finishing Existing Basements

Finishing an existing basement is preferred to constructing an addition when adding conditioned space to a home (Figure 5.31a and Figure 5.31b). In general, finishing a basement uses fewer materials and improves the efficiency of a home. To ensure proper performance, the renovation must include plans to fix any moisture issues, provide adequate heating and air conditioning for the new space, address combustion and radon safety, and appropriately frame and insulate against foundation walls.

Figure 5.31a Before a thorough renovation, this unconditioned basement was a source of moisture problems and served limited purposes.

Figure 5.31b This is the same basement, but after a renovation. The homeowners were able to add living space while using materials efficiently and improving the overall performance of the home.

Summary

In choosing a foundation, making the right decisions help make a building greener. Building smaller footings and walls that take advantage of the existing strength of the soil, using piers instead of walls, or installing prefabricated foundations can provide opportunities to reduce material use. Effective management of soil gases, water vapor, and bulk moisture help keep the house healthy and durable. When you understand how these choices fit into the whole-house system, you are well on your way to creating a green and sustainable building.

Review Questions

1. What foundation type has no radon risk?
   a. Conditioned crawl space
   b. Unconditioned basement
   c. Slab-on-grade
   d. Piers
2. What building materials may contain CCA?
   a. CMUs
   b. Pressure-treated wood
   c. ICFs
   d. Steel studs
3. What foundation type is best suited to a flood-prone site?
   a. Slab-on-grade
   b. Basement
   c. Crawl space
   d. Pier
4. Which item below does not reduce the amount of concrete used in a foundation?
   a. Frost-protected shallow foundation
   b. ICFs
   c. Prefabricated foundation
   d. Using thin brick or stone veneer
5. Which foundation type does not include integrated insulation?
   a. AAC
   b. ICFs
   c. Precast concrete
   d. Poured-in-place concrete
6. Which is not a benefit of prefabricated foundations?
   a. Reduces job-site waste
   b. Reduces material required
   c. Provides improved earthquake resistance
   d. Shortens construction schedule
7. Which of the following techniques helps to eliminate moisture movement between the ground and concrete foundation walls?
   a. Exterior insulation
   b. Interior insulation
   c. Capillary break on top of footing
   d. ICFs

8. Which foundation is the best choice when there are large trees close to the house?
   a. Slab-on-grade
   b. Basement
   c. Crawl space
   d. Pier
9. When should crawl spaces be vented?
   a. Only when they cannot be kept dry
   b. In all dry climates
   c. In all wet climates
   d. In all warm climates
10. When should crawl spaces not be conditioned?
   a. In earthquake-prone areas
   b. In humid climates
   c. In dry climates
   d. In flood-prone areas

Critical Thinking Questions

1. For a home located in Atlanta, Georgia, that is not within a high-risk flood zone, what foundation type is best and why?
2. What is the "greenest" foundation type?
3. Describe the process of constructing a conditioned crawl space.
4. What is fly ash, and why is it considered green?

Key Terms

autoclaved aerated concrete (AAC), 128
capillary break, 137
chromated copper arsenate (CCA), 134
coal fly ash, 127
concrete, 127
concrete masonry unit (CMU), 128
conditioned crawl space, 137
damp-proofing, 135
drainage mat, 135
footing, 126
foundation drainage, 135
foundation walls, 125
frost-protected shallow foundations, 127
grade beam, 132
ground granulated blast furnace slag, 127
hydrostatic pressure, 135
insulated concrete form (ICF), 130
permanent wood foundations (PWF), 133
pier foundation, 130
Portland cement, 127
pozzolan, 127
precast concrete, 133
prefabricated foundations, 132
radon, 140
radon ventilation, 140
retaining wall, 126
soil gas, 140
vapor barrier, 137
waterproofing, 135

Additional Resources

American Coal Ash Association: http://www.acaa-usa.org/

Autoclaved Aerated Concrete Products Association: http://www.aacpa.org/

Advanced Energy crawl space resources: http://www.crawlspaces.org/

U.S. EPA radon resources: http://www.epa.gov/radon/

Form-a-Drain: http://www.certainteed.com/

Frost-protected shallow foundations: http://www.toolbase .org/Technology-Inventory/foundations/frost-protected-shallow-foundations

Insulated Concrete Forms Association (ICFA): www.forms.org/

Permanent Wood Foundations: http://www.toolbase.org/Technology-Inventory/Foundations/wood-foundations

Structural Insulated Panel Association (SIPA): http://www.sips.org/

Superior Walls: http://www.superiorwalls.com/

TOOLBASE TECHSPECS: Frost-Protected Shallow Foundations: http://www.toolbase.org/pdf/techinv/fpsf_techspec.pdf

6

Floors and Exterior Walls

The materials and methods selected for construction of the floors and walls of a house have a direct effect on durability, energy efficiency, and resource efficiency. While traditional wood stick framing is used for most new and remodeled homes, there are alternatives that should be considered when building green. Every system has both positive and negative impacts on the building performance, the project budget, and the environment. This chapter deals with the structural elements of floors and walls. Interior and exterior finishes, windows, water-resistive barriers and other issues will be addressed in later chapters.

LEARNING OBJECTIVES

Upon completion of this chapter the student should be able to:

- Describe the different floor and wall structure types and their relationship to green building principles
- Describe the elements of advanced framing and the benefits of employing these techniques
- Demonstrate how to determine the resource efficiency in terms of materials used in floor and wall construction
- Describe the process of evaluating an existing structure to determine its suitability for renovation

Green Building Principles

 Energy Efficiency

 Resource Efficiency

 Durability

Introduction to Floors and Walls

Residential construction in the United States is dominated by wood framing with cavity insulation; however, there are many alternative systems available to the green builder. Options include walls with integrated insulation, such as insulated concrete forms (ICFs), straw bale, or structurally insulated panels (SIPs), as well as wood or concrete panels or full modular construction. Each method can have a place in a green home, and within each method there are options available that affect the sustainability of a project.

Stick-framed walls consist of horizontal and vertical wooden members. With balloon framing, the studs are continuous from the foundation sill to the top wall plate (Figure 6.1). In the early 1900s, most balloon framing was replaced with platform framing. Platform frame construction is composed of single-story walls that are erected on constructed floor decks or platforms (Figure 6.2). Traditionally, all framing occurred on-site; today, individual assemblies or even entire homes are manufactured in factories. Panelized framing consists of wall,

---

balloon framing a system of wood-frame construction, first used in the 19th century, in which the studs are continuous from the foundation sill to the top wall plate.

sill first horizontal wood member resting on the foundation supporting the framework of a building; also, the lowest horizontal member in a window or door frame.

platform framing a method of wood frame construction in which walls are erected on a previously constructed floor deck or platform.

panelized framing consists of wall, floor, ceiling, and roof panels constructed in a controlled environment and delivered to the site ready for installation.

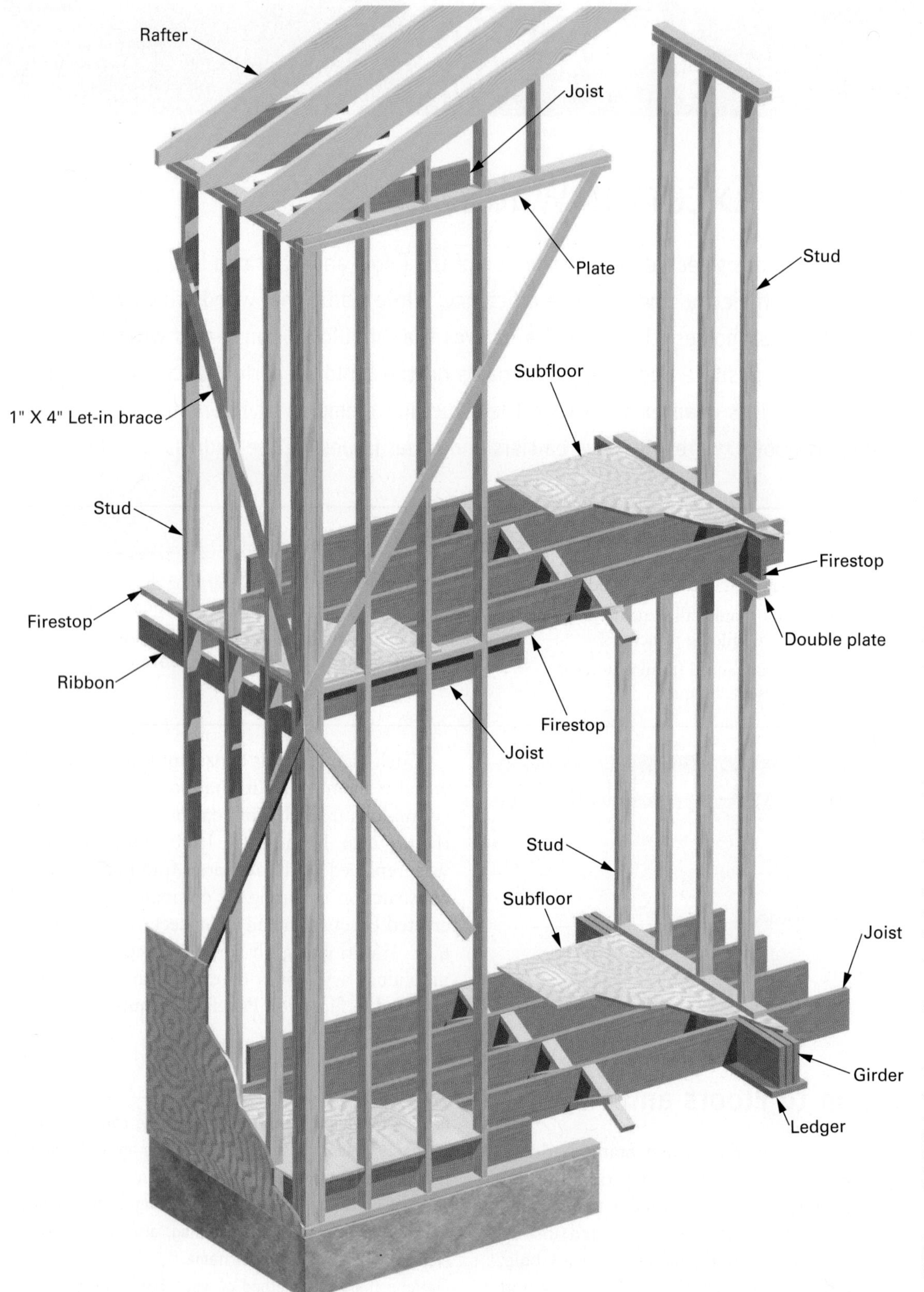

Figure 6.1 Balloon-frame construction.

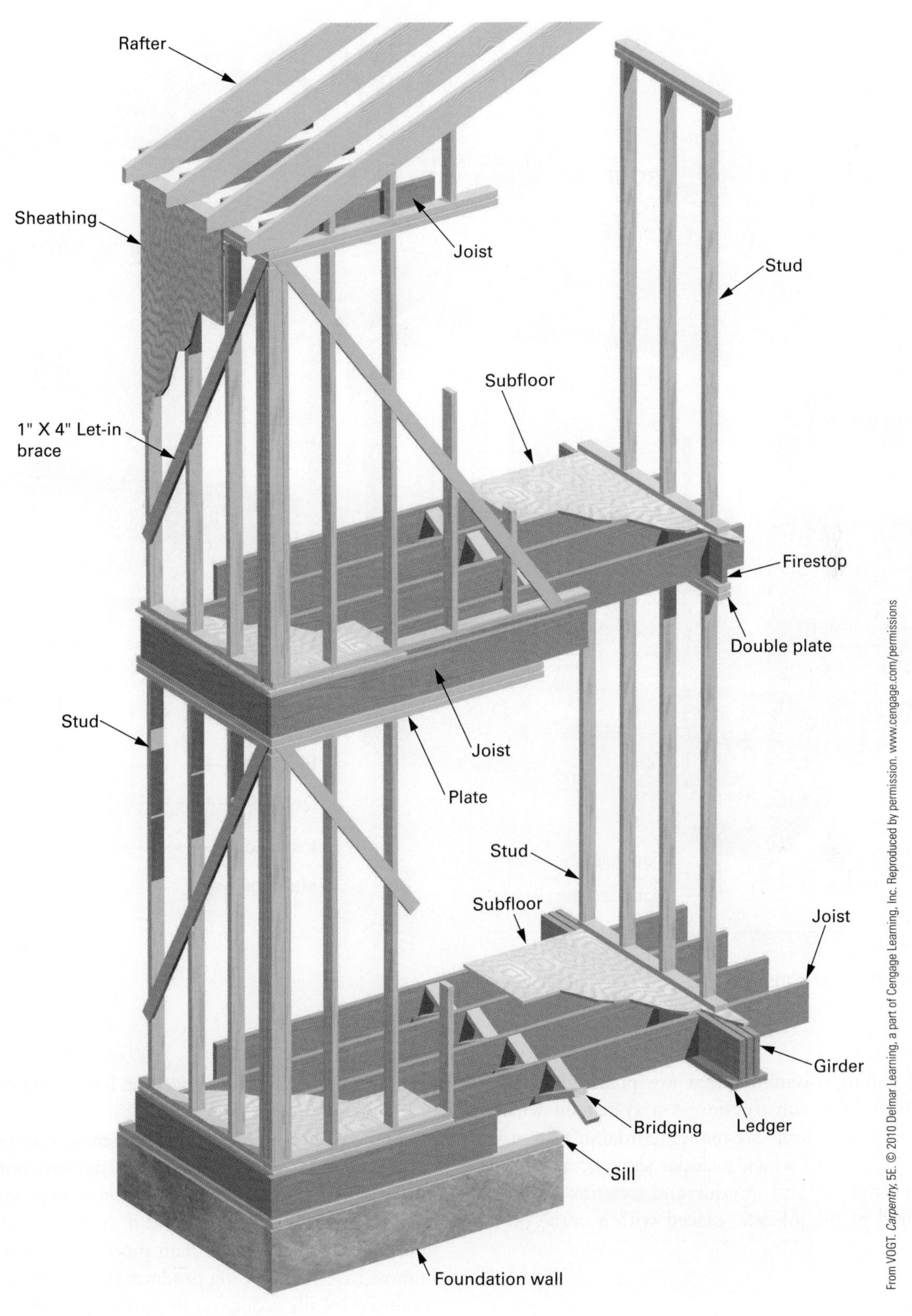

Figure 6.2 Platform-frame construction.

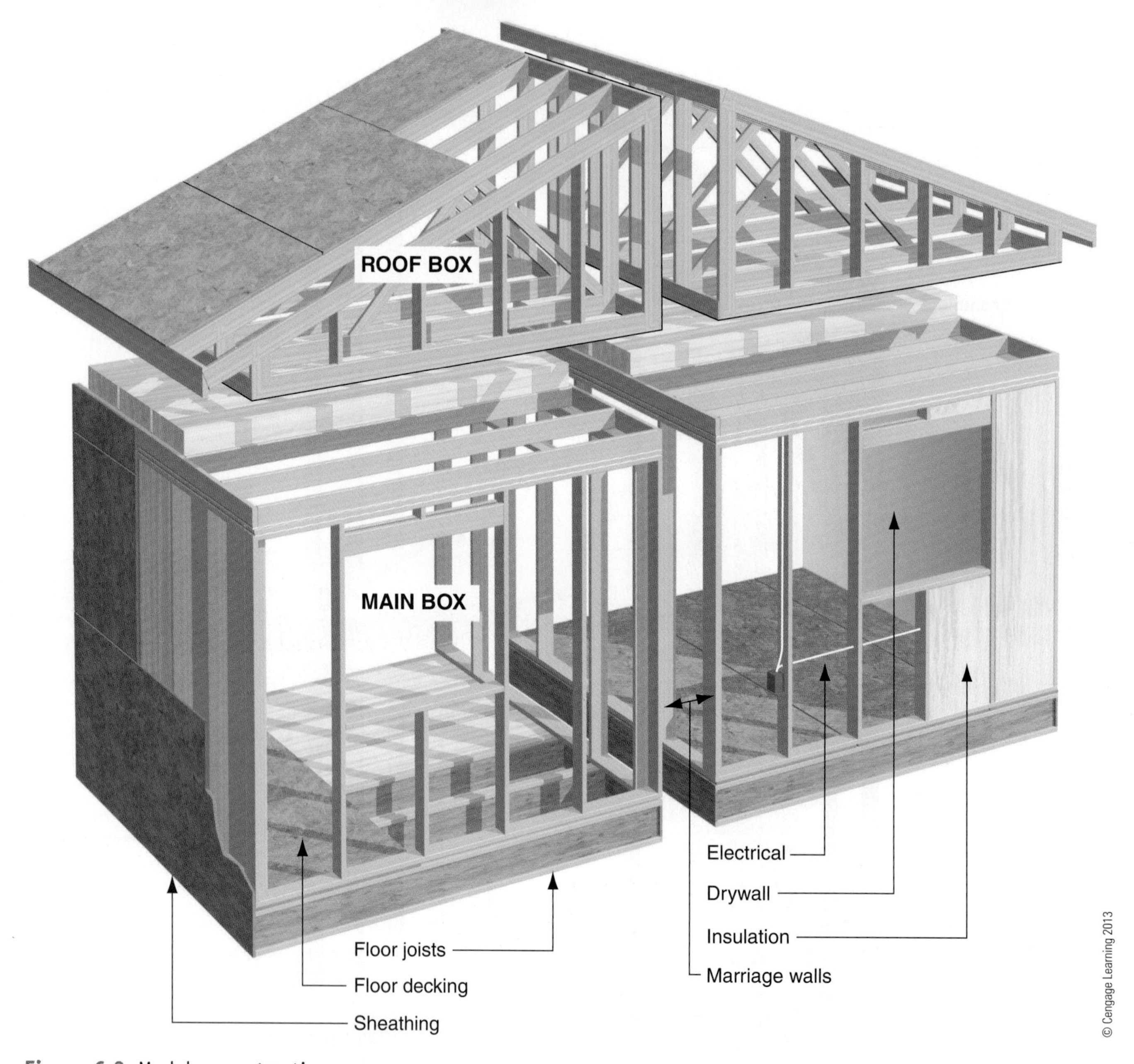

Figure 6.3 Modular construction.

floor, and ceiling assemblies that are preassembled in a factory. Modular construction is a system in which whole sections of a home are manufactured in a factory. The sections, each of which includes some or all of the mechanical systems and interior and exterior finishes, are delivered to the job site, placed with a crane on a foundation, and attached together. Required finish work is completed on-site (Figure 6.3).

In the United States, wood is the most common structural material used for residential construction and is available almost everywhere. Less common products may not be as readily accessible, and if available, may be significantly more expensive than more standard materials. In most cases, purchasing products that are extracted and produced locally reduces transportation energy and associated air pollution. Key decisions that affect the type of floor and wall structural systems include material availability and cost, access to an adequate skilled labor force, site accessibility, and local climate considerations.

---

modular construction the practice of factory fabrication of complete sections of a house including floors, walls, ceilings, mechanical systems, and finishes, which are delivered by truck to the job site where they are placed on a foundation and finished in place.

Skilled Labor Force

Successful green building may require different methods for trades in the field, ranging from minor technique changes for framers to entirely new skills for such materials as ICFs or SIPs. When you select an uncommon product or method, make sure that the available craftspeople have the skills needed to execute the job or are willing to acquire them. A severe lack of skilled tradesmen may lead you to consider panelized or modular construction, reducing the amount of on-site work required.

Site Accessibility

The characteristics of a particular lot helps guide the selection of construction techniques. Some methods, such as panelized and modular construction, require storage space for preassembled parts and access for large equipment that may not be available on all building sites. Sites with limited access may require on-site assembly as material is delivered in small orders.

Climate

While wood-framed construction predominates and is used in almost all climates, there are alternatives that should be considered, particularly where wood is susceptible to moisture or termite infestation (Table 6.1 and Table 6.2). In climates with hurricane and flooding risks, many builders avoid wood floors and walls completely, substituting concrete and masonry products that are better able to withstand both impact and water damage. Keep in mind that you can combine different construction methods in a single house, such as framed floors with SIP walls, or a concrete slab floor combined with stick-framed or ICF walls.

Climate impact, availability of materials, and the skill sets of available personnel are key factors when selecting floor and wall structural systems. Upfront construction costs must be weighed against long-term maintenance and utility expenses. Each system poses distinct advantages and drawbacks.

Table 6.1 Floor Selection Chart

Floor System	Pros	Cons
Stick framing	Inexpensive Materials and skilled labor readily available	Susceptible to moisture
SIPs	Strong Energy efficient	Susceptible to moisture Materials and skilled labor less readily available

Table 6.2 Wall Selection Chart

Wall System	Pros	Cons
Stick framing	Inexpensive Materials and skilled labor readily available	Susceptible to moisture
SIPs	Strong Energy efficient	Susceptible to moisture Materials and skilled labor less readily available
ICFs	Strong Energy efficient Not affected by moisture	Materials and skilled labor less readily available High cost High embodied energy

Wood

As an abundant resource in this country, wood has been used for a building material for many generations. Before the advent of long-range transportation, wood was scarce in areas where trees did not grow, leaving the southwestern United States and other such areas to rely on traditional adobe, brick, stone, and rammed-earth construction methods. Buildings constructed of stacked logs were the first wood homes, followed by post-and-beam construction, often with masonry infill. With the advent of mechanical milling equipment in the early 19th century, small-dimensional lumber became widely available, and stick-frame methods began to emerge. Evolution of stick-frame construction has been modest and gradual over the past several hundred years with continued fine tuning in recent decades to improve building efficiency and reduce material use.

Grown naturally, wood is renewable, recyclable, and compostable, and it can be used without additives and little processing. All wood products, however, are not created equal. Wood products are drawn from a wide variety of raw sources.

Timber Harvesting

The three primary sources of wood are old-growth, second-growth, and plantation forests. Old-growth refers to forest or woodlands having a mature or overly mature ecosystem that is more or less uninfluenced by human activity. Second-growth are forests that have

---

old-growth a forest or woodland area having a mature or overly mature ecosystem that is more or less uninfluenced by human activity.

second-growth a forest or woodland area that has regrown after the removal of all or a large part of the previous stand by cutting, fire, wind, or other force; typically, a long enough period will have passed so that the effects of the disturbance are no longer evident.

regrown after the removal of all or large part of the previous stand by cutting, fire, wind, or other force; a long-enough period will have passed so that the effects of the disturbance are no longer evident. Conventional timber harvesting from old- and second-growth forests destroy wildlife habitat, may lead to soil erosion, and release greenhouse gas emissions (GHG). Timber harvested in forests that are not sustainably managed can reduce the forest's ability to remove greenhouse gases from the atmosphere, and cleared lands are often burned, which releases additional greenhouse gases.

Today, most of the wood used in construction comes from managed timber plantations. **Tree plantations** are areas where trees are managed like agricultural crops. Plantations use short timber rotations, contain only one or two species, and rely on the intensive use of herbicides, pesticides, and fertilizers. Mature monoculture plantations often have little undergrowth and minimal ecologic diversity. In fact, from a wildlife habitat standpoint, such tree plantations are little different from an acre of corn.

Although tree plantations have their drawbacks, they are a valuable alternative to harvesting old-growth and act as a carbon sink. A **carbon sink** is any natural storage reservoir, such as trees and oceans, that removes carbon from the atmosphere. Trees and timber store the carbon until the wood burns or decomposes. By using the wood for timber in homes, the carbon is effectively removed from the atmosphere for many years (Figure 6.4).

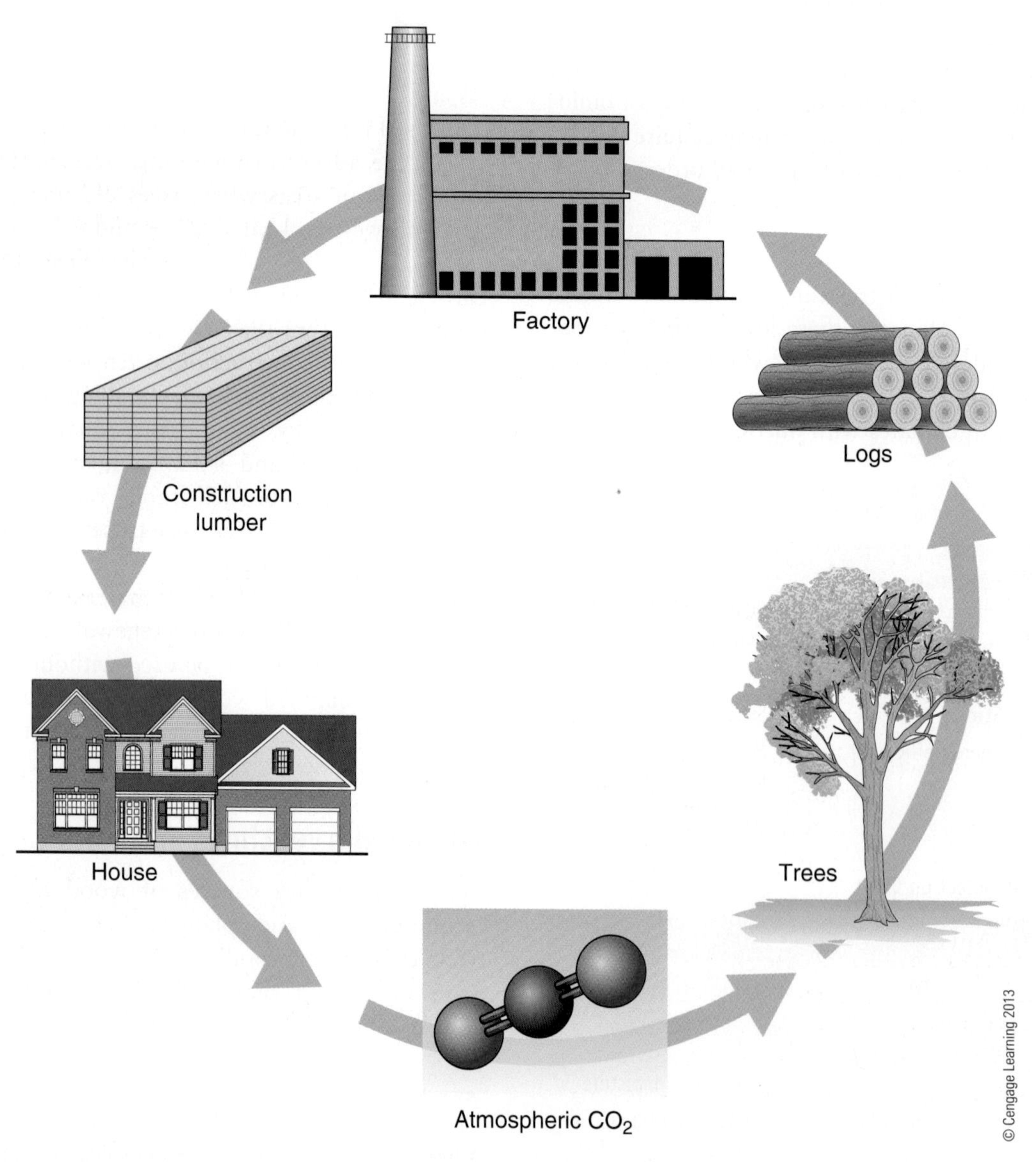

Figure 6.4 Timber is a carbon sink and helps to remove greenhouse gases from the atmosphere.

---

tree plantation an actively managed crop of trees that, unlike a forest, contains one or two tree species and provides little wildlife habitat.

carbon sink an environmental reservoir that absorbs and stores carbon, thereby removing it from the atmosphere.

Timber Certification

Numerous programs exist to certify sustainable forestry practices. Sustainable forest management includes forestry management practices that maintain and enhance the long-term health of forest ecosystems while providing ecologic, economic, social, and cultural opportunities for the benefit of present and future generations. Programs evaluate forestry practices on the basis of social, economic, and environmental criteria. Like green building certification programs, third-party verification is essential to the success of sustainable forestry programs. The two most common certification programs are offered through the Sustainable Forestry Initiative (SFI) and Forest Stewardship Council (FSC). Both certifications require third-party chain-of-custody verification throughout the timber harvesting and manufacturing process. Although these certification programs have seen tremendous growth in the last decade, only about 10% of global forests are certified, most of which are located in North America.

Good for you. Good for our forests.℠

The SFI program was launched in 1994 by the U.S. forest industry. Today, SFI Inc. is an independent, nongovernmental, nonprofit organization responsible for maintaining, overseeing, and improving a sustainable forestry certification program for North America. SFI has certified 125 million acres.

---

sustainable forest management is forestry management practices that maintain and enhance the long-term health of forest ecosystems while providing ecologic, economic, social, and cultural opportunities for the benefit of present and future generations.

Sustainable Forestry Initiative (SFI) a nonprofit organization responsible for maintaining, overseeing, and improving a sustainable forestry certification program.

Forest Stewardship Council (FSC) is an independent, nongovernmental, not-for-profit organization established to promote the responsible management of the world's forests.

FSC was established in 1993 as an independent, nongovernmental, not-for-profit organization to promote the responsible management of the world's forests. FSC is widely regarded as the most stringent, most responsible forestry certification program and the only program accepted within the LEED Rating Systems. FSC is internationally represented in over 50 countries.

Tropical hardwoods, defined as wood harvested in countries between the Tropics of Cancer and Capricorn, should always be sustainably harvested or avoided altogether whenever possible. Tropical forests are often poorly managed, and deforestation is a significant threat to local habitat and indigenous peoples. Long transportation distances also contribute to the overall negative environmental impact of tropical hardwoods.

Climate Impact

Wood is one of the most "climate-friendly" materials due to its lower greenhouse gas footprint. Compared with concrete and metals, the manufacturing of lumber produces significantly fewer greenhouse gas emissions. Life cycle analysis shows that most of wood's emissions are connected with the end-of-life treatment due to generally poor rates of recycling and the associated landfill emissions. By comparison, metal and concrete are inert and release zero landfill emissions.

Engineered Wood Products

A piece of wood may have natural defects that occur as a tree grows or during the wood drying process. Natural defects include knots, check, wane, insect holes, and pitch pockets (Figure 6.5). Engineered products, on the other hand, offer both resource and energy efficiency

Figure 6.5 Examples of common wood defects.
From SPENCE/KULTERMANN. *Construction Materials, Methods and Techniques,* 3E. © 2011 Delmar Learning, a part of Cengage Learning, Inc. Reproduced by permission. www.cengage.com/permissions

benefits for builders. **Engineered lumber** is wood that is manufactured by bonding together wood strands, veneers, lumber, or fiber to produce a stronger and more uniform composite. Engineered products are used in floor, roof, and wall systems. **Trusses** are engineered lumber products consisting of wood or wood and metal members used to support roofs or floors (Figure 6.6a and Figure 6.6b). Trusses replace site-built framed construction. **Open-web floor trusses**, one particular type of floor assembly consisting of dimensional lumber joined by metal connector plates, uses fewer materials than larger single structural members while providing equivalent strength.

Beams, studs, and **joists** can be made from peeled, chipped, or short pieces of wood that are assembled with adhesives (Figure 6.7). Engineered products make use of small, fast-growing, low-value trees, such as aspen and pine, thereby reducing the need to harvest large, mature trees for building materials. These products generally provide equivalent or higher strength while using less material than the solid lumber they replace. Some traditional and common engineered products include **plywood**, oriented-strand board (OSB), laminated beams, and finger-jointed studs. Other examples of engineered

engineered lumber is wood that is manufactured by bonding together wood strands, veneers, lumber or fiber to produce a stronger and more uniform composite; also known as *manufactured wood product.*

truss an engineered lumber product consisting of wood or wood and metal members used to support roofs or floors that reduces the amount of lumber required to support a specific load.

open-web floor truss an engineered assembly of dimensional lumber with metal connector plates that replaces a larger single structural member with one using fewer materials while providing equivalent strength.

joist horizontal framing members used in a spaced pattern that provide support for the floor or ceiling system.

plywood a piece of wood made of three or more layers of veneer joined with glue, and usually laid with the grain of adjoining plies at right angles.

Figure 6.6a Wood trusses are made from 2× stock joined with metal gusset plates.

Figure 6.6b Open-web floor trusses are made from 2× stock joined with metal gusset plates.

Figure 6.7 This I-joist is composed of flanges made of solid or composite lumber and a composite wood web.

Figure 6.8 Laminated veneer lumber is made by bonding together layers of wood veneer with an exterior adhesive.
From SPENCE/KULTERMANN. *Construction Materials, Methods and Techniques*, 3E. © 2011 Delmar Learning, a part of Cengage Learning, Inc. Reproduced by permission. www.cengage.com/permissions

lumber include laminated strand lumber made of small strips of wood assembled into beams and posts, and laminated veneer lumber, made of sheets of thin wood assembled in a manner similar to plywood (Figure 6.8).

Engineered products are lighter, straighter, and stronger than the traditional materials they replace and are much less likely to warp or split. Their added strength allows for longer spans, reducing the amount of intermediate beams required and providing additional material savings. From an energy-efficiency standpoint, engineered products are often thinner and more widely spaced than the materials they replace, allowing more room for insulation. In particular, open-web trusses save material by using short lengths of small dimensional lumber instead of larger solid boards. A 1996 case study by the Structural Building Components Association and National Association of Home Builders (NAHB) determined that open-web trusses used 25% less raw material than solid lumber in the floor structure and required 68% less labor to install.

Engineered wood products are designed as systems. The layout, sizing, connectors, installation requirements, and size and location of allowable cuts must follow suppliers' specifications and be carefully managed on-site. Professionals who are not familiar with engineered products can make mistakes if they rely on prior experience with solid lumber rather than follow the design and installation details provided with the materials.

Engineered Wood Product Chemicals and Adhesives

Interior-grade adhesives are made of urea formaldehyde, which releases significant amounts of formaldehyde, a known carcinogen. Exterior-grade adhesives are made from phenol formaldehyde, a more moisture-resistant adhesive that releases much less formaldehyde than urea formaldehyde and is a better choice for a green home. Alternate adhesives include methyl diisocyanate (MDI), a polyurethane binder that, while quite toxic during manufacture, is extremely stable when cured with virtually no off-gassing after installation.

Extremely chemically sensitive people may need to have their buildings constructed of solid lumber products with no chemical additives, avoiding even such common engineered products as plywood and OSB.

Solid Milled Lumber

Prior to the development of engineered wood products, solid lumber was the only material available for wood framing. Solid lumber is harvested from larger and older trees and creates more waste than engineered products made from smaller and faster-growing trees. Traditional solid lumber can be part of a sustainable home, particularly if it is made from reclaimed materials, is locally harvested, or is certified as being cut from a sustainably managed forest.

Moisture and Pests

Wood-framed structures are susceptible to water damage from both bulk moisture (e.g., rain or flooding) and excessive vapor that can condense inside walls, floors, and roofs. Effective moisture management, a critical component in green building, is necessary to ensure the structural integrity and durability of wood-framed buildings. Excessive moisture promotes mold growth and termite infestations, and it can lead to structural failure if not corrected. Wood that is exposed to changes in relative humidity expands and contracts, creating gaps in critical air seals and causing cracks in drywall and other interior finishes.

To avoid termite infestations, a sill plate of wood treated to resist termites must be installed where there is contact with concrete or masonry. There should also be a capillary break to eliminate moisture transmission from the foundation to the wood floor structure (Figure 6.9). Several chemicals are used to protect wood from termites, including alkaline copper quaternary (ACQ) and borates; the U.S. Environmental Protection Agency has also recently approved a new manufacturing process in which sodium silicate is infused into wood and heated to create a physical termite barrier (TimberSIL®). Design elements that can help control pests include establishing the finish exterior grade at least 12" below any wood framing, avoiding any poured-in-place concrete slabs adjacent to wood-framed floors, and providing access for visual inspections of termite activity in all basements and crawl spaces. Spray-on borate treatments, such as Bora-Care®, are applied to the bottom 3' of the framing before insulation is installed (Figure 6.10) and are less toxic than traditional soil treatments.

Courtesy of The Dow Chemical Company

Figure 6.9 A foam plastic gasket below the bottom plate acts as a capillary break and provides additional air sealing. The blue roll of foam gasket is visible on the left.

Courtesy of Nisus Corporation

Figure 6.10 Borate is a termite treatment option that is less toxic than traditional pest control chemicals. Borate treatments are generally applied to the bottom 3' of framing.

Wood Treatment

Manufacturers are continuing to develop processes and chemicals to reduce or eliminate mold, rot, and pest infestations in wood framing materials. Recent developments include chemical coatings and heat treatment.

Available chemical applications include BluWood® (which factory-treats all wood framing), sheathing, and engineered wood products with a borate-based product for pest and mold resistance as well as a proprietary moisture-resistant coating (Figure 6.11). BluWood provides a 30-year limited warranty against mold, rot, and pests; however, it is not approved for sill plates or masonry and concrete contact applications. A similar product, MOLD-RAM®, is applied on-site after framing is complete and can be combined with a borate termite treatment (Figure 6.12).

Courtesy of Viance

Figure 6.11 Wood frame building components can be treated in the factory with mold inhibitors, fungicides, and insecticides. Products are typically colored to allow for easy identification on-site.

Figure 6.12 Framing is treated with MOLD-RAM®, a proprietary mold-preventative product, and a borate termite treatment.

---

sheathing boards or sheet material that are fastened to joists, rafters, and studs and on which the finish material is applied.

Did You Know?

Wood Treatment Products Are Only Part of the Solution

Wood treatments can be effective in protecting a building from moisture and other damage during construction. Installed framing exposed to rain is susceptible to mold, rot, and pests—vulnerabilities that can be reduced by the application of chemical additives to the wood. While generally not considered to be toxic, these chemicals make scrap wood unsuitable for grinding into mulch, eliminating the opportunity for recycling. Regardless of how much these treatments will reduce mold or structural deterioration, they are not substitutes for high-quality building practices, such as effective wall and foundation moisture management, designs that divert water away from buildings, and details that help reduce termite infestations. If a building is not designed and built to remain dry and pest-free, no amount of wood treatment will keep it healthy and durable.

Heat treatment, or thermal modification, involves heating wood without oxygen to improve decay resistance and dimensional stability. The process removes unstable compounds, kills organisms, and modifies the structure of the wood, reducing its ability to absorb moisture. In doing so, the wood becomes more resistant to pests, mold, and fungus. Materials like Blu-Wood, however, are not a substitute for approved concrete contact-treated products. Heat-treated wood is still establishing its place in the U.S. market. The most likely applications will be siding, decking, and other exterior finishes, but we may see it in structural products as the technology develops.

Wood Framing

Residential wood framing uses techniques that have been developed and perfected over several generations, allowing for fast and inexpensive construction of new homes. Newer techniques have helped to reduce material use, save money, and improve thermal performance without sacrificing structural integrity.

Conventional Wood Framing

Standard residential wood framing is simple, inexpensive, and prone to material waste and energy inefficiency. In most of the United States, wood framing materials are relatively inexpensive while labor costs are high. This cost relationship tends to lead builders toward using more materials quickly with minimal regard to waste and energy efficiency. Rather than selecting headers, beams, and posts sized for actual structural loads, designers and builders usually specify and install standard sizes that frequently exceed minimum load requirements. (Figure 6.13).

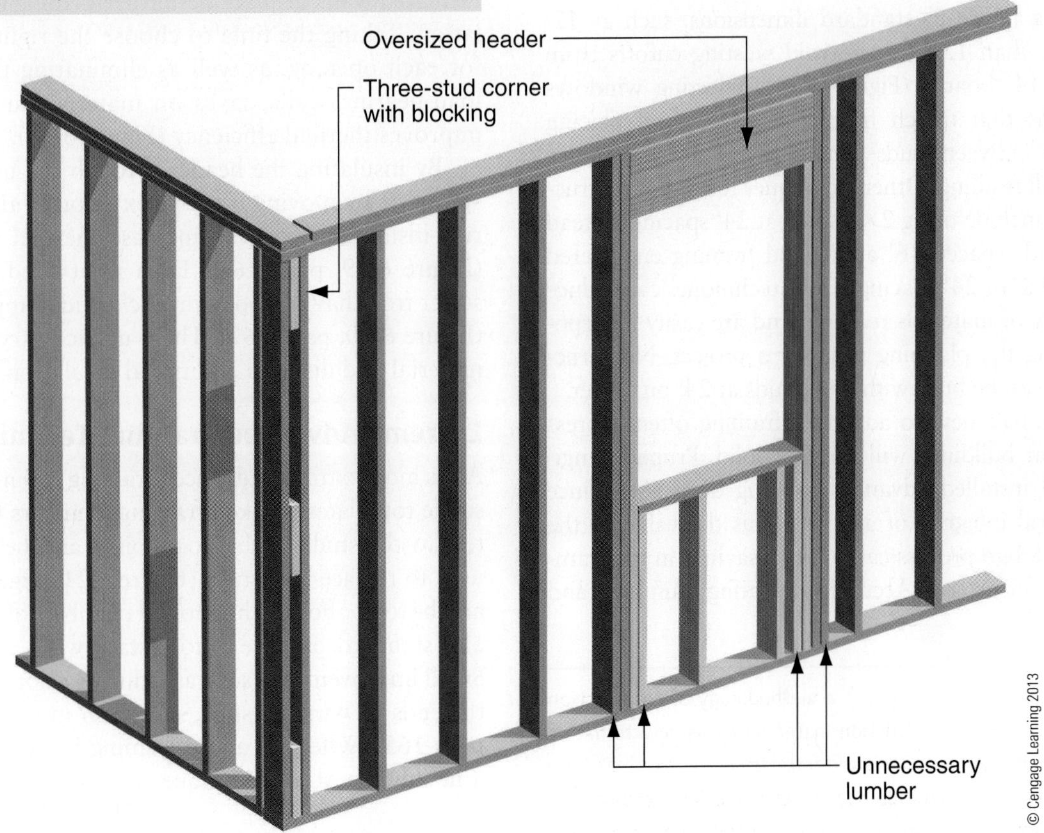

Figure 6.13 Conventionally framed homes contain excess framing, which wastes money and resources and hurts the home's overall performance.

Advanced Framing

In the 1960s, optimum value engineering (OVE), also known as advanced framing, was developed under a U.S. Department of Housing and Urban Development (HUD) initiative to decrease material use, and its associated costs, in wood-framed buildings. Because the reduction in material provides more space for insulation, advanced framing has been embraced by the green building movement. Many advanced framing techniques are simple, cost effective, and energy efficient; however, their adoption has been slow and inconsistent due to the building industry's slow pace of change. Advanced framing requires extra thought and care, in both the planning and the construction stages of a project.

Building professionals often incorrectly believe that more structure is better—if one stud is good, then two studs are better, and three studs must be great. Overbuilding uses more materials and costs more money than required. Regardless of the materials and methods chosen to construct a building, right-sizing structural members to their actual loads saves money and improves building efficiency. Advanced framing does not sacrifice safety or strength. Instead, it provides the appropriate amount of structure required while avoiding excess material use and allowing for the maximum amount of insulation.

Benefits of Advanced Framing

Designing a house to standard dimensions, such as 12' long rather than 12'6", can avoid wasting cutoffs from dozens of 14' boards (Figure 6.14). Selecting windows with widths that match framing modules and placing them to fit between studs eliminates the need for extra studs in wall framing. Other techniques for building structural walls include using 2×6 studs at 24' spacing instead of 2×4 studs spaced 16' apart, and framing engineered floors at 19.2' or 24' spacing. These techniques can reduce the quantity of materials required and are easily incorporated during the planning stage of a project. Nonstructural walls can be built with 2×4 studs at 24' on center.

Professionals new to advanced framing often express concern that buildings will not feel solid. Properly engineered and installed advanced framing does not reduce the structural integrity of a house, plus the value of the integrated design process can provide savings in the framing phase of a project. Accurately locating plumbing and heating, ventilating, and air conditioning (HVAC) lines to minimize cuts in structural members can eliminate the need to double-up framing members that are weakened by cuts and holes for pipes and ducts, allowing for more space for insulation and a reduction in material and labor costs.

Planning for Advanced Framing

To be successful, advanced framing techniques must be managed carefully in the field. Such details as open two- or three-stud corners (Figure 6.15) and ladder-framed T walls (Figure 6.16) are simple to construct, but framers unfamiliar with these techniques may fail to utilize them. Traditional builders are often reluctant to reduce the amount of lumber in their homes, wary of change from well-established traditions and experience.

A common practice in residential construction is the use of double 2×10 headers at all door and window openings, often with 2×4s nailed to the top and bottom with a space in the middle (Figure 6.17). This design is wasteful because the top and bottom boards provide no structural value. The double 2×10 is frequently larger than needed for most of the openings, and the space between them does not provide any insulation value, particularly when a smaller header would be adequate (Table 6.3, page 162). Also, structural headers are frequently installed in non-load-bearing walls. These practices waste material, and oversized headers installed in insulated walls displace better-performing thermal insulation. Taking the time to choose the right size header for each opening, as well as eliminating them in non-load-bearing walls, saves on materials and labor and improves thermal efficiency (Figure 6.18).

By insulating the header through the use of a foam spacer or by moving it to the exterior to allow for interior insulation, you can increase thermal performance (Figure 6.19, page 162). Even right-sized headers can suffer from having too many jack studs supporting them (Figure 6.20, page 162). These unnecessary studs waste material and displace additional insulation.

Extreme Advanced Framing Techniques

At its most extreme, advanced framing techniques include single top plates, stacked framing members from the rafters to the studs to the floor joists, and header hangers used to replace jack studs (Figure 6.21, page 164). Plates are the top or bottom horizontal members of a wall frame, and standard practice is to install two 2×4 top plates. Small improvements, such as reducing cripple studs under the ends of window stills, save even more (Figure 6.22, page 165). When advanced framing is incorporated into a new home, these techniques can reduce total material

(Continued on page 162)

---

optimum value engineering (OVE) a methodology of construction designed to conserve construction materials by using alternative framing methods; see also *advanced framing*.

advanced framing a methodology of construction designed to conserve construction materials by using alternative framing methods; see also *optimum value engineering (OVE)*.

plate top or bottom horizontal member of a wall frame.

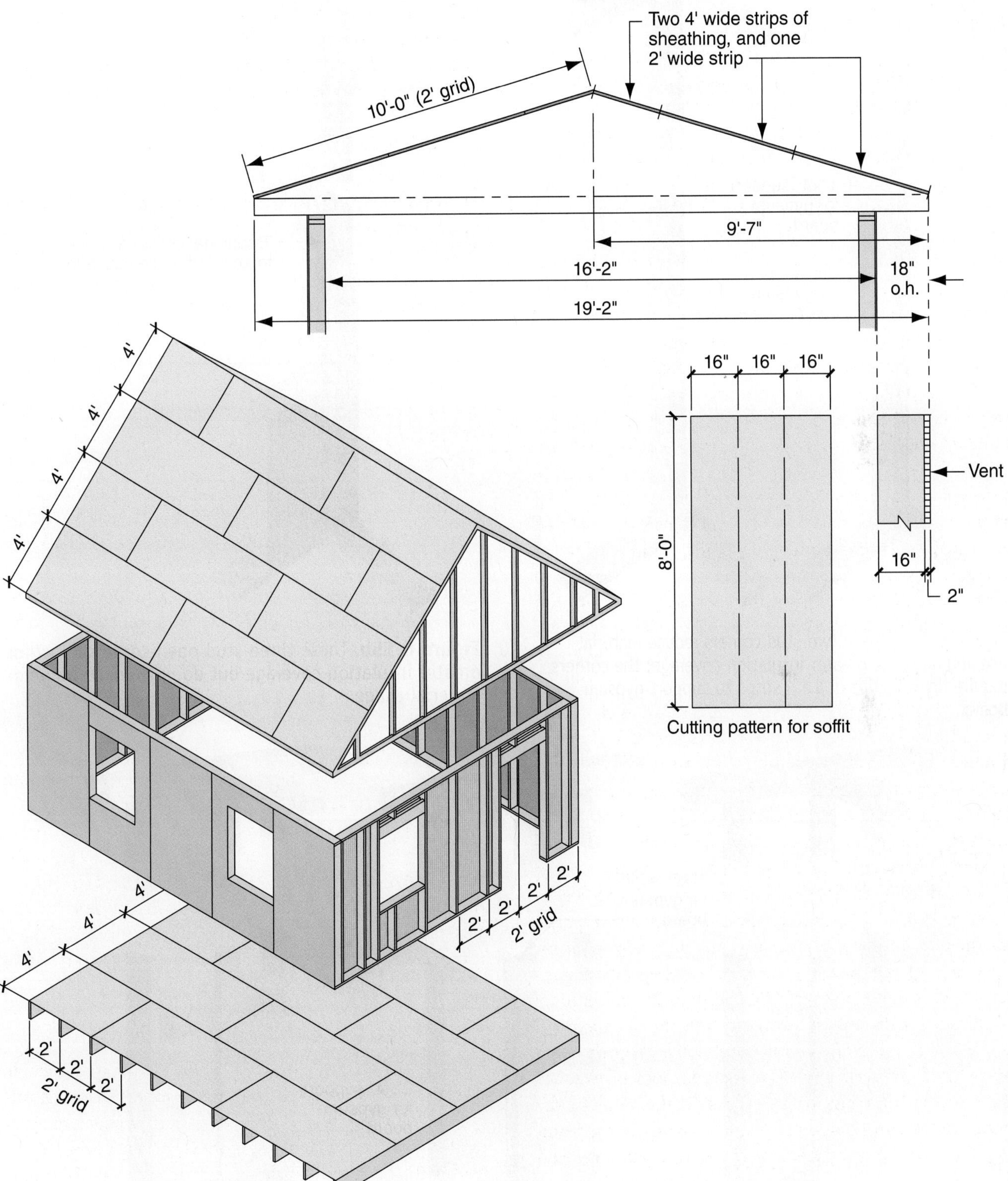

Figure 6.14 Lay out and cut framing and sheet goods to take advantage of the full dimension of the material and to reduce job-site waste.

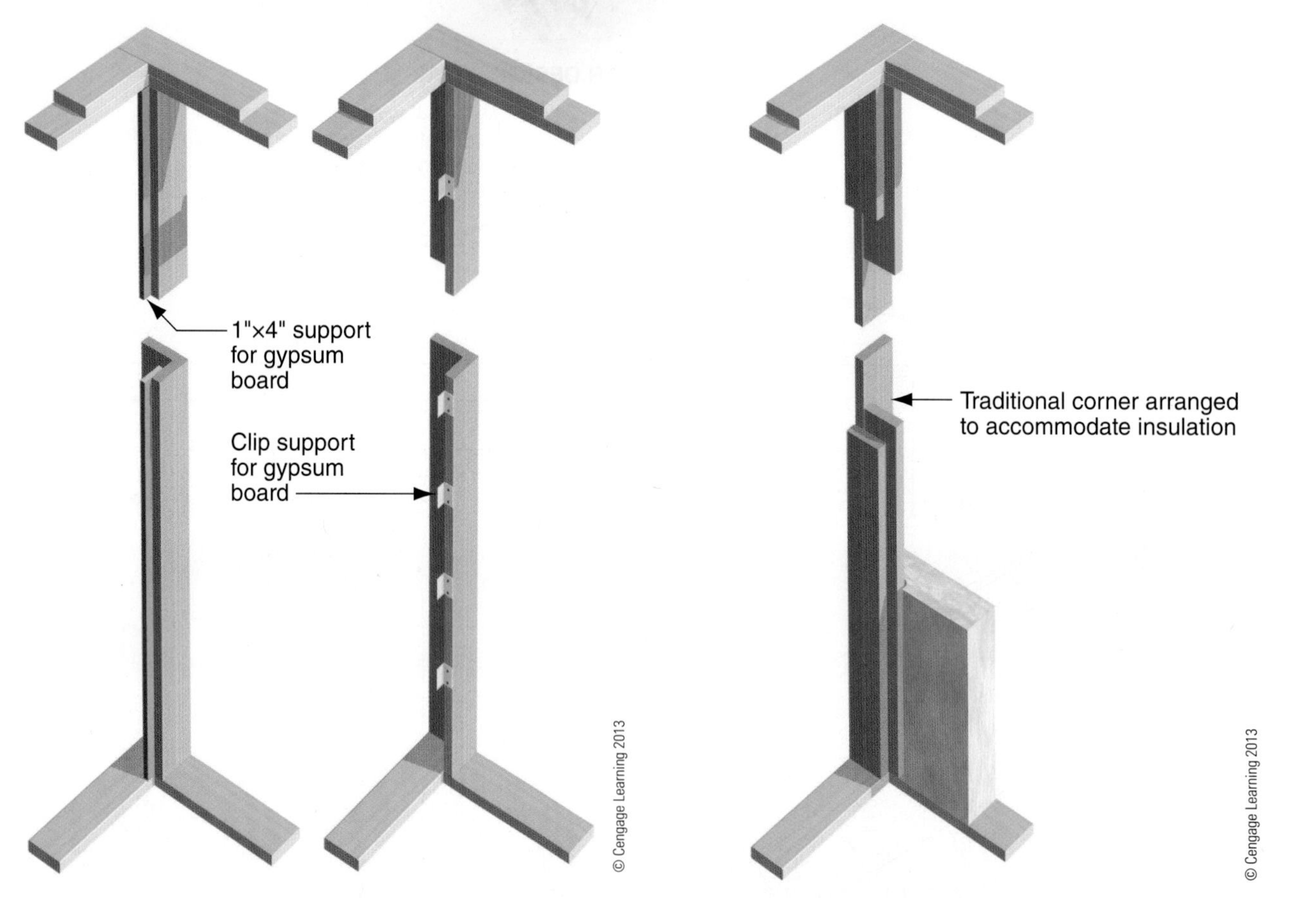

Figure 6.15a These two-stud corners reduce material use and allow for greater insulation coverage; the corners require drywall clips or 1×4 strips to support gypsum board.

Figure 6.15b These three-stud open corners allow for greater insulation coverage but do not reduce framing material usage.

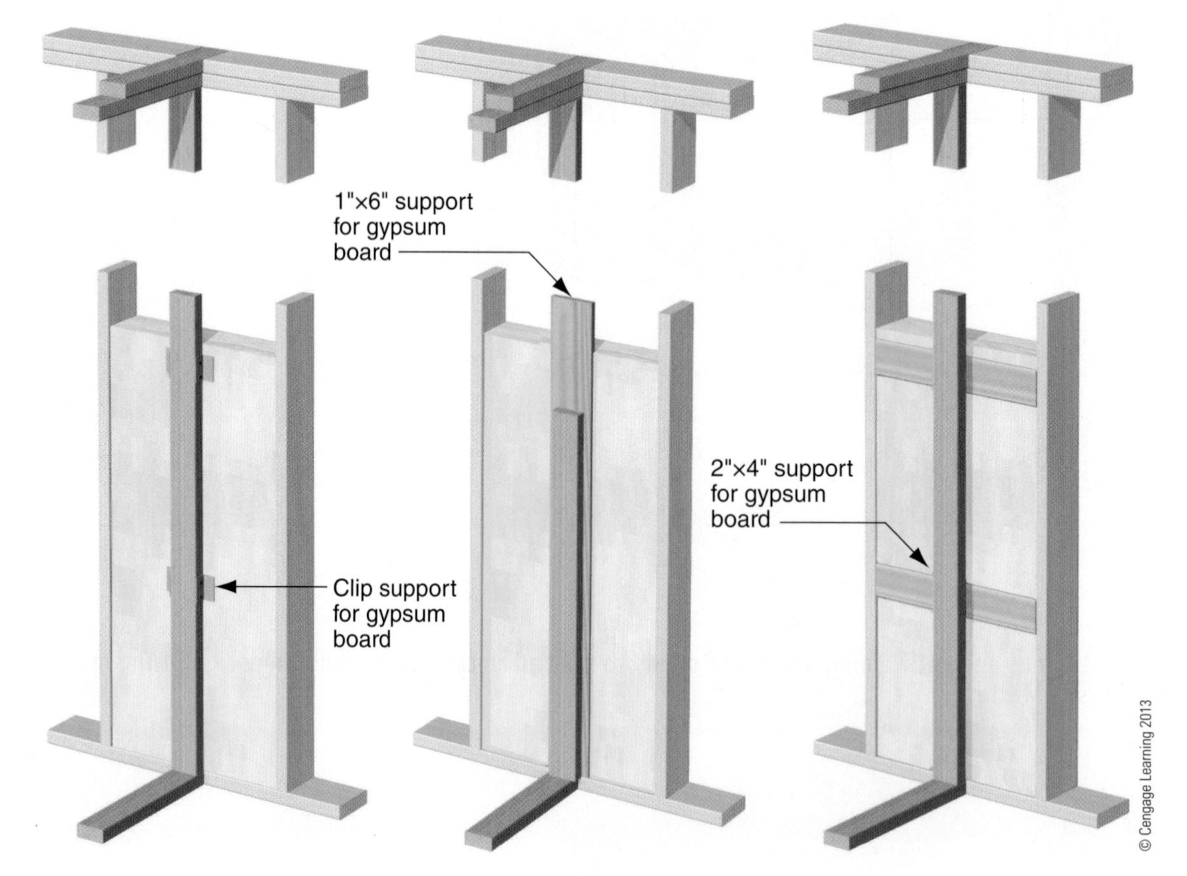

Figure 6.16 Ladder T-walls reduce material use and allow for greater insulation coverage.

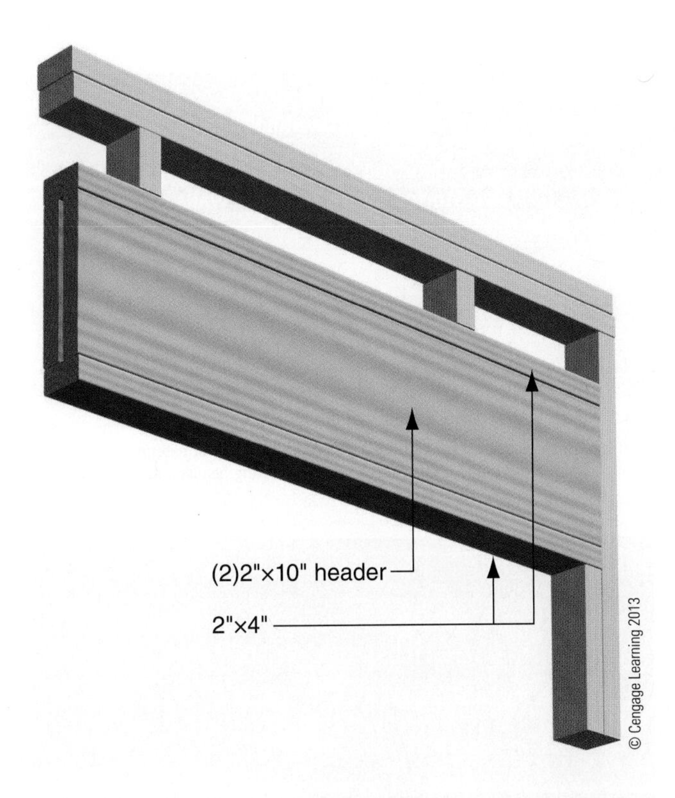

Figure 6.17 Headers made from double 2×10s are often oversized for actual loads, and the practice of connecting them with 2×4s at top and bottom wastes material and reduces available space for insulation.

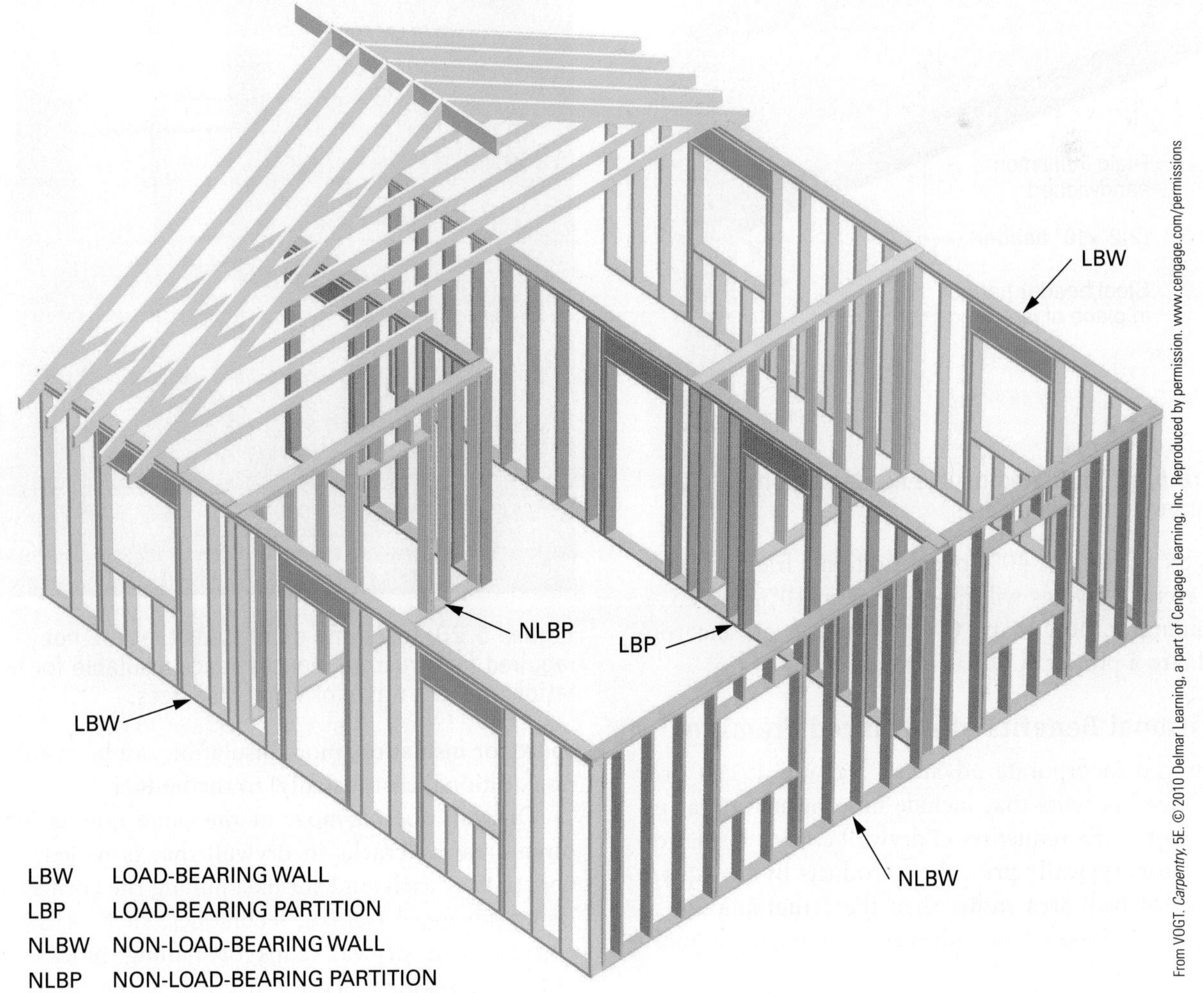

Figure 6.18 Headers are only located in load-bearing walls.

Table 6.3 Header sizing chart

Maximum Spans for Two-Inch Double Headers (in feet)[a] (Derived from Table 602.6 of the 1995 International Code Council's One- and Two-Family Dwelling Code)				
Header Size	**Supporting Roof Only**	**Supporting One Story Above**	**Supporting Two Stories Above**	**Not Supporting Walls or Roofs**
2×4	4	0	0	b
2×6	6	4	0	b
2×8	8	6	0	10
2×10	10	8	6	12
2×12	12	10	8	16

[a]Also applies to nominal 4-inch single headers. Based on No. 2 lumber with 10-foot tributary loads. Not to be used where concentrated loads are supported by headers.

[b]Load-bearing headers are not required in interior or exterior nonbearing walls., Single flat 2-inch-by-4-inch members may be used as headers in interior or exterior nonbearing walls for openings up to 8 feet in width if the vertical distance to the parallel nailing surface above is not more than 24 inches. For such nonbearing headers, no cripples or blocking are required above the header.

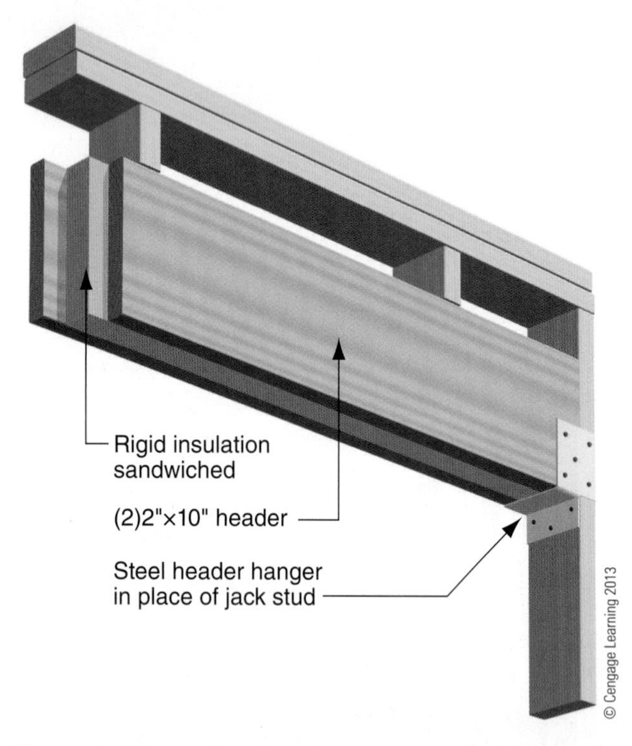

Figure 6.19 Insulated double header with rigid insulation board in the middle.

Figure 6.20 Installing extra studs that are not required for structure reduces space available for insulation and is a waste of wood resources.

costs by as much as 40% over traditional framing methods; average savings will range between 10% and 25%, depending on how many OVE techniques are incorporated into a project.[1]

Additional Benefits of Advanced Framing

When you incorporate advanced framing, it can provide "free" benefits that include more insulation at no extra cost and a reduction in drywall cracks. Insulation contractors typically price their products by the square footage of wall area rather than the actual amount of product they install. Since advanced framing leaves more space for insulation, more insulation can be installed at no additional cost (usually) to the builder.

Drywall doesn't move at the same rate as lumber, often creating cracks in drywall that is nailed to corner studs as each stud shrinks, pulling the corner apart. Using advanced framing techniques, such as two-stud corners with drywall clips or nailing blocks, helps reduces cracking since the drywall is not attached to framing that moves in opposite directions at the corner.

[1]Baczek, S., Yost, P., and Finegan, S. *Using Wood Efficiently: From Optimizing Design to Minimizing The Dumpster*. Boston, MA: Building Science Corporation, 2002.

The Pros and Cons of Steel Framing

Wood is much more common than metal when it comes to framing houses, but light-gauge steel studs outperform their wooden counterparts in a number of categories, such as recycled content, rot resistance, and reduced job-site waste. Steel framing has two major drawbacks however—high cost and much lower energy performance.

While there is much to be said in favor of steel framing, the key drawback is its poor thermal performance. Steel conducts heat much more effectively than wood, and even though the web is far thinner in cross section than 2× material, heat loss is significant. The California Energy Commission claims that a steel stud conducts 10 times as much heat as a wood stud.

This thermal bridging drastically lowers the performance of insulation in wall cavities. A 2001 study by the Oak Ridge National Laboratory (ORNL) found that thermal bridging in a conventionally framed wood wall lowers the performance of cavity insulation by 10%; in a steel-frame wall, performance drops by up to 55%. Simply making wall cavities deeper to compensate for this thermal short-circuiting is ineffective.

The ORNL research, however, looked at a number of possible solutions and found that steel-stud walls can perform as well as or better than similar wood-framed walls. The most common fix is to wrap the exterior of the building with rigid foam insulation, which provides a thermal break for the steel framing. A steel-framed wall wrapped with foam performs better than a steel-framed wall without foam—but a wood-framed wall wrapped with foam performs better still.

Other options cited in the ORNL report included using spacers to isolate the sheathing from the studs, and using foam-covered steel studs; both techniques interrupt heat conduction through the steel frame to the outdoors. Perhaps the best fit for steel framing is in mild climates, such as Hawaii, where steel's thermal performance is much less of an issue. In locations like this, the need to ship materials from great distances, along with moisture and insect pressures, give durable, lightweight steel framing an inherent advantage over wood.

Source: Reprinted with permission from *Green Building* Advisor. http://www.greenbuildingadvisor.com/green-basics/steel-studs

Stick Framing and Thermal Bridging

One disadvantage of wood-framed construction is thermal bridging (refer to Chapter 4, Insulation and Air Sealing) that occurs wherever low R-value framing members, such as studs and rafters, transfer heat between the interior and exterior at a much higher rate than thermal insulation. One method of reducing heat transfer is to install insulated wall sheathing, which acts as a thermal break at the framing. The most common insulated sheathing product is rigid polystyrene foam, although glass and mineral fiber alternatives are available. Another method to limit thermal bridging is to build double-framed walls (Figure 6.23) with a thick layer of insulation filling the entire cavity (refer to the "From Experience" sidebar in Chapter 4).

While eliminating thermal bridging is desirable in all climates, it provides the most benefit in colder regions that have the biggest temperature differential (referred to as Delta T or ΔT) between the inside and outside. Most hot climates have few days over 90°F or 100°F, and even on those days, the ΔT is usually between 15°F and 20°F because summer interior temperatures are usually maintained between 72°F and 78°F. Cold climates often have a ΔT between 30°F and 75°F as the outside temperatures are often well below freezing while interior temperatures are kept between 68°F and 72°F.

Wall Sheathing

Stick-framed walls all require sheathing on the exterior face to serve as a base for a water-resistive barrier (WRB) and the finished siding materials. The WRB is the moisture-resistant barrier between the exterior wall finish and the structure that, when combined with flashing, directs water down and away from the structure (WRBs and siding will be addressed in detail in Chapter 9). Many early stick-framed structures had no sheathing; instead, the siding was attached directly to the studs. The first structural sheathing was made of individual boards, usually placed diagonally to provide resistance against racking.

Structural Wall Sheathing

Plywood, the original engineered building material, is made from thin layers of wood veneer that are glued together. Plywood began to replace solid sheathing following World War II and is still in use today, but other modern materials make up most of the exterior sheathing currently used in new home construction.

Oriented-strand board (OSB) is the most common structural wall sheathing currently in use. An engineered material made of rectangular wood strands that are

---

oriented-strand board (OSB) an engineered lumber product that is often used as a substitute for plywood in the exterior wall and roof sheathing.

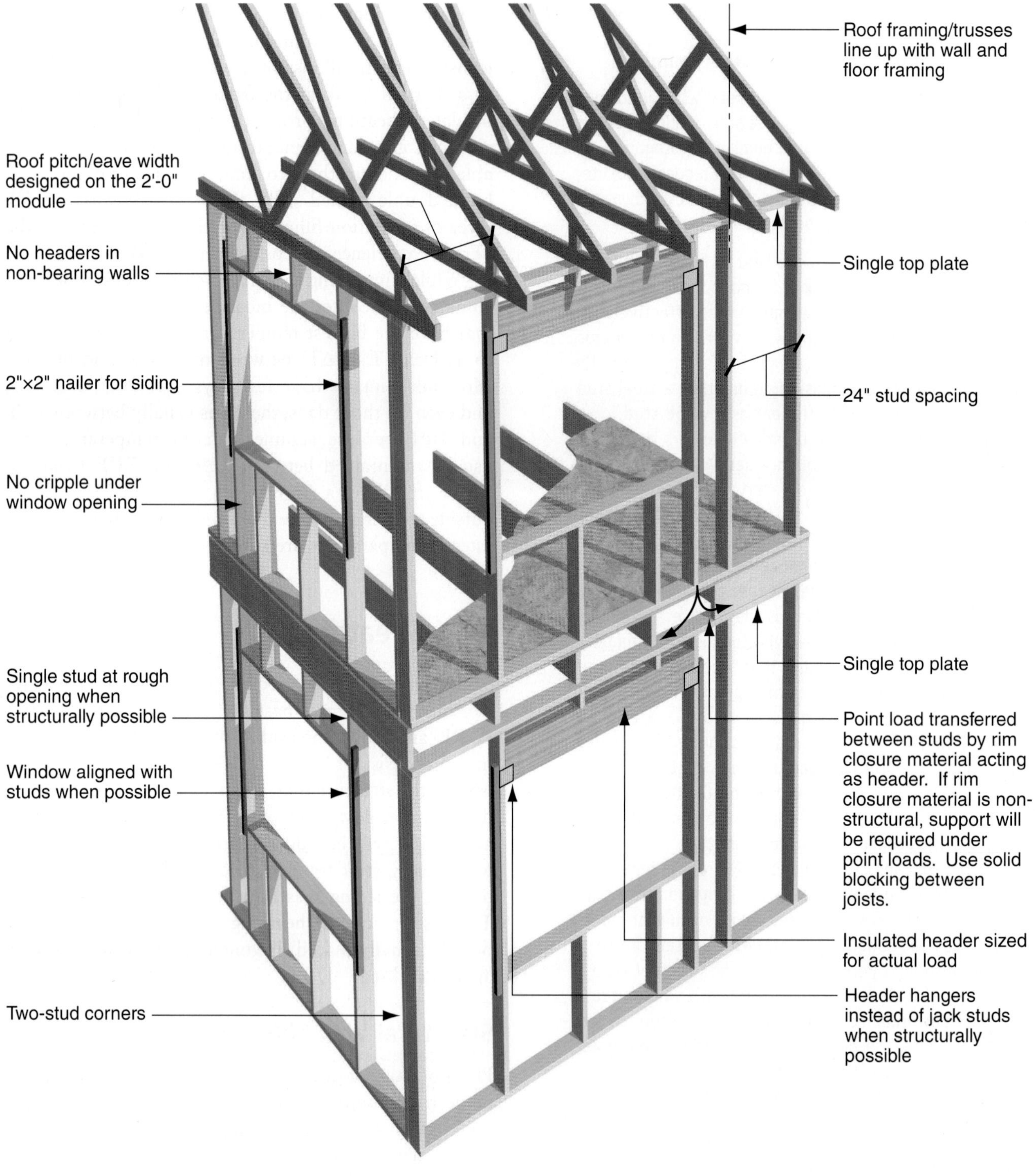

Figure 6.21 Advanced framing techniques.

arranged in cross-oriented layers and pressed into sheets, OSB wall sheathing is the same material that is used in the webs of I-joists. The manufacturing technique used to produce OSB is closely related to other engineered wood products.

I-joist a structural building component consisting of a wide vertical web of OSB with engineered wood flanges at the top and bottom. I-joists can be used for floor, roof, and wall framing.

Many different structural sheathing materials are available for residential construction. Gypsum-based products include GP DensGlas Silver™, an inorganic, mold-resistant, noninsulative structural sheathing (Figure 6.24). Magnesium oxide (MgO) boards, which are water- and fire-resistive and can be used as both interior and exterior finishes, are not common in the United States but are gaining acceptance (Figure 6.25). MgO products include DragonBoard and Magnum

Figure 6.22 Under this window, the two outside cripples are unnecessary.

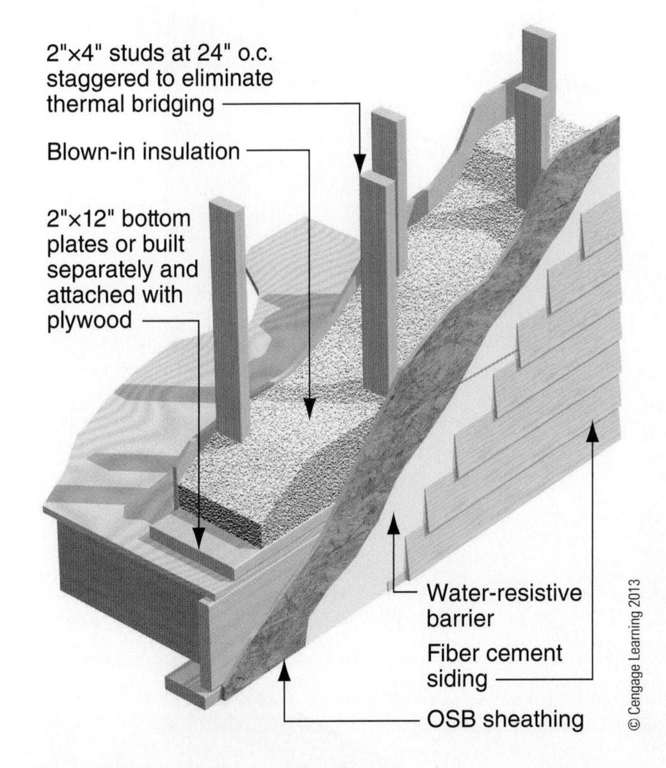

Figure 6.23 Double-stud walls contain two parallel 2×4 walls. The studs are staggered to eliminate thermal bridging.

Board®. Dow's structural insulated sheathing (SIS)(see Chapter 4), and Huber ZIP System® walls are relatively new products that are expanding their market shares in residential construction. These two products can serve both as water-resistive barriers and as structural sheathing. Other structural sheathings, unless designed with an integrated moisture-resistant facing, must have a separate WRB applied to keep bulk moisture out of the building structure.

Figure 6.24 Gypsum-based structural sheathing.

Figure 6.25 Magnesium oxide (MgO) board used for exterior wall sheathing and roof decking.

Insulated Sheathing

With the exception of SIS and similar products, insulated sheathing does not provide structural integrity to walls and must be combined with OSB, plywood, or recessed diagonal bracing to maintain the building structure. As discussed in Chapter 4, insulated sheathing is available in polystyrene, glass fiber, and mineral fiber alternatives and provides a thermal break at all framing members, helping to reduce heat gain and loss through walls.

Insulated Sheathing in Combination with Structural Sheathing

One option for using insulated sheathing is to install structural sheathing at only the corners of the structure where needed, which is typically less than 25% of the wall area (Figure 6.26a). If 0.5" sheathing is desired on the entire house, insulated sheathing is installed flush with the structural sheathing on the remaining walls (Figure 6.26b). Other options include using 1" or thicker insulated sheathing on all walls except the

Figure 6.26a Exterior sheathing is replaced with rigid insulation wherever structural requirements permit.

corners, with 0.5" structural sheathing where a layer of insulated sheathing is installed over the structural panels in a thickness to bring it flush with the rest of the wall (Figure 6.26c). Finally, if the wall is built with recessed (or let-in) diagonal bracing, a full layer of insulated sheathing can be installed on all framed walls.

Fiberboard

Another traditional wall sheathing, fiberboard, also known as blackboard or buffaloboard, is available under the trade names of QuietBrace® by Temple-Inland and Stedi-R® by Georgia-Pacific Corporation, among others (Figure 6.27). Fiberboard is made of wood fibers (with

fiberboard is an engineered lumber product used primarily as an insulating board and for decorative purposes, but may also be used as wall sheathing.

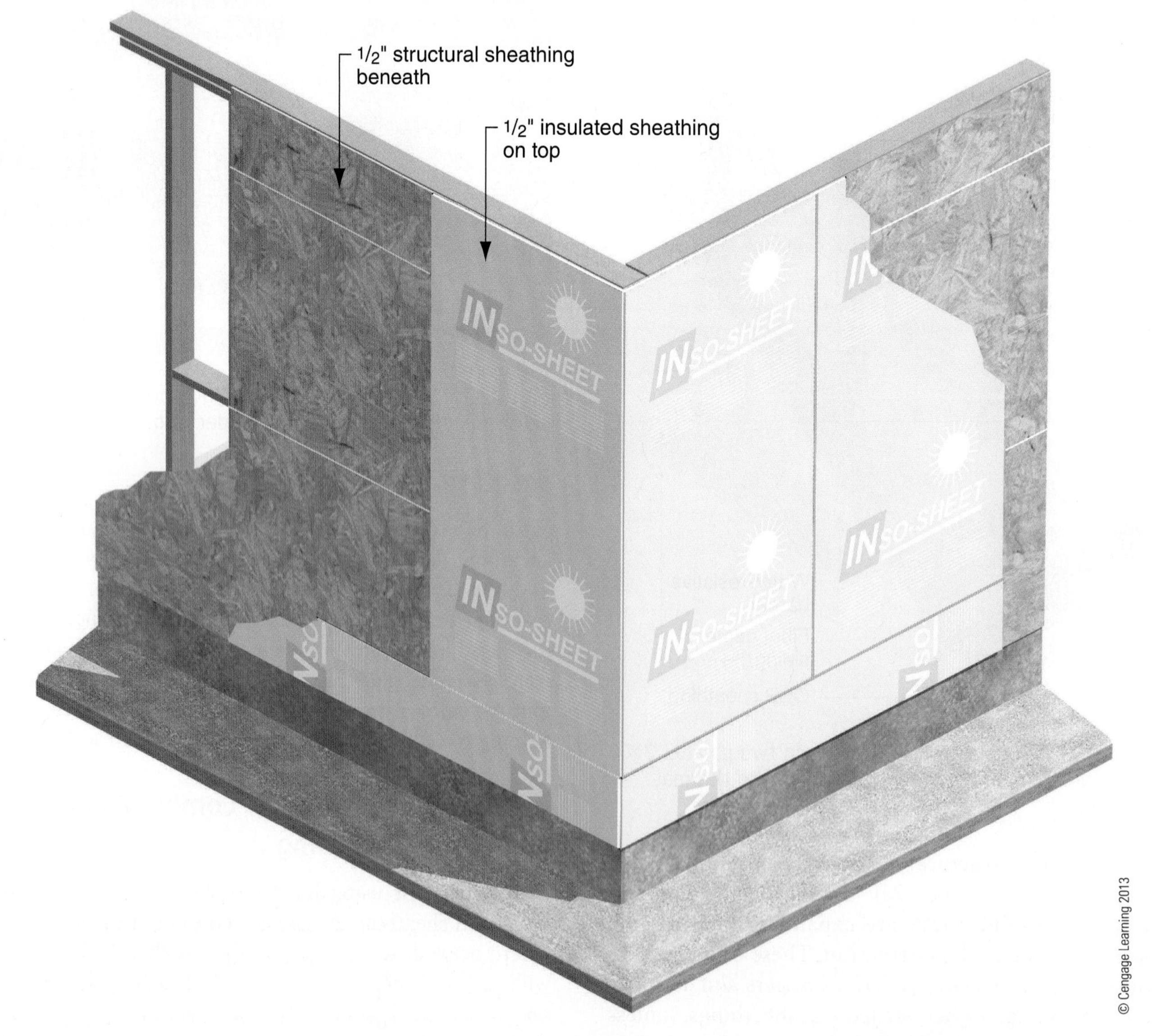

Figure 6.26b The walls are completely wrapped in 0.5" structural sheathing and 0.5" insulated sheathing over top.

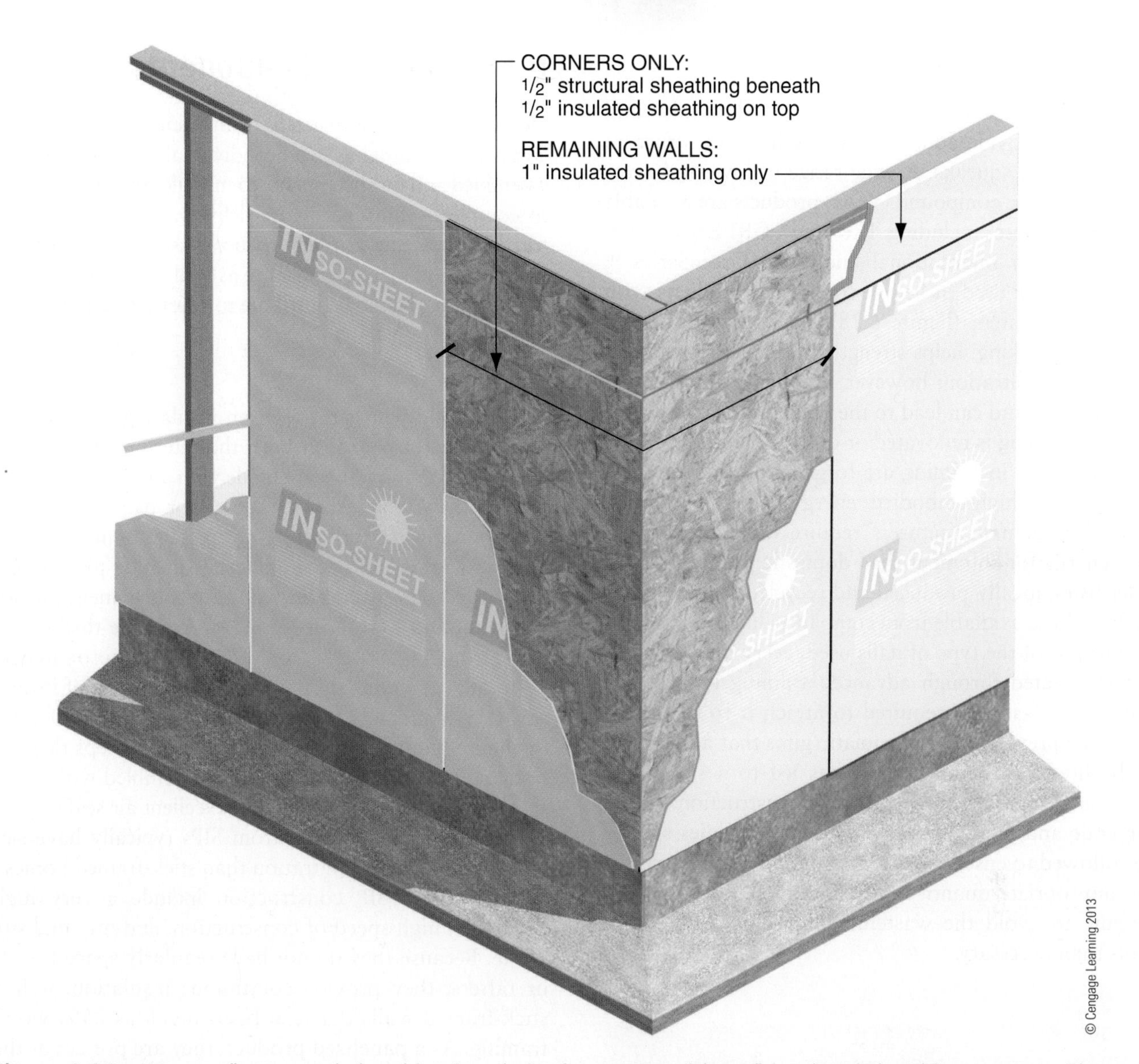

Figure 6.26c Here, 0.5" structural sheathing is only in the corners with 0.5" insulated sheathing over top. The rest of the wall contains 1" insulated sheathing to bring it flush with the corners.

Figure 6.27 A home sheathed with fiberboard.

as much as 85% recycled content) and formed with binders into sheets; an asphalt-based water-resistive coating may also be applied. Available in structural or nonstructural panels, fiberboard is a suitable low-cost alternative to plywood and OSB, with a higher R-value and sound-deadening properties.

Floor Decking

Structural floor decking products include plywood, OSB panels, and, less commonly in modern construction, solid lumber. Improved adhesives in panel products provide excellent moisture resistance, maintaining structural integrity and dimensional stability after wetting and drying events during construction.

Adhesives and Fasteners

Adhesives are commonly applied between floor joists and decking to provide additional strength to the structure and to reduce squeaks. A wide range of low- and zero-volatile organic compound (VOC) products are available for this purpose, including Titebond® GREEN*choice*™ from Franklin International and OSI® GreenSeries™, and should be used in lieu of traditional products with high VOC content (Figure 6.28). The use of adhesives reduces squeaking, helps strengthen a structure and can reduce air infiltration; however, they make disassembly very difficult and can lead to the creation of more waste when a building is renovated or deconstructed.

Nails used in framing are forged from molten steel and have a high-embodied energy. Nails purchased from foreign manufacturers require more transportation energy for shipment than domestic products. Consider using locally produced and recycled content nails, such as those available from companies like Maze Nails. Regardless of the type of nails used, every piece of lumber eliminated through advanced framing reduces the number of fasteners required to attach it to the structure. The prevalence of pneumatic guns that allow nails to be fired easily and quickly has led to a significant overuse of fasteners in wood frame construction. Building code and manufacturers' fastener schedules should be followed to ensure structural integrity and determine the appropriate quantity of nails for each application, helping to avoid the wasteful practice of using more nails than necessary.

Figure 6.28 A wide range of low- and zero-VOC adhesives and sealants are available for construction projects.

Manufactured Components

Homes can be constructed from factory-manufactured components, ranging from individual panels that are assembled and finished on-site to modules that are partially to completely finished before delivery. Factory manufactured components can help reduce waste, improve energy efficiency, and manage moisture as well or better than site-built systems, helping contribute to a home's sustainability.

SIPs

SIPs are solid panels made of an insulating core bonded to a structural board on both the interior and exterior. The most common SIPs are made with an expanded polystyrene (EPS) core with OSB exterior panels, although polyisocyanurate (polyiso) and polyurethane SIPs are also available. Alternative materials include plant-based structural boards and non-wood exterior panels, including metal and MgO (Figure 6.29). Available thicknesses range from 4" to 12.25", with R-values of approximately 14 to 58, depending on the thickness and type of insulation core. SIPs can be used for floors, walls, and roofs, and the limited number of structural members keeps thermal bridging to a minimum. Panels are assembled with glued splines and screws, providing an excellent air seal (Figure 6.30). Homes constructed from SIPs typically have significantly lower air infiltration than stick-framed homes.

Benefits of SIP construction include a very high R-value, a high speed of construction, and minimal site waste. Because they do not have regularly spaced studs or rafters, they provide continuous insulation, unlike stick-framed walls that can be as much as 25% wood framing. As a panelized product, they are pre-cut at the factory to the proper size, including all door and window openings. When planned properly, SIP buildings have virtually no job-site waste in the framing stage.

Structural Considerations

While SIPs have inherent structural capabilities, load-bearing points must be determined in the design stage and, where required, posts are factory-installed to carry beam loads through the walls to the foundation.

Figure 6.29 SIPs with MgO board on the interior and exterior.

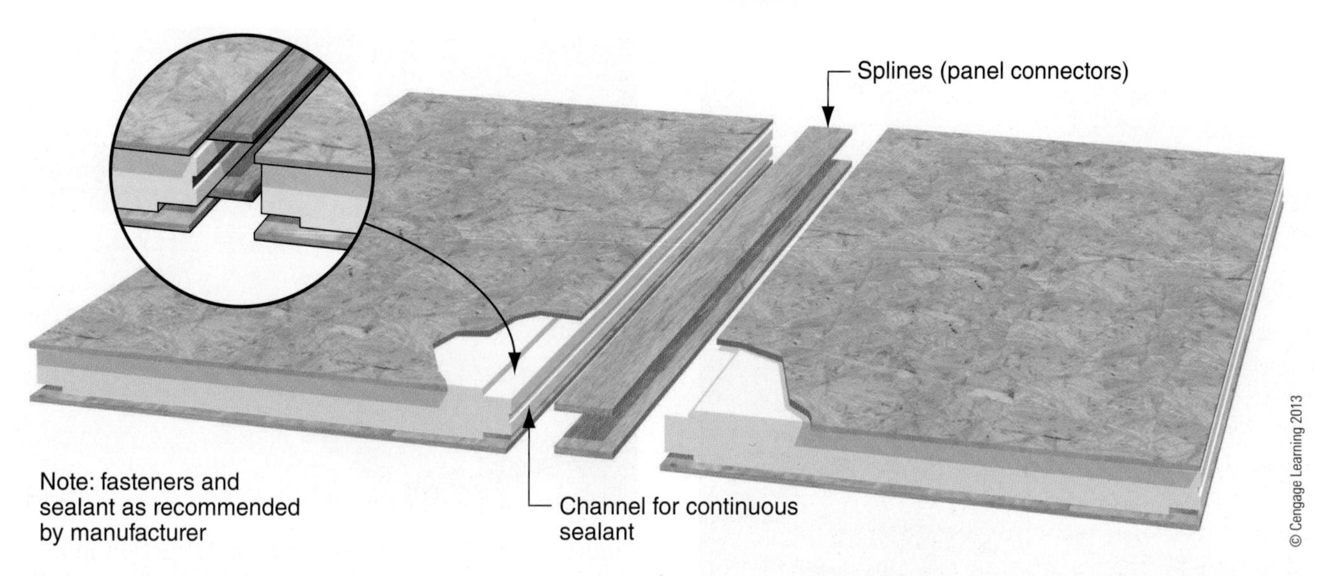

Figure 6.30 SIP wall assembly using splines as connectors. SIPs must be carefully air sealed to prevent air leakage and moisture problems.

Cautions and Challenges As a panelized product, SIPs demand careful and thorough construction planning. The foundation must be built accurately and sized to fit the panels to avoid any major site adjustments that would add both cost and time to the process. Plans must account for the cost and accessibility of the equipment needed to lift the panels into place. While the panels contain precored channels for wiring and plumbing, electricians and plumbers without experience in SIP projects will have to learn new installation techniques. Finally, there have been studies, particularly in cold climates, where poorly sealed joints between panels can cause excessive moisture buildup that promotes mold growth and, in some cases, causes structural problems with the panels.[2] As with any wood construction, homes constructed with SIPs require careful air sealing and the installation of a complete and effective WRB to ensure high performance.

Panelized Construction

Panelized homes combine techniques of both traditional framing and mass production to reduce costs, improve quality, shorten the construction cycle, and minimize weather damage to structural components. Panelized homes typically consist of factory-built wall, roof, and floor sections that are assembled on-site. Advantages of panelized construction include quality control, completeness of the framing package, speed of assembly, and waste reduction. Product quality is managed through the factory process where panels and other components are assembled, cut, and packaged in climate-controlled conditions, limiting the effects that weather and unknown field conditions have on the project. A panelized house package arrives with all the panels, blocking, fasteners, and miscellaneous lumber required for assembly. These systems minimize wasted time and money associated with running to the store for missing parts throughout the framing process. Floor, roof, or wall sections are delivered to the site and lifted into place with cranes. They are then assembled by field crews, saving time and practically eliminating site waste by reducing the amount of cutting required on-site.

Components of panelized homes can include trusses, I-joists, SIPs, precast concrete, and stick-framed walls. Wall panels are almost always factory-assembled for panelized projects, although floors and roofs may be delivered either assembled or with precut trusses and sheathing ready for field assembly (Figure 6.31).

Modular Construction

Modular homes are built with modules or "boxes" that are constructed and substantially finished in a factory, then delivered to the job site where they are assembled (Figure 6.32 and Figure 6.33). Modular homes are different than manufactured housing, commonly referred to as mobile homes or trailer homes. Compliant with federal HUD building standards, **manufactured homes** are typically fully finished in the factory and delivered on a permanent steel chassis. Manufactured homes are not always installed on a permanent foundation.

[2]Lstiburek J. SIPA Technical Report: Juneau, Alaska, Roof Issue, 2001.

manufactured homes buildings that are fully finished in the factory and delivered on a permanent steel chassis.

(A)

(B)

Figure 6.31 (A) Wall panels being assembled in a factory and (B) installed on-site.

In comparison, modular homes are partially finished in the factory, delivered on truck beds, conform to local building codes, and are installed on permanent foundations. Both modular and manufactured homes can meet green building standards.

Figure 6.32 Home modules under construction in a factory.

Figure 6.33 Home module installation on-site.

Modular homes are typically built using standard wood-frame construction methods. Multiple boxes can be assembled into homes of almost any size, including multi-unit buildings (Figure 6.34a and Figure 6.34b). The level of factory completion ranges from as low as 50% to as much as 90%, with the remaining work being completed on-site. At a minimum, the boxes are framed, insulated, wired, plumbed, and sheathed, and drywall, windows, and doors are installed. Additional work that may be completed in the factory includes the installation of cabinets, tile, flooring, and siding, depending on the details of the particular project. Foundations, utility connections, siding, exterior trim, and roofing are completed on-site. A typical modular home is manufactured in one week and, after delivery, is assembled and dried-in on-site within one or two days. According to NAHB, approximately 10% of all single-family homes in the United States are modular construction.

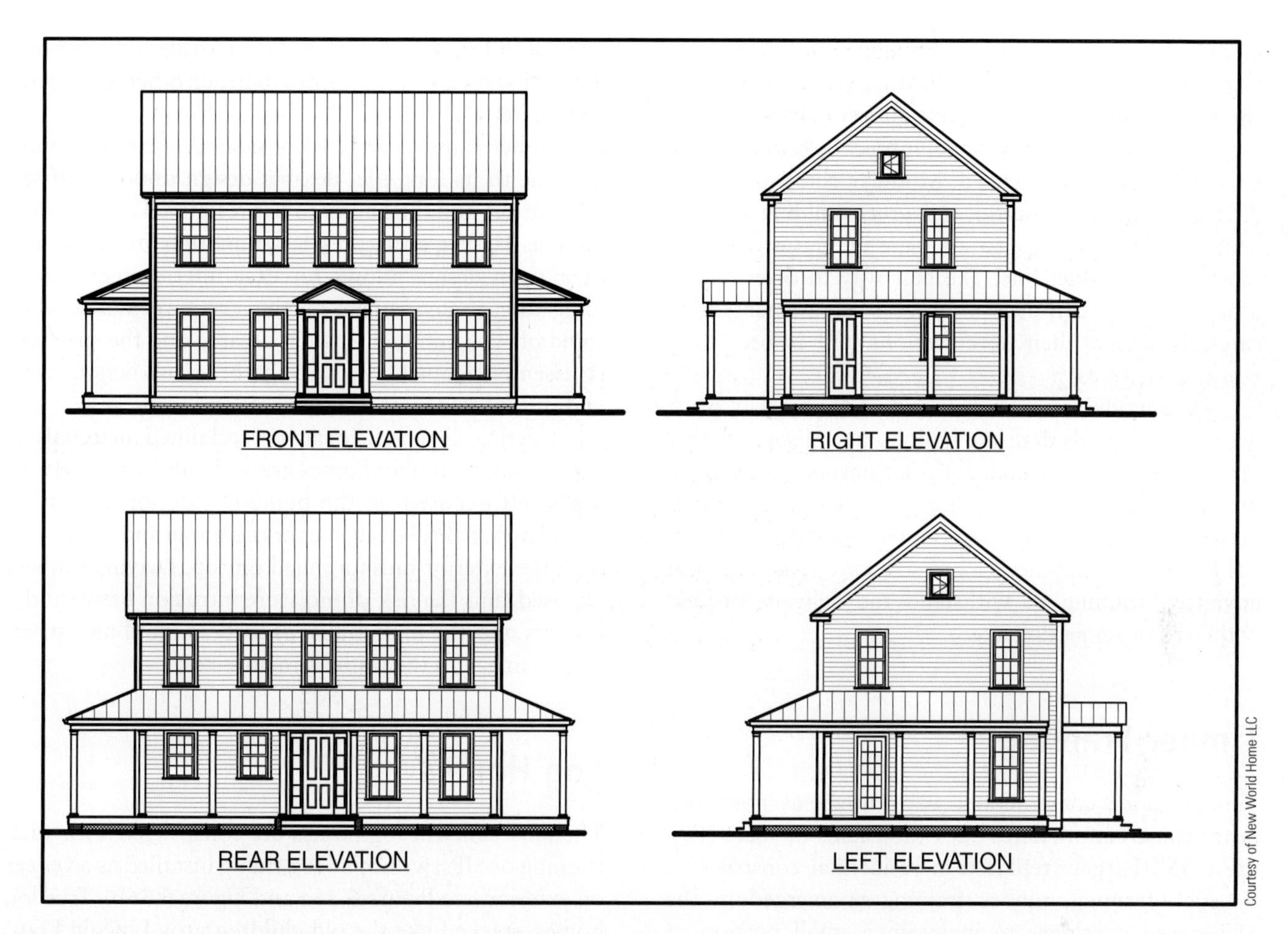

Figure 6.34a Once assembled, modular homes look no different from traditional site-built homes.

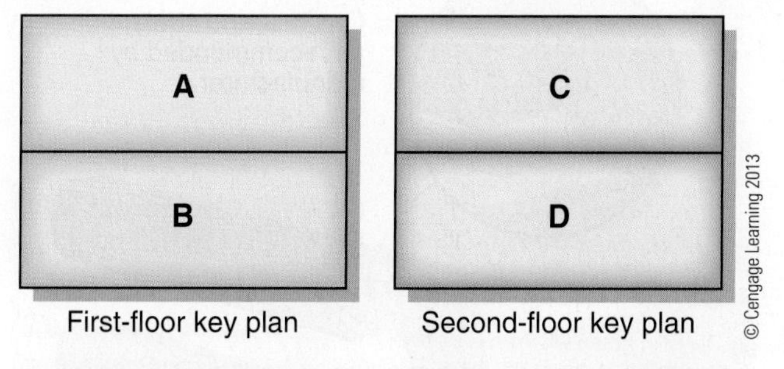

Figure 6.34b This home consists of four modules or boxes that are shipped to the construction site. Each letter corresponds to a specific module. The attic and roof are integrated into the second-floor modules, and the porches are built on-site.

Modular manufacturers offer a wide range of products, including standard designs with no available options, basic design packages with a wide range of options, custom designs, and construction to customers' own plans. Homes range in size from small cottages to mansions of almost any size. High-quality modular homes are often indistinguishable from site-built homes.

Modular Homes as a Green Building Strategy

Modular homes help minimize site-generated waste, and sophisticated manufacturing processes can optimize material use in the factory. They are much less susceptible to moisture damage and associated mold problems than site-built homes because the boxes are constructed in a factory and the site assembly time is only days instead of weeks or months to dry-in. Quality control is generally excellent; compared with site-built homes, modular home construction is less likely to be affected by such uncontrolled factors as weather, labor shortages, and missing materials. Many modular companies design their homes to meet the criteria of ENERGY STAR, LEED for Homes, the National Green Building Standard, and other local green programs, providing the builder with documentation of available program points for certification.

Modular homes can be green, but the builder or owner must confirm that they are being built to the desired specifications. Since the boxes arrive on-site prefinished, there is no opportunity to inspect insulation and thermal bypasses on-site; this must be done in the factory. Factories may perform their own pre-drywall inspections; however, just as third-party inspections for site-built homes provide the best quality assurance, having an independent professional inspect boxes in the factory is highly recommended. Modular companies have different levels of commitment to green building, ranging from little or no interest to complete commitment across their entire product line. The best way to determine if a manufacturer can meet a project's requirements is to carefully review their specifications and inspect their work in the factory.

While modular homes help reduce waste, they often use more materials than site-built homes. Since each box must have a bottom and top for transport, multistory buildings have separate ceilings and floors, and walls between adjacent boxes are double thickness. Individual boxes also may need more structure than site-built advanced framing to withstand the delivery process without sustaining damage.

Timber Framing

Timber frame construction, also known as post-and-beam construction, dates back thousands of years (Figure 6.35). Largely replaced in residential construction by stick framing, timber framing re-emerged in the 1970s and continues to maintain a small portion of the construction market. This type of construction uses large timber posts supporting beams and rafters, usually assembled using mortise and tenon joints, although metal plate connectors may also be used. Timber framing requires specialized skills and heavy equipment to install the structure. Most timber framed homes leave the structure exposed as a decorative element on the interior, filling the spaces between structural members with SIPs, stick framing, straw bale, or other alternative wall materials.

Timber framing can be considered a green technology due to its long life, flexible design options, off-site fabrication, and the ability to use reclaimed and certified lumber. Timber structures last hundreds of years, and their open design allows for easy interior renovations with little, if any, impact on the structure. When fabricated off-site minimal waste is created and the structure is assembled quickly, reducing the construction cycle and the risk of water damage before dry-in.

Selecting certified, local, or reclaimed materials is key to making timber homes green. Timbers are not normally left exposed on the building exterior; instead, a complete WRB is installed over the frame and infill walls with the exterior finish applied on top. Leaving timbers exposed runs the risk of moisture intrusion between the timbers and the infill walls, as well as mold and structural damage to the timbers themselves.

Figure 6.35 An example of a timber-framed home. The structural elements are larger posts, beams, and girders. Connections are typically bolted-type joints.

Log Homes

The most efficient log homes are constructed with stick framing or SIPs, with split-log siding installed as a veneer on exterior (see Figure 6.36a and Figure 6.36b). True log homes, stacked like the old children's toy Lincoln Logs,

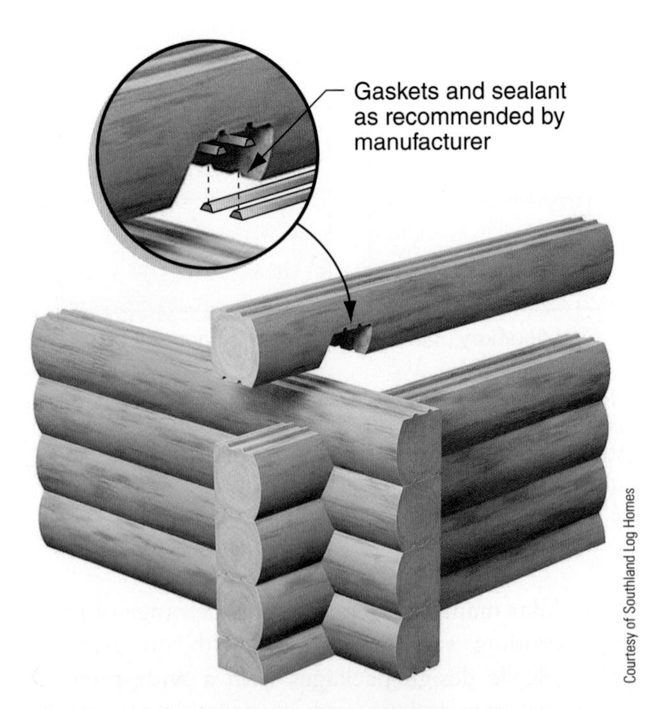

Figure 6.36a Saddle notch logs contain double tongue-and-groove white pine logs with a flush corner system in which logs are notched from the bottom as opposed to the side.

Figure 6.36b A home built using a saddle notch log has the appearance of a traditional western log home. Logs extend just beyond their intersection and create a sealed corner that makes a solid visual statement.

can be considered green if they are made from sustainable wood and are fully air sealed and insulated. Unlike timber frame homes that are sealed and insulated on the exterior, the solid logs are left exposed on the exterior. Solid log homes that are also exposed on the interior provide only a minimal amount of insulation, and air sealing all the joints between logs can be a challenge. Installing a full layer of insulation and air sealing on the interior of a log home will dramatically improve the efficiency; however, the aesthetic value of the exposed logs is lost.

Because the wood is exposed on the exterior, water management is critical to creating a durable log home. The foundation must be designed to keep the bottom of the walls above grade to prevent termite infestation and to keep water splashing on the ground from damaging the logs. Large overhangs help protect the log walls from additional water damage. Logs can shrink as much as 0.5" per 1" of thickness, so allowances must be made for window and door openings. Sealants must also be able to withstand these movements to maintain an effective air seal at openings. Log homes do not have a water-resistive barrier, so the individual logs must be finished and maintained regularly with a durable wood sealer. The joints between logs must also be filled with a flexible sealant that will hold up against weather and wood shrinkage and expansion.

Unit Masonry

A construction strategy that has been around for centuries, solid masonry is resistant to fire, termites, storms, and moisture damage. Masonry can serve as thermal mass to help maintain interior temperatures; however, traditional materials provide very little insulation. Masonry walls should be sealed on the exterior but are hygroscopic (meaning they can take on and give off some amount of water vapor in response to humidity changes). Consequently, they are not easily damaged from moisture intrusion. Since some rain entry through the face sealing is expected, unit masonry walls serve as a storage reservoir and allow both bulk moisture and vapor to move in and out of the wall as conditions change.

In cooling and heating climates, masonry buildings must be insulated. Traditional concrete masonry unit (CMU) construction is insulated on either the interior or the exterior surface, usually with rigid foam. Open cells in the block can be filled with insulation, but this allows for thermal bridging at the webs of individual blocks. Niagara NRG™ blocks are two-piece interlocking blocks sandwiched around a layer of foam that provides continuous insulation (Figure 6.37). These insulation options for CMU walls can pose a problem in some climates by creating an undesired vapor diffusion retarder in the wall structure. As moisture in walls may dry to either the interior or exterior (depending on temperature, humidity, and heating and cooling systems), vapor diffusion retarders should be avoided in most climates to prevent trapping moisture in the wall and creating condensation (see Chapter 2 for more information).

One masonry product that provides both moisture management and insulation is autoclaved aerated concrete (AAC). This lightweight cement product is filled with tiny air bubbles that provide a high insulation value. Commonly known as Hebel block, AAC was developed in the 1920s and became popular in Europe, but it has not yet achieved deep market penetration in the United States. AAC is available in blocks and as large panels, which are stacked with a thin mortar to build walls. The product can be cut, routed, and

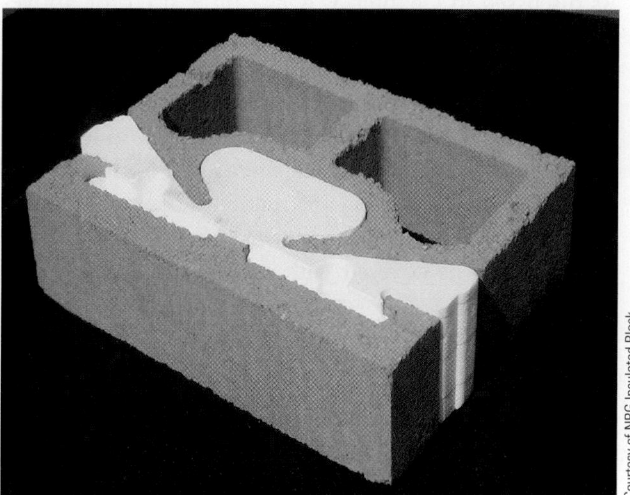

Figure 6.37 An NRG™ block.

Figure 6.38 AAC can be easily cut on-site without specialized equipment.

shaped with power tools, and blocking and framing can be nailed to completed walls (Figure 6.38). Resistant to moisture intrusion due to tight mortar joints, AAC is not affected structurally from moisture that does enter the wall, and it can help temper humidity by absorbing and releasing vapor from the interior. It can also serve as thermal mass to help manage interior temperature swings.

Walls are typically reinforced with grouted vertical steel reinforcement through holes that are drilled into the blocks and with bond beams at openings and across the top of walls. AAC construction must have an exterior finish applied, such as stucco or siding, and the interior can be covered with stucco, clay plaster, or drywall.

ICFs

As discussed in Chapter 5, ICFs can be used for above-grade walls as well as for foundations. With a high R-value and no thermal bridging, ICFs are not subject to termite or water damage, and they are storm resistant. ICF homes are both energy efficient and very quiet; however, they have high embodied energy due to the amount of concrete used in their construction. ICFs can be made from virgin or recycled foam or from mineralized wood, and they can be covered with a wide range of exterior finishes.

Concrete Slab Floors

Slab-on-grade floors can provide thermal mass for passive solar applications and can reduce material use by serving as the finished floor when protected with a sealant or other coating. Slab floors are appropriate for traditional wood-framed walls, ICFs, SIPs, and alternative systems, such as straw bale and rammed earth; they make an excellent choice for homes with hydronic radiant heating. More information about slab floors can be found in Chapter 5 (Foundations) and Chapter 16 (Renewable Energy).

Pre-Cast Concrete Walls

Pre-cast concrete products, such as Superior Walls® (see Chapter 5), can also be used for above-grade walls. The panels can be left exposed for a contemporary look or have almost any type of exterior finish applied. They have many of the same attributes as unit masonry and ICFs, including storm and sound resistance, but use fewer materials. While heavy equipment is required to set the walls in place, installation time is much faster.

Alternative and Natural Wall Construction Methods

Many ancient construction methods using natural materials are used in green building. Considered natural because few, if any, heavily processed materials are required, these methods include straw bale, light straw clay, cob, rammed earth, and adobe. These approaches are typically used only for walls; floors are generally slab-on-grade (although other methods can be used). Some of these construction materials are not directly addressed in the building code, so special approval from local building officials may be required for their use.

Straw Bales

Straw bales are made of wheat straw, an agricultural by-product that is normally burned or composted (Figure 6.39). Straw bale walls can be used as infill in a timber frame structure or as complete structural walls. The stacked bales are normally covered on both sides with clay plaster or stucco, creating a very airtight structure. Due to their thickness, straw bales have R-values as high as 33, and the surface coating provides some thermal mass. They are very susceptible to moisture that can cause them to mold or rot, so standard moisture-protection methods (e.g., deep overhangs and keeping the bales above grade) should be employed. Straw bale infill walls are installed after the roof is in place, protecting them from rain. The use of straw bale structural walls

Courtesy of David Eisenberg, Development Center for Appropriate Technology

Figure 6.39 Although straw bale construction is not commonly used, it does provide many benefits.

is generally limited to dry climates due to the danger of rain events before the roof is installed, which could cause structural failure and mold in the bales.

Cob

Cob is a combination of natural clay and straw that is mixed together and built in a free-form style into structural walls. Light straw clay, similar but lighter weight and looser than cob, is installed between forms so it retains its shape while it dries. Light straw clay is generally not considered to be a structural product, so it must be installed between timbers or other structural members (Figure 6.40).

Rammed Earth

Rammed earth walls use locally gathered soil, often combined with some Portland cement that is compacted into removable forms with the aid of hand tools or power machinery (Figure 6.41a). Once set, rammed earth walls are very durable and can be plastered, oiled, or finished on the interior with traditional materials. Old automobile tires may also be used as permanent rammed earth forms. Rammed earth has a low R-value, but the overall thickness of the walls provides some tempering effects due to the thermal mass. Exterior or interior insulation can be installed to improve efficiency (Figure 6.41b).

Courtesy of Brian "Ziggy" Liloia

Figure 6.40a The home builder applies cob by hand to his home in the Dancing Rabbit Ecovillage in northeastern Missouri. The cob is a mixture of sand, clay, and straw that is mixed by foot and applied by hand.

Courtesy of Brian "Ziggy" Liloia

Figure 6.40b The completed home contains a vegetated roof and deep overhangs.

---

cob a combination of natural clay and straw that is mixed together and built in a free-form style into structural walls.

rammed earth an ancient form of construction in which soil and additives, such as straw, lime or cement, are placed within forms in multiple 6- to 8-inch thick layers and compacted to create a structural mass wall.

Courtesy of Solum Rammed Earth Builders Ltd. (www.solumbuilders.ca)

Figure 6.41a Interior view of rammed earth walls.

Courtesy of Solum Rammed Earth Builders Ltd. (www.solumbuilders.ca)

Figure 6.41b Rammed earth walls with integral insulation being installed. This builder installs the rigid insulation, rebar, electrical conduit, and any other items in the wall during construction. The soil is mixed on-site and shoveled into the forms. An 8" layer of soil, or lift, is placed into the form and then pneumatically compacted. Once that layer is firmly compacted, another fresh layer is shoveled on top and the process continues until we get to the top of the wall.

These alternative wall construction techniques have several things in common:

- Fire resistance
- Labor intensiveness
- Low material costs (clay and sand can often be obtained directly from the job site)
- Lack of widespread acceptance by building code officials
- Need for special approval to use
- Natural aesthetic variations (which may be attractive to some people and unappealing to others)

The extensive labor requirements generally relegate these techniques to owner-built or craftsman-built homes. The average contractor who is looking to reduce labor costs will often shy away from non-standard methods such as these. Where a specific method is not addressed in the local building code, professional engineering drawings and calculations may be required to obtain approval from building officials.

Green Remodeling Considerations

Green remodeling projects must evaluate and address deficiencies in existing floors and walls, making necessary improvements to the existing structure before undertaking renovations and additions. The choice of materials and techniques for additions to existing buildings should be made on the basis of the key principles of energy efficiency, durability, resource efficiency, and compatibility with the existing structure.

- *Energy Efficiency*: Select wall and floor systems that allow for insulation that exceeds the levels in the existing structure.
- *Durability*: While all structural systems should be installed to be as durable as possible, remodeling projects demand that those systems, at a minimum, do not diminish the durability of the existing structure and preferably enhance it.
- *Resource Efficiency*: Reusing existing structural elements is inherently resource-efficient because fewer new materials are required for construction and less waste is produced. Required improvements should be made with the minimum amount of materials necessary. Additions should take into account all techniques described earlier in this chapter to reduce material use.
- *Compatibility of additions with existing structure*: While most any structural system can work for an addition, some factory assemblies (such as SIPs or modular construction) require superior advanced planning and do not allow for much site adjustment. With the proper planning, panelized and SIP walls can save site time, thereby helping to limit weather exposure—particularly in the case of second-story additions in which the roof over finished space may be removed for a period of time. Regardless of the methods chosen for constructing an addition, connections between existing and new areas must be structurally sound, resistant to exterior moisture, well insulated, and air sealed effectively.

FROM EXPERIENCE

Alternative Building Materials: How to Raise and Train Your Local Building Official

Bruce King, Structural Engineer Here at the start of the 21st century, building in North America is at once a wide-open canvas full of possibilities and a narrowly constricted system governed by rigid regulations and myopic worldviews. If you are living in the United States, you have inexpensive and stunningly easy access to an array of building materials never before seen in history. Want Italian marble for your Spokane kitchen? No problem! Douglas fir framing lumber from British Columbia for your Tucson shopping mall? We do it every day!

But if you would like to build with the materials most familiar to our not-that-distant ancestors, you may have some work to do to persuade your engineer, subcontractors, building official, lender, insurer, and maybe even neighbors to accept something "different." The Romans built their empire with concrete made of lime plaster and volcanic mud—concrete far better than what we use today—but you would likely get a lot of hard questions about your "experimentation" if you tried to do the same. Straw bale building was invented in Nebraska in the late 1800s, and many of those homes are still here. Nevertheless, if you try and build with bales today, you will likely still get a lot of questions about durability. And if you should try to build with adobe or rammed earth anywhere but New Mexico, you may get folks nervous about the (yikes!) return of the dreaded mud hut.

Do not freak out, and do not give up. If you have decided to build with a structural material other than concrete, steel, masonry, or wood (the entire known universe to most structural engineers), you probably have a lot of good testing and experience to find and read—and bolster your case. Do your homework, be patient when folks are resistant or even sometimes hostile, and listen to their concerns. Better still, make them your allies! If you are like the hundreds of people I have had the pleasure of working with, you are not trying to be different or contrarian; you are simply trying to find a better way to build. When people recognize that, they will usually work with rather than against you. Have your facts straight, be patient, and know that you are part of an ever-growing group of people working to make buildings safer, healthier, and smarter.

Ecological Building Network:
http://www.ecobuildnetwork.org/

Leading Home Network:
http://leadinghome.net

Green Building Press:
http://www.greenbuildingpress.com

Bruce has been studying, writing about, designing, and building with alternative materials for 20 years. Initially focused mainly on so-called natural building, such as adobe, straw bale, rammed earth, and bamboo, Bruce pursued a path that led to disaster relief work in Haiti and other areas in which builders use whatever is locally available. In the modern urban context, that means considering alternative uses for waste plastic, shipping containers, barbed wire, fishing net, clay, concrete rubble, and so forth. Ultimately, it also means using and guiding the best resource of all, human ingenuity, to create safe, effective, and affordable shelter anywhere from Port-au-Prince to Portland.

Summary

Selecting the appropriate structural system for walls and floors for a particular project involves consideration of the climate, availability of materials, and the skill sets of available personnel. Material options range from standard wood framing to high-efficiency systems (e.g., SIP and ICF construction) to alternative methods (e.g., rammed earth and straw bale). Each structural system has unique characteristics, such as first cost, durability, skill requirements for installation, energy efficiency, and resistance to pests and moisture damage. Matching the properties of a particular system with the needs of a project is key to creating a sustainable project.

Review Questions

1. Advanced framing can save as much as ___% of the wood required to frame a house.
   a. 10%
   b. 25%
   c. 40%
   d. 75%
2. Which of the following is not a feature of advanced framing?
   a. Lumber savings
   b. Increased space for insulation
   c. Higher cost
   d. Reduced waste
3. Which of the following techniques is not considered advanced framing?
   a. Two-stud corners
   b. Right-sized headers
   c. 2×4 studs at 16" on center
   d. Single top plates
4. Which assembly has the most thermal bridging?
   a. 2×4 wall framing with 0.5" rigid foam sheathing
   b. 2×6 wall framing with OSB sheathing
   c. Steel stud wall with OSB sheathing
   d. 8" SIP panel
5. Which of the following is not an engineered wood product?
   a. FSC certified wood stud
   b. I-joist
   c. Laminated beam
   d. Open-web truss
6. Which wall sheathing is nonstructural?
   a. OSB
   b. Plywood
   c. Fiber-reinforced gypsum
   d. Rigid foam
7. Which of the following structural system is most susceptible to termites?
   a. ICFs
   b. CMU
   c. Slab-on-grade
   d. Stick framing
8. True or false: Stick-framed projects treated with a mold inhibitor don't require the same amount of moisture management as traditional framing.
   a. True
   b. False
9. Wood framing is commonly treated with Borate to prevent which of the following?
   a. Fire
   b. Material expansion
   c. Termites
   d. Thermal bridging
10. Which of the following is not a characteristic of timber framing?
    a. Provides wide-open flexible spaces
    b. Requires standard carpentry skills
    c. Can be constructed from reclaimed wood
    d. Can be used in combination with SIPs, stick framing, or straw bale

Critical Thinking Questions

1. Compare and contrast different wall structures and their implications on cost, durability, and thermal performance.
2. Consider how you can incorporate advanced framing into both new construction and renovation projects, and explain how it affects the cost of construction and effectiveness of insulation.
3. How would you go about specifying and obtaining approval for an alternative wall structure, such as straw bale or rammed earth?
4. In areas subject to severe termite infestation, what materials would you consider for floor and wall structures for a project with a modest budget?
5. Explain the benefits of engineered wood products over conventional lumber.

Key Terms

advanced framing, 158
balloon framing, 147
carbon sink, 152
cob, 175
engineered lumber, 154
fiberboard, 166
Forest Stewardship Council (FSC), 153
I-joist, 164
joist, 154
manufactured homes, 169
modular construction, 150
old-growth, 151
open-web floor truss, 154
optimum value engineering (OVE), 158
oriented-strand board (OSB), 163
panelized framing, 147
plate, 158
platform framing, 147
plywood, 154
rammed earth, 175
second-growth, 151
sheathing, 156
sill, 147
sustainable forest management, 153
Sustainable Forestry Initiative (SFI), 153
tree plantation, 152
truss, 154

Additional Resources

Forest Stewardship Council: http://www.fsc.org/

Sustainable Forestry Initiative: http://www.sfiprogram.org/

Solar Energy International: http://www.solarenergy.org

Advanced Wall Framing (six-page fact sheet by the U.S. Department of Energy, 2001): http://www.nrel.gov/docs/fy01osti/26449.pdf

Roofs and Attics

This chapter presents different methods for making the roof and attic of a home green by exploring different options available for the location of the thermal envelope at the top of a building and how these decisions, along with roof design and roofing material selection, play an important role in keeping the home healthy, dry, and energy-efficient. The impacts of insulation and roofing materials on a home's energy and resource efficiency will also be addressed. Finally, we'll discuss roof and attic decisions that should be considered while renovating, remodeling, or adding on to an existing home.

LEARNING OBJECTIVES

Upon completion of this chapter the student should be able to:

- Describe how roofs and attics affect building performance
- Explain how to specify the appropriate roof and attic details for a house
- Differentiate between vented, conditioned, and semi-conditioned attics
- Describe how a radiant barrier works in a roof assembly
- Explain why ice damming occurs in cold climates and how it can be prevented
- Describe how to create an effective thermal envelope at the ceiling line of a house

Green Building Principles

Energy Efficiency

Resource Efficiency

Durability

Indoor Environmental Quality

Homeowner Education and Maintenance

Sustainable Site Development

The Role of Roofs and Attics in Green Homes

The purpose of a roof is to keep elements out of the interior and deflect rain off the exterior walls of the house as much as possible. Together with the exterior walls and basement or foundation, the attic acts as a buffer between the conditioned space and the exterior. The boundary can be the ceiling, gable ends, the roofline, or a combination of all three. From a building science standpoint, attics, and roofs play a role in managing heat, air, and moisture flows in the house (Figure 7.1). The choice of finish roofing materials not only affects resource efficiency but also may impact the site and surrounding community by raising the temperature around the home (heat island effect).

Problems arise when roofs and attics do not effectively manage rain events or have improper flashing that allows bulk moisture into the structure. An ineffective building envelope will also allow heat, air, and vapor to move in and out of conditioned space. Solutions to these problems require considering the roof and attic as part of the whole house system, creating simple and effective designs that shed water, and incorporating a complete and comprehensive insulation and air barrier at the building envelope. Each of these solutions will be addressed in detail throughout this chapter.

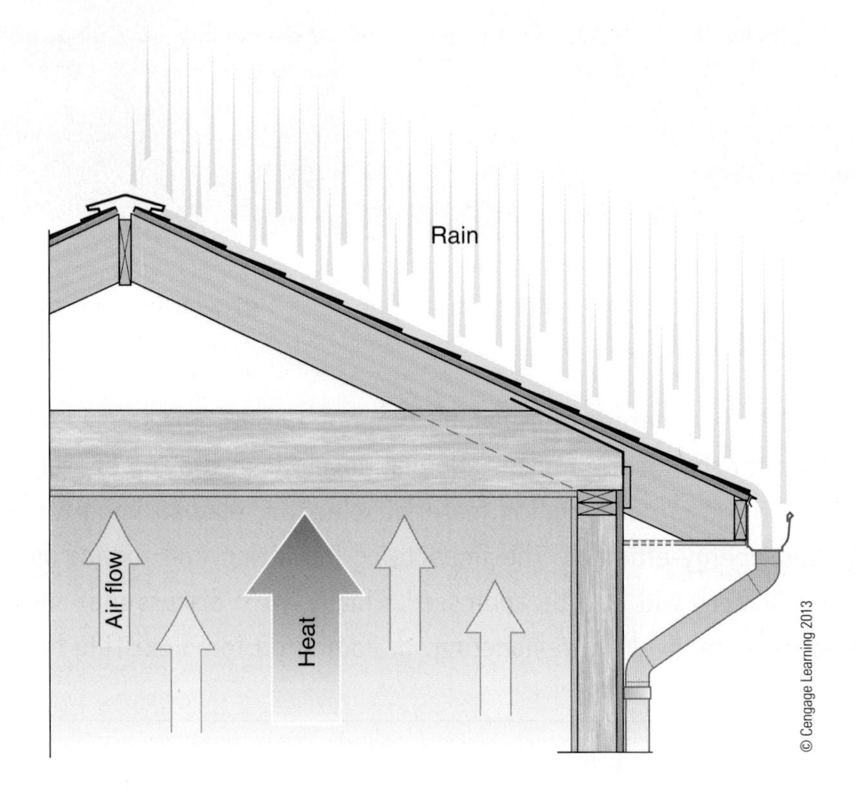

Figure 7.1 Roofs control the flow of moisture into a home, and attics provide a buffer area for heat transfer.

Types of Roofs and Attics

Roof types are generally classified by their slope, from steep to low slopes. Roof slopes are described by their rise (vertical height) over their run (horizontal length). A 6:12 slope means that the roof rises 6" for each 12" of horizontal run. A steep-slope roof is anything above 2:12 (Figure 7.2a). Low-slope roofs, less than or equal to 2:12, are sometimes inaccurately referred to as flat roofs. With some having as little as 0.25:12 slope, low-slope roofs must be fully waterproofed at every seam and joint for protection during periods when water is standing on the roof surface (Figure 7.2b). In comparison, steep-slope roofs do not need to be waterproofed to the same extent because they are designed to shed water down their slope quickly enough to prevent leakage to the interior.

Roofs on homes can be steep- or low-sloped, or a combination of both. The design and type of roof determine its ability to shed rain effectively, to keep the structure dry, and to direct water away from exterior walls and into gutters or ground features that either collect it or divert it away from the building foundation.

roof slopes the angle of a roof, described by the rise (vertical height) over the run (horizontal length).

steep-slope roof a roof whose angle is more than 30° (2:12).

low-slope roof a roof angle or pitch that is 30° (2:12) or less.

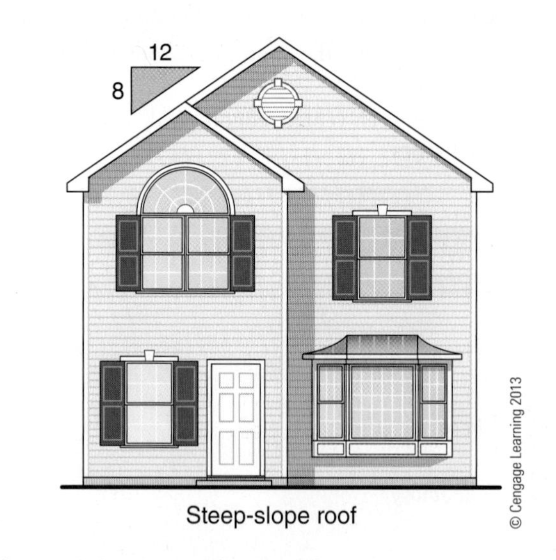

Figure 7.2a Steep-slope roof.

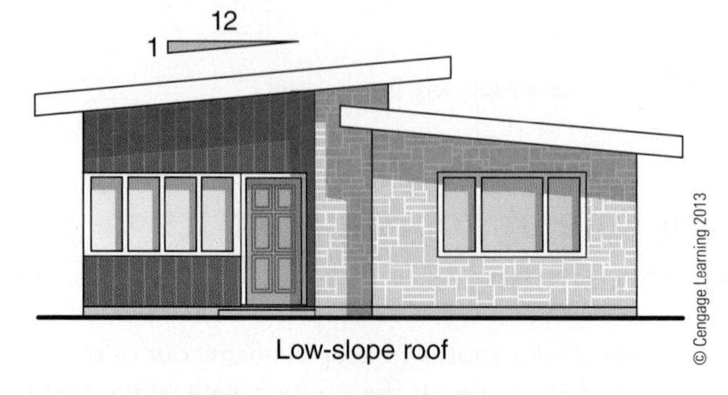

Figure 7.2b Low-slope roof.

Gutters are a wood, metal, or plastic trough used at the roof edge to carry off rainwater and water from melting snow.

The steeper the roof slope, the faster water moves off of it and the roof finish itself lasts longer; however, they require more roofing material than shallower roofs. Steep slopes can use a wider range of materials, including asphalt or wood shingles, clay or cement tiles, or metal. In climates with heavy snow, steeper roofs better distribute the weight of the snow and allow for smaller structural members than shallow roofs.

Low-slope roofs are generally limited to asphalt, rubber, or plastic-based products that can be sealed together on-site into a waterproof membrane. Metal roofing panels that are seamed and soldered together are also suitable for low-slope applications, but they can be very costly and are therefore not common in residential construction. One benefit of low-slope roofs is their suitability for vegetated coverings that can reduce runoff, extend roof life, and provide thermal insulation.

The Effects of Roof Design on Green Homes

The design and dimensions of a roof can have either positive or negative impacts on the sustainability of a building from the standpoint of efficient material use and durability. The roof shape and complexity affect the amount of material required for construction as well as the amount of waste produced. By designing to 2'0" modules for both framing and sheathing, limiting the number of different roof planes, and selecting simpler over more complicated designs, you will reduce waste from cutoffs—particularly the angle cuts required at hips and valleys (Figure 7.3a, Figure 7.3b, and Figure 7.3c). More complex roof designs also make ventilation more difficult.

Sheds and gables are the most economical roof designs, particularly when designed to fit modular material measurements (e.g., 16" or 24") that reduce excess waste from lumber cutoffs (Figure 7.4). Shed roofs are roofs that slope in one direction only. Gable roofs slope in two directions. Hip roofs covering a whole house have a minimum of four planes, and those covering porches or additions have at least three. While cutoffs from hip roofs can often be reused, they produce more waste than shed and gable roofs. Dormers are structures that project outward from a sloping roof to form another roofed area that provides a surface for the installation of windows and additional interior space. Dormers can provide day-lighting, natural ventilation, and additional living space. If they are not considered when calculating building material measurements, dormers can create significant amounts of waste materials.

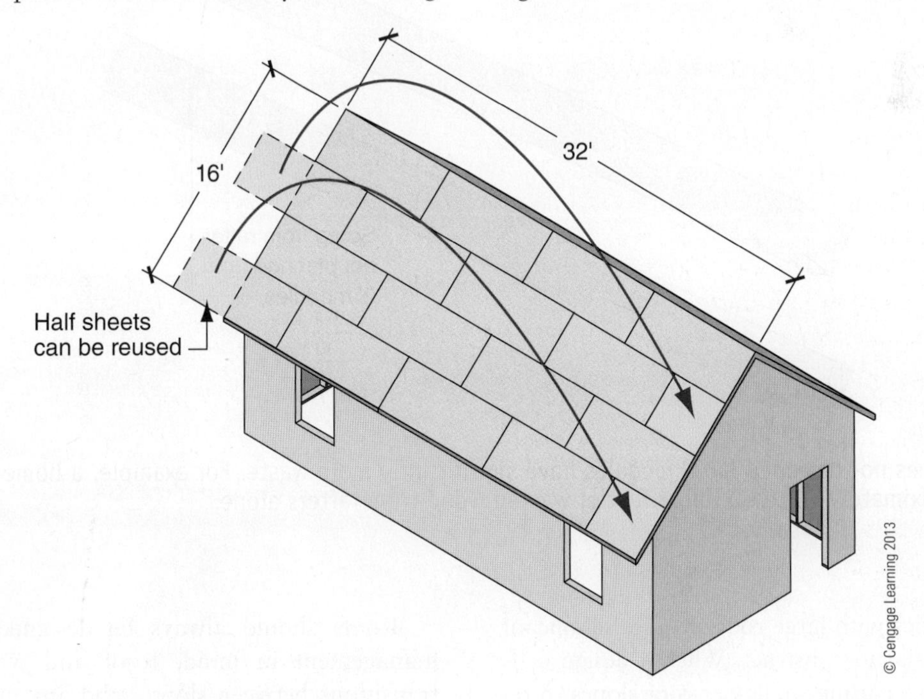

Figure 7.3a Roof designed for 2' modules.

gutter a wood, metal, or plastic trough used at the roof edge to carry off rainwater and water from melting snow.

shed roof a type of roof that slopes in one direction only.

gable roof a type of roof that slopes in two directions.

hip roof a three or four-sided roof having sloping ends and sides.

dormer a structure that projects out from a sloping roof to form another roofed area to provide a surface for the installation of windows.

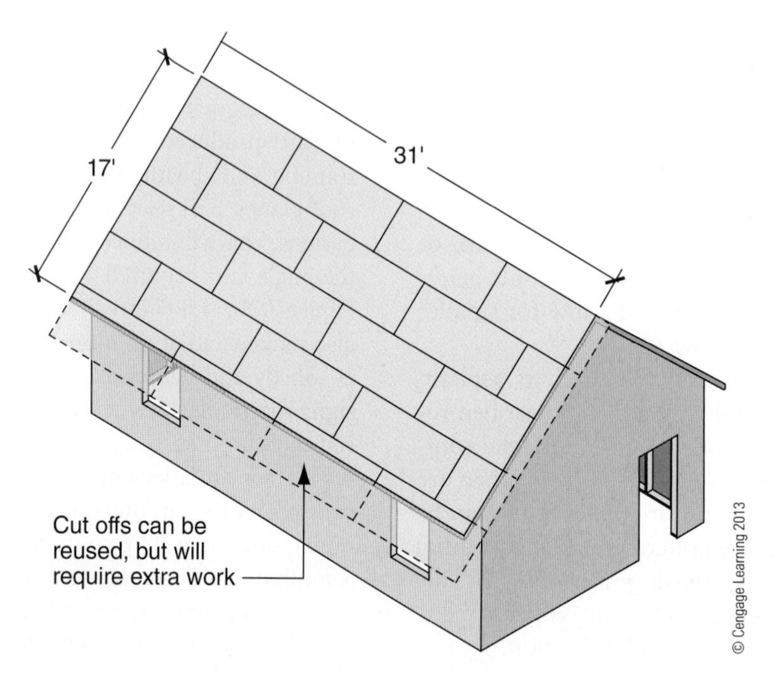

Figure 7.3b Traditionally, roofs have not been designed to maximize material efficiency.

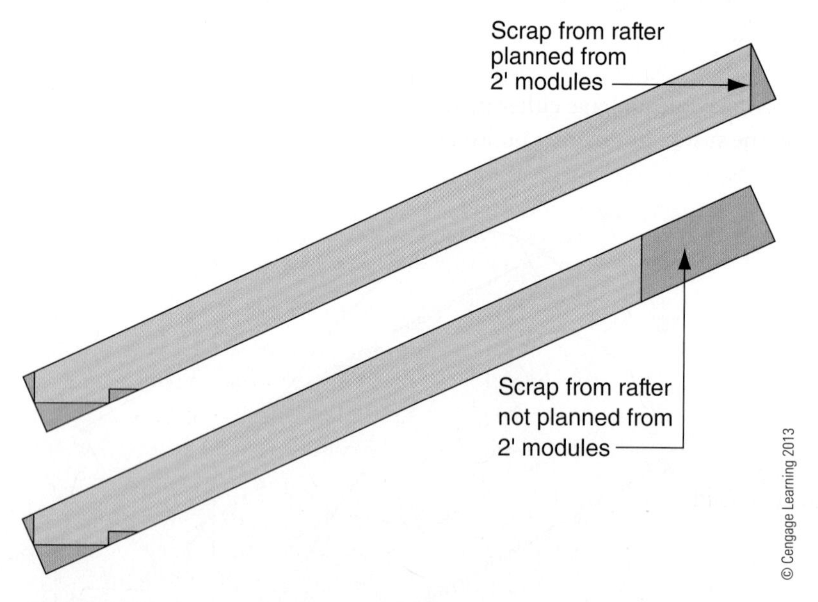

Figure 7.3c Homes not designed for 2' modules have significantly more waste. For example, a home with 40 rafters would have approximately 40 to 60 linear feet of wasted wood from rafters alone.

Unfinished attics with large roofs require significant amounts of material to construct. When a design calls for an attic, consider using shallower roof slopes to reduce the attic volume. This approach saves both framing and roofing materials, particularly when the local climate and precipitation allow it. When considering different plans, creating finished living space under the roof can allow for a smaller building footprint and significant material savings.

Roofs should always be designed with moisture management in mind. Roof and wall intersections, transitions between slopes, and any penetrations (e.g., chimneys and skylights) must be planned to carry the maximum rainfall off the roof without backing up and creating leaks. Areas that can collect leaves should be easily accessible for regular cleaning (Figure 7.5). Avoid narrow spaces between vertical walls that can collect snow and cause leaks. A combination of simple roof

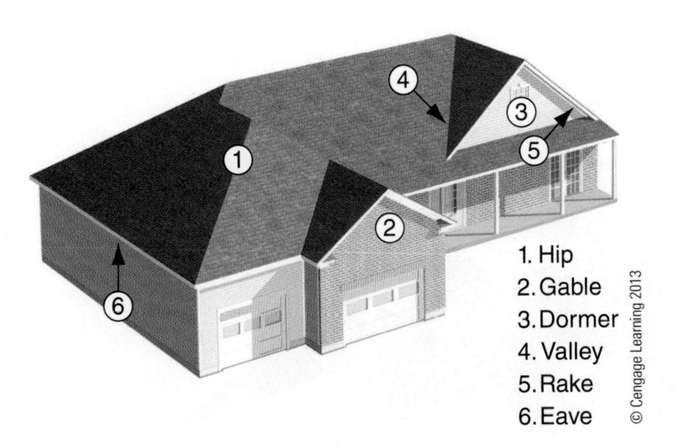

Figure 7.4 Roof terminology.

Figure 7.5 The likelihood of leaf and snow accumulation is high where these two gables come together. Any areas with narrow spaces should be avoided whenever possible or, at the very least, made easily accessible for cleaning.

designs and careful workmanship provides the best opportunity for effective moisture management and resource efficiency.

Regardless of the roof type, attics can be conditioned finished space or left unfinished. An unfinished attic may be fully conditioned, semi-conditioned, or unconditioned, with each type having different levels of efficiency and indoor environmental quality (Table 7.1). A finished space has both space conditioning and final interior finishes. Conditioning refers to either directly or indirectly supplying heating and cooling. In each case, proper location and installation of insulation and air sealing creates an effective building envelope. Making the appropriate decisions on roof and attic design and construction are key components in creating a sustainable and healthy building.

Roofing Basics

The roof structure consists of the framing members to which sheathing is attached, and the finish roofing material installed on top of the sheathing. Structural insulated panels (SIPs) can be used for roof structures as an alternative to or combined with traditional framing. Flashing and edging materials complete the total assembly to keep the house dry (Figure 7.6). The interior finishes can be attached to flat or sloped ceilings that are separate from the roof framing or, as is common for vaulted ceilings, directly to the underside of the roof structure. Wood is the most common material used to create the sloped plane. The vertical wall framing, however, may be made of wood studs and plates, SIPs, or alternative systems, such as insulated concrete forms (ICFs), autoclaved aerated concrete (AAC), and other non-wood systems.

Roof Structural Systems

Wood roofs may be framed with solid lumber, engineered lumber, heavy timber, or a combination of any of these. Open-web trusses are a common and efficient method of framing the roof and the ceiling at the same time (Figure 7.7a and Figure 7.7b). Trusses are available in a variety of designs that include open storage space and vaulted ceilings. When the area immediately under the roof is designed for living space, the attic floor and roof are almost always separate structural systems rather than open-web trusses. As shown in Figure 7.7b, most truss designs do not provide usable attic space.

Table 7.1 Attic Assembly Selection

Potential Asset	Unconditioned Attic	Semi-Conditioned Attic	Fully Conditioned Attic
Additional living space	No	No	Yes
Additional storage space	Limited to items not affected by temperature and relative humidity swings	Yes	Yes
Mechanical equipment inside the building envelope	No	Yes	Yes

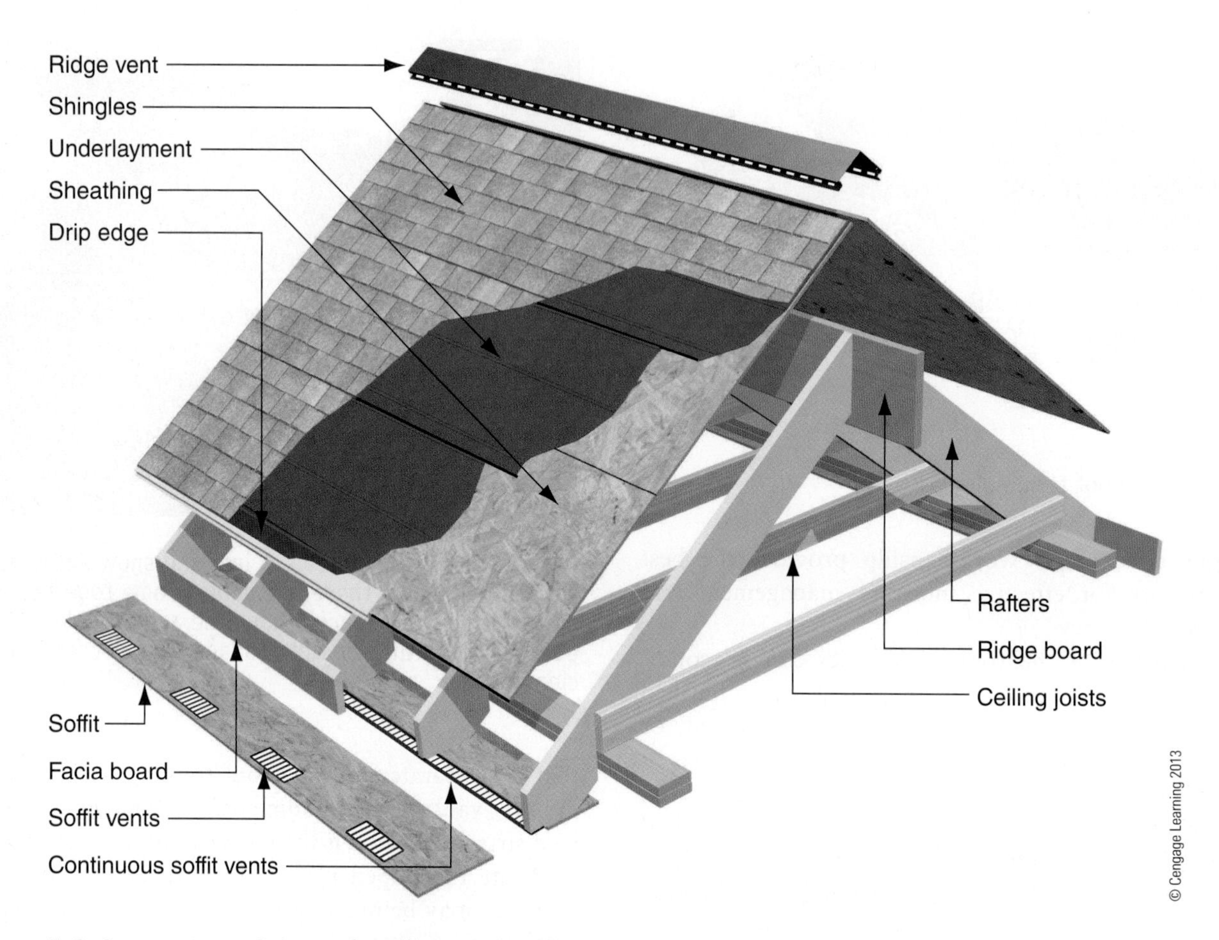

Figure 7.6 Components of the roof, ceiling, and soffits in an unconditioned attic.

Alternative structural systems for roofs include SIPs and steel. SIPs are prefabricated in the factory and lifted into place by cranes, creating a fully insulated roofline (Figure 7.8). SIPs may be used to provide the entire roof structure or be used in conjunction with timber framing. Steel-framed roofs are more common in commercial construction but are also suitable for homes. When framed over unconditioned attics, steel framing does not create the same thermal bridging issues as it does in walls, as discussed in Chapter 4.

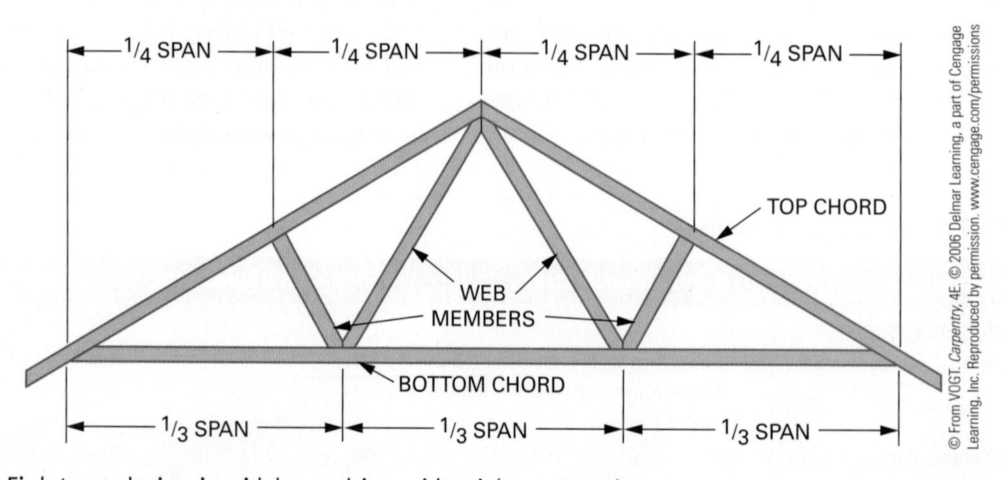

Figure 7.7a The Fink truss design is widely used in residential construction.

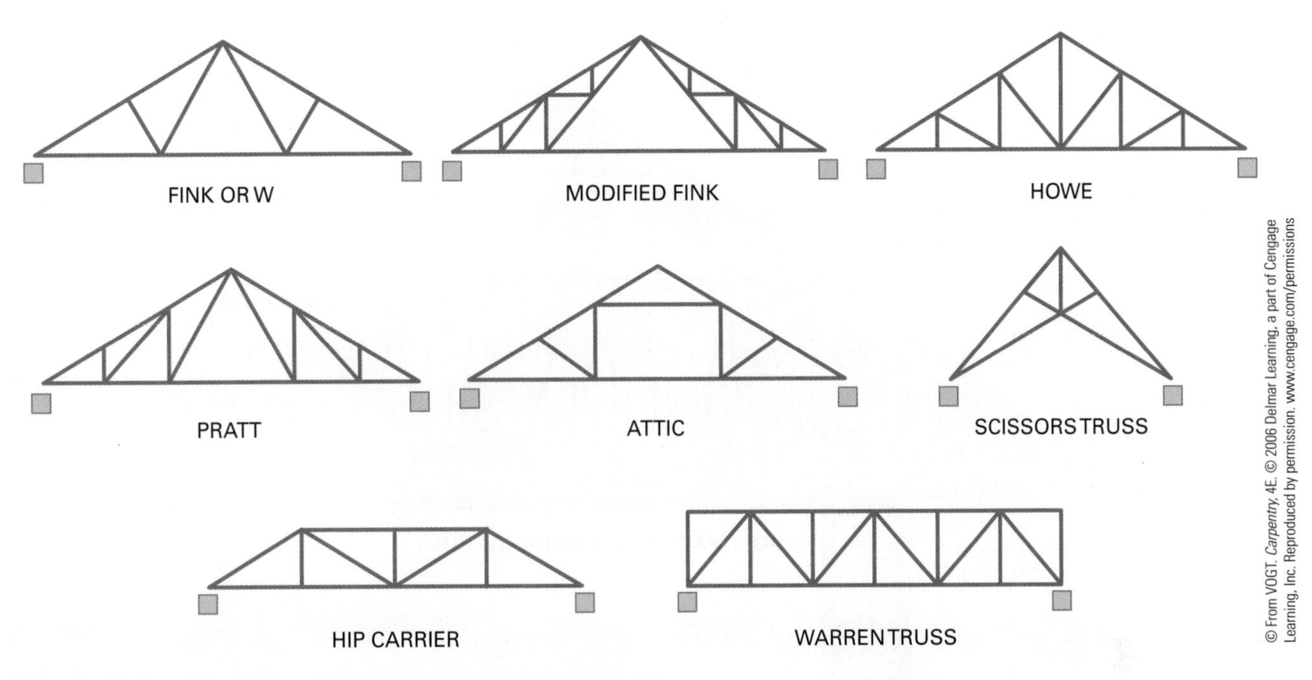

Figure 7.7b Various truss designs for special requirements.

Figure 7.8 SIPs may be used for walls, floors, and roofs. They are assembled without the traditional wood framework.

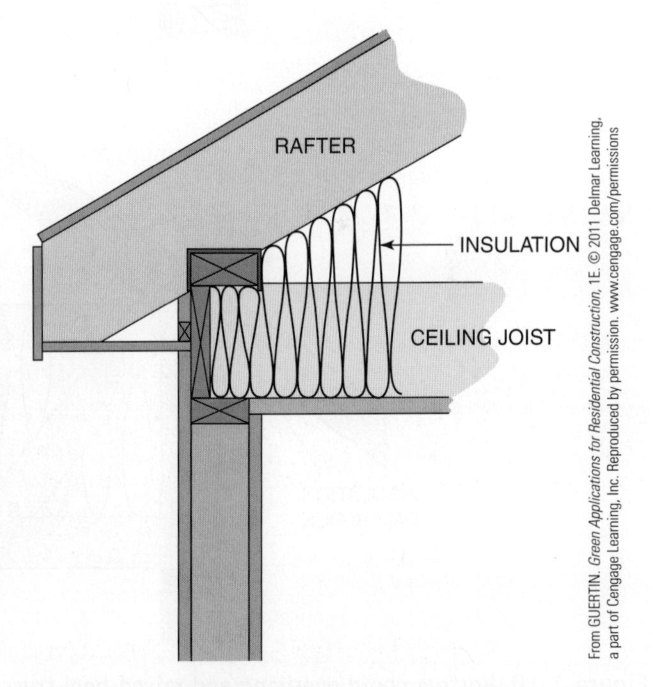

Figure 7.9 Resting the rafters on top of the ceiling joists instead of the wall plate creates more space for insulation near the edge of the roof.

Roof designs over unconditioned attics should allow for full-depth insulation at eaves to provide for full insulation depth at the edges of the roof. This can be accomplished with raised-heel trusses or, if site-framed, by resting the rafters on top of the ceiling joists instead of alongside them. The rafters may rest on a plate laid across the top of the ceiling joist or directly on top of the ceiling joists if a band joist is installed at the end of the joists (Figure 7.9). The vertical space for insulation is then increased by the height of the ceiling joists. **Raised-heel trusses**, also called **energy trusses**, are roofing trusses designed to span an area and provide adequate space for full-depth attic insulation across the entire area (Figures 7.10).

---

raised-heel truss a roofing truss designed to span an area and provide adequate space for full-depth attic insulation across the full area; see also *energy truss*.

energy truss a roofing truss designed to span an area and provide adequate space for full-depth attic insulation across the full area; see also *raised-heel truss*.

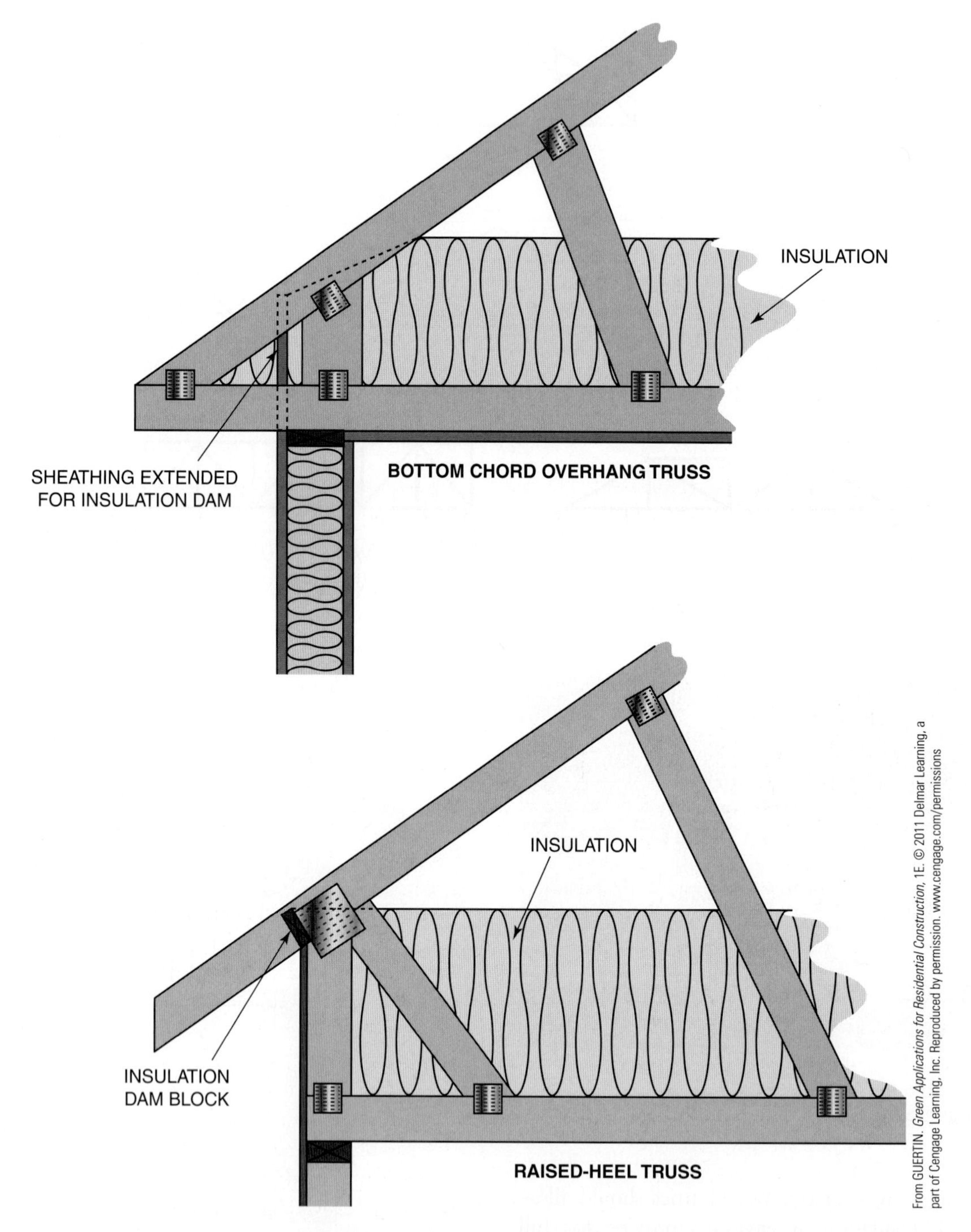

Figure 7.10 Bottom chord overhang and raised-heel truss designs allow full-height insulation to extend over the top of the exterior wall. Built-in or added insulation dams prevent insulation from falling into the soffit.

Roof Decking

Roof decking provides a structural base for the finish roofing material. Oriented-strand board (OSB) and plywood are the most common roof deck materials used in homes. Conditioned and semi-conditioned attics can be created using spray polyurethane foam (SPF) or rigid insulation combined with structural decking (Figure 7.11a and Figure 7.11b). Alternate materials, such as Homasote® structural fiberboard and Tectum roof panels, can serve as both structural sheathing and insulation and are manufactured with recycled or rapidly renewable materials. Rarely does the structural

---

roof decking the wooden or metal surface to which roofing materials are applied.

Figure 7.11a SPF along the roofline and exterior gable walls eliminates the need to insulate the vaulted ceiling and attic knee walls.

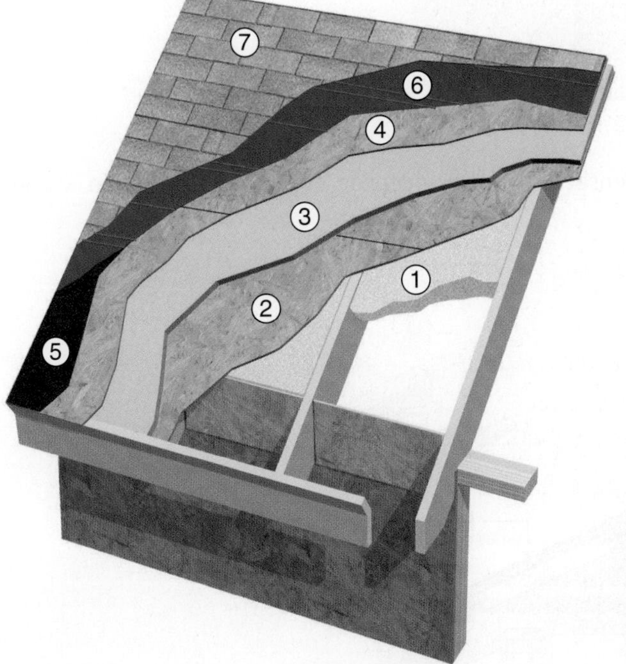

Figure 7.11b Installing rigid foam insulation above the roof deck is an alternative method to creating an insulated roofline.

sheathing provide adequate insulation on a sloped roof, so additional cavity insulation is usually required. Some of these alternate materials can serve as the finished surface when exposed to the interior above exposed rafters (Figure 7.12).

Figure 7.12 Exposed Tectum roof decking. Tectum manufactures roof decking products with nominal thickness of 1.5" to 10" and thermal resistance up to R-5/inch.

Roof Ventilation

Building codes generally require ventilation of attics or beneath the roof decking to help remove moisture and extend the life of asphalt roofs, although unvented roofs are now accepted in most building codes. Modern ventilation standards are largely based on research from the 1930s by the U.S. Forest Products Laboratory and a 1942 document from the Federal Housing Authority (FHA).[1] These studies were limited in scope and modern research has disproven many of the assumptions. Studies have shown that most traditional methods of roof ventilation are not as effective at removing moisture as intended, and they do not have a significant effect on the life of the roofing material. Additionally, condensation risks are largely eliminated by the proper air sealing of the attic ceiling (Figure 7.13a, Figure 7.13b, and Figure 7.13c). Determining whether or not to ventilate a roof depends on a number of factors, including the climate, local building codes, roofing material, insulation type and location, and the location of heating, ventilating, and air conditioning (HVAC) equipment.

Roof ventilation is generally required by building codes when an attic is unconditioned or when air-permeable insulation is installed between the rafters. The best practice for roof ventilation incorporates low-intake vents at the soffits, baffles wherever insulation comes in contact with the roof decking, and either ridge vents or high vents at the gables (Figure 7.14). This arrangement allows for warm air rising in the attic to be drawn out of the high vents by convection and replaced by outside

[1] Rose, William B. *Water in Buildings: An Architect's Guide to Moisture and Mold.* John Wiley & Sons, Inc, 2005.

air coming in the low vents. If the ceiling plane is not fully air sealed, high winds can overcome this convection and draw conditioned interior air into the attic, creating drafts and reducing energy efficiency. In areas of high rain and wind, bulk moisture can be drawn or blown into vents, creating problems in wall or roof structures. Proper installation of eave baffles (also known as positive-ventilation chutes) that extend over the top of

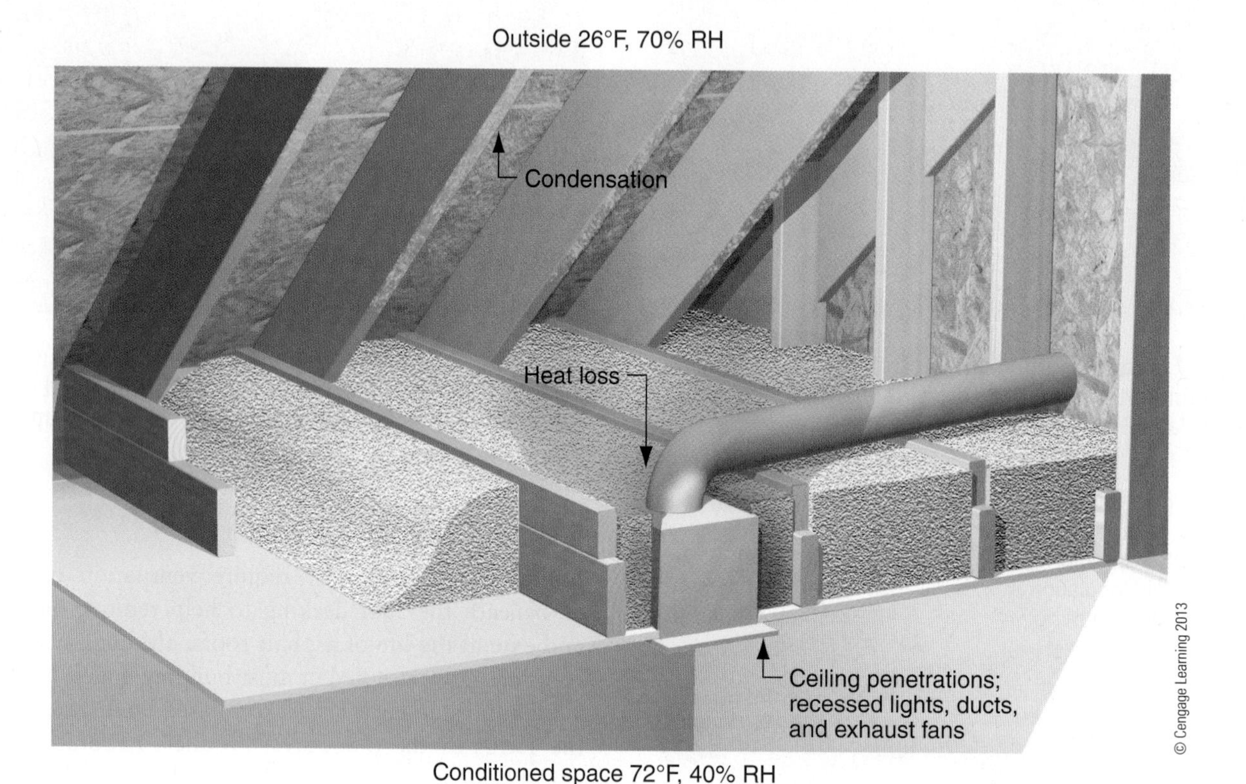

Figure 7.13a Relatively warm humid attic air can form condensation on the underside of the roof decking in winter.

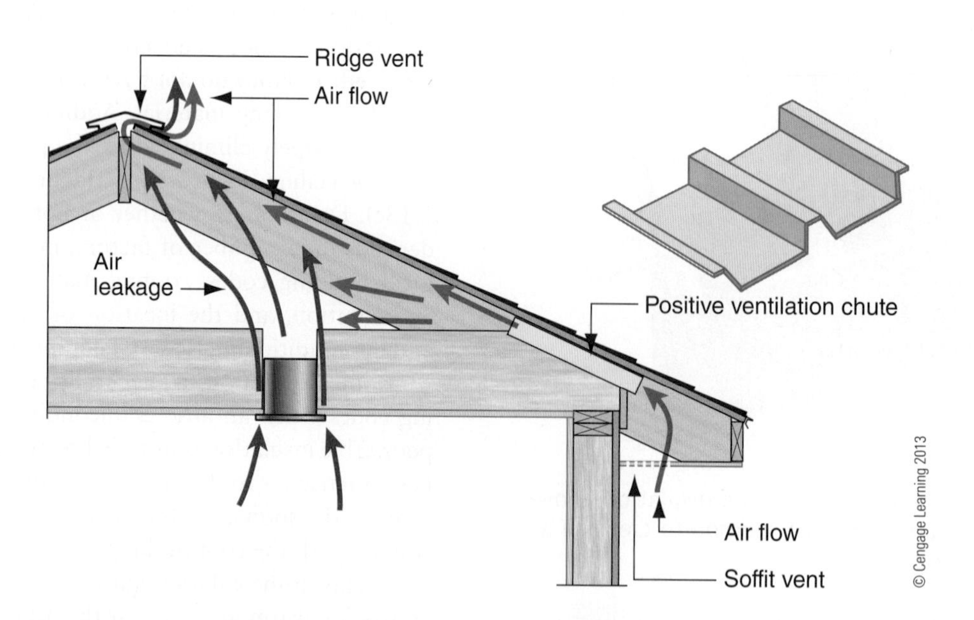

Figure 7.13b Attic ventilation removes air leaking from the house into the attic before condensation forms.

eave baffles materials that prevent attic insulation wind-washing by directing soffit air flow over attic insulation; also known as *positive-ventilation chutes*.

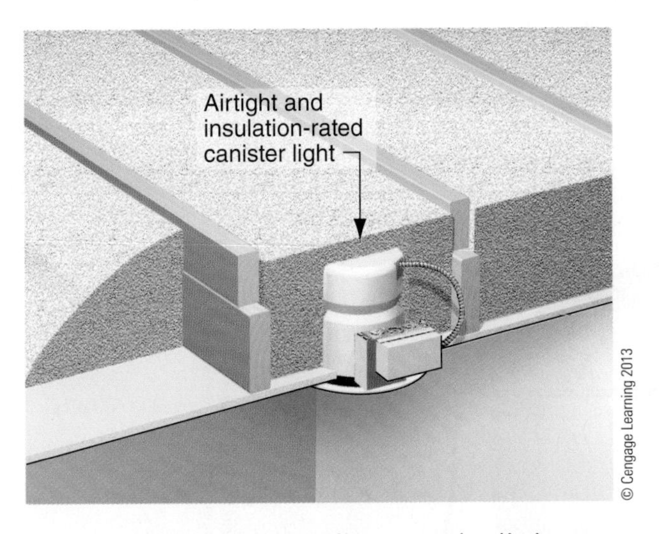

Figure 7.13c Proper air sealing not only eliminates condensation risks in the attic but also improves the overall efficiency of the home.

the ceiling insulation can help prevent "wind washing" in which air is allowed to blow through loose insulation, degrading its thermal performance (Figure 7.15a and Figure 7.15b).

Many manufacturers offer electric roof vents that are tied to the traditional home electric supply, but solar-powered units are also available. Research has shown that electric vents do not provide significantly better ventilation than passive systems; in fact, they can depressurize the attic space enough to draw conditioned air into the attic, decreasing the overall energy efficiency of the building (Figure 7.17, page 194). The energy saved by power vents is generally offset by the cost of the power to run them, making these units even less effective. Solar-powered units, while not requiring line voltage to operate, generally do not provide enough benefit to offset their cost. In all cases, when power roof vents are installed, the

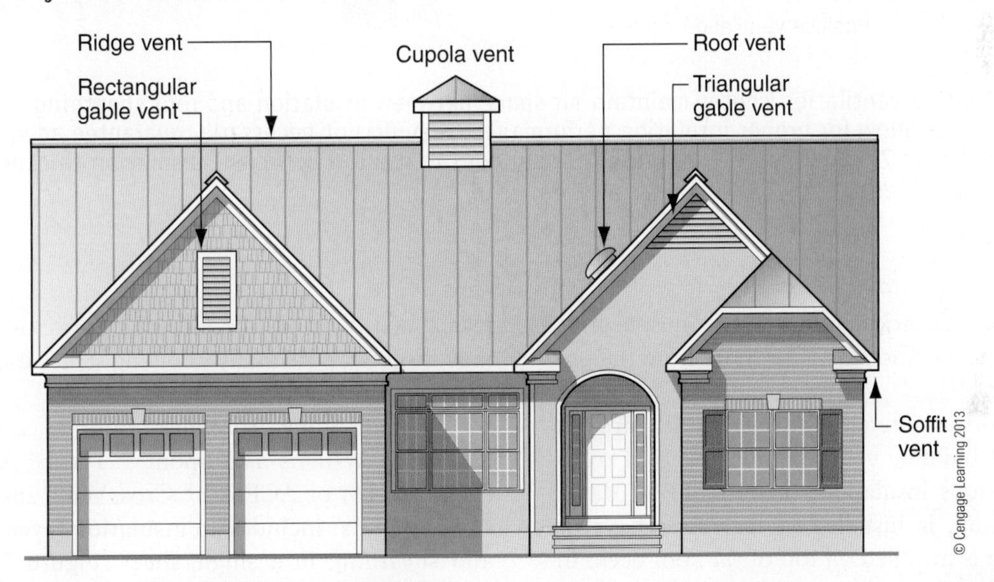

Figure 7.14 Common attic ventilation strategies.

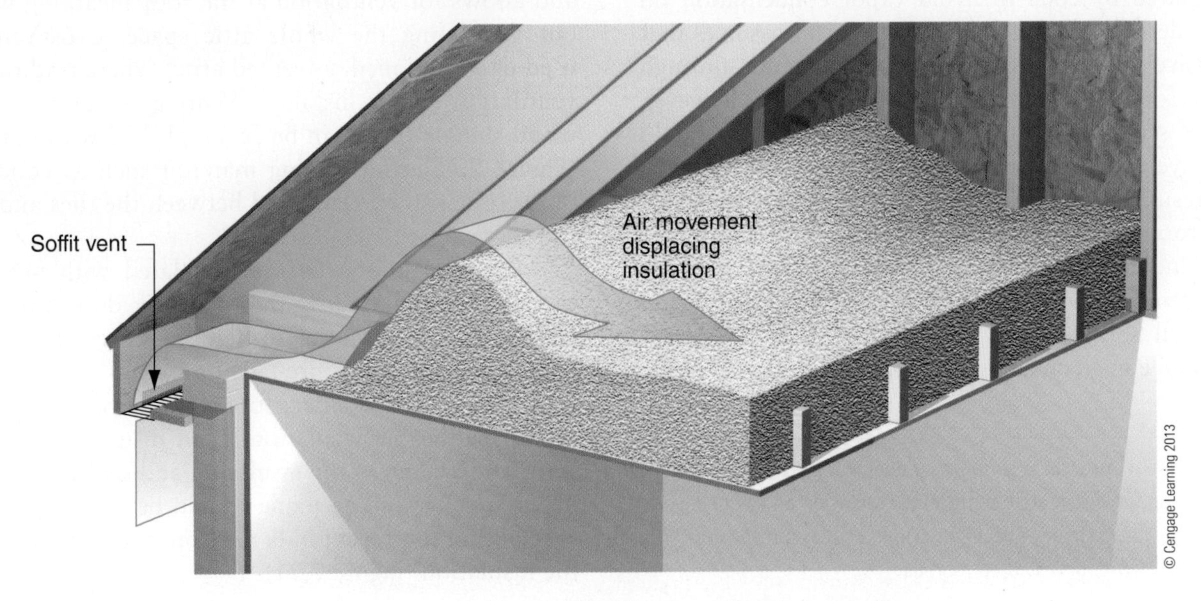

Figure 7.15a Wind washing significantly reduces the effectiveness of attic insulation.

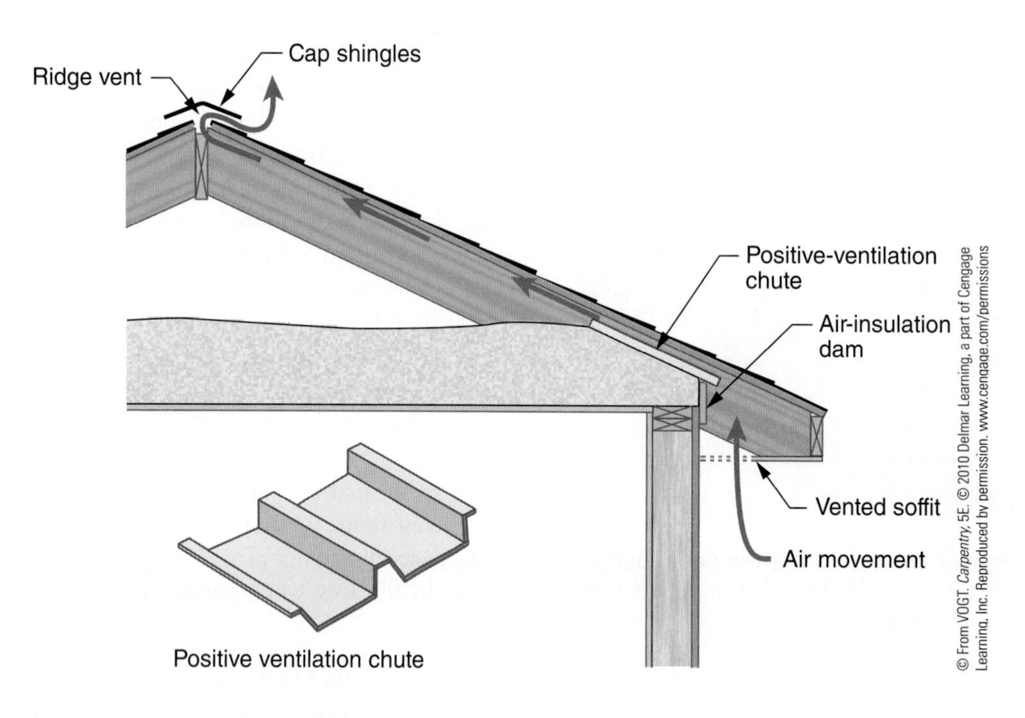

Figure 7.15b Positive-ventilation chutes maintain air space between insulation and roof sheathing. Although eave baffles allow for proper insulation performance, they do not necessarily guarantee adequate insulation depth. Figure 7.9 and Figure 7.10 present techniques that allow sufficient insulation thickness and prevent wind washing.

air barrier between the attic and the interior must be fully sealed to keep the conditioned air separate from the unconditioned attic.

Unvented roofs, also referred to as hot roofs, are not ventilated. Most building codes permit unvented roofs when air-impervious insulation, such as SPF or flash-and-batt insulation, is installed in the rafter bays or when rigid foam is installed on top of the roof deck. In cold climates, closed-cell SPF or other vapor retarder is required by code to avoid vapor condensation on the underside of roof deck in winter. This assumes that excessive vapor inside the house will migrate through to the roof deck without the vapor retarder; however, tightly sealed and properly ventilated homes should not have these conditions. Excess vapor inside a home can result in condensation forming on the underside of a roof in cold climates; however, this is most likely an indication of larger problems in construction or management of the structure. Warmer climates can use open-cell SPF. SIP roofs use closed-cell foam, which is acceptable in all climates. Open-cell foam can allow for faster identification of roof leaks that closed-cell foam may disguise, potentially allowing structural damage to develop undetected.

Venting along rooflines can be accomplished with such systems as Dupont™ Tyvek® AtticWrap™ (Figure 7.18) or ACFoam® CrossVent® insulated roof sheathing that includes an insulation layer, vent space, and sheathing in a single sheet (Figure 7.19). AtticWrap is an alternative to traditional attic ventilation and allows for ventilation at the roof sheathing without ventilating the whole attic space. CrossVent is used in conditioned, unvented attics where traditional ventilation is not installed. Venting under metal or wood shingle roofs can be accomplished with battens (Figure 7.20), and roofing material such as concrete or clay tile can be ventilated between the tiles and the roof deck.

Low-slope roofs that are insulated with SPF below or rigid foam above the roof deck do not require ventilation because the insulation serves as an air barrier, keeping moisture-laden air from reaching the roof deck where condensation may occur. When using air-permeable insulation in an insulated roofline, ventilation between the insulation and the roof deck is necessary. The ceiling plane should be air sealed with special care to prevent bulk air from traveling through the insulation (Figure 7.21).

---

unvented roof an attic assembly that does not contain ventilation.

hot roof an unvented attic containing insulation on the underside or directly above the roof decking; also known as a *cathedral attic, conditioned attic*, or *insulated roofline*.

FROM EXPERIENCE

Preventing Water Intrusion through Roof Ventilation

Mike Guertin is a custom homebuilder and remodeler with more than 25 years of experience specializing in energy- and resource-efficient construction. He is a contributing editor to Fine Homebuilding *magazine and an advisor to http://www.greenbuildingadvisor.com.*

Mike Guertin Conventional roof design relies on roof ventilation to help reduce attic temperature and carry away moisture vapor. But in order to incorporate ventilation into roofs, we have to put holes in them. Holes leak water—probably not during light rain, but add a stiff breeze or heavy wind, and water is bound to get inside. Almost any vent—a ridge vent, a roof vent, a gable vent, or even a soffit vent will leak under the right conditions.

I use four strategies to reduce the chance that water will get into roofs. The method I choose depends on several factors, including where the house is located, the design of the roof, the design of the house, the insulating and air sealing systems, the local building code, and the building official enforcing it.

1. Skip roof vents altogether, and you don't have to worry about holes. Rather than insulating and air sealing at the ceiling level, insulate the roof rafters or trusses at the sheathing. Moving the insulation from the ceiling line to the roofline brings the attic within the thermal and air control layers of the house. The 2009 International Residential Code has specific conditions that must be met, depending on whether the insulation is air-permeable or air-impermeable.
2. Skip the roof vents and install the thermal and air control layers at the ceiling level. For this strategy to work, the air sealing must be meticulous, a good vapor retarder has to be installed, and the insulation must cover the ceiling evenly from exterior wall to exterior wall. You will have to get approval from your building official to use this method because it is not prescriptively permitted under building codes.
3. Ventilate the attic with indirect paths so air can flow but water is less likely to leak (Figure 7.16). It is hard to configure ordinary ridge vents or roof vents to resist water, but gable vents can be used for exhaust ventilation and protected to block water. Cover conventional louvered gable vents with an additional face cover that extends below the bottom of the vent. This can be a separate box or incorporated into the cladding. An opening at the bottom lets air flow down and out but helps resist wind-driven water and snow. Rather than using conventional vented soffit panels or strip soffit vents, install fascia or frieze board–mounted vents that incorporate a vertical channel. The vertical element provides a baffle to water.
4. Ventilate the attic with special wind-resistant vents. Some ridge vent manufacturers have internal baffle designs or flaps that close when conditions are windy to resist water intrusion. Use these exhaust vents along with tortured-path soffit intake vents.

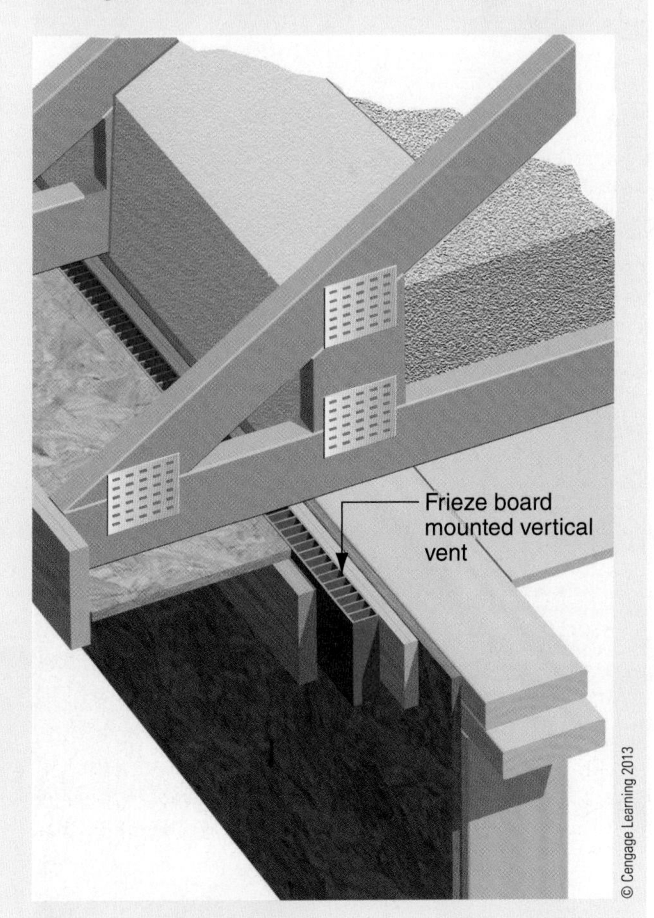

Figure 7.16 This soffit vent design can help prevent wind-driven rain from entering the roof structure.

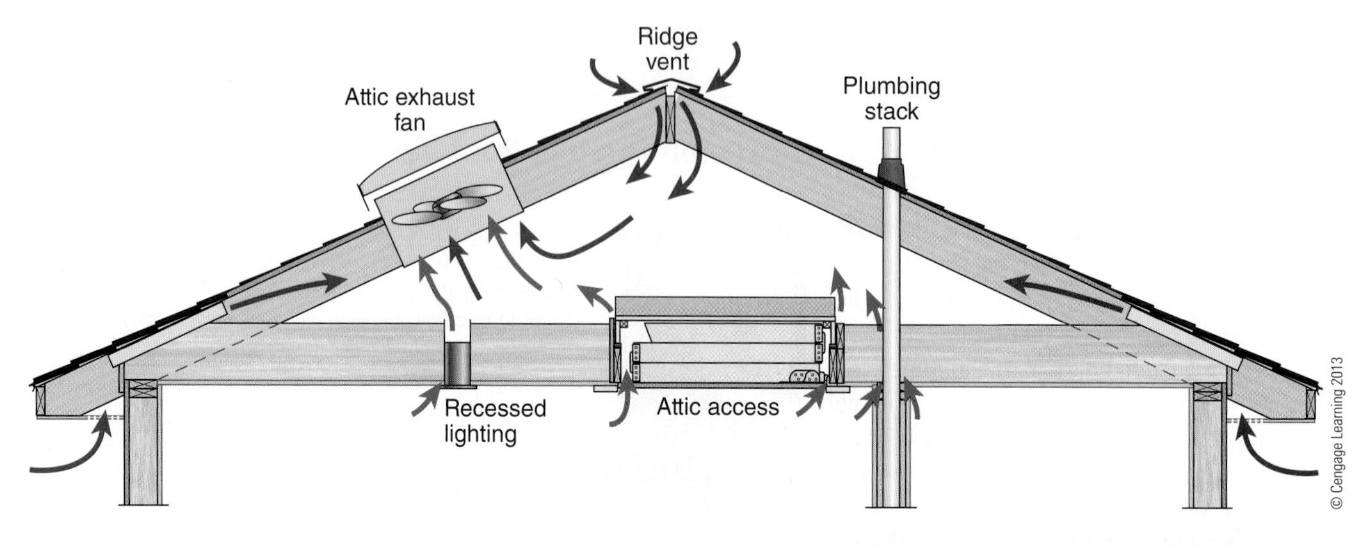

© Cengage Learning 2013

Figure 7.17 Electric attic ventilators can draw air from the house into the attic and potentially cause backdrafting of combustion appliances.

Courtesy of Huber ZIP Panels

Figure 7.18 Tyvek® AtticWrap™ is a system that provides attic ventilation while allowing for a sealed, semi-conditioned or conditioned attic.

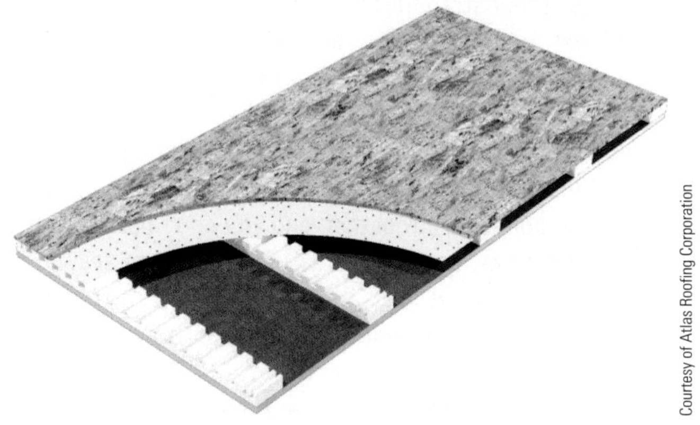

Courtesy of Atlas Roofing Corporation

Figure 7.19 The ACFoam® CrossVent® RB product contains vent spacer strips that separate OSB from the polyisocyanate insulation board. To improve performance, a radiant barrier is applied to underside of the OSB surface during the manufacturing process.

Combustion Equipment in Attics

In many climates, furnaces and sometimes water heaters are located in attics. Often this equipment is atmospherically vented ("open combustion"), using the area around the heater for makeup air for burning the fuel. Open-combustion equipment should not be located in a conditioned or semi-conditioned attic unless it is placed in a sealed combustion closet to isolate the burners from the ambient air. Failure to do so can result in backdrafting of carbon monoxide, which can combine with the air supply in the house and lead to health problems. In addition to the danger that carbon monoxide poses to occupants, backdrafting may cause water vapor from burning fuel to remain in the attic space, increasing the potential for condensation in cold weather.

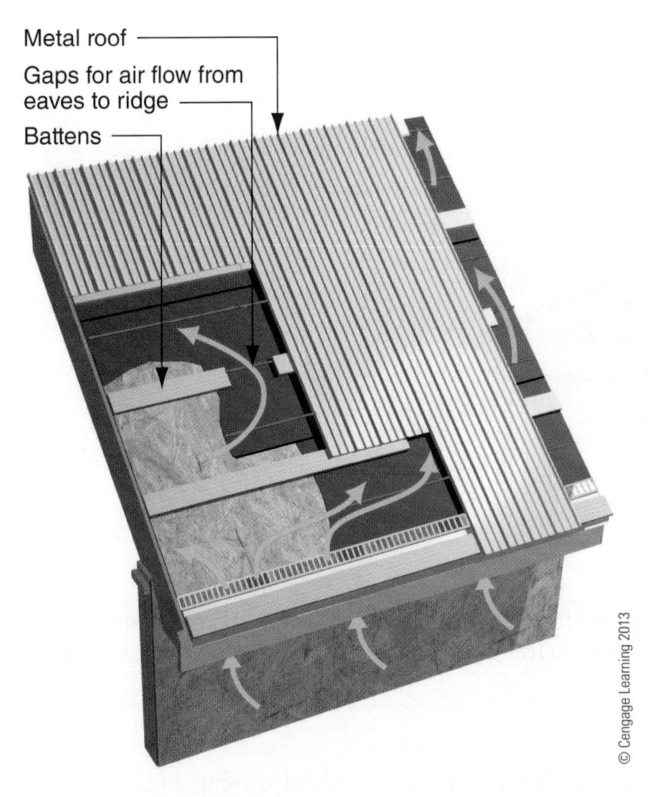

Figure 7.20 Battens provide an air space below the metal roofing.

Ice Dams

Steep-slope roofs in cold climates can suffer from ice dams. Ice dams occur when ice forms on an overhang and causes water buildup behind it to back up under roofing materials (Figures 7.22). Ice dams occur when snow accumulates on roof surfaces that, due to interior air and heat movement, reach temperatures above 32°F. When this happens, snow on the warmer, upper roof melts and refreezes on the colder sections, creating an ice dam at the lower edge of the roof. Above the ice dam, snow is warmed by interior air that reaches the roof deck and continues to melt, backing up and leaking in between the shingles.

The warm roof conditions that cause ice dams are caused by convection of warm air from the living space into the attic, conduction heat loss through the roof structure, and HVAC duct leakage. Ice dams can be effectively eliminated through careful management of air and duct leakage in vented attics or through the use of insulated rooflines. To reduce interior water damage, building codes often require the installation of a waterproof membrane at the bottom end of roofs in areas prone to ice dams; while this may be a good practice, it is not a substitute for proper control of air and duct leakage.

ice dam ice that forms at the eave of sloped roof, causing water build-up behind it to back up under roofing materials.

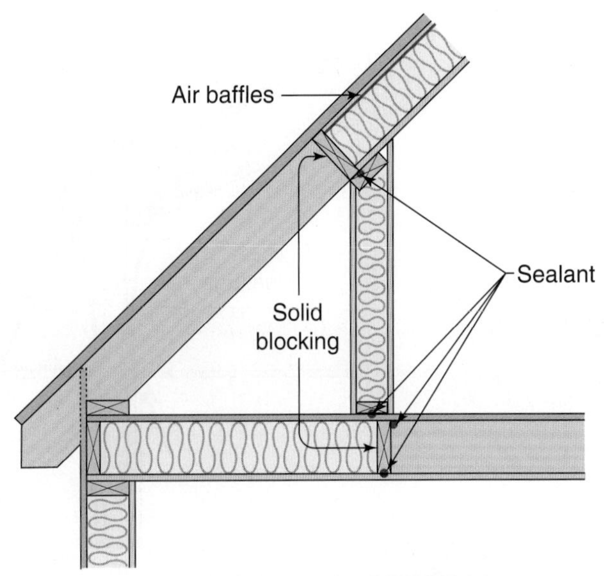

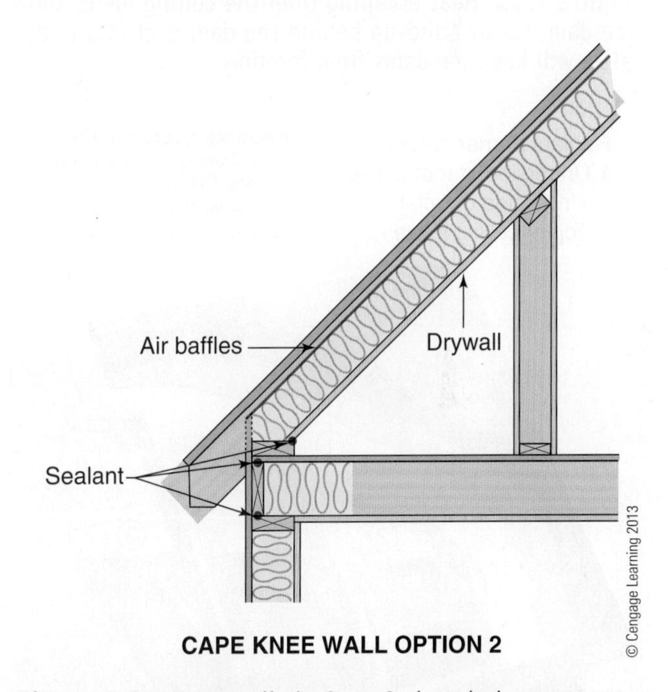

Figure 7.21 Knee walls in Cape Cod–style homes must be framed, air-blocked, and insulated to prevent air from leaking through the insulation.

Radiant Barriers

Vented attics and rooflines can benefit from radiant barriers, particularly in hot climates. A radiant barrier is a thin layer of reflective material, either applied directly to the roof sheathing or installed as a separate film attached to the rafters, that faces into the attic space (Figure 7.23). This metallic layer prevents heat from radiating into the attic space, reducing temperatures on hot days by

radiant barrier a material that inhibits heat transfer by thermal radiation; commonly found in attics.

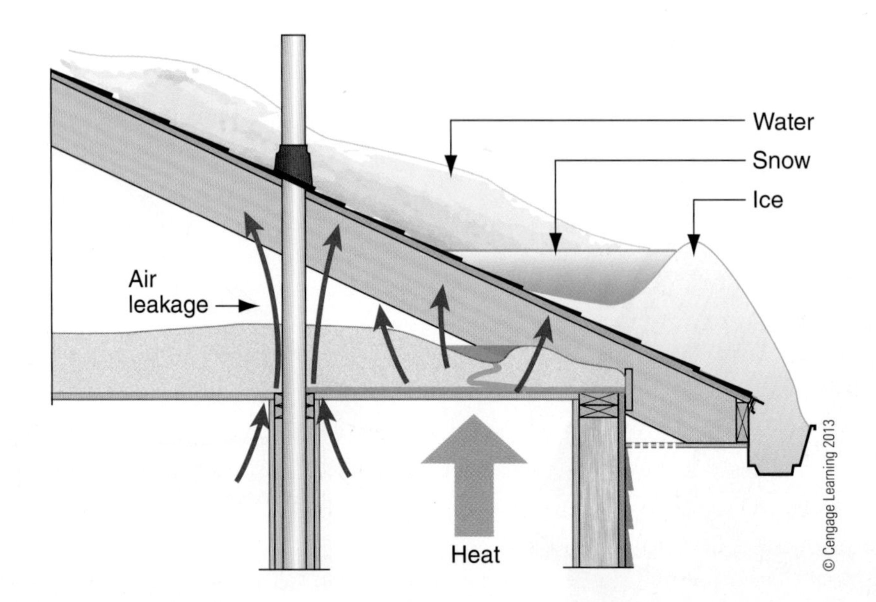

Figure 7.22 Heat escaping from the ceiling melts snow, and water flows to the overhang where it freezes into an ice dam. Water build-up behind the dam backs up under the roofing material. A properly constructed and ventilated attic will keep ice dams from forming.

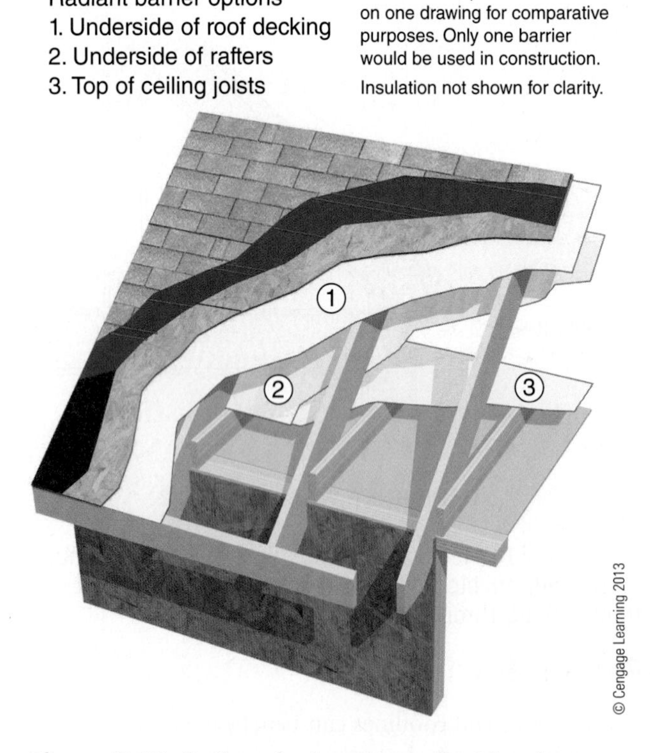

Figure 7.23 Options for locating radiant barriers.

between 17% and 42%, according to the U.S. Department of Energy. Since these barriers only reduce radiant heat, they do not serve their purpose when other materials touch them and allow heat to move into the adjacent material through conduction. Radiant barriers can be installed in insulated rooflines, reducing heat buildup in the roof structure with as little as 0.75" between the barrier and the insulation below. Metal roofing can also serve as a radiant barrier when a vent space is left between the roofing and the decking. While radiant barriers do reduce attic temperatures, when a house is constructed with adequate insulation and air sealing at the ceiling and no HVAC equipment or ducts located in the attic, there is little, if any, effect on the temperature of the conditioned space or overall energy efficiency of the home.

Radiant barriers may also be placed on top of ceiling insulation to increase its performance; however, it does not reduce attic temperatures. This method also allows dust to settle on the radiant barrier, reducing its effectiveness.

Regional Considerations

Roof design and construction decisions can vary between different climates.

In cold climates, roofs should be designed to avoid ice dam formation and to keep vapor from condensing on the underside of cold roof decks. In hot climates, radiant barriers can be used in vented attics containing HVAC systems.

In climates where hurricane and tornado resistance is a concern, design roofs to resist uplift by installing extra hurricane clips where the roof meets the wall and select roofing that will resist high winds. Where hail is a problem, choose a roofing material that is impact-resistant. In regions at risk for wildfires, select roofing that is resistant to fire, and avoid vented roofs that can draw in flames through soffit vents. In coastal areas that endure fierce storms, consider sealed and insulated rooflines to avoid problems with wind-driven rain entering through vents.

Properties of Roofing Materials

The choice of the roof finish for a green home depends on many factors, including shape, durability, the recycled content and recyclability, embodied energy, reflectivity, and availability of both labor and materials. Since the primary purpose of roofing is to keep water out of the house, selecting products that can be installed properly by available labor is key. Regardless of the sustainability of a particular material, the building will not be durable or healthy if poor installation allows water to leak inside.

Durability

When considering the lifetime of the building and the number of times a particular roof finish must be replaced, selecting a material with the longest life reduces both the amount of materials required for replacement and the waste created by removal. More durable products have a higher initial cost, but their total lifetime costs, including repair and replacement, are usually lower than less expensive products.

Recycled Content

Many manufacturers are now incorporating recycled content into their roofing products, particularly metal and rubber. Reducing the amount of virgin materials helps lower the overall environmental impact of a product.

Recyclability

With the exception of certain plastics, most roofing materials are recyclable; however, the availability of resources to accept materials for recycling varies widely by region. Recyclability becomes more critical with materials that have shorter lives. Roofing that must be replaced every 20 to 30 years will create much more waste than one that lasts centuries.

Embodied Energy

The amount of energy required to extract raw materials, manufacture the finished product, and deliver it to the job site are important considerations (Table 7.2). Keep in mind, however, that a material with a higher embodied energy that only requires replacement every 100 years may be a better choice than another with lower embodied energy that requires replacement every 25 years.

Table 7.2 Embodied Energy of Roof Materials

Roofing Material	Btu/ft^2
Clay tile	23,760
Concrete tile	3,520
Slate	11,440
Coated steel	15,840
Aluminum, virgin	48,400
Aluminum, recycled	2,640
Shingles, asphalt	25,080
Shingles, wood	Very low
Membrane structures	High

Source: Green Building Press. http://www.greenbuildingpress.co.uk/archive/sustainable_roofing.php.

Reflectivity

Albedo, or solar reflectance, is the fraction of electromagnetic energy that an object or surface reflects. Albedo includes the visible, infrared, and ultraviolet wavelengths on a scale of 0 to 1. An albedo value of 0.0 indicates that the surface absorbs all solar radiation, and a value of 1.0 represents total reflectivity. More reflective roofs reduce energy used for cooling as well as the urban heat island effect, making reflective roofing a key strategy in green building (Table 7.3).

Table 7.3 Solar Reflectance of Roofing Materials

Roofing Material	Initial Solar Reflectance	3-Year Solar Reflectance
Polyvinyl chloride		
White	0.87	0.61–0.81
Gray	0.67	NA
Thermoplastic polyolefin		
White	0.79	0.69
Gray	0.46	0.43
Ethylene propylene diene terpolymer (EPDM)		
White	0.76	0.64
Black	0.06	0.07
Modified bitumen, white	0.27–0.75	0.28–0.62
Fiber cement	NA	NA
Rubber and plastic	0.25–0.4	NA
Slate	NA	NA
Wood	NA	NA
Metal	0.6–0.7	0.6–0.7
Tile	0.15–0.6	NA
Composition shingles		
White	0.25–0.27	0.26–0.29
Dark	0.04–0.15	NA

Source: Data from Cool Roof Rating Council (http://www.coolroofs.org) and manufacturers' information. NA: data not available.

albedo the ratio of electromagnetic energy that an object or surface reflects.

solar reflectance is a decimal number less than one that represents the fraction of light reflected off the roof; see also *albedo*.

The Cool Roof Rating Council (CRRC) classifies roofing products according to their solar reflectance as well as their thermal emittance. The thermal emittance is the ability of the roof surface to radiate absorbed heat. Both the solar reflectance and thermal emittance of roofing are measured on a scale from 0 to 1, from low to high (Figure 7.24). Roofs with higher solar reflectance and thermal emittance measurements are cooler, allowing less thermal energy to be conducted into the building, potentially lowering air conditioning costs, particularly if HVAC equipment is located in an unconditioned attic. Although this effect may also cause a slight increase in heating costs in the winter, the cumulative effect is beneficial in terms of both energy use and environmental impact.

A study by the Lawrence Berkeley National Laboratory (LBNL) determined that the use of cool roof technologies provide annual cooling energy savings of between 10% and 50%, and peak demand savings of up to 30%. The LBNL Cool Roof Calculator allows you to compare roofing strategies and calculate the estimated annual savings in energy costs. Cool roofs do provide the greatest savings in locations with large cooling loads, but there is little to no energy penalty for using cool roof technologies in heating-dominated climates. The amount of heat provided by solar energy absorbed through roofing is generally quite small, and the gains from reducing peak cooling load can be significant.

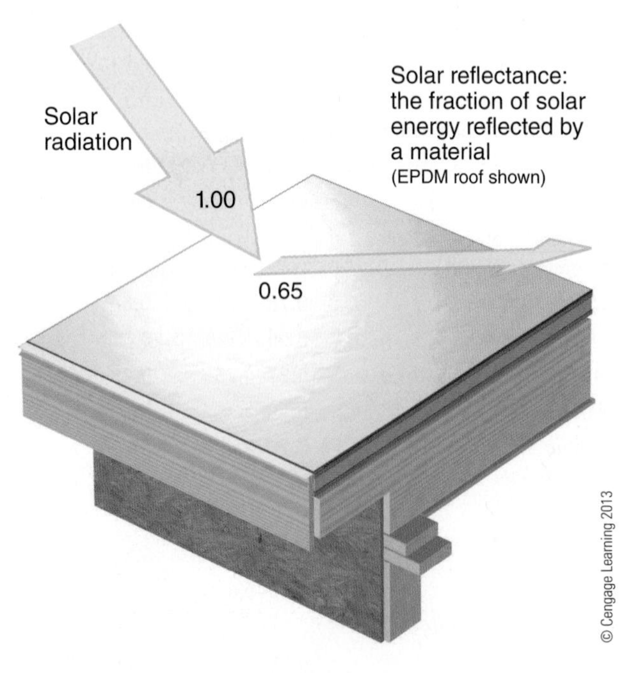

Figure 7.24 Solar reflectance.

Cool Roof Rating Council (CRRC) an independent, nonprofit organization that maintains a third-party rating system for radiative properties of roof-surfacing materials.

thermal emittance a decimal number less than one that represents the fraction of heat that is re-radiated from a material to its surroundings.

Emerging research indicates that the use of reflective coatings over insulated low-slope roofs in regions with clear night skies, such as the American southwest desert, may result in moisture problems.[2] Nighttime radiant cooling can lead to the chilling of surfaces exposed to the night sky with indications that the roof sheathing may remain colder than the outdoor air temperature all winter long. One possible solution is to insulate the topside of the roof sheathing, though this is typically more expensive than fiberglass batts in the roof cavity.

An alternate measurement for cool roofs is the Solar Reflectance Index (SRI). SRI measures a material's ability to reject solar heat and thereby stay cool. Typical values range from 0 to 100, representing low to high ability to reject heat, but some materials may fall outside that range. SRI is defined so that a standard black material (reflectance 0.05, emittance 0.90) is 0 and a standard white (reflectance 0.80, emittance 0.90) is 100. SRI is calculated using the Solar Reflectance and Thermal Emittance ratings that are based on ASTM Standard E 1980. The reflectance of roofs can change over time, and the CRRC addresses this by including both new and aged ratings for roofing materials.

In addition to CRRC ratings, roofing materials can qualify for an ENERGY STAR label based on their reflectivity. ENERGY STAR qualified roofs must meet minimum initial and aged reflectivity ratings (Table 7.4). At this writing, ENERGY STAR does not include emissivity as a factor for their roof label.

Steep-Slope Roofing Materials

Steep-slope roofing materials are available as shingles, tiles, or panels. Designed to shed water, each plane of a steep-slope roof does not need to be fully watertight;

Table 7.4 ENERGY STAR Labeled Roofing

Roof Type	Initial Solar Reflectance Rating	3-Year Aged Solar Reflectance Rating
Steep slope	≥ 0.25	≥ 0.15
Low slope	≥ 0.65	≥ 0.5

[2] Rose, William B. "White Roofs and Moisture in the US Desert Southwest." *ASHRAE*, 2007.

Solar Reflectance Index (SRI) measure of a material's ability to reject solar heat, thereby staying cool; typical values range on a scale from 0 to 100, from low to high ability to reject heat.

instead, it sheds the water as it flows down the surface. Combined with flashing at walls and penetrations, and underlayment installed on the roof deck below, the roofing finish keeps water out of the structure by placing all gaps above the highest level at which water will accumulate during the most severe rain events (Figure 7.25). Proper design, installation, and maintenance are critical to a complete and dry roof system.

Material Choices

Material choices for steep-slope roofing includes standard composition shingles, concrete and clay tiles, metal, slate, plastic, rubber, fiber cement, and wood. The properties of each material (including durability, recycled content and recyclability, embodied energy, and reflectivity) have an effect on the overall sustainability of a project.

Composition Shingles

Composition shingles made of asphalt and glass fiber are the most popular steep-slope roofing material, comprising approximately 60% of the residential market. Wide availability, low cost, and simple installation methods that require no special skills or tools account for their popularity. Composition shingles are available in a variety of styles and thicknesses—a thicker shingle generally means a longer lifespan, although roof slope, local climate, and other factors also play a role in durability. Typically lasting between 20 and 30 years, composition shingles have the shortest life of all steep-slope roofing products. Few composition shingles are made with recycled content, and although they are recyclable, most shingles are discarded instead of recycled when they are removed for replacement. Owens Corning is developing a national recycling program for their shingle products in an effort to increase the amount of recycled waste. With petroleum-based asphalt representing as much as one third of their content, composition shingles have a relatively high embodied energy. Accounting for more frequent replacement than other materials, they have greater overall environmental impact than alternative products. Composition shingles generally have low solar reflectance ratings as well, ranging from 0.04 to 0.15 for black and from 0.25 to 0.27 for white. Manufacturers are now offering medium-colored roofs with solar reflectance close to that of white roofs.

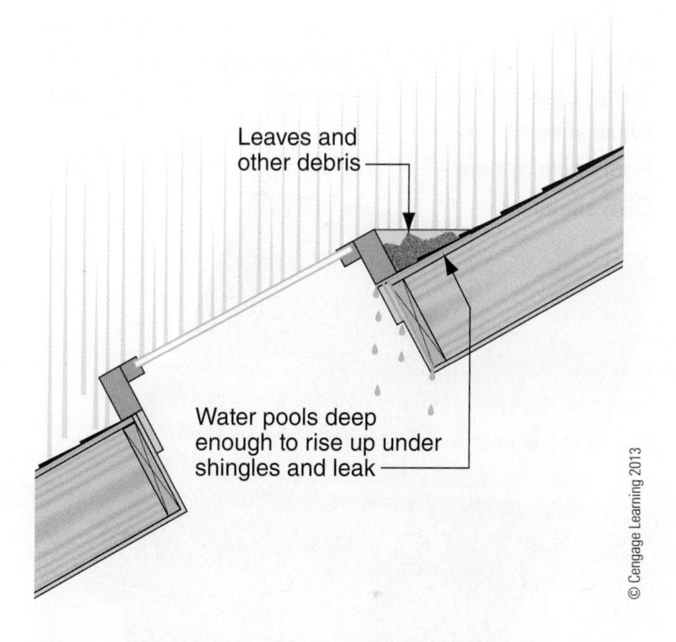

Figure 7.25 Even proper skylight installation cannot overcome poor maintenance. Flashing is installed to keep water out during normal rain events, however if debris is allowed to collect, water can back up higher than the flashing.

Roof Tiles

Roof tiles date back thousands of years and are a common material in many climates. Originally made only of clay, tiles are now also made of concrete that can be designed to look like wood shingles, slate, or clay tiles. Both types of tile are extremely durable, although clay has a reputation of holding its color better than concrete. Individual tiles can be removed and replaced numerous times with practically an indefinite lifespan. Tile roofs have little, if any recycled content; however, they can be ground into aggregate and recycled when no longer reusable. Many clay tile roofs that are hundreds of years old are still functional. Simple to manufacture using locally available materials, clay tiles do require firing that consumes significant amounts of energy; however, their long life and reusability makes them a good option for green buildings. Although they are manufactured out of Portland cement, a high–embodied energy product, concrete tiles do not require high-temperature firing. The concrete versions weigh less than their clay counterparts, which makes them a suitable alternative. Tile roofs generally have initial solar reflective ratings in the range of 0.4 to 0.6, with some white tiles rated as high as 0.8.

Metal Roofing

Metal roofs can be made of galvanized or painted steel, stainless steel, copper, and aluminum, in both sheet and shingle designs. The most common metal roof design is referred to as standing seam. A standing seam

---

composition shingles made of asphalt and glass fiber; the most popular steep-slope roofing material.

standing seam a roof assembled from metal panels with vertical seams that are snapped or crimped together to form a seal.

roof is assembled from metal panels with vertical seams that are snapped or crimped together to form a seal (Figure 7.26). Metal roofing can last 50 years or more, requiring little maintenance.

Steel roofing is available with galvanized or factory-applied permanent coatings that help eliminate rust problems. Uncoated steel roofs require regular recoating to protect from rust. Copper roofing and zinc coating used on galvanized steel can leach chemicals from roofs into waterways where it can be toxic to marine animals. Metal roofing is available with as much as 100% recycled content, and the waste is generally fully recyclable. The manufacture of metal roofing requires significant amounts of energy, but the recycled content and its long life make it an appropriate choice for a green building. Metal roofs' initial solar reflective ratings, the reflectance at the time of installation, are in the range of 0.6 to 0.7. Solar reflectance can diminish as roofs age. Cool roof coatings in medium and dark colors are available that provide very high reflectance ratings equivalent to light colors.

Wood Roofing

Wood shingle and shake roofing can be made from many species, but cedar is the most commonly available. A major challenge with wood roofs on green buildings is the limited availability of sustainably certified materials. The best source for wood shingle roofs, all-heart vertical-grain cedar, comes from old-growth forests that are not a sustainable supply source. Alternative materials include treated pine and oak. Wood shingles, with the exception of those made from pressure-treated pine, can be recycled as can any other wood product. Wood shingles have very low embodied energy, limited to the fuel for harvesting, cutting, and delivering the material. Selecting locally harvested shingles from other than old-growth forests can be a sustainable roofing choice. The CRRC does not provide solar reflectance ratings for wood shingles; however, a report from the U.S. Environmental Protection Agency lists solar reflectance ratings between 0.4 and 0.55 for uncoated wood shingles and shakes.

Wood shingles should be able to dry evenly to avoid warping and cracking. Traditionally installed over stripping or gapped sheathing that allowed for ventilation and drying, modern wood roofs are usually installed on solid sheathing. Using a system that provides ventilation, such as Cedar Breather® (Figure 7.27), allows for even drying and extends the shingle life.

Slate Roofing

Slate is among the most durable roofing materials available. Cut in thin sheets from natural stone, it can last over 100 years and, like clay and concrete tiles, it can be removed and reinstalled and ground and recycled at the end of its life. Salvaged and scrap slate of sufficient thickness can even be reused for flooring. A solid stone, slate is not made with recycled content, but salvaged slate is often available for new buildings and renovation projects. The embodied energy in slate is limited to the extraction, cutting, and delivery of the material. The CRRC does not provide solar reflectance ratings for natural slate.

Figure 7.26 Standing-seam metal roofs are durable, can reflect heat, and are available made with recycled content.

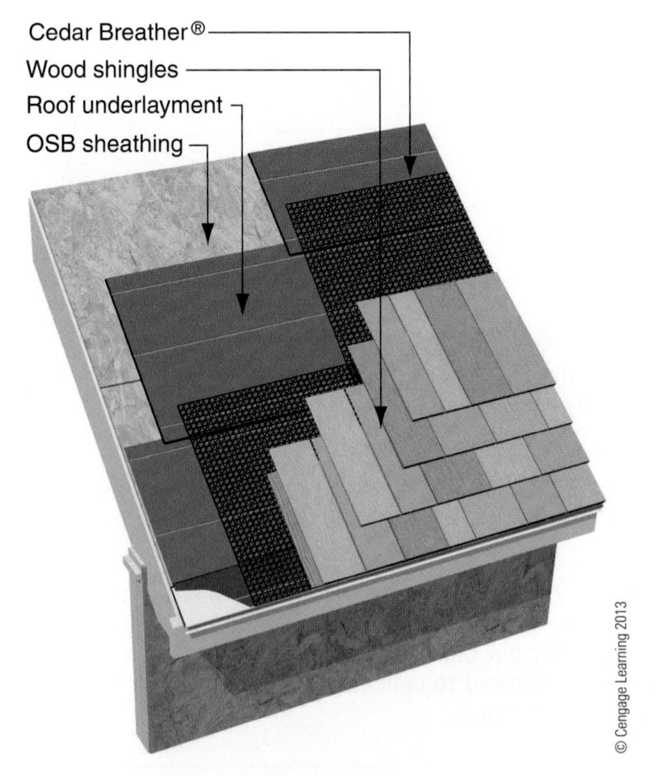

Figure 7.27 Cedar Breather® installation with cedar shingles.

Rubber and Plastic Roofing

Several companies make roof shingles of plastic or rubber, some of up to 100% recycled content. These products are made to look like natural slate or wood shingles and typically have warranties of up to 50 years. Although shingles that are made from recycled materials can be considered sustainable, it is not easy to recycle installation scrap or to recycle the old roof when it is replaced. These products have a moderate level of embodied energy based on the transportation of raw materials and energy used in the manufacturing process. The initial solar reflective ratings of these shingles range from approximately 0.25 to 0.4.

Fiber Cement Roofing

Fiber cement shingles made of cement, sand, clay, and wood fiber are a durable choice for steep-slope roofs, with many carrying warranties of up to 50 years. They are available in designs that replicate wood, slate, and tile roofs. These products were once manufactured with asbestos until it was removed from building products in the 1980s. Most of these products do not use significant amounts of recycled material in their manufacture, nor are they easily recycled. They have a relatively high level of embodied energy due to the Portland cement content and transportation costs for both raw and finished materials. Solar reflectance ratings for fiber cement shingles are similar to plastic and rubber roofing.

Roof Underlayment

Most steep-slope roofing requires a water-resistant underlayment be installed directly on the roof decking. Traditionally, this underlayment has been roofing felt, although new substitute products are lighter weight and provide better slip-resistance for workers on the roof. In addition to providing a backup moisture barrier, underlayment provides temporary protection from weather, allowing the house to be "dried in" before the roofing is installed. One of the newer-generation underlayment products, Huber ZIP System® consists of OSB sheathing with an integrated waterproofing material that is taped at the joints after installation. This technique reduces the chance of damage or swelling to the panels and eliminates the need for a separate moisture barrier underlayment to keep the building dry before the roof is installed (Figure 7.28).

Roofing Felt

Roofing felt, also referred to as tar paper, is made from sheets of paper or felt that are impregnated and coated with tar to provide a water-resistant coating for protecting roofs as well as exterior walls.

Courtesy of DuPont™

Figure 7.28 The Huber ZIP System® provides roof decking and underlayment in a single layer by incorporating a waterproof finish on the OSB and a durable, waterproof tape to seal between panels.

Synthetic Underlayments

Substitutes for roofing felt include polypropylene, olefin, and fiberglass fabric that are inherently water-resistant, eliminating the need for tar coatings. They are lighter than roofing felt, are more resistant to tearing, provide longer lasting temporary protection before the finish roof is installed, and are available with special surfaces that provide more traction for workers on the roof.

Waterproof Membranes

Bituminous membranes, such as Grace Ice & Water Shield® from W.R. Grace, are adhesive-backed sheets that provide superior moisture resistance in critical areas on roofs. Their flexibility makes them excellent for complicated roof details where they will be covered with finish roofing or metal flashing because the membranes will deteriorate if left exposed after construction. One unique property of these membranes is their ability to seal around nails, which makes them a particularly effective underlayment on roofs below 3:12. Since these membranes are very effective at keeping moisture from penetrating them, they should only be installed on completely dry roof decking because they will slow down the drying process to the exterior.

Flashing

Steep-slope roofs require metal flashing to direct rain away from potential points of entry, such as side walls, skylights, and chimneys where it can be directed off the roof and away from the structure. Most flashing remains partially exposed after the roof is complete, so aluminum, copper, galvanized steel, or other durable sheet metal is typically used.

Step and Kick-Out Flashing

All flashing must be designed and installed to provide a barrier to water entry that is always higher than the water level on the roof. Typical flashing is installed at least 4" above the roof level to provide adequate protection in heavy rains. In addition to proper flashing, maintenance and regular cleaning of the roof is necessary to keep a house dry.

Step flashing is installed on walls behind the water-resistant barrier (WRB) and is interlaced with the roof shingles (Figure 7.29). Siding and other finish materials must be raised above the roof to allow for water to flow freely (see Chapter 9 for additional details).

Wherever a roof terminates in a sidewall, kick-out flashing must be installed to direct water away from the wall and back onto the roof (Figure 7.30). Kick-out flashing is one of the most overlooked critical details in roof flashing. Kick-out flashing is installed at the bottom of a roof slope that is adjacent to a wall, preventing roof rainwater from washing down the wall and from getting behind the wall cladding. It can be fabricated of sheet metal on-site or off, or manufactured pieces of metal or plastic can be integrated with the sidewall flashing.

Edge and Rake Flashing

Edge flashing should be installed at the eaves to protect the edge of the roof decking from damage and to direct water off the roof and into gutters. Underlayment must be installed on top of the edge flashing so that any moisture that gets under the roofing is directed over the flashing (Figure 7.31). Rake flashing should be installed to protect the sides of roof, directing any water running off the side of the roof away from the decking and trim. A drip edge is essential for these flashings. The drip edge is a metal strip that extends beyond the other parts of the roof and is used to prevent water from wicking back under the flashing by capillary action and from running directly down the surface of exterior trim.

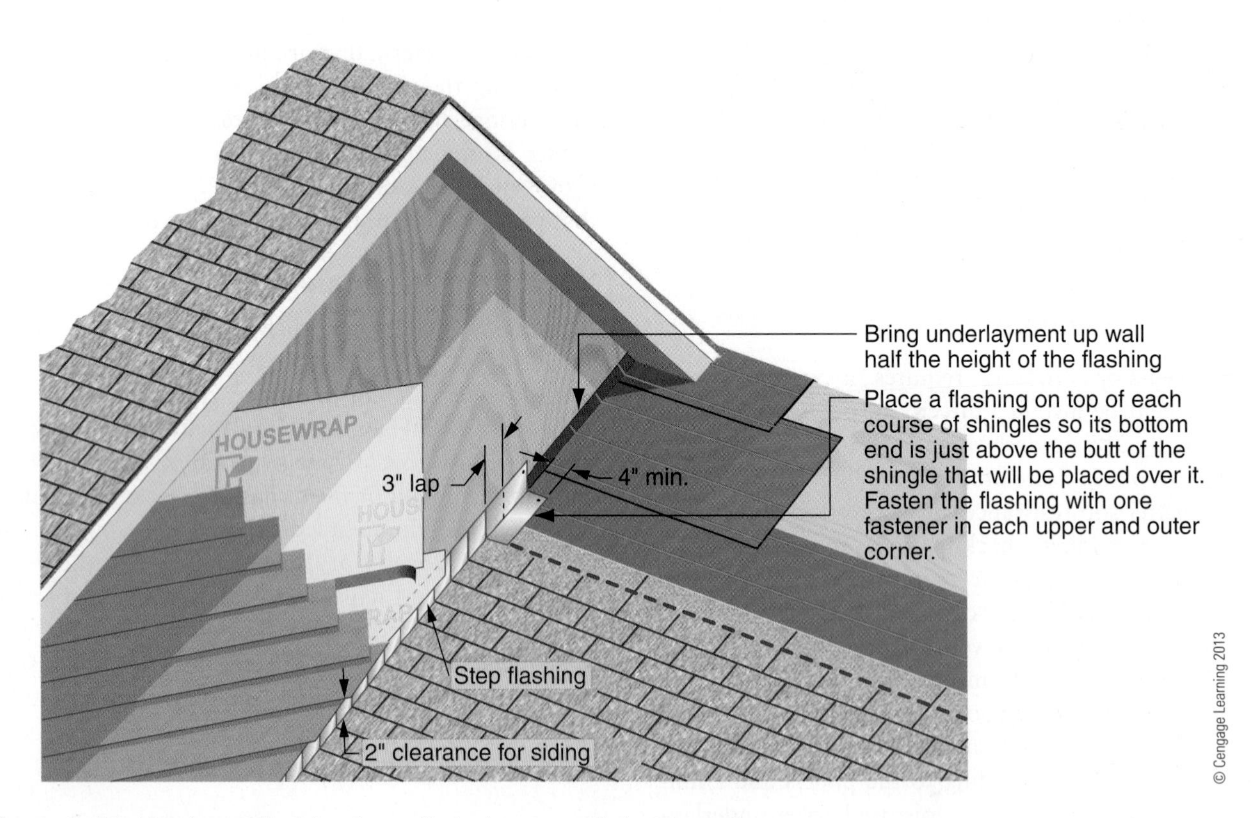

Figure 7.29 Metal step flashing is applied where a roof abuts a wall.

step flashing individual metal pieces installed behind the water-resistive barrier (WRB) and interlaced with the roof shingles.

kick-out flashing a flashing piece installed at the bottom of a roof slope that is adjacent to a wall, preventing roof rain water from getting behind the wall cladding material and WRB.

drip edge a metal strip that extends beyond the other parts of the roof and is used to direct rainwater away from the structure.

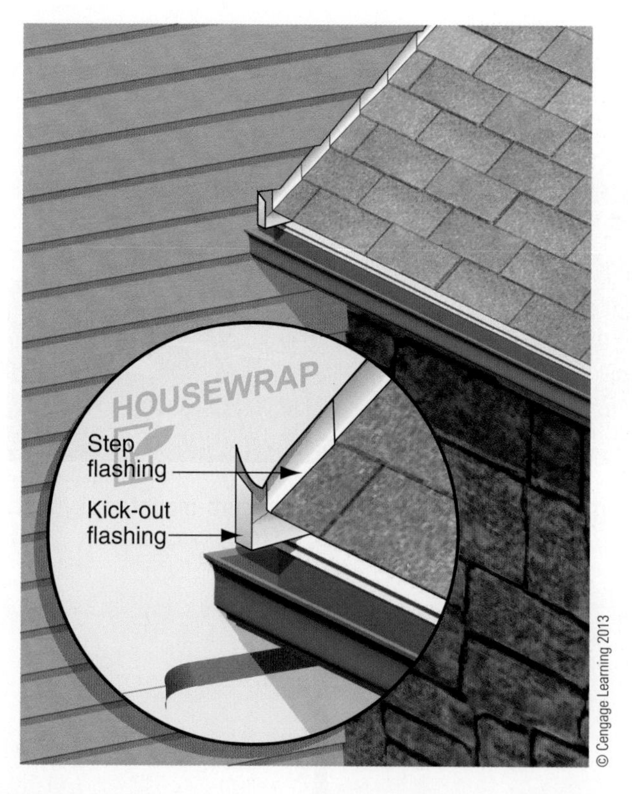

Figure 7.30 Metal kick-out flashing directing water away from the wall and into the gutter.

Chimney Flashing

Chimney flashing poses several challenges and must be detailed and installed carefully to keep water from entering the structure around them. Chimneys are typically made of either solid masonry or wood framing with siding or stucco applied as a finish. All chimneys should have a cap that directs rain away from the top and off onto the roof (Figure 7.32). Solid masonry chimneys may accomplish this with a sloped concrete top, but a secondary cover is highly recommended to keep moisture from getting into the structure and to reduce the amount of water that runs down the sides. Any roof that slopes toward a chimney face must have a cricket or saddle that directs water away from the chimney to the roof slope (Figure 7.32). Crickets can be framed and roofed or constructed completely of metal. Chimney sidewall flashing is done in the same manner as previously described for building sidewalls. For solid masonry chimneys, flashing is placed against the masonry surface and counter flashing is inset into the wall, covering the top of the wall flashing to keep water from running behind the wall flashing.

Skylight Flashing

Many skylight manufacturers now provide high-quality flashing systems with their products that ensure an excellent seal against rain. VELUX® provides an integrated flashing system with their units that includes a labor and

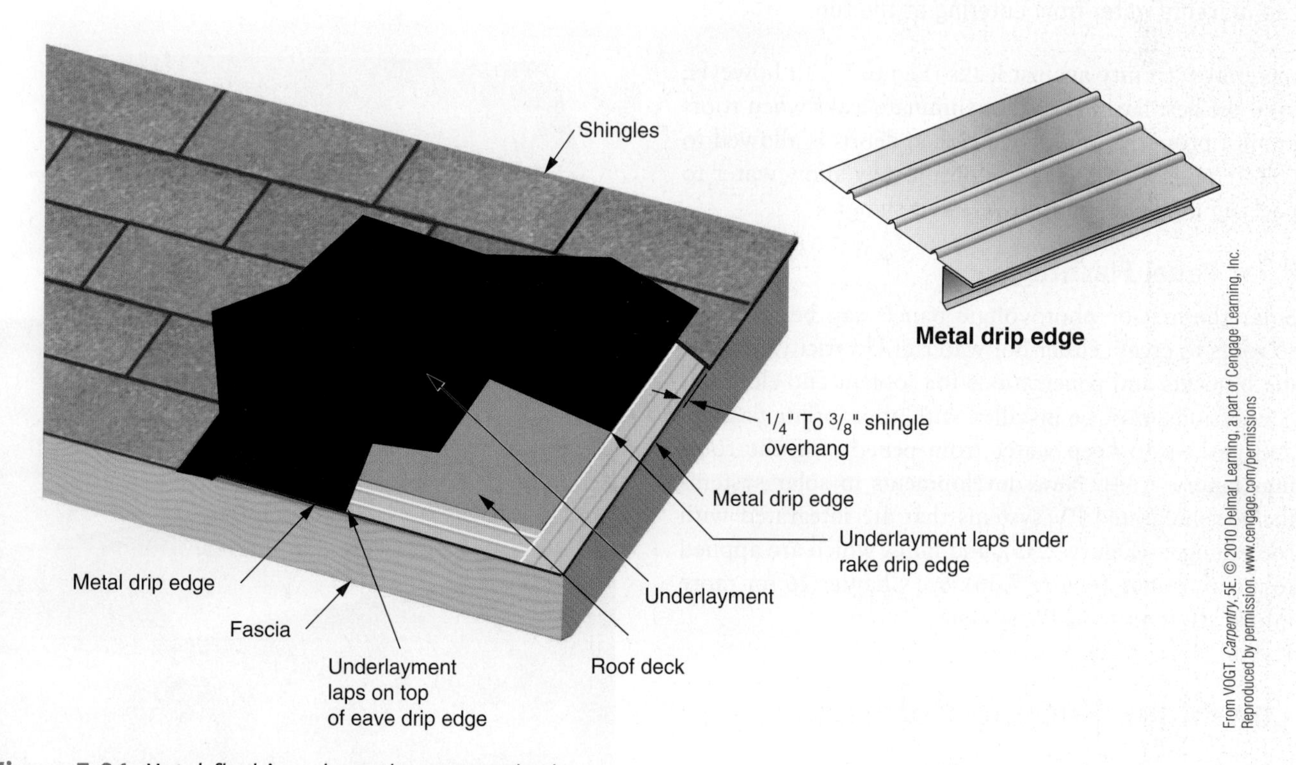

Figure 7.31 Metal flashing along the eave and rake.

cricket a small, false roof built behind a chimney or other roof obstacle for the purpose of shedding water; also called a saddle.

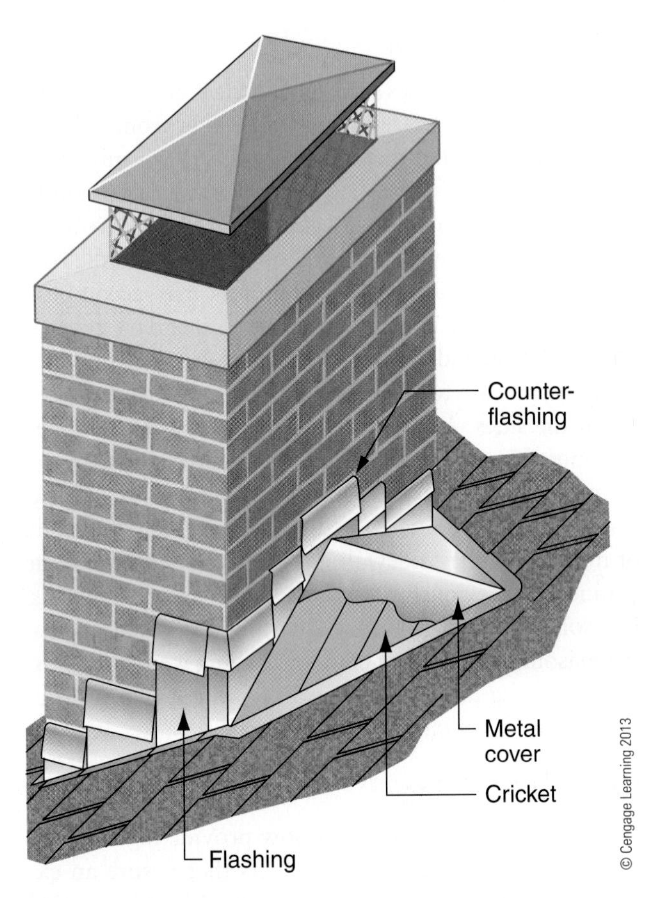

Figure 7.32 A cricket is built to prevent the accumulation of water behind the chimney. The chimney cap prevents water from entering at the top.

Figure 7.33 Step flashing around curb-mounted skylight.

material warranty against leaks (Figure 7.33); however, even the best flashing cannot eliminate leaks when roofs are not properly maintained. When debris is allowed to collect against roof penetrations, it may allow water to back up high enough to leak under shingles.

Solar Panel Flashing

Solar thermal or photovoltaic panels can be mounted to roofs to create either hot water or electricity. Mounting brackets and penetrations for coolant and electrical connections must be installed with appropriate flashing and gaskets to keep water from penetrating the roofing (Figure 7.34). New developments in solar systems include integrated PV systems that are integrated with roof shingles (Figure 7.35), and films, which are applied to metal roofing (Figure 7.36). See Chapter 16 for more information on solar PV systems.

Low-Slope Roofing Materials

Low-slope roofs use less framing and finish materials than steep slopes; however, they are more complicated to install, have a shorter life, require more maintenance, and are more prone to leaks. Fewer material choices are

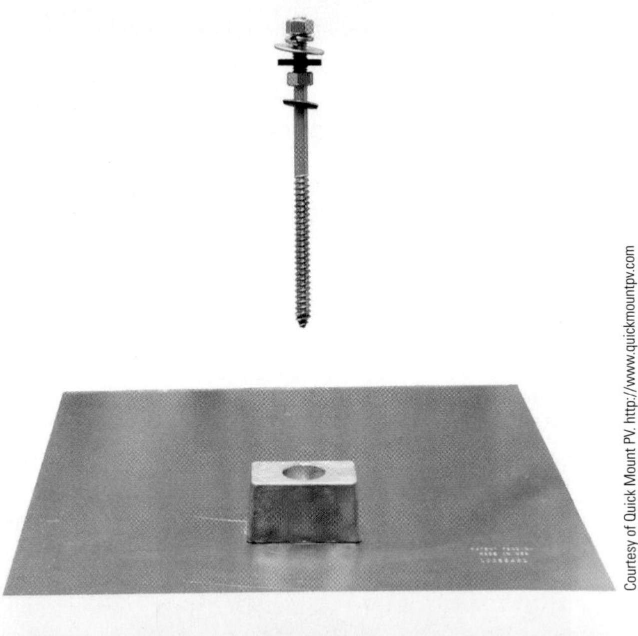

Figure 7.34 Brackets for solar panels must be properly flashed with the roof finish to eliminate the leaks.

Courtesy of Dow Chemical Company

Figure 7.35 Solar PV systems, like DOW™ POWERHOUSE™ Solar Shingle, can be integrated into the roofing shingle.

Courtesy of Fabral

Figure 7.36 Thin film solar PV on standing seam roof.

available for low-slope roofs, and most have little recycled content. They do, however, provide opportunities for outdoor living space and vegetated roofs that steep-slope roofs do not. Because low slopes shed rain more slowly than steep slopes, water is able to back up higher on the roof, often creating temporary pools of standing water. This design factor requires low-slope roofs to be completely watertight, like a swimming pool or shower pan, to keep the house dry. Most low-slope roofs contain plastic, petroleum, or coal tar products that are difficult to recycle. These systems also require high heat to install and release volatile gases in the process.

Low-slope roofs should be designed with as high a pitch as practical and with as much free drainage area as possible (Figure 7.37). **Parapets** are low walls at the edge of a roof, terrace, balcony or other structure. Where it is not possible to drain directly to one side and scuppers or internal drains are required, design the drain capacity for the maximum expected rainfall and allow for overflow locations that direct water away from the structure (Figure 7.38). **Scuppers** are openings in the side of a building or wall, such as a parapet, that allow water to flow out. **Internal drains** are openings in the surface of a low-slope roof that lead to downspouts placed inside the building structure to remove water from the roof.

Figure 7.37 Low-slope roof with as much slope as possible.

Figure 7.38 Low-slope roof with parapet walls and scupper drains.

---

parapet a low wall at the edge of a roof, terrace, balcony, or other structure.

scupper an opening in the side of a building, such as a parapet, that allows water to flow outside.

internal drains are openings in the surface of a low-slope roof that lead to downspouts placed inside the building structure to remove water from the roof.

Downspouts, also called conductors or leaders, are vertical members used to carry water from the gutter downward to the ground.

Traditional low-slope built up roofs are made of hot mopped asphalt combined with four to five layers of roofing felt that together provide a waterproof barrier over the roof deck. Older built-up roofs were made of coal-tar pitch without layers of felt. Both types may be covered with a layer of gravel or a reflective coating to protect the surface from damage and deterioration. While built-up roofs are used on some multifamily homes, they are rarely used in single-family construction, having been replaced by more easily installed alternative membrane products.

Mineral-surfaced roll roofing, essentially the same material used in composition shingles in large sheet form, is sometimes used for low-slope roofs, but it has a limited lifespan as the primary roof covering on low slopes. Durable substitutes to built-up roofing include single-ply membranes made of plastics or modified bitumen. Torch-down modified bitumen is a rolled roof material with a heat-activated adhesive. Single-ply membrane roofing comes in sheets that are attached to the roof deck and seamed together with mechanical fasteners, heat, or chemical solvents. Membranes can be fully adhered or loosely attached with a stone ballast to keep them in place.

Single-ply membranes are typically applied over rigid foam insulation that can be sloped to assist with drainage. It can also be applied directly to the roof deck if insulation is installed below. Membranes are wrapped up walls and attached to metal flashing to provide a complete water seal. Membrane roofs can be made of polyvinyl chloride (PVC), thermoplastic polyolefin (TPO), ethylene propylene diamine monomer (EPDM), or modified bitumen.

PVC

PVC roofing is composed of plastic waterproofing combined with a fiber reinforcement layer. Sheets can be either mechanically fastened or fully adhered to the roof deck with seams heat-welded together. PVC scrap is recyclable, but little if any post-consumer materials are actually recycled, particularly in the residential sector. White, the standard color, has an initial solar reflectivity rating of 0.87. Flexible PVC contains pthalates, a chemical that is considered by many to be a long-term health risk to humans and animals.

TPO

TPO was developed in the 1980s as an alternative to PVC roofing due to environmental concerns. It is attached to roof decks with adhesives or mechanical fasteners, or it can be loose-laid with ballast. TPO may be made with a small amount of post-industrial recycled content, but the material is not recyclable. Available in white, tan, and gray, TPO carries initial solar reflectance ratings of up to 0.79

EPDM

EPDM is a rubber roofing membrane that, while primarily found in commercial construction, can also be used in residential projects. A common substrate for green roofs, EPDM can be mechanically fastened, fully adhered, or loose-laid with ballast. It is not made with recycled content and is not recyclable. The standard color is black, although a white coating can be applied at the factory or in the field to give an initial solar reflectance rating in the range of 0.76.

Modified Bitumen

Bitumen, also known as asphalt or tar, is modified with plastic or rubber to enhance its flexibility, ultraviolet (UV) light resistance, and workability. Typically combined with reinforcing layers of fiberglass or polyester, the final product is a flat roofing membrane shipped in rolls for installation on the job site. Its performance is similar to that of built-up roofing with lower material and energy use. Modified bitumen roofing is attached to the roof deck by heat welding, with applied adhesives, or as a self-sticking membrane. It is not made of recycled materials although waste and removed material may be recycled through the same process as composition shingles. Modified bitumen can have a granular finish or a smooth surface that can receive a cool roof coating. Solar reflectance ratings are in the range of 0.28 for dark colors up to 0.75 for white.

Field-Applied Coatings

Roof coatings can be applied to many low-slope roofing products to provide higher solar reflectance, provide extra durability, and extend the life of older roofs. Some coatings are available with low levels of volatile organic compounds.

---

downspout a vertical member used to carry water from the gutter downward to the ground; also called a conductor or leader.

torch-down modified bitumen a rolled roof material with a heat-activated adhesive.

single-ply membrane roofing material that comes in sheets, which are attached to the roof deck and seamed together with mechanical fasteners, heat, or chemical solvents.

ethylene propylene diamine monomer (EPDM) a single-ply membrane consisting of synthetic rubber; commonly used for flat roofs.

Green Roofs

Sometimes referred to as living roofs or vegetated roofs, **green roofs** employ special plants in a growing medium that covers a membrane roof. Green roofs provide additional insulation, reduce sound transmission, and can extend the life of the roofing membrane by protecting it from UV light and damage. The vegetation helps to reduce roof temperature thereby minimizing the heat island effect as well as helping to reduce the amount of stormwater runoff. These benefits do come with a higher initial price tag due to added costs for both the green roof materials and additional structural reinforcement that may be required for support.

Green roofs are classified as either *intensive* or *extensive*. **Intensive vegetated roofs** use deep layers of soil and can support shrubs and small trees. **Extensive vegetated roofs** use a thin layer of a special growing medium (usually placed over a drainage mat) and require special low-growing, short-rooted plants, such as sedum. Plants must be selected for full sun exposure and the specific climate where they are installed. Intensive roofs can support foot traffic, but they require extra structural support for the extra weight of soil and rain accumulations. Extensive roofs require little, if any, additional roof structure, but they are not designed for foot traffic (Figure 7.39). Both types of roofs require root-resistant roof layers installed over the roof membrane, or the membrane itself must be designed and installed to resist root penetration.

Green roofs are popular in Europe and are continuing to gain a presence in the United States. Although they can contribute to the sustainability of a home, they should not be considered ahead of improving building performance through window orientation, insulation, air sealing, and high-performance mechanical systems.

Figure 7.39 The Eden House in Atlanta, Georgia, was designed to showcase numerous green building techniques. The majority of the roof space is covered with an extensive green roof of small sedums.

Metal

Flat-seam soldered metal roofs are a sustainable solution for low-slope roofs; however, they are complicated to install properly and can be very expensive.

Gutters and Downspouts

Gutters and downspouts help to keep water coming off the roof from soaking the ground around the house, reducing the possibility of foundation leaks. They also reduce splashing at ground level by keeping water off the sidewalls, which helps reduce decay of siding and trim. Gutters should be installed to carry the flow of the heaviest rain without overflowing, and the downspout exits should be extended at least 5' away from the foundation (Figure 7.40). Gutters can be made of aluminum, galvanized steel, copper, or vinyl. When gutters are installed on a metal roof, the gutter material should be compatible with the roof metal to avoid galvanic corrosion that occurs between different materials.

Gutters must be maintained regularly to remove leaves and other debris to keep them from clogging. Gutters that are filled with debris can be worse than having none at all—backed-up gutters can overflow and soak the foundation, damage the roof structure, and exacerbate ice dams in cold climates.

Effective gutter guards that keep debris out and allow water to enter are a good alternative to regular cleaning. Several manufacturers of high-quality gutter guards, such as Gutter Helmet®, offer lifetime warranties against clogging.

---

green roof a roof that is partially or completely covered with vegetation and a growing medium, planted over a waterproofing membrane; also known as a *vegetated* or *living roof*.

intensive vegetated roofs a type of green roof containing deep layers of soil that can support shrubs and small trees.

extensive vegetated roofs a type of green roof that uses a thin layer of a special growing medium (usually placed over a drainage mat) and requires special low-growing, short-rooted plants, such as sedum.

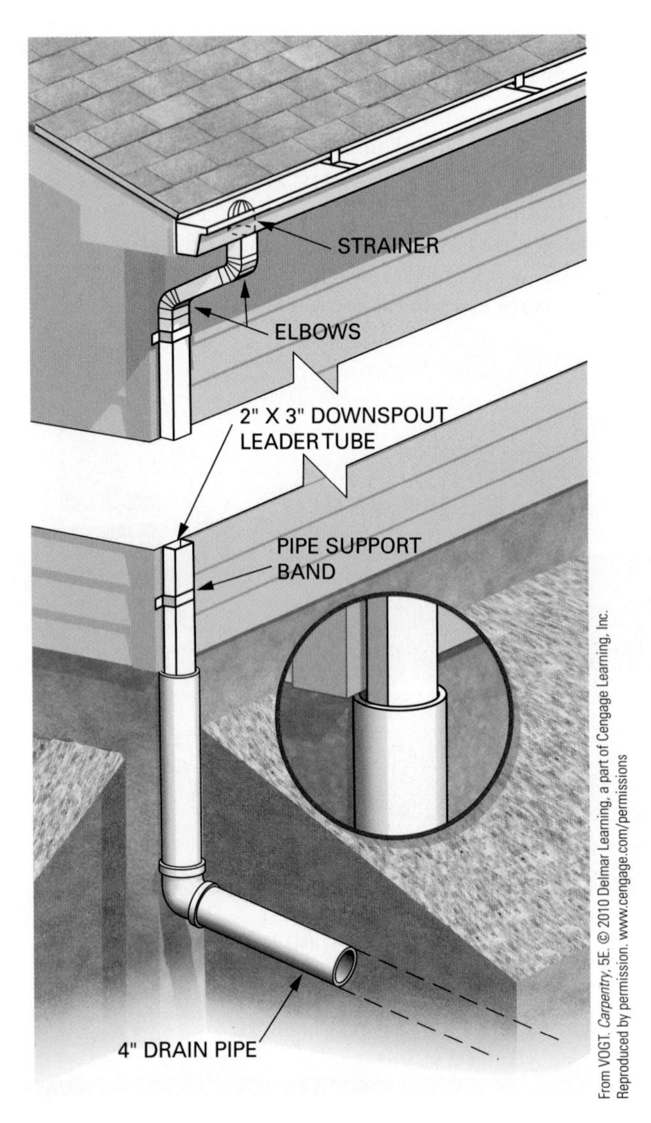

Figure 7.40 Gutter downspout run into drain pipe to direct away from site.

Rainwater Collection

Rainwater collection is a very effective and moderately priced path to water efficiency. When considering rainwater collection for irrigation or potable use, the choice of roofing materials will have an effect on the quality of the water. Metal roofs provide the cleanest water (generally free of debris or toxic chemicals), whereas composition shingle roofs can release toxic chemicals into runoff (see Chapter 15 for more information).

Remodeling Considerations

When undertaking a remodeling project, the roof and attic provide excellent opportunities to create a sustainable project. If the project includes expansion into an existing attic area, start with a thorough evaluation of the existing structure and roofing material. In some cases, the existing ceiling joists will be capable of supporting the new floor loads, saving the cost of a new floor system. Regardless of whether or not a new floor system is installed, the exterior edges of the floor structure should be completely insulated and air sealed before covered with decking to complete the thermal envelope.

The depth of existing rafters will dictate the amount of insulation that can be installed along the roofline. If the roofing is to be replaced, consider adding rigid insulation on top of the roof decking; if the existing roof is to be retained, you may furr down rafters to allow for additional insulation below the roofline.

When adding insulation in ceilings, any old knob-and-tube wiring should be replaced to avoid potential fire hazards. Old ceiling insulation that will be disturbed should be tested for any asbestos content; if asbestos is present, use a certified abatement contractor to remove and dispose of the material properly.

Existing attics may have decking on the floor that compresses insulation. This decking should be removed, and additional framing should be added to allow for proper depth insulation. After the insulation is installed, the decking can be replaced on top of the new, deeper ceiling structure.

Evaluation of roofs and attics includes visual inspections, blower door testing, and infrared cameras. Trained thermographers using infrared cameras can help identify minor water leaks that remain in the structure, as well as insulation gaps and thermal bypasses.

Summary

Deciding the type of roof structure, thermal envelope location, and roofing materials must be considered with respect to the climate, the overall building design, budget, available materials and trade contractor skills, resource efficiency, recycled content, embodied energy, and the health of installers and the final occupants.

The more complicated the roof structure, the more difficult it becomes to direct water away from the structure and off the house. Avoid designs that require complex flashing details. Install all flashing to direct water away from the structure, and never rely on caulking or other sealants to keep out the water; instead, the materials should be layered to shed water under all conditions.

Make sure that the roof or attic creates a complete thermal envelope at the ceiling plane or at the roofline, or both. Gaps in the thermal envelope reduce energy efficiency and air quality.

Review Questions

1. What is the most critical component of roof design?
   a. Insulation
   b. Ventilation
   c. Water management
   d. Roof pitch
2. Which of the following situations would suggest using a conditioned attic?
   a. Simple ceiling
   b. Sealed combustion furnace in attic
   c. Open combustion furnace in attic
   d. Raised-heel trusses
3. Which of the following situations would suggest not using an unconditioned attic?
   a. Sealed combustion furnace in attic
   b. Raised-heel trusses
   c. Air source heat pump in attic
   d. Multiple recessed can lights
4. Which system does not provide for a conditioned attic?
   a. SIPS
   b. Rigid foam on roof deck
   c. Insulation on ceiling with ridge and soffit vents
   d. Spray foam on roofline
5. Which is not true of radiant barriers?
   a. They should be in direct contact with insulation
   b. There should be at least 0.75" of space next to the foil
   c. The foil surface should always face inward
   d. They should not installed with unvented insulated rooflines
6. Which of the following roofs has the highest initial solar reflectance according to the Cool Roof Rating Council (CRRC)?
   a. White Polyvinyl Chloride
   b. White EPDM
   c. Slate
   d. White Composition Shingles
7. Which of the following do not help eliminate ice dams?
   a. Duct sealing in basement
   b. Waterproof membrane on roof
   c. Air sealing at ceiling and knee walls
   d. Eliminating thermal bridging in roof
8. What type of flashing must be installed where a roof eave terminates into a two-story wall?
   a. Saddle
   b. Kick-out
   c. Step
   d. Rake
9. What solar reflectance is most appropriate in a cold climate?
   a. Low
   b. High
   c. Medium
   d. Solar reflectance does not matter in cold climates
10. Which is the most effective method to ventilate an unconditioned attic?
    a. Power roof vent with soffit vents
    b. Solar-powered roof vent with gable end vents
    c. Ridge vents with soffit vents
    d. Ridge vents with gable end vents

Critical Thinking Questions

1. In what situations would you consider constructing an unvented roof assembly?
2. Compare and contrast different roof designs and how their applications differ in hot, moderate, and cold climates.
3. What strategies would you incorporate into a house to avoid ice dams?
4. Identify critical locations for insulation and air sealing in unconditioned attic structures.

Key Terms

albedo, 197
composition shingles, 199
Cool Roof Rating Council (CRRC), 198
cricket, 203
dormer, 183
downspout, 206
drip edge, 202
eave baffles, 190
energy truss, 187
ethylene propylene diamine monomer (EPDM), 206
extensive vegetated roofs, 207
gable roof, 183
green roof, 207
gutter, 183
hip roof, 183
hot roof, 192
ice dam, 195
intensive vegetated roofs, 207
internal drains, 205
kick-out flashing, 202
low-slope roof, 182
parapet, 205
radiant barrier, 195
raised-heel truss, 187
roof decking, 188
roof slopes, 182
scupper, 205
shed roof, 183
single-ply membrane, 206
standing seam, 199
solar reflectance, 197
Solar Reflectance Index (SRI) , 198
steep-slope roof, 182
step flashing, 202
thermal emittance, 198
torch-down modified bitumen, 206
unvented roof, 192

Additional Resources

Cool Roof Rating Council (CRRC):
http://www.coolroofs.org

U.S. Department of Energy (DOE) Cool Roof Calculator:
http://www.roofcalc.com

Green Roofs for Healthy Cities North America (GRHC):
http://www.greenroofs.org

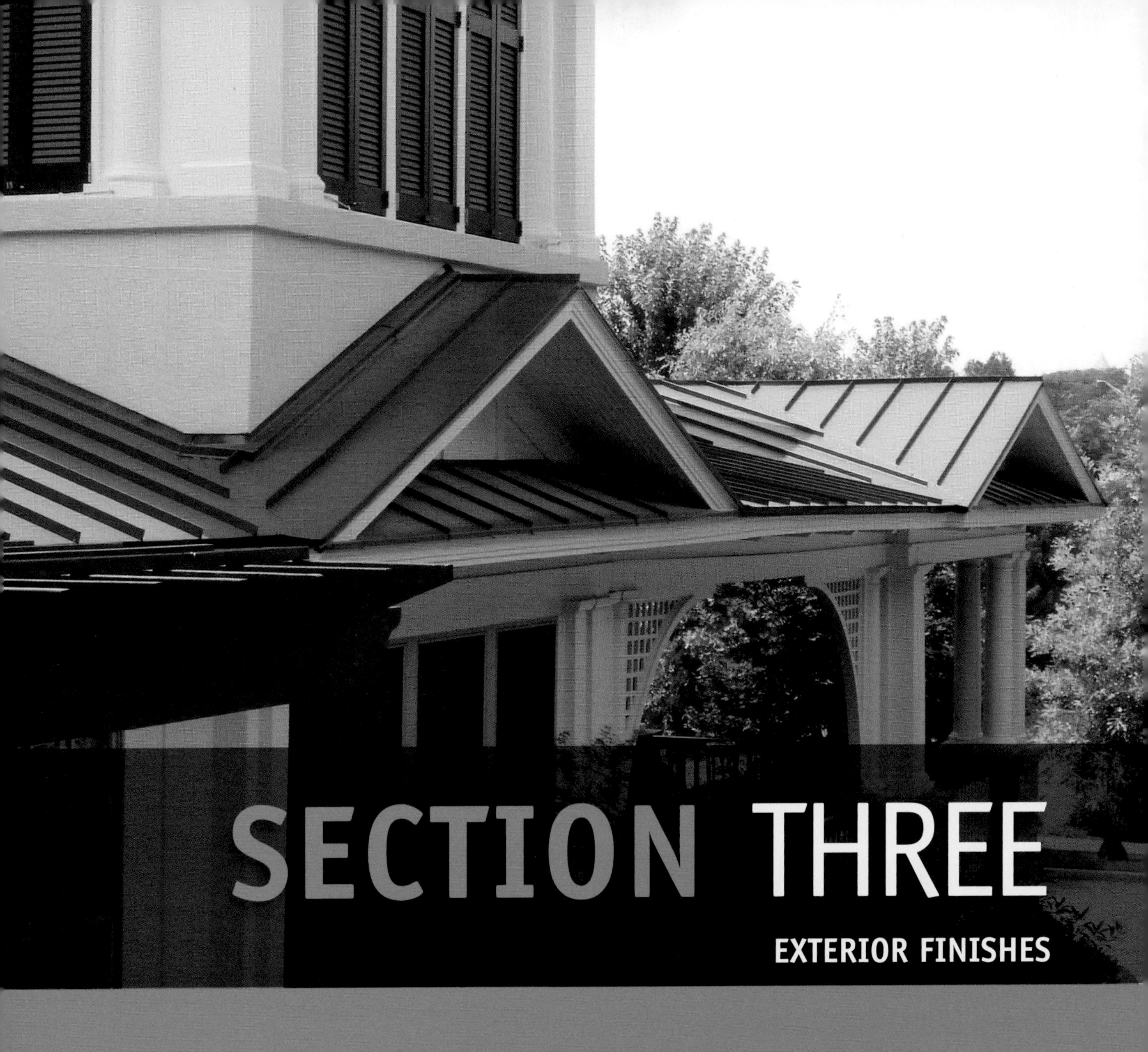

SECTION THREE

EXTERIOR FINISHES

Fenestration

Fenestration describes all the products that fill openings in a building envelope, including windows, doors, and skylights that allow air, light, people, or vehicles to enter. In this chapter, we will discuss how windows, doors, and skylights offer both benefits and challenges for green homes. They can provide daylighting, ventilation, emergency egress, and "free" heat in the winter. Challenges include the energy penalties from allowing heat to escape in the winter and enter in the summer. All fenestration must be installed and flashed properly to keep damaging moisture out of the structure. Proper selection, location, and installation of windows, doors, and skylights are some of the most critical elements of green homes.

LEARNING OBJECTIVES

Upon completion of this chapter the student should be able to:

- Describe fenestration's effects on energy efficiency
- Describe the performance ratings of windows, doors, and skylights
- Convert U-factors into R-values
- Describe the pros and cons of various window and door materials
- Describe the different types of windows, doors, and skylights
- Describe proper window, door, and skylight installation
- Explain how to select the most appropriate products for a specific project

Green Building Principles

 Energy Efficiency

 Resource Efficiency

 Durability

Indoor Environmental Quality

Types of Fenestration

Fenestration describes all the products that fill openings in a building envelope and allow air, light, people, or vehicles to enter. Fenestration types differ by their operation, glass, and frame. Window sashes may be operable or fixed and may slide vertically or horizontally, or swing in or out. Glazing refers to the glass in fenestration. For example, windows may be single, double, or triple glazed. The window or door frame is the fixed portion of windows and doors that is connected to the building structure. The choices available for each of these criteria can have either a positive or negative effect on the performance of a green home. The choice of fenestration used in a project is affected by the climate, building design, architectural style, and solar orientation.

Fenestration Selection

Making appropriate decisions in the selection of fenestration are critical to creating a sustainable home. These decisions include the size, location, shading, glazing,

---

fenestration describes all the products that fill openings in a building envelope, including windows, doors, and skylights that allow air, light, people, or vehicles to enter.

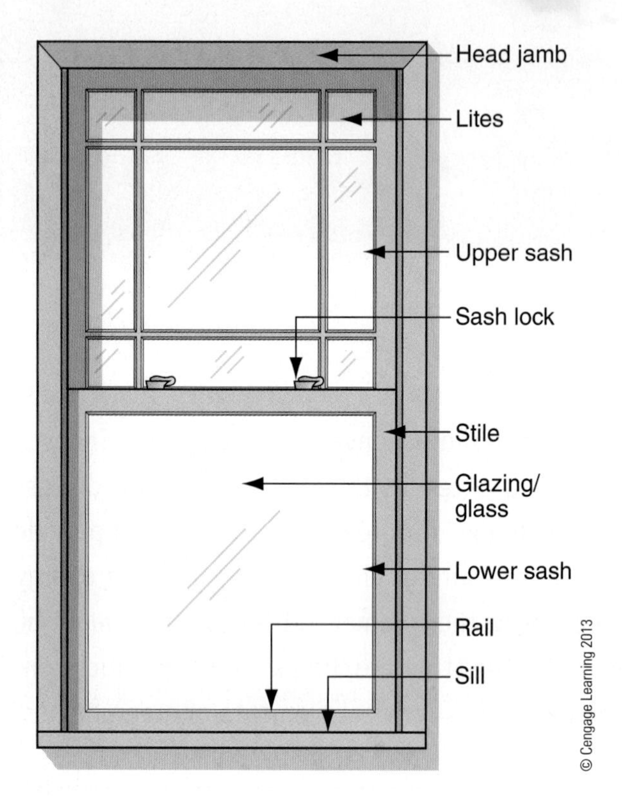

Figure 8.1 Anatomy of a double-hung window.

sash, and frame materials. Additional considerations include the unit operation, hardware and weatherstrip, grilles or muntins, and installation method (Figure 8.1).

Understanding the multiple purposes of windows, doors, and skylights is key to choosing the correct products for each project. They are all capable of providing natural light, passive cooling and outside air through ventilation, heat through **solar gain**, and views to the outside. **Daylighting** is the use of natural light to supplement or replace artificial lighting. Daylighting can reduce the amount of electricity required to light home interiors, but if too much heat enters during the cooling season, or too much exits during the heating season, it will more than offset all the lighting energy saved. Improperly installed windows, doors, and skylights will allow air and moisture to enter the home, wasting energy and reducing durability and indoor environmental quality.

Natural light has been shown to improve productivity in offices and schools and reduce recuperation time in hospitals. We can assume that it will have similar positive effects on residential occupants. Passive cooling and ventilation, when properly managed, help reduce the energy needed to operate ventilation and air-conditioning equipment. Providing pleasing exterior views helps make a home more desirable, reducing the probability of major renovation or demolition and replacement.

The size, location, and style of windows, doors, and skylights are primarily a design decision, affecting both the interior and exterior elevations of a house. When these design decisions are made in the absence of the house as a system concept, the home's efficiency and comfort will inevitably suffer. We will discuss this in more detail later in the chapter.

The Effect of Fenestration on Efficiency and Comfort

The energy efficiency of fenestration is a product of radiant heat gain, convection, heat conduction, and air leakage. In cold climates, heat gain can reduce the need for additional energy; in hot climates, the same heat gain can increase that need. Understanding and comparing the energy efficiency of different windows, doors, and skylights is critical to making the most appropriate choices for a project.

Fenestration in general, and windows in particular, affect efficiency and comfort through four specific principles: radiation, convection, conduction, and air leakage.

- *Radiation* is the movement of heat as infrared energy through glass. Most radiant heat originates outside and enters the building, although smaller amounts of heat radiate outward in cold weather.
- *Convection* occurs in cold weather when warm interior air loses heat as it comes in contact with the colder glass surface, causing the cooler air to sink down toward the floor. This movement draws warmer air toward the glass, creating air currents that reduce interior comfort. Note that this type of convection is a form of uncontrolled air movement that differs from the intentional forced convection used for heating or cooling a home discussed in Chapter 2.
- *Air leakage*, which is directly related to the convection of Chapter 2, causes energy loss through gaps between sashes, frames, and other components. Air leakage affects efficiency, moisture vapor control, and comfort—particularly in cold climates where drafts are more apparent.
- *Conduction*, as discussed in Chapter 2, is the direct transfer of heat through a solid. In the case of fenestration it moves through the sash, frame, and the glass at different rates. The rate of conduction through the entire unit, along with radiation, convection, and air leakage, determine the U-factor. Conduction heat loss in cold climates generally exceeds the heat gain in warm climates (refer back to the ΔT discussion in Chapter 6).

---

solar gain the heat provided by solar radiation.

daylighting the use of natural light to supplement or replace artificial lighting.

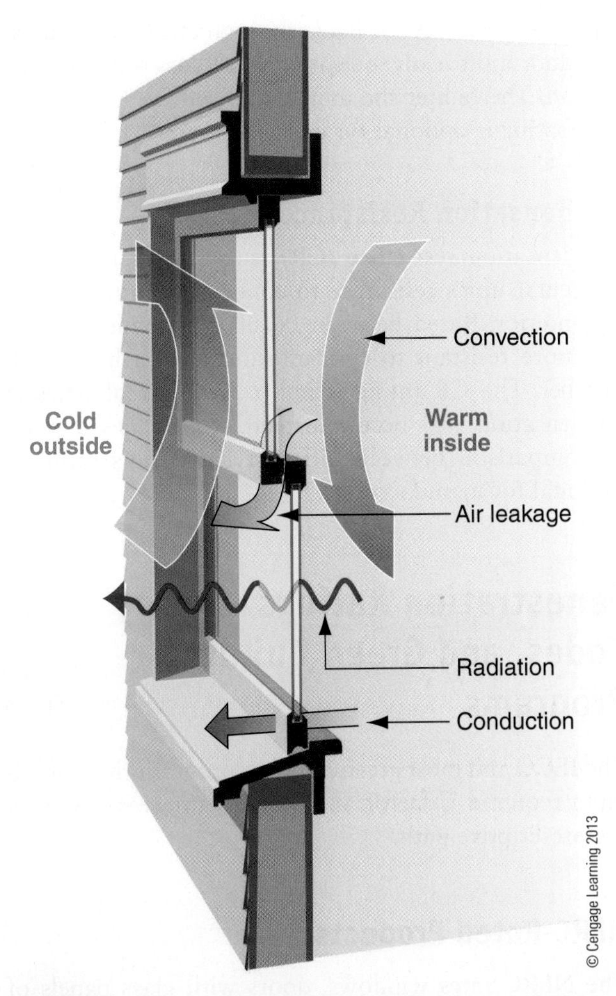

Figure 8.2 Heat flows through windows by radiation, convection, air leakage, and conduction.

Figure 8.2 provides examples of radiation, convection, air leakage, and conduction through windows. Air leakage is important in all climates although radiation is most critical in warm climates, and conduction is most important in cold climates. These climate differences are reflected in the ENERGY STAR and green building program criteria for windows (see Table 8.2).

National Fenestration Research Council (NFRC)

The **National Fenestration Rating Council (NFRC)** rates and certifies most fenestration products. These ratings are used by energy codes to identify which products meet the minimum requirements, and by ENERGY STAR to determine which products are entitled to use their label.

Following the energy crisis of the 1970s, the fenestration industry in the United States began to develop many new efficient technologies for windows, doors, and skylights. As is often the case with new products, many manufacturers' claims exceeded their actual performance. As consumer complaints increased, the government intervened to investigate deceptive claims in the industry.

A group of industry leaders met in 1989, founding the National Fenestration Research Council (NFRC) to develop verification and performance standards for their products. The NFRC is a nonprofit organization that administers an independent energy performance labeling system for fenestration products. Their goal is to provide information for industry professionals and individuals to use in comparing products and making purchasing decisions. Other groups utilizing these ratings include government and utility energy efficiency programs and building officials who work in code development and enforcement. In addition to providing manufacturers with a point of reference to compare their products with their competition, the NFRC ratings provide objective criteria for them to use in marketing efforts.

Required criteria for a window, door, or skylight to be rated by the NFRC are the U-factor, solar heat gain coefficient (SHGC), and visible transmittance (VT). Additional optional ratings are air leakage (AL) and condensation resistance (CR). Both the required and optional ratings include the entire unit (glazing, sash, and frame) and are listed on the window label (Figure 8.3).

Figure 8.3 A typical window label from the National Fenestration Rating Council (NFRC).

National Fenestration Research Council (NFRC) a nonprofit organization that administers a uniform, independent rating and labeling system for the energy performance of windows, doors, skylights, and attachment products.

Required NFRC Ratings

Manufacturers who choose to provide NFRC ratings for their products are required to include the U-factor, SHGC, and VT on their labels. These ratings provide critical information regarding energy efficiency and how clear the glass appears.

U-Factor

The U-factor, also referred to as the U-value, is the measurement of the heat loss of a particular window or door assembly, including the glazing, sash or slab, and frame. U-factors usually range between 0.20 and 1.20. A lower U-factor indicates a higher resistance to heat flow. As we discussed in Chapter 2, the U-factor is the mathematical inverse of the R-value and vice versa. For example, a U-factor of 0.33 is the same as R-1.

Solar Heat Gain Coefficient

Solar heat gain coefficient (SHGC) is the measurement of how much of the sun's radiation a window or door will block. The SHGC describes the fraction of solar radiation that is released to the interior. SHGC is defined by a number between 0 and 1. A lower SHGC indicates less heat is transmitted through the unit.

Visible Transmittance

Visible transmittance (VT) is a measurement of how much light comes through glazing. VT is defined by a number between 0 and 1; the more light admitted, the higher the VT.

Optional NFRC Ratings

In addition to the previously discussed criteria, optional NFRC ratings are available for manufacturers who wish to include them to provide their customers. Air leakage and condensation resistance ratings provide consumers with additional information to help in their purchasing decisions and to differentiate products from the competition.

Air Leakage

Air leakage (AL) is a measurement of the total amount of air leakage, equivalent to the total cubic feet of air passing through a square foot of window area in one minute (cfm/ft^2). Air leakage through cracks in window and door units leads to heat gain and loss through infiltration. The tighter the unit, the lower the AL number. This rating is optional for manufacturers.

Condensation Resistance

Condensation resistance (CR) is a measurement of a particular unit's resistance to condensation forming on the interior. Rated between 0 and 100, windows that are more resistant to condensation have a higher CR number. The CR rating is not a predictor of whether condensation will occur; instead, it provides a point of comparison between different units. This rating is optional for manufacturers.

Fenestration Ratings, Energy Codes, and Green Building Programs

The IECC and most green building programs have minimum required U-factor and SHGC ratings when using the prescriptive path.

NFRC-Rated Products

The NFRC rates windows, doors with glass panels of 410 in^2 or more, glass skylights, and tubular daylighting devices (TDD), which are often referred to as solar tubes or sun tunnels. The NFRC also provides ratings for SHGC of applied window films and dynamic glazing products. Examples of dynamic glazing (DG) products include glass that changes properties electronically via an electric current, and glazing with blinds between glass layers that control light and heat.

2009 International Energy Conservation Code (IECC)

The prescriptive path of the 2009 International Energy Conservation Code (IECC) requires minimum U-factor and SHGC ratings based on climate zone (Table 8.1).

---

solar heat gain coefficient (SHGC) the fraction of solar radiation admitted through glazing.

visible transmittance (VT) a measurement of how much light comes through glazing.

air leakage (AL) a measurement of the total amount of air leakage, equivalent to the total cubic feet of air passing through 1 ft^2 of window area per minute (cfm/ft^2).

condensation resistance (CR) a measurement of a particular unit's resistance to condensation forming on the interior.

tubular daylighting devices (TDD) a cylindrical skylight with a reflective tube to provide daylight to interior rooms.

dynamic glazing (DG) products either glass that changes properties electronically via an electric current, or glazing with blinds between glass layers that control light and heat.

Table 8.1 Prescriptive Fenestration Requirements

Table N1102.1 Insulation and Fenestration Requirements by Component[a]											
Climate Zone	Fenestration *U*-Factor	Skylight[b] *U*-Factor	Glazed Fenestration SHGC	Ceiling *R*-Value	Wood Frame Wall *R*-Value	Mass Wall *R*-Value[k]	Floor *R*-Value	Basement[c] Wall *R*-Value	Slab[d] *R*-Value and Depth	Crawl Space[c] Wall *R*-Value
1	1.2	0.75	0.35[j]	30	13	3/4	13	0	0	0
2	0.65[i]	0.75	0.35[j]	30	13	4/6	13	0	0	0
3	0.50[i]	0.65	0.35[e,j]	30	13	5/8	19	5/13[f]	0	5/13
4 except Marine	0.35	0.60	NR	38	13	5/10	19	10/13	10, 2 ft	10/13
5 and Marine 4	0.35	0.60	NR	38	20 or 13 + 5[h]	13/17	30[f]	10/13	10, 2 ft	10/13
6	0.35	0.60	NR	49	20 or 13 + 5[h]	15/19	30[g]	10/13	10, 4 ft	10/13
7 and 8	0.35	0.60	NR	49	21	19/21	30[g]	10/13	10, 4 ft	10/13

[a] *R*-values are minimums. *U*-factors and solar heat gain coefficient (SHGC) are maximums. R-19 batts compressed in to nominal 2 × 6 framing cavity such that the *R*-value is reduced by R-1 or more shall be marked with the compressed batt *R*-value in addition to the full thickness *R*-value.

[b] The fenestration *U*-factor column excludes skylights. The SHGC column applies to all glazed fenestration.

[c] The first *R*-value applies to continuous insulation, the second to framing cavity insulation; either insulation meets the requirement.

[d] R-5 shall be added to the required slab edge *R*-values for heated slabs. Insulation depth shall be the depth of the footing or 2 feet, whichever is less, in zones 1 through 3 for heated slabs.

[e] There are no SHGC requirements in the Marine Zone.

[f] Basement wall insulation is not required in warm-humid locations as defined by Figure N1101.2 and Table N1101.2.

[g] Or insulation sufficient to fill the framing cavity, R-19 minimum.

[h] "13 + 5" means R-13 cavity insulation plus R-5 insulated sheathing. If structural sheathing covers 25% or less of the exterior, R-5 sheathing is not required where structural sheathing is used. If structural sheathing covers more than 25% of exterior, structural sheathing shall be supplemented with insulated sheathing of at least R-2.

[i] For impact-rated fenestration complying with Section R301.2.1.2, the maximum *U*-factor shall be 0.75 in zone 2 and 0.65 in zone 3.

[j] For impact-resistant fenestration complying with Section R301.2.1.2 of the *International Residential Code*, the maximum SHGC shall be 0.40.

[k] The second *R*-value applies when more than half the insulation is on the interior.

Source: *2009 International Residential Code*, © 2009. Washington, DC: International Code Council. Reproduced with permission. All rights reserved. http://www.ICCSafe.org

These are the minimum ratings required to meet the energy code; however, green buildings should aim for higher performance standards.

ENERGY STAR

The ENERGY STAR program of the U.S. Environmental Protection Agency has a labeling program for windows, doors, and skylights that is based on NFRC ratings. ENERGY STAR-labeled fenestration, while slightly better than the requirements of the 2009 IECC, cannot be described as very high-performance products, particularly in colder climates. Green homes should aim for significantly lower U-factors than ENERGY STAR or 2009 IECC in cold climates, and, in most applications, higher SHGC in warm climates. The prescriptive requirements of the higher tiers of green building programs can provide guidance when selecting fenestration.

LEED for Homes and National Green Building Standard (NGBS) Ratings

LEED for Homes and the National Green Building Standard (NGBS) specifications have set their own minimum U-factor and SHGC ratings in the prescriptive paths of both programs. Table 8.2 compares the ENERGY STAR, LEED for Homes, and the NGBS ratings.

Table 8.2 Fenestration Rating Comparisons by Climate Region

	North		North/Central		South/Central		South	
Standard	U-Factor	SHGC	U-Factor	SHGC	U-Factor	SHGC	U-Factor	SHGC
Windows and glass doors								
2009 IECC	0.35	NA	0.35	NA	0.50	0.30	.65/1.2	0.30
ENERGY STAR	0.30	NA	0.32	0.40	0.35	0.30	0.60	0.27
LEED								
Good	0.35	NA	0.40	0.45	0.40	0.40	0.55	0.35
Enhanced	0.31	NA	0.35	0.40	0.35	0.35	0.55	0.33
Exceptional	0.28	NA	0.32	0.40	0.32	0.30	0.55	0.30
NGBS								
Enhanced 1	0.30	NA	0.30	NA	0.35	0.30	0.45	0.30
Enhanced 2	0.25	NA	0.25	NA	0.35	0.25	0.45	0.25
Skylights								
2009 IECC	0.60	NA	0.60	NA	0.65	NA	0.75	NA
ENERGY STAR	0.55	NA	0.55	0.40	0.57	0.30	0.70	0.30
LEED H	LEED does not address skylights in the prescriptive path.							
NGBS	NAHB uses the same standards for skylights as for windows.							
Enhanced 1	0.30	NA	0.30	NA	0.35	0.30	0.45	0.30
Enhanced 2	0.25	NA	0.25	NA	0.35	0.25	0.45	0.25

Note: The LEED for Homes and National Green Building Standard (NGBS) specifications listed above are for the prescriptive certifications paths. Both programs also have performance paths that allow for greater flexibility. The ENERGY STAR specifications are for window products and not the new homes program.

Using NFRC Ratings

Understanding NFRC ratings allows you to compare the performance of different windows and provides you the information needed to make the best choice for your project. Cold climate fenestration should have the lowest U-factor available. The costs involved in selecting windows with extremely high performance ratings should be weighed with the climate-based needs of the building. In a warm climate, purchasing windows with a higher U-factor may provide savings that could be directed to other, more effective efficiency measures.

In warm climates, lower SHGC fenestration helps reduce cooling loads where glass is exposed to direct sunlight during the cooling season, reducing the amount of heat allowed into the building. In cold climates and wherever passive solar heating is part of a design, higher SHGC glass is preferred to allow heat to enter the space. Cold climate windows with very low U-factors and high SHGC can actually perform better than the surrounding walls when the energy gain from the glass is taken into account.

Some manufacturers will list a high U-factor at the center of glass. This number does not indicate the overall efficiency of the entire unit, which is calculated as a weighted average of the center of the glass, the edge of glass, and the sash and frame. This is the U-factor listed on the NFRC label.

Passive solar designs often require that windows with different SHGC ratings be installed on different faces, allowing more heat in on the south side and less on the east and west. Windows with ratings other than those considered standard often are only available by special order. Even when different window ratings are available, installers may mix up windows, installing differently rated units in wrong locations. One solution is to design different size units for each different rating, making it impossible to install them in the wrong place.

Very High Performance Windows

Very high performance windows are available with U-factors below 0.15, equivalent to an R-value of 7. Some European and Canadian companies produce windows with even lower U-factors, although these are not readily available in the United States. Note that VT may go down to below 0.30 in some of these windows, potentially requiring more electric lighting and making them appear dark to the occupants.

Did You Know?

Are Windows Really Leaky?

Properly installed and fully intact windows are rarely as leaky as the typical homeowner suspects. Windows tend to feel drafty, not because air is exchanging with the exterior through leaks, but because convective currents are forming in front of the poorly insulated glazing. In the winter, warm interior air is drawn to the cold window and falls as it cools. As the cool air is heated, it is once again attracted to the cool window surface and creates a loop (Figure 8.4). Highly insulated windows and insulated drapes help prevent heat loss through convection.

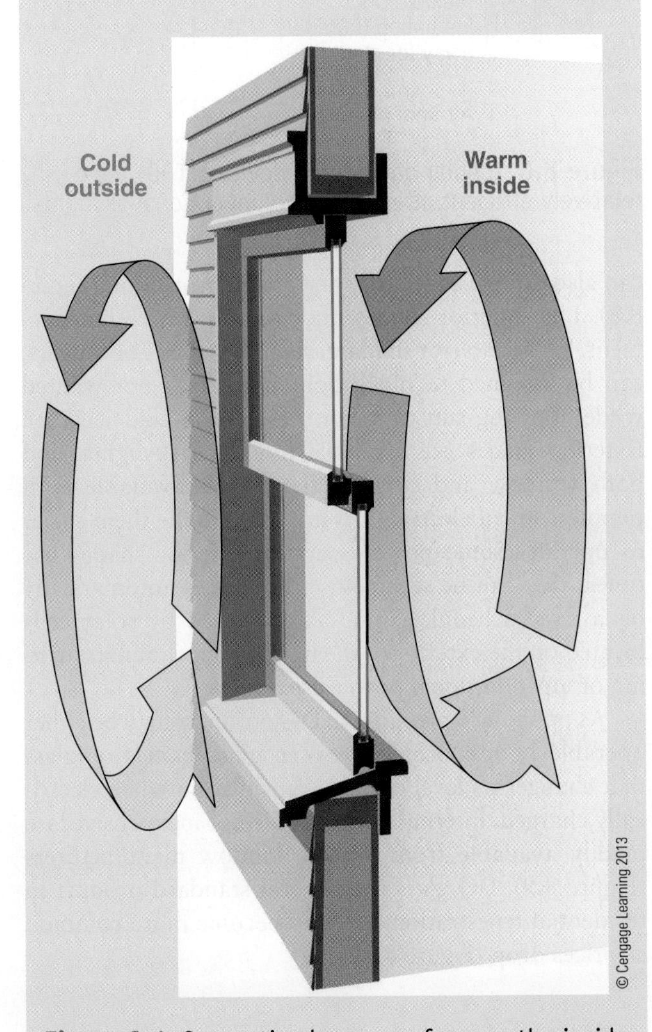

Figure 8.4 Convective loops can form on the inside and outside of highly conductive glazing. On the interior of the glass, air cools as it transfer its heat to the cold glass. As the air cools, it becomes less dense and falls. The falling air is then replaced by warmer air and creates a continuous loop of moving air. A similar process happens on the exterior as air is heated by the glass and rises. These processes may reverse in the cooling system.

Fenestration Size, Location, and Shading

The size, location, and shading of building fenestration all have a significant effect on energy efficiency. Poorly located windows can lead to overheating in hot weather and excess heat loss in cold weather. Carefully designed fenestration can reduce energy use most of the year in many climates.

Fenestration Size

Since fenestration, in most cases, is less efficient than the walls in which they are installed, from an energy standpoint, smaller is better. While a building with no windows or doors will have the least heat gain and loss, it would not be a very pleasant place to live. Green homes should strike a balance between too much and too little fenestration. Determining the appropriate sizes of windows, doors, and skylights, and how they provide natural light and radiant heat, is critical to making the right decisions for these key components of all buildings.

Location

Locate windows, doors, and skylights to take advantage of radiant heat and natural light where desired and to avoid it where it will not provide appropriate benefits. As discussed in Chapter 3 (window location and shading section), south-facing windows can be effectively shaded with modest-sized overhangs that provide heat in cold weather while blocking the sun when extra heat is not needed. North-facing windows get no direct sunlight, so they will not allow radiant heat in at any time of the year. In cold climates, limit fenestration on the north side to avoid excess heat loss. East-facing windows provide direct morning sun, which is desirable in all but the hottest climates. West-facing windows allow late afternoon sun to enter, which can overheat the interior in most climates and create significant amounts of glare. Late afternoon sun is generally uncomfortable from both a temperature and visibility standpoint, leading most homeowners with large expanses of west-facing windows to keep their blinds drawn much of the time. Minimizing the size and quantity of east- and west-facing windows is a good practice to follow. See Chapters 3 and 16 for additional information about passive solar design strategies.

Operable units can also be used to create thermal chimneys for passive ventilation. High windows or operable skylights, combined with lower-level open windows or doors, can create natural currents that draw warm air up and out to be replaced by cooler air from the outside. Lower-level openings can be located to take advantage of seasonal prevailing winds, providing no-cost cooling when weather conditions allow (Figure 8.5).

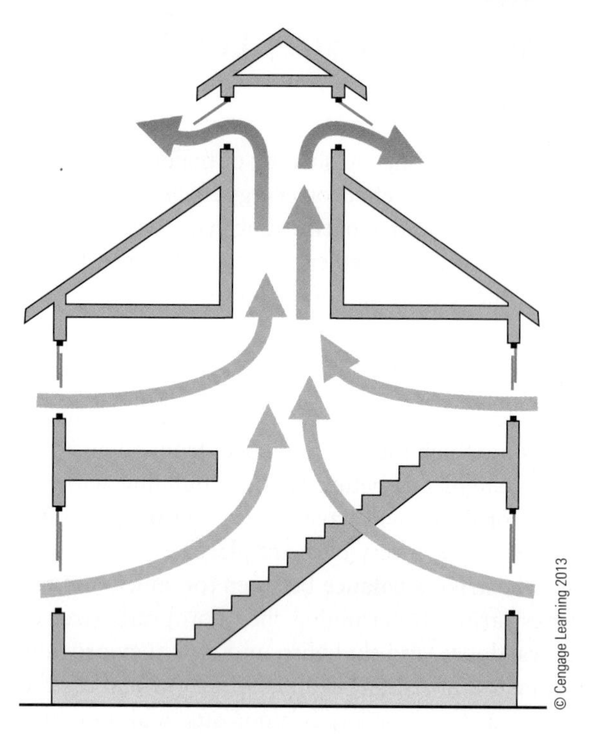

Figure 8.5 Intentional thermal chimneys provide passive ventilation and cooling. Warm air exits the home through high windows or skylights and is replaced by cooler air through lower windows.

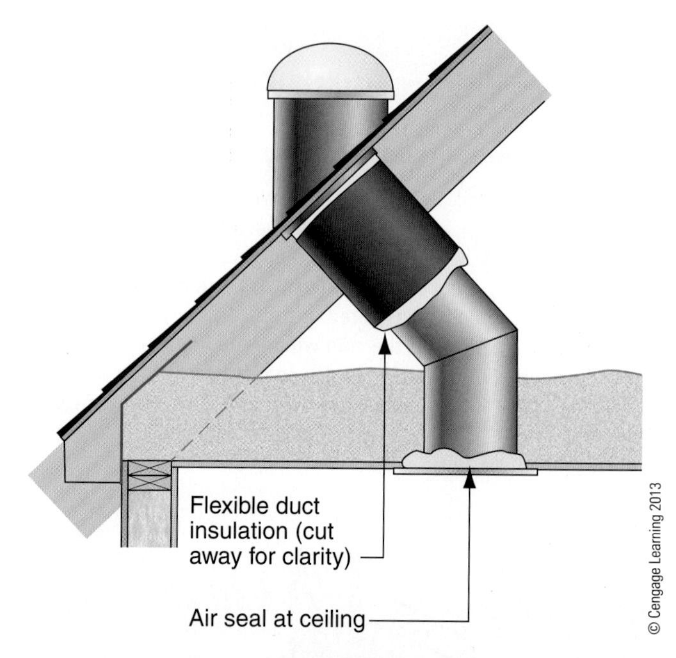

Figure 8.6 Tubular daylighting devices (TDD) are a relatively efficient alternative to conventional skylights.

Compared with conventional nonwest-facing windows, skylights and TDDs can allow in significantly more radiant heat due to their orientation. Unless located on a north-facing roof, skylights and TDDs can allow in significant amounts of heat as well as create glare from direct sunlight. They can also create an energy penalty through heat loss in cold weather. Despite these issues, small skylights and TDDs can provide excellent daylighting opportunities and help to reduce electricity use, offsetting some of the extra heat gain or loss (Figure 8.6).

Shading

Shading fenestration, whether through permanent overhangs or operable devices, helps to control solar gain when extra heat is not needed. External shading that keeps sunlight from hitting glass eliminates infrared radiation from entering and heating the interior (Figure 8.7). Interior shading does not effectively keep excess heat out of the interior.

Window shading with overhangs is addressed in Chapter 3. Other methods to protect from radiant heat include interior and exterior operable shades, fixed exterior shading, and DG systems. Interior shading requires active management—opening and closing the shades as needed. Even with active management of interior shading, solar gain through the glass is not completely eliminated because the shading still allows the sun's heat to penetrate into the building.

External shading, while often costly, can provide partial or complete shading from unwanted sun. Some units can also double as security and storm protection (Figure 8.8). Like interior shades, they require active management. Fixed exterior shading, such as trellises or louvers, can be designed to block light when it is not wanted while allowing sun to reach the glass in cold weather. Exterior shades are available for some skylights, and both window and skylight shades are available with remote control electric operators that make them easier to operate. Some power-operated exterior shades use timers that can be set to close the shades automatically on a fixed schedule. Some are powered by solar cells located on the exterior of the unit, which eliminates wiring or any additional power use.

As previously mentioned, DG products may be either operable blinds located between glass panels or glass that changes its level of light transmission when electrically charged. Internal blinds, while not inexpensive, are readily available from several window manufacturers (Figure 8.9). DG glass is not yet a standard product in residential fenestration but may become more common as prices drop (Figure 8.10).

Glazing

Single-thickness glass was common as recently as the 1980s. **Insulated glass**, originally invented in the 1940s, was not generally available until building codes first

---

insulated glass a window unit made up of at least two panes separated by a sealed space that is filled with air or other gases.

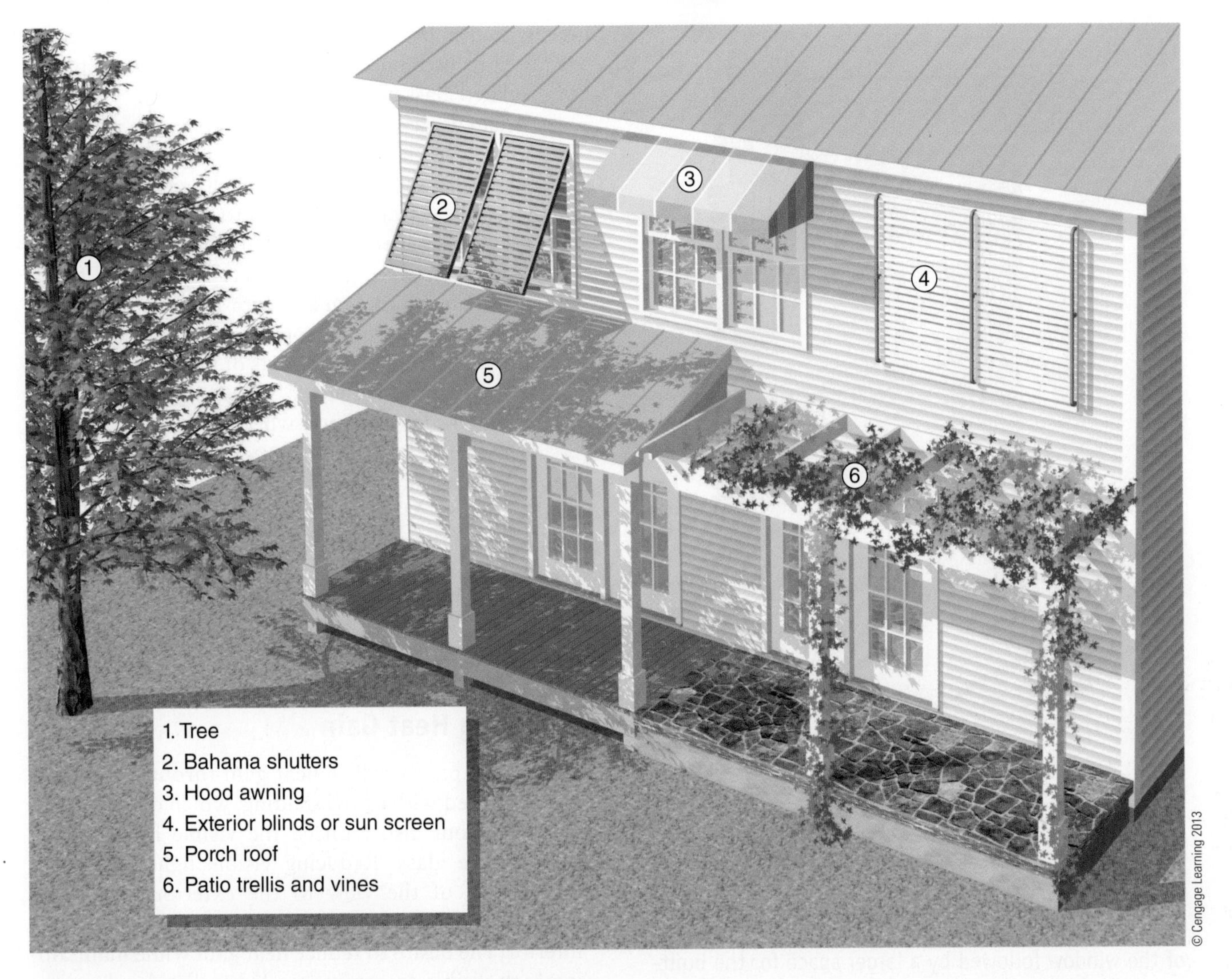

Figure 8.7 Options for shading windows from the exterior.

Figure 8.8 Rolling shutters may provide solar shading, security, and storm protection.

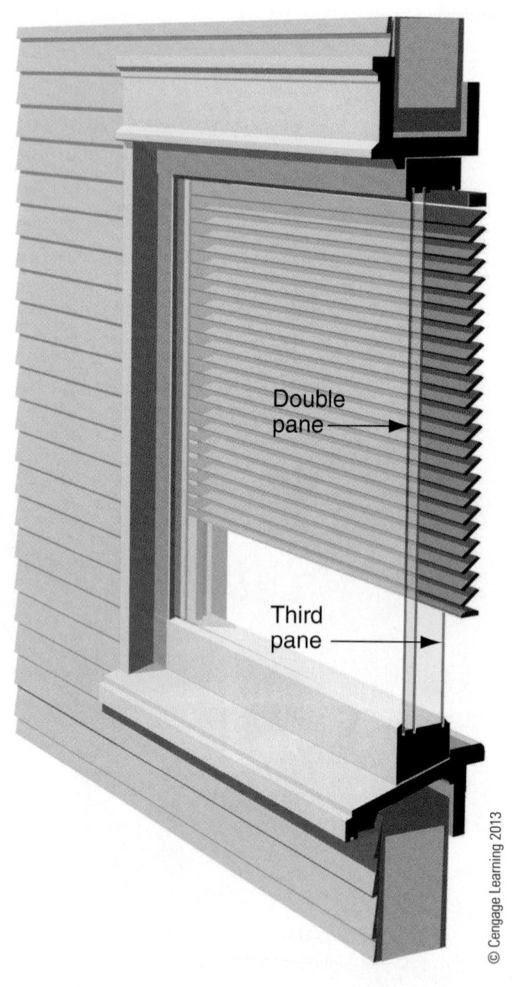

Figure 8.9 Window with internal blinds. The window has a traditional insulated double pane on the exterior of the window followed by a larger space for the built-in blinds. Lastly, there is another piece of glass that encloses the blind system.

recognized it in the 1970s; by the 1990s, most codes required its use. Since its introduction, the performance of insulated glass has advanced significantly.

Insulated glass consists of a combination of two or more sheets of glass separated by spacers with a pocket of dead air between the layers, which reduces the transmission of heat from the interior to the exterior. Argon and krypton are inert gases that are commonly added to the air space between glass panes to lower the U-factor. Most insulated glass is made with two sheets of glass, but triple glazing is more common in extreme cold climates; triple-glazed glass consists of three sheets of glass with two separate air spaces.

Insulated glass spacers, when made of metal, can create a thermal bridge that reduces the overall efficiency of the entire unit and causes condensation in cold climates. Most manufacturers now offer warm edge spacers that have a built-in thermal break, reducing the amount of heat conducted through them (Figure 8.11). Warm edge spacers improve both the U-factor as well as the condensation rating—both important issues in cold climates.

Reducing Heat Gain

Early attempts to reduce heat gain through glass was accomplished with tinted shading, which did cut down solar gain but also reduced the visible light coming through the glass. Reducing visible light diminishes the quality of the view to the exterior as well as increases the amount of electricity required to light the interior. The desire to reduce heat gain while maintaining high visible transmittance led to the development

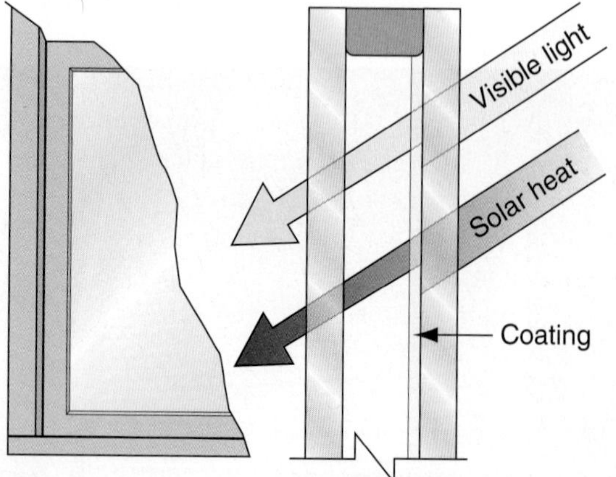

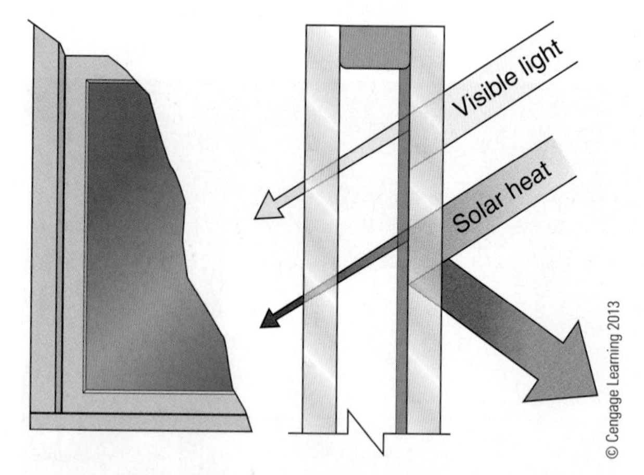

Figure 8.10 Electronically tintable glass, for use in building windows and skylights, can be switched from clear to darkly tinted at the click of a button, or programmed to respond to changing sunlight and heat conditions.

argon an inert gas commonly added to the air space between glass panes to lower the U-factor.

krypton an inert gas commonly added to the air space between glass panes to lower the U-factor.

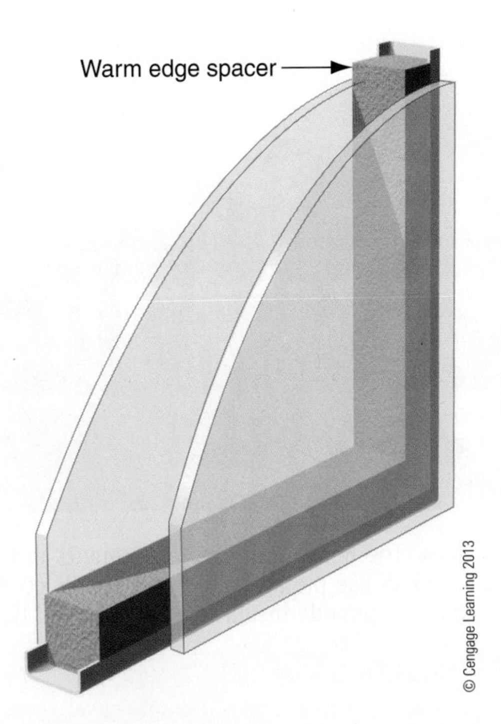

Figure 8.11 A warm edge spacer is a nonconductive material between the panes of glass.

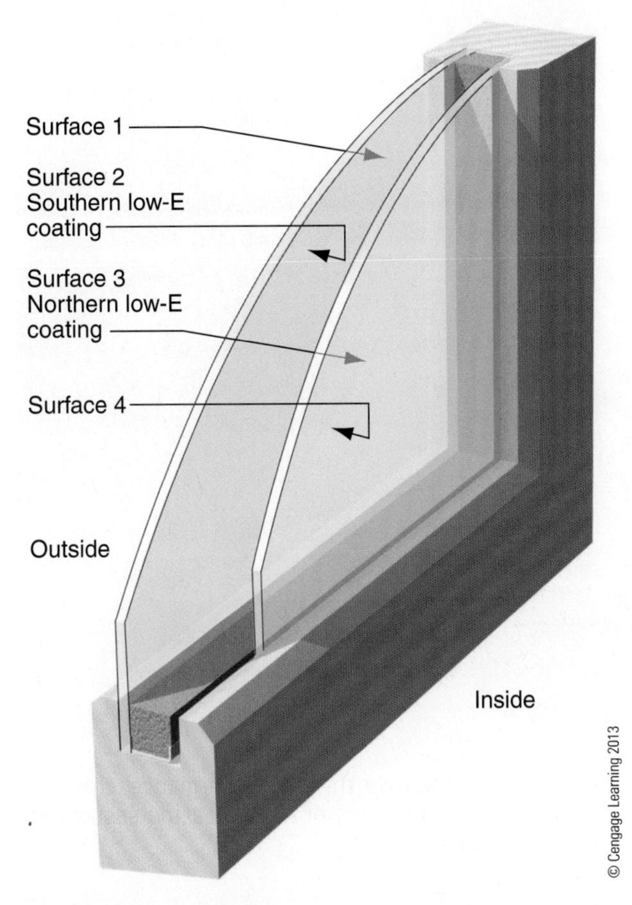

Figure 8.12 Low-E coatings are placed within the window according to climate zone to maximize solar heat rejection or internal heat retention.

of new forms of glass coatings. Spectrally selective glass makes use of low-emissivity (low-E) coatings that block much of the infrared energy and ultraviolet light while allowing in most of the visible light. With low-E coatings, a microscopic layer of metal is applied to the glass surface to act as a radiant barrier (similar to a radiant barrier in an attic, as described in Chapter 7), reducing the amount of infrared energy that penetrates through the metallic surface. In hot climates, low-E coatings reflect as much of 90% of the infrared heat while allowing visible light to penetrate from outside. In cold climates, the same infrared radiant heat from the interior is reflected back, keeping the house warmer.

Location of Low-E Coatings

Low-E coatings are fragile, so they are applied to one or more of the inside surfaces of the glass before it is assembled into an insulated panel. The coating works best when it is on the warmer side of the window, reflecting the heat outside during hot weather and keeping heat inside when the weather turns cold. Glazing surfaces are numbered from the outside to the inside. Southern low-E glass (designed for warmer climates) has the coating on surface two, the inside surface of the outside pane of glass (Figure 8.12), to allow less heat penetration. Northern low-E glass (designed for colder climates) has the coating on surface three, the outside surface of the inner pane, to increase solar gain. Some double-glazed units have the coating on both surfaces. Regardless of the number of low-E coatings, the best way to compare the performance between different products is to refer to the NFRC label.

Low-E Films between Glass

While triple-glazed glass provides benefits in cold climates, it is heavier and thicker than double-glazing and requires heavier sashes to hold it securely. Lighter alternatives with similar performance use layers of low-E film instead of the middle layer of glass. Sometimes referred to as suspended film (SF) technology, up

---

spectrally selective glazing a coated or tinted glass with optical properties that are transparent to some wavelengths of energy and reflective to others.

low-E coating a microscopic layer of metal applied to the glass surface that acts as a radiant barrier, reducing the amount of infrared energy that penetrates through the metallic surface.

Courtesy of Serious Materials, Inc.

Figure 8.13 To improve thermal performance, this window has two thin layers of low-E coating suspended between the panes of glass.

to two additional layers of low-E films are available for extremely high performance (Figure 8.13).

VT Ranges of Low-E Glass

To most people, glass with 60% VT appears clear. Below 50% VT, glass can start to look dark. Personal reactions to spectrally selective glass vary; to some people, a particular type of low-E glass may seem colored or dark although others may see the glass as clear.

Super Insulated Glazing Systems

Translucent glazing panels are available filled with aerogel, a translucent insulation. These panels, typically made of fiberglass, can have U-factors as low as 0.05 (R-20), making them as efficient as many walls. At this writing, these panels are primarily used in commercial applications but are expected to expand into the residential market as material costs come down and additional product lines become available (Figure 8.14).

Storm-Resistant Glazing

In some regions that are at risk for hurricanes or tornadoes, building codes require or recommend special storm-resistant glazing. Such glazing is made of

Courtesy of Kalwall Corporation

Figure 8.14 The Bearwood Road Apartments in the United Kingdom use prefabricated wall panels containing Nanogel® to provide insulation and diffused lighting.

laminated glass similar to automobile windshields. When installed in standard window frames, this thicker glass causes a reduced air space in the insulated glazing panel, which can reduce the U-factor of the unit. This can significantly reduce the overall building efficiency, particularly in cold climates. Alternatives include special window frames that are designed for thicker glazing; however, this may only be available by special or custom order. When impact-resistant glazing is necessary, installation of storm shutters with standard windows that have higher performance ratings may be a more cost-effective approach.

Decorative Glazing

While not recommended in high-performance homes, up to 15 ft^2 of uninsulated glass is allowed per the requirements of the 2009 IECC and the prescriptive path of the NGBS. This type of glazing is typically leaded or stained glass used in entry doors or accent windows. To avoid excessive heat loss or gain, decorative glass should be installed together with a clear insulated glass panel that has climate-appropriate efficiency ratings.

Applied Window Films

Window films can be applied to existing windows to reduce the SHGC, which can be a cost-effective way to improve their performance without replacement. NFRC ratings on window film provide guidance as to the level of improvement that can be expected by applying window film (Figure 8.15).

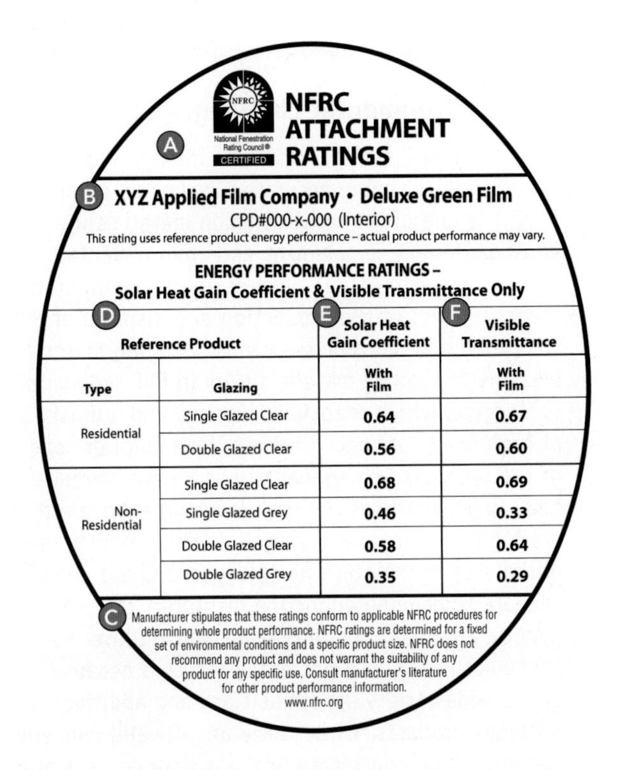

NFRC ATTACHMENT RATINGS

National Fenestration Rating Council® CERTIFIED

XYZ Applied Film Company • Deluxe Green Film

CPD#000-x-000 (Interior)

This rating uses reference product energy performance – actual product performance may vary.

ENERGY PERFORMANCE RATINGS – Solar Heat Gain Coefficient & Visible Transmittance Only

Reference Product		Solar Heat Gain Coefficient	Visible Transmittance
Type	Glazing	With Film	With Film
Residential	Single Glazed Clear	0.64	0.67
	Double Glazed Clear	0.56	0.60
Non-Residential	Single Glazed Clear	0.68	0.69
	Single Glazed Grey	0.46	0.33
	Double Glazed Clear	0.58	0.64
	Double Glazed Grey	0.35	0.29

Manufacturer stipulates that these ratings conform to applicable NFRC procedures for determining whole product performance. NFRC ratings are determined for a fixed set of environmental conditions and a specific product size. NFRC does not recommend any product and does not warrant the suitability of any product for any specific use. Consult manufacturer's literature for other product performance information. www.nfrc.org

(A) This mark indicates that the product's energy performance has been rated and certified in accordance with NFRC's certification process.

(B) This area is reserved for the name of the manufacturer and the product.

(C) This space provides details about NFRC's rating procedures.

(D) Consumers, building officials, and others should use the information in the **Reference Product** columns to choose the glazing system that most closely matches the product on which the film is applied.

(E) **Solar Heat Gain Coefficient** (SHGC) measures how well a product blocks heat from the sun. SHGC is expressed as a number between 0 and 1. The lower the SGHC, the better a product is at blocking heat gain. Blocking solar heat gain is particularly important during the summer cooling season and in southern climates.

(F) **Visible Transmittance** (VT) measures how much light comes through a product. VT is expressed as a number between 0 and 1. The higher the VT, the higher the potential for daylighting.

Courtesy of NFRC.org

Figure 8.15 NFRC's Window Film Energy Performance Label.

Glass Block

Glass block is often specified for high-security areas where light is desired but a view is not needed. Glass block walls and windows should be limited in size due to their low insulation value. Manufacturers state U-values ranging from 0.50 to 0.60; however, glass block is not NFRC rated, so these figures should not be used to directly compare with rated fenestration.

Window Sashes, Door Slabs, and Frames

Window sashes and doors slabs, the frames into which they are set, and glazing strategies are all key components of high-performance fenestration. The glass must be held securely in place with no air or water leakage, and a tight air seal must be maintained between the sash or slab and the frame. In the case of operable units, the mechanism must be able to open and close thousands of times without failing.

Overhead Doors

While garages must always be outside of the building envelope, many homeowners use them for workshops and may condition the space. For this reason, overhead doors should be as efficient as possible, although comparisons of different units are difficult to make. Overhead doors used for garages typically do not have NFRC or other independent ratings, but some manufacturers make claims of certain R-values for their products. Many of these claims overstate the performance because they only address center-of-panel ratings rather than ratings for the entire unit, which would include air leakage. Regardless of the insulation value, air leakage is the most critical element in overhead doors, and finding units with high-performance air sealing is a challenge. Attached garages, particularly in cold climates, can benefit from doors that are insulated and installed with effective weatherstripping to help keep the space warm during cold weather.

Sashes, slabs, and frames should conduct as little heat and air as possible for the unit to be energy efficient. As mentioned earlier, the U-factor takes into account the entire unit, so even high-performance glazing will not be efficient if installed in a poorly performing sash and frame.

---

slab a single door panel, excluding the jamb, hinges, threshold, and door hardware.

sash a structure that holds the panes of a window in the window frame.

Window Sashes

Traditionally, most windows were built from wood, and many today still are. Other common materials include vinyl, fiberglass, aluminum, and steel. Many wood windows are available with aluminum or vinyl exterior cladding to improve durability and reduce maintenance costs.

Wood, vinyl, and fiberglass are poor conductors, so they all help improve thermal performance. Metal is a good conductor, so any metal window should have a thermal break in the sash as well as the frame to reduce heat transfer. Some manufacturers offer vinyl and fiberglass units with foam-filled cavities, providing additional thermal resistance (Figure 8.16).

Wood windows are generally aesthetically pleasing; however, they require regular maintenance and, if not adequately protected from moisture, can suffer from premature deterioration. Vinyl windows are efficient, moderately priced, and durable, but there are concerns about toxicity in the manufacturing process and their limited recycling options (see sidebar above, *Environmental Challenges of Vinyl*). Vinyl expands and contracts more than other materials, creating the potential for air and water leaks when adjacent materials move at different rates.

Courtesy of Serious Materials, Inc.

Figure 8.16 Double-pane window with one layer of suspended film and a foam insulation in the cavity.

Environmental Challenges of Vinyl

Used for pipes, siding, weatherstripping, and windows, vinyl (or, more accurately, polyvinyl chloride [PVC]) is one of the most common materials in construction. Both scientific and environmental groups have expressed concerns about the environmental impact of the production and disposal of PVC that should be considered when specifying materials. One of the raw materials used in PVC production is chlorine, which is toxic to humans and animals at high levels of concentration. Vinyl chloride, one of the processed components, is a known carcinogen—particularly at the high levels of exposure that are typical in manufacturing. Soft PVC products use phthalates to maintain flexibility, and these chemicals are considered endocrine disrupters that cause genetic abnormalities in humans and wildlife. Post-consumer PVC is also not easily recycled because of the wide variety of formulations and additives in different products. While these are all valid concerns and should be considered in the overall context of a home's environmental impact, they will have a limited effect on the home's occupants or builders.

Fiberglass Windows

Fiberglass windows, a more recent entrant into the market, are an interesting compromise between wood and vinyl. Fiberglass is more durable than wood while avoiding the toxicity problems of vinyl manufacturing. Fiberglass also offers the benefit of being made from the same raw materials as the glazing, therefore it tends to expand and contract at the same rate; this reduces the possibility for the seal between glass and frame to separate. While vinyl and fiberglass windows provide durability on the exterior, they offer a less desirable finish on the interior. Some models are available with applied wood finishes on the interior for a more aesthetically pleasing finish.

Exterior Cladding

Many manufacturers offer premium wood windows and doors that are covered (i.e., clad) in aluminum, vinyl, or, in some cases, bronze. Clad windows and doors offer the warmth and sustainability of wood with a more durable exterior finish. All clad finishes help resist deterioration,

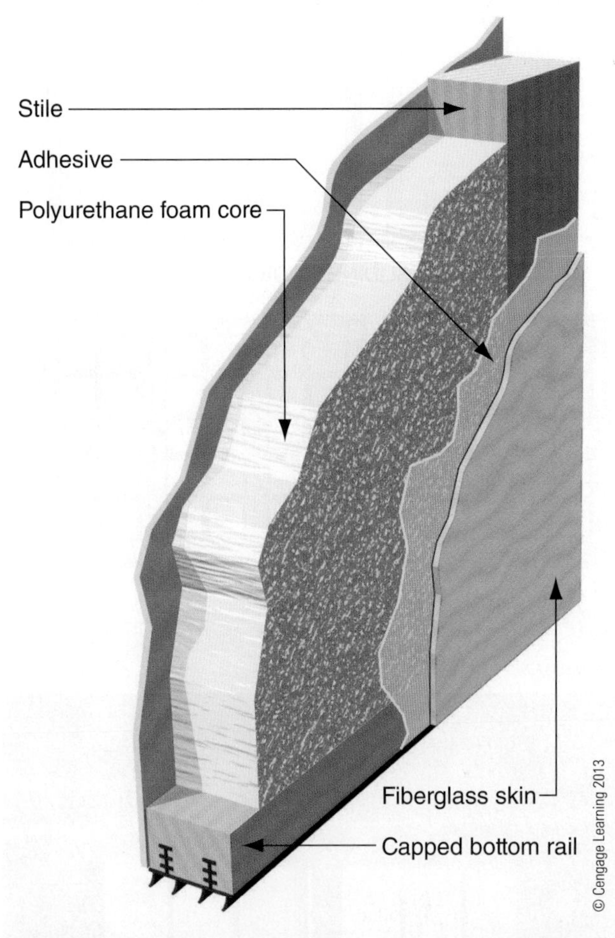

Figure 8.17 Door slabs can be insulated to provide additional thermal resistance.

but recycling at the end of life is more difficult. Vinyl cladding has the same toxicity issues as vinyl window frames.

Most skylight frames are made of aluminum, vinyl, or wood with applied cladding. The same considerations for windows and doors apply in selecting these products.

Door Slabs

Door slabs can be made from steel, plastic, and fiberglass or solid or engineered wood, with or without a cladding. Although not always as aesthetically pleasing as wood and clad doors, the most durable and efficient door slabs are made of fiberglass with a polyurethane foam filling (Figure 8.17). Door slabs can be ordered solid opaque or with glass panels that vary from a small openings to almost the entire door.

Operable and Fixed Fenestration

Fenestration can be fixed or operable, with several different methods of operation. Hinged units are referred

Figure 8.18a Casement windows swing outward.

Figure 8.18b Awning windows are often used in stacks or in combination with other types of windows.

to as casement or awning windows, depending on whether or not they swing from the side or the top (Figure 8.18a and Figure 8.18b). Hinged doors are described

casement window a side hinged window that swings open to interior or exterior.

awning window an operable window with a sash hinged at the top that swings outward.

as swinging. Sliding windows can be horizontal sliders or, if vertical, single-hung or double-hung, referring to whether or not one or two sashes operate (Figure 8.19a and Figure 8.19b). Sliding doors, often referred to as patio doors, are a common product (Figure 8.20). Folding exterior doors provide the opportunity to create a very wide opening, effectively bringing the outside and inside together in appropriate climates (Figure 8.21).

Courtesy of Andersen Windows, Inc.

Figure 8.19a Windows with horizontal sliding sashes.

Courtesy of Andersen Windows, Inc.

Figure 8.19b Double-hung window with vertical sliding sashes that also tilt for easy cleaning.

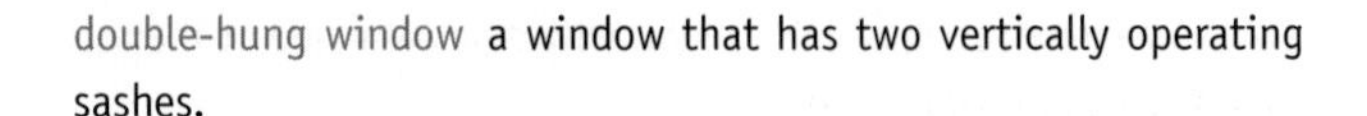

double-hung window a window that has two vertically operating sashes.

Courtesy of Andersen Windows, Inc.

Courtesy of Andersen Windows, Inc.

Figure 8.20 Two or three doors usually are used in sliding or swinging type patio door units.

Hinged Units

Hinged units provide a better air seal than sliders, provided that the weatherstrip and locking systems are of equivalent quality. This will generally be reflected in different U-factors and air leakage ratings on the NFRC

© Cengage Learning 2013

Figure 8.21 Wide expanses of operable doors, such as these folding units, can open the interior directly to the exterior, expanding the available living space in moderate weather.

Courtesy of Velux

Figure 8.22 Operable skylights provide daylighting and ventilation.

labels. Hinged units also provide as much as much as 100% of their total area for ventilation, although sliding units can only open up to 50% of their total area. Operable skylights are typically hinged at the top, providing a small opening at the bottom for ventilation (Figure 8.22).

Electrically Operated Units

Some casement and awning windows and operable skylights are available with electric operators, which are helpful when using difficult-to-reach windows for ventilation. Occupants are more likely to take advantage of natural ventilation when high windows and skylights can be opened and closed by pressing a button instead of manually operating each one individually. Electrically operated skylights are available with rain sensors that automatically close the unit to keep the house dry. Fixed units must have the same quality seals against air and water infiltration as operable ones.

Screens

Screens are necessary to promote natural ventilation in climates where insects are a problem. If there are no screens on operable units, then occupants will be less likely to open them if bugs will be flying in the house. Aluminum screening can be considered a more environmentally friendly product than flexible fiberglass, which is coated with soft PVC that is manufactured with harmful phthalates (see sidebar, *Environmental Challenges of Vinyl*).

Some homeowners may be concerned about security when leaving windows open for ventilation. Window screens that are wired into a home security system or security bars can provide protection against intruders while still allowing for natural ventilation.

Hardware and Weatherstrip

Quality hardware and weatherstrip on windows and doors ensure smooth operation and a tight seal against air and moisture. Casement and awning windows and operable skylights typically use crank-and-arm-type hardware, often with latches that are manually snapped shut to secure the sash in the frame. Double- and single-hung windows use balances that incorporate springs to counterbalance the weight of the sashes, making them easier to raise and lower. Latches secure them in the closed position, compressing the weatherstrip for a tight seal. Sliding doors and windows have tracks that can carry the weight of the sash or door slab, with latches that seal them tightly when closed. Folding exterior doors employ special heavy-duty track-and-wheel systems that allow the panels to fold flat when open. Many high-quality doors have three-point latching systems that provide a tighter air seal than traditional single-point door locks (Figure 8.23).

Hardware typically is included with the window, door, or skylight, although some double-hung windows and standard doors will accept standard hardware purchased separately. The hardware should be durable, close the unit securely, and come with a long warranty to ensure long-term performance. In very cold climates, metal hardware may create a thermal bridge through the unit, increasing the possibility of interior condensation.

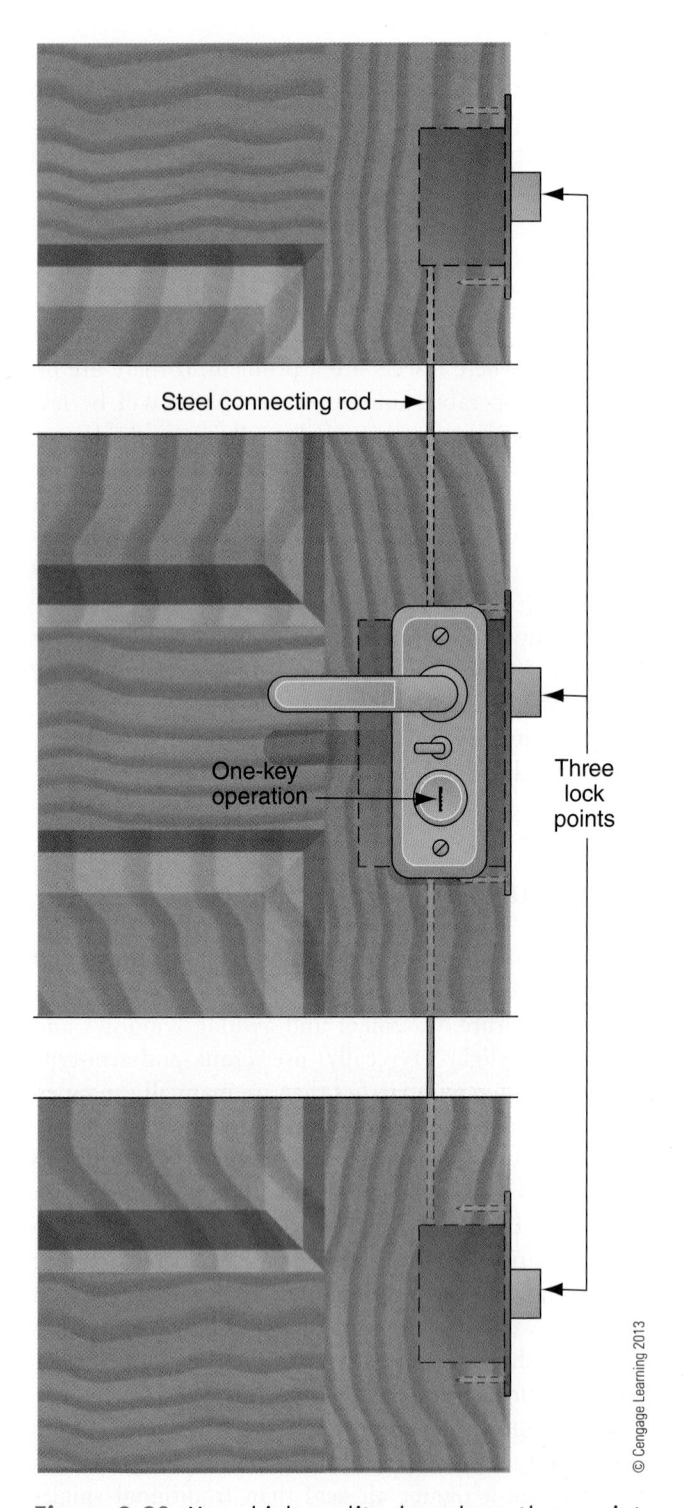

Figure 8.23 Many high-quality doors have three-point latching systems that provide a tighter air seal than traditional single-point door locks.

Most weatherstrip is made of PVC, which provides a long-term flexible air and moisture seal. Alternative products include felt, foam, and mohair. The best sealing systems employ double or triple sets of weatherstrip (Figure 8.24). Units with the best NFRC Air Leakage ratings will generally have the highest quality hardware and weatherstripping.

Grilles

Many designers and owners prefer grilles or muntins, the bars that create the look of individual panes of glass in a sash, in new windows that mimic the look of old windows with small individual panes or lites. The original purpose of grilles was to allow for large windows when large sheets of glass were unavailable or unaffordable. With glass available in almost any dimension, grilles are now exclusively an aesthetic decision.

Grille options available in modern fenestration includes true divided light (TDL), with individual panes of glass separated by grilles; simulated divided light (SDL), with grilles permanently installed over of a large pane of glass; removable grilles installed over a large pane of glass; and grilles between glass (GBG), with grilles permanently installed between two layers of insulated glass (Figure 8.25). TDL insulated glass is an option available from some manufacturers, but each individual pane of glass has spacers around the perimeter and the dividers are not insulated, so TDL fenestration is less efficient than other options. SDL can be combined with GBG to provide a look very close to TDL. Grilles that are applied to the interior, exterior or both sides of the glass have no impact on the efficiency of the unit; however, when GBG are included, the same conduction problems that occur at the edges of insulated glass happens wherever the internal grilles are installed. The selection of nonconductive spacers is critical to high-performance windows with GBG.

Window, Door, and Skylight Installation

No matter the quality of a window, door, or skylight, improper installation, flashing, or air sealing will prevent the unit from functioning effectively and will degrade the overall home performance.

Moisture Sealing

Windows and doors are installed in two types of walls: those with drainage planes or those without, which are generally referred to as storage reservoir walls (for

---

grille the bars that divide the sash frame into smaller lites or panes of glass.

muntins the actual bars that comprise a grille and divide the sash frame into smaller lites of glass.

true divided light (TDL) window or doors in which multiple individual panes of glass or lites are assembled in the sash using muntins.

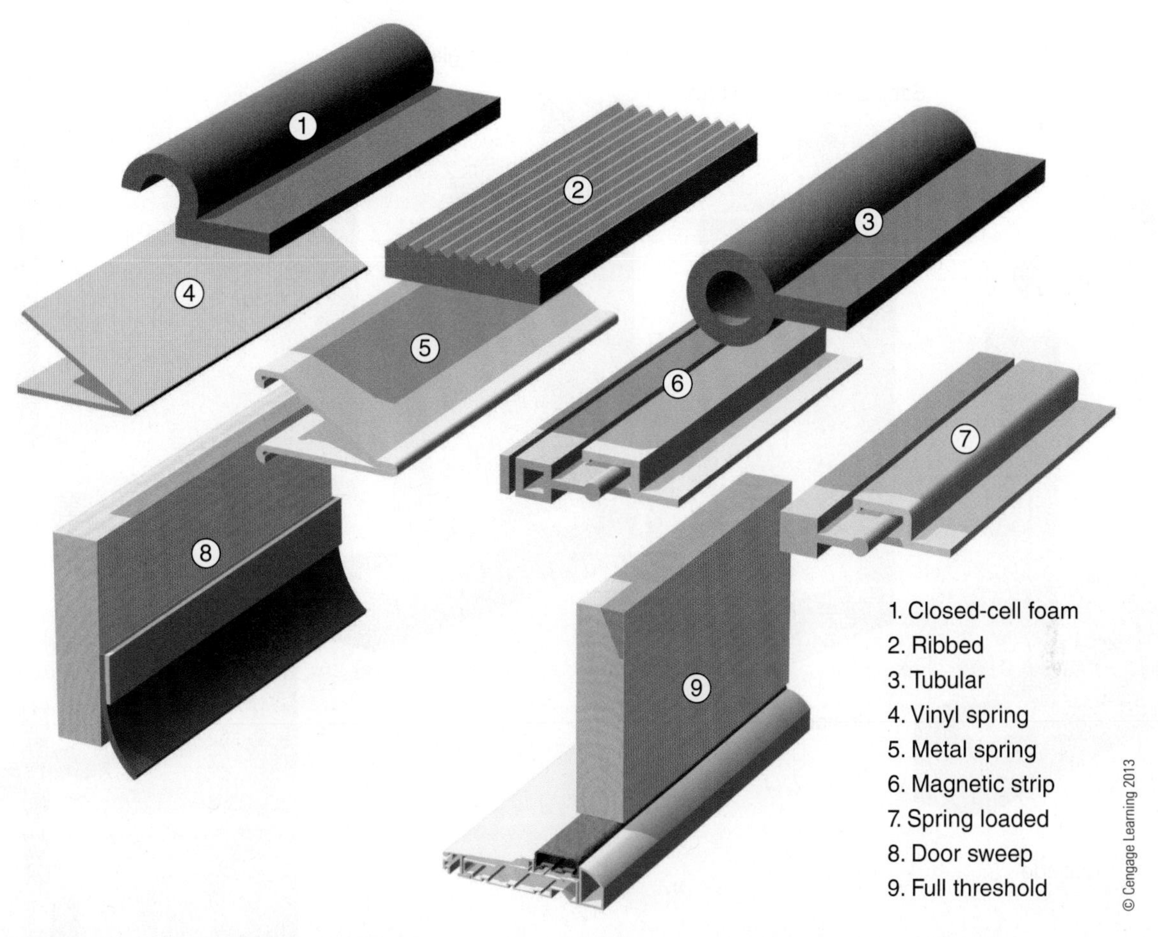

Figure 8.24 Various types of weatherstripping.

more information, see Chapter 6). For drainage plane installation, each unit must be fully integrated with the water-resistive barrier (WRB), directing all exterior moisture away from the wall structure. Drainage plane installations incorporate pre-formed or site-fabricated sill pans, side flashing, and heat flashing.

Windows and doors arrive on the job site with integrated nailing fins or flanges or shop-applied exterior casing (Figure 8.26). Alternatively, exterior trim may be applied on-site. Both types require sill flashing; however, the side and top flashing differs between them (see Step-by-Step instructions in Figure 8.27 and Figure 8.28).

Storage reservoir wall installation does not usually use flashing because openings are designed with a preformed seat to drain water to the exterior. The units are face-sealed to the wall surface (See Figure 8.29).

Skylight and TDD manufacturers typically provide integrated roof flashing with their units, with different products available for specific roofing types (see Chapter 7). When skylights and TDDs are installed, they must be integrated into the thermal envelope to prevent thermal bypasses (see Chapter 4 for more information).

Windows in Showers

Although many designs incorporate windows in shower and tub areas to provide light, they should be avoided in the design process to prevent problems that occur from regular wetting due to showering. Repeated wetting can damage not only the window itself but the entire wall below, which can fail if the window is not properly installed. First, wood windows should be avoided in shower areas; vinyl or fiberglass units are better alternatives for this wet environment. Other alternatives include solid panels, such as Kalwall systems or glass block; however, the latter is not very energy efficient, with estimated U-factors in the range of 0.50 to 0.60. Second, the window must be integrated into the interior drainage plane behind the tile wall to keep all water out of the wall structure. Unless this is done properly, the wall will quickly suffer severe water damage. Finally, if the shower area cannot be avoided, the window should be placed as high as possible to limit the amount of water hitting it during showers.

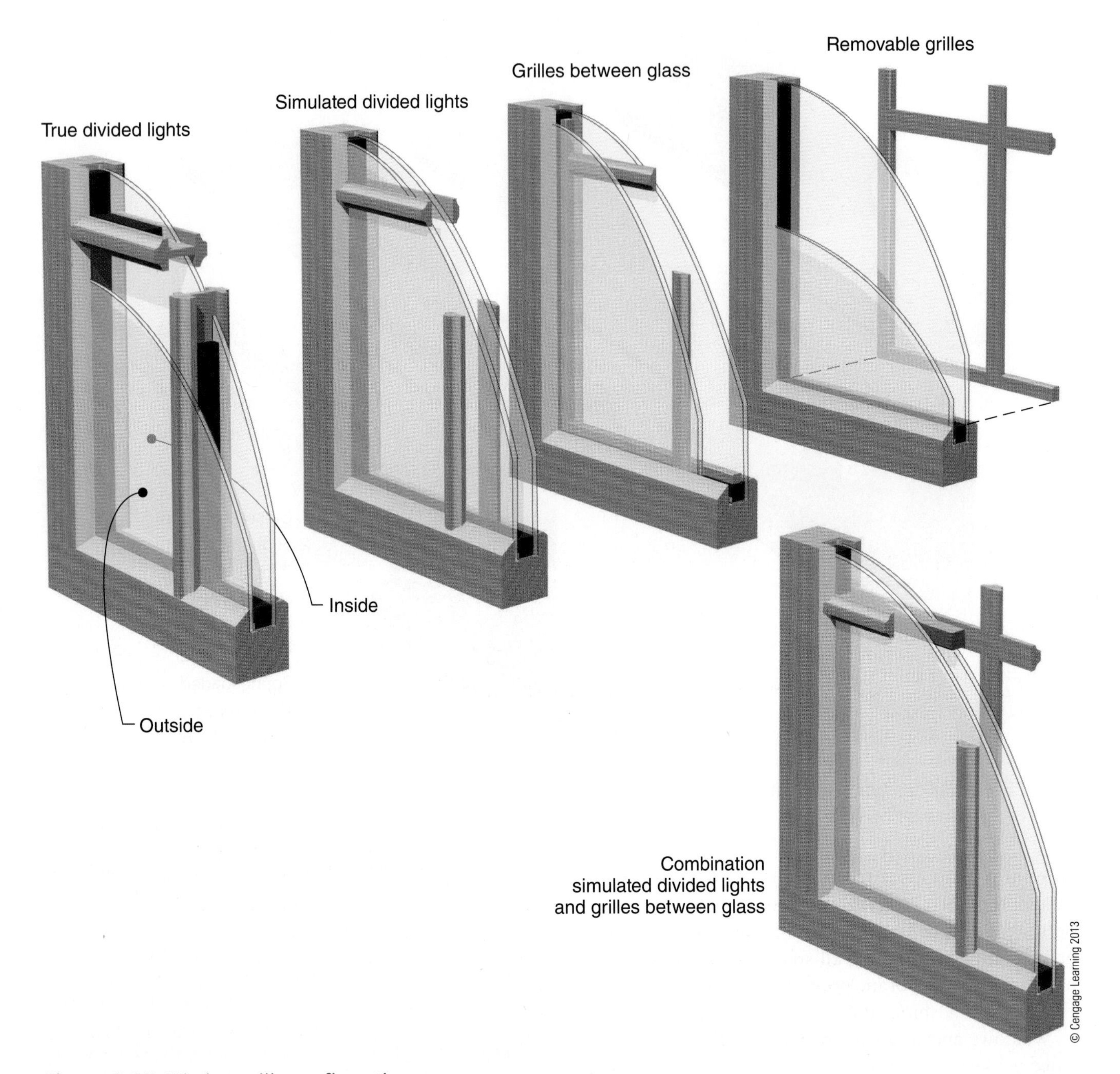

Figure 8.25 Window grille configurations.

Air Sealing around Windows, Doors, and Skylights

After windows, doors, and skylights are installed, the space between the frame and the structure must be sealed against air infiltration. This is typically accomplished with a low-expansion spray foam insulation designed specifically for this purpose (Figure 8.30). Standard high-expansion spray foam should not be used for air sealing because it can force window and door frames together, causing damage to the units. Caulk and other sealants are acceptable; however, fiberglass and mineral wool insulation, products often used for this purpose, do not provide the required air seal. Regardless of the type of sealant used, it must provide a complete air seal, be flexible enough to withstand material expansion and contraction, and be durable enough to maintain its seal for the life of the structure.

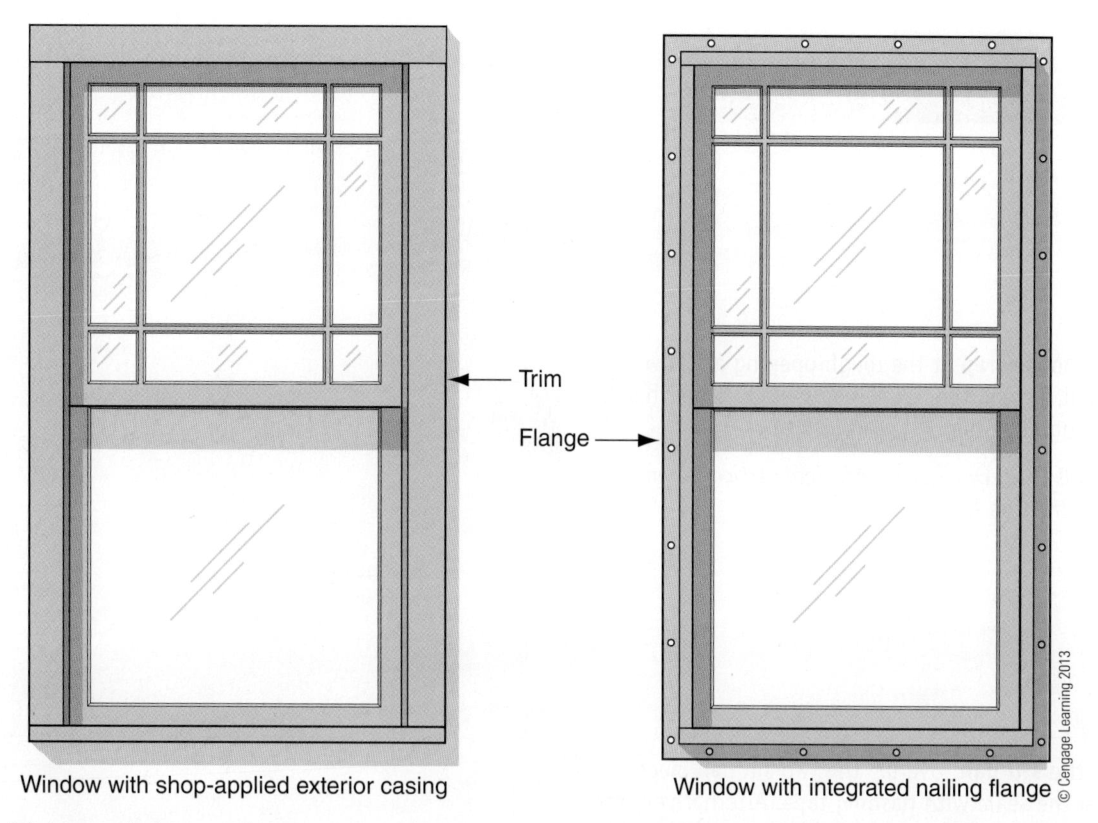

Figure 8.26 Most windows and doors arrive on the job site with integrated nailing fins or flanges or with shop-applied exterior casing.

Did You Know?

Fenestration as a Sound-Control Strategy

When high-quality fenestration is installed properly with careful air sealing, it reduces the amount of sound that enters the house. This quality can be a great benefit for homes close to busy roads or near airports that suffer from excessive noise. As part of a high-performance whole-house strategy, windows, doors, and skylights can help insulate a home from exterior noises.

Fenestration Management

As discussed earlier in this chapter, moveable interior and exterior shades must be manually opened and closed when necessary to effectively manage heat gain and loss. In addition, windows should be open to provide ventilation during moderate weather, reducing the need for heating and air conditioning. Many homeowners do not bother, preferring to keep the house closed up much of the year and use heating, ventilating, and air conditioning (HVAC) systems more than necessary. Homeowner training, combined with easily accessible operable windows, is a key component of sustainable building. Part of this training is to educate occupants to turn off HVAC systems when windows and doors are open for ventilation. Advanced electronic home management systems use alarm sensors installed on operable units to automatically turn off entire HVAC systems or zones when windows and doors are open. While systems like these are not a substitute for proper home management, they are available for homeowners willing to invest in them.

Sustainable Material Use

The amount of material used in fenestration is modest compared with that used in other phases of construction, and the energy savings from installing high-quality units will likely more than offset the embodied energy used to make and transport them. Still, some manufacturers offer wood sashes and frames that are harvested sustainably, and tropical woods that are not certified as sustainable should be avoided for windows and doors.

---

operable window a window with movable sashes.

STEP-by-STEP Figure 8.27 **Window with an integrated nailing fins or flanges installation in a non-reservoir wall.**

Step 1.

Cut the housewrap at the rough opening as shown. At the sill, extend the cut 4" horizontally on each side of the rough opening.

Overlap all housewrap seams 6". Seal vertical seams with construction tape.

1

Step 2.

Install the sill pan. Overlap the two sill pan pieces, and seal the seam with flashing tape. Alternatively install flexible flashing tape across sill and up sides 6".

Layer the vertical sides of the sill pan behind the housewrap, and layer the front face of the sill pan over the housewrap.

Ensure that the end dam on the sill pan isn't cracked or broken.

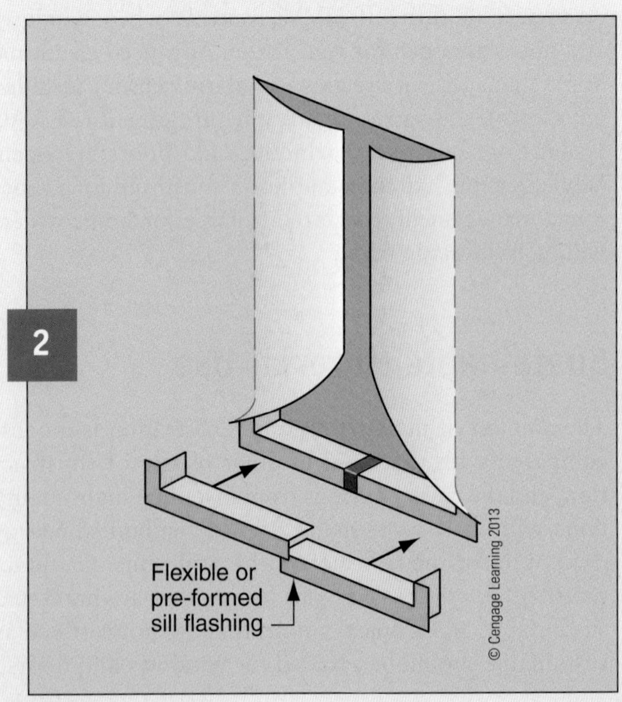

Step 3.

Fold the housewrap into the rough opening, and fasten it to the interior stud face. Seal the seams where the housewrap was cut with construction tape.

At the head, make two 6" 45° cuts in the housewrap.

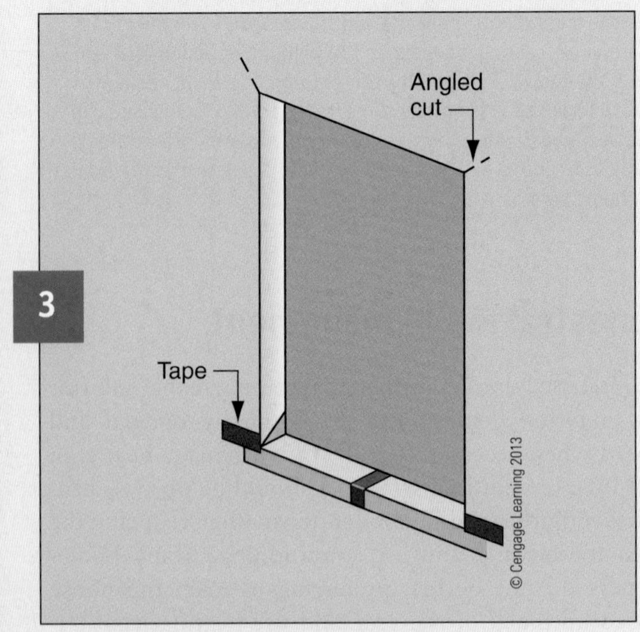

STEP-by-STEP **Figure 8.27 (Continued) Window with an integrated nailing fins or flanges installation in a non-reservoir wall.**

Step 4.

Fold the housewrap up away from the rough opening, and temporarily tape it out of the way with construction tape.

Apply a continuous bead of caulk to the head and jambs. Install the window in the rough opening.

Make sure caulk is applied only to the head and jambs. Leave the sill uncaulked, so that moisture can drain out and away from the rough opening.

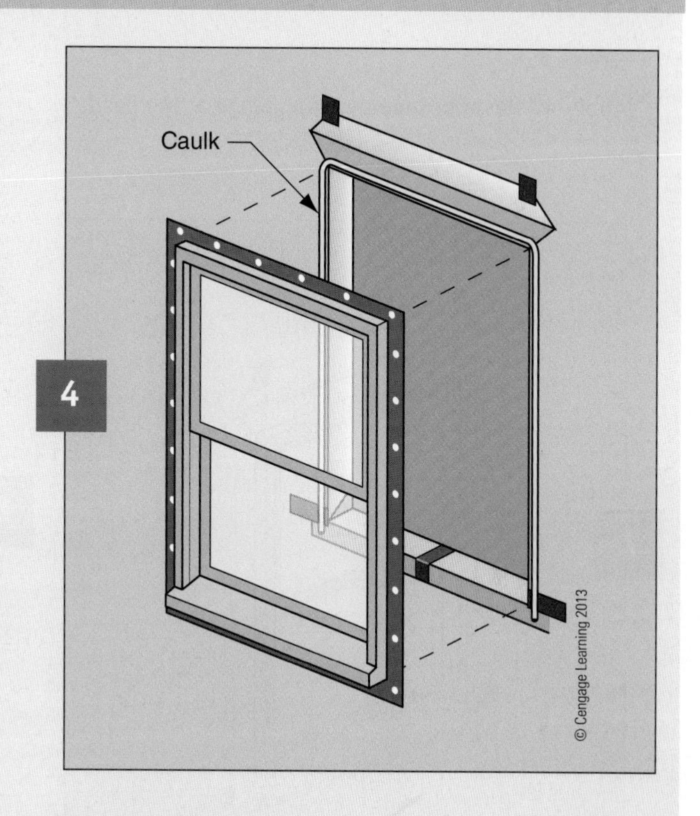

Step 5.

Shim the window to achieve an even, continous space between the window frame and the rough opening. Hold the window in place by nailing one of the top corners. Check the window for plumb, level, and square. Confirm that the window opens and closes smoothly without sticking.

Check for square by measuring the diagonals. The window is square when the measurements are equal.

Fasten the window to the frame per the manufacturer's instructions.

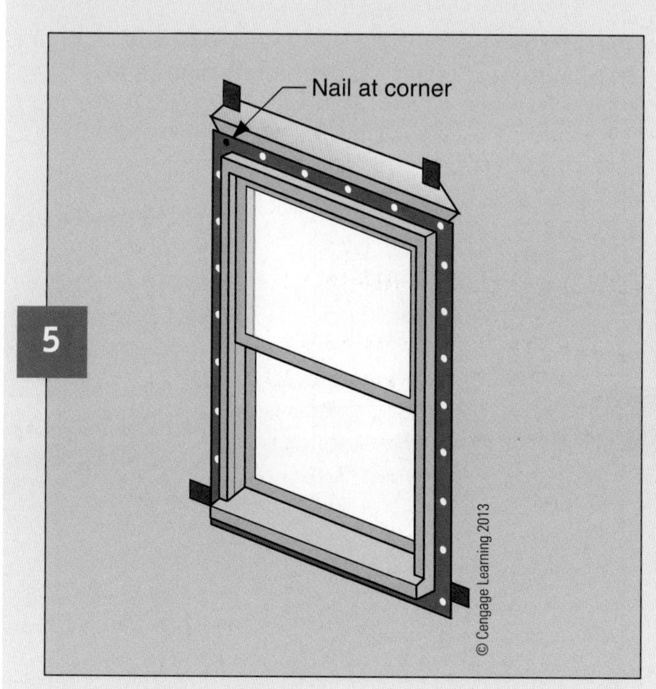

Step 6.

Apply 4" flashing tape over the jamb flanges.

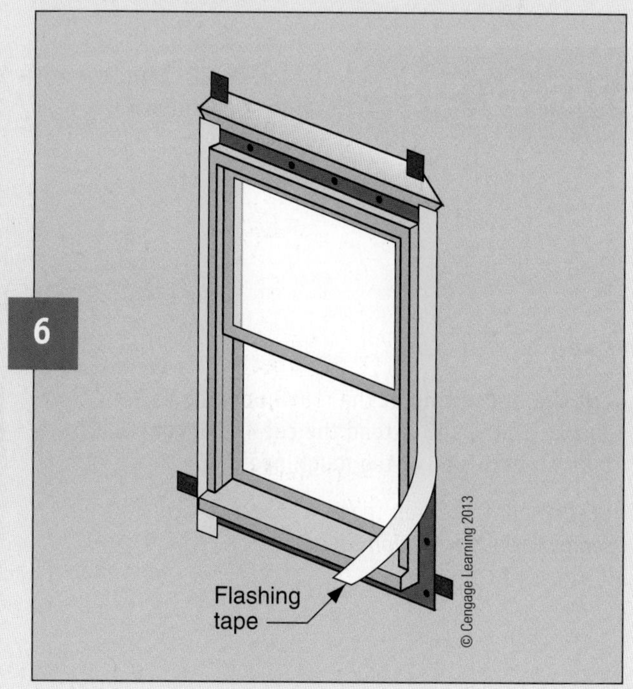

(Continued)

STEP-by-STEP

Figure 8.27 **(Concluded) Window with an integrated nailing fins or flanges installation in a non-reservoir wall.**

Step 7.

Apply 4" flashing tape over the flange at the head.

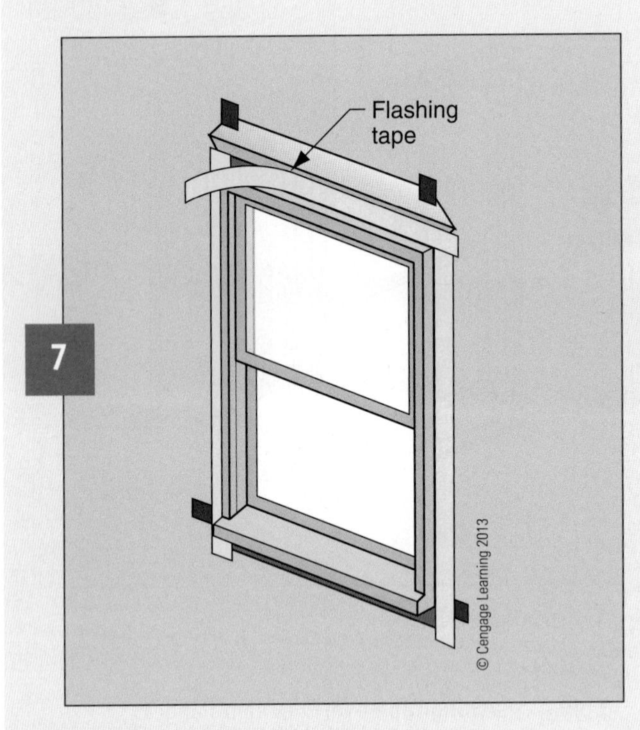

Step 8.

Unfold the housewrap, and seal the 45° cuts with construction tape.

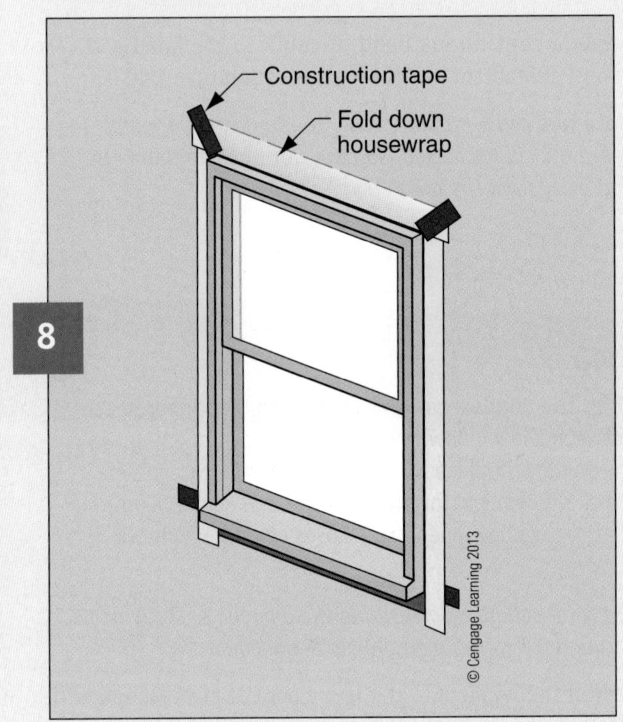

STEP-by-STEP

Figure 8.28 **Window with a shop-applied exterior casing installation in a non-reservoir wall.**

Step 1.

Cut the housewrap at the rough opening as shown. At the sill, extend the cut 4" horizontally on each side of the rough opening.

Overlap all housewrap seams 6". Seal vertical seams with construction tape.

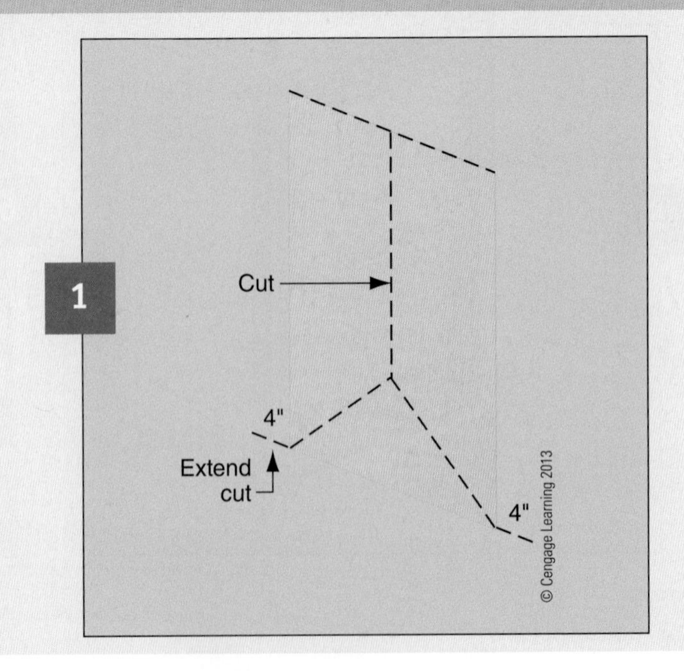

STEP-by-STEP Figure 8.28 **(Continued) Window with a shop-applied exterior casing installation in a non-reservoir wall.**

Step 2.

Install the sill pan. Overlap the two sill pan pieces, and seal the seam with flashing tape. Alternatively install flexible flashing tape across sill and up sides 6".

Layer the vertical sides of the sill pan behind the housewrap, and layer the front face of the sill pan over the housewrap.

Ensure that the end dam on the sill pan isn't cracked or broken.

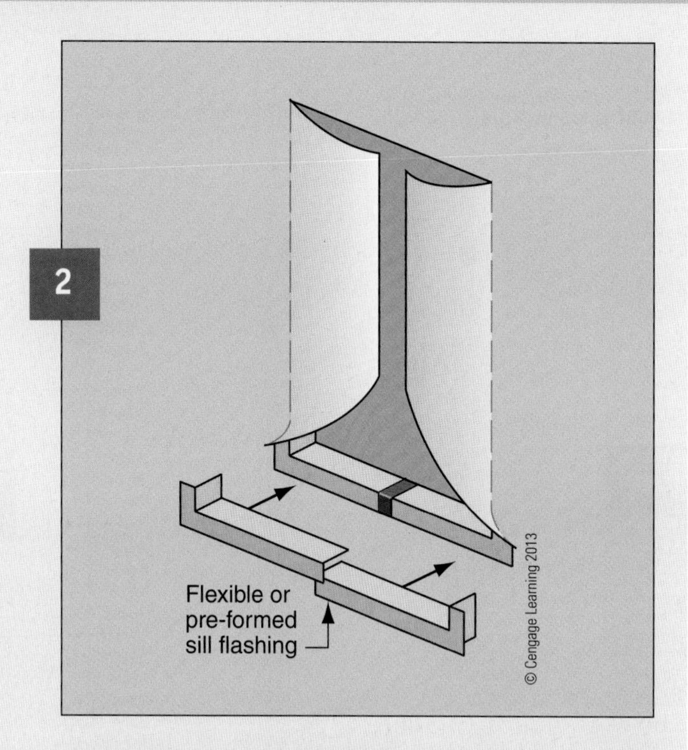

Step 3.

Fold the housewrap into the rough opening, and fasten it to the interior stud face. Seal the seams where the housewrap was cut with construction tape.

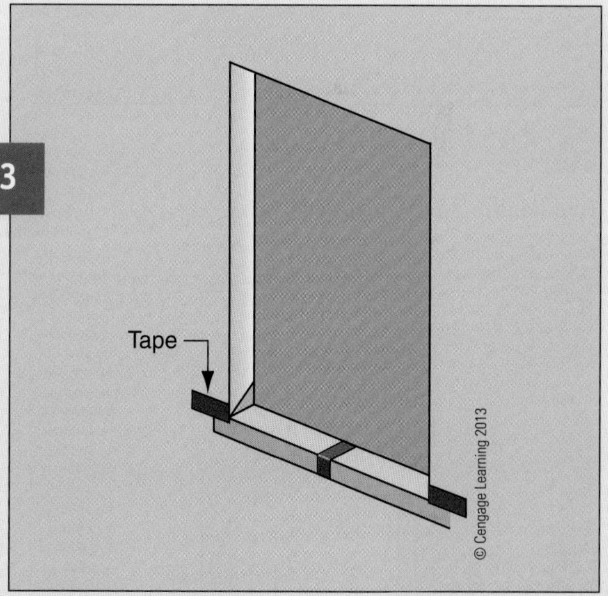

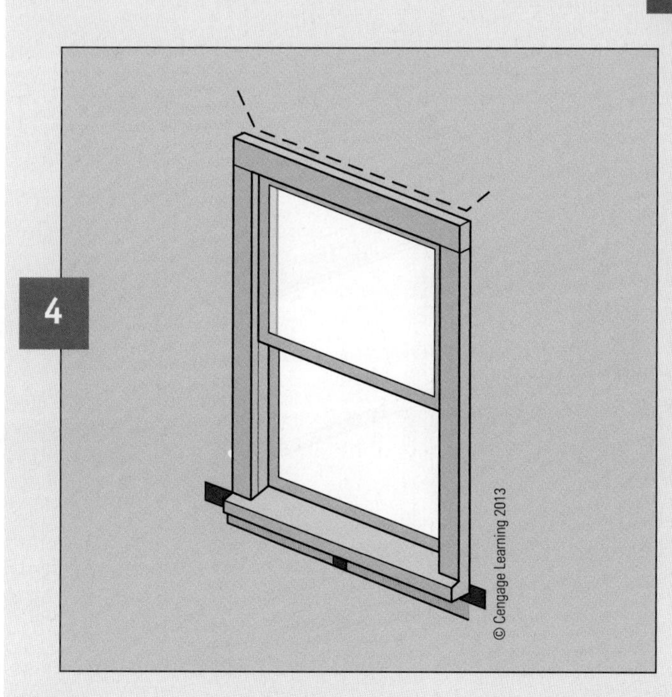

Step 4.

Install window plumb, level, and square per manufacturer's instructions.

(Continued)

STEP-by-STEP Figure 8.28 **(Concluded) Window with a shop-applied exterior casing installation in a non-reservoir wall.**

Step 5.

Cut housewrap at head and temporarily fold up or tuck under.

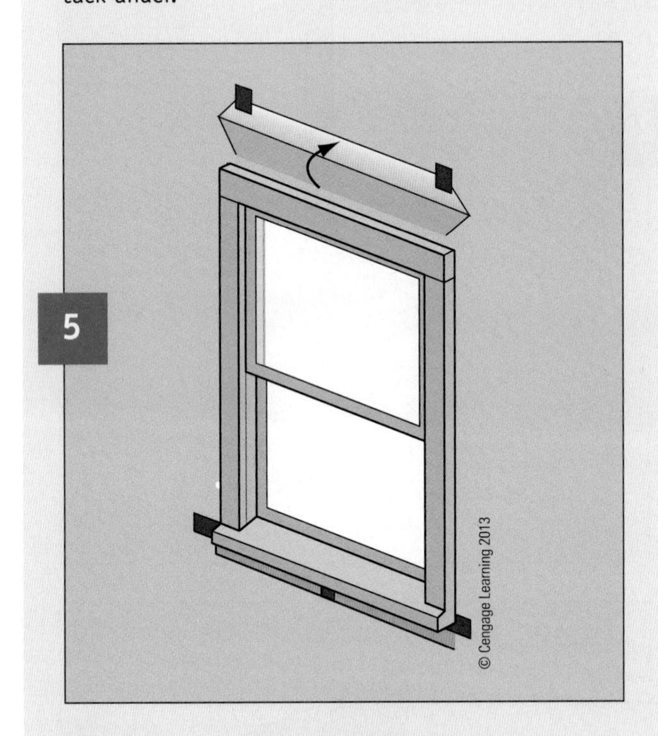

Step 6.

Install a drip cap/head flashing.

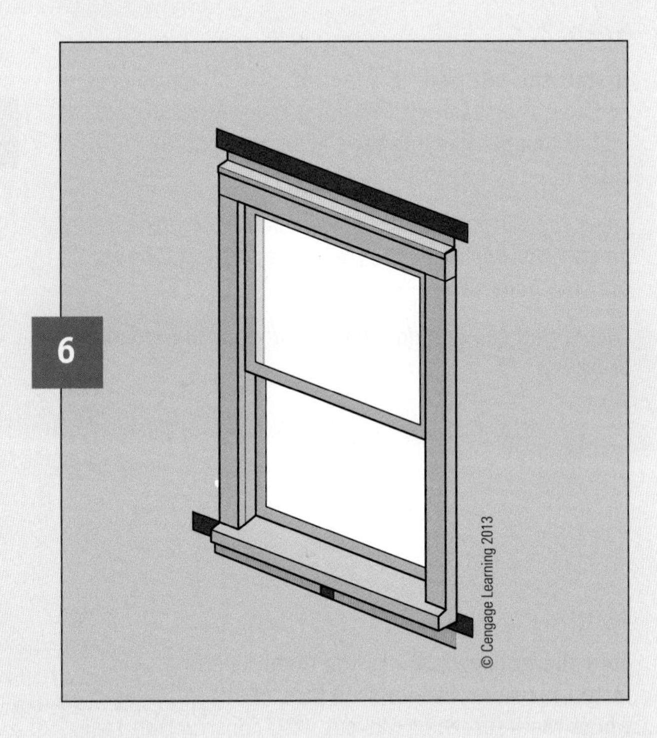

Step 7.

Fold housewrap down at head.

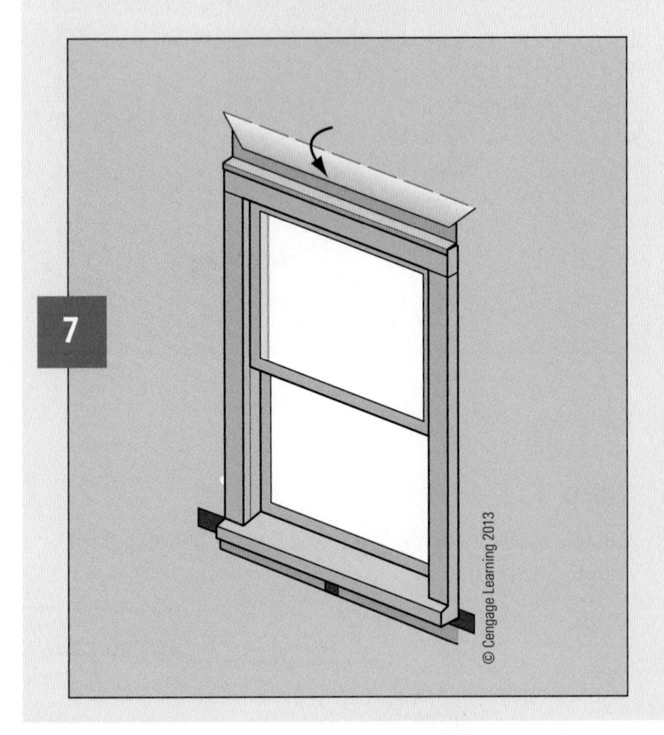

Step 8.

Tape head.

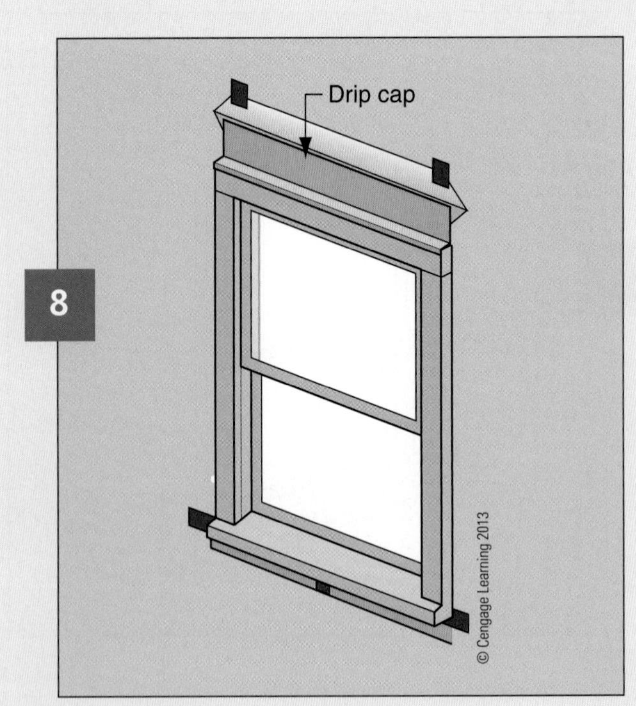

STEP-by-STEP **Figure 8.29 Window installation in a reservoir wall.**

Concrete masonry wall.

Step 1.

Install pre-cast tiered sill or cast in place sill. Seal to sill with a liquid applied waterproof sealant.

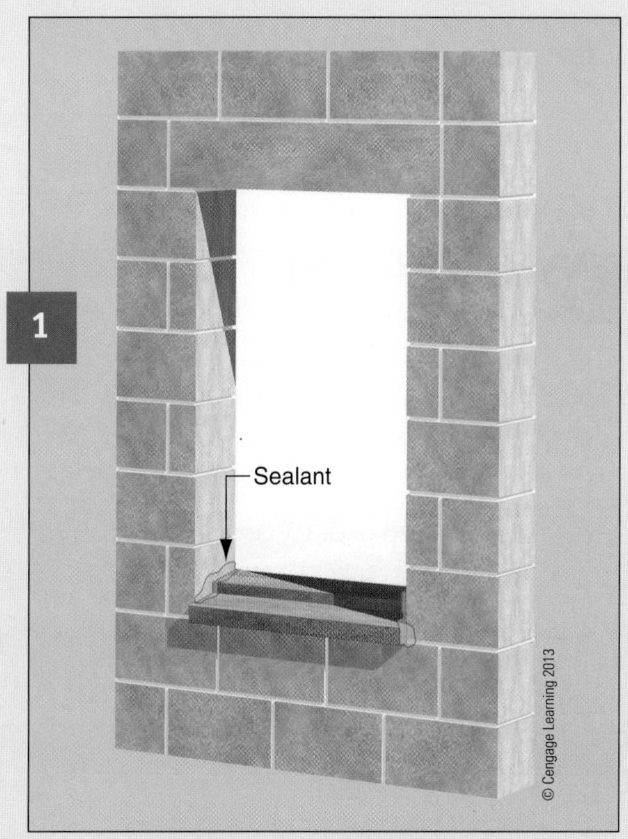

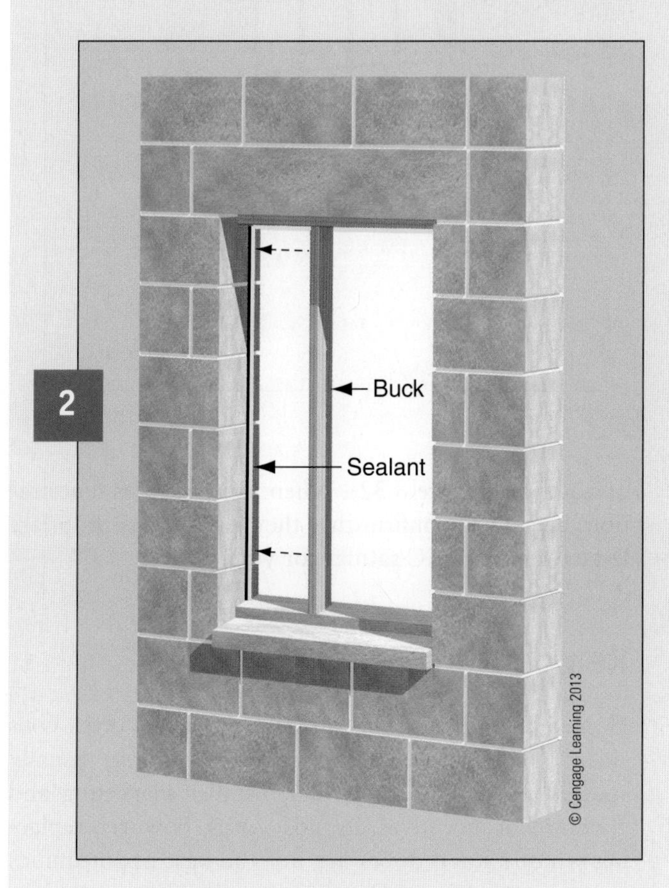

Step 2.

Install wood bucks over sealant.

(Continued)

STEP-by-STEP Figure 8.29 **(Concluded) Window installation in a reservoir wall.**

Step 3.

Apply sealant at the jambs and head.

Step 4.

Install window per manufacturer's specifications and apply sealant over exposed wood bucks.

Step 5.

Install exterior finish with continuous bead of sealant at jambs and head. Seal window to the stucco.

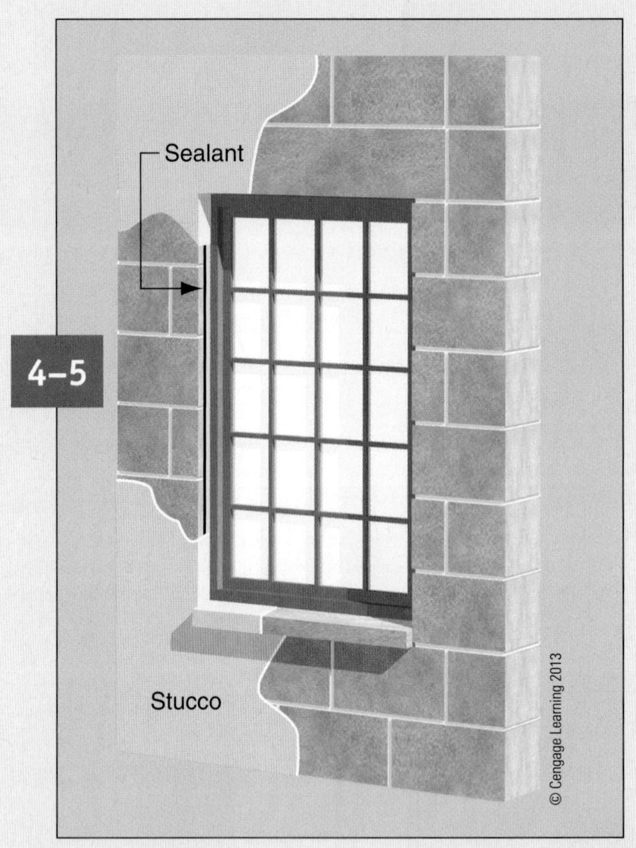

Windows are not easily recycled at the end of their useful life, so purchasing the highest quality units possible is of utmost importance. Correct installation and proper maintenance will ensure long service life. Quality manufacturers offer long warranties of 10 years on insulated glass and even longer on hardware and frames.

Ideally, old, inefficient windows should not be used in new homes or renovations, but sales on odd lots of unused high-quality new windows are not uncommon. These are typically units that were ordered or manufactured incorrectly and are often available at extremely low prices. A minor renovation or addition could be designed around surplus windows and doors of this sort, or they can be used as design features in a new home or addition (Figure 8.32). When using surplus fenestration, be sure to confirm that they have the appropriate U-factor and SHGC ratings for your project.

Remodeling Considerations

Homeowners looking to improve efficiency often consider window and door replacement first, usually spurred by manufacturer and installer marketing and government incentives. In most cases, however, replacing windows and doors are not the best opportunities for energy savings. Determining whether to replace or repair existing fenestration is just one piece of a

Courtesy of GREAT STUFF™ Insulating Foam Sealant

Figure 8.30 Window rough opening sealed with a low-expanding urethane foam.

whole-house evaluation. Understanding the extent of existing deficiencies in insulation air sealing, mechanical systems, and moisture management will allow for setting appropriate priorities for improvements, with fenestration being only one of many items to be considered. Replacing windows and doors without addressing other issues can create more problems than it solves if not done in the context of the whole house as a system.

Inspect Existing Units

Proper evaluation of existing fenestration is critical to making the best decisions in an existing house. Questions to ask include the following:

- What is the condition of wood frames, sashes, or door slabs? Are they deteriorated; and, if they are, are they beyond repair?
- Do window sashes and door slabs operate smoothly, close securely, and make a tight air seal?
- Does the hardware securely seal the unit? Is there thermal bridging through the window frame?
- Is the glass insulated and the unit well air sealed?
- How much solar gain comes through the glazing?
- Is the unit properly flashed and air sealed to the structure?

The answers to these questions can help you make appropriate choices regarding improving and replacing existing windows, doors, and skylights. Once you have determined whether to repair or replace existing fenestration, you have a wide range of options from which to choose.

Full Unit Replacement

When replacing complete window and door units in existing walls, it is critical to determine if there is a WRB and if it is possible to integrate the new units

Daylighting with Interior Windows and Glass Doors

Homes are often designed with closets, bathrooms, and other accessory spaces that have no windows. Many of these dark spaces are used infrequently and for just a few seconds at a time, but they require you to turn on a light to see. By adding just a small amount of natural light, you can eliminate the need for electric lighting for these brief visits during the day. Installing glass doors or windows between these rooms and other rooms with natural light can save electricity and provide a nice decorative element. These are appropriate locations for antique or decorative leaded glass pieces (Figure 8.31).

© Cengage Learning 2013

Figure 8.31 Skylights and decorative glass provide natural daylighting.

Photo by Jim Heafner. www.HaefnerPhoto.com

Figure 8.32 The Nautilus House project used surplus windows to reduce costs and provide the windows with a new life.

and flashing effectively into it. Many older wood-framed homes do not have a complete WRB. In this case, the best practice is to remove all of the siding, corner to corner and sill to soffit; install a new WRB; install new windows and doors; and replace the siding. This may seem extreme, but the risk of long-term moisture damage leading to catastrophic failure is too big to ignore. Unit replacement will involve removal and replacement of interior and exterior trim, providing easy access for air sealing between the unit and rough framing.

Improving Existing Fenestration

Options for window improvements other than full-unit replacement include sash and balance replacement and insert replacement units. Glazing in existing sashes may also be replaced. All of these methods retain the frame, interior, and exterior trim, reducing the costs and time involved. These methods neither affect any existing flashing, nor do they provide any moisture protection if none exists. Since the frame remains in place when any of these methods is employed, you may need to remove interior trim to provide an effective air seal. Pockets that formerly contained sash weights may also need to be filled with insulation. While this is a best practice, the project team should consider whether or not this additional work is appropriate for a particular project.

Installing storm windows over existing units is another alternative that will provide a certain measure of additional air sealing, thermal resistance, and if available with a low-E coating, a reduction of solar gain. Other options to reduce solar gain include an applied window film and exterior solar screens, which replace traditional insect screens while reducing solar gain by as much as 70% (Figure 8.33).

Jalousie windows have operable parallel glass, acrylic, or wooden louvers set in a frame. This style of window provides a marginal air seal (Figure 8.34), so jalousie windows should be replaced in any conditioned areas.

Doors

Swinging doors can be replaced in the existing frame, or the entire unit can be replaced, similar to the previously described methods. Door weatherstripping and thresholds can be replaced to provide a better air seal if the slab and frame are in good condition. Sliding doors are difficult to repair; if they do not operate or seal effectively, complete replacement should be considered.

JoshHobbs.com Solar Screen Services

Figure 8.33 Solar screens can significantly reduce solar gain into a home.

---

jalousie a window with operable parallel glass, acrylic, or wooden louvers set in a frame.

Figure 8.34 Jalousie windows tend to be very leaky.

Figure 8.35 Construction waste materials can be reused in numerous creative ways. Old windows are common canvases for artists.

Skylights

Old skylights are unlikely to be very efficient, but they are difficult to replace without causing roof leaks. They are best replaced at the same time as the roofing material. If existing units are to remain, consider adding window film or exterior shading to reduce solar gain. Ensure that the thermal envelope is complete with no bypasses at the skylight well, and a complete air seal is essential between the unit and the roof.

Recycling Removed Fenestration

There are limited opportunities to recycle old windows and doors. If the glass can be removed from the sashes, you may be able to locate a recycling center that will accept it. Occasionally, local artists may be interested in using old sashes and doors for craft projects and are willing to pick them up from the job site, saving this waste from entering landfills (Figure 8.35).

Summary

Selecting the appropriate fenestration is critical when creating a high-performance home. The house should be viewed as a system to guide decision making with regard to climate, building orientation, unit efficiency, frame materials, and proper installation method. For a green home, always buy windows, doors, and skylights of the highest possible quality, then install them properly and maintain them so they will perform well for the life of the structure.

Review Questions

1. Which of the following is not a benefit of windows, doors, and skylights?
   a. Daylighting
   b. Fresh air
   c. Insulation
   d. Passive cooling
2. Which of the following is not part of the NFRC rating system?
   a. R-value
   b. Air leakage
   c. Solar heat gain coefficient
   d. U-factor
3. What is the purpose of a low-E coating?
   a. Keep heat out in cold weather
   b. Keep heat in during hot weather
   c. Reduce visible light passing through glass
   d. Reduce infrared light passing through glass
4. In a cold climate, low-E coating should be applied to which glass surface?
   a. Outside of exterior pane
   b. Inside of exterior pane
   c. Inside of interior pane
   d. Outside of interior pane

5. What is the most important rating for fenestration in hot climates?
   a. SHGC
   b. U-factor
   c. VT
   d. Condensation
6. What NFRC ratings are best for south-facing windows in a cold climate?
   a. High U-factor, low SHGC
   b. Low U-factor, low SHGC
   c. High U-factor, high SHGC
   d. Low U-factor, high SHGC
7. Which is the best face to maximize windows in a cold climate?
   a. North
   b. South
   c. East
   d. West
8. Which of the following is the purpose of warm edge glass spacers?
   a. Increase VT
   b. Decrease heat loss
   c. Decrease VT
   d. Increase SHGC
9. Which of the following is appropriate for air sealing around windows and doors?
   a. Fiberglass insulation
   b. Mineral wool insulation
   c. Durable, flexible caulk
   d. High-expansion spray foam
10. Which of the following is the most cost-effective method to reduce solar gain through an existing windows?
   a. Installation of interior shutters
   b. Replacement with high U-factor windows
   c. Installation of solar screens
   d. Replacement with high-SHGC sashes

Critical Thinking Questions

1. Discuss the different U-factor and SHGC requirements for cold, moderate, and hot climates.
2. When designing a new house in a hot climate, what are the best strategies to control solar gain?
3. When designing a new house in a cold climate, what are the best strategies to take advantage of solar gain?

Key Terms

Additional Resources

National Fenestration Research Council (NFRC): http://www.nfrc.org

ENERGY STAR: http://www.energystar.gov

Exterior Wall Finishes

Exterior wall finishes, such as siding, brick, or stucco, and associated exterior trim, serve as a home's first line of defense against the elements and are part of an architecturally appealing design. Although exterior wall finishes are viewed as a home's raincoat, the moisture management system behind those finishes is what keeps most homes dry. Effective moisture management is essential to creating a durable, high-performance home.

Decisions regarding exterior finish and moisture management must take into consideration the climate, the type of wall structure, sustainable material use, and personal preferences. In this chapter, we will discuss how to select the right exterior finish systems for a green home.

LEARNING OBJECTIVES

Upon completion of this chapter, the student should be able to:

- Discuss the purpose and elements of a water-resistive barrier system
- Describe the purpose of a vented rainscreen and how to create one
- Explain how rain is kept out of the wall cavity in a wood-framed home
- Describe the sustainable attributes of different exterior finish materials

Green Building Principles

 Resource Efficiency

 Durability

 Indoor Environmental Quality

Introduction to Exterior Finishes

Exterior wall finishes comprise the materials applied to the vertical wall surfaces, the cornices and other exterior trim, and the water management system behind them. The choice of water management systems depends on the type of wall structure. While most homes use wood framing, some are made of alternative systems, such as structurally insulated panels (SIPs), insulated concrete forms (ICFs), steel framing, and other products. Each wall type is classified as either a storage reservoir or a non-reservoir system. **Storage reservoir walls** are capable of absorbing water and drying out without damage to the structure. **Non-reservoir walls** cannot absorb water without risking structural damage.

The most common wall types are non-reservoir systems, such as stick framing or SIPs. Exterior water must be managed and drained away from the structure. According to both the International Residential Code and good building practices, non-reservoir walls must have a water-resistive barrier (WRB) installed beneath the finish material to keep bulk moisture out of the wall structure.

---

storage reservoir walls wall assemblies that can absorb water without risking structural damage.

non-reservoir walls wall assemblies that cannot absorb water without risking structural damage.

Hygroscopic materials, like reservoir systems, readily attract and retain moisture. Reservoir systems, such as ICFs or autoclaved aerated concrete (AAC), are able to absorb and release moisture without causing structural damage. They do not need a separate WRB to keep water out of the structure; however, careful detailing of window and door openings is critical to keeping rain out of the interior spaces.

Wall Finish Materials

Exterior wall finish materials can be made of solid or engineered wood, wood fiber composites, cement fiber, vinyl, metal, masonry, or stucco. Any of these materials can be applied to either reservoir or non-reservoir wall systems, although some materials are more appropriate for one wall type over another.

Cornices

The cornice is the combination of the soffit, the horizontal surface that projects out from the exterior wall; fascia, the vertical trim between the soffit and roof; and associated trim that covers roof overhangs from the edge of the roofing material to the top of the wall finish (Figure 9.1). Cornice finishes include wood, cement fiber, cellular polyvinyl chloride (PVC), vinyl, and metal. Other trim on homes includes corner boards, columns, pilasters, brackets, and window or door casings. Homes with masonry or stucco exterior finishes without roof overhangs may not have cornices.

Selecting Exterior Finishes

Exterior finishes in most homes are selected for their aesthetic appeal. When planning a green home, the project team should also consider durability and sustainable extraction or harvesting of materials.

For many years, exterior finish materials were limited to wood, brick, stone, and stucco. Now, many products mimic the looks of these traditional materials while also offering improved durability, lower cost, reduced material use, or greater recycled content. This range of options allows the design team to select a style first, and then choose a specific material and installation method to achieve the desired look. Typical exterior finish styles include lap siding, shingles, shakes, vertical paneling, brick, stone, and stucco.

Keeping Bulk Water Out

Understanding how water gets into above-grade walls is key to designing systems that effectively keep it out. Rain is the primary type of bulk water intrusion into

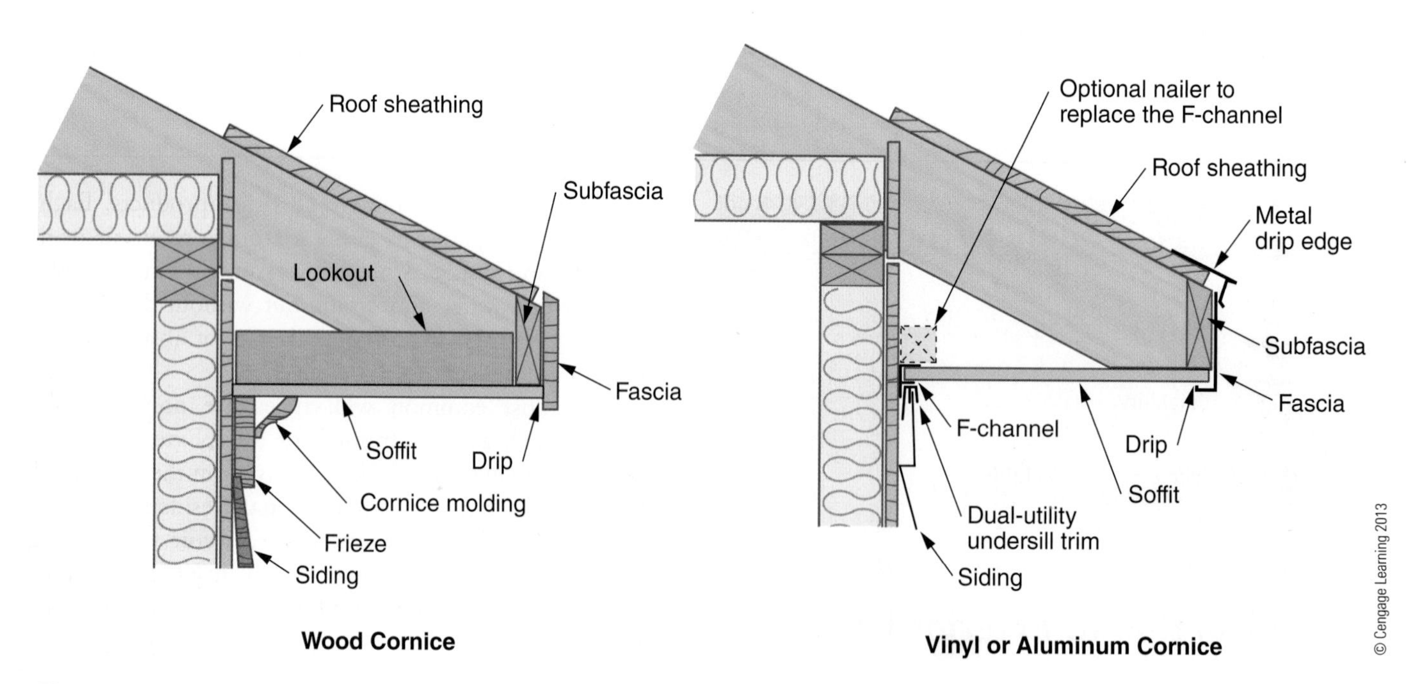

Figure 9.1 Traditional cornice details.

hygroscopic materials that readily attract and retain moisture.
cornice the entire finished assembly where the walls of a structure meet the roof; sometimes called *eaves* or *soffits*.
soffit a horizontal surface that projects out from an exterior wall.
fascia the vertical trim between the soffit and roof.

above-grade walls—the more it rains, the greater the importance of having an excellent moisture management system. That being said, water management is still important in locations that experience very little rainfall. The driving forces behind water entry are gravity, capillary action, momentum, surface tension, and wind pressure.

Gravity is the most common cause of water intrusion and one of the easiest to manage. Water always takes the path of least resistance and travels from areas of high concentrations to low. Through proper installation of the WRB and flashing, the use of roof overhangs, and appropriate water-shedding details on exterior finishes, water can be directed away from the home. WRB are bulk water-shedding surfaces that protect the home's sheathing and prevent water intrusion.

Capillary action sucks water into porous materials and through small cracks in the exterior finish (Figure 9.2). An air space between the exterior wall finish and the WRB provides a capillary break that helps to keep the water from entering the wall structure. This gap is referred to as a **rainscreen** and allows pressure equalization to prevent rain from being forced into the home (Figure 9.3 and see below for more information).

Rain can work its way through a wall by its own *momentum*, which is controlled by eliminating direct paths into the structure through the use of sloped

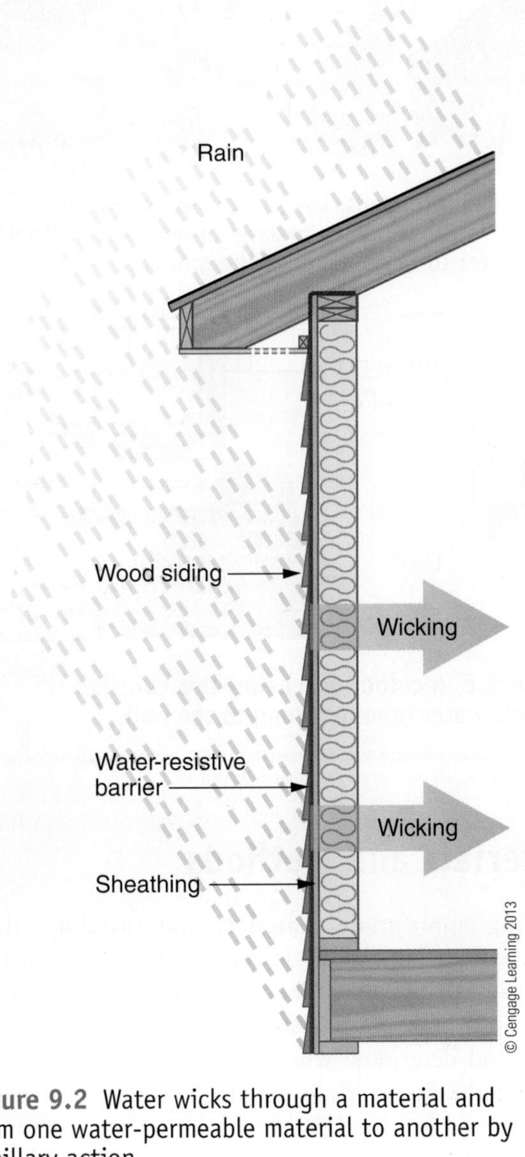

Figure 9.2 Water wicks through a material and from one water-permeable material to another by capillary action.

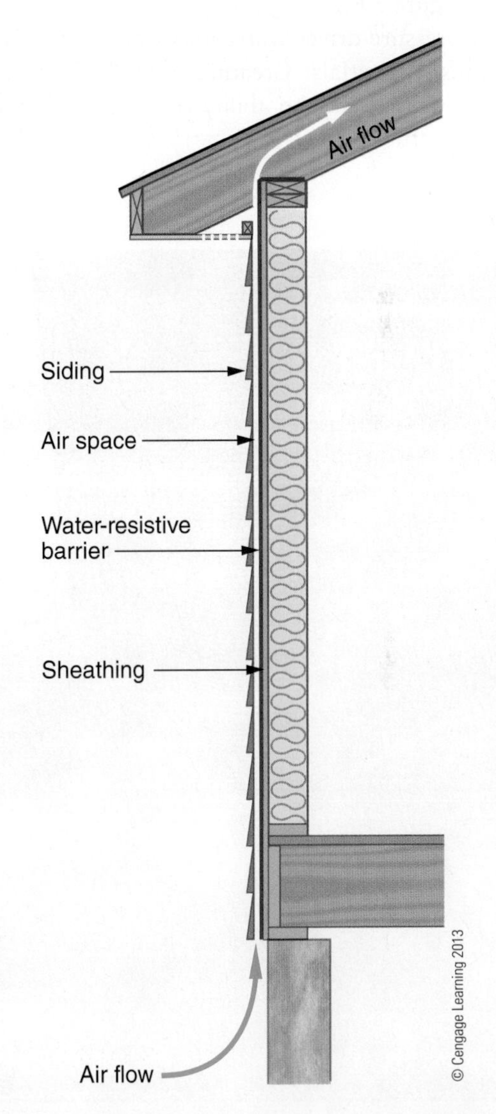

Figure 9.3 A rainscreen is an air space between the exterior wall finish and the WRB. This gap prevents the migration of water from the exterior to the interior of the wall.

---

rainscreen a method of constructing walls in which the cladding is separated from the water-resistive barrier by an air space that allows pressure equalization to prevent rain from being forced inside.

windowsills and similar features (Figures 9.4). Drip caps are horizontal molding or flashing installed over the frame for a door or window to direct water away from the frame (Figure 9.5).

Window installation details are covered in Chapter 8, and Figure 8.28 shows the installation of a drip cap.

Water can flow around material and into wall structures through *surface tension.* A drip groove is a cutout on the underside of the projection intended to prevent water from traveling beyond it and back to the face of the wall (Figure 9.6).

Wind pressure drives water into walls through gaps in the finish materials. Creating a pressure-equalized rainscreen will reduce the ability of water to penetrate the wall structure (Figure 9.3).

Figure 9.4 A sloped window sill directs water away from the home.

Figure 9.5 Drip caps, the small metal or PVC projection above the top edge of exterior windows and doors, allow water to fall directly to the ground.

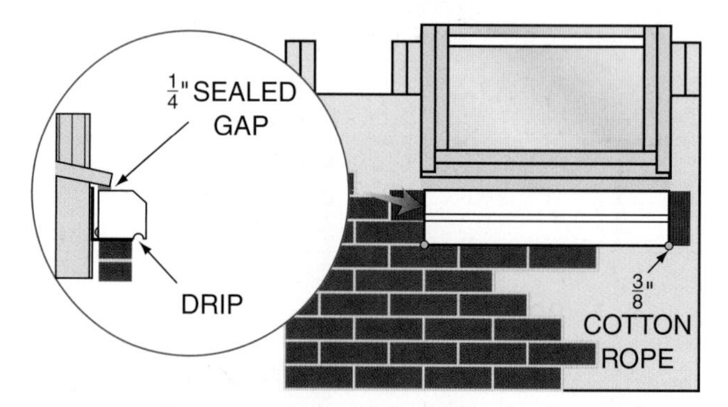

Figure 9.6 A cutout drip below this concrete sill prevents water from returning to the wall.

Materials and Methods

Selecting appropriate materials and installing them correctly are the critical elements of high-performance exterior wall finishes. When selecting these materials, the project team must identify the desired style of the house and determine which material is best suited to creating the appropriate details. The material should be evaluated for durability and appropriateness for the climate. The WRB, rainscreen design, and installation

---

drip cap a horizontal molding or flashing installed over the frame for a door or window to direct water away from the frame.

drip groove a cutout on the underside of the projection intended to prevent water from traveling beyond it and back to the face of the wall.

details should also be factored into material selection to ensure careful management of the entire installation process.

Considerations in Selecting a Finish Material

Exterior finish selection is primarily an aesthetic decision, usually controlled by the designer or building owner. Options include lap siding, shingles or shakes, brick, stone, stucco, or panels. Many of these styles can be created with any of a number of different materials and products.

Lap, shingle, shake or panel siding is available in solid or engineered wood, wood fiber composite products, vinyl, aluminum, and fiber cement. Brick finishes are available in standard veneer, thin veneer, and fiber cement panels. Stone can be cut in a quarry or manufactured. Stucco can be made with traditional sand and cement or with acrylic resins applied directly to the wall or over foam insulation or fiber cement panels.

Each of these materials has different features and benefits, offering varying levels of durability and embodied energy. Recycled content, recyclability, and sustainable material use may vary greatly from product to product. Appendix A, located at the end of this chapter, summarizes the environmental benefits and drawbacks to each siding option.

Durability

Considering the lifetime of a building and the number of times a finish must be refinished or replaced, the selection of a material with the longest life can reduce not only the amount of materials and labor required for refinishing or replacement but also the waste created by these activities. Some more durable products may have a higher initial cost, but their total lifetime cost (including maintenance, repair, and replacement) may be lower than less expensive products. In areas with a high risk of termite or other pest infestations, the use of materials and methods that resist pest damage should be considered in material selection decisions.

Embodied Energy

The amount of energy required to extract raw materials, to produce the finished product, and to deliver it to the job site are important considerations. Keep in mind, however, that a material with a higher level of embodied energy that only requires replacement every 100 years may be a better choice than another with lower embodied energy but requires replacement every 25 years.

Recycled Content

Few exterior finish products incorporate recycled materials in their manufacture; they are limited primarily to fly ash used in cement fiber products and recycled materials in manufactured stone veneer. Brick, stone, and wood can all be reclaimed for reuse from old buildings that are being demolished.

Recyclability

With the possible exceptions of brick, stone, and wood that can be reused or ground up, most exterior finish materials are not easily recycled. Thus, selecting products with the longest life helps reduce the overall environmental impact.

Sustainable Material Content

Selecting products that are harvested, extracted, and processed with the fewest hazardous chemicals will reduce the environmental impact on workers in the factories where the products are manufactured, as well as the occupants of homes where they are installed. Careful product choices can also benefit other individuals who may be affected by manufacturing plant emissions.

Exterior Finish Material Options

Exteriors can be finished with traditional materials, such as wood, brick, and stone; with more modern products, such as vinyl, wood fiber composites, cultured stone, and fiber cement; and with less common products, such as metal panels. Comparing and choosing among the available options requires an understanding of the properties of each, which we explore in the following sections.

Wood

Wood has been used for siding and exterior trim for hundreds of years; when installed and maintained properly, it can last the life of a house. Wood siding is available in many styles, including lap, shingles, shakes, and even bark (Figure 9.7). Finishes can be smooth or rough textured, and the wood can be stained or painted.

Wood is very desirable from a design standpoint and is easily worked with standard tools and skills; however, it does require regular maintenance. The frequency of maintenance depends on the climate and installation methods. Wood from old-growth trees is the most durable but is not readily available today. Such wood usually costs more and is less sustainable than material harvested from faster growing, younger trees. Although solid wood has no recycled content, some siding and

---

cultured stone a cast masonry unit made of cement and various additives that simulates the look of natural stone.

Figure 9.7 The poplar tree's bark is usually a waste byproduct from the timber mill process, but it can also be used for siding a home.

trim is finger-jointed from smaller scraps of wood, which makes better use of resources. Unpainted and untreated wood can be recycled, but other wood scraps must be discarded. Wood production is generally sustainable, particularly when harvested from certified forests (see Chapter 6 for a discussion on sustainable forestry programs). Wood is very efficient to produce, and its production generates little waste because most mill scraps are used in engineered wood products or as a biofuel.

Wood Fiber Composites

Siding and exterior trim can be made from wood fiber and glues that are combined under heat or pressure to create a dense, smooth, and durable material. Hardboard siding and other early products had durability problems, which led to lawsuits and their ultimate removal from the marketplace. Newer products, such as MiraTEC® from CMI and SmartSide® from Louisiana-Pacific, require less maintenance than solid wood and are more durable than their predecessors, with the latter product offering a 50-year substrate warranty. Composites make efficient use of sustainable new growth trees and waste from other processes. Similar to exterior-grade plywood and oriented-strand board (OSB), most of these composites are produced with resins that contain phenol formaldehyde, which does not have the same negative health effects of urea formaldehyde. Due to the use of these resins in their manufacture, however, these products are not recyclable. Wood fiber composites have a higher embodied energy level than that in solid wood, but the level is still lower than that in fiber cement or brick, which use significant amounts of heat in their manufacture.

Composites are heavier than wood, so they have increased shipping and handling costs. However, most are available pre-primed or pre-finished, which reduces time and costs involved in on-site painting. Working with composites is similar to working with wood, but strict installation requirements must be upheld to maintain their warranty. Some composites are not suitable for miter or scarf joints or countersunk nails, which can make for less attractive installations than solid wood. Finally, composites are only available in limited lengths (generally 16' or 20'), which can leave a large volume of cutoff waste if the building design and product selection are not coordinated carefully.

Fiber Cement

Fiber cement siding and trim, the successor to asbestos siding, has commanded a significant segment of the market. Made from Portland cement, sand, wood fibers, and clay, fiber cement is extremely durable and dimensionally stable; some products carry warranties of up to 50 years. It is very resistant to both fire and rot. The amount of recycled content from water waste and fly ash varies among manufacturers, but waste siding and trim material are not recyclable. While it is an extremely durable product, fiber cement has high embodied energy because of its Portland cement content as well as the transportation costs of wood fiber sourced from overseas. James Hardie is the dominant manufacturer of fiber cement products, although Nichiha USA, CertainTeed, and other companies now offer competing products.

Fiber cement provides a look similar to wood and composites with some of the same installation limitations of composite products, such as limited use of miters and countersinking of nails. In addition to wood looks, fiber cement is available in panels with brick, stucco, and stone-look finishes. Most fiber cement siding and trim boards are between 5/16" and 3/4" thick, and new-generation 5/8" siding can be mitered for a more traditional look. Available siding and shingle design finishes include smooth, wood-grain, and beaded, and 4' × 8' panels with smooth or stucco finishes can be used to create a board-and-batten look or other architectural details (Figure 9.8). Soffit panels are available smooth, beaded, and vented (Figure 9.9). Most fiber cement is

Figure 9.8 Fiber cement siding is available as 4' × 8' panels and lap siding.

Figure 9.9 Fiber cement soffits are available in a variety of styles to fit any project.

installed with nails or screws, although Nichiha USA offers brick, block, and stone-look panels installed with a proprietary metal clip system that provides a built-in rainscreen (Figure 9.10).

Fiber cement contains crystalline silica, a known carcinogen. Breathing excessive amounts of silica dust—created from cutting fiber cement products—can cause silicosis, a potentially fatal lung disease. Product manufacturers have specific safety recommendations for dust collection and respirator usage that should be followed to protect workers and others in the work area (Figure 9.11).

Brick

Originally used in houses as the structural wall, brick was laid in as many as three or four layers to provide a very durable (if not well-insulated) wall structure. Modern brick is installed as a solid veneer on top of a footing, as a thin layer attached directly to the wall structure, or with a proprietary fastening system (for more information, see Chapter 5). In addition to the materials required to create a thick brick veneer, additional concrete is required for the footing to support its weight. Thin brick uses less material and does not require the extra concrete footing. Most brick is very durable and requires little maintenance; however, some manufacturers are producing bricks with applied decorative finishes that are not designed to withstand power washing or acid cleaning and may not have the same long-term durability of solid-color bricks. While most brick is made from virgin clay, some styles are available with recycled content such as those from Green Leaf Brick. Salvaged brick from demolished buildings can be reused in new homes and renovations (Figure 9.12), and scrap brick can be ground into aggregate (Figure 9.13). Brick has a high level of embodied energy from the high temperature firing required to manufacture it, but its overall life cycle analysis is good because it lasts indefinitely with little or no maintenance.

Because of its weight, transporting brick uses significant amounts of fuel. To keep embodied energy levels down, many brick manufacturers are now using alternative fuel sources to fire the kilns, such as methane from landfills and wood waste. The most sustainable brick is locally produced and as lightweight as practical for the application.

Stone

Like brick, stone originally constituted the entire wall structure in some buildings. Modern stone is installed as a solid veneer or as a thin layer adhered to the wall structure (Figure 9.14). Thick stone veneers, like their brick counterparts, use more material than thin veneers

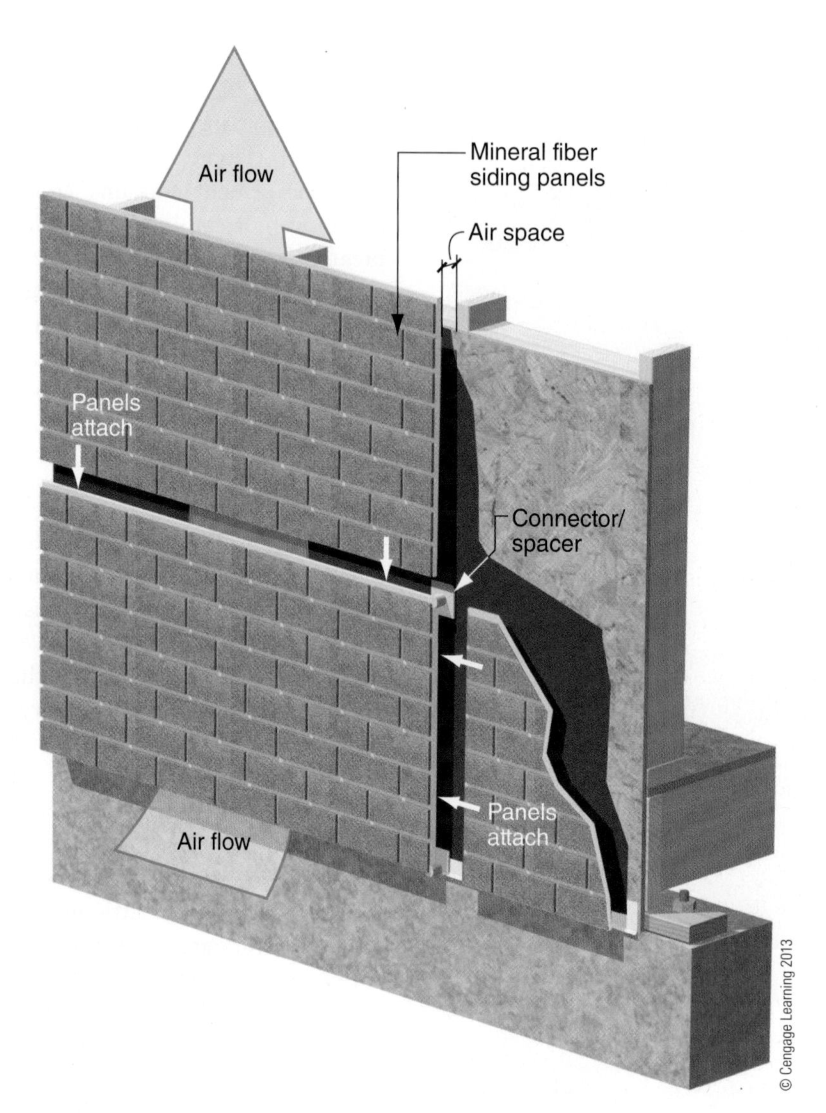

Figure 9.10 Fiber cement siding products are available in panels that are installed with metal clips. The clips hold the siding away from the housewrap, creating a vented rainscreen.

and require an additional concrete footing for support; however, natural stone has a low embodied energy and is a durable, low-maintenance finish material.

Stone is available in solid form, either cut from a quarry or as a cast cement product referred to as cultured stone. Both types are very durable, although some cultured products have a surface coating that can chip. Cultured stone may be made with some recycled content, usually pre-consumer.

Natural stone salvaged from demolition projects is usually suitable for reuse on new construction or renovations. Both natural and cultured stone scraps can be ground into aggregate. Natural stone quarries cause tremendous environmental impact by removing vegetation and topsoil and generating large volumes of waste products that require disposal. Large equipment operation in quarries also produces air pollution. The use of cultured stone reduces these impacts but carries the high energy demand and other downsides of producing Portland cement. Using reclaimed or local natural stone that is extracted in a responsible manner is a sustainable material choice.

Stucco

Stucco finishes have long been created with a traditional three-coat process or a one-coat system. Three-coat systems are a traditional, nonproprietary mix that uses Portland cement, sand, and water applied to a total thickness of approximately 1" over a plastic or metal lathe (Figure 9.15). One-coat stucco, which actually has two layers (a base coat and a finish), adds fibers and proprietary chemicals to the mix and is also applied over lathe (Figure 9.16).

A newer alternative to the one-coat or three-coat systems, the **exterior insulated finish system (EIFS)** is a synthetic stucco finish applied over foam insulation. EIFS originally was designed as a water-resistive

---

exterior insulated finish system (EIFS) a synthetic stucco finish applied over foam insulation.

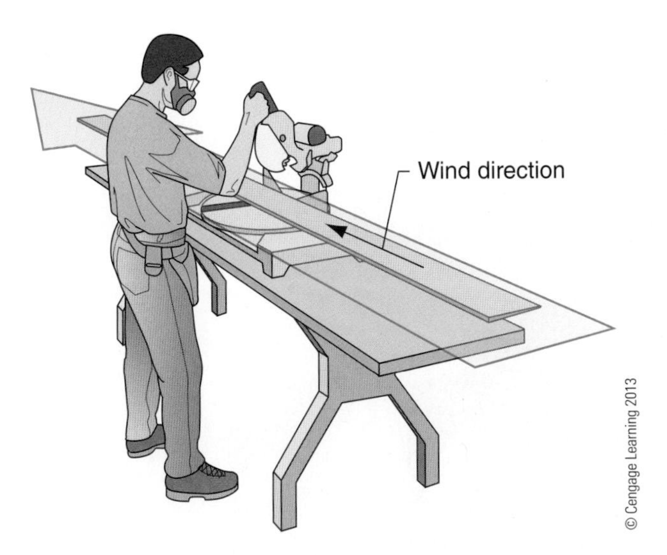

Figure 9.11 Always follow manufacturer's guidelines when cutting fiber cement siding. Set up the cutting station so that the wind carries dust away from the saw operator. In addition, the operator should wear a NIOSH-approved respirator, and the debris should be collected in such a way as to prevent sending dust particles into the user's breathing area.

Figure 9.12 Stacking salvaged brick on a demolition site.

Figure 9.13 Brick can be ground and reused as aggregate or for landscaping.

Figure 9.14a Stone is installed as a solid veneer over a WRB.

Figure 9.14b A thin layer of stone veneer is adhered to the wall structure.

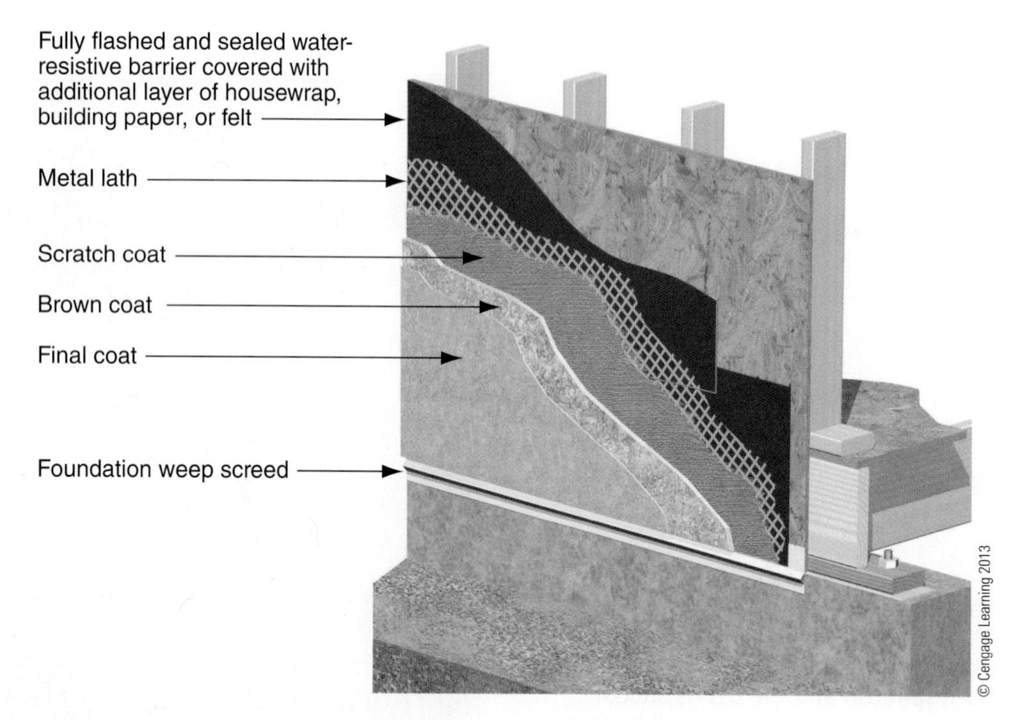

Figure 9.15 A three-coat stucco system consists of a scratch coat, brown coat, and finish coat. The first layer is the scratch coat embedded in metal lathe. This provides a strong base for the system. The brown coat is then applied to create an even surface for the finish coat. The finish coat is applied last, creating the decorative finish on the wall surface.

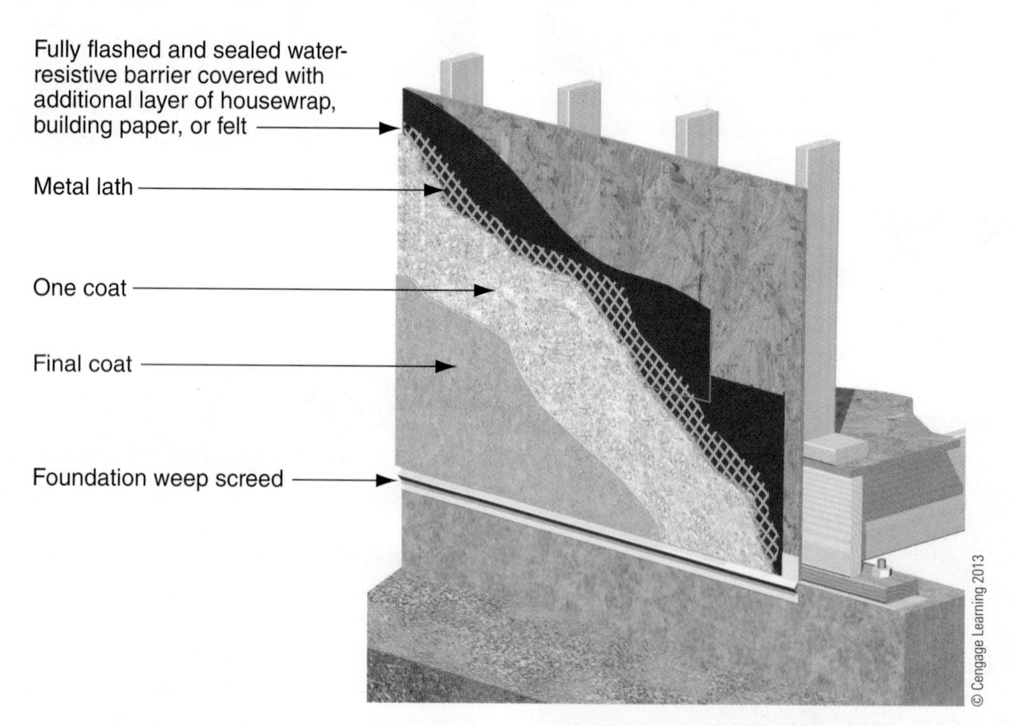

Figure 9.16 One-coat stucco consists of a blend of Portland cement, sand, fibers, and special chemicals.

barrier system without a drainage plane, but it led to water damage and mold in wood-framed buildings, which resulted in multiple lawsuits and a redesign of the system (Figure 9.17a). Although EIFS received most of the blame for these problems, both three-coat and one-coat stucco finishes can also create water problems when installed over wood framing without effective drainage behind them (Figure 9.17b). EIFS finishes can

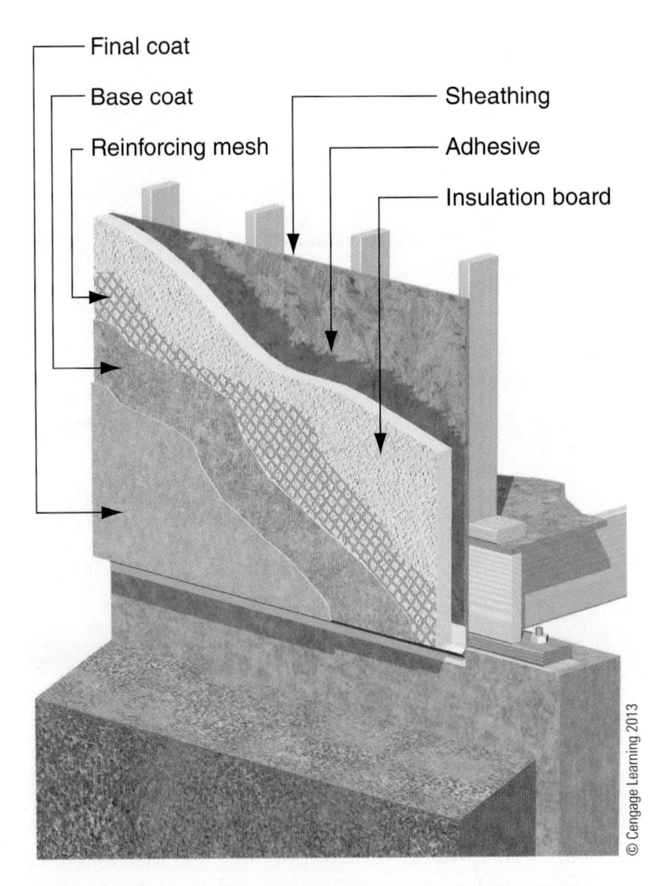

Figure 9.17a Early EIFS systems did not provide a drainage plane, which resulted in numerous moisture problems. Any moisture that penetrated behind the insulation board was unable to drain or dry to the exterior.

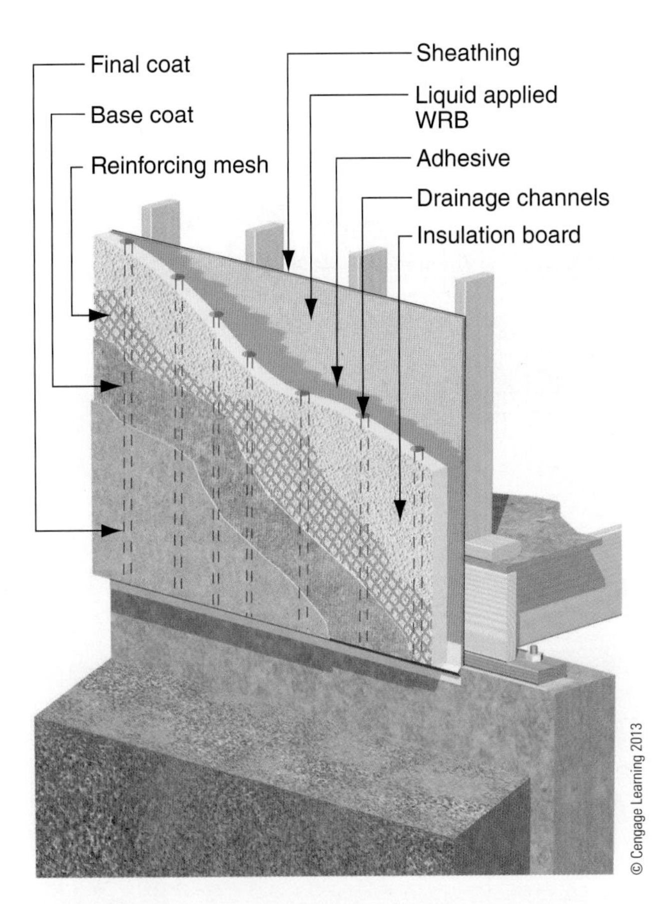

Figure 9.17b The Dryvit® Infinity EIFS provides a watertight membrane and an insulation board with drainage channels to capture, control, and discharge any incidental moisture that may enter the system.

also be applied in a single coat over masonry like traditional stucco. Stucco is durable, although cement-based finishes are susceptible to surface cracking and require cleaning or repainting periodically. Stucco and EIFS can be applied with integrated colors that eliminate the need to repaint them.

Stucco is not made with recycled materials, and it is not recyclable at the end of its life. The Portland cement used in traditional stucco has a high level of embodied energy. EIFS does not have as much embodied energy, but the plastics and other additives have environmental impacts that offset the elimination of cement from these mixtures.

Lime-based stucco is less common but is used by some straw bale builders, traditionalists, and natural builders who prefer its finish and because it has a lower embodied energy than cement-based masonry stucco.

PVC Siding and Trim

PVC siding, commonly known as vinyl siding, has commanded a significant share of the residential market due to its low price and easy installation. Available in patterns that mimic smooth or wood-grain lap siding, shingles, and shakes, vinyl siding provides a low-maintenance exterior finish that carries the benefit of an integrated rainscreen (Figures 9.18). Although PVC is considered to be a durable material, vinyl is susceptible to damage from severe cold or heat or by heavy wind. PVC siding may even be vulnerable to melting from sunlight reflecting off low solar heat gain coefficient (SHGC) windows. Some styles are available with integrated insulation which, when installed and sealed carefully against the building sheathing, can provide additional R-value (Figure 9.19).

PVC trim materials are available in many designs, including flat stock, moldings, and panels (Figure 9.20). PVC is a durable, easily workable substitute for wood trim, proving particularly useful in locations that are subject to extreme weather conditions where wood might not last. It does, however, expand and contract more than other products, which must be taken into account during installation to avoid large gaps or boards pushing away from the structure.

PVC is not typically made with recycled content. Installation scraps can be recycled, but collection centers are limited, and demolition waste is generally not recyclable.

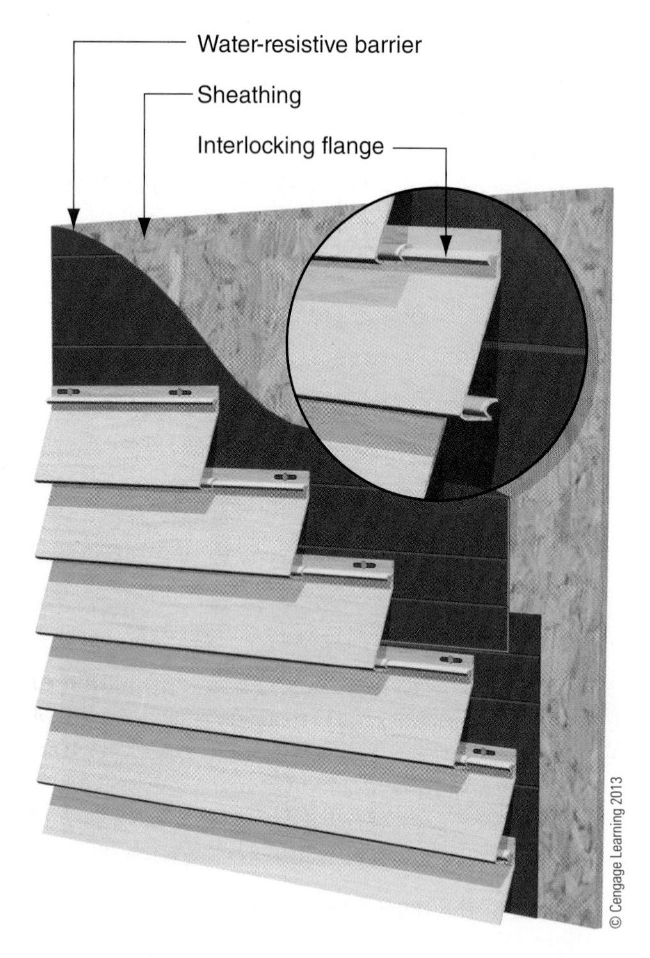

Figure 9.18 Vinyl and aluminum siding often have a built-in air space behind them. Although not originally intended to provide a rainscreen, the air gap is actually quite effective at allowing the assembly to dry.

Figure 9.19 Vinyl siding with integrated insulation.

Figure 9.20 PVC trim materials.

The primary concerns with PVC as a building material are the toxicity problems in manufacturing, which are addressed in detail in Chapter 8. PVC also releases toxic fumes when burned in disposal or a house fire. PVC has relatively low embodied energy.

Metal

The original residential metal siding was aluminum, which has lost much of its market share to vinyl and fiber cement in recent years. Aluminum siding is available in traditional designs with smooth and wood-grain finishes. Corrugated metal panels are also used for siding in both traditional and contemporary projects (Figure 9.21).

Figure 9.21 Metal siding is durable and available with recycled content.

Courtesy of Pacific Columns, Inc.

Figure 9.22 Fiberglass columns are nearly indistinguishable from their wood counterparts and are resistant to moisture damage.

Metal siding is extremely durable, but it is subject to denting and scratching. Many metals are made with high recycled content, and it is generally easy to recycle scraps and demolition waste. The embodied energy of metals is relatively high; however, the long life and recyclability of the material make it a suitable choice for a green home.

Fiberglass

A limited amount of exterior trim is available made of fiberglass—most notably round and square columns that are usually provided with plastic composite bases and caps (Figure 9.22). Fiberglass is a suitable substitute for wood columns, which require regular maintenance and are susceptible to water damage. Fiberglass is not made with recycled content and is not easily recycled; however, it does not raise the same concerns in manufacture as PVC, and it has relatively low embodied energy.

Paints and Finishes

Many exterior materials, such as wood, wood composites, fiber cement, and some stucco, must have finishes applied for durability and aesthetics. Cedar, cypress, and some other woods can be finished with clear or semitransparent stain, but most other wood products require paint finishes. Wood, composites, and fiber cement are available with factory-applied primers and, in some cases, final finishes.

The primary environmental concern with paints and stains is the release of volatile organic compounds (VOCs) during the drying process. Many paints and stains are available with low or zero VOC content, and some exterior stains are made from nontoxic and biodegradable oils. Paint made with recycled content is also available, although it is not very common.

Mineral silicate paints are used on masonry and stucco and create a chemical bond with the substrate. These paints are made from natural inorganic compounds, have zero VOC content, are naturally mold resistant, and can last 30 years or more without repainting.

Exterior finishes that are durable and contain the fewest VOCs and solvents are the best choices for a green home.

Design for Durability and Ease of Maintenance

Any exterior materials that have an applied finish, such as wood, fiber cement, or painted stucco, will require maintenance. These materials are affected by water, freezing temperatures, excessive heat, sunlight, wind, hail, and storms that create flying debris. Each has specific installation requirements or recommendations that extend life and reduce the frequency of maintenance. In addition to proper installation, the incorporation of such design elements as overhangs can extend the life of finish materials by improving site drainage and protecting walls from excessive rain and sun exposure. Providing easy access to all exterior walls and cornices will also increase the likelihood that the homeowner will perform the necessary maintenance.

Avoiding Design Pitfalls

Exterior finish details, such as roofs that terminate into walls and masonry veneer walls above roofs, can interfere with the creation of a complete WRB. There are proven methods to manage bulk moisture with these designs; however, they are complicated and are prone to failure when not installed and maintained properly. Best design practices will prevent the creation of these conditions in a green home. When these design elements cannot be avoided, moisture management systems at these areas must be both well designed and properly executed in the field.

Roofs terminating in walls require kick-out flashing to be installed and properly layered with the WRB and roof flashing to keep water from running inside the wall structure. All siding materials should be installed with a minimum 2" clearance from roofing materials (or whatever distance is recommended by the manufacturer). Installing siding or stucco tight against roofing can lead to premature failure of both the finish and the roofing; see Chapter 7 for more information.

Masonry that is installed above roofs, such as above a bay window or porch roof in a brick wall, must have proper flashing integrated with the WRB. Weep holes should direct water away from the interior, without allowing any water to penetrate the interior below the brick veneer (Figure 9.23).

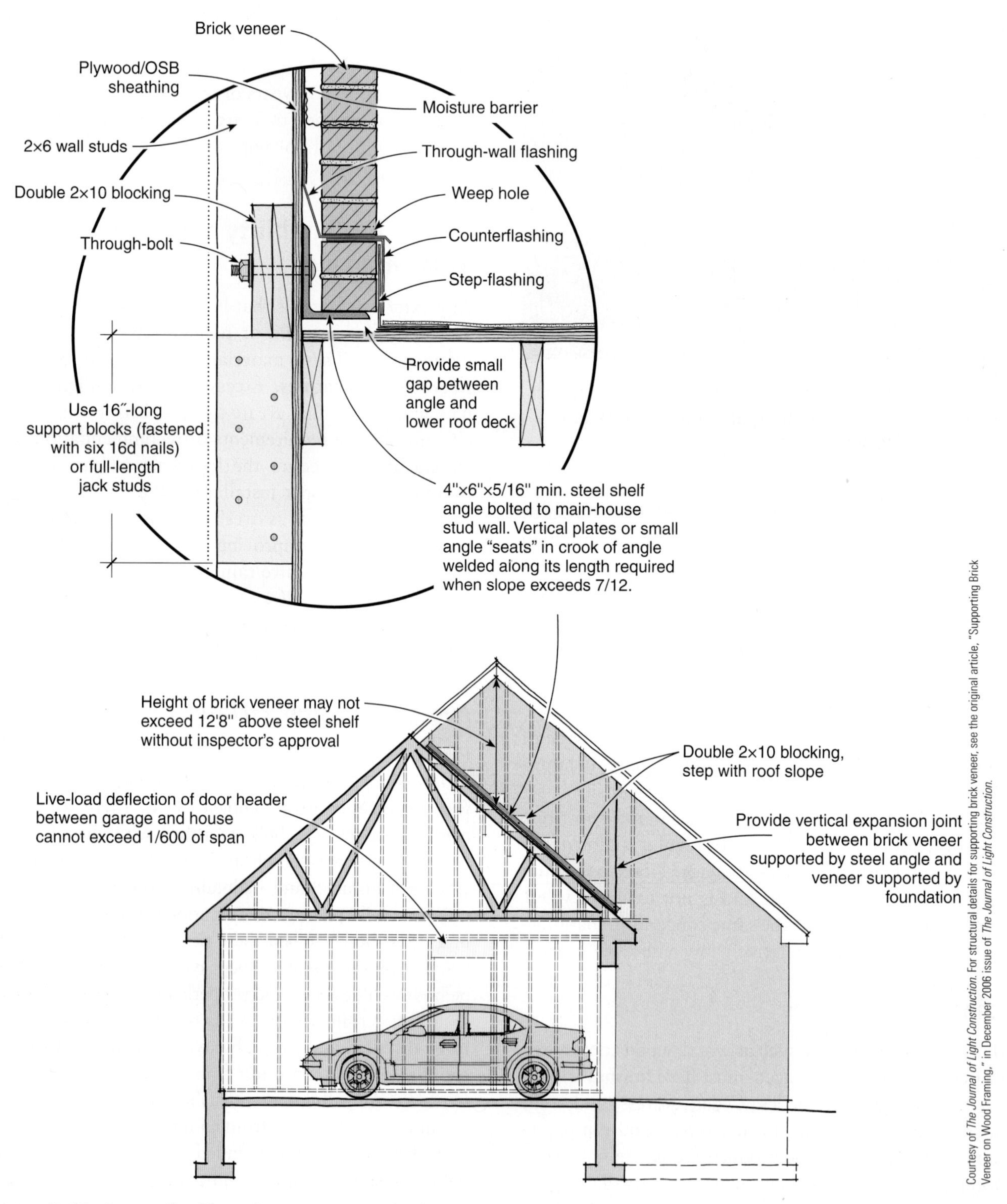

Figure 9.23 Proper flashing of a masonry clad wall above a roof.

Planning Exterior Finishes

Stick-framed homes using optimum value engineering techniques (OVE) may lack structural backing for some siding details, particularly corner boards and large window casings; these issues should be addressed in the design development stage of the project. Appropriate nailers should be used, and structural sheathing should be installed to provide a suitable nailing surface (Figure 9.24).

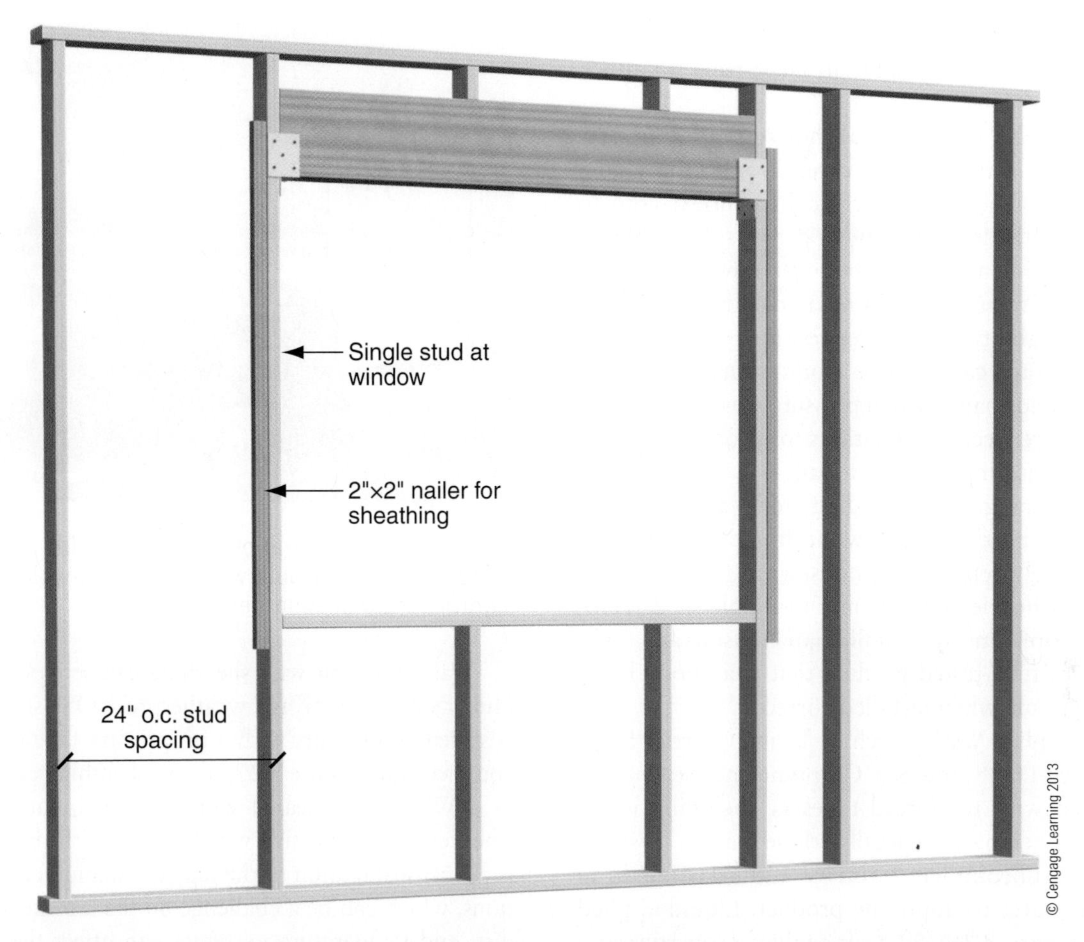

Figure 9.24 When using advanced framing, ensure that adequate support is provided for siding materials.

Did You Know?

ENERGY STAR Version 3 Moisture Management Requirements

The new ENERGY STAR Version 3 standards incorporate moisture management best practices. Among these new specifications are a builder-verified Water Management System Checklist. As discussed in Chapter 2, some features of energy-efficient homes are less forgiving to moisture and must be kept dry. The new standards help to ensure water does not enter the home through the roof, walls, and foundations.

Building modules should be designed to coordinate with available sizes of finish materials. For example, if exterior siding is only available in 16' lengths, design the building in 2' or 4' modules to avoid layouts that require cuts at just over the module, which would leave large quantities of waste.

Moisture Management Systems

Effective moisture management in nonreservoir walls consists of a WRB, flashing, and, as a best practice, a vented rainscreen. The WRB can be an applied sheet, a self-sticking membrane, or a liquid-applied coating. Flashings may be a combination of metal, rigid or flexible plastic, and self-sticking tapes. Vented rainscreens provide an air space between the exterior finish and the wall sheathing that helps keep the wall dry.

Storage reservoir walls do not need a WRB; however, openings in these walls must be properly detailed and sealed to eliminate water intrusion to the interior. Window and door sills should have a back dam at the base, a drip groove at the top, and sealant applied between frames or casing and the wall material (see Chapter 8).

WRB Materials

WRBs must be installed on all nonreservoir walls to keep bulk moisture out of the structure and to prevent structural damage and mold growth. The most common WRB is housewrap, which is available from several manufacturers under a variety of names, including DuPont's Tyvek® HomeWrap® and Pactiv's GreenGuard® RainDrop®. As described in Chapter 4, housewraps are synthetic WRBs that are designed to shed bulk moisture and allow vapor passage, thereby keeping rain out of the structure while allowing any water vapor in the wall to dry out. Housewrap is lightweight, durable, and available in large rolls for quick installation.

Proper shingle lapping of all horizontal joints, correct overlap of vertical joints, careful joint taping, and well-chosen fasteners are all critical for housewrap to perform as intended. Housewraps are susceptible to tearing from wind and should be covered with finish siding within the manufacturer's recommended timeframe to ensure reliability. The water-resistant properties of some housewraps can degrade from natural tannins in cedar siding and soaps used in pressure washing. Review all manufacturer recommendations for product compatibility when selecting these materials.

Asphalt-impregnated building felt, a traditional water-resistive barrier used in wood-frame construction, has been largely replaced by housewrap (Figure 9.25). Felt is much heavier and is only available in 3' rolls, requiring more time to install than housewraps. Felt, however, is a time-tested product that is a good choice for a green home when installed correctly.

Liquid-applied WRBs, such as Tremco Barrier Solutions' Enviro-Dri™ and Sto Corporation's StoGuard® are installed with reinforced tapes at sheathing joints and rough openings. Generally, these barrier systems require that subcontractors receive authorization from the manufacturer to apply the product. Liquid-applied products (Figure 9.26) offer air sealing properties and a seamless WRB that cannot tear like housewrap or felt. Although still relatively uncommon in single-family residential construction, they have gained popularity on multifamily projects. As with all WRBs, they must be carefully coordinated with all openings in the wall for proper flashing.

Figure 9.26 Liquid-applied WRBs provide continuous protection against bulk water intrusion while offering significant air sealing benefits.

Water-resistant wall sheathing can serve as the WRB. Huber's ZIP System® is a weatherproof OSB sheathing that is sealed at all joints with a proprietary tape to provide a complete seal (Figure 9.27). Foam sheathing can also serve as a WRB when sealed with an appropriate joint tape. The key to an effective weather seal with these products is proper installation of the tape per manufacturer instructions, which can be a challenge on the job site where rain, dirt, and temperature extremes can affect the quality of installation. When installed correctly, the tape provides an excellent rain barrier; however, if any tape should fail, the lack of shingle lapping may allow water to penetrate the structure. This pitfall can be avoided by installing a piece of flashing at horizontal joints before the tape is applied (Figure 9.28).

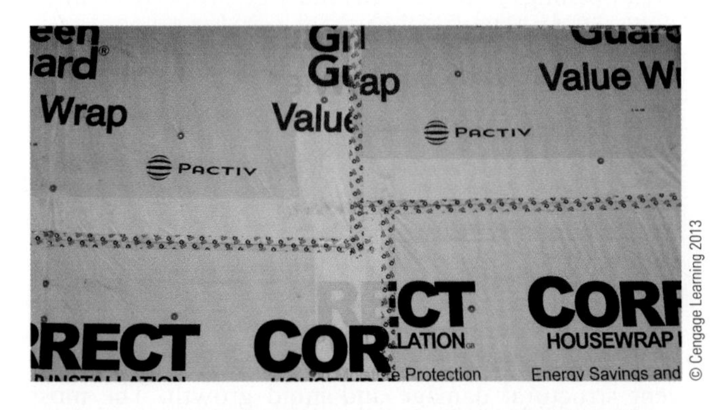

Figure 9.25 Housewrap is installed in "shingle-lap" fashion, with higher courses lapped over lower courses by at least 2". Vertical seams should have an overlap of at least 6". The housewrap is secured firmly in place by fastening along the vertical studs with large-headed or plastic-cap nails or 1" crown staples. All seams are sealed with a manufacturer-approved tape.

Figure 9.27 Huber's ZIP System® wall and roofing panels have integrated WRBs.

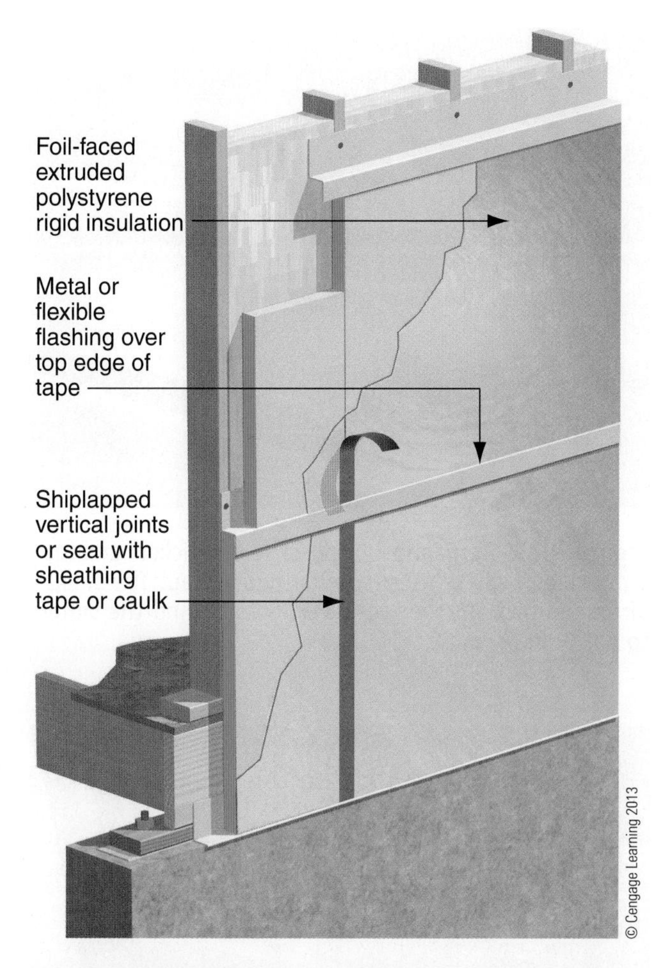

Figure 9.28 Foam board insulation may be used as a WRB. Unlike housewrap, this material requires that all horizontal seams be flashed before applying tape to direct water away from the wall.

Did You Know?

Controlling Vapor Drive

Exterior finishes that absorb large amounts of water when heated by the sun (e.g., stucco and brick) can cause water vapor to be driven into nonreservoir wall structures. WRBs are designed to allow water vapor to pass through them, so the wall system must be designed to limit vapor drive into the wall. Since most vapor is transported via air movement, wall sheathings that are less permeable to air (e.g., OSB and foam) will reduce the amount of air and vapor that is driven into the wall. Sealing joints between panels will also help to control this vapor passage. Brown board and other permeable sheathings should be avoided.

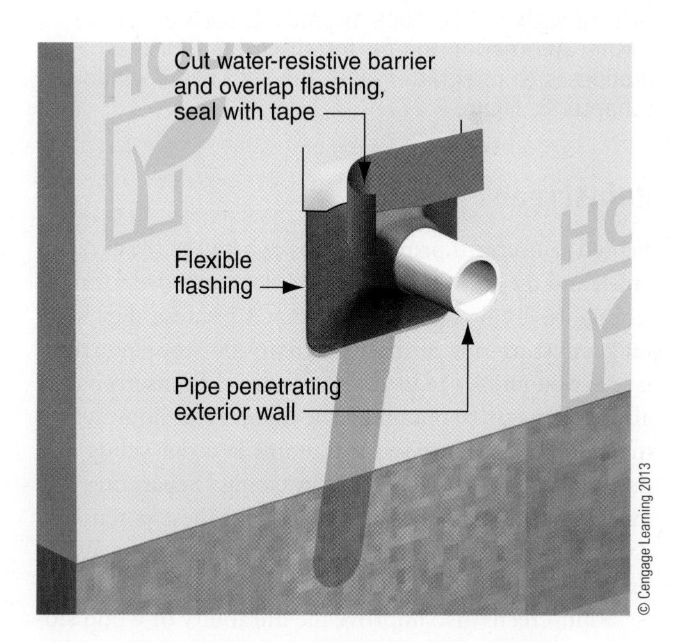

Figure 9.29 Penetrations through the exterior sheathing can be sealed with flexible tapes.

Flashing Materials

To maintain the integrity of the WRB, all penetrations must be flashed, including windows, doors, pipes, wires, and ducts. Windows and doors must have flexible sill flashing or pre-formed sill pans, as well as side and **head flashing**, that are fully integrated with the WRB. Pipes, exhaust and mechanical ventilation ducts, and wires can be sealed with flexible tape (Figure 9.29) or with such products as Quickflash® panels (Figure 9.30).

All mechanical penetrations must be installed so that they can be both properly flashed and trimmed without trapping moisture. Dryer vents and air intakes installed in thin finishes can be a challenge to flash properly. Preformed flanges must be carefully set so they integrate properly with the finished surface of the wall. To simplify the installation of vents on walls with thin exterior finishes, consider relocating them to roofs or foundations where they are easier to install and flash properly.

head flashing the flashing over a projection, protrusion, or window opening.

Planning for Future Needs

Changes to pipes or wires after a house is built are difficult to anticipate, and any holes cut into the siding and sheathing after completion are prone to leakage and almost impossible to properly flash. You may consider installing flashed conduits through the wall that are capped and insulated. When a new pipe or wire needs to be installed in the wall, a pre-planned opening can prevent potential moisture problems in the wall.

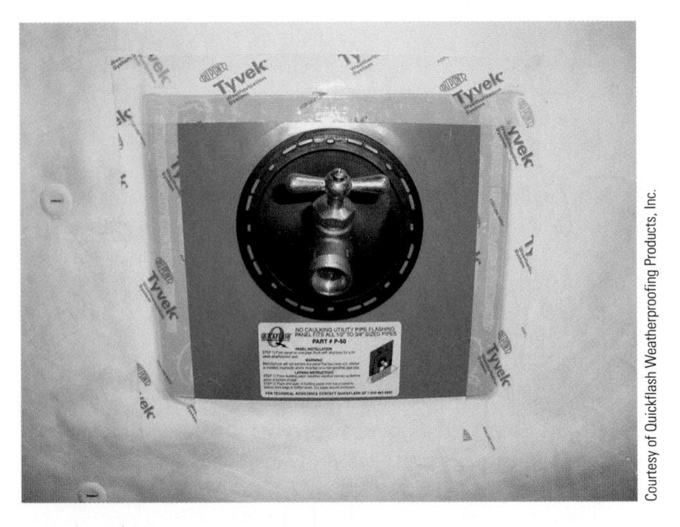

Figure 9.30 Exterior penetrations can be sealed and flashed with flexible tape or products such as Quickflash®. Another option for installing the Quickflash product is to integrate it with the WRB like a window (Chapter 8, Figure 8.27).

Figure 9.31 Benjamin Obdyke's Home Slicker® creates a rainscreen when installed over housewrap. The small air gap allows any moisture that gets behind the siding to easily drain out.

Rainscreens

Vented rainscreens provide air space between the exterior finish and the sheathing. They are recommended for non-storage reservoir walls in all climates because they keep bulk moisture out of the wall cavity by stopping capillary action and by reducing vapor drive. Rainscreens are also effective in maintaining the integrity of housewraps that can be compromised by tannins in cedar siding and detergents used for pressure washing. Separating the exterior finish from the WRB helps keep these potentially damaging chemicals from adversely affecting the WRB. Rainscreens are not required for storage reservoir walls.

Rainscreens also improve the durability of wood siding by keeping the temperature and relative humidity the same on both the interior and the exterior surfaces. When wood is installed without a rainscreen, temperature and humidity differences can lead to premature deterioration or paint failure.

Rainscreens can be created by installing products manufactured specifically for the purpose, such as Benjamin Obdyke's Home Slicker® or Cosella-Döerken's DELTA®-DRY (Figure 9.31); by applying furring strips to the wall (Figure 9.32); or by simply creating a ventilated air space behind brick veneer or vinyl siding (Figure 9.33). Rainscreens should be vented at both the top and bottom of the wall to allow vapor to escape up, and condensed water to drain out the bottom. The top vents should not open into the soffit because this can allow moisture to condense and cause damage in the roof structure. Screens should be installed over vent spaces at both the top and bottom of the wall to keep insects out of the wall cavity.

Figure 9.32 Wood strips are applied on top of the WRB to provide a rainscreen (vented airspace) behind the siding.

Rainscreens with Thick Masonry Veneer

Thick masonry veneers of brick and stone must have drainage space behind them to keep both bulk moisture and vapor out of the wall structure. Brick absorbs large amounts of rain, and when heated by the sun, vapor from the brick is driven inward toward the house. The air space between the brick and the WRB allows the rain that works its way through the brick to drain down and out of the wall, thereby limiting the amount of vapor that is driven into the wall. It is not uncommon for mortar droppings to fill in some of the space behind brick veneer, which can cause excess moisture buildup between the brick and the wall sheathing. Using

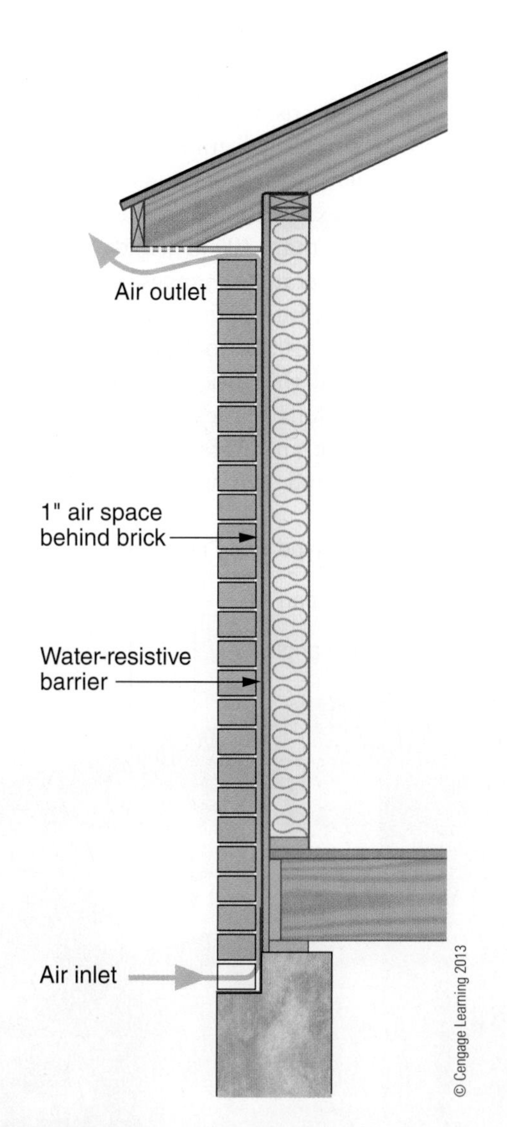

Figure 9.33 A clear air space behind brick veneer is an effective rainscreen.

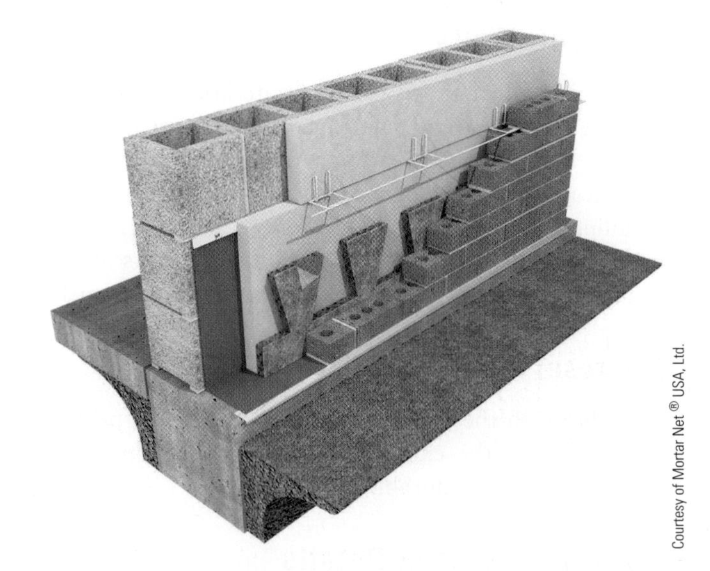

Figure 9.34 Mortar Net® is designed to collect mortar droppings and prevent clogged weep holes in masonry cavity construction.

a rainscreen material designed for brick, such as Mortar Net® (Figure 9.34), can help maintain the necessary air space.

Most stone is less porous and than brick, but rain can still work its way to the back side of the veneer through mortar and gaps between stones. Using a manufactured rainscreen, such as DELTA®-DRY or Home Slicker, will provide the necessary space for drainage behind stone.

Rainscreens with Stucco and Thin Masonry

Stucco and thin masonry are adhered directly to the wall sheathing without the same air space as thick veneers; if not designed to effectively manage rain, they can allow water into non-reservoir walls. These materials absorb large amounts of water that can work its way through the finish, either as bulk moisture seeping through porous materials or as vapor driven through walls that have been heated by the sun. Thin masonry veneer is installed over a base coat, which, like stucco, adheres to the building paper or felt to which it is applied. In doing so, the veneer eliminates the moisture-repellant properties of that material and leaves the home with little, if any, moisture protection. Installing a second layer of building paper, felt, or housewrap over the primary WRB allows the stucco to adhere to a surface that remains separate from the WRB, providing space for drainage between both layers.

Non-absorbing Materials

Exterior finishes that do not absorb water, such as PVC, metal, and fiber cement, will not have the same durability or moisture penetration problems as wood siding or masonry if installed without a rainscreen. Nonetheless, maintaining an air space helps water drain out before it gets into the wall cavity. The best practice is to install a vented rainscreen behind all exterior finishes on non-reservoir walls, particularly when building in areas with heavy rainfall.

Finish Installation Details

Correct installation of moisture management systems will keep water out of the structure and extend the life of the house. Careful design and installation of exterior finishes provide the best assurance that they will be durable and low maintenance.

Reservoir Walls

Reservoir walls that are able to dry to both the interior and the exterior are not damaged by moisture and can employ face sealing instead of a WRB behind the finish. A critical component of face-sealed walls is the proper detailing and installation at windows and doors, including sloped sills with back dams, capillary breaks at the head, and sealant between the casing and the wall finish.

Nonreservoir Walls

In addition to the WRB and vented rainscreen, proper installation of the exterior finish material helps shed water and increase durability.

Wood Installation Details

The most durable wood for exterior finishes is old-growth heartwood like redwood and cedar; however, these products are becoming increasingly rare and, if available, are generally not harvested sustainably. More sustainable, new-growth wood is less resistant to water damage, so it must be detailed and installed to repel water.

All exterior joints should be designed to shed rain. To keep rain out of wood trim, horizontal joints should be limited, and the tops of all trim where water can collect should be sloped. Flashing should be installed over window heads and column caps, and window and door sills should have properly positioned drip grooves. Hollow columns and posts should be primed on the interior surfaces and at both ends, and the tops and bottoms should be vented. Finally, all bare wood should be primed thoroughly before installation, including the backs and ends of all boards. Apply durable sealants at joints between siding and trim; however, caulking the gaps at the bottom of lap siding will prevent the water from draining out, causing damage to the wood. Finally, keep all wood trim high enough above grade to protect from insects and splashing rain.

Composite, PVC, and Metal Installation Details

Manufactured products, such as wood fiber composites, PVC siding and trim, and aluminum siding, all have very specific manufacturer's installation instructions. Following these recommendations generally provides for the most durable installation and ensures that warranties will be honored. Nail types and installation depth, gaps required between boards, clearance to differing materials (e.g., roofing and stone), and priming and painting specifications should all be followed. Reviewing manufacturer's instructions with the product installers before they start can help ensure proper installation.

Masonry Installation Details

Solid brick and stone veneer must have flashing and weep holes at all window heads and at the bottom of walls that drain above the finished grade (Figure 9.35a and Figure 9.35b). While uncommon in single-family residential construction, expansion joints are recommended for brick walls wider than 20' in length to allow for movement due to temperature changes. The Brick Institute of America and the International Masonry Institute can provide additional installation specifications.

Cultured stone and thin brick should be installed according to the manufacturer's instructions, taking special care to provide an appropriate drainage plane between the veneer and the wall sheathing. Where

Figure 9.35a Window flashing is fully integrated with the WRB to direct water away from the structure.

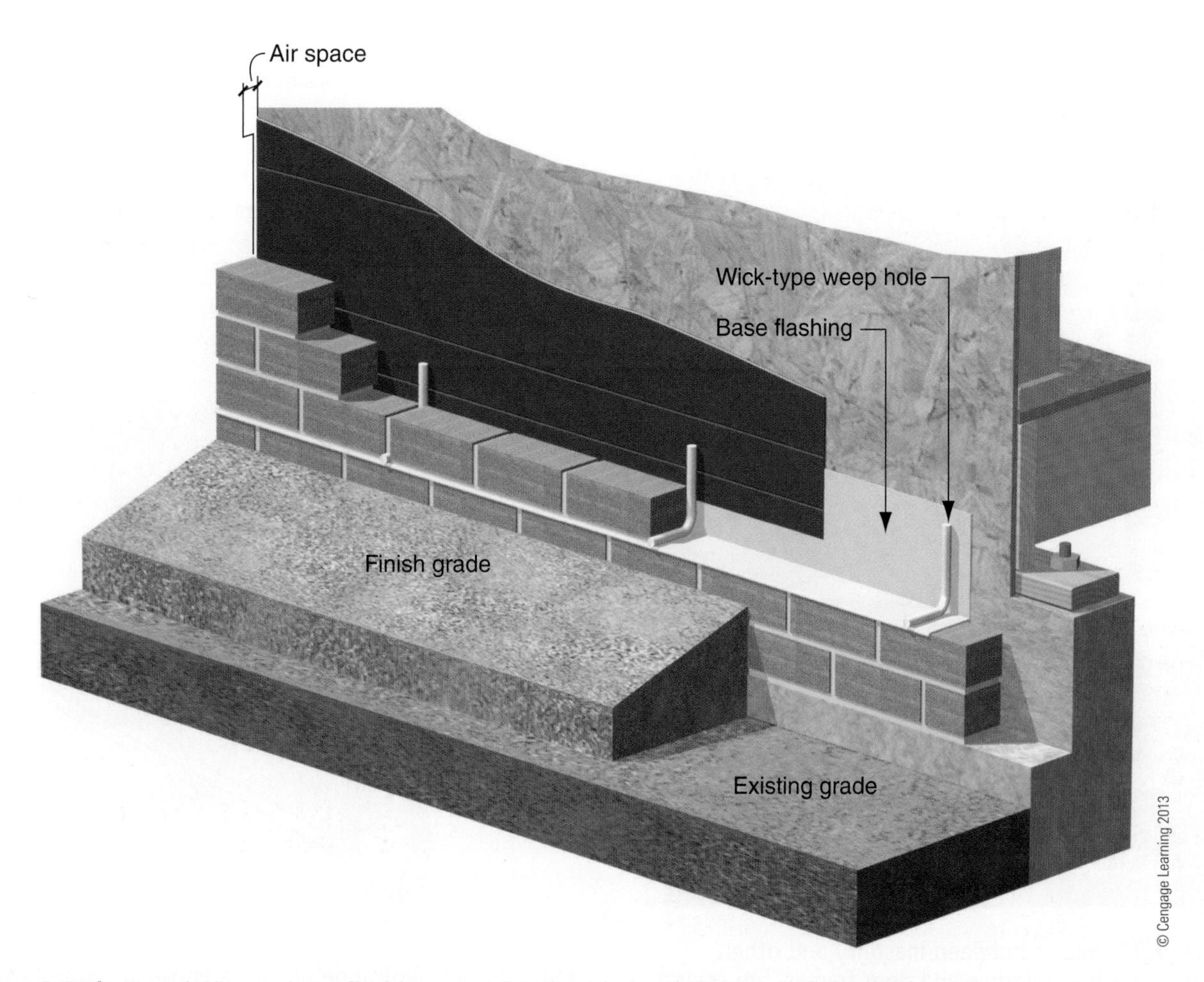

Figure 9.35b Foundation or base flashing must be placed above the level of the final grade.

masonry meets other materials (e.g., window and door frames), the materials expand and contract at different rates, creating small gaps that must be sealed where water can enter. Caulk is not necessary, however, where water will shed off wood or stucco onto the surface of masonry (Figure 9.36).

Stucco Installation Details

As discussed previously, masonry stucco must have at least two layers of WRB to provide a rainscreen in front of the wall sheathing. Weep screeds, a metal or plastic flashing at the base of walls, allow moisture to drain away from the building structure. Control joints should be installed at critical movement points and between large spans, and casing bead should be applied at windows, doors, and dissimilar materials to provide clean edges for sealants and limit cracking (Figure 9.37). Stucco should be caulked around all windows and doors.

weep screed a metal or plastic flashing at the base of walls that allows moisture to drain away from behind masonry and stucco.

EIFS and resinous stucco should be carefully installed according to manufacturer's instructions, and a complete rainscreen should be applied on nonreservoir walls to prevent moisture penetration.

Managing the Process

Effective on-site quality control is critical to effective moisture management and a durable wall finish. Every penetration in the WRB must be properly flashed before the exterior finish is installed. In addition to doors and windows, penetrations include pipes, wires, ducts, and deck ledgers. A common practice is for builders to install siding or brick while mechanical contractors are still completing their interior work. Unfortunately, these contractors then, after the exterior finish is installed, cut holes that are not properly flashed to the WRB. Best practices include delaying the start of the exterior finishes until all the mechanical work is complete and all penetrations flashed to the WRB.

Review all materials, methods, installation details, and sequencing with craftspeople on the job site before they begin their work, and schedule regular inspections with them to confirm that the work meets the project

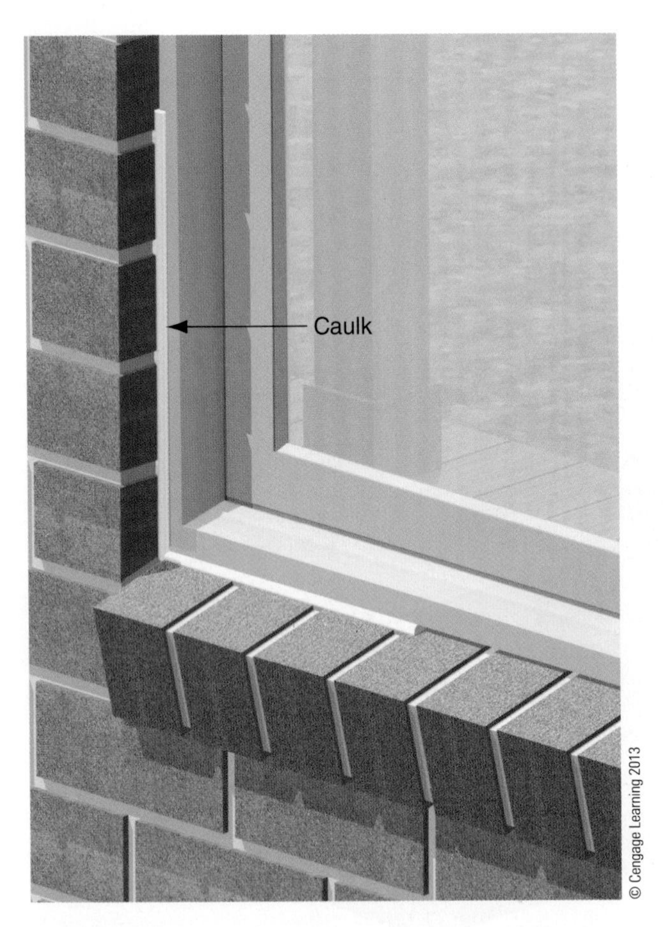

Figure 9.36 Seams between masonry and other materials, such as window and door frames, are sealed with caulk. Select the sealant based on the types of adjoining materials.

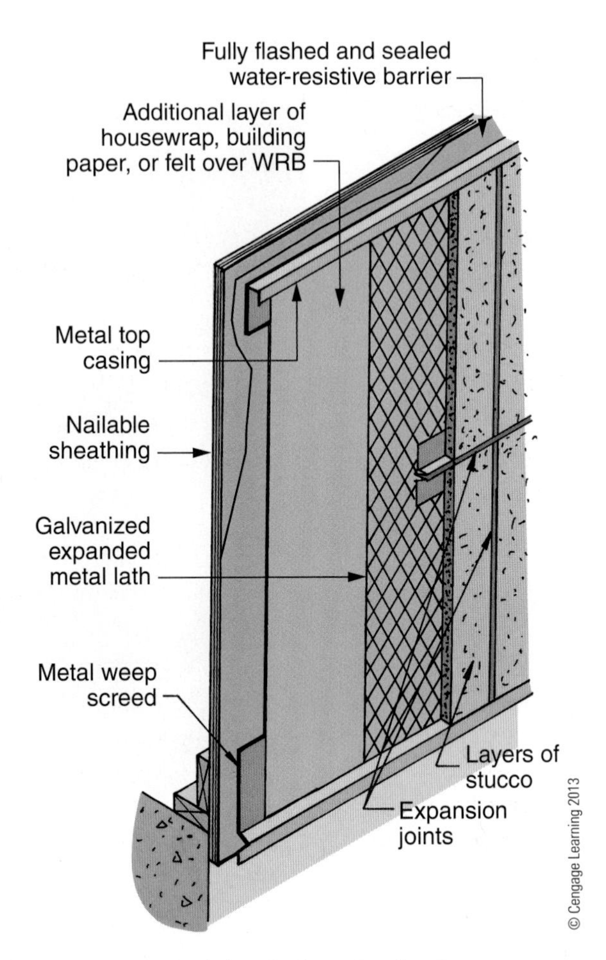

Figure 9.37 Typical details for a Portland cement stucco installation.

requirements. When something goes wrong or work proceeds too far for proper quality control inspections, stop the process and make the required corrections before proceeding.

Inspecting the Work

The old expression "You only get what you expect when you inspect" is particularly true in construction. Someone on the project team should inspect the entire WRB and all flashing work for proper shingle lapping to shed water, for correct horizontal overlaps, and for quality tape and flashing seals. Reverse lapping of WRBs is not uncommon, and only an effective quality control and inspection process can ensure that it is done correctly every time.

Rainscreens should be inspected both during installation and after completion to ensure that they have continuous ventilation from bottom to top and that appropriate vents are installed in the exterior finish material. Insect screens should also be properly installed.

All exterior finish material should be inspected after completion to ensure that no flashing was damaged during installation. All weep holes, control joints, and clearance to adjacent materials should also be inspected, and the correct use of all sealants should be confirmed.

Frequent Mistakes in Applying Exterior Finishes

- Reverse-lapped water-resistive barrier: housewrap, felt, or flashing installed backward so water flows into wall structure
- Missing masonry weep holes
- Weep holes placed below grade
- Lack of vents at top and bottom of vented rainscreen
- Top of vented rainscreen open to soffit
- No back or end priming on wood siding and trim
- No flashing at mechanical penetrations
- No rainscreen behind stucco, thin brick, or stone
- Accumulation of mortar droppings in space behind brick and stone, filling up rainscreen

FROM EXPERIENCE

Water Testing Windows

Peter Yost, Director of Technical Services, Building Green LLC. One really great way to make sure that window flashing is done correctly is to performance-test the installation (or at least threaten to). You can include language in your Scopes of Work for the window installer that states the following: "We reserve the right to spray-test each window installation to assess water leakage after the window installation is complete but prior to installation of the insulation." This means you can see the leaks, and the flashing can be corrected.

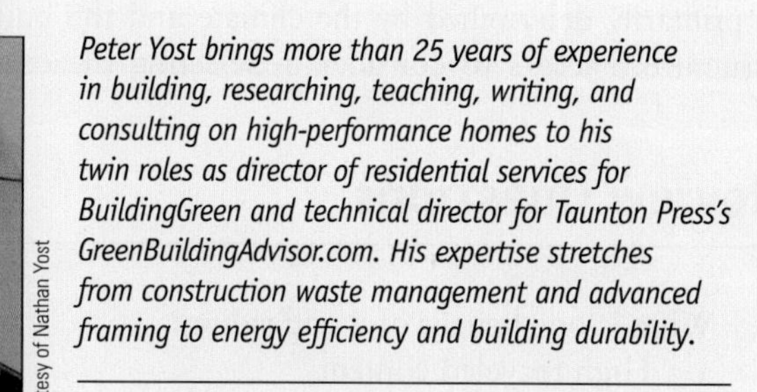

Courtesy of Nathan Yost

Peter Yost brings more than 25 years of experience in building, researching, teaching, writing, and consulting on high-performance homes to his twin roles as director of residential services for BuildingGreen and technical director for Taunton Press's GreenBuildingAdvisor.com. His expertise stretches from construction waste management and advanced framing to energy efficiency and building durability.

In Figure 9.38, we see spray-testing of a window installation. Anytime we can make simple but effective performance testing a part of the process, everybody wins (ultimately, if not immediately).

Courtesy of Johnson Construction Consulting, Portland, Oregon

Figure 9.38 Testing a window in a masonry wall for a leak.

Remodeling Considerations

Remodeling that affects exterior finishes must be done in consideration of the existing moisture management systems on the wall and with an understanding of how any changes will affect the ability of the wall to repel water and transmit vapor. Attention must be paid to existing finishes that may contain asbestos or lead paint. Any changes or removal of asbestos or lead paint must be done in accordance with all applicable local, state, and federal regulations.

Whenever changing or repairing exterior finishes or adding windows and doors to an existing wall, it is critical to determine if a WRB exists, and if so, whether or not it is complete and properly lapped. When there is no WRB, or when the existing one is compromised, best practices suggest that no changes should be made to the exterior finishes unless they are removed corner to corner and bottom to top in order to install a complete WRB. Changing exterior finishes on an existing building without addressing moisture management may create significant liability issues for design and construction professionals, particularly in climates with significant amounts of rainfall that can damage walls.

Summary

Although choosing an exterior finish is primarily an aesthetic decision, the water-resistive barrier system is primarily determined by the climate and the building structure. Factors to consider in selecting the exterior finish include the amount of recycled content, its recyclability, its embodied energy, its hazardous material content, and its effect on the health of the installers.

Review Questions

1. What is one benefit of wood siding?
   a. High recycled content
   b. Durable
   c. Low energy to manufacture
   d. Low maintenance
2. Which of the following is not a benefit of fiber cement siding?
   a. Recycled content
   b. Durable
   c. Low energy to manufacture
   d. Low maintenance
3. Which water-resistive barrier system is not appropriate for a wood-framed wall?
   a. Building felt
   b. Housewrap
   c. Liquid-applied WRB
   d. Face-sealed OSB
4. Select the siding material that cannot be recycled.
   a. Unpainted wood
   b. Fiber cement
   c. Brick
   d. Vinyl
5. Which exterior wall finish requires a double layer of felt or housewrap?
   a. Stucco
   b. Brick
   c. Wood siding
   d. Mineral fiber siding
6. Which of the following is not an acceptable WRB?
   a. Housewrap
   b. OSB
   c. Felt paper
   d. EPS insulation
7. What are the benefits of a rainscreen?
   a. Increases wall R-value
   b. Improves wall durability
   c. Reduces radiant heat gain
   d. Prevents termite entry
8. When properly installed, WRB provide all of the following benefits except:
   a. Thermal resistance
   b. Moisture control
   c. Air infiltration reduction
   d. Drainage plane
9. Which of the following is a common mistake with exterior finishes that causes moisture problems?
   a. Reverse-lapped water resistive barrier (WRB)
   b. Weep holes above grade
   c. Back priming of wood siding and trim
   d. Rainscreen behind stucco and stone
10. Which of the following is not an issue that must be addressed when replacing existing or installing new windows and doors in existing walls during remodeling projects?
    a. Lead and asbestos content in materials to be removed
    b. Determining whether or not there is a WRB on the existing wall
    c. U-factor and SHGC of windows being removed
    d. Existence of a vented rainscreen behind exterior finish

Critical Thinking Questions

1. What is the greenest siding material and why?
2. Explain four ways that bulk water can enter a home through the above-grade walls.
3. What are the pros and cons of using traditional stucco vs. EIFS?

Key Terms

cornice, 246
cultured stone, 249
drip cap, 248
drip groove, 248
exterior insulated finish system (EIFS), 252
fascia, 246
head flashing, 261
hygroscopic, 246
nonreservoir walls, 245
rainscreen, 247
soffit, 246
storage reservoir walls, 245
weep screed, 265

Additional Resources

Building Enclosure Council: http://www.bec-national.org/
Brick Industry of America: http://www.gobrick.com/
International Masonry Institute: http://www.imiweb.org/
Vinyl Siding Institute: http://www.vinylsiding.org
Western Red Cedar Lumber Association (WRCLA): http://www.cedar-siding.org

Appendix A: Environmental Impacts of Siding Materials

Siding Product	Environmental Advantages	Environmental Disadvantages
Solid wood	Low energy to manufacture; good durability if maintained	Most durable wood from old-growth forests; in most areas requires regular recoating with pain or sealer
Wood composites	Made from low-value trees	Petrochemical binders required; questionable durability
Vinyl	Minimal maintenance required	Toxins from manufacturing and disposal; moderate durability
Stucco, traditional	Durable; minimal maintenance required	High quality of cement makes it energy intensive; polluting to manufacture
Stucco, synthetic	Durable; minimal maintenance required; low embodied energy	Toxins from manufacturing and disposal; moderate durability
Fiber cement	Durable; low-maintenance requirements	Somewhat energy intensive to manufacture due to cement content and transportation of imported wood fibers
Brick	Durable; low-maintenance requirements; often locally manufactured	Energy intensive to manufacture
Cultured Stone	Durable; low-maintenance requirements	Energy intensive to manufacture

Source: *Adapted from http://www.buildinggreen.com. Used with permission.*

Outdoor Living Spaces

Outdoor structures, such as porches, decks, and patios, are an important feature of a home. They provide living space that does not require heating and cooling, and they offer an opportunity to interact with neighbors and the environment. Decks and patios are uncovered, although porches have roofs that provide protection from the elements. Porch roofs can provide window shading to reduce solar gain in hot climates, but decks provide an opportunity to enjoy nice weather in cooler climates. The design and construction of outdoor spaces can affect the energy efficiency, durability, indoor environmental quality, and community as a whole—either positively when done properly, or negatively when not.

LEARNING OBJECTIVES

Upon completion of this chapter, the student should be able to:

- Explain how outdoor spaces can add to the sustainability of a home
- Identify sustainable materials used in decks and porches
- Describe appropriate methods to sustainably build decks and porches
- Describe how to properly attach decks and porches to homes to properly manage bulk moisture
- Describe how to determine the location of prevailing breezes

Green Building Principles

 Energy Efficiency

 Resource Efficiency

 Durability

 Indoor Environmental Quality

Exterior Living Spaces

In addition to serving as traditional entrances and storage areas, outdoor living spaces can provide space for a wide range of other uses, including entertaining, dining, working, relaxing, and cooking (Figure 10.1). Outdoor spaces that often adjoin the home include decks (a flat-floored roofless area), patios (a paved or unpaved roofless area), and porches (covered structures attached to the front or rear entrance of a building). Depending on the particular design of a house, the surrounding land, and the climate, one or more porches, decks, or patios can provide economical, efficient living space for much of the year. In most climates, a smaller house with a variety of outdoor spaces can be more useful and efficient than a larger house with few decks and porches.

Structures at or near grade can benefit from built-in seating, railings, walls, or plantings that help define the space better than a design that leaves all sides fully

deck a raised, roofless structure adjoining a house.

patio an on-grade outdoor space for dining or recreation that adjoins a home and is often paved.

porch an unconditioned open structure with a roof attached to the exterior of a building that often forms a covered entrance or outdoor living space.

Figure 10.1 Outdoor spaces can be an efficient way to increase the home's living area.

open. Outdoor spaces can be compact with just enough space for a grill and a cook, or as large as a great room with tables and chairs, couches, and even a porch swing. Detached gazebos, screen rooms, and decks can serve as private retreats separate from a house.

When designing outdoor spaces, employ the same resource efficiency principles that you would utilize inside a home—minimize corners, angles, and curves; design to building modules; and avoid details that require extensive maintenance and regular repair and replacement. Plan to diminish the effect of construction on existing mature trees and natural habitats. Look for ways to minimize impervious surfaces, and manage surface water to limit erosion. Finally, if future plans may entail enclosing a deck or patio into conditioned space, avoid construction features that would be costly to change later, such as enlarged footings and roof insulation.

Porches

A porch is a structure attached to the exterior of a building that often forms a covered entrance or outdoor living space. Porches have roofs that provide protection from the rain and sun and, if screened, insects. Properly designed porch roofs can shade windows to reduce solar gain, which is particularly useful for east- and west-facing windows that are difficult to shade due to low sun angles (Figure 10.2). They can, however, shade windows from desired sunlight, making rooms darker and requiring additional electric lighting. Striking the correct balance of shaded and unshaded windows is important. Particularly in cold climates, the avoidance of porch roofs allows for solar gain on windows where it will provide benefits.

In hot and wet climates, porches create outdoor spaces that are usable much of the year. Roofs also protect windows and walls from solar gain and excess moisture. Porches can be oriented to capture prevailing winds

Figure 10.2 Porches provide additional living space and provide window shading.

in warm climates or protect from them in cold climates. Constructing full- or partial-height walls can help protect from strong winds in cooler weather. Wind direction can be observed on-site, or regional wind information is available from the National Oceanic and Atmospheric Administration (NOAA). This information can be used to determine where to locate porches for the best use of breezes. NOAA is the U.S. federal agency that conducts research and gathers data about the global oceans, atmosphere, space, and sun.

Small entry porches help protect entrance doors and provide cover from weather for occupants and visitors (Figure 10.3). If the porch cannot be made wide enough to be useful for sitting, make it only large enough to provide weather protection for the entry door. Although some homes have very narrow porches that are practically useless for sitting and serve only as decoration, this design should be avoided except when used to shade windows from excessive solar gain. An appealing option where there is no room for a porch or deck is a French balcony, which is a set of doors with an exterior rail that can open the interior space to the outside (Figure 10.4). These are a common sight in many foreign countries.

National Oceanic and Atmospheric Administration (NOAA) the U.S. federal agency that conducts research and gathers data about the global oceans, atmosphere, space, and sun.

French balcony a set of doors with an exterior rail that can open the interior space to the outside.

Figure 10.3 This small porch protects the rear entrance to the home and provides shelter for guests and occupants. The home also features a full-size usable front porch.

Figure 10.4 French balconies are a cost-effective way to provide daylighting and ventilation.

Decks and Patios

A deck is a flat-floored roofless area adjoining a house that is constructed above grade, most commonly from wood or a synthetic wood substitute. A patio is an outdoor space for dining or recreation that adjoins a home and is often built directly on grade from stone, tile, brick, or concrete.

Decks and patios are very enjoyable when occupants can take advantage of sun on moderate days where they might feel cool without the radiant heat. Trellises, pergolas, or even umbrellas can make an open deck or patio more comfortable on hot days.

Most decks require regular maintenance, including sealing, cleaning, and repairs. Masonry and concrete patios are more durable but still require cleaning. Decks should be located 2' or more above grade; concrete patios can also be elevated on steel structures, but they are more economical when built directly on grade. Concrete or masonry patios or porch floors have a higher initial cost and environmental impact than decks, but their minimal maintenance needs and increased durability make them a more sustainable solution than wood-framed floors. Patios are discussed further in Chapter 11, "Landscaping and Paving."

Upper-Level Porches and Balconies

Second-floor decks and porches are desirable features for private outdoor living; however, they must be carefully detailed to keep out the rain when built over interior space. Unless they are perfectly sealed, the structure is very vulnerable to deterioration. Typical moisture management strategies include a waterproof membrane similar to that used for low-slope roofing (see Chapter 7), with wood-framed or tile floors installed above (Figure 10.5). Membranes with a walkable surface, such as Duradeck™, provide waterproofing without the need for a separate finish above. Regardless of the system used, all penetrations in the membrane for posts, rails, and other connections must be designed to keep all water out of the structure. This can be accomplished by installing steel braces through the framing, flashed to the membrane, and then connecting the braces to the posts and rails. Proper detailing of second-floor outdoor living space should not be undertaken unless all the details to manage moisture have been thoroughly addressed.

Materials and Methods

Porches are essentially decks or patios with columns and a roof. Starting at the bottom, we will consider available materials and methods for construction of each component of porches and decks.

On-Grade Patio and Porch Floors

Concrete patios and porch floors are almost identical to slab foundations. The key differences are their smaller footings and slopes for drainage. (See Chapter 5 for details on slab foundations.) Concrete slabs can be stained, left natural, or finished with brick, stone, or tile.

In areas that are susceptible to termites, separation between poured concrete slabs and wood framing is important to avoid infestations and structural damage. Exterior slabs should be placed a minimum of 6" below

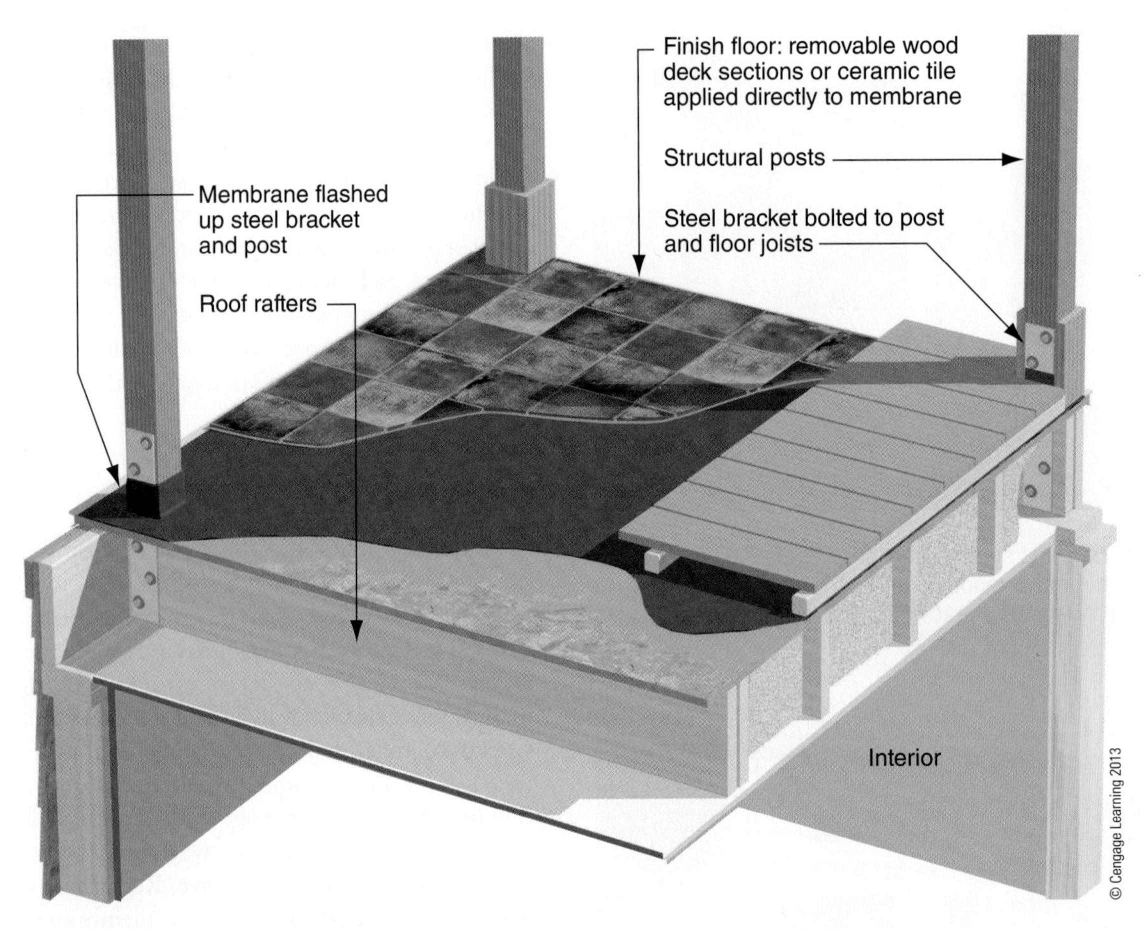

Figure 10.5 Decks and porches installed over finished space must have a comprehensive moisture barrier installed below the walking surface, including careful sealing around all penetrations for posts and rails.

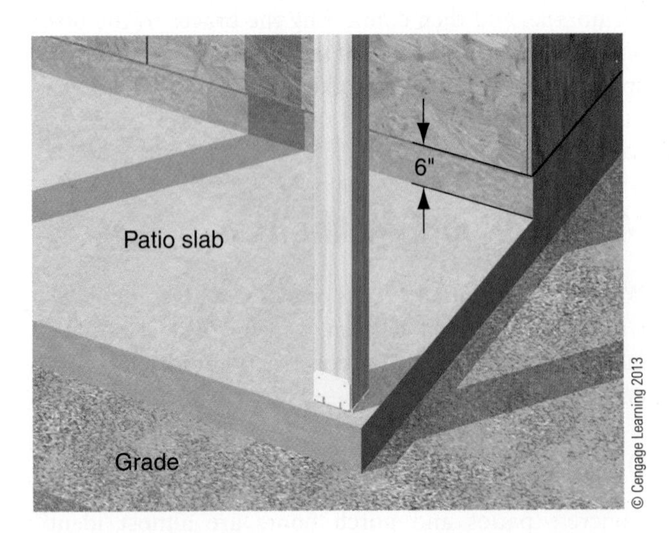

Figure 10.6 Placing exterior slabs a minimum of 6" below the bottom of the floor structure prevents termite entry.

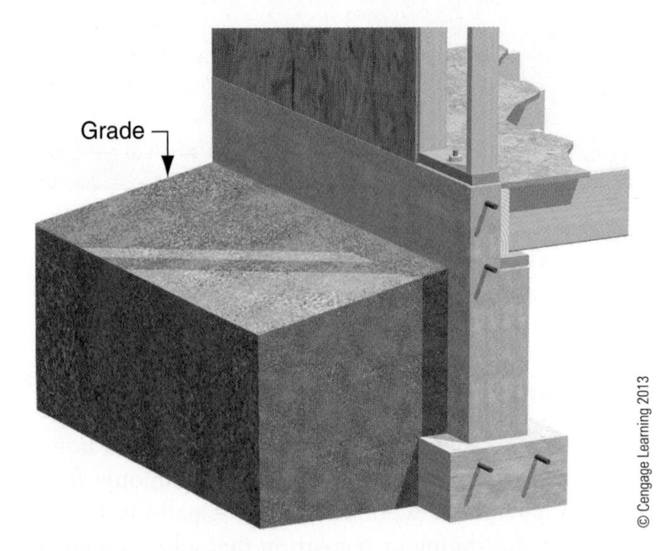

Figure 10.7 Extending the poured foundation wall to the bottom of the wall structure helps prevent termite entry.

the bottom of the floor structure or any wood framing (Figure 10.6), or the foundation wall should extend above the floor framing to the bottom of the wall structure (Figure 10.7).

Concrete and masonry porch floors have higher embodied energy than wood framing but are more durable than wood-framed floors and decks, making them a more sustainable solution—particularly when built on

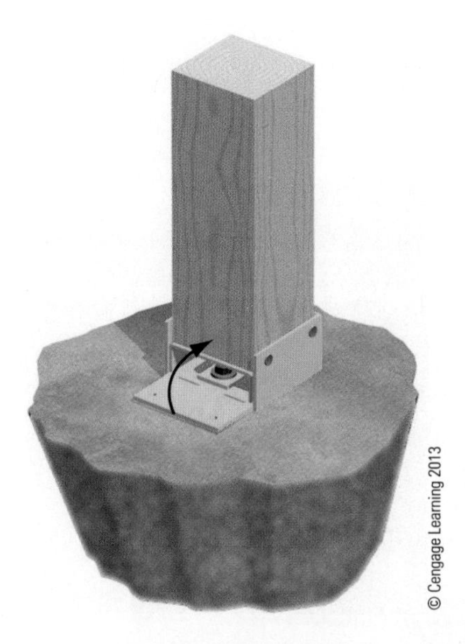

Figure 10.8 A post base prevents the wooden posts from coming in contact with the concrete footing and ground.

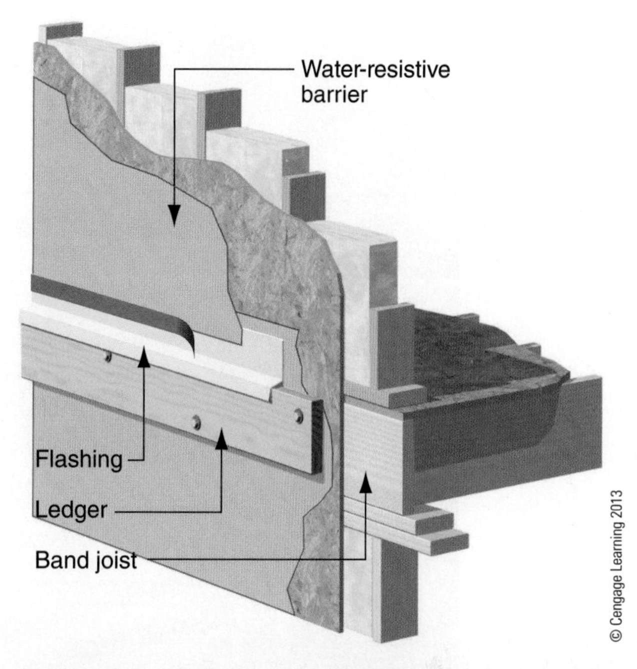

Figure 10.9 Wall drainage planes are often interrupted when decks or porches are attached. By properly flashing the ledger board to the WRB, the wall drainage plane is left intact.

grade. Adding recycled aggregate and fly ash in the concrete mix or using reclaimed masonry can help to reduce the environmental impact of construction.

Above-Grade Decks and Porches

Porch floors and decks built above grade are constructed like pier foundations. Each structural corner and intermediate bearing points must have a footing to support a post. Footings can be poured-in-place concrete, prefabricated piers installed with drive pins, or helical piers (see Chapter 5). On top of the footings, such materials as wood, steel, or masonry posts can be used to support the floor above. Wood posts should not be buried in the ground; rather, they should be separated from the footing with a post base (Figure 10.8). Steel posts should be primed and painted to protect from rust and kept above the dirt. Masonry columns of block, brick, or stone require more materials and labor but are the most durable.

Building the Floor Structure

Above-grade structures are normally framed with pressure-treated lumber for durability, although recycled plastic structural lumber is available in some regions in limited quantities. One of the most critical components in porch and deck construction is to maintain the wall drainage plane. This can be accomplished either by carefully flashing the ledger board to the water-resistive barrier (WRB) (Figure 10.9) or by keeping the structure separate from the house, supported on posts at the house as well as at the perimeter (Figure 10.10). A third option is to install standoffs or blocks that allow the ledger to be bolted to the house while leaving an air space between the board and the wall finish. The standoffs must be flashed into the WRB, which, in some cases, may be easier than flashing the entire ledger or keeping the structure completely separate (Figure 10.11).

Where decks and porches are designed with double and triple beams, consider installing a flexible self-sealing flashing over the top of these beams to keep water out of the joints between members (Figure 10.12). Even pressure-treated lumber is susceptible to deterioration when it is not allowed to dry out. The same self-sealing flashing can be installed on the top of every joist in open decks, providing additional protection against water intrusion and extending the life of the structure.

Due to the corrosive properties of pressure-treated lumber on metals, the fasteners must be approved for use with the specific treated material being used. Stainless steel, hot-dipped galvanized steel, or other approved fasteners, post bases, and joist hangers are key to creating a durable structure.

Decks are typically built flat to allow water to drain through the spaces between boards. Porches and decks with solid flooring should be built with a slight slope to keep water from standing on the floor (Figure 10.13).

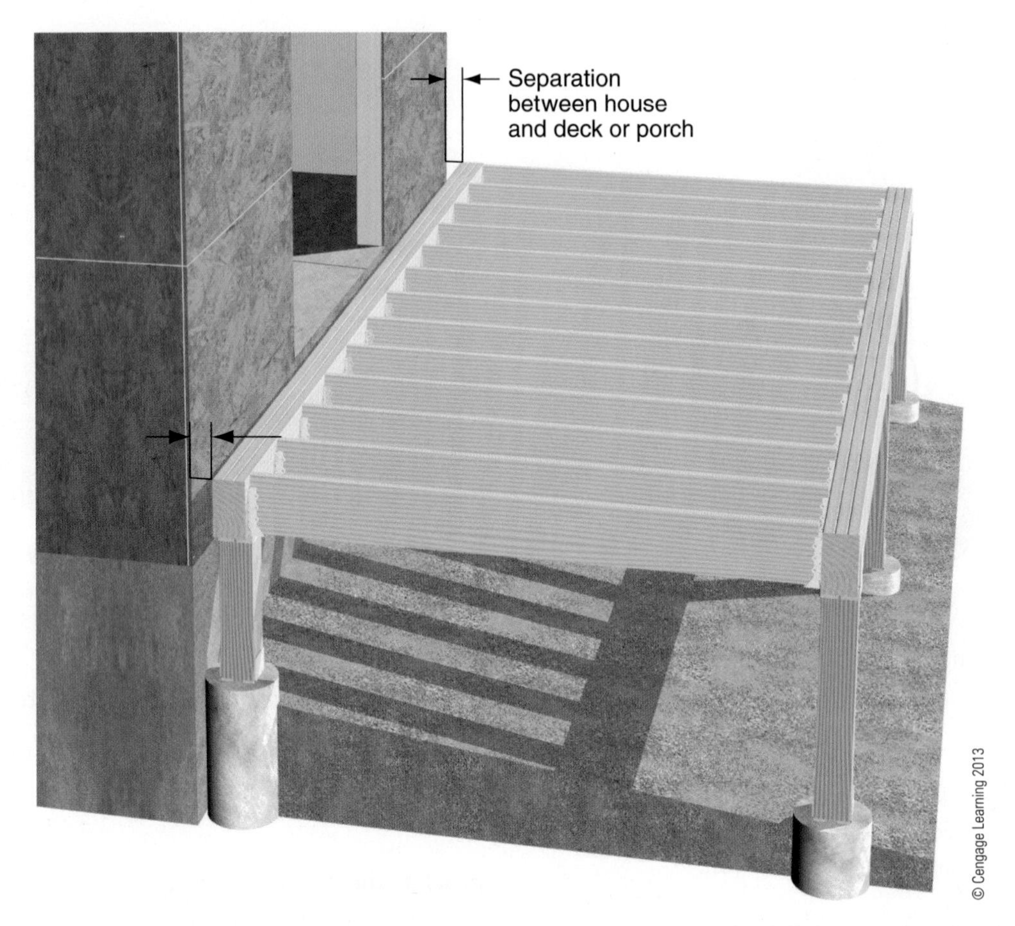

Figure 10.10 Another option to maintain wall drainage is to construct a freestanding deck that is completely separate from the house.

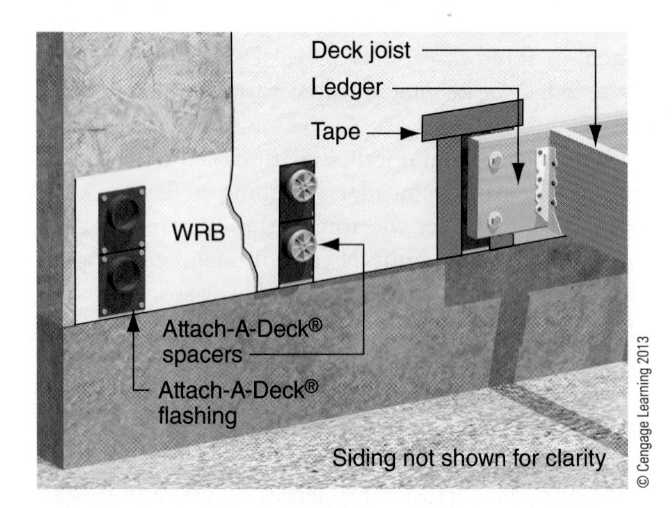

Figure 10.11 Standoffs or blocks, such as Attach-A-Deck®, allow the ledger to be bolted to the house while leaving an air space between it and the wall finish.

Figure 10.12 Installation of flexible self-sealing flashing.

Elevated concrete porch floors can be constructed on steel posts and beams, providing a long-lasting, low-maintenance structure (Figure 10.14). This type of floor should include a waterproof membrane to protect the steel from rust over the long term. Just like the previous example utilizing a waterproof membrane, this deck should also slope away from the house.

Decking and Trim

Decking is the flooring material installed over the supporting deck structure. For aesthetic reasons, many decks and porches have trim around the perimeter to cover the structure. Porch floors are typically covered with tongue-and-groove (T&G) boards with no gaps between them. However, if a porch is not going to be enclosed with screens, spaced square- or eased-edge (the slight bevel helps to compensate for minor variations in plank heights) decking may be used.

A wide variety of decking and trim materials is available, including wood, wood composites, plastics, and aluminum. Wood suitable for exterior use can be pressure treated or thermally treated to resist pests and withstand weather exposure, or a naturally rot-resistant species can be used.

Pressure-Treated Lumber

Pressure-treated lumber decking, typically made from southern pine, is fairly durable and inexpensive but susceptible to cracking and warping. Due to the chemicals used in its production, it must be handled and disposed of carefully. The most common use for pressure-treated lumber is for spaced deck boards, although T&G flooring suitable for porches is also available. Since pressure-treated lumber is prone to expansion, T&G installations should be primed on all sides and ends, and all joints should be sealed upon installation to minimize movement.

Thermally Treated Lumber

Thermally treated lumber, a relatively new entrant into the decking market, is produced by exposing wood to very high temperatures and steam, which transforms it into a product that is not affected by insects or decay. This process also minimizes shrinkage, cracking, and swelling. Thermally treated wood can be machined, handled, and disposed of like standard lumber. Due to the high heat required in treatment, it has high embodied energy; however, its enhanced durability and absence of treatment chemicals make it a sustainable choice for decking. Thermally treated wood is available in patterns for spaced deck boards (Figure 10.15).

Wood Composites

Wood composites are made from a combination of wood fiber and plastics, with the amount of recycled content varying by manufacturer. Most composites are available in open decking boards, although T&G patterns are available from some manufacturers. Composites are rot-resistant, safe to handle, and very durable; however, due to their wood content, they must be cleaned regularly to prevent mold formation. Colors can also fade

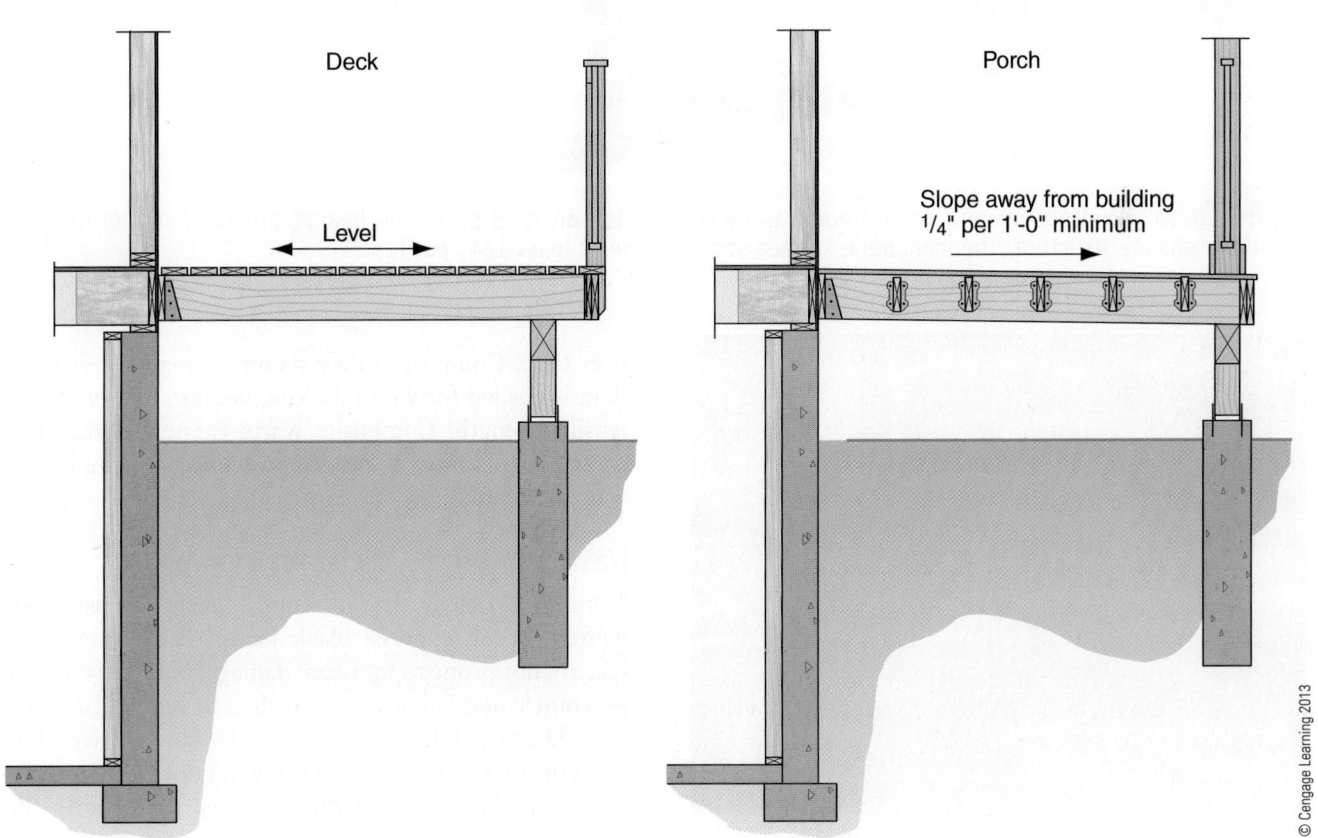

Figure 10.13 Decks may be constructed level because they allow water drainage through slats. Porches and decks with solid flooring, on the other hand, should be built with a slight slope to drain water away from the home.

---

thermally treated lumber produced by exposing wood to very high temperatures and steam, transforming it into a product that is not affected by insects and decay.

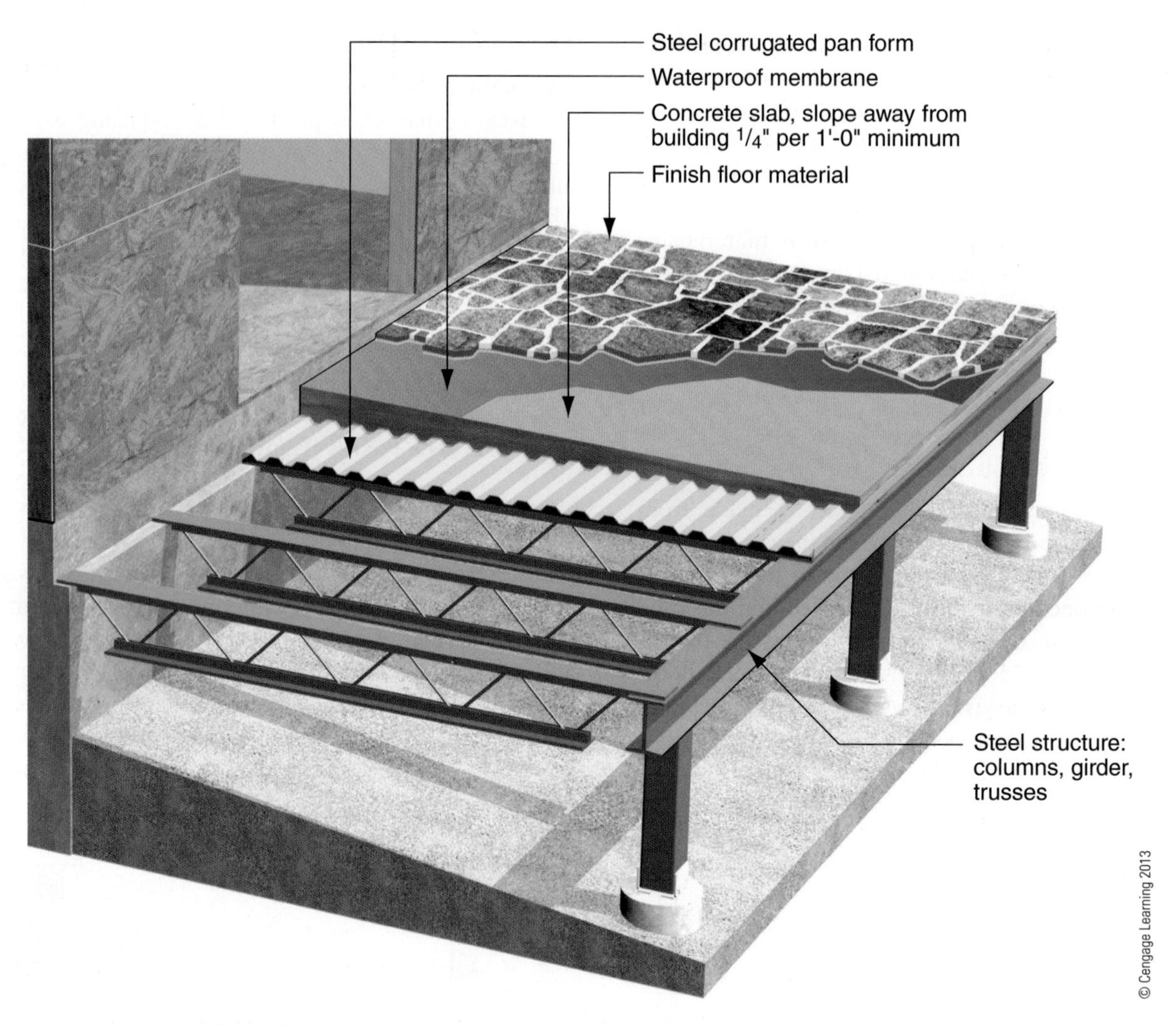

Figure 10.14 Elevated concrete porch floors can be constructed on steel posts and beams, providing a long-lasting, low-maintenance structure. The membrane surface should slope at least 1/4" per foot.

Figure 10.15 Thermally treated wood begins as standard lumber and then undergoes a heat-based, chemical-free process that changes compounds in the wood. The result is a wood product that is resilient to rot, decay, and the elements.

over time. Composites may require closer joist spacing than is needed for wood decking because of their lower tensile strength. Composite waste cannot be recycled, so careful planning is needed to reduce the amount of waste generated.

Naturally Decay-Resistant Lumber

Naturally decay-resistant woods, such as redwood, western cedar, cypress, black locust, Pacific yew, and Ipe, are not prone to moisture damage. Naturally decay-resistant wood is less common than in years past. Only the old-growth heartwood of cedar and redwood is resistant to rot; new-growth sapwood does not have the same properties. Use of these woods, particularly the

---

naturally decay-resistant wood wood species, such as redwood, western cedar, cypress, black locust, Pacific yew, and Ipe, that are not prone to moisture damage.

heartwood, should be avoided unless it has been certified as sustainably harvested by the Forest Stewardship Council (FSC). Ipe is a South American hardwood that is very heavy and naturally resistant to insects and rot. Like any tropical wood, Ipe should only be used in a green home if FSC-certified.

Plastic decking is available from both virgin and recycled materials and is available in square- and eased-edge deck boards. Resistant to cracking, warping, and rot, plastic decking materials can be worked with standard woodworking tools. Some products are available reinforced with fiberglass for added strength. Because plastic decking expands and contracts from temperature changes more than most products, this characteristic must be taken into account during installation. Some of these materials may be recyclable, but locating facilities that will accept surplus material may prove challenging.

Aluminum decking is available with a durable powder coating finish, and interlocking styles can provide a waterproof covering on a porch or deck. Due to its high cost and somewhat sterile appearance, aluminum has not captured much of the residential market. Aluminum has relatively high embodied energy unless made from recycled content; however, its low maintenance, durability, and recyclability make the product a sustainable choice.

Planning for Repair and Replacement At some point, most decking materials will need repairs or replacement due to damage or deterioration. To simplify future removal, these materials should ideally be installed with screws or hidden fastening systems (Figure 10.16). Posts and columns should be designed so that decking can be replaced without their removal to avoid additional work and waste over the lifetime of the home.

Figure 10.16 Whether using pre-grooved deck board, grooving your deckboards on the job site, or using solid-edge deckboards, there are hidden fastening systems available.

Ipe a South American hardwood that is very heavy and naturally resistant to insects and rot.

Finishing Most decking does not require finishing, but, with the exception of Ipe, wood decking should be sealed regularly to extend its life. Paint can be an appropriate finish in dry climates but is not durable in wet climates. Use finishes with little to no volatile organic compound (VOC) whenever possible.

Stairs and Landings

Most above-grade decks and porches will require stairs for yard access. Stairs and landings can be either wood-framed or solid masonry. Although wood-framed stairs are finished with the same materials and methods as decks, masonry stairs usually have a veneer finish over a poured concrete or concrete masonry unit structure. Masonry stairs have higher embodied energy than wood but last longer and require less maintenance over their lifetime. To prevent premature deterioration, masonry stairs must be constructed to keep water away from any adjacent wood structures.

Columns, Posts and Rails, and Trellises

Porches generally have columns that hold up beams, which, in turn, hold up the roof. Patios, decks, and porches have posts and railings, and decks often have trellises to provide shade.

Columns are vertical structural members made of wood, wood composites, plastic, or fiberglass. Hollow wood and composite columns should be fully pre-primed on all sides before they are assembled, with ventilation at the top and bottom to allow for drying and extending their life. Both hollow and solid wood columns should be set up on bases to protect the bottoms from water damage. Plastic columns can be prefabricated or site-built; however, unlike fiberglass columns, they are not usually load-bearing and must have a wood or steel post inside for support. Wood columns require regular maintenance, whereas plastic and fiberglass are essentially maintenance-free products. Wood base and cap trim is susceptible to deterioration and should be designed to shed water by sloping horizontal areas and by using cap flashing to minimize water entry. Formed plastic caps and bases are very common, providing an extremely durable, low-maintenance trim product.

Posts and rails serve both decorative and safety purposes on decks and porches and are fabricated on-site from solid lumber or assembled from prefabricated parts made of wood composites, plastics, fiberglass, or metal. Wood rails, particularly on open decks, require regular maintenance and replacement due to exposure to the elements. While metal, fiberglass, and other alternative products may have higher initial environmental impacts than wood, they can provide a more durable product that requires less frequent replacement—making them good choices for a green home. All railings, particularly those made of wood, should have sloped or curved tops to shed water, as should intermediate posts.

Partial or full-height walls used as wind or privacy screens on porches should only be finished on one side, or, if two-sided, they should be constructed like hollow wood columns with ventilation at the top and bottom to allow them to dry out.

Trellises are structures that can provide partial shade, a framework for vines to grow, and a sense of enclosure to decks and patios. Usually made from interwoven pieces of wood, bamboo or metal, trellises must be constructed to withstand the elements to ensure maximum durability. Similar to deck ledgers, they must be properly flashed to the WRB where they attach to the walls of the house. Support columns should be of rot-resistant materials or fully flashed at the top and vented at top and bottom. Overhead beams with multiple members should be separated with spacers or flashing over the assembly to keep out water (Figure 10.17). Single beams should have metal flashing or sloped tops to shed water, and all exposed wood should be sealed regularly to extend life.

Roofs

Porch roofs do not differ from building roofs except that they are normally not insulated (please refer to Chapter 7 for construction details). If there is a possibility that a porch may be enclosed into living space in the future, insulating the roof or ceiling may be worthwhile. You may also consider using radiant-barrier roof decking in hot climates to keep the porch cooler. Large overhangs help keep out rain and protect the columns, rails, and floor from damage. As described in Chapter 3, an overhang is a soffit, porch, or other projection that extends beyond a window or entranceway.

A critical element in porch construction is maintaining the building envelope where the roof meets the house. Where a first-floor roof intersects a two-story wall, care must be taken to prevent a thermal bypass

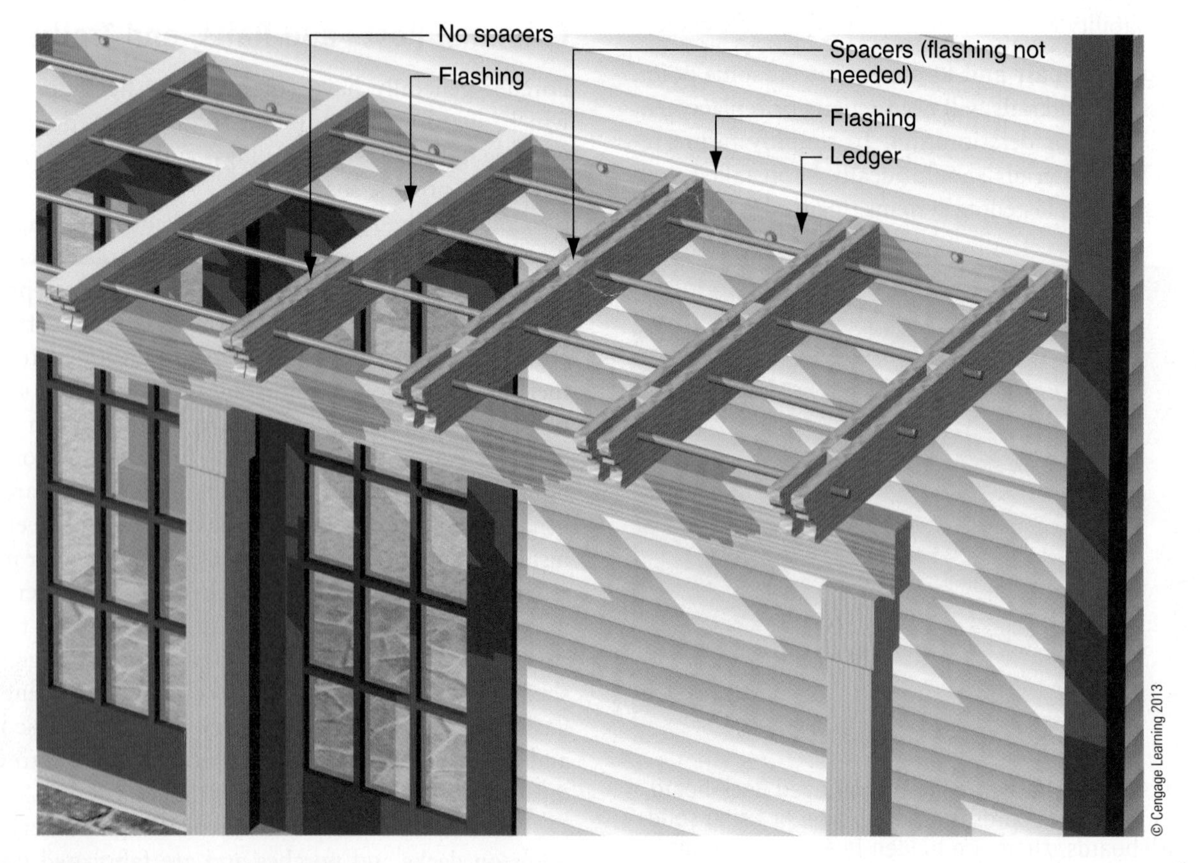

Figure 10.17 Overhead beams should have multiple members separated with spacers or flashing over the assembly to keep out water.

trellis a structure used to provide shading or support for climbing plants; usually made from interwoven pieces of wood, bamboo, or metal.

from the wall into the porch roof volume (Figure 10.18). Porch roofs that are an extension of the main house roof may require a wall that aligns with the house wall or insulation on the roofline, ceiling, and any gable ends to maintain a complete thermal envelope (Figure 10.19).

Skylights or sun tubes on porches bring in extra light without the energy penalty when used inside. They can brighten both the porch itself as well as the windows and doors that are shaded by the porch roof.

Screening

In climates with flying insects, screening can significantly extend the amount of time that a porch is useful to homeowners. Almost any porch with a solid floor and a roof can be screened; however, advanced planning of the screen location and installation methods is critical to a successful project. Screened porches with large doors to interior spaces can make a house feel much larger during moderate weather without the inconvenience of screens on individual doors. In cold climates, storm panels can be installed to extend the usefulness of a porch through additional seasons.

Screening a porch typically requires framed panels between columns that reduce the spans to fit available screen dimensions. The screens themselves can be set in metal frames that are installed with screws or clips, or they can be installed with snap-in tracks, such as SCREENEZE® (Figure 10.20) or Screen Tight™.

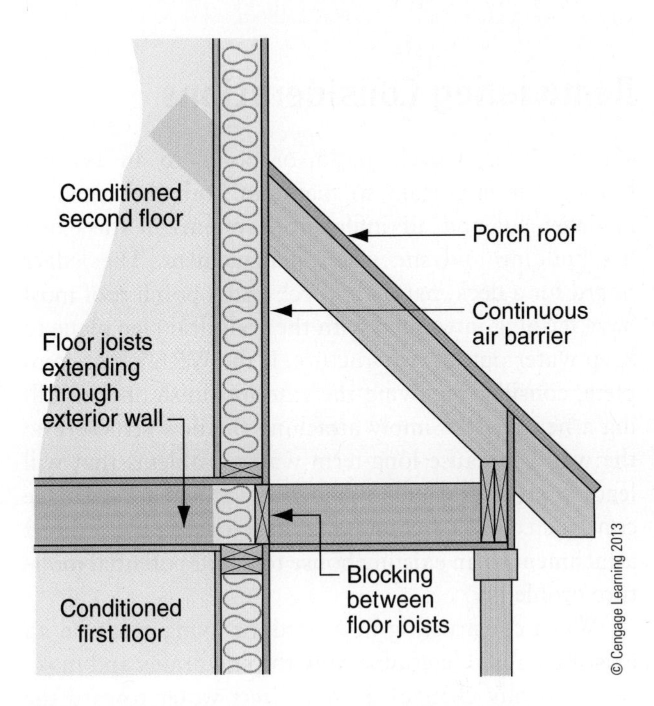

Figure 10.18 The home must be isolated from unconditioned spaces like porches. Common thermal bypasses are open-band joists.

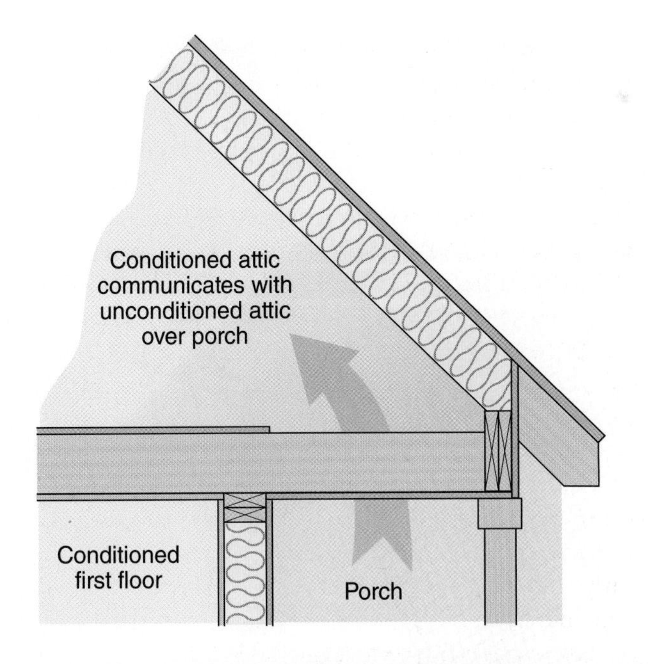

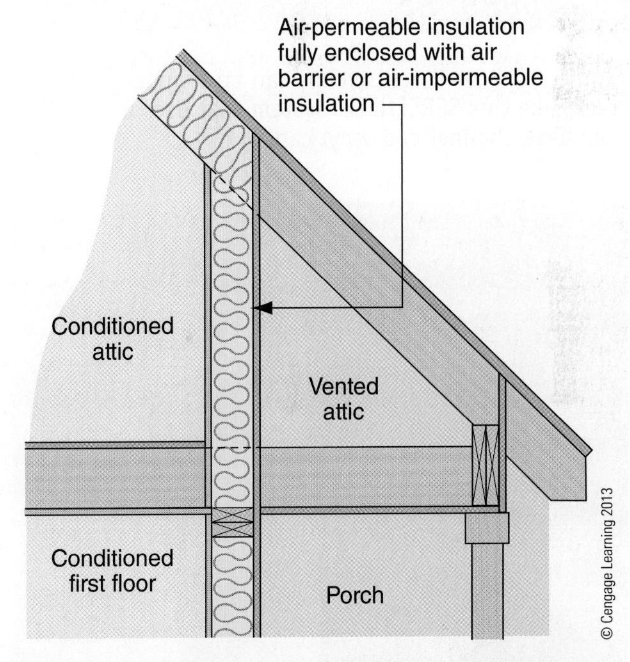

Figure 10.19 When insulating rooflines to create "conditioned" attics, careful attention must be paid to fully enclose the building envelope (top image). A special dividing wall (attic knee wall) may need to be constructed between conditioned attic space and unconditioned attic space, such as over porches (bottom image).

Removable panels or track systems allow for easy replacement of damage screens. Screens can also be stapled in place and trimmed with lattice strips, though stapled screens are more labor-intensive to replace because the lattice strips require periodic replacement or repainting. Screens can also be manually or electrically operated, opening and closing as desired for protection from insects.

Courtesy of SCREENEZE®

Figure 10.20a Porch screens can be snapped into a track, like this SCREENEZE® system that uses an aluminum base channel and vinyl cap.

© Cengage Learning 2013

Figure 10.20b The SCREENEZE® system installed.

Flexible plastic-coated fiberglass screen material is the most economical and common product in residential construction. Metal screens made of aluminum, stainless steel, and copper are also available. Fiberglass screens, like all flexible plastics, typically contain pthalates and VOCs, although some products are available with GREENGUARD℠ certification as low-emitting materials.

Outdoor Kitchens and Fireplaces

Cooking outdoors in hot weather can reduce the air-conditioning loads. Outdoor kitchens can range from a single grill to a full kitchen with ovens, refrigerators, and sinks. The cleanest cooking fuel is natural gas, followed by natural lump charcoal. Avoid using charcoal briquettes and lighter fluid as these products are made with chemicals that are released into the air when burned.

Open wood or gas fireplaces can degrade the air quality when inside the home but, when installed outside, become a focal point for outdoor living while avoiding air quality issues. Fireplaces can be masonry or metal structures that burn gas or wood logs. Natural draft fireplaces may need taller-than-normal flues to counteract wind that may affect its ability to vent properly. Avoid putting fireplaces on wood-framed porches for safety. When using metal fireboxes, flash them carefully to eliminate moisture intrusion and allow for ventilation of any chases built around the firebox.

Remodeling Considerations

When adding a deck, patio, or porch to an existing house, it is important to maintain both the moisture management and air infiltration systems in the existing building and site water management. The ledger board for a deck, patio, or porch and a porch roof must have flashing integrated with the wall drainage plane to keep water out of the structure. If the WRB is not complete, consider removing the exterior finish and installing a new WRB. Simply attaching the new structure to the wall can cause long-term water problems that will lead to structural damage in the wall. A deck can be constructed as a freestanding structure with no direct attachment to an existing house to avoid potential moisture problems.

When constructing new outdoor living areas on an existing house, determine how the site drains and make sure that any changes do not direct water toward the house or create erosion problems in the yard.

Summary

Decks, patios, and porches can provide comfortable and efficient living space that takes advantage of views and fresh air. When properly constructed, they can improve the durability of the structure by protecting walls from water intrusion and reduce energy use through window shading and lower heating and air conditioning demands. Incorporating outdoor living spaces into a project are an important part of a sustainable design.

Review Questions

1. What is the best way to attach a deck to an existing house?
   a. Install a ledger board directly to the siding or brick
   b. Install posts and a structural ledger not connected to house wall
   c. Remove siding and install ledger directly to wall structure
   d. Construct the deck at grade level
2. Which of the following is not an important consideration in construction of a porch roof?
   a. Creation of a complete thermal envelope where the roof meets the house
   b. Flashing at intersection with exterior walls
   c. Integration of the roof with the wall WRB
   d. Roof reflectivity
3. In a hot, wet climate, porches can provide which of the following benefits?
   a. Allow extra light into rooms that open onto the porch
   b. Provide space for sunbathing in nice weather
   c. Provide cooling shade and protection from rain
   d. Increase ability of siding to dry to the exterior
4. In a cool, dry climate without flying insects and a house floor set at 48" above grade, which outdoor living space is an appropriate choice?
   a. Patio
   b. Deck
   c. Screened porch
   d. Outdoor fireplace
5. Which of the following is an efficient solution to providing light to rooms adjacent to covered porches?
   a. High-efficiency electric light fixtures
   b. Skylights or sun tunnels
   c. Reflective floor and ceiling paint
   d. Low porch ceilings
6. Which of the following is the most critical element of second-floor porches and balconies when constructed over finished space below?
   a. Ceiling height
   b. Floor finish material
   c. Waterproofing flooring and connections for posts
   d. Integration of floor structure to house wall
7. Which of the following details provides for easier porch and deck flooring replacement?
   a. Screw fasteners
   b. Nail installation
   c. Painted finishes
   d. Glue and nail application
8. Which of the following strategies is most appropriate to prevent termite damage to decks and porches?
   a. Install pressure-treated wood posts in holes to support structure
   b. Use naturally rot-resistant wood, such as cedar or redwood, for structure
   c. Keep all earth at least 6" below the level of wood framing
   d. Slope the grade toward foundation below decks and porches
9. Which of the following materials would be the least sustainable choice for a deck?
   a. Virgin cellular polyvinyl chloride
   b. FSC-certified Ipe
   c. Recycled content aluminum
   d. Recycled content plastic decking
10. Which of the following is the most appropriate material for a porch floor that is close to the ground (approximately 4" above grade) in a mixed-cold climate?
    a. Pressure-treated floor framing with recycled content decking
    b. Slab below grade
    c. Pressure-treated floor framing with FSC-certified Ipe decking
    d. Slab on grade

Critical Thinking Questions

1. Discuss the key decisions in deciding between porches, decks, and patios.
2. What benefits can a porch provide that a deck or patio cannot?
3. Compare the strengths and weaknesses of porches, decks, and patios from a durability standpoint.

Key Terms

deck, 271
French balcony, 272
Ipe, 279
National Oceanic and Atmospheric Administration (NOAA), 272
naturally decay-resistant wood, 278
patio, 271
porch, 271
thermally treated lumber, 277
trellis, 280

Additional Resources

Fine Homebuilding: http://www.finehomebuilding.com/pages/build-a-deck.asp

http://www.finehomebuilding.com/Design/Porches-and-Patios/93396.aspx?channel=2

Landscaping

This chapter presents strategies for creating sustainable sites that minimize negative environmental impacts and decrease energy and water consumption. The landscape includes grading, plants, paving, walls, fences, storm water management, irrigation, and water features. Landscaping plays an important role in the local heat island effect and in storm water management, and is an important component in passive solar and energy efficiency strategies. Rainwater collection, irrigation, and renovation strategies are additional issues addressed in this chapter. Vegetated roofs, which are covered in Chapter 7, can play an important part in the landscape and overall stormwater management plan.

LEARNING OBJECTIVES

Upon completion of this chapter, the student should be able to:

- Describe the effects of landscaping on energy consumption
- Identify methods of reducing water consumption through landscaping decisions
- Explain the influence that landscaping has on the heat island effect
- Describe efficient irrigation systems

Green Building Principles

- Energy Efficiency
- Resource Efficiency
- Durability
- Water Efficiency
- Reduced Community Impact
- Homeowner Education and Maintenance
- Sustainable Site Development

Landscape Planning

The landscape comprises all outdoor features of the home, including natural and built elements. The landscape is divided into two types of features: softscape and hardscape (Figure 11.1). The softscape is the vegetated elements of a landscape, including plants and soil. The hardscape is the nonvegetated elements of a landscape, including paving, walkways, roads, retaining walls, street amenities, fountains, and pools. Landscaping serves as both the introduction to a house and the buffer between it and the neighborhood. From large suburban and rural sites to tiny urban plots, softscapes and hardscapes are an important part of green construction and renovation. When properly designed and constructed, they help contain storm water runoff, allowing water to infiltrate into the ground and recharge aquifers, thereby reducing the flow of pollutants into waterways and the need for additional storm sewer capacity. Selection and location of climate appropriate plants can help reduce both water and energy use.

---

landscape all outdoor features of the home, including natural and built elements.

softscape the vegetated elements of a landscape, including plants and soil.

hardscape the nonvegetated elements of a landscape, including paving, walkways, roads, retaining walls, street amenities, fountains, and pools.

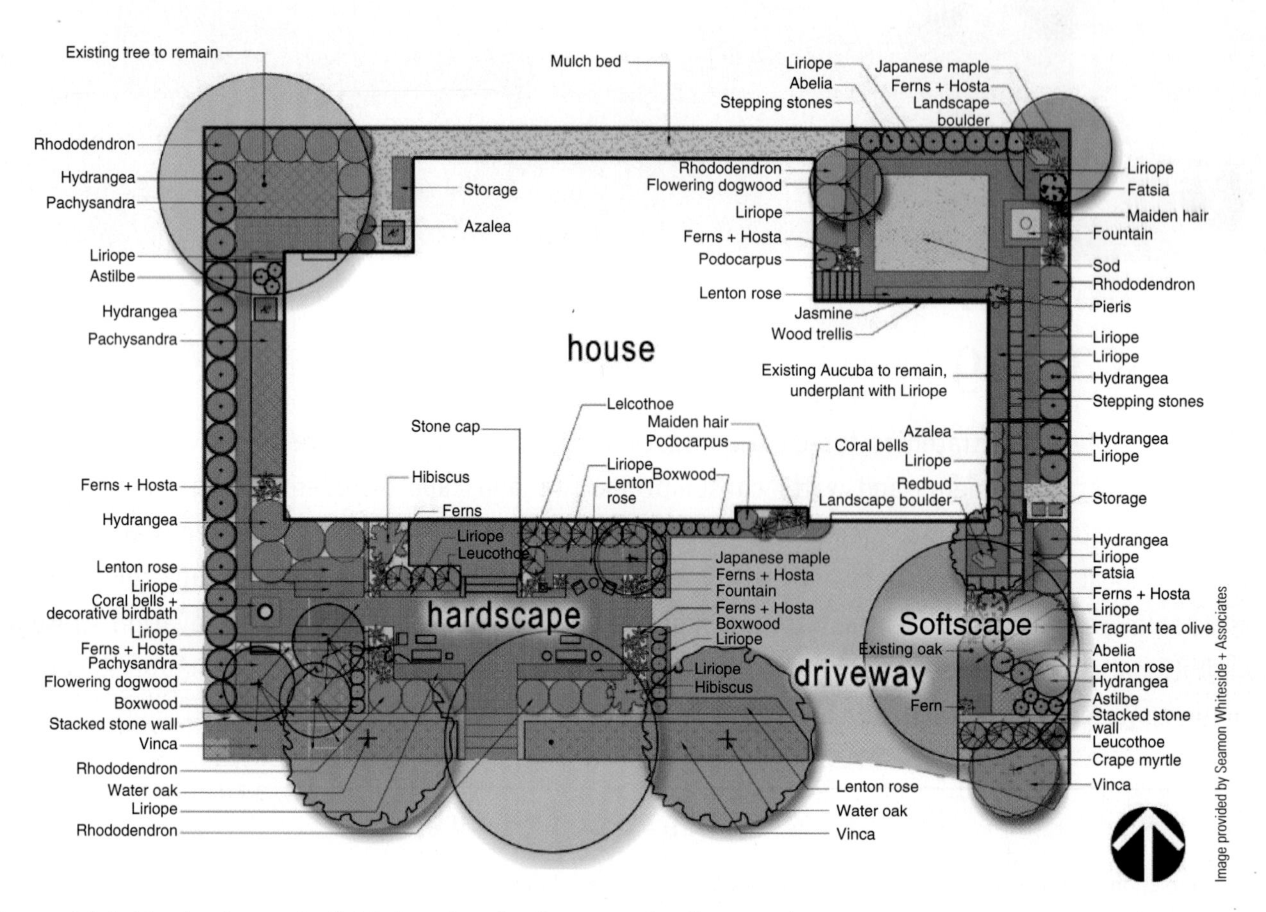

Figure 11.1 The landscape is divided up into hardscape and softscape elements.

As discussed in Chapter 3, landscape planning should be an integral part of the design from the start. The beginning of the design process should include a site survey to identify any sensitive areas, such as mature trees, wildlife habitats, existing native plants, and adjacent wetlands. It should also identify invasive plants to be removed and existing or potential storm drainage problems. Invasive species are region-specific, non-native plants that tend to spread aggressively. The building and site work should be planned to minimize any work in the critical root zones of mature trees to remain in place. Existing established native plants should be identified and incorporated into the design or relocated off site. Development near wetlands and waterways should be avoided, and existing buffer zones should be maintained or new ones created to protect these sensitive areas from disturbance during and after construction.

Landscaping can create private areas or outdoor "rooms" that work together with decks, porches, and patios to provide unconditioned living space. Outdoor spaces can be defined by any combination of fences, walls, trellises, and plants. Fences and walls can have climbing vines and vertical plantings to provide a view, privacy, fragrances, and even food for the occupants.

Traditional front yards are public or semi-public places that, when designed effectively, encourage community interaction and leave rear yards for private spaces. Compact urban developments may not have rear yards, using space behind homes for alleys and garages. Sites without rear yards may benefit from semi-private front yards set up as courtyards that provide partial separation from the street.

Water and fire can serve as key decorative components of outdoor spaces. Manmade ponds, fountains, and waterfalls provide visual and aural stimulus, making yards more enjoyable places to relax. Outdoor fireplaces, chimineas, and fire pits provide additional focal points and gathering spots on chilly evenings.

Landscape Components

Landscaping with plants and manmade structures provides visual interest and structural support for buildings and grade changes. Most traditional landscapes are neither resource nor water efficient. They can also have a detrimental effect

---

invasive species non-native plants that tend to spread aggressively.

on the environment through storm water runoff and erosion. Sustainable alternatives are readily available and can be implemented on a site with proper planning.

Landscaping methods and materials include, but are not limited to, the following:

- *Patios* provide enjoyable outdoor living spaces that take advantage of the sun to provide warmth on moderate days.
- *Paving* is both a method and material and is required for most patios, driveways, and walkways.
- *Retaining walls* are used to transform slopes into level areas for buildings and yards, or to maintain tree roots near construction zones.
- *Fences* and *freestanding walls*, along with vegetated trellises and rows of trees, provide privacy and windbreaks while helping to define the limits of outdoor living spaces.
- *Softscapes*, such as lawns, ground cover, shrubs, and trees, are all part of traditional landscaping work. They can provide permeable surfaces, reduce runoff, provide windbreaks, and help reduce local temperatures by absorbing heat from the sun.
- *Grading*, while necessary for most construction, must be done properly to help manage storm water runoff both during and after construction.
- *Rainwater collection systems* help reduce runoff and can provide free water for irrigation as well as some interior uses.
- *Rain gardens* are depressed, vegetated areas that collect surface runoff from impervious surfaces. Once collected, the water may infiltrate the surface and return to the groundwater supply or evaporate into the atmosphere.
- *Irrigation systems* help establish and maintain plants on-site.
- *Water features*, such as pools, ponds, and fountains, can be used for recreation and to attract desirable wildlife while providing stimulating sights and sounds to outdoor living areas.

Impervious Paving

Traditional paving surfaces, such as concrete, asphalt, pavers, and stone, can create excess storm water runoff by creating impervious surfaces. Such surfaces can also add to the heat island effect. Every impervious surface on a site increases the required capacity of municipal storm water systems and allows contaminants, such as oil, gasoline, insecticides, and fertilizers, to run into waterways.

Sustainable Uses of Impervious Paving

Concrete, asphalt, stone, and other traditional impervious surfaces can be part of a sustainable site design by limiting their area and by directing runoff to such features as dry wells, rain gardens, or pervious paving areas that keep the water on-site. Sustainable alternatives to standard concrete are discussed in Chapter 5, Foundations. Reclaimed masonry, cast pavers with recycled content, and locally quarried stone are good alternatives.

Pervious Paving Where paving is required, the use of water-pervious materials as part of an installation will retain much of the storm water on-site, allowing it to drain through the ground and filter out pollutants, thereby reducing the amount of water entering storm sewers. Unlike traditional materials, pervious paving allows water to enter the ground below. Depending on site conditions and the paving design, pervious paving may provide enough storage capacity to eliminate retention ponds, swales, and other rain catchment requirements. Pervious paving can be concrete, asphalt, loose-laid pavers, or plants grown in structural grids.

Pervious Concrete and Asphalt Pervious concrete combines small, uniform aggregate with water and Portland cement; no sand or fine aggregates are added in the mix (Figure 11.2). The result is paving with voids throughout that allow water to penetrate. The concrete is placed over a gravel sub-base of at least 12" that provides a

Figure 11.2 Pervious concrete contains large voids and a deep sub-base that allows water to easily penetrate. Pervious concrete is easily used for sidewalks and driveways.

---

rain garden a depressed vegetated area that collects surface runoff from impervious surfaces and allows infiltration into the groundwater supply or return to the atmosphere through evaporation.

pervious paving paving materials that allow water to enter the ground below.

pervious concrete a method of concrete paving that allows water to enter the ground below.

storage reservoir for rainwater, allowing it to percolate into the ground (Figure 11.3). Some installations may include a filter fabric below the sub-base to reduce the chance of silt build-up. Pervious concrete installations should be engineered for the specific site and soil and installed by an experienced contractor.

Pervious asphalt is installed with the same sub-base and a similar open structure in the finish layer as pervious concrete (Figure 11.4). While not as durable as concrete paving, pervious asphalt is flexible, making it more resistant to cracking. The surface can also be recoated and paved over without removing the original layer.

Permeable and Porous Pavers Solid paver blocks can be installed as a permeable surface so that water flows between the blocks into a gravel sub-base, similar to pervious concrete. Pavers are installed between curbs with gravel or sand (instead of mortar) between each block

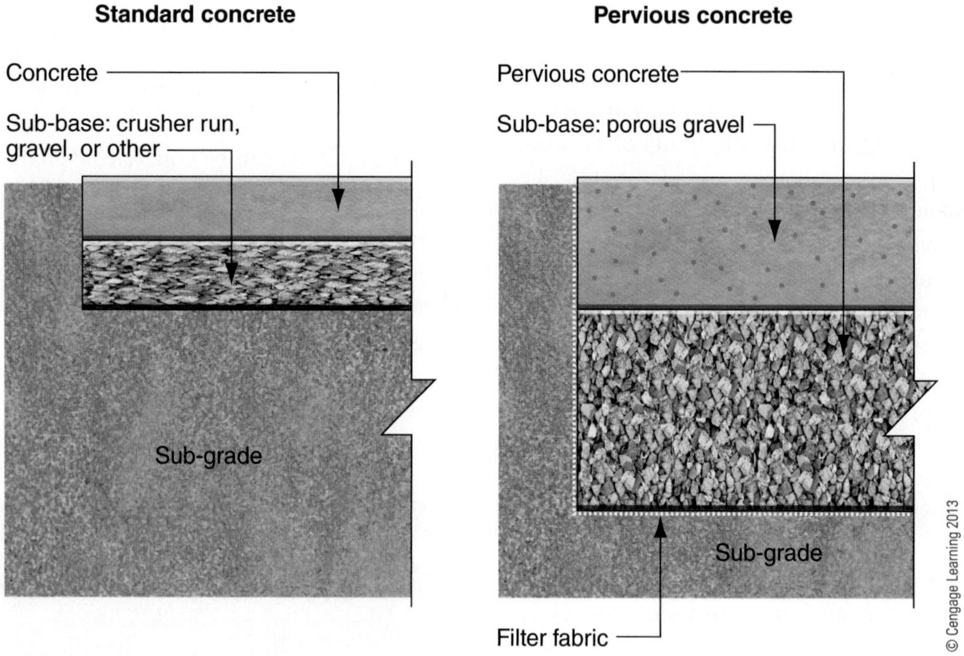

Standard concrete

Concrete

Sub-base: crusher run, gravel, or other

Sub-grade

Pervious concrete

Pervious concrete

Sub-base: porous gravel

Sub-grade

Filter fabric

Figure 11.3 Typical cross-section of pervious concrete. The pervious concrete surface layer (15% to 25% voids) and the sub-base (20% to 40% voids) provide stormwater storage.

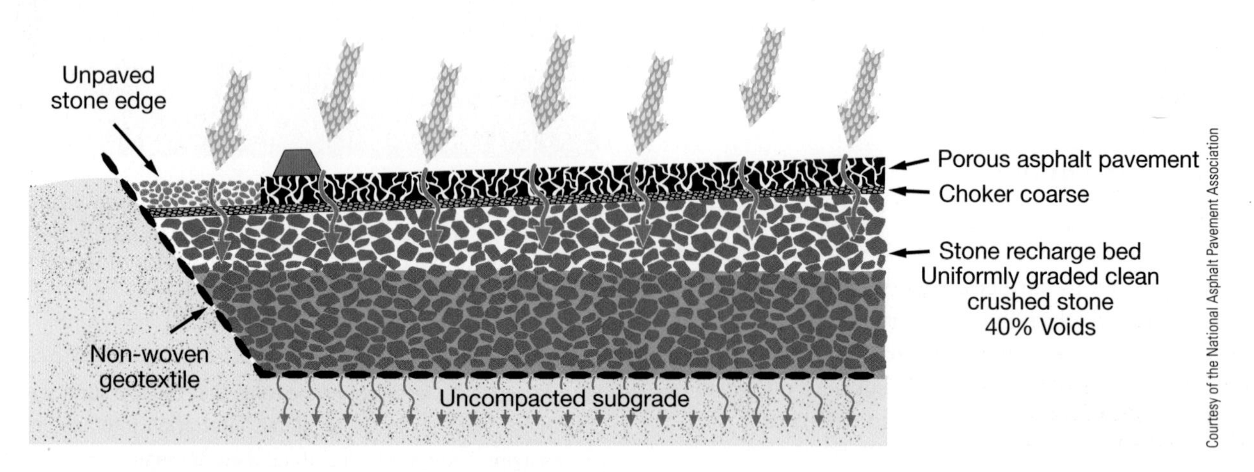

Figure 11.4 Typical cross-section of porous asphalt paving.

---

pervious asphalt a method of asphalt paving that allows water to enter the ground below.

for drainage. Pavers can be made of concrete, brick, or stone. Porous pavers have openings between and within the pavers that are filled with vegetation or gravel, allowing water to drain (Figure 11.5).

Another porous paving option is a plastic grid, strong enough to support vehicle traffic, which is normally installed over a gravel sub-base (Figure 11.6).

Heat Island Effect of Paving Hardscapes can contribute to the heat island effect by absorbing heat from the sun during the day and heating the air as it is released

Figure 11.5 Paving stones are a pervious paving option that allows water to easily migrate to the ground.

Figure 11.6 Plastic grid structures are available to transform vegetated surfaces into porous pavement. They provide load-bearing strength while protecting vegetation root systems from deadly compaction. This allows grass areas to perform the functions of asphalt or concrete pavement but with the aesthetics of a lawn.

porous pavers pavers with openings between and within the pavers that are filled with vegetation or gravel, allowing water to drain through to the ground.

late in the day, offsetting the normal evening cooling process. The more reflective a surface is, the less heat it absorbs and releases. Reflectivity, or albedo, is measured as a decimal between 0 and 1. The higher the number, the more reflective the material. As described in Chapter 7, the solar reflective index (SRI) is an alternative measurement that ranks materials between 0 (low reflectivity) and 100 (high reflectivity). Materials with an SRI of 29 or higher are generally considered reflective and contribute less to the heat island effect than darker paving materials. Table 11.1 compares the albedo, emittance, and SRI of various paving materials.

Asphalt and other dark-colored paving materials have the lowest SRI, concrete and light-colored pavers have the highest. Paving that is partially shaded from the sun also reduces the heat island effect.

Retaining Walls

Retaining walls are used to support cuts in the grade and the structures above them, as well as to help eliminate erosion on steep slopes. While working with the existing grades is preferable to reduce the need for retaining walls, it is often not possible to do so. Some sites may require retaining walls to turn a sloped area into a suitable building site while maintaining existing natural features on a level area that otherwise would be suitable for construction. Stone, concrete, concrete masonry units (CMUs), interlocking concrete blocks, and wood or recycled plastic beams are some of the available materials for retaining walls.

Table 11.1 Solar Reflectance (Albedo), Emittance, and Solar Reflective Index (SRI) of Select Material Surfaces

Material Surface	Solar Reflectance	Emittance	SRI
Acrylic paint, black	0.05	0.9	0
Acrylic paint, white	0.8	0.9	100
Asphalt, new	0.05	0.9	0
Asphalt, aged	0.1	0.9	6
"White" asphalt shingle	0.21	0.9	21
Concrete, aged	0.2–0.3	0.9	19–32
Concrete, new (ordinary)	0.35–0.45	0.9	38–52
Concrete, new white	0.7–0.8	0.9	86–100

Sources: Levinson, R., and Akbari, H. "Effects of Composition and Exposure on the Solar Reflectance of Portland Cement Concrete," Lawrence Berkeley National Laboratory, Publication No. LBNL-48334, 2001; Pomerantz, M., Pon, B., and Akbari, H. "The Effect of Pavements' Temperatures on Air Temperatures in Large Cities," Lawrence Berkeley National Laboratory, Publication No. LBNL-43442, 2000; Berdahl, P., and Bretz, S. "Spectral Solar Reflectance of Various Roof Materials," *Cool Building and Paving Materials Workshop*, Gaithersburg, MD, July 1994; Pomerantz, M., Akbari, H., Chang, S.C., Levinson, R., and Pon, B. "Examples of Cooler Reflective Streets for Urban Heat-Island Mitigation: Portland Cement Concrete and Chip Seals," Lawrence Berkeley National Laboratory, Publication No. LBNL-49283, 2002; Heat Island Group, Lawrence Berkeley National Laboratory. http://concretethinker.com/solutions/Heat-Island-Reduction.aspx.

Concrete, masonry, and most stone walls require a concrete footing to support them (Figure 11.7). Large stone walls, interlocking block walls, and walls made of wood or plastic ties often do not require a separate footing (Figure 11.8). Retaining walls are expensive and materially intensive to build—and even more expensive to repair or replace—so the wall must be engineered to support the structural load as well as the hydrostatic pressure created by the earth it is supporting. Properly sized footings or wall structure and adequate drainage behind the wall are critical to long-term performance.

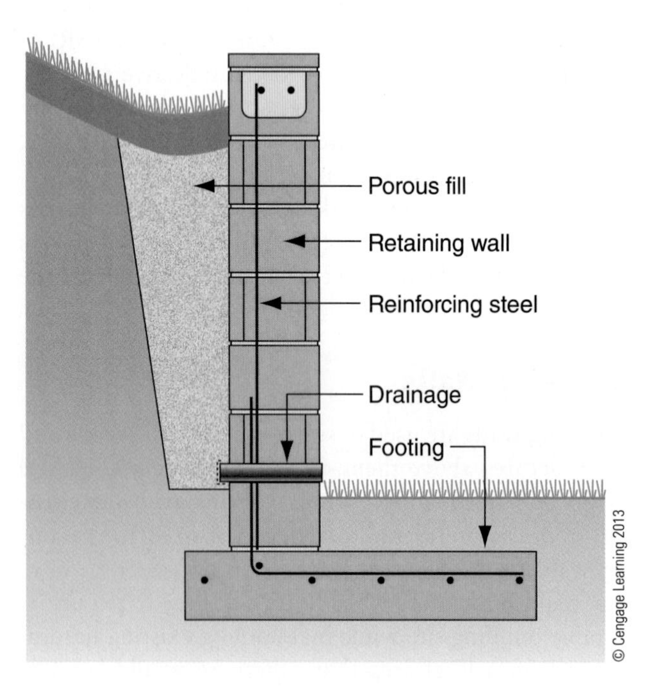

Figure 11.7 Concrete footings provide structural foundation to concrete, masonry, and most stone walls.

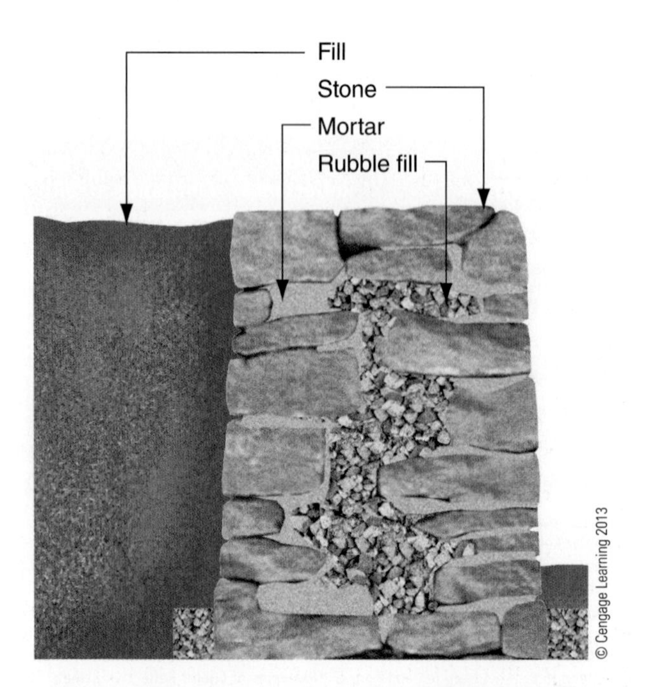

Figure 11.8 Separate footings are typically not required for large stone walls, interlocking block walls, and wood or plastic ties.

Common methods and materials for retaining walls include:

- Concrete, interlocking blocks, and CMUs all have high embodied energy, but they are durable and require very little maintenance. In some cases, they are the only practical option available for large walls.
- Stone walls can be solid or veneered over concrete or CMUs. Brick is typically veneered over a structural wall.
- Recycled and new railroad ties, pressure-treated timbers, recycled plastic ties, and other wood products are suitable for retaining walls (Figure 11.9). Consider the toxicity of any preservative and its impact on the site before selecting a product. Naturally decay-resistant woods that are sustainably harvested or recycled are suitable choices, particularly in areas with low risk of termites.
- Old concrete slabs broken into small pieces and installed like stone are another option for constructing retaining walls. Often referred to as "urbanite," slabs can be removed and recycled on the same site, saving the cost of disposal and new material.

Fences and Freestanding Walls

Made from wood, metal, stone, brick, or CMUs, fences and walls can range from a short picket fence to delineate a front yard to a tall wall for security and privacy. They can serve as windbreaks that reduce heat loss in winter, and they may be able to provide shade for east- and west-facing glazing, reducing overheating in the summer. Suitable material choices include reclaimed stone or brick, recycled plastic, and sustainably harvested wood.

Softscapes

Plantings set the scene for a home and provide an introduction to visitors. Creating landscapes that have the least impact on the site, including using existing grades and local, drought-tolerant plants help limit erosion, minimize water required for irrigation, and reduce the need for fertilizers, pesticides, and regular maintenance. Drought-tolerant plants are those species that are able to survive in environments with little available water or moisture. Local agricultural offices generally provide lists of these plants. Flowers provide visual interest and fragrance, whereas edible plants provide fresh, healthy low-cost food.

Figure 11.9 Retaining walls may be made from recycled and new railroad ties, pressure-treated timbers, recycled plastic ties, and other wood products.

Although invasive species are non-native, not all non-native plants are considered invasive. A full list is available at the local Cooperative Extension System office or through state agencies. The U.S. Department of Agriculture maintains a list of resources at http://www.invasivespeciesinfo.gov/.

Shrubs and trees should be native species, located to provide shade for windows and paving where desired. Each plant should be positioned far enough away from the house to leave a minimum of 24" between the mature plant's leaves and the home's structure. This distance will help to reduce moisture intrusion and pest infestation in the structure.

Soils

Soil health can be assessed through chemical and biological testing. Soils may be amended to improve fertility, drainage, structure, pH, and microbe levels. Typical amendments include compost, sand, clay, gypsum, and peat moss. Some construction wastes may even be reused on-site as soil amendments. Gypsum from drywall and untreated wood are often ground and applied on-site. Care should be taken with fertilizers, however; research indicates excessive fertilizers may actually encourage undesirable plants. Soils that are compacted during construction should be aerated before replanting to provide a better base for new plants, allowing for better drainage and reducing runoff.

Lawns

Lawns require chemical fertilizers, weeding, watering, and regular mowing. Minimizing or eliminating lawns saves water and energy and reduces the amount of fertilizers that pollute waterways. Alternatives include native groundcovers and drought-resistant, slow-growing, regionally appropriate grasses, such as sheep fescue and Bermuda.

Ground Cover, Shrubs, Decorative and Edible Plants, and Vines

Native ground cover plants can be used as sustainable substitutes for turf grass that reduce the need for regular maintenance, irrigation, and fertilizers. Shrubs provide visual interest and help define outdoor areas. Decorative plants can include both perennial and annual flowers, vegetables, fruit trees and bushes, and herbs. Select native and noninvasive species and rotate them annually for healthier soil and plants. Vines planted on a trellis, fence, or freestanding wall can attract wildlife while providing fragrance and food, shade and cooling for a building, and a focal point for a view.

Trees

Healthy trees are a valuable resource, providing shade, erosion control, and tangible value—studies show that trees can increase property values up to 25%. Begin a project by evaluating all the trees on the site and determining which ones should remain, depending on the building location and their condition. Locate buildings and other improvements as far away from the most desirable trees. Deciduous trees on the south side of a house can provide shade in warm weather and allow for solar gain when cold. Evergreens on the east and west help shade from hot sun year-round and can provide windbreaks in cold climates.

Tree roots should be protected from equipment by placing protective fencing at the edge of the drip line to minimize damage and compaction of soil. Locate new trees so that, as they mature, they will provide shade where desired and their roots and branches will not interfere with any structures.

Xeriscaping

Xeriscaping is a landscaping technique that uses drought-tolerant plants to minimize the need for water, fertilizers, and maintenance. More diverse than lawns, xeriscaped sites are generally less vulnerable to insects and other threats. When designed and installed properly, they require no irrigation, mowing, or fertilizers (Figure 11.10).

Invasive and Native Plant Species

Every climate has had invasive plant species introduced, many of which have no natural enemies to prevent them from taking over landscapes and growing uncontrollably. Through the removal of any existing invasive plants and avoiding the installation of new ones, a more sustainable yard can be created. Plants that are native to the region have evolved to thrive in local conditions with minimal irrigation and generally do not grow uncontrollably as invasive species can. Most local agricultural extension offices can provide lists of invasive plants and native and drought-tolerant species that are well suited to the growing area.

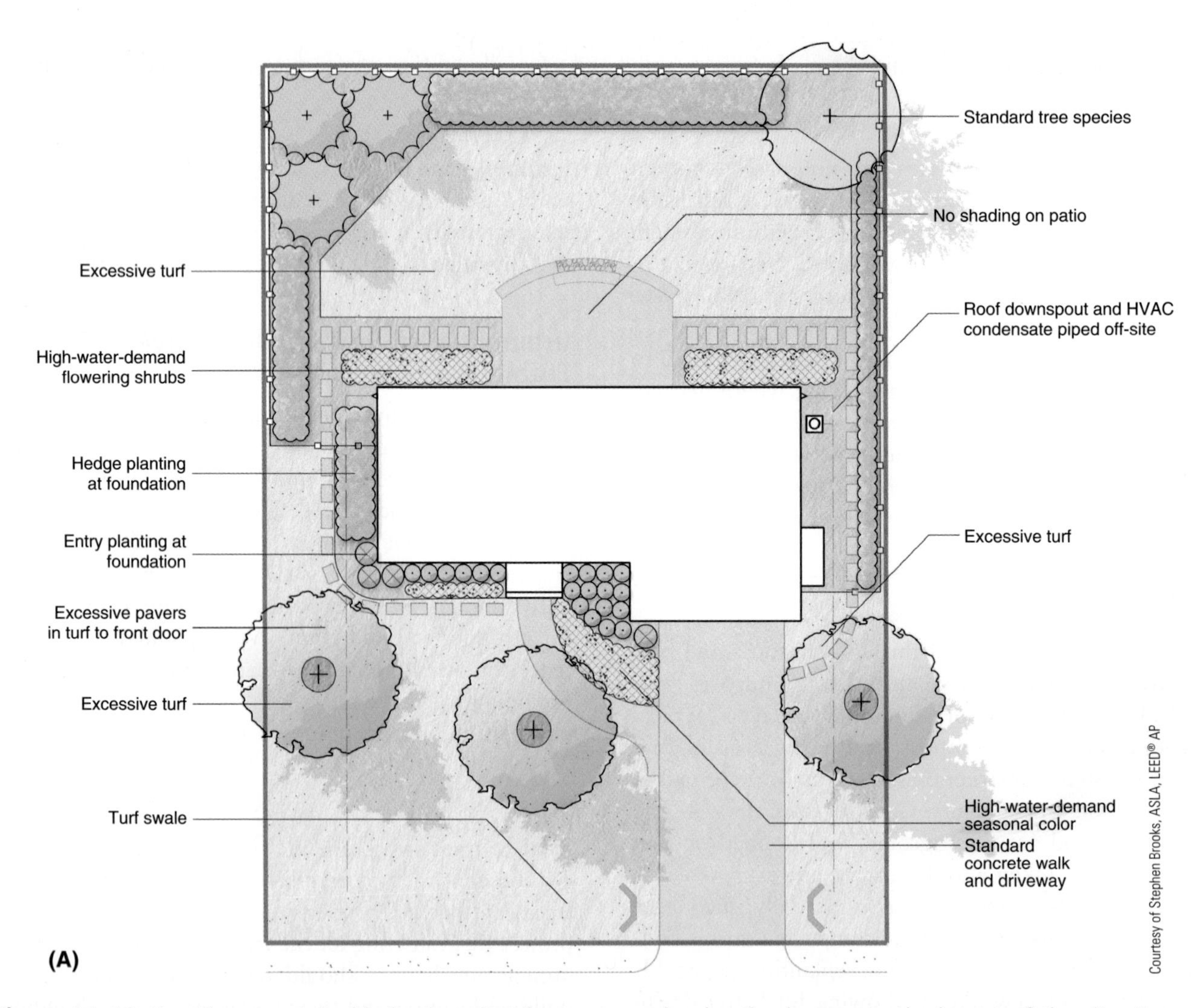

Figure 11.10 Creating a sustainable landscaping does not require drastic changes to the layout of the site. Here we see that through relatively simple modifications, the landscape can be greatly improved. Image (A) shows the home with a standard landscape plan, and image (B) is a sustainable plan.

---

xeriscaping a landscaping technique that uses drought-tolerant plants to minimize the need for water, fertilizers, and maintenance.

drought-tolerant species a tree or plant that is able to grow and thrive in arid conditions.

Composting

Composting yard waste on-site reduces the impact on landfills and can provide healthy, natural fertilizer with minimal effort. Most kitchen waste can be added to yard compost, further reducing waste going to landfills. Including a dedicated location for composting on a site can help encourage the practice (Figure 11.11).

Integrated Pest Management

Pesticides used in residential landscapes to control insects and other pests can pose serious health risks to people and animals exposed to them. Integrated pest management (IPM) is a strategy to first limit the use of pesticides and, only when necessary, use the least hazardous products sparingly. IPM strategies include setting "action thresholds," which define levels of infestation below which nothing is done; rotating plants; and selecting pest-resistant plants. IPM can be equally or more effective and less expensive than regular use of pesticides in controlling insects.

Grading

By using as much of the existing grade as possible, site developers can disturb less soil and reduce the impact on trees while lowering costs. Less new topsoil is required, and existing soil may require less conditioning. The retention of existing grades and plants also helps to minimize erosion and the need for new replacement plants.

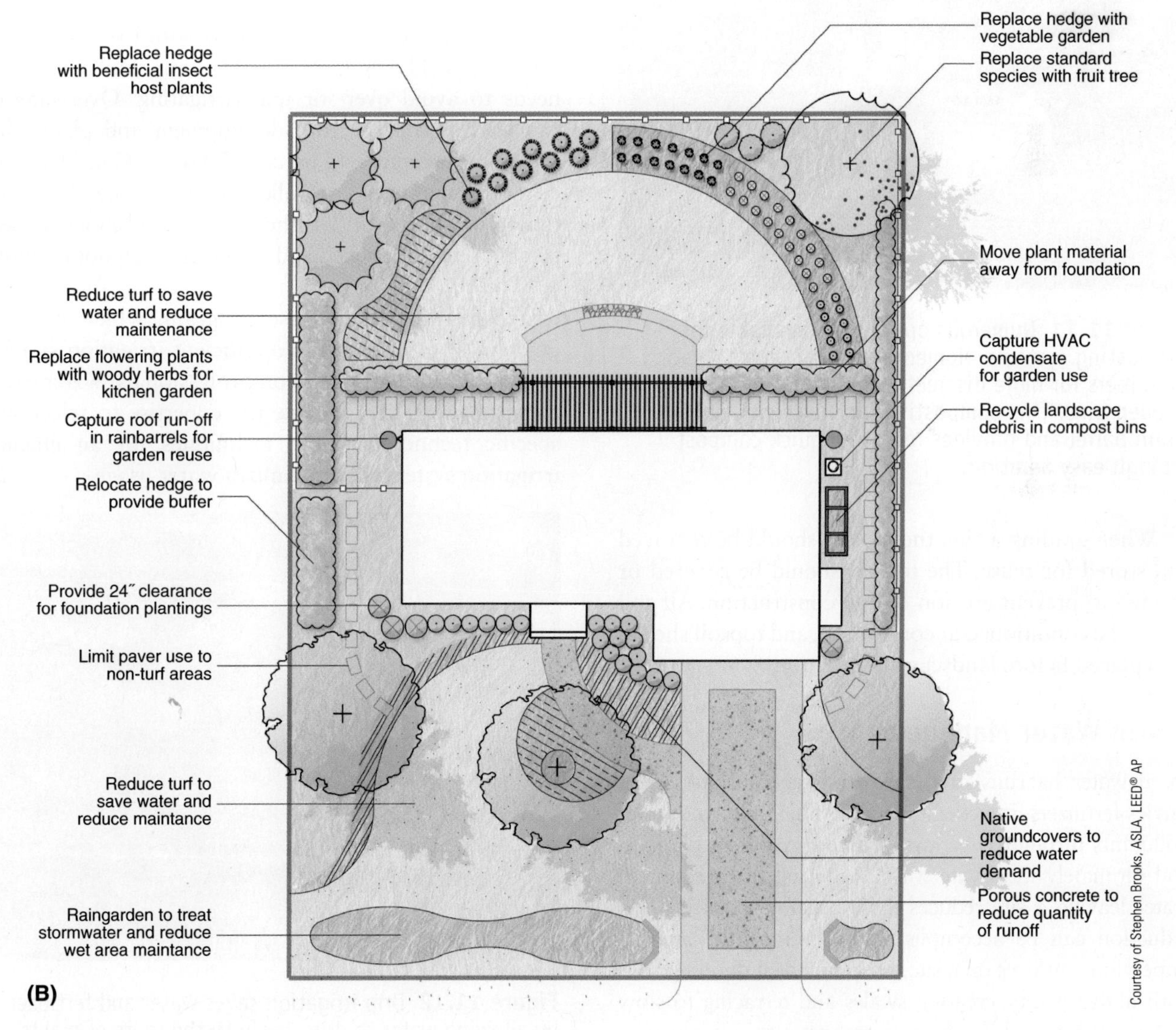

Figure 11.10 (Concluded)

integrated pest management (IPM) a strategy to first limit the use of pesticides and, only when necessary, use the least hazardous products sparingly.

Figure 11.11 Numerous options are available for composting systems. Homeowners may select plastic containers for more discreet composting, or open-air wooden bins. This composting bin looks similar to a rain barrel and provides relatively quick compost through easy aeration.

When grading a site, the topsoil should be removed and stored for reuse. The topsoil should be covered or planted to prevent erosion during construction. All soil should be conditioned at completion, and topsoil should be replaced before landscape installation.

Storm Water Management

Storm water that runs off sites, down streets, and into sewers carries fertilizers, insecticides, vehicle fluids, silt, and other pollutants through sewer systems and treatment plants—and ultimately into waterways. Reducing the amount of water leaving a site reduces these negative impacts. This reduction can be accomplished by limiting the area of impervious surfaces on a site, providing well-draining soils with native plants, creating swales and terracing to slow water velocity, and directing storm water to rain gardens, dry wells, and rainwater collection systems. Water that is allowed to percolate through the soil recharges groundwater, filtering out pollutants in the process.

When incorporating underground pipes for storm water management, use rigid rather than flexible corrugated pipe, and install easily accessible cleanouts. Rigid pipe is more durable and not as susceptible to damage. Incorporate overflow outlets to avoid backups during heavy rain events.

Rainwater Collection

Rainwater can be collected for reuse in irrigation as well as for some interior uses, reducing the demand on potable water supplies. This strategy also helps cut water bills. Rainwater collection systems are discussed in detail in Chapter 15, Plumbing.

Irrigation

Where irrigation is required to establish or maintain landscaping, use underground and drip irrigation systems rather than broadcast sprayers. Underground systems provide water directly to the roots with less evaporative loss (Figure 11.12). Group plants with similar watering needs to avoid over- or underirrigating. Overwatering can lead to shallow root development and plants that are more susceptible to pests and disease. Consider using evapotranspiration controllers that manage irrigation timing through local weather data or soil moisture sensors that override scheduled watering when not required.

The U.S. Environmental Protection Agency has established the WaterSense program, which provides guidelines for installation of efficient irrigation systems and certification of irrigation professionals. WaterSense-certified professionals have the expertise to select site-specific technologies and techniques for an efficient irrigation system that minimizes water usage.

Figure 11.12 Drip irrigation saves water and fertilizer by allowing water to drip slowly to the roots of plants, either onto the soil surface or directly onto the root zone, through a network of valves, pipes, tubing, and emitters.

WaterSense the U.S. Environmental Protection Agency–administered program that provides guidelines for installation of efficient irrigation systems and certification of irrigation professionals.

FROM EXPERIENCE

Sustainable Landscapes

Stephen Brooks, ASLA, LEED AP, Director of Landscape Architecture, Pond l Ecos. As our population increases, so does our demand of our natural resources. Many regions have taken dramatic measures to conserve and protect through policies that control when and how much a resource can be used. Many policies start at a regional level, with most focusing on entities as small as the individual homeowner. Depending upon the region of the country, the at-risk resources vary somewhate; almost universally, however, water is the resource that is now controlled through some level of policy. Much of the policy focuses on the interior and exterior of the residence; principles for both are very simple to incorporate.

An example is the typical newly built residential home. Most have a small landscaping budget that is generally put toward non-native showy plants that create the most visual appeal to help sell the house. Much of the lot is covered in lawn, which is the most water-intensive material found in a home's landscape and is usually placed on heavily compacted soils that discourage any infiltration. With a little coordination among the builder, homeowner, and designer, the average builder's landscaping budget can be used to provide a more sustainable landscape.

Courtesy of Stephen Brooks

Stephen Brooks directs the Design & Construction Studio at Ecos, training and mentoring his staff through hands-on project management. Since joining the firm in 2000, he has incorporated his design and technical expertise into a variety of projects, including such large-scale technical construction projects as the Georgia Institute of Technology's Klaus Advanced Computing Building and the Heart of Lancaster Regional Medical Center.

The first step is to understand both the macro and micro conditions of the site. The macro conditions deal with natural climate factors that cannot be readily manipulated, such as sun angles, shade patterns, wind directions, humidity levels of an area, soil moisture levels, and heat gain of surrounding surfaces. Micro conditions relate to factors that are unique to growing areas on the site, such as topography and soil quality. Ideally, the soil should be rich, loose, and organic to provide nutrients for plants and reduce the need for fertilizers that deplete soil moisture.

To create a sustainable landscape, canopy trees should be positioned to provide shading for the structure as well as any surfaces that may retain heat. Evergreen trees can be used to block winter winds and provide necessary visual screening. By reducing the size of the lawn and increasing the size of mulched planting areas, the amount of fertilizers and water needed for a property can be significantly reduced. Instead of selecting the typical non-native flowering shrubs and groundcover, choose native alternatives.

Other design solutions include capturing stormwater runoff for irrigation, utilizing vegetables and herbs in lieu of traditional ornamental plants, and recycling yard debris as compost to help regenerate the soil's nutrients. Porous surfaces can also be used to help reduce runoff.

Rainwater is an excellent alternative to potable water for irrigation and can be used for both underground and broadcast watering. Gray water collected from inside the house can be used for underground irrigation, subject to local restrictions. Gray water reuse systems are discussed in detail in Chapter 15, Plumbing.

Water Features

In general, pools, fountains, and ponds are unnecessary for the overall performance of the home and should be avoided. When they are included in a project, they should be designed to be as energy- and water-efficient as possible.

Solar pool heating systems reduce or eliminate the use of fossil fuels for heating. Traditional pool pumps are single-speed, designed for the highest load conditions (including spa jets and other features) and using more energy than necessary when less power is required. Variable-speed pumps that adjust to the required load require less power to run. Pumps can also be set on timers to run only when necessary. Larger-diameter pipes with wide sweeps instead of tight elbows and larger filters reduce pressure, allowing pumps to work more efficiently.

Freshwater swimming pools must be purified with added chlorine for health and safety. Salt-water pools generate chlorine from salt added to the water or to the

---

gray water nonpotable water reclaimed from sinks, baths, and washing machines that may be used to flush toilets and for irrigation.

filter equipment; while they eliminate the need to handle chlorine directly, they do require regular balancing of water chemicals. The chemicals in both chlorine and salt-water pools are corrosive to equipment and finishes, requiring additional maintenance and regular repairs.

An alternative to these purification systems are natural pools that filter the water by running it through pipes with titanium plates (to reduce algae growth) or through constructed wetlands. The latter, a manmade treatment system that uses a combination of plants and soils to improve water quality, can serve as a landscape design feature near the pool or be hidden from view. Water is pumped through the landscaping, which filters out impurities that are retained to nourish the plants.

Ponds and fountains can employ captured rainwater instead of potable water, and water features can be created from required detention ponds on the site. Circulation pumps for ponds and waterfalls can employ solar photovoltaic power, reducing or eliminating energy use to operate them.

Special Uses

Projects may include areas with special landscaping requirements, such as a basketball or tennis court, a turf grass sports field, a parking lot, flower gardens, or a wildlife sanctuary. Like all paving, the area for sports courts should be kept to a minimum, and they should drain to site features that manage the water on-site. Fields should use native, drought-resistant grasses that are maintained with a minimum amount of fertilizer and irrigated to promote deep root growth. Vegetable and flower gardens are appropriate designs for a sustainable site, provided they avoid invasive species and irrigation minimizes potable water use.

The National Wildlife Federation (NWF) provides guidelines for creating wildlife habitats within yards of any size. Breeding grounds for wildlife can be restored in developed areas by planting shrubs that provide cover for nests and food to support local fauna. NWF has a program to certify individual wildlife habitats, which are listed on a national registry (Figure 11.13).

Courtesy of http://www.northseattlesarah.com

Figure 11.13 The National Wildlife Federation (NWF) certified this yard as wildlife habitat.

Remodeling

Remodeling provides opportunities to improve the structure of the home as well as the landscape. Replacing plants with drought-tolerant species, upgrading irrigation systems, adding rainwater collection or gray water reuse systems, and replacing hardscape elements can reduce energy and water consumption.

When planning additions and renovations, design to minimize the impact on existing landscaping and mature trees. Working within the existing footprint of a house instead of expanding the foundation eliminates the negative impacts of grading and erosion on the job site (Figure 11.14). When plans include removal of existing native plants, consider donating them to individuals or local plant rescue organizations. Major landscape updates should incorporate the green building principles and use water as efficiently as possible to minimize negative environmental impacts.

Figure 11.14 This home's footprint remained unchanged despite a substantial renovation. Such design decisions as building up rather than out prevent site disturbances.

Summary

Site finishes play several important roles in a home. They provide the frame through which a building is viewed and approached by the public. They also have an effect on energy efficiency, water efficiency, impact on the community, and building durability. Making appropriate selections during the design and construction process are key to a sustainable site that improves the building and adjacent property.

Review Questions

1. Which of the following is not generally considered a part of a sustainable landscape design?
   a. Storm water control
   b. Aquifer recharge
   c. Native and drought-tolerant plants
   d. Impervious paving
2. Which of the following is considered to contribute to the heat island effect?
   a. Asphalt
   b. White concrete
   c. Gray concrete
   d. Turf grass
3. Which of the following is not a component of pervious concrete?
   a. Gravel base
   b. Filter fabric
   c. Sand
   d. Portland cement
4. Which of the following plants are the most sustainable?
   a. Native, drought-tolerant ground cover
   b. Turf grass
   c. Non-native ground cover
   d. Non-native, drought-tolerant vines
5. What is the primary purpose of retaining walls?
   a. To support the house foundation
   b. To provide a level surface for construction
   c. To create privacy
   d. To promote water conservation
6. Which of the following are not benefits of evergreen trees?
   a. They provide shade in summer and allow for sun in winter
   b. They can provide year-round shade on west- and east-facing windows
   c. They provide wind breaks
   d. They can increase the value of a property
7. Which strategy does not improve storm water management?
   a. Rain gardens
   b. Concrete paving
   c. Swales
   d. Terracing
8. What is not a result of overwatering?
   a. Shallow root growth
   b. Reduced disease resistance
   c. High water bills
   d. Stronger root systems
9. Which of the following programs certifies irrigation systems and professionals?
   a. ENERGY STAR
   b. LEED for Homes
   c. WaterSense
   d. The National Green Building Standard
10. What strategies are appropriate for pools and water features in a sustainable landscape (choose all that apply)?
    a. Single-speed pumps
    b. Constructed wetland filter system
    c. Solar heating
    d. Use of gray water

Critical Thinking Questions

1. Discuss different options for paving that reduce runoff and the heat island effect.
2. How does the existing landscape affect decisions on house location and size?
3. Review the benefits of minimizing traditional turf grass on lots.

Key Terms

drought-tolerant species, 292
gray water, 295
hardscape, 285
integrated pest management (IPM), 293
invasive species, 286
landscape, 285
pervious asphalt, 288
pervious concrete, 287
pervious paving, 287
porous pavers, 289
rain garden, 287
softscape, 285
WaterSense, 294
xeriscaping, 292

Additional Resources

National Asphalt Pavement Association: http://www.hotmix.org

Portland Cement Association: http://www.cement.org

National Ready Mixed Concrete Association: http://www.nrmca.org

SECTION FOUR

INTERIOR SYSTEMS

CHAPTER 12: Interior Finishes

Interior Finishes

This chapter covers material selection and installation methods for walls, floors, ceilings, and permanent fixtures, such as cabinets and counters. The manufacturing process, transportation energy, and in-place emissions of volatile organic compounds have an effect on the environment as a whole as well as the building's indoor environmental quality. There is rarely a perfect answer to the question of what materials to use. Understanding the tradeoffs between different products and how to make the best decision for your particular project is key to selecting interior finishes that are sustainable and durable but do not diminish indoor environmental quality. Interior finish issues in remodeling projects will also be discussed.

LEARNING OBJECTIVES

Upon completion of this chapter, the student should be able to:

- Select interior finish materials appropriate to green homes
- Explain how interior finish materials contribute to the sustainability of a project
- Describe how finish selection and installation affect indoor environmental quality and the sustainability of a home

Green Building Principles

- Energy Efficiency
- Resource Efficiency
- Durability
- Indoor Environmental Quality
- Reduced Community Impact

Types of Interior Finishes

Interior materials include floor, wall, and ceiling finishes; millwork, including doors, trim, and cabinets; counters; paint, stains, and clear coatings; and the methods used to install and apply these products in a house.

Floor Finishes

Floors usually have a decorative material applied on top of the structural subfloor; the most popular surfaces include carpet, wood, sheet vinyl, and tile. Concrete slabs can be stained or polished for a floor finish.

Wall and Ceiling Finishes

Drywall is the most common wall and ceiling finish for most homes, particularly wood-framed structures. **Drywall**, also referred to as gypsum board, is a construction material used for finished wall and ceiling surfaces and is made of kiln-dried gypsum that is pressed between fiberglass mats or paper facings. Buildings with integrated insulation, such as autoclaved aerated concrete or insulated concrete forms, can have plaster or stucco applied directly to the structure for interior wall finishes, although most interior walls and ceilings in these buildings will be

---

drywall a construction material used for finished wall and ceiling surfaces; made of kiln-dried gypsum pressed between fiberglass mat or paper facing; also referred to as *gypsum board*.

drywall. Plaster is a mixture of lime or gypsum mixed with sand and water that hardens into a smooth solid and is used to cover walls and ceilings. Both traditional plaster and gypsum veneer plaster, which consists of two finish coats applied over a gypsum board substrate, can be applied directly to masonry walls. Alternatively, plaster can be applied over wood or metal lath, both of which are premium products that may not be readily available in some markets. Wood panels and trim for wall finishes are usually used as accents in such rooms as libraries and dens.

Millwork

Millwork refers to the wood and composite products used to create doors, trim, cabinets, and similar finishes in homes. Traditionally made from solid wood, millwork now incorporates engineered materials, plywood, and composite panels for cabinets, as well as metals, plastics, and other nontraditional materials.

Counters

Counters can be made of almost any material; traditional choices are natural stone or wood. Alternatives include composite materials made of crushed stone and recycled glass cast with binders, cast concrete, and ceramic tile. Ceramic tile is a thin surfacing unit composed of various clays fired to hardness.

Paints, Stains, and Clear Coatings

Most finish materials require a coating to increase their durability and reduce maintenance. Walls and ceilings can be painted or coated with integrally colored clay plasters. Wood millwork can be painted or finished with a clear coating, with or without a stain. Wood floors are normally finished with a clear coating or a penetrating sealer, with or without a stain.

Selecting Interior Finishes

Selecting sustainable interior finishes is important in a green project, but keep in mind that green building is about the process, not the products. With the exception of the most toxic materials, almost any product can be part of a green home if installed and maintained properly.

The following criteria should be considered in selecting finish materials:

- Durability
- Recycled content
- Recyclability when replaced
- Toxicity in manufacture and on the site
- Embodied energy
- Environmental impact and working conditions in manufacturing
- Local extraction and processing
- Renewable resources used
- Suitability to the climate

Rarely does one product meet all of these criteria. Material selection usually requires striking a balance among them. Although one product may meet most of your criteria but could have high-embodied energy, an alternative choice may be more toxic in the manufacturing process—no single answer is correct. Understanding the differences between products and making an informed decision is an important responsibility of the project team.

Interior Finish Colors

Light-colored and reflective interior materials help reduce the amount of lighting needed for both task and general purposes. Dark surfaces absorb light and increase the need for artificial lighting, which creates waste heat, adds to the cooling load, and requires electricity for operation. Choosing lighter surfaces can contribute to the home's overall energy efficiency by reducing both lighting and cooling loads in a home.

Durability

Given the lifetime of a building and the number of times a finish must be refinished or replaced, the selection of a material with the longest life reduces both the amount of materials and labor required for refinishing or replacement and the waste created by these activities. Some more durable products may have a higher initial cost, but their total lifetime cost, including maintenance, repair, and replacement, may be lower than that of less expensive products. Solid wood floors can be refinished many times and last 100 years or longer before requiring replacement. Ceramic tile and stone, when installed properly over a durable substrate, can last several hundred years. The initial costs of long-life materials are higher than those for such products as carpet, sheet vinyl, or linoleum, but the latter require replacement every five to ten years—ultimately costing more and creating more waste than the more durable materials.

---

plaster a mixture of lime or gypsum with sand and water that hardens into a smooth solid; used to cover walls and ceilings.

millwork refers to the wood and composite products used to create doors, trim, cabinets, and similar finishes in homes.

ceramic tile a thin surfacing unit composed of various clays fired to hardness.

Recycled Content

Waste generated by consumers and manufacturers may be recycled and is accounted for separately in products. The total recycled content is the amount, usually expressed as a percentage, of recovered material that would otherwise be discarded in the waste stream but is instead introduced as raw materials in the production process. Post-consumer recycled content is the portion of a product that is reclaimed after consumer use. In general, post-consumer content is more variable because of the added costs of collection, sorting, and processing. Post-industrial recycled content is the portion of a product that contains manufacturing waste material that has been reclaimed. This is also referred to as pre-consumer recycled content. Interior finish materials that may contain recycled content include counters, tile, drywall, and carpet. Reclaimed wood beams can be milled into flooring and other trim.

Recyclability

Most interior finish materials are not easily recycled. Wood and drywall are easily recycled when unfinished but are difficult or impossible to recycle after paint or other finishes are applied. Most plastics are difficult or impossible to recycle. Products that can be recycled include ceramic tile, stone, metals, and masonry. Selecting products with the longest life helps reduce the overall environmental impact.

Embodied Energy

The amounts of energy required to extract raw materials, produce the finished product, and deliver it to the job site are important considerations. Keep in mind, however, that a material with a higher embodied energy that only requires replacement every 100 years may be a better choice than another that has lower embodied energy but requires replacement every 25 years.

Toxicity

Manufactured products often contain volatile organic compounds (VOCs) that can be harmful to humans, such as urea formaldehyde (UF), benzene, toluene, and other chemicals. VOCs from interior finish materials are released into the indoor air, creating unpleasant and sometimes toxic odors, with both short- and long-term health effects for workers and occupants. Selecting products with low and zero VOC content is important to creating a healthy home. Selecting products that are harvested, extracted, and processed with as few hazardous chemicals as possible reduces the environmental impact on workers in factories, individuals who may be affected by manufacturing plant emissions, and occupants of homes where they are installed. Material Safety Data Sheets (MSDS), available for most products that fit the U.S. Occupational Health and Safety Administration definition of hazardous, identify potentially dangerous content, exposure limits, safe handling instructions, proper clean-up protocol, and other factors to consider in selecting and using products (Figure 12.1).

Environmental Product Declarations

Environmental product declarations (EPDs) are an evolving process to provide consumers with comparative information on product performance based on specific product attributes. Independent consultants prepare these declarations for manufacturers to address their product's life cycle assessment, performance characteristics, toxicity, and other factors, all in compliance with rules created by the International Standards Organization (ISO). The ISO is a nonprofit and the world's largest developer and publisher of international standards. EPDs are common in Europe but are only beginning to make inroads in the American market.

Product Certifications

There are a number of certification organizations that evaluate products for their sustainable attributes. They range from rigorous and objective third-party evaluations to marginally useful certifications that are little more than marketing tools for manufacturers. First-party certification refers to a single company developing their own rules, analyzing their own performance, and

---

recycled content the amount of pre- and post-consumer recovered material introduced as a feedstock in a material production process, usually expressed as a percentage.

post-consumer recycled content the portion of a product that is reclaimed after consumer use.

post-industrial recycled content the portion of a product that contains manufacturing waste material that has been reclaimed; also called *pre-consumer recycled content*.

Material Safety Data Sheets (MSDS) are documentation available for most products that fit the Occupational Safety and Health Administration definition of hazardous; they identify potentially dangerous content, exposure limits, safe handling instructions, clean-up protocol, and other factors to consider in selecting and using products.

environmental product declarations (EPDs) the quantified environmental data for a product with pre-set categories of parameters based on the ISO 14040 series of standards, but not excluding additional environmental information.

first-party certification when a single company develops its own rules, analyzes its performance, and reports on its compliance.

Material Safety Data Sheet

Bamboo Flooring Products

Teragren LLC
12715 Miller Road N.E. Suite 301
Bainbridge Island, WA 98110

Emergency Phone: (206) 842-9477
Additional Information: (800) 929-6333
Email: ann@teragren.com

This Material Safety Data Sheet (MSDS) applies to Teragren Synergy strand bamboo flooring products.

1. Product Identification

Product	Manufacturing Location
Unfinished Wheat, Chestnut, and Java Flooring	USA Headquarters - Bainbridge Island, WA.
Prefinished Wheat, Chestnut, and Java Flooring	USA Headquarters – Bainbridge Island, WA.

Synonyms: Bamboo Flooring System

2. Hazardous Ingredients/Identity Information

Name	CAS#	Percent	Agency	Exposure Limits	Comments
Bamboo [1]	None	94-95	OSHA OSHA ACGIH ACGIH	PEL-TWA 15 mg/m3 PEL-TWA 5 mg/m3 TLV-TWA 3 mg/m3 TLV-STEL 10 mg/m3	Total dust Respirable dust fraction Respirable dust fraction Inhalable particles
Phenol-formaldehyde resin solids[2]	None	4-5	OSHA OSHA ACGIH	PEL-TWA 0.75 ppm PEL-STEL 2 ppm TLV- Ceiling 0.3 ppm	Free gaseous formaldehyde Free gaseous formaldehyde Free gaseous formaldehyde
UV Finish[3] Polymerized polyurethane	None	0-1	OSHA ACGIH	PEL-TWA None TLV-TWA None	None None

1 Bamboo is a member of the grass family which has distinct anatomical differences from that of wood. Therefore, bamboo would be regulated as an organic dust in a category known as "Particulates Not Otherwise Regulated" (PNOR), or Nuisance Dust by OSHA. The ACGIH classifies dust or particulate in this category as "Particulates Not Otherwise Specified".
2 Contains less than 0.02% free formaldehyde.
3. For pre-finished flooring

3. Hazard Identification

Appearance and Odor: A matrix of natural/blonde or caramel colored interlocking bamboo fibers bonded with phenol-formaldehyde resin having a slightly aromatic odor.

Figure 12.1 The first page of an MSDS for bamboo flooring. This particular product is 94% to 95% bamboo, and the rest resins. Note that UF is not among the resins.

reporting on their compliance. Third-party certification, on the other hand, involves a review and confirmation by a nonaffiliated, outside organization that a product meets certain standards. Determining which certifications are useful in selecting products is difficult and is probably will not become any easier as more programs enter the market.

The most robust product certifications come from third-party organizations that only evaluate products and are fully independent from individual manufacturers (or industry organizations supported by manufacturers). The most robust third-party organizations are

third-party certification a review and confirmation by a nonaffiliated, outside organization that a product meets certain standards.

FROM EXPERIENCE

Life Cycle Assessment

Cindy Ojczyk, LEED-AP, Vice President, Verified Green, Inc. Three simple words—life cycle assessment (LCA)—represent a complicated process that seeks to make the total environmental impact of any product transparent to all users. It is the measurement, in absolute numbers, of all the inputs needed and outputs created in the extraction of raw materials, manufacturing, transportation, installation, use, maintenance, and disposal of a "product." The "product" in reference can be a singular item, such as a galvanized nail, or it can be a complex system, such as a building that is composed of many subproducts, including the galvanized nail.

Manufacturers use the LCA process to meet regulatory compliance or to inform product design, manufacturing, packaging, and disposal. They also use LCA as a competitive marketing tool. Architects, engineers, and building owners use LCA information to inform building design and material selection.

The International Organization for Standardization (ISO) has created a global framework and principles for conducting an LCA that can be found in the ISO 14000 series of standards. The broad range of products and building types and their varying costs make an LCA complicated to perform, so the actual methods for collecting, analyzing, and reporting data are left to the individuals charged with the task. The ISO framework provides LCA guidance through four phases:

Phase 1: Define the goal and scope of the LCA

- Define the purpose of the study and which phases of the product life cycle to study
- Identify the assumptions that must be made due to gaps in availability of information
- Determine how results will be reported

Phase 2: Perform a life cycle inventory

- Define the flow diagram of all processes in the life cycle
- Calculate the amount of all energy, water, and raw materials used
- Calculate all air emissions, water emissions, solid waste, and other releases

Phase 3: Perform a life cycle impact assessment

- Determine the impact categories against which data from the life cycle impact assessment will be evaluated. Although European researchers use up to 12 impact categories, in the United States there are 11 categories that are considered including global warming potential, smog, over fertilization (eutrophication) of water, acidification of air, and human health impact
- As ISO 14000 provides a framework rather than specific requirements, the decision as to which of the impact categories to include falls to the individual or group performing the assessment. The assessment may include as few as one or up to 12 impact categories.
- The ISO standard has a computer modeling methodology within their framework that is used to determine the impact of the measured emissions, the intended goal being that if all assessments follow the same protocols, then the results will be able to be meaningfully compared between products.

Phase 4: Report and interpret results

- Include limitations of the study so transparency of the process and results is readily evident
- Report results following the guidelines of ISO

The green building industry is moving toward a time at which LCA will inform every design and construction decision, but the limited availability of data makes a complete building analysis impossible. However, LCA information can inform parts of the design and material selection process through the use of LCA computer-based tools that are currently available.

Architects and builders would be wise to begin familiarizing themselves with LCA now since it will become an important component of design and building in the future. The U.S. Environmental Protection Agency (EPA) has created a reference called *Life Cycle Assessment: Principles and Practice*, which further expands upon the information provided here.

Resources

U.S. EPA: http://www.epa.gov/NRMRL/lcaccess/lca101.html
ISO: http://www.iso.org/iso/iso_14000_essentials
National Institute of Standards and Technology: http://www.bfrl.nist.gov/oae/software/bees
Athena Sustainable Materials Institute: http://www.athenasmi.org/tools/impactEstimator/index.html

Cindy Ojczyk is actively engaged in the residential green building industry as both an educator and a practitioner. She works with Verified Green, a consulting and training business, and is the owner of Simply Green Design, an interior design firm. She was a key author of MN GreenStar and the consultant to the first silver-level LEED home in Minnesota.

approved by the American National Standards Institute, verifying their objectivity. Second-party certifications are often prepared by separate entities that are supported by industry groups or individual companies. First-party evaluations are typically manufacturer's statements of content, such as those normally found in an MSDS.

Currently, most product certification programs evaluate only one or a few individual attributes, as opposed to a full LCA. Most certification programs still only review a small selection of the available market, and most of the products listed are for commercial rather than residential use. In the future, we will likely see LCA and EPD evaluations of more products, both residential and commercial, including structural and exterior finish materials.

Even though the scope of certification information is currently limited, product reports from third-party organizations can be useful in evaluating products for a green home. See Table 12.1 for a list of third-party organizations that provide product certification.

No single certification should be expected to provide a comprehensive evaluation of any product. The project team should evaluate products with all available criteria, using the information collected to make as informed a decision as possible.

Materials and Methods

Selecting interior materials is typically an aesthetic decision, controlled by the designer or building owner. When selecting products, always consider their durability, recycled content and recyclability, sustainability, embodied energy, and toxicity. Most finishes are available with or without appropriate green attributes; the project team must take responsibility for reviewing products and selecting the best ones for their project on the basis of these criteria.

Floors

Applied floor finishes can be made of solid or engineered wood, bamboo, cork, vinyl, linoleum, carpet, tile, brick, stone, or concrete. Each material, finish, and installation method has specific green attributes that affect the project.

Passive solar applications (see Chapter 3) use thermal mass to store heat. Projects that require thermal mass may want to specify tile, masonry, or concrete flooring in the design phase for this purpose.

Wood Floors

Wood floors can be classified by material, finish, and installation type. They are either solid or engineered which are often produced with a finish layer applied over multiple layers of wood or composite material. Engineered flooring is made up of multiple layers of wood that are glued together as one board. Although solid wood has lower embodied energy due to the limited about of processing, engineered products can make more efficient use of raw materials.

Selecting wood that is certified as sustainably harvested, locally produced, or reclaimed are appropriate choices for green homes. Avoid tropical woods unless they are harvested sustainably (see Chapter 6 for a further discussion on sustainable forestry practices). Engineered products should use adhesives with no added UF.

Wood floors, whether solid or engineered, can be either pre-finished or finished on-site after installation. Most engineered floors are delivered pre-finished. Site

Table 12.1 Product Certifications

Wood Products	Emissions	Multi-Attribute Programs
Forest Stewardship Council (FSC): http://www.fsc.org	Greenguard Environmental Institute: http://www.greenguard.org	Scientific Certification Systems Sustainable Choice (carpet, others in future): http://www.scscertified.com
Sustainable Forestry Initiative (SFI): http://www.sfiprogram.org	Scientific Certification Systems Floor Score (non-textile flooring): http://www.scscertified.com	SMaRT Consensus Sustainable Product Standards: http://www.sustainableproducts.com Green Seal: www.greenseal.org
American Tree Farm System (ATFS): http://www.treefarmsystem.org	Scientific Certification Systems Indoor Advantage (furniture): http://www.scscertified.com	EcoLogo/Environmental Choice: http://www.ecologo.org
Canadian Standards Association Sustainable Forest Management System (CSA): http://www.csa-international.org		Pharos Project: http://http://www.pharosproject.net/

---

second-party certification when an industry or trade association fashions its own code of conduct and implements reporting mechanisms.

engineered flooring a product that is made up of multiple layers of wood and glued together as one board.

finishing involves sanding and applying clear coatings or penetrating sealers in place. Solid wood floors can be sanded and refinished multiple times; some engineered products can be refinished once or twice, but others are not suitable for refinishing and will require replacement when damaged or worn. Site-applied finishes with zero or low VOC content are the healthiest for occupants (see the section on "Paint and Coatings" later in this chapter for details on VOC content in site-applied finishes), but pre-finished floors help reduce the amount of dust contamination and VOC exposure on the job site.

Solid wood floors are installed directly to the floor structure with nails, glue, or a combination of the two. Engineered floors can be fastened to the subfloor or installed as floating floors in which each board is attached to the adjacent board but only attached to the structure at the edges of the room. Floating floors allow for easier replacement of damaged boards and often have a thin layer of padding beneath them, providing a softer finish underfoot (Figure 12.2). Floating floors are often more suitable for concrete slab installations than nailed or glue-down installations. Follow manufacturers' and trade associations' recommended instructions regarding subfloor moisture content, installation methods, and suitability for a particular application for the most durable finish.

All wood floors are susceptible to water damage, so they should not be used in high-moisture areas, such as basements that may flood, bathrooms, and laundry rooms. Also, the finish on wood flooring in entry areas and kitchens may wear faster, requiring more frequent refinishing.

Bamboo and Cork Floors

Bamboo flooring is one of the most common "green" flooring products available. Although most bamboo is imported from China, its durability and short growth cycle make it a good consideration for a green home. Bamboo is typically assembled into pre-finished boards from thin strips using adhesives (Figure 12.3); look for products with low or zero UF content in the adhesives. Bamboo flooring can be installed with mechanical fasteners, glued down, or installed as a floating floor. Bamboo is generally delivered with a factory-applied finish. Look for products that can be refinished several times for maximum life.

Cork flooring is made from the outer bark of the cork oak tree, which is harvested without killing the trees; the trees grow new bark about every 10 years (Figure 12.4). Cork is durable, provides a cushion underfoot, and has antibacterial properties, but it is often processed with binders and adhesives that contain UF. Flooring is available in tiles that are glued directly to the subfloor or as veneers on engineered panels that are glued down or installed as a floating floor (Figure 12.5). Look for products that are finished with polyurethane or penetrating oils that can be refinished for maximum life. Some products may be made with a polyvinyl chloride (PVC) wear layer on top of the cork, which introduces less desirable plastics into the home and eliminates the possibility of refinishing the floor.

Linoleum and Vinyl Floors

Linoleum, invented in the mid 1800s, was a common flooring material for kitchens, baths, and hallways until it was largely replaced by vinyl beginning in the mid 1900s. **Linoleum**, made of linseed oil, cork and wood particles, and a fabric backing, is a very durable flooring material that is experiencing a recent resurgence, particularly in green building projects (Figure 12.6).

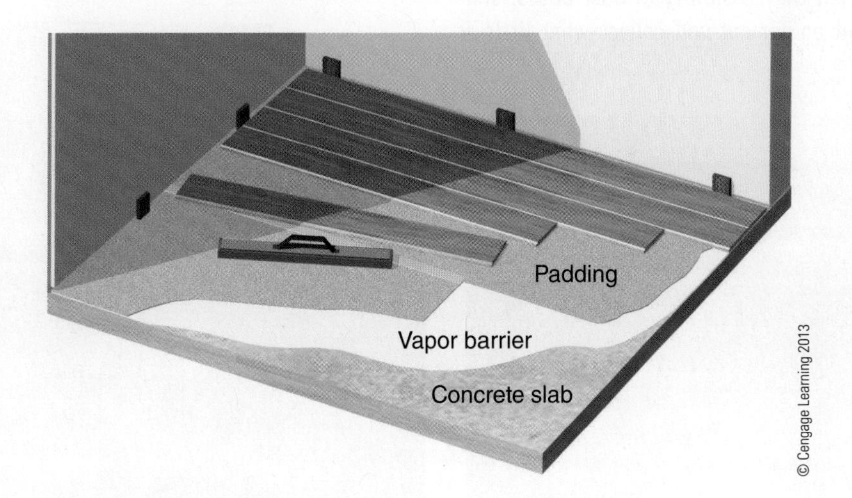

Figure 12.2 Floating floor installation.

---

linoleum durable flooring materials made of linseed oil and cork and wood particles.

FROM EXPERIENCE

Greening the Restoration of Wood Floors

Michael Purser, President, The Rosebud Co. We always wondered how many times wood floors could be sanded before they reached a dead end. Since most of my wood floor refinishing work takes place on old wood floors, I had to deal with this issue many years ago. The challenge was the same: get the old finish off and make the floors look great, but do it without any loss of wood. This meant traditional sanding methods (the powerful drum/belt sanders and edgers) were out. These machines are designed with one thing in mind—speed! And with that speed comes excessive loss of wood.

Courtesy of Michael Purser

Michael Purser is a second-generation wood-flooring contractor. He started The Rosebud Co. in 1973 in Atlanta, Georgia. He has written extensively on wood floor products, restoration, and the greening of a dirty building trade

The most obvious method was to remove the old finish by chemical means, but the most popular products were too toxic and dangerous to use. One option was to use chemical cleaners that contained *N*-methyl-2-pyrrolidone. Long used in hand cleaners, this product proved to be effective at removing many of the older clear coatings (and some contemporary finishes) in an environmentally responsible manner. This became my cornerstone as I developed Passive Refinishing®. It has opened the door for me on some high-profile restoration projects and extended the life of these old floors indefinitely.

Passive Refinishing® produces little if any loss of wood, and since the surface is wet during much of the process, there is no dust. When the possibility of dust arises, simple dust containment equipment will collect what little is created. In terms of environmental issues, this option does not foul the air, create hazardous by-products, or send clouds of dust wafting through the home.

When chemical removal is not an option, other, less aggressive sanding equipment will remove old coatings and not result in substantial wood loss. This secondary sanding equipment lacks the speed of the powerful drum sanders and uses copious amounts of sandpaper but leaves most of the wood intact. Slower to use than traditional equipment, these sanders are not as popular with most flooring contractors who see "getting in and out quickly" as the greater priority to removing less wood. Much of this equipment has evolved over the last 15 years and was initially regarded as fine sanding equipment, but its role is gradually changing.

Courtesy of Michael Purser

Wood floor before restoration.

Courtesy of Michael Purser

Wood floor after restoration.

(Continued)

Many thought that old wood floors present us with challenges that were going to be expensive and difficult to resolve, but that does not have to be the case. The options of Passive Refinishing® and nontraditional sanding approaches give us environmentally safe, economical, and reasonable options to tearing out an old floor. The wood-flooring trade has come a long way. Once considered the dirtiest trade imaginable and an air quality nightmare, the industry has done a complete turnaround. VOC-compliant finishes, effective dust containment, and new equipment that meets the demands of old floors are rewriting our future.

© Teragren LLC

Figure 12.3a Bamboo flooring with vertical grain.

© Teragren LLC

Figure 12.3b Bamboo flooring with horizontal grain.

Courtesy of iStock Photo

Figure 12.4 Cork is a sustainable building material because it is a renewable resource. The cork tree is not harmed during harvesting.

Courtesy of USFloors Inc.

Figure 12.5 Cork flooring.

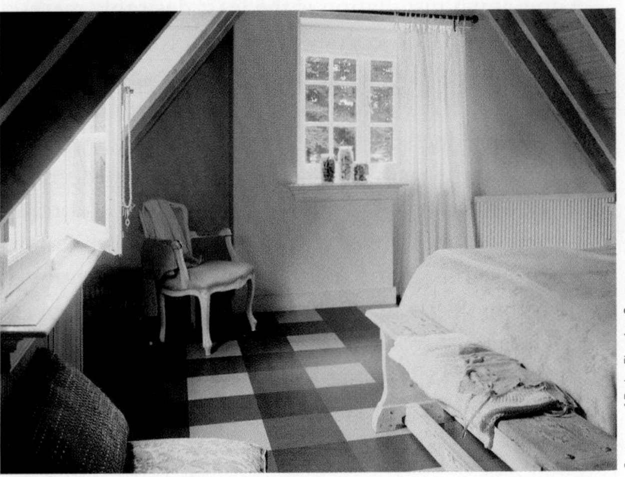
Courtesy of Forbo Flooring Systems

Figure 12.6 Linoleum flooring.

Manufactured with fewer toxic materials than vinyl, linoleum does not off-gas any toxic fumes, although it does have an odor that dissipates after installation. Linoleum is installed directly to the subfloor with an adhesive, and it should have a protective coating applied during installation and reapplied regularly to maintain the finish.

Vinyl flooring became popular due to its wide range of patterns and low maintenance. Like all PVC products, however, it uses toxic nonrenewable chemicals in its manufacture, is difficult to recycle, off-gasses VOCs, and releases hazardous fumes when burned. Flexible vinyl also contains phthalates, chemicals known to cause health problems, particularly in young children. For these reasons, vinyl flooring should be avoided in green homes.

Carpet

In general, wall-to-wall carpet is not the best choice for a green home. Most carpets, padding, and adhesives off-gas many harmful chemicals. Carpet collects dust, dirt, and chemicals tracked in from outdoors, and it absorbs moisture, interior odors, chemicals, and smoke. Even the most benign carpet installations must be cleaned frequently to maintain healthy indoor air. Carpet cleaning uses energy, water, detergents, and solvents. Hard-surface floors need cleaning less frequently and require fewer resources to do so.

When carpet must be part of a project, consider using area rugs over hard-surface floors. Rugs can be removed and cleaned regularly and do not require pads, adhesives, or physical fasteners for installation.

For wall-to-wall installations, select carpets made of natural materials, such as wool, cotton, or sisal, a fiber created from leaves of the agave plant. Less-expensive options include carpet made from recycled plastic, which is suitable for low- to moderate-traffic areas that can be recycled at the end of its life. Installation options include gluing directly to the subfloor, laying loose tiles, and mechanical fastening.

Glue-down installation is the least desirable method because the carpet becomes difficult to remove and recycle, and the adhesives used may have high VOC content. Carpet tiles are easily replaced when damaged, and many products are available that have high recycled content and are easily recycled (Figure 12.7). Mechanically fastened carpet is attached at the edges of a room and stretched tight over a pad. Stretched installations use the least amount of adhesive and are easily removed for recycling. Pads made of jute or other natural fibers are preferable to foam, which often contains brominated flame retardants. **Brominated flame retardants** belong to a group of chemicals that consist of organic compounds containing bromine and are suspected of causing long-term health problems. Adhesives used in seaming carpet together should have low or zero VOC content.

Figure 12.7 Carpet tiles are available in a variety of colors and patterns that may be easily mixed and matched.

The **Carpet and Rug Institute (CRI)** is a nonprofit industry association that created the second-party certification programs, Green Label and Green Label Plus. These programs provide guidelines for VOC content and off-gassing of carpet and padding materials.

Tile, Stone, and Brick

Hard-surface flooring, such as tile and masonry, is durable and attractive, and can provide thermal mass for passive solar designs. Sustainable material choices include tile and brick with recycled content and local production. Stone can be reclaimed or locally extracted. Tile and brick have high embodied energy due to the heat used in production; however, their durability and low maintenance make them a good choice for a green home.

Most tile is made either with a durable glaze finish or from solid porcelain, both of which require minimal maintenance. Brick, stone, and unglazed porous tiles are susceptible to staining and should be sealed on installation and at regular intervals afterward.

Tile, brick, and stone can be installed over concrete or wood floors. Because installations over concrete are susceptible to telegraphing cracks in the slab through the finish floor, consider using a **decoupling membrane**,

---

brominated flame retardants a group of chemicals that inhibit the spread of fire and consist of organic compounds containing bromine.

Carpet and Rug Institute (CRI) a nonprofit industry association that created second-party certification programs, Green Label and Green Label Plus, which provide guidelines for VOC content and off-gassing of carpet and padding materials.

decoupling membrane a flexible plastic sheet that is placed between ceramic tile and the subfloor to provide strength and crack resistance.

a flexible plastic sheet that is placed between the tile and the concrete with an adhesive. This membrane creates a kind of shock absorber that reduces the potential for cracks in the finish floor (Figure 12.8). Tile and masonry floors on wood structures can be installed over a thick layer of mortar, referred to as a mud base, over a cement or gypsum-based 1/4" or 1/2" backer board, or over a decoupling membrane (Figure 12.9). Select the most durable installation method based on manufacturer's recommendations for the finish material and the structure. Although these floors have high embodied energy, their long service life helps to reduce the overall negative environmental impacts.

Showers and Tubs Showers and tubs are typically finished with ceramic tile or stone walls with matching material on shower floors. Shower and bath wall finishes must be installed either on a masonry or gypsum substrate that will withstand moisture, or over a waterproof membrane, such as Schluter®-KERDI from Schluter Systems (Figure 12.12). Tiling directly on moisture-resistant drywall or other moisture-sensitive materials will lead to premature failure and water damage to the underlying structure. Shower floors must have a fully waterproof shower pan membrane integrated with a drain to keep the structure dry. Traditionally made of lead, most shower pans are now made from flexible plastic or rubber sheets that are similar to low-slope membrane roofing materials. Careful installation and leak testing are critical to building an effective shower. Options to tile and stone showers and tub surrounds include prefabricated fiberglass and acrylic units that are available in one piece or as multi-part systems.

Figure 12.9 Tile installation in a bathroom over a framed floor. The Schluter®-DITRA product is used as a decoupling membrane. The decoupling membrane anchors the tile to the wood subfloor, eliminating the need for cement board or mortar underlayment. The overall result reduces weight, materials, and likelihood of cracking.

Figure 12.8 Tile installation over a concrete slab. The Schluter®-DITRA product is used as a decoupling membrane. The decoupling membrane anchors the tile to the concrete slab and eliminates the telegraphing of cracks from concrete to tile.

Concrete

Using a structural concrete slab as the finished floor eliminates the work and materials required for an applied surface. Concrete slabs can be finished with a wide range of options, including stains, acid etching, polishing, and clear sealers (Figure 12.13). Finishes can be applied at the time the concrete is poured, which requires careful protection during construction, or at the end of the project. Selecting finish materials with low or zero VOC content will ensure better IEQ.

Walls and Ceilings

Wall and ceiling finishes include drywall, plasters, paneling, wallcovering, tile, brick, and stone. The choice of material is affected by the wall structure, how the space is used, aesthetic requirements, and green attributes of each specific product.

Did You Know?

Occupant Behavior and Indoor Environmental Quality

One of the best ways to maintain high indoor environmental quality (IEQ) in a house is to not wear outdoor shoes inside the home. Shoes worn outdoors can track in significant amounts of dirt, dust, mud, chemicals, animal excrement, and other pollutants that can be transferred to the floors—particularly carpet. Frequent cleaning and vacuuming can minimize the impact of these pollutants, but keeping them off the floors entirely is the best strategy. Providing a comfortable place to remove and store shoes at each entry encourages keeping outdoor shoes outdoors (Figure 12.10). Another alternative, more common in commercial buildings, is to use a **walk-off mat**, an abrasive floor mat installed over a shallow receptacle that catches debris as it is scraped off shoes (Figure 12.11). Walk-off mats are typically 3' to 4' long and are installed at each entry door to help reduce the amount of pollutants that are tracked into a home.

Figure 12.10 Mudroom benches with built-in shoe storage encourage removing shoes at the entrance.

Figure 12.11 Recessed walk-off mats include space to collect dirt and debris.

Drywall

By far the most common interior wall finish in homes, drywall is inexpensive and reasonably durable while serving as an excellent substrate for paint finishes. Most manufacturers use recycled content in the paper facing. Synthetic gypsum, a waste product of coal plant flue scrubbers, may be used for some or all of the panel core; however, concerns have been raised about possible health effects from mercury and other toxins that it may contain.

Drywall's biggest enemy is moisture. When wet, the paper facing on drywall can develop mold quickly, requiring the costly and disruptive process of removal and replacement. Highly mold-resistant paper and fiberglass facings should be used in any areas subject to moisture damage, such as bathroom wet areas and basements. Moisture-resistant drywall, commonly referred to as greenboard, provides only minimal resistance to moisture and should not be used in wet areas or as a substitute for non-paper-faced drywall or tile backer board (Figure 12.14).

One product currently under development is drywall with phase-change beads embedded in them that increase the thermal mass of the material. Phase-change beads absorb and release energy as temperature changes, helping to manage room temperatures in passive solar projects.

Drywall is installed with screws or nails, and adhesives. Drywall adhesive, when installed in continuous beads, provides an air seal between the boards and framing, reducing air infiltration. Adhesives with zero or low VOC content should be selected to ensure highest indoor air quality. See Chapter 4 for more information about how to reduce air leakage through the airtight drywall approach (ADA).

Unpainted drywall scraps can serve as beneficial soil amendments for plants. Gypsum is hydrated calcium sulfate, and is often marketed as a soil conditioner for improving soil structure (tilth). Both calcium and sulfur are essential plant nutrients. The need for these nutrients

---

walk-off mat an abrasive floor mat installed over a shallow receptacle that catches debris as it is scraped off shoes.

Figure 12.12 Here the Schluter®-KERDI product is installed directly to paperless drywall. This product provides a waterproof underlayment for wall tile and eliminates the need for cement-based backer board.

Figure 12.13 Structural concrete slabs can be stained, sealed, and polished to produce an attractive finished floor.

depends on the crop, the soil type, the existing soil supply, and the contribution from other sources. Grinding gypsum waste and disposing of it on-site can improve soil conditions and reduces waste going to landfills.

Plaster

Plaster, once a standard interior wall finish, has been largely replaced by drywall in residential construction. Gypsum veneer plaster is used in some areas, and clay plasters have emerged as a sustainable interior finish material. Gypsum veneer plaster is a one- or two-coat finish that creates a very hard, durable surface for walls and ceilings (Figure 12.15). Plaster can be pigmented before application or painted after it is applied.

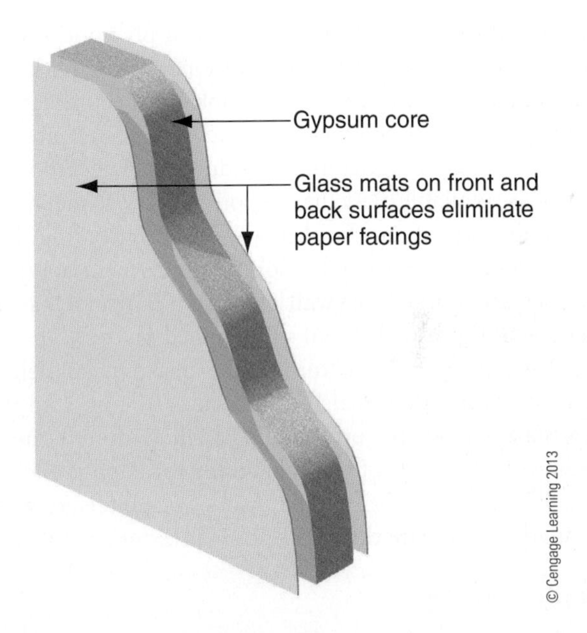

Figure 12.14 Non-paper-faced drywall.

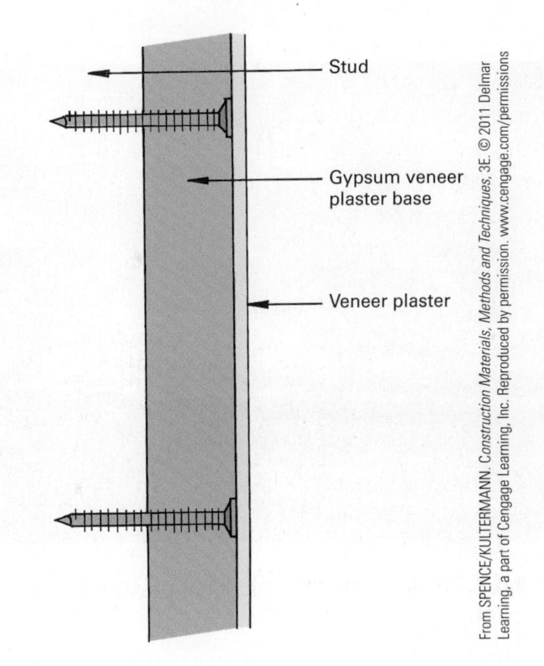

Figure 12.15 Veneer plaster is a single-coat plaster applied over a gypsum veneer plaster base.

Clay plaster is made with natural clays, aggregates (such as recycled marble dust), and pigments. Applied over gypsum or masonry, clay plaster is intended to be a final finish and requires no painting (Figure 12.16). Both gypsum and clay plasters can provide some level of humidity control—absorbing moisture when humid, releasing it when dry. This characteristic may reduce cooling loads in mixed and humid climates by reducing the amount of dehumidification required.

Paneling

Walls and ceilings can be finished with plywood sheets with trim at the seams or, for additional detail, with flat and tongue-and-groove boards. Look for sheet materials with no added UF in glues and binders, such as exterior grade plywood or agricultural products like wheatboard, which is available with hardwood veneers. Paneling is generally stained and finished with a clear coating, but painting may be preferred in some design schemes. When finishes are painted, drywall or plaster can serve as the main wall finish with detail trim applied on top.

When using the wall finish as the air barrier, paneling must be thoroughly sealed at all joints to complete the thermal envelope (Figure 12.17). Paneling often requires significant amounts of wood blocking for nailing. Blocking should be installed so that it does not interfere with insulation or create thermal bridges in exterior walls.

Wallcovering

Wallcoverings, printed or textured sheets that are glued to walls, were originally made from paper or natural fibers; however, most are now made from vinyl or PVC. Vinyl wallcoverings should be avoided for two reasons. First, they act as a vapor barrier that traps moisture on the wall surface behind the vinyl, particularly in buildings with air conditioning (Figure 12.18). Second, like all PVC products, they off-gas VOCs, and they are not easily recycled. Look for wallcoverings made of paper, grasses, and other natural materials that are not vapor barriers and have low VOC content.

Courtesy American Clay Enterprises

Figure 12.16 A painter installs clay plaster over gypsum board.

wallcovering a covering on a wall, such as vinyl or paper wallpaper.

Tile, Stone, and Brick

Tile and masonry wall finishes are used primarily in bathroom and kitchen areas for durability, although they are also seen in entries, wine cellars, and other rooms for decorative purposes. Brick is available in a thin veneer style, reducing both weight and bulk while providing the look of solid brick. In areas that are not in direct contact with water, tile and masonry finishes can be installed directly to drywall.

Millwork

Solid wood, the original millwork material, is becoming more scarce and costly. Consequently, its use is generally limited to areas where it will have a natural finish. Wood veneer applied to cabinet panels, doors, and molded trim is a more sustainable use of hardwoods than solid material (Figure 12.19). Millwork from reclaimed material, locally harvested wood, and certified sustainably harvested wood are good choices for a green home. Many millwork products are available in composite products made of wood and agricultural waste with binders. Look for products made with no added UF for best indoor environmental quality. Unpainted solid wood products can be ground up and recycled as mulch. Composites offer a more efficient use of resources but are generally not recyclable, so waste must be discarded. Regardless of recyclability, waste can be reduced by designing to standard modules, carefully ordering to avoid mistakes, and installing products correctly.

Doors

Doors can be made of solid wood, finger-jointed wood, composition cores with wood veneer, molded medium-density fiberboard (MDF) or other composites. Veneered or molded-panel doors are available in solid-core or hollow-core styles. The interiors of solid-core doors are filled completely with wood or composition board, providing a solid, heavy product. Hollow-core doors are available with flush or molded panels assembled to a wood frame, with cardboard or similar material used to keep the panels separated while significantly reducing the amount and weight of material required for construction. Hollow-core doors use less material and are lighter to ship than solid-core designs. Styles include flush, raised panel, flat panel, and glass panel.

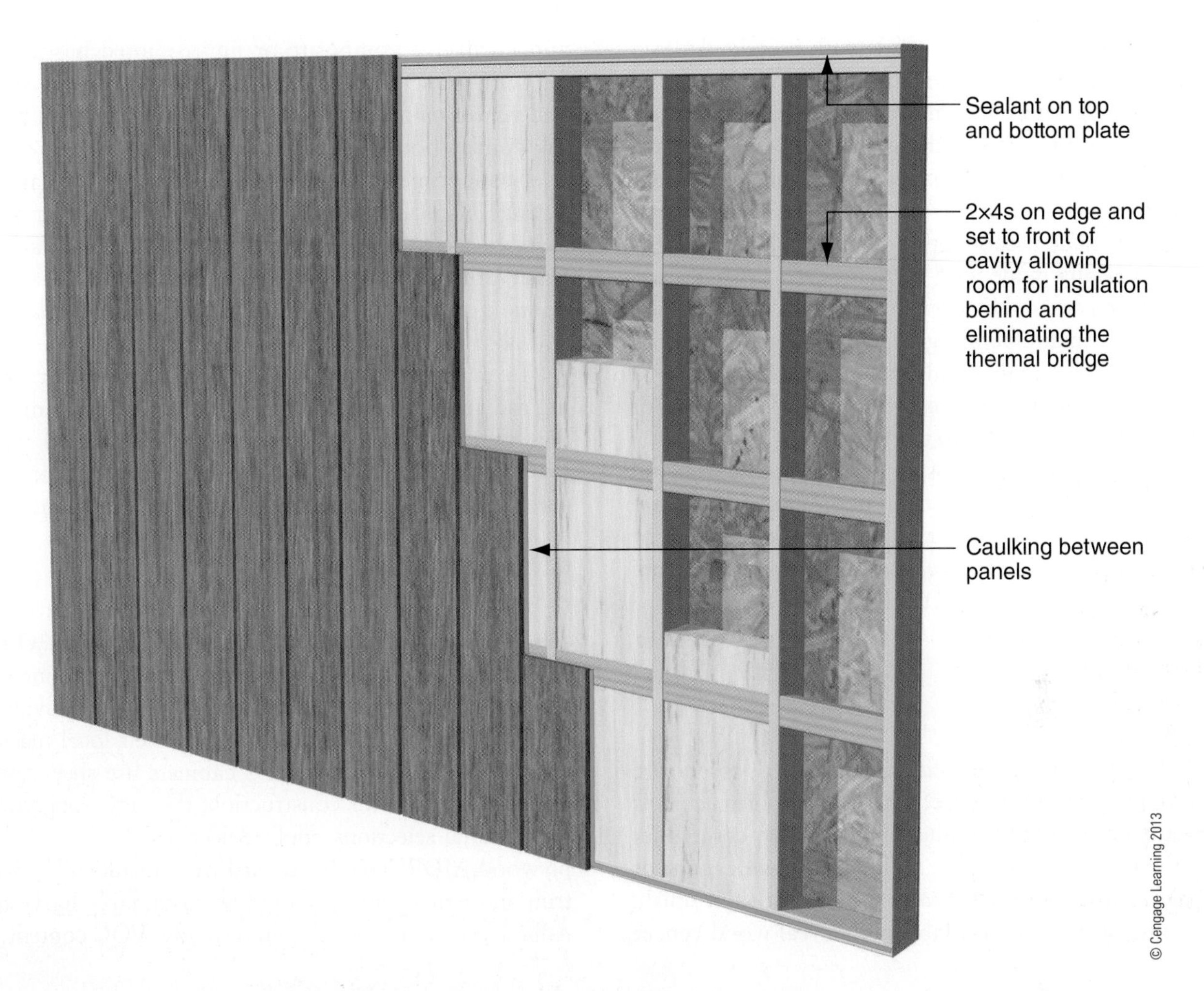

Figure 12.17 Air sealing wall paneling so that it may be used as an air barrier.

Figure 12.18 Vinyl wallpaper is a vapor barrier that may trap moisture and encourage mold growth. To make matters worse, the wallpaper adhesive is generally an excellent food source for mold. In this home, the vinyl wallpaper had to be removed due to extensive mold growth.

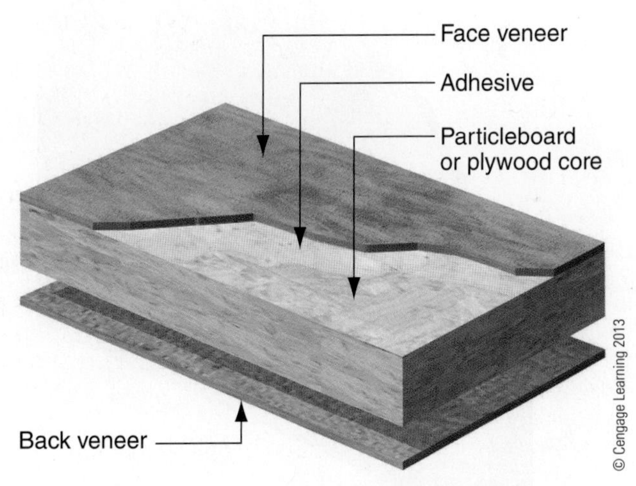

Figure 12.19 When using wood veneer products, always specify UF-free products. Particleboard often contains UF binders.

Look for doors that are made with no added UF in glues or binders. Many standard composite doors will have high UF content, although some manufacturers now produce products without the resin. Wheatboard and other agricultural products are similar to wood-based composites that are made with farm waste and are usually made without any added UF. UF-free materials are typically made with urethane resins that do not off-gas when cured; however, these alternative resins are very toxic in manufacturing, increasing risks to factory worker health. Any materials that contain UF should be coated with a low-toxic finish to reduce the amount of formaldehyde released into the home.

Consider using composite or finger-jointed wood instead of solid wood for trim. Stain-grade doors are available with wood veneer applied over composite or finger-jointed wood (Figure 12.20). Doors made from solid wood and veneers, particularly those from tropical regions, should be certified as coming from sustainably harvested forests.

Trim

Like doors, trim can be solid, finger-jointed, composite material, or plastic (Figure 12.21). Composite trim is typically MDF that contains UF; however, companies like Sierra Pine offer UF-free moldings in a wide range of profiles. Most composite trim is made for a paint finish, but some profiles are available with a real wood veneer applied over a composite or finger-jointed base, offering stain-grade trim while using less solid wood. Wood and veneer trims, particularly those created from tropical woods, should be certified as sustainably harvested.

Plastic trim, both virgin and recycled, is available in a range of profiles; some even has a wood grain finish that can accept stain. Flexible plastic trim is available to finish around curved walls and arched-top windows and doors. Trim is typically installed with nails or screws, and occasionally with adhesives.

To minimize environmental impact, consider limiting the amount of decorative trim, and select trim profiles that produce the least waste. Engineered wood products with no added UF are appropriate choices for green homes, as are adhesives with low VOC content.

Cabinets

Cabinets are available as stock, semi-custom, and full custom products from large and small manufacturers. Often, cabinets are shipped from a distant state or country to the job site. Independent cabinet shops and individuals typically produce custom cabinets in their local market, either in a shop or on-site. As cabinets use sheet goods and running trim for construction, the same suggestions for material selections apply. Select panel materials like plywood, MDF, and wheatboard with no added UF. Solid trim and panel veneers should be sustainably harvested. Adhesives and finishes should have low VOC content.

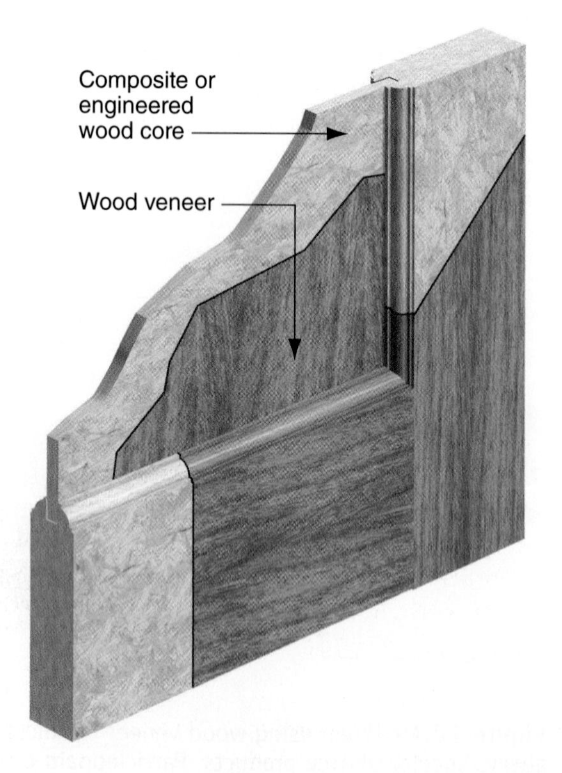

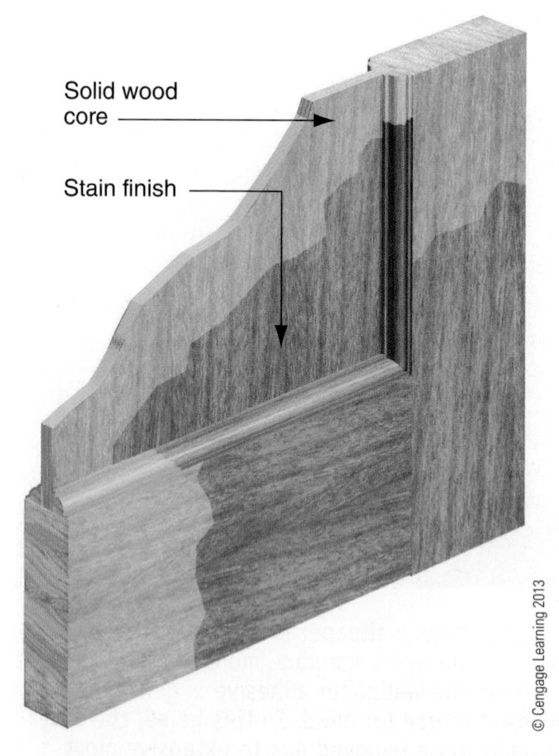

Figure 12.20 Doors are available from solid wood (right) and composite or engineered wood (left).

Figure 12.21 Visually, there is no difference between trim made from composite materials and that made from solid wood.

Many cabinets are made with particleboard panels, which often have a high UF content. Consider selecting products with low UF content panels, or seal all faces of cabinet panels with low-emission sealers before assembly. Most cabinets are finished at the factory, many with highly toxic finishes. Selecting products with low-VOC finishes reduces their impact on the air quality on site. Site-finished cabinets should use zero- or low-VOC paints and coatings.

Purchasing locally manufactured cabinets reduces the amount of transportation energy required. Long-haul deliveries of cabinets add to both the cost and environmental impact of the project.

Shelving

Closet shelving has evolved from a single shelf and rod installed in each bedroom into sophisticated systems that have spawned an entire industry. Closet storage specialists provide fully outfitted storage systems that range from an assembly of stock products to fully custom installations with drawers, multi-tiered clothes rods, and shoe racks. These companies also outfit wine cellars, basement and garage storage areas, offices, and entertainment rooms.

Closet systems are typically made from vinyl-coated wire shelving, plastic-covered particleboard, wood-veneered MDF and plywood, and solid wood trim. Shelving systems made of solid wood, plywood, and MDF can be site-built by trim carpenters.

Regardless of the type of system chosen, recommendations for materials and finishes are the same as for all millwork: avoid vinyl and plastic finishes, select panel products with no added UF, and use low-VOC finishes and sustainably harvested wood. Closet systems are often installed on brackets, making them easily removable for reuse following renovations or demolition.

Counters

Counters provide one of the best places to incorporate sustainable materials in a home. Traditionally, counters were made from natural stone and wood. Solid surface counters are manufactured from acrylic polymers, stone composites, recycled paper and wood pulp, and recycled glass. Stainless steel and ceramic tile provide durable, attractive finishes for counters.

Stone

Solid stone counters go through a minimal amount of processing (limited to cutting and polishing) before installation. Stone is very durable; however, it is also porous and should be sealed periodically. Stone counters are most often made of granite, although marble, slate, and soapstone are also used. The most important issue in stone counters is where it is quarried. Shipping stone requires significant amounts of transportation energy, adding to the overall environmental impact of the material. Selecting stone that is extracted locally reduces that impact. There has been some concern that granite counters may emit radon into a house; however, there is limited evidence that this creates a health hazard in a well-ventilated home.

Solid Surface

Dupont™ Corian®, one of the first manufactured solid-surface countertop materials, is a cast acrylic polymer that is very durable and nonporous. Available in dozens of colors and patterns, Corian can be worked and formed with carpentry tools. Numerous manufacturers produce solid-surface products that are similar to Corian,

---

solid surface a manufactured product typically used for countertops that emulates stone, created by combining natural minerals with resin and additives.

Figure 12.22 PaperStone® is made from 100% post-consumer recycled paper or cardboard with non-petroleum-based phenolic resins. The final product is VOC free and emits no radon gases.

Figure 12.23 Countertop of recycled glass.

providing a wide range of styles and colors. These types of acrylic counters have very low VOC emissions, are very durable, and are not porous, but they are not recyclable, and they use many chemicals in their manufacture. Acrylic solid-surface counters are usually installed on top of a wood composite substrate for support. A substrate with no added UF should be used, and low-VOC adhesives and sealants specified for installation.

Newer solid-surface materials include stone composites that are comprised of over 95% quartz crystals cast with acrylic resins, colors, and binders. These counters are nonporous, low emitting, and extremely durable. Stone composites are cut from solid slabs and polished with abrasive finishing equipment.

Recycled content solid surface counters include such products as PaperStone®, which is made with recycled paper and water-based resins (Figure 12.22). It contains no formaldehyde and, like acrylic counters, is worked with traditional carpentry tools. Richlite® is a similar product that is made with recycled paper and phenolic resins.

Counters of recycled glass are one of the more recent entries into the market (Figure 12.23). These counters consist of approximately 85% recycled glass (much of which is post-consumer), combined with cement binders and coloring. They provide a durable, VOC-free surface with a unique look and high recycled content. Glass counters are worked like stone and stone composites.

Laminate

Plastic laminate is an economical material made of layers of resin-impregnated kraft paper with a decorative layer compressed under pressure and heat to create a veneer for counters. Some laminates are available made with recycled paper content. Typically adhered to a particleboard substrate, traditional laminate counter assemblies have high UF and VOC content. Newer substrates made without added UF and water-based cements ensure that these counters do not degrade indoor air quality.

Laminates are not easily recycled and are susceptible to scratches and cracks, but they are among the least expensive counter options available. When budget is important, the right laminate counter is a suitable choice for a green home.

Concrete

Concrete counters are another alternative to solid surface and stone counters (Figure 12.24). Although they use high-embodied energy Portland cement, the overall amount is small; consequently, concrete is a reasonably sustainable choice for counters. Concrete counters are durable; however, they must be sealed to reduce staining. Like all concrete, these countertops can also crack. They can be cast in place or fabricated off-site, and fly ash can be substituted for some of the cement.

Courtesy of Trueform Concrete, LLC. www.trueformconcrete.com

Figure 12.24 Concrete is a very versatile material and may be used for countertops, sinks, vanities, tabletops, and other applications.

Stainless Steel

Stainless steel counters, a mainstay in commercial kitchens, are durable and nonporous, and can use recycled steel in their manufacture. Generally made to order, they are susceptible to scratches and dents, which can become part of the character of the finish or damage it, depending on your point of view. When stainless steel is bonded to a substrate, it should have no added UF and use low-VOC adhesives.

Tile

Ceramic tile makes a durable, attractive counter that can mimic the look of solid stone at a lower cost. As with any counter, select a substrate with no added UF, and consider locally manufactured and recycled content tiles. Keep in mind that the grout required between tiles can absorb stains and bacteria, making it a less healthy choice for food preparation areas.

Paints and Coatings

Oil-based paints and clear finishes, long preferred by many paint and floor finish professionals, provide excellent performance but have high levels of VOCs and other hazardous chemicals. Newer water- and acrylic-based finishes with lower VOC content provide performance similar to that of traditional oil-based finishes.

Paint

The key features to look for in an interior paint are its durability against abrasion and cleaning, how effectively it covers, and its VOC content (Figure 12.25). Select the highest quality paint with the longest warranty for best performance. Water-based paints are less toxic and clean up with water, oil-based paints are solvent based and must be cleaned with mineral spirits or turpentine.

Courtesy of Benjamin Moore

Figure 12.25 A paint can stating that the product contains zero VOCs.

Lighter colors and flat finishes have the lowest VOC content, which increases as the tint and gloss increase. Higher gloss paints tend to be more durable and are able to better withstand repeated cleaning; however, they often have higher VOC content and require more surface preparation because they accentuate defects more than flat finishes. Recycled-content paint is available in some areas in limited colors, most with low VOC content.

Paint is applied by brush, roller, or sprayer. Spray applications should be done with all recommended personal protective equipment, including respirators and eye protection. Ducts, filters, and other equipment should also be protected from damage by paint droplets. Paint cleanup should be done in buckets or sinks—never by dumping waste paint and rinse water on the ground or into storm drains.

Stains and Clear Finishes

Most stains and clear finishes have higher VOC content than water-based paints; therefore, certification agencies and green building programs allow for higher limits in these products (Tables 12.2 and 12.3). Oil-based stains, varnishes, and polyurethanes have traditionally provided the most durable finish; however, many new water-based products offer similar durability with lower VOC content and easier cleanup.

Waterborne polyurethane finishes, suitable for wood trim and floors, are available in one- and two-part systems; the two-part approach is generally more durable. Waterborne finishes dry more quickly than oil-based finishes, so they require experience to achieve the best finish. When using stains and sealers, compatibility with the final finish is critical for maximum performance.

GreenSeal VOC Limits

Green Seal is a nonprofit organization that certifies the performance and VOC contents of many products, including paint and clear finishes. The following charts list the maximum limits allowable for their certification, providing a benchmark for selecting finishes for a green home.

Table 12.2 Green Seal Paint and Opaque Finish VOC Limits

Product Type	VOC Level (g/L)
Topcoat, flat	50
With colorant added at point of sale	100
Topcoat, non-flat	100
With colorant added at point of sale	150
Primer or undercoat	100
With colorant added at point of sale	150
Floor paint	100
With colorant added at point of sale	150
Anticorrosive coating	250
With colorant added at point of sale	300
Reflective coating, wall	50
With colorant added at point of sale	100
Reflective coating, roof	100
With colorant added at point of sale	150

Table 12.3 Green Seal Stain and Clear Finish VOC Limits

Coating Type	VOC Content as Applied (g/L)
Varnishes	350
Varnishes, conjugated oil	450
Lacquer	550
Lacquer, clear brushing	680
Shellacs, pigmented	550
Shellacs, clear	730
Stains	250
Sealers	200
Sealers, waterproof	250

Natural oils and waxes with zero VOC content are another alternative for clear finishes, but they require regular re-application and buffing to maintain their finish. Wood stains made from plant products with no solvents or VOCs are also an available option.

Remodeling Considerations

Remodeling projects provide opportunities for recycling and reuse but may also be rife with challenges surrounding existing finishes that may include lead or asbestos. Materials that can be reused (on-site or off) or repurposed include cabinets, counters, wood flooring, doors, and trim. Cabinets can be reused in the same location, reinstalled in a different location of the house, donated, or sold. When reused in place, cabinet surfaces can be painted or refinished, or new doors and drawer fronts can be installed on existing frames for a less expensive and more resource-efficient project. Solid surface and stone counters may be able to be reworked for use, sold, or donated. Unique interior trim profiles and doors can be salvaged and reinstalled, saving the time and cost to locate special-order materials to match. Existing wood flooring can be removed and used to patch in for repairs, usually providing a better match than new material.

Before beginning any renovation work, confirm that there is no lead paint or asbestos that will be disturbed. Lead paint is common in homes built prior to 1978, and EPA regulations require training and certification for contractors and their workers, lead-safe work practices in all houses containing lead paint, and documentation and record keeping of all work. Asbestos can be found in sheet vinyl and vinyl composition tile and adhesives; in wall, ceiling, and pipe insulation; and in duct sealants, plaster, and some drywall. All potential asbestos sources should be tested prior to beginning work, and any contaminated materials should be removed by abatement professionals. Asbestos falls under the jurisdiction of state or local governments who license inspectors and contractors for testing and removal.

Cover and protect unaffected areas and equipment (particularly duct systems) throughout the renovation process to avoid damage and contamination that would need to be remedied later.

Look for the least-destructive methods to repair and refinish existing surfaces to remain after renovation. Repairing and repainting plaster walls and ceilings may be a suitable substitute for demolition and replacement. Refinishing existing floors is preferable to replacement, provided that proper cautions are taken for dust protection and low-VOC finishes are used.

Summary

In selecting finish materials for green homes, attributes between choices should be carefully compared to make the best decision for a particular product. Material selection can be very subjective. For example, very chemically sensitive clients may opt for zero-VOC finishes over durability or sustainable manufacturing.

Some more durable products may have other less sustainable attributes, but that does not necessarily disqualify them from use in a project. If a product with higher VOC content or higher embodied energy offers longer life and lower maintenance, it may still be a suitable choice for a particular project. Understanding the specific requirements of the project and the clients will help you make the right choices for each home.

Review Questions

1. Which product certification type provides the most objective evaluation of green attributes?
   a. First-party
   b. Second-party
   c. Third-party
   d. Fourth-party
2. Which of the following flooring finishes can provide thermal mass for passive solar projects?
   a. Wood
   b. Carpet
   c. Stone
   d. Linoleum
3. When in doubt, which of the following is the best guide to use when installing a new material?
   a. Subcontractor experience
   b. Manufacturers' instructions
   c. Personal experience
   d. Web search results
4. Which of the following is a potential problem with wood paneling?
   a. Poor durability
   b. Difficulty maintaining the air seal at the building envelope
   c. Low-VOC applied finishes
   d. Potential to create a vapor barrier in the wall
5. Which of the following is *not* a potential problem with vinyl wallcovering?
   a. It uses paper and grasses in manufacture
   b. Moisture can condense on wall surfaces in cooling climates
   c. It off-gases VOCs
   d. It is not recyclable
6. Which of the following criteria do not apply to solid wood trim?
   a. Sustainably harvested
   b. Added UF content
   c. Recyclability
   d. Location of forest and mill
7. Which of the following flooring materials is the least sustainable choice for a green home?
   a. Recycled plastic carpet installed with jute pad and low-VOC glues
   b. Ceramic tile
   c. Forest Stewardship Council–certified oak with waterborne polyurethane finish
   d. Structural concrete slab sealed with low-VOC finish
8. Which of the following substrates should *not* be used behind tile in a shower?
   a. Paperless gypsum backer board
   b. Cement-based backer board
   c. Moisture-resistant drywall or greenboard
   d. Mud base
9. Which of the following finishes tend to have the highest VOC content?
   a. Water-based polyurethane
   b. Acrylic latex flat finish paint
   c. Clear sealers
   d. Varnish
10. Which of the following interior trim materials is the best choice for a green home?
    a. PVC
    b. Solid wood
    c. MDF
    d. Finger-jointed wood

Critical Thinking Questions

1. What resources would you consider using to determine the most appropriate interior finishes for a green home?
2. How would you rank the following criteria from most to least important in choosing interior finishes for a green home? How might they vary from one project to another?
   a. Recycled content
   b. Recyclability
   c. Low VOCs
   d. No added UF
   e. Life span
   f. Location of harvesting and manufacture

Key Terms

brominated flame retardants, 310
Carpet and Rug Institute (CRI), 310
ceramic tile, 302
decoupling membrane, 310
drywall, 301
engineered flooring, 306
environmental product declarations (EPDs), 303
first-party certification, 303
linoleum, 307
Material Safety Data Sheets (MSDS), 303
millwork, 302
plaster, 302
post-consumer recycled content, 303
post-industrial recycled content, 303
recycled content, 303
second-party certification, 306
solid surface, 317
third-party certification, 304
walk-off mat, 312
wallcovering , 314

Additional Resources

EPA Renovation Repair and Painting requirements for lead paint:
http://www.epa.gov/lead/pubs/renovation.htm

Building Green, *Green Building Product Certifications: Getting What You Need*, 2011:
https://www.buildinggreen.com/ecommerce/certifications-report.cfm?

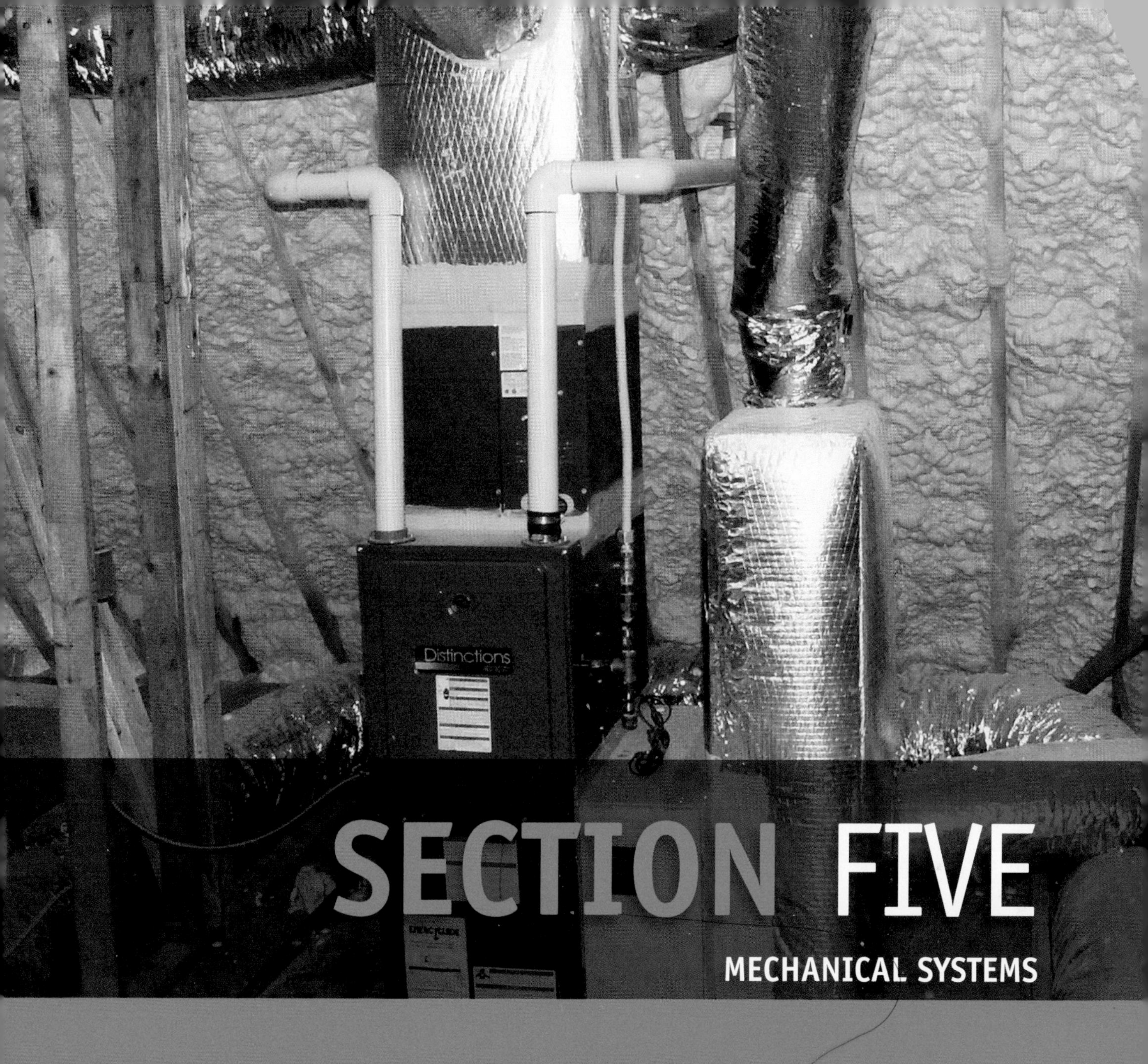

SECTION FIVE

MECHANICAL SYSTEMS

13

Heating, Ventilation, and Air Conditioning

Heating, ventilation, and air conditioning (HVAC) systems control the indoor environment by heating, cooling, circulating, and filtering the air and managing humidity. Proper equipment selection, installation, and maintenance are essential for comfortable and healthy indoor environments. This chapter presents the different types of HVAC systems available for use in homes and explains their respective pros and cons. Climate, building type, and homeowner preference are key factors in system selection and design. Ventilation, control systems, filtration systems, combustion safety, fireplaces, and remodeling considerations are addressed.

LEARNING OBJECTIVES

Upon completion of this chapter, the student should be able to:

- Explain different types of heating systems
- Explain different types of cooling systems
- Discuss different types of distribution systems
- Describe different types of ventilation systems
- Describe different efficiency ratings and how they affect HVAC equipment
- Calculate ventilation requirements by using standards from the American Society of Heating, Refrigeration, and Air Conditioning Engineers (ASHRAE)
- Describe proper ductwork installation
- Describe the components of an HVAC system and their contributions to efficiency

Green Building Principles

Energy Efficiency

Indoor Environmental Quality

Durability

Homeowner Education and Maintenance

The Building Envelope

As we discussed in Chapter 4, whenever a house is heated or cooled, the quality of the building envelope has a significant effect on the need for air treatment and, in turn, the size and capacity of the heating, ventilation, and air conditioning (HVAC) systems. Inadequate insulation or air sealing will result in the structure requiring higher capacity heating or cooling equipment as well as a greater degree of humidity management to maintain desired comfort levels. It stands to reason that energy savings arise from a more efficient building envelope. To specify, design, and install an HVAC system properly, the efficiency of the building envelope must be taken into account. When the envelope in a new building is deficient in design, or an existing building exhibits envelope defects, these issues should be corrected before sizing and installing HVAC equipment.

A carefully designed and constructed house that is managed properly by the homeowner may require so little additional heating and cooling that a central HVAC system is not required, or the central system can be much smaller than in a standard code-built home. Many industry professionals are surprised to hear that modern energy-efficient homes do not always require traditional heating and cooling systems. Super-efficient homes may be heated with what would otherwise be considered supplemental systems, such as solar radiation, wood stoves, space heaters, or fireplaces. When designing a home, the first priority is to minimize the heating and cooling loads as much as possible.

Heating and Cooling Systems

Most homes require space conditioning, defined as heating, cooling, or both, depending on the local climate. The Internal Residential Code (IRC) requires that dwellings in cold climates must have heating equipment that can maintain indoor temperature at a minimum of 68°F, although there is no code requirement for cooling equipment in any climate.

The key components of HVAC systems are the equipment used to create the heating or cooling, the fuel sources, and the method used to distribute the heating and cooling throughout the house. Heating and cooling are delivered by convection through air or by radiation. Individual systems that serve multiple areas of the home are referred to as *whole-house systems,* and those that serve only sections or single rooms are called *local or non-distributed systems.* Either of these can provide heating only, cooling only, or both, depending on the system type and climate requirements. Fuels can be fossil fuels (e.g., gas or oil), electricity, or wood, or, in some cases, solar energy can be employed. Selecting the most appropriate distribution system and equipment, combined with proper design and installation, are all critical to creating an effective and efficient HVAC system.

Whole-House Systems

Whole-house systems use centrally located equipment that heats and cools either air or water, which is then distributed throughout the house through ducts or pipes. Depending on the climate and the total load on the home, systems may provide heating only, cooling only, or both. Most systems that provide both heating and cooling will use the same equipment for distribution; in some cases, however, heat may be supplied by one system and cooling by another.

Local and Non-Distributed Systems

Local or non-distributed systems can provide heating, cooling, or both to a room or open area. Individual rooms or areas within a home (as well as entire homes) that are small, tightly sealed, and well-insulated can be heated with a centrally located heater or cooled with a single air conditioner. Alternatively, a heat pump can provide both heating and cooling.

Equipment

HVAC equipment includes furnaces and boilers that heat water or air; heat pumps, which are devices that transfer heat between a fluid and the outside air, the ground, or water; and air conditioners, which transfer heat in the same manner as heat pumps. Furnaces and boilers provide heat only. Heat pumps provide both heating and cooling, and air conditioners provide cooling only.

Heating Systems

Heating systems consist of fuel-fired and electric furnaces, boilers, and heat pumps, providing heated air or water for distribution to the home. Furnaces use fans to blow heated air through ducts, supplying conditioned air throughout a home. Boilers heat water, which can be used for either radiant heat or forced-air systems. Heat pumps are combined with an air handler that conditions the air, which is then distributed through ducts. Heat pumps can also heat water that is either used directly for radiant heat or transferred to the air for distribution through ducts.

Furnaces

Furnaces burn fuel or use electricity to create heat, which is then transferred to the air for distribution to heat a home. Fuel-fired furnaces utilize a heat exchanger to prevent the introduction of combustion gases into the conditioned air stream (Figure 13.1). Some amount of heat is always lost through the fuel combustion process, therefore oil and gas furnaces are not as efficient as electric furnaces. Electric furnaces use resistance to convert electricity into heat. Electric resistance heating is very efficient, converting 100% of the site energy (electricity) into heat. Electric systems, however, can be relatively expensive compared with fossil fuel heating systems, and central electricity generation and transmission is inefficient (see Chapter 1). Furthermore, most electricity

---

furnace an appliance fired by gas, oil, or wood in which air is heated and circulated throughout a building in a duct system.

boiler piece of heating equipment designed to heat water (using electricity, gas, or oil as a heat source) for the purpose of providing heat to conditioned space or potable water.

heat pump a heating and cooling unit that draws heat from an outdoor source and transports it to an indoor space for heating purposes or, inversely, for cooling purposes.

air conditioner a home appliance, system, or mechanism that dehumidifies and extracts heat from an area.

electric furnace a heating system that uses electric resistance to convert electricity into heat.

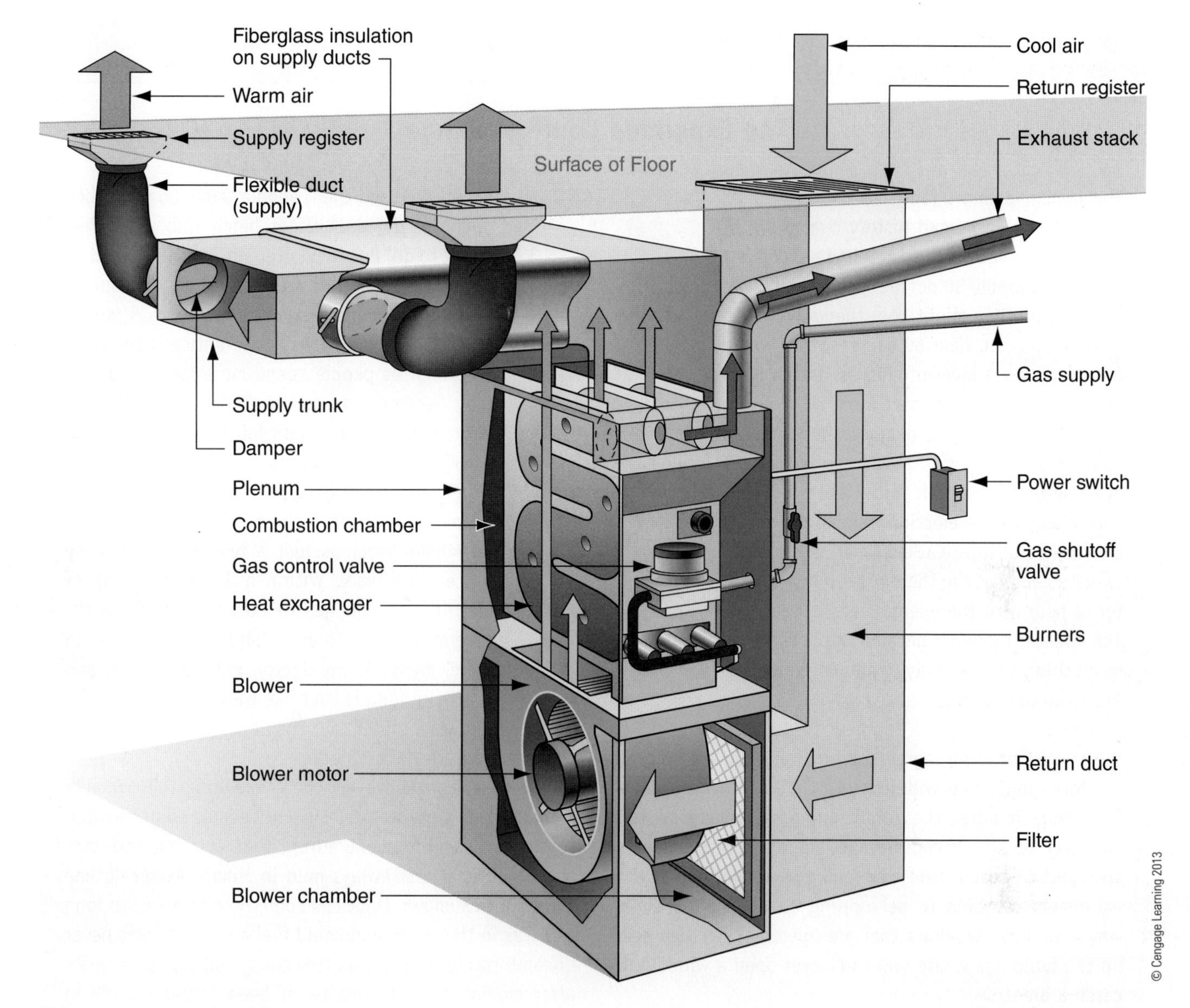

Figure 13.1 Gas-fired furnace.

produced in the United States is generated at plants that use fossil fuels—so while there is no on-site pollution from using electricity, there is pollution at the power plant. For these reasons, most green building programs and ENERGY STAR do not allow electric resistance heating to be used as the primary heat source.

Boilers

Boilers produce hot water that is either heated only when needed or heated and stored in an insulated tank where it is used for space heating and domestic hot water (Figure 13.2). Storage tanks typically use heat exchangers, which are devices that transfer the heat from the stored water to the domestic hot water or radiant heat systems. This system keeps potable water separate from heating water to avoid any contamination. Boilers can be fueled by gas, oil, wood, wood pellets, or, less frequently, electricity. Wood-fired boilers are typically installed on the exterior of a home. These are most common in extremely cold climates—mostly in rural areas where there is both room for the exterior installation and a steady supply of firewood available. Hot water can also be used to heat forced-air distribution systems. Boilers can be combined with solar pre-heating, which reduces the amount of energy required by the boiler to heat the water, thereby reducing energy consumption. Water heaters can be used instead of boilers in moderate climates.

Heat Pumps

Heat pumps are heating and cooling systems that draw heat from an outdoor source, such as the air, the ground, or a body of water, and transport it to an indoor space for

FROM EXPERIENCE

The Expanded Comfort Range

The human comfort range has shrunk to its smallest size in human history over the past half-century. Our ancestors had a comfort range of probably 30 degrees Fahrenheit. Near 90 degrees, they might cool themselves with a hand-held fan. Near 60 degrees, they would put on an extra layer of clothes. Today, however, there are "thermostat wars" all over the United States over 2 degrees. Don't laugh: you likely have participated in some of them at some point yourself. And President Jimmy Carter lost his re-election campaign in part because he famously asked Americans to wear sweaters and cut the thermostat down in winter to help with the energy crisis of that day. The sweater therefore became the only article of clothing to ever play a role in ending an American presidency.

Courtesy of Steve Mouzon, Architect & Urbanist

Steve Mouzon is an architect, urbanist, author, and photographer. He founded the New Urban Guild and is a principal in Mouzon Design, a residential and planning firm in Miami, Florida.

Ask any mechanical engineer to describe the impact of a 30-degree comfort range vs. a 2-degree comfort range. She will tell you that a 2-degree comfort range requires the conditioning equipment to run basically all the time, because outdoor temperatures are almost never within that 2-degree range. And if the equipment is going to be running almost all the time, why even have windows that are operable? So they seal up the buildings where you can't ever open a window to catch a breeze.

A 30-degree range, on the other hand, means that there are several months per year when the air outside is within the comfort range at least part of the day. So if the building is designed cleverly enough, it can condition itself for most of the year in many places, requiring mechanical conditioning only in more extreme weather.

How do we expand the human comfort range again, getting it back close to where it has been for almost all of recorded human history? President Carter's approach of telling us what we ought to do is no more likely to work now than it did then. People rarely do what they ought to do, and resent being told what they ought to do. But they often do what they want to do. So what's the most effective way of ensuring that people want to expand their comfort range?

The best known way is to entice them to go outdoors. As people spend more time outdoors, they become more acclimated to the local environment and need less full-body conditioning when they return indoors.

My own experience provides a good example. I moved to Miami in the fall of 2003. My home on Miami Beach is just a few blocks from my office, so I walk. Within a 10-minute walk of my office, I can get to dozens of restaurants, several grocery stores, a hardware store, a drug store, my bank, my doctor, my accountant, and lots more. And it isn't like walking alongside the highway, either . . . they are highly interesting walks through beautiful places.

Because I walked everywhere, cranking the car only a couple times per week, I quickly became so acclimated to the local environment during that first fall and winter, which is almost always mild in Miami. As springtime turned into summer, I noticed something strange: so long as I was in the shade and could feel a breeze, I was never uncomfortable. That is still true today, almost seven years after moving here: I have never been uncomfortable in Miami so long as there's a breeze in the shade . . . in a place where the basketball team is named "the Heat," and unaccustomed tourists sweat profusely.

The difference between running the mechanical conditioning equipment all the time and cutting it off several months of the year is so big that it dwarfs any equipment efficiency increases we could hope for in the near future.

Excerpted from The Original Green: The Mystery of True Sustainability *by Steve Mouzon; Miami, FL: New Urban Guild, 2010.*

heating purposes or, conversely, for cooling purposes. In the cooling season, heat pumps remove heat from indoors and release it to the outside, which acts as the heat sink.

Heat pumps are able to run in both directions, extracting heat from the cold air outside and pumping it into the interior of the home to heat, as well as extracting heat from warm air inside and pumping it outdoors to cool the occupied space ((Figure 13.3a and Figure 13.3b). Heat pumps achieve great efficiencies by moving ("pumping") heat rather than creating heat from a fuel source. Heat pumps do not primarily create heat; rather, they move it from one location to

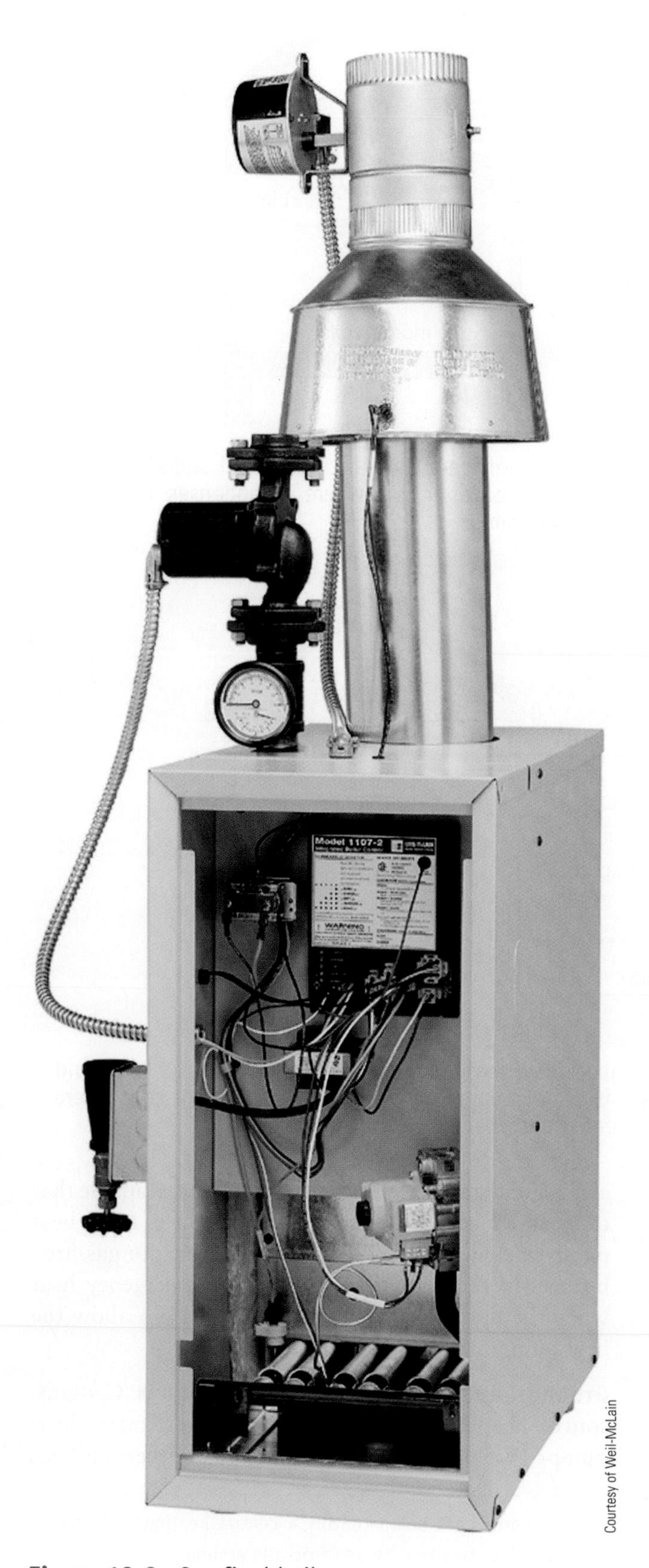

Courtesy of Weil-McLain

Figure 13.2 Gas-fired boiler.

another. Some heat that is created through the operation of motors is used when heating, but that same heat must be offset when in cooling mode.

Heat pumps are available with distribution systems (ductwork) or non-ducted systems, such as one-piece units similar to window air conditioners or ductless mini-splits, which have separate exterior and interior sections (Figure 13.4).

Air-Source Heat Pumps A heat pump consists of a compressor and two coils made of copper or aluminum tubing (one located inside and one outside) that are surrounded by aluminum fins to aid heat transfer (Figure 13.5). A refrigerant, a chemical that transfers heat as it changes from a liquid to a gas and back to a liquid, flows back and forth between the indoor and outdoor coils. The compressor is a mechanical pump that increases the pressure and temperature of vapor refrigerant and is typically located outside. In the heating mode, liquid refrigerant in the outside coil extracts heat from the air and evaporates into a gas. The indoor coil, now acting as a condenser, releases the heat from the refrigerant as it condenses back into a liquid.

Heat pumps also cool by working in reverse from the heating mode. In cooling mode, the liquid refrigerant on the inside coil extracts heat from the indoor air and evaporates into a gas. A *reversing valve*, near the compressor, can change the direction of the refrigerant flow for cooling as well as for defrosting the outdoor coil in winter. Now the indoor coil acts as an evaporator coil, changing the refrigerant from a liquid to a gas. The evaporator is the system component responsible for performing the actual cooling or refrigerating of the occupied space. The outdoor coil serves as a condenser, releasing the heat from the refrigerant into the outside air and condensing back into a liquid.

Air to water heat pumps use the same technology to transfer heat to and from water, which can then be used in hydronic systems to condition a home.

In cold weather, heat is removed from the outside air and transferred to the refrigerant in the compressor, then transferred to air through the heat exchanger

---

ductless mini-splits compact, wall-mounted air-conditioner or heat pump connected to a separate outdoor condensing unit via refrigerant lines.

refrigerant a chemical that transfers heat as it changes from a liquid to a gas and back to a liquid.

compressor a mechanical pump that uses pressure to change a refrigerant from liquid to gas.

condenser the component in a refrigeration system that transfers heat from the system by condensing refrigerant.

evaporator the system component responsible for performing the actual cooling or refrigerating of the occupied space.

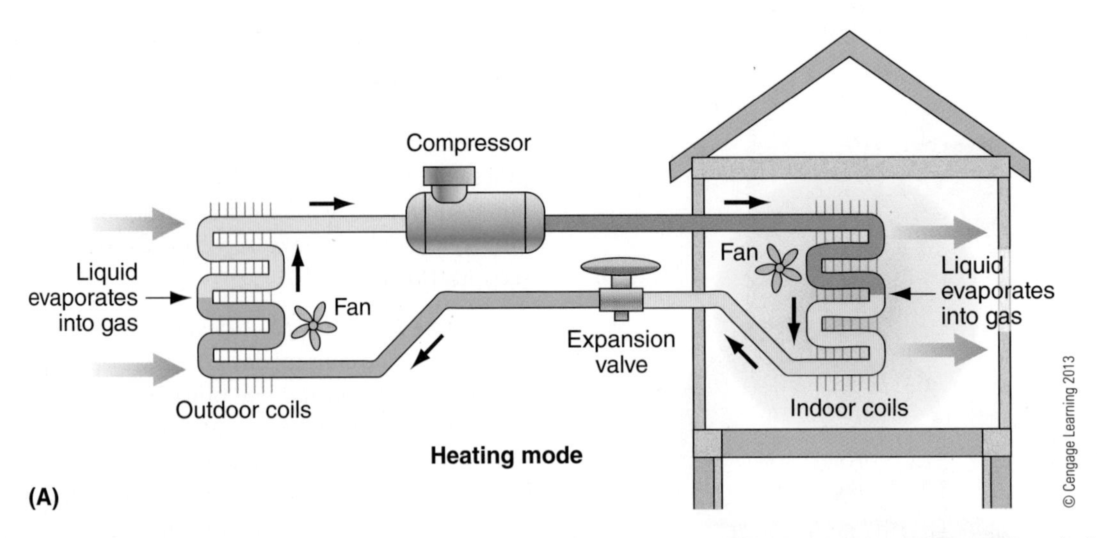

Figure 13.3a In cooling mode, an air-source heat pump evaporates a refrigerant in the indoor coil and absorbs heat from the home's air. The refrigerant is then compressed and sent to the outdoor coil, where it condenses at high pressure. At this point, it releases the heat it absorbed earlier in the home.

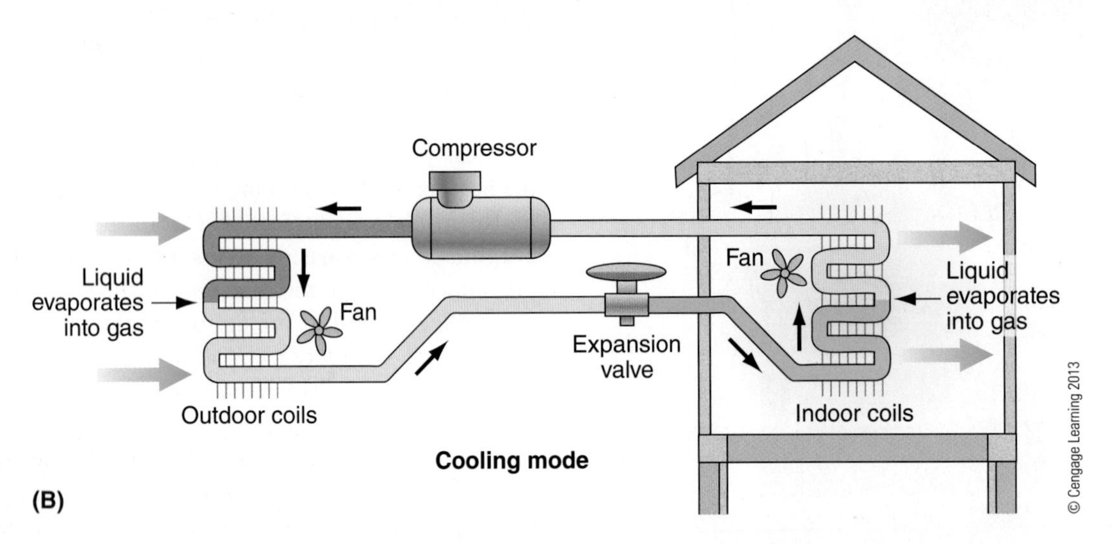

Figure 13.3b In heating mode, an air-source heat pump evaporates a refrigerant in the outdoor coil; as the liquid evaporates, it pulls heat from the outside air. The refrigerant is then compressed and sent to the indoor coil, where it condenses at high pressure. At this point, it releases the heat it absorbed earlier in the home.

for distribution in the house. In hot weather, heat is removed from interior air and transferred to the refrigerant. Heat is then removed from the refrigerant and transferred to the outside air. Air-source heat pumps are rated to provide a specific amount of heat at a specific outdoor air temperature, which is typically 47°F. As the outdoor temperature drops, the amount of heat that can be removed from the air decreases and the efficiency goes down. Most standard heat pumps can only efficiently remove heat from air that is approximately 40°F, although low temperature heat pumps are available that can remove heat from air at 0°F and lower. Most heat pumps include either electric resistance heat or gas-fired burners in the air handler to provide emergency heat when the outside temperature is too cold to allow the heat pump to remove heat from the air.

Ground- and Water-Source Heat Pumps Ground-source heat pumps (GSHPs), and water-source heat pumps (WSHPs), also referred to as geothermal heat

air-source heat pump a heating and cooling system that consists of a compressor and two coils made of copper tubing, one located inside and one outside, and surrounded by aluminum fins to aid heat transfer.

ground-source heat pump (GSHP) a central heating and cooling system that pumps heat to and from the ground.

water-source heat pump (WSHP) a heating and cooling unit that exchanges heat between the ground or water and the home interior; also referred to as geothermal heat pumps.

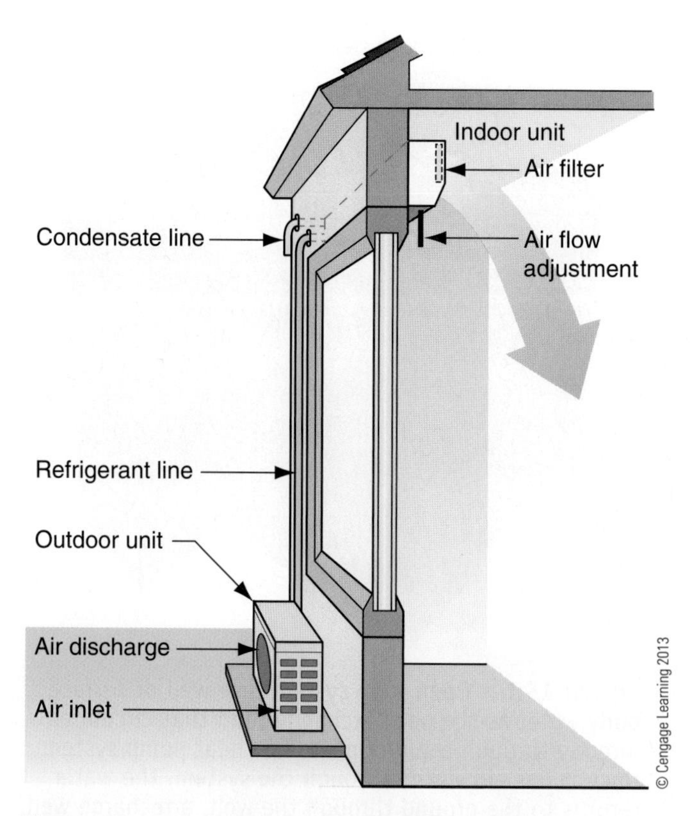

Figure 13.4 A single zone mini-split heat pump. Multiple indoor units can be run off one exterior unit and operated by remote control.

pumps, exchange heat between the ground or water and the home interior, using it to heat or cool a house with forced-air or hydronic distribution systems. GSHPs use loops of pipe that are buried in the ground and filled with a refrigerant, which transfers heat to and from the interior. Loops can be vertical wells or horizontal loops set in shallow trenches. WSHPs can have either open or closed loops (Figure 13.6a). Open loops, also known as "pump and dump" systems, draw in well or lake water and exchange heat, dumping the water back into the ground or lake to recapture or release heat (Figure 13.6b). Closed systems use refrigerant-filled loops placed in a lake or other water body.

Cooling Systems

Cooling systems are referred to as air conditioning (AC), a process that cools and dehumidifies air. In dry climates, evaporative cooling is an alternative system that cools the air but does not remove humidity.

Air Conditioning

Air conditioning cools indoor air by transferring indoor heat to a refrigerant, which then moves it to the exterior (Figure 13.7). Humidity is removed from the warm air

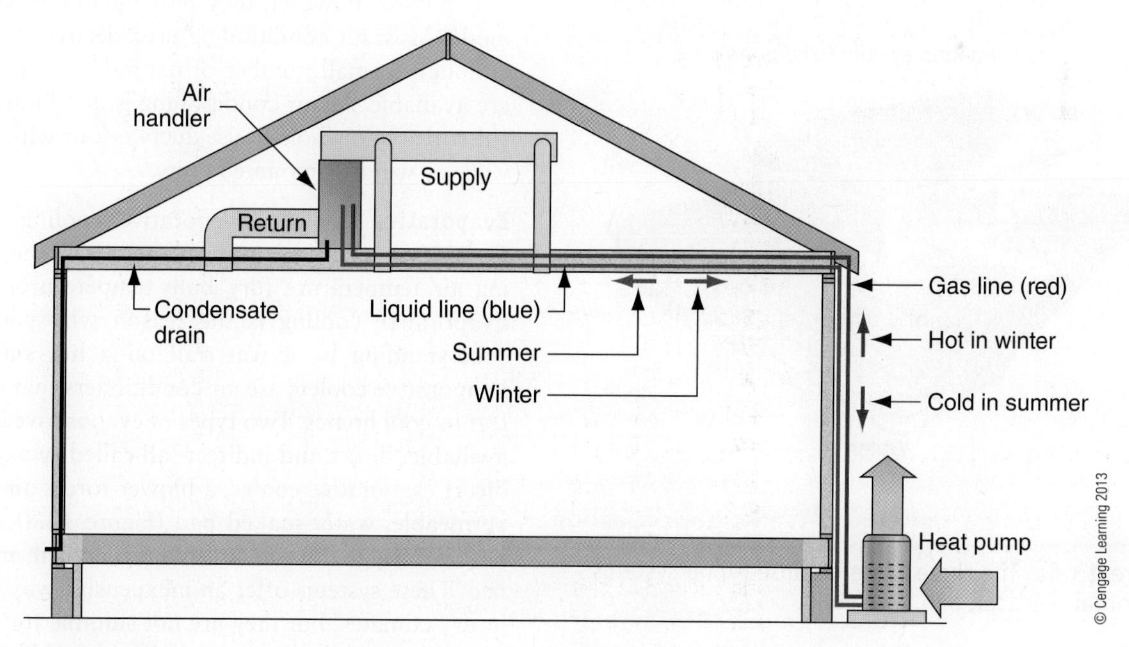

Figure 13.5 Heat pumps are air conditioners that are able to provide space heating and cooling.

---

air conditioning (AC) the process of cooling indoor air by transferring indoor heat to a refrigerant, which then moves it to the exterior.
evaporative cooling a means of temperature reduction that operates on the principle that water absorbs latent heat from the surrounding air when it evaporates.

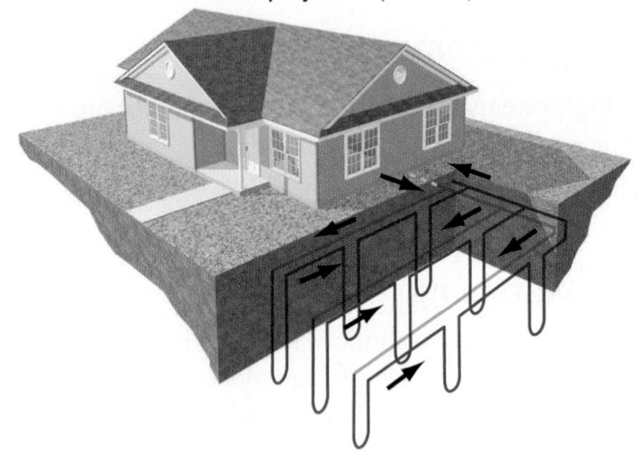

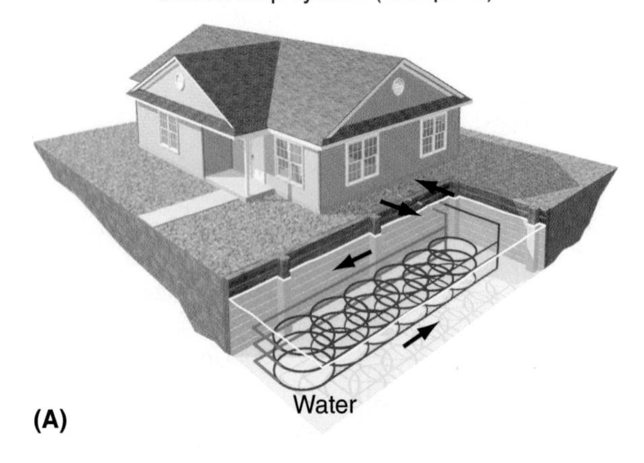

Figure 13.6a The three types of closed-loop systems: horizontal, vertical, and lake/pond.

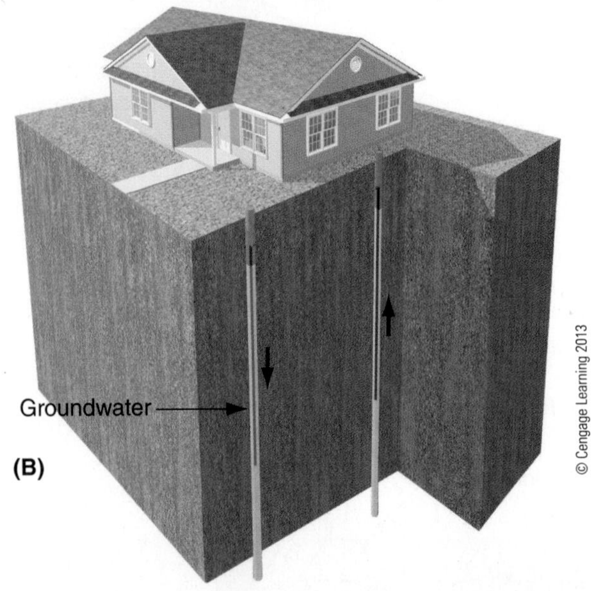

Figure 13.6b Open-loop systems use well or surface body water as the heat exchange fluid that circulates directly through the ground-source heat pump system. Once it has circulated through the system, the water returns to the ground through the well, a recharge well, or surface discharge.

as it passes over the cooler evaporator coil. The moisture is removed from the air as water is drained to the exterior. The compressor and condenser are normally combined in an outdoor component called a condensing unit.

Air conditioning uses the same technology as air-source heat pumps; however, they only operate in the cooling mode. Most air conditioning uses electricity to operate, although a small number of natural gas-fueled systems are available. All air conditioning is distributed with air, either through whole-house ductwork or with individual wall- or window-mounted units.

Evaporative Cooling Evaporative cooling works on the principle that as water evaporates, the surrounding air temperature (dry bulb temperature) decreases. Evaporative cooling is the reason why you will feel cool standing by a waterfall on a hot summer day. Evaporative coolers are air conditioners that use this effect to cool homes. Two types of evaporative coolers are available: direct and indirect (all called two-stage). In a direct evaporative cooler, a blower forces air through a permeable, water-soaked pad (Figure 13.8). As the air passes through the pad, it is filtered, cooled, and humidified. These systems offer an inexpensive way to cool air in dry climates, but they are not suitable for humid climates because they increase indoor humidity instead of reducing it. Direct evaporative coolers are commonly referred to as *swamp coolers* due to the algae growth that was seen on early models.

---

condensing unit the portion of a refrigeration system where the compression and condensation of refrigerant is accomplished.

direct evaporative cooler a device that reduces the temperature of air by passing it through water-soaked pads.

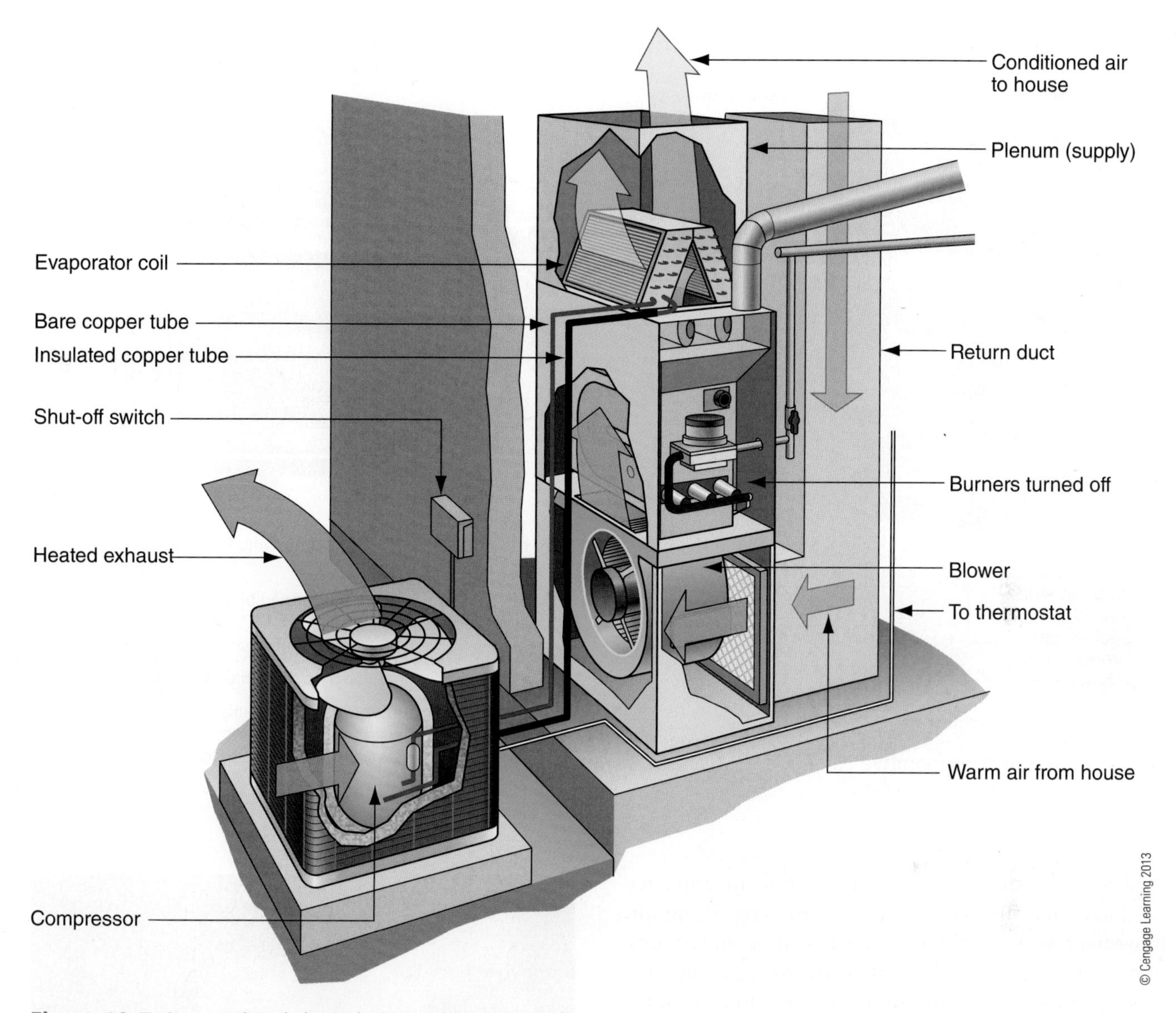

Figure 13.7 Conventional ducted air conditioner.

An indirect evaporative cooler has a secondary heat exchanger, which prevents humidity from being added to the airstream that enters the home (Figure 13.9). The air is first sent through a heat exchanger that is cooled by evaporation on the outside. Next, the pre-cooled air passes through a water-soaked pad and picks up humidity as it cools. Since the air is pre-cooled first, less humidity is added to the air. Although indirect evaporative coolers do not increase the relative humidity of the air entering the home, they do not provide any dehumidification benefits either; therefore, they are not recommend for homes in humid climates.

indirect evaporative cooler similar to direct evaporative cooling but uses some type of heat exchanger to prevent the cooled moist air from coming in direct contact with the conditioned environment.

Evaporative coolers may use as little as one fourth of the electricity required to run an air conditioner and do not require refrigerants, but they do use significant amounts of water to operate.

Alternative Cooling Strategies

In moderate climates, and during cool, dry periods in hot and mixed climates, air movement can substitute for air conditioning, saving both energy and money. Air movement can be achieved by opening low and high windows to bring in outside air and take advantage of the stack effect in multi-story homes. This passive cooling technique is commonly referred to as a thermal chimney and is covered in Chapter 8 (see Figure 8.5). Whole-house fans can be used in conjunction with opening windows to bring in outside air when the temperature and relative humidity are lower than that inside. Ceiling fans can be

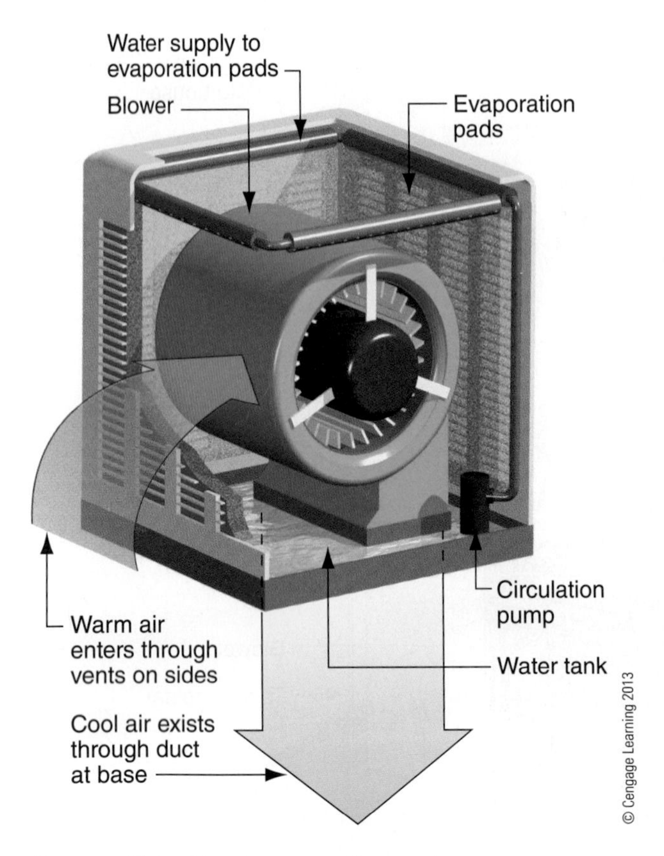

Figure 13.8 Direct evaporative cooler.

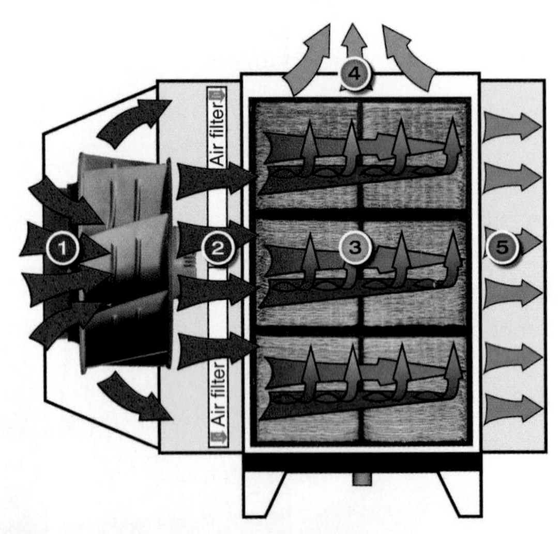

used to cool individuals sitting beneath them, reducing or eliminating the need for air conditioning. Using air movement as a substitute for air conditioning requires homeowners to actively participate by opening and closing windows, turning fans on and off, and adjusting thermostats.

Distribution

HVAC can be delivered by convection through the air, by radiation, by passive solar heating, or through natural ventilation. Air delivery systems can provide both heating and cooling. Radiant systems typically provide only heat, but radiant cooling is appropriate in certain dry climates. Passive solar design (addressed in Chapters 3 and 16) provides only heat, and natural ventilation can offer cooling during some seasons in moderate climates.

Air, or convection delivery of HVAC, is typically distributed through forced-air systems that use blowers and ducts to deliver hot or cold air to all rooms. Alternatives include non-ducted systems that blow conditioned air from individual room units, and natural convection from single-source heaters, such as wood

Figure 13.9 Indirect evaporative cooler.

stoves (Figure 13.10). **Radiant heating** uses hot water circulating through pipes embedded in the floor or through radiators, or electric-resistant heat installed in floors or individual radiators to provide heat. Steam

---

radiant heating a heating system in which the heating source (electric resistance or hot water) is installed under the finish flooring or individual radiators.

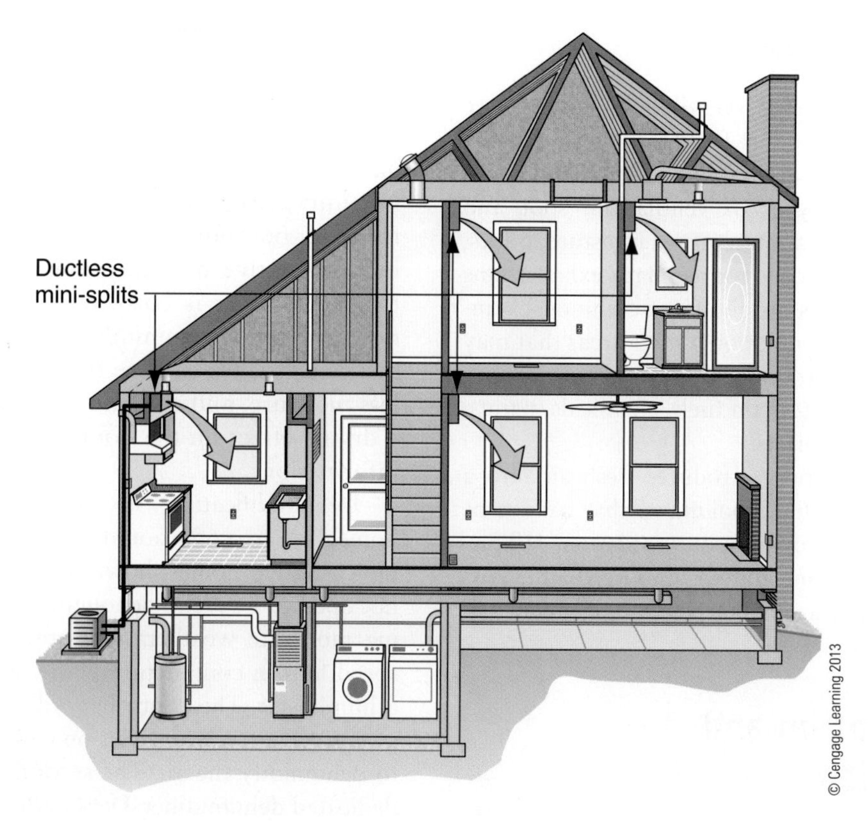

Figure 13.10 A ductless mini-split is one example of a non-ducted distribution system. Other systems include fireplaces and hydronic systems.

heat systems use individual radiators. When the outdoor temperature and humidity levels are comfortable, natural ventilation can provide free cooling.

Air Distribution

Most residential systems in the United States are air distribution systems that use ductwork to deliver conditioned air throughout the house. Ductwork may be made from sheet metal, plastic pipe, plastic flexible duct ("flex duct"), and fiberglass duct board. While such building components as stud and joist cavities may be code-approved for return air ducts, they are not recommended for high-performance homes and are prohibited in some green building programs and energy codes. The most basic distribution systems, such as window air conditioners, have integrated blowers and no ductwork.

Radiant Distribution

Radiant heat is delivered through hydronic or electric resistance systems (Figure 13.11). Hydronic, or water systems, include wall- or baseboard-mounted radiators, in-floor tubing, or a combination. Electric resistance heat can also be installed in-floor or as radiators. Electric resistance heat is generally not considered a sustainable solution for whole-house heating, but it may be suitable for a super-insulated home that requires minimal heating, particularly if it uses a renewable power source.

hydronic a space-conditioning system that circulates heated or cooled water through wall- or baseboard-mounted radiators, in-floor tubing, or a combination.

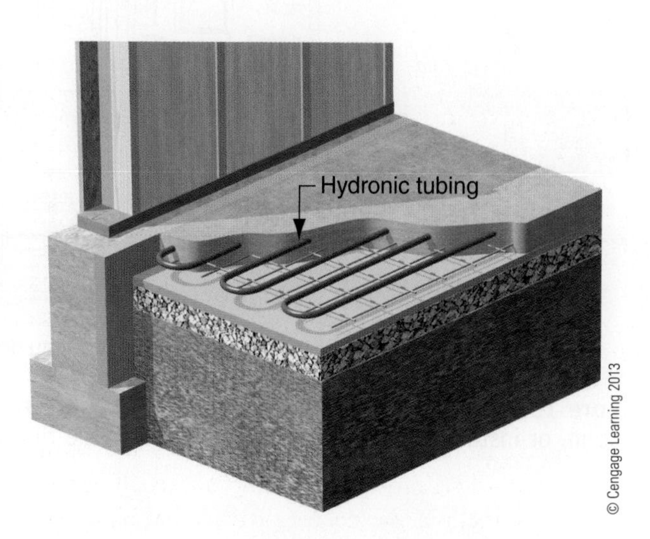

Figure 13.11 A hydronic, or radiant, floor piping system awaiting encasement in concrete.

Ventilation Systems

As discussed in Chapter 2, ventilation systems are designed to remove pollutants from the interior of the home and supply outside air to dilute pollutants. Homes typically require two types of ventilation: spot and whole house. Spot ventilation removes moisture, odors, and pollutants directly at the source. Bath exhaust fans and kitchen range hoods are the most common examples of residential spot ventilation. Other areas that may benefit from spot ventilation include laundries, garages, and storage rooms that contain fuels, chemicals, paints, cleaners, and other chemicals.

Whole-house ventilation introduces fresh air into a tightly sealed home to dilute pollutants that cannot be removed completely through spot ventilation. Whole-house ventilation can be integrated with forced-air HVAC equipment, or it can be a completely separate system.

Dehumidification and Humidification

Supplemental humidification systems are typically attached to central HVAC systems at the air handler. Humidification systems vary, but most introduce water vapor directly into the supply air stream. A well-sealed home rarely requires supplemental humidification, especially in cooling-dominated climates. Homes with excessively dry air in moderate climates typically have very leaky building envelopes that allow any humidity created by occupant activities to migrate to the dryer outdoor air through exfiltration, leaving the indoor relative humidity lower than desired. When this air leakage is corrected, it often eliminates the need for indoor humidification. Using the psychrometric chart, we can see that cold air is able to hold less moisture, and cold air infiltrating is likely having a drying effect on the home (see Chapter 2 for more information).

Dehumidification is normally accomplished in homes by air conditioning systems; in well-insulated, tightly sealed homes, however, the air conditioner may not operate frequently enough to remove humidity during moderate weather. One option is to use variable-speed fans in conjunction with controllers that contain a humidistat. The controller will call for the air conditioner to turn on and run on low speed, which works to dehumidify the air. The second option is to install a dedicated dehumidifier. Dehumidifiers remove humidity without cooling the air by using the primary duct system, a secondary distribution system, or a free-standing unit (Figure 13.12). Dehumidifiers in high-performance

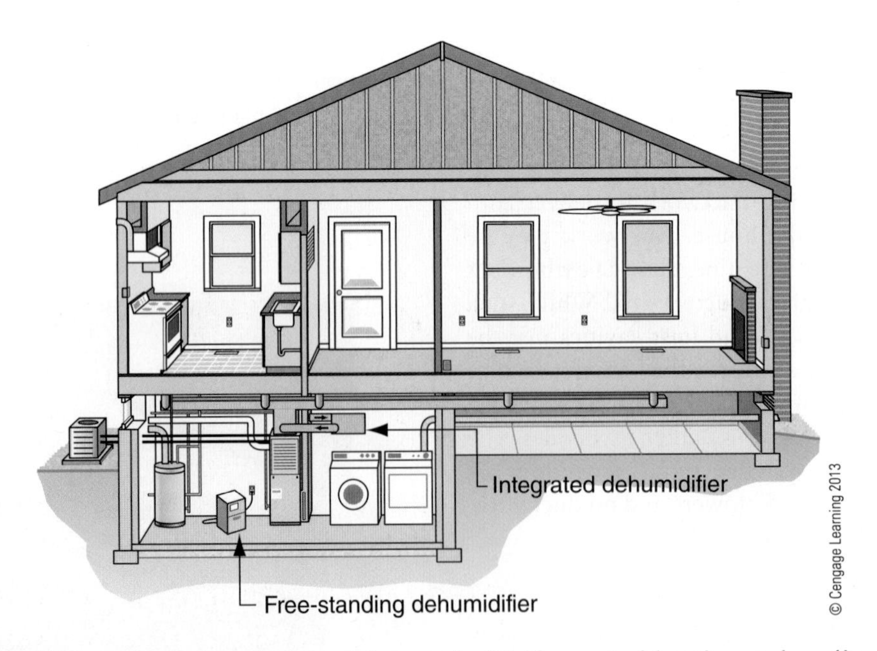

Figure 13.12 Dehumidifiers may be stand-alone, integrated with the central heating and cooling distribution system, or installed with their own duct system. The first two options are presented here.

---

spot ventilation the mechanical process of removing moisture, odors, and pollutants directly at the source.

homes can keep the indoor relative humidity below 50% during spring and fall seasons when the air conditioning may be used infrequently but the humidity is high. They are also useful in removing moisture from basement areas that may be damp and require very little air conditioning throughout the year.

Filtration Systems

Filtration removes pollutants from the air and is normally installed as part of a forced-air HVAC system. When designing a home for good indoor environmental quality, the following steps should be taken:

- Avoid introducing pollutants into the home; examples include excess moisture, dirt from street shoes, household chemicals and solvents, pet hair, and carbon monoxide from combustion appliances and attached garages
- Remove pollutants at the source through spot ventilation, such as bath and kitchen vent fans
- Dilute pollutants through whole-house ventilation
- Remove remaining pollutants with filters

Filtration is the least effective means of pollutant removal because it typically occurs after pollutants are distributed throughout the home. The three primary types of filters are mechanical, pleated, and electronic. The most common type used in homes, **mechanical air filters** use synthetic fibers, fiberglass, or charcoal to remove particulates. **Pleated air filters** are more effective than other mechanical air filters because they contain more fiber per square inch than mechanical filters. **Electronic air filters** use electricity to attract smaller molecules, such as smoke, mold and pet odors, to metal fins. The efficiency of removing particulates increases with mechanical and pleated filters as they become dirty because smaller and smaller particulates are captured in the increasingly fine openings. The effectiveness of electronic filters decreases over time without cleaning because the metal fins become ineffective when dirty. Filtration systems may be installed as separate stand-alone equipment, at the air handler unit, or at the return register (Figure 13.13).

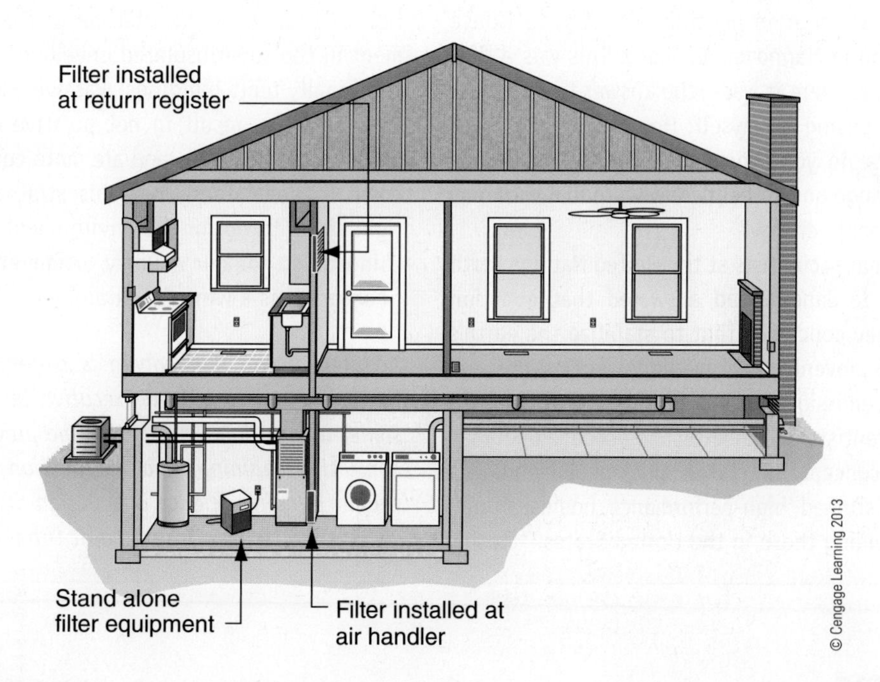

Figure 13.13 The three most common locations for filters. Single-room filters, often called "room cleaners," are not as effective as whole-house systems.

mechanical air filters filters that use synthetic fibers, fiberglass, or charcoal to remove particulates; most common type of filter used in homes.

pleated air filters high-efficiency paper mechanical air filters that contain more fiber per square inch than disposable fiberglass filters.

electronic air filters use electricity to attract smaller molecules, such as smoke, mold, and pet odors, to metal fins.

FROM EXPERIENCE

Passive Houses in America

Katrin Klingenberg, Executive Director, Passive House Institute US In 2000, I was working for a successful architecture firm in Chicago, famous for high-end commercial and urban architecture. Many of the firm's projects made use of innovative energy efficiency and natural-ventilation strategies, such as double-glazed walls for high-rises. However, the measured results of those strategies were rather disappointing. As it turned out, they reduced overall building energy consumption by only about 30%.

Then, in 2000, the United States withdrew from the Kyoto Protocol.

This action prompted me to search even harder for a way to leapfrog from conventional architecture to a carbon-neutral and economically feasible approach to building, without getting bogged down in incrementalism. It inspired my search for a path that would lead away from dependence on an oil-based economy and toward the creation of a global economy based on sustainable energy and equity.

That year, I set foot in my first-ever Passive House at the World Expo in Hannover, Germany. This was a life-changing moment. There it was—the answer to the question I had been posing for myself: How much energy can you afford to use in your home if your goal is to mitigate climate change and to help create a global economy based on sustainable energy?

In 1992, climate scientists at the United Nations Earth Summit in Rio de Janeiro had answered that question theoretically. They concluded that to stabilize the earth's atmosphere and prevent global warming, energy use and greenhouse gas emissions must be cut by a factor of 10. The building scientists and architects who developed the Passive House concept had taken this recommendation to heart. They studied high-performance homes around the world—including those in the United States—closely examining passive solar and superinsulation techniques, to arrive at an answer that had been unthinkable until then. Yes, we can build homes that use 80% less energy—for a price people can afford—by employing efficiency techniques. We do not need to rely on costly active-solar technologies.

We chose to build a prototype in Urbana, Illinois. Constructed in 2002–2003, it was the first house built in the United States using the specific practices, technologies, and energy- modeling tools developed by the Passivhaus Institut in Germany to achieve the Passive House standard. It was designed with two primary goals in mind: first, reduce operational energy use by a factor of 10, and second, follow cradle-to-cradle principles as closely as possible.

The experience of building the Smith House has been setting the bar for my designs ever since. I soon thereafter made a decision to not to work on any projects other than Passive Houses. As we have gotten more practice over the past seven years, we can now build Passive Houses that, over the lifetime of the additional investment in the superinsulated envelope, cost less than conventionally built buildings! Passive Houses afford energy independence, result in net positive economic benefits, promote good health, and are more comfortable. We have come to understand that this strategy is not only the right thing to do for the environment but also the smart thing to do for our country, communities, and personal finances. It is a win-win all around.

Architect Katrin Klingenberg is co-founder of the Passive House Institute and the Executive Director and Lead Designer at e-co lab. For PHIUS, she provides Passive House consulting, training, and certification U.S.-wide. She has taught Building Science and Design Studios at the University of Illinois in Chicago and at Urbana-Champaign.

Duct Systems

The most common HVAC delivery method is a forced-air duct system. Standard duct systems move air at low velocity through large-diameter ducts (Figure 13.14a). High-velocity systems, most often used for retrofit applications where space for large ducts is limited, deliver air at high speed through small-diameter ducts (Figure 13.14b). By their very nature, high-velocity systems must be well-designed and leak-free to operate effectively. High-velocity systems will be discussed in the remodeling section at the end of the chapter.

Duct System Design and Installation

Forced-air systems must provide enough air flow to adequately distribute heating or cooling to all rooms in a home. They must also provide equal amounts of supply

© Cengage Learning 2013

Figure 13.14a Conventional HVAC system with a flex duct distribution system.

Courtesy of Unico System

Figure 13.14b High-velocity distribution systems use small diameter ducts. Courtesy of Unico System Inc.

and return air flow, both in individual rooms as well as in the entire system. When the amount of supply and return air is unbalanced (more of one than the other), the home or individual room pressures are affected. For example, a bedroom with 85 CFM of supply air but no ducted return will have a positive pressure with respect to the outdoors when the door is closed. Positive pressure results in exfiltration (forcing air out of the home), which forces conditioned air out of the house through cracks and gaps. Negative pressures promote infiltration, bringing unconditioned air from the exterior in through those same gaps.

Heating and cooling equipment efficiencies will also suffer as a result of undersized return air flow, which can lead to high static pressure that reduces a system's ability to effectively condition a space. **Static pressure** refers to the pressure inside the duct system and is an indication of the amount of resistance to air flow within the system. Static pressure is usually measured in inches of water column (IWC) or Pascals (Pa).

The best way to ensure pressure balance is to install supply and return ducts in each major room with the exception of bathrooms and kitchens which receive only supply ducts to avoid drawing excess moisture and contaminated air into the HVAC system. Some systems are designed with a single large central return register. Central return systems create major imbalances in rooms when doors are closed, pressurizing rooms with supplies and depressurizing the room with the return. Systems with central returns can be effectively balanced by installing **jumper ducts**, which are small sections of duct installed in ceilings that allow air to flow between rooms, or **transfer grilles**, which are holes cut in walls to allow air to flow between rooms (Figure 13.15a and Figure 13.15b). Undercutting the door to allow air flow may also be a cost-effective way to balance systems if enough free air space can be created between the floor and the door when it is closed.

Ducts should be run through the conditioned space whenever possible, avoiding unconditioned attics, exterior walls, vaulted ceilings, crawl spaces, basements, and garages. Keeping ducts in conditioned space (Figure 13.16) reduces the extra load on the HVAC system and limits the opportunity for pollutants, such as mold, dust, and dirt, to enter the ducts and disperse throughout the house. Keeping ducts out of garages is particularly important to avoid introducing carbon monoxide and other pollutants inside the building envelope.

Residential duct systems generally follow two basic designs: trunk-and-branch systems and spider ducts. **Trunk-and-branch duct systems** have large ducts (trunks) installed through the center of the house with smaller ducts (branches) run off the trunks to supply individual rooms (Figure 13.17). **Spider duct systems** have large supply trunks connected to remote mixing boxes with small individual runs of ductwork to individual rooms (Figure 13.18a and Figure 13.18b). Both systems employ **plenums**, which are rectangular

---

static pressure the pressure inside the duct system and an indicator of the amount of resistance to air flow within the system; usually measured in inches of water column (IWC) or Pascals (Pa).

jumper ducts small sections of duct installed in ceilings that allow air to flow between rooms.

transfer grilles vents placed in walls to allow air to flow between rooms.

trunk-and-branch duct systems large ducts (trunks) installed through the center of the house with smaller ducts (branches) that run off the trunks to supply individual rooms.

spider duct systems individual runs of ductwork directly from the furnace or air handler to individual rooms.

plenums rectangular boxes attached to the furnace that receive heated or cooled air which is then distributed to the trunks and individual ducts.

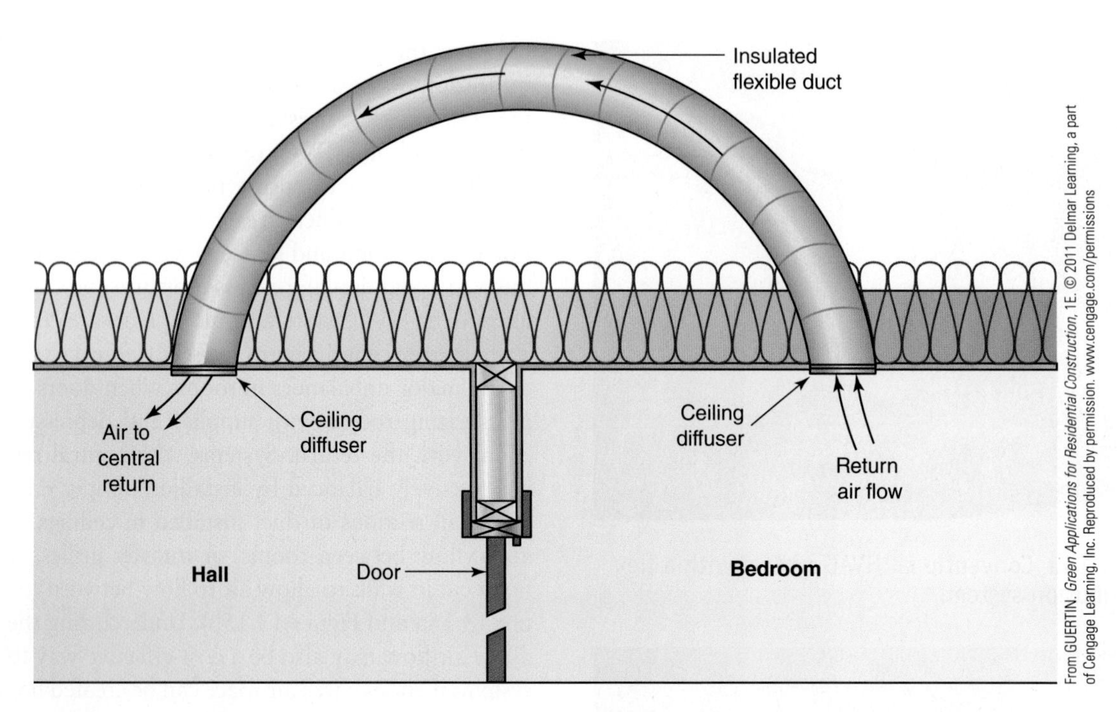

Figure 13.15a Jumper ducts provide a pathway for air behind closed doors to access central returns.

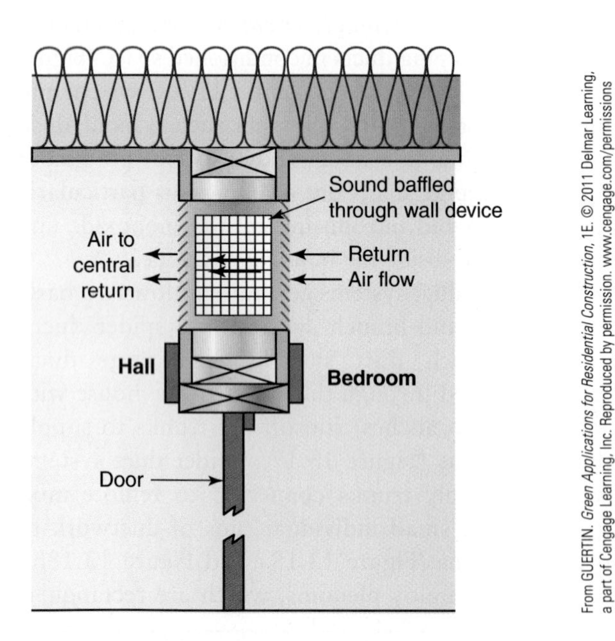

Figure 13.15b A through-the-wall vent provides return pathway from closed rooms to a central return. The sound baffle reduces noise transmission.

boxes attached to the furnace that receive heated or cooled air which is then distributed to the trunks and individual ducts. Trunk-and-branch systems generally use less material and provide a more even supply of air; however, spider systems can work effectively when designed and installed properly. The shortest and straightest duct runs provide the best performance. In a radial duct system, the branch ducts that deliver conditioned air to individual supply outlets are connected directly to a small supply plenum.

Most residential ductwork is either round or rectangular and made from rigid metal, fiberglass duct board, or plastic flexible ducts (Figures 13.19a, b, and c). For maximum performance, flexible ducts must be installed fully extended in straight runs with adequate support to avoid sagging, using sweeping turns or elbows instead of tight turns, and it must not be compressed to fit in available building cavities. Metal ducts are less prone to the same installation problems as flexible ducts. Metal ducts also have a smoother interior, creating less resistance to air flow than flex duct, which can allow for smaller diameter ducts that provide equivalent air flow. The building code allows the use of building cavities (e.g., walls or floors) for return ducts, but this strategy is not recommended for high-performance homes because these areas are difficult to seal and insulate effectively (Figure 13.20a and Figure 13.20b).

Duct Insulation

Most flexible duct material comes with attached foil-faced fiberglass insulation. Duct board combines the duct structure and insulation in a single material, and metal-ducts are wrapped with separate insulation, usually foil-faced

---

radial duct system a distribution system where the branch ducts that deliver conditioned air to individual supply outlets are connected directly to a small supply plenum.

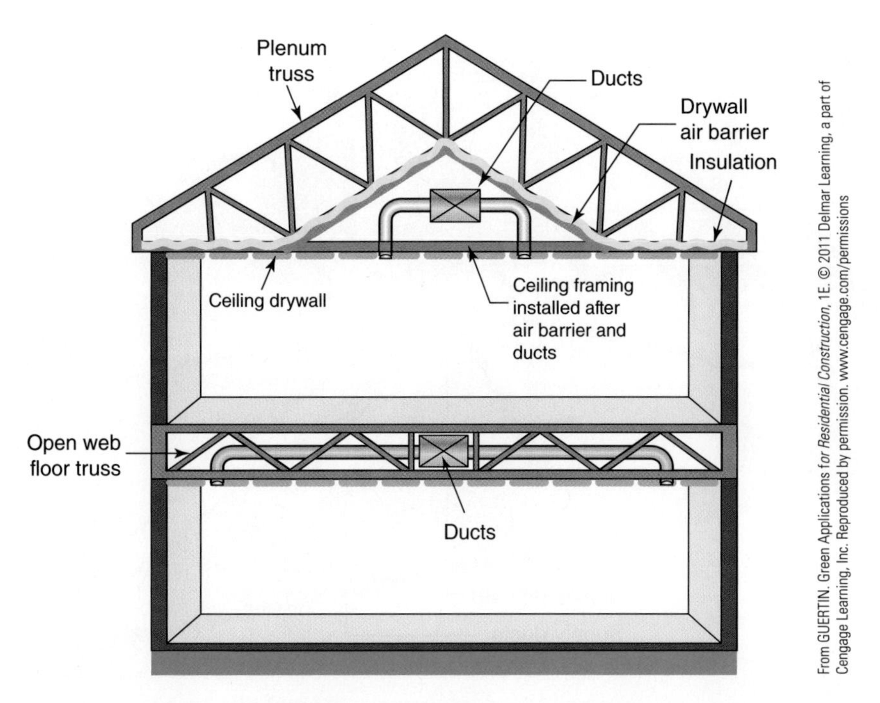

Figure 13.16 Ductwork can be run within the conditioned space through special lateral chases created using plenum trusses or open-web floor trusses. Insulating the roofline also brings the attic within the building envelope.

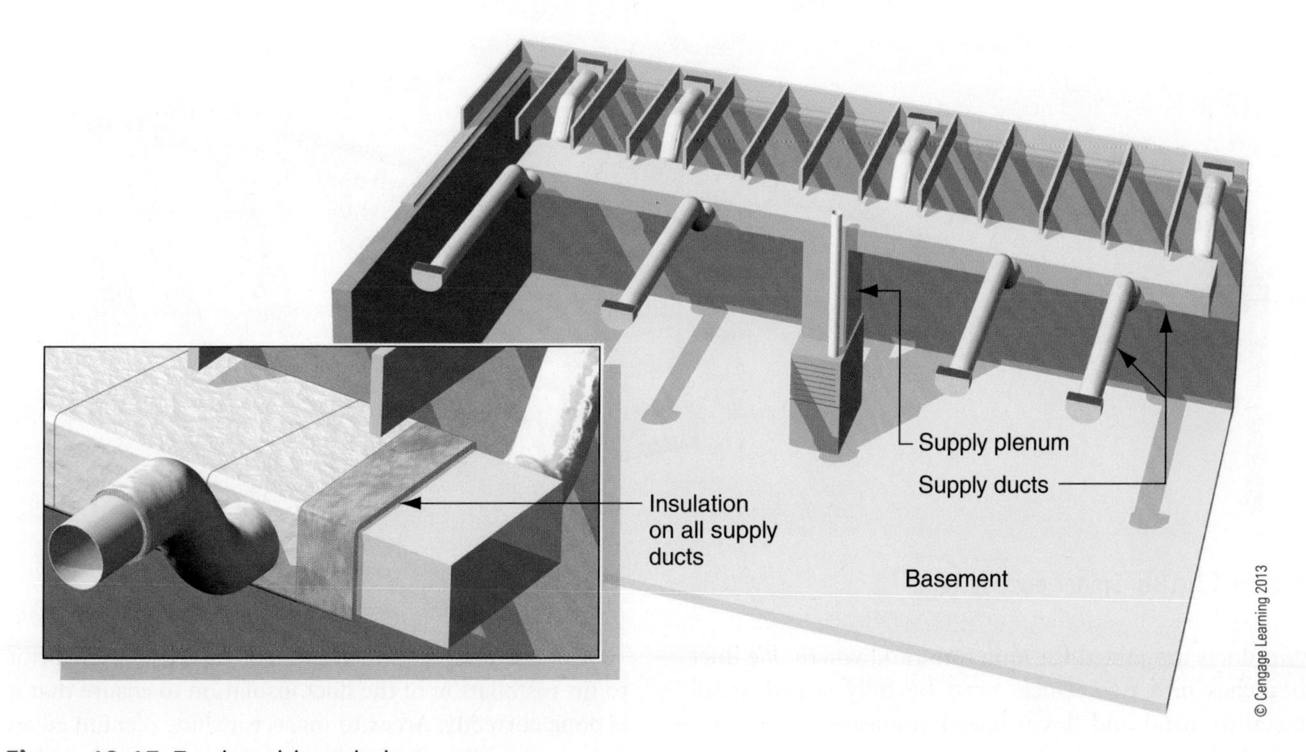

Figure 13.17 Trunk-and-branch duct system.

fiberglass, after assembly. Ducts must be insulated when located in unconditioned space; insulation is not required when installed in conditioned space, but most ductwork in residential projects is insulated.

Duct Sealing

Forced-air systems should be fully ducted using code-approved materials. Use metal collars and connectors at connections, such as take-offs at plenums, where

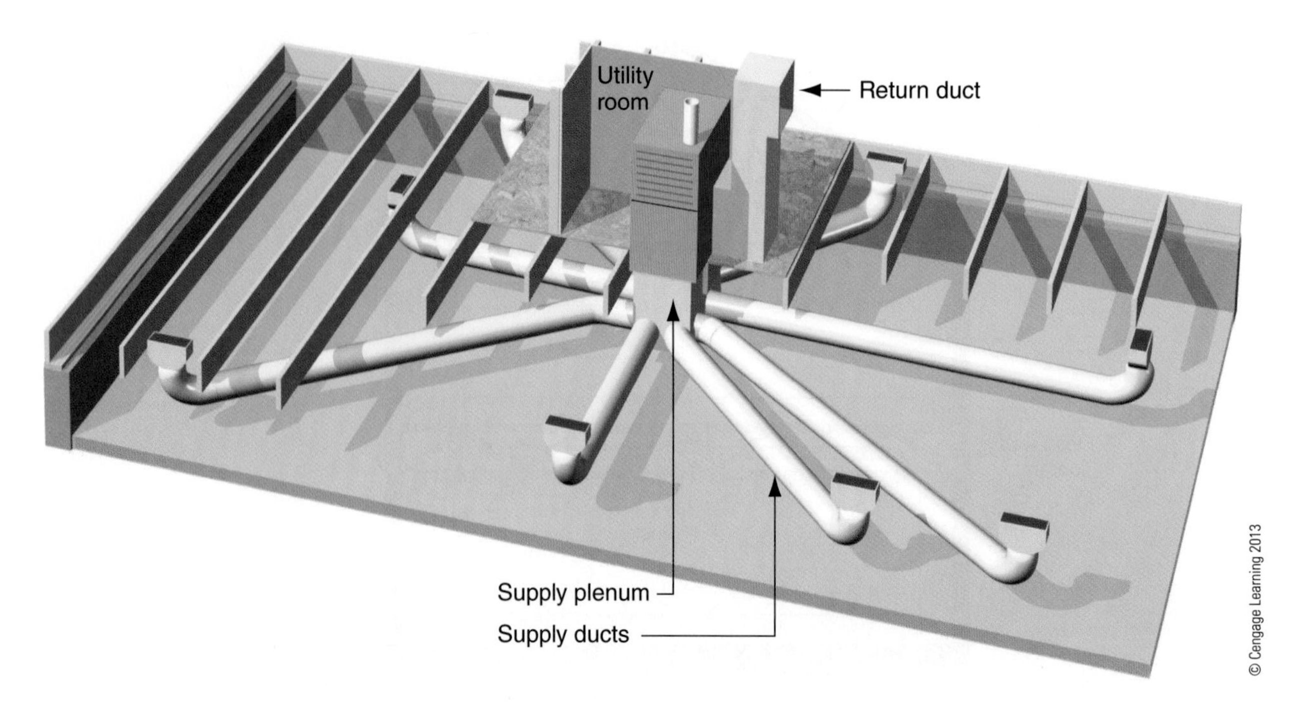

Figure 13.18a Radial duct system.

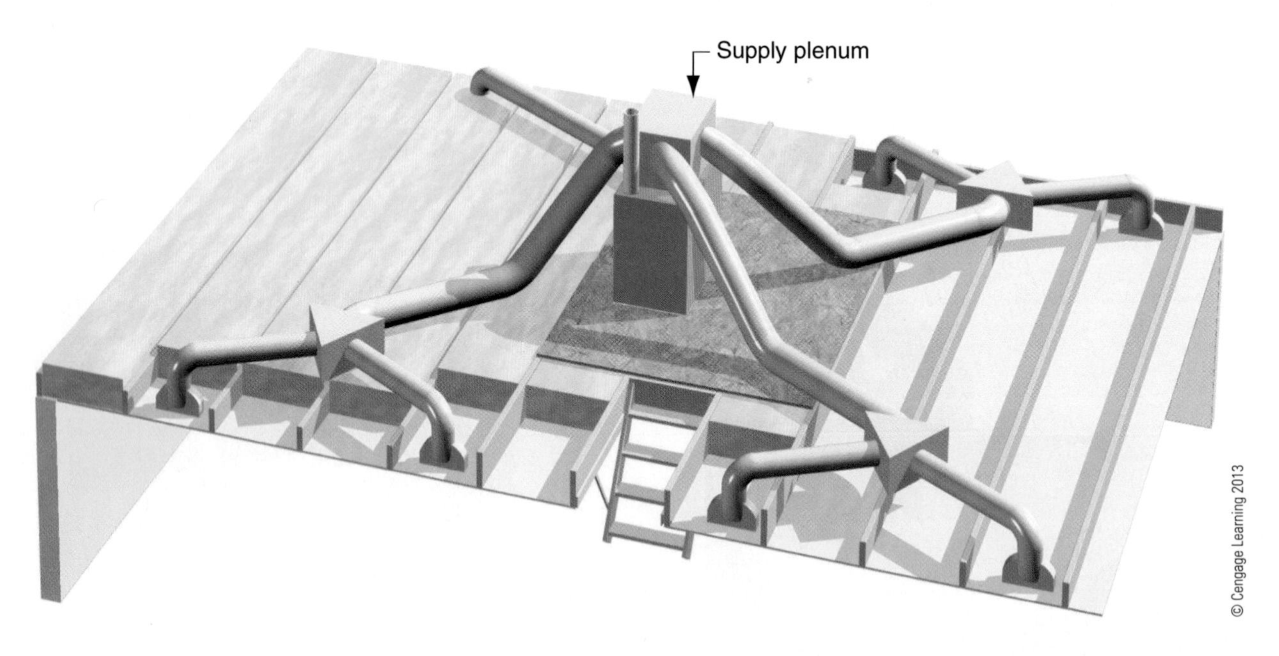

Figure 13.18b Spider duct system.

two ducts are joined for long runs, and where one duct branches into two. Ducts must be fully sealed at all metal-to-metal and flex-to-metal connections to eliminate leaks and maximize the system performance. For best results, ducts must be sealed at all seams with mastic or mastic tape (Figure 13.21).

Foil-tape labeled approved according to UL181 requirements is code-approved for duct sealing, but it is difficult to install it effectively in the field and is best avoided in favor of mastic or mastic tape in a green home. Duct tape is not designed for sealing ducts and is not appropriate for use in HVAC installations.

Duct sealing should be thoroughly inspected prior to the installation of the duct insulation to ensure that it is done correctly. Areas to inspect include plenum edges (where the plenums meet the furnace or air handler) as well as the seams in the units themselves, ducts and trunk lines where they connect to plenums, all linear seams in metal ducts, joints between sections of metal duct, joints between flex duct and metal pipes, and all seams in elbows and boots (Figure 13.22). Ducts are often sealed at the foil face of the insulation, which does not provide an adequate air seal and is not a substitute for sealing at metal and flex connections (Figure 13.23).

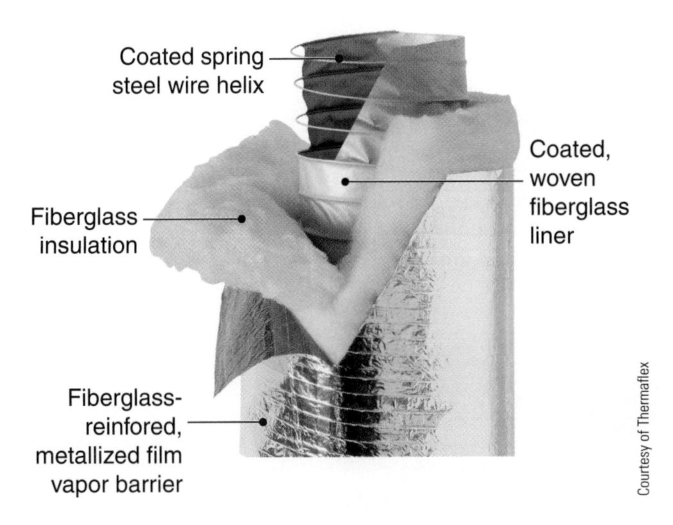

Figure 13.19a Flex duct consists of an inner liner with wire supports, fiberglass insulation, and an outer liner.

Figure 13.19b Ducts manufactured from compressed fiberglass with a foil backing.

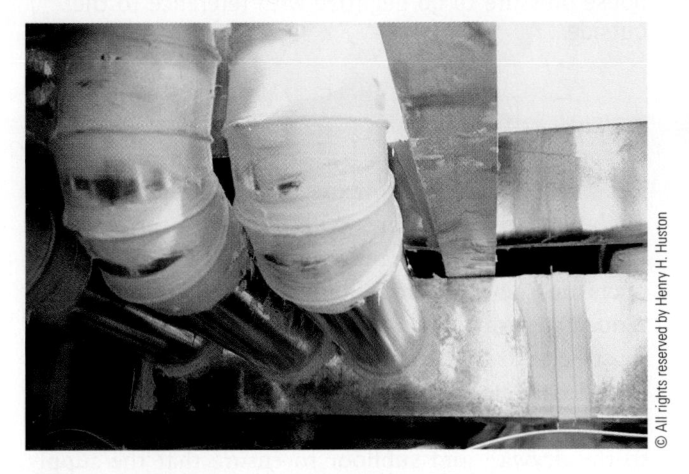

Figure 13.19c Metal ducts sealed with mastic.

Figure 13.20a A panned return using a wall cavity. Using building cavities as ducts, like this panned wall return, are not recommended.

Figure 13.20b A panned return using the floor joists. Sheet metal is used to block off between the joists as well as to create the "pan" on the bottom.

Figure 13.21 A technician applies mastic to a joint between two metal ducts.

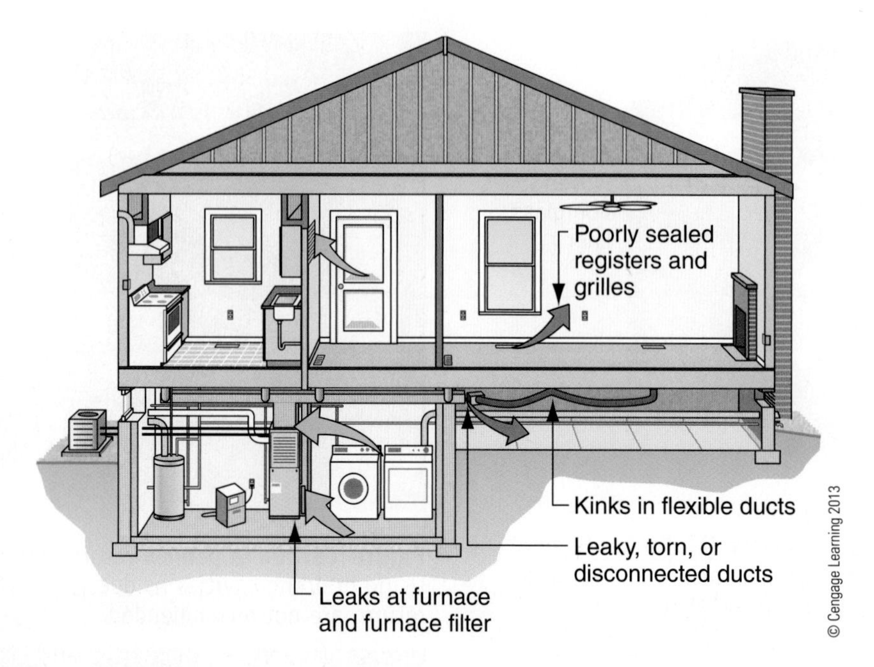

Figure 13.22 Common duct leaks.

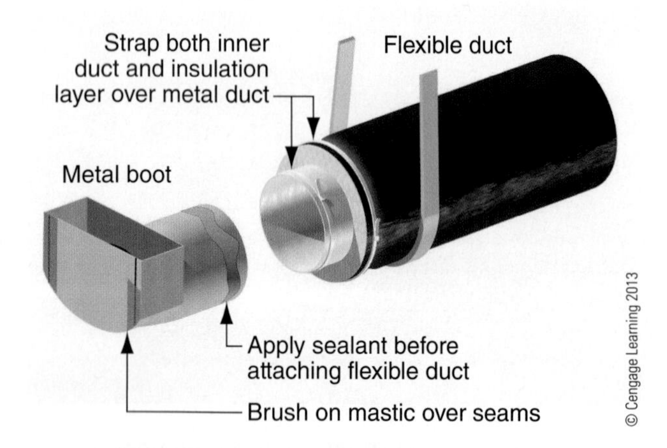

Figure 13.23 All joints and connections should be sealed throughout the duct system.

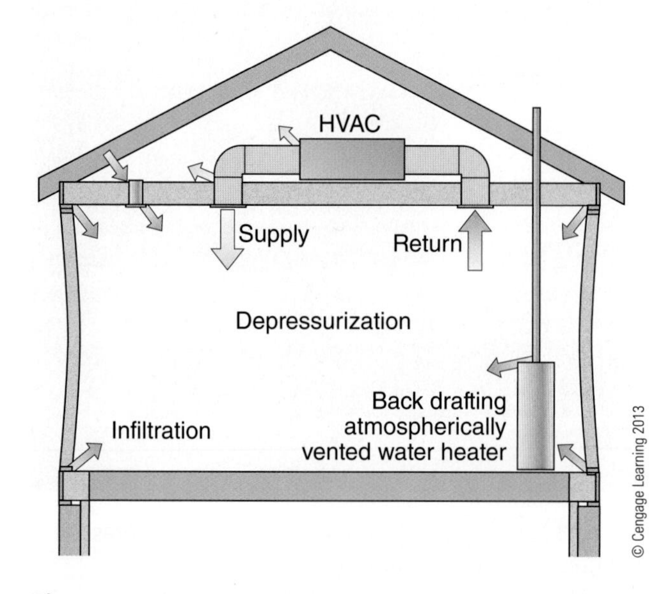

Figure 13.24 Predominant supply leaks cause the house pressure to go negative with reference to the outside.

Poorly sealed ductwork allows conditioned air to leak both inside and outside the building envelope, reducing system efficiency and creating pressure imbalances. Supply ducts that leak more than return ducts create negative pressure as the returns draw more air back into the system than the supplies deliver (Figure 13.24). Negative pressure can draw air in through leaks in the building envelope; this air may be hot or cold and contain excess humidity, dirt, dust, pollen, combustion gases, and other pollutants. Similarly, leaky return ducts create positive pressure, which can force air out through envelope leaks, forcing moisture into wall cavities and wasting energy (Figure 13.25).

Duct Register Location

Traditionally, duct system design locates supply registers near the exterior walls (usually near windows) and return registers at the interior. This design was based on the principle that homes had significant heat gain and loss through inefficient exterior walls and windows. High-performance homes with minimal heat gain and loss at exterior walls and windows can locate supply ducts on the interior to allow for shorter duct runs, reducing energy loss and saving on material. Duct boots must be sealed where they penetrate the building envelope according to the 2009 International Energy Conservation Code (IECC). As good practice, seal all boots to the drywall and subfloor to ensure that the supply

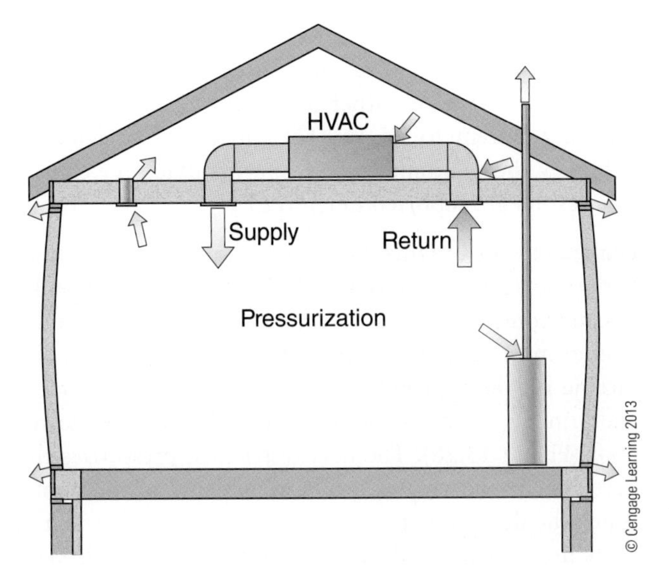

Figure 13.25 Predominant return leaks cause the house pressure to go positive with reference to the outside.

air is delivered effectively and return air is not drawn from unintended areas, such as semiconditioned attics or basements (Figure 13.26).

Duct Leakage Testing

Ducts should be tested for leakage with such equipment as the Minneapolis Duct Blaster® or Retrotec DucTester. Ducts are tested either during construction (at the rough-in stage) or after the home is completed. There are pros and cons to both approaches. Testing duct leakage at the rough-in stage allows for easy repair of defects that may be impossible to locate after completion. Testing at the final stage, on the other hand, will allow for identification of ducts that were disconnected during construction and for instances where the drywall is covering supplies or returns.

Duct testing equipment consists of a calibrated fan, manometer, and pressure hoses. The setup consists of turning off the HVAC system and sealing all registers

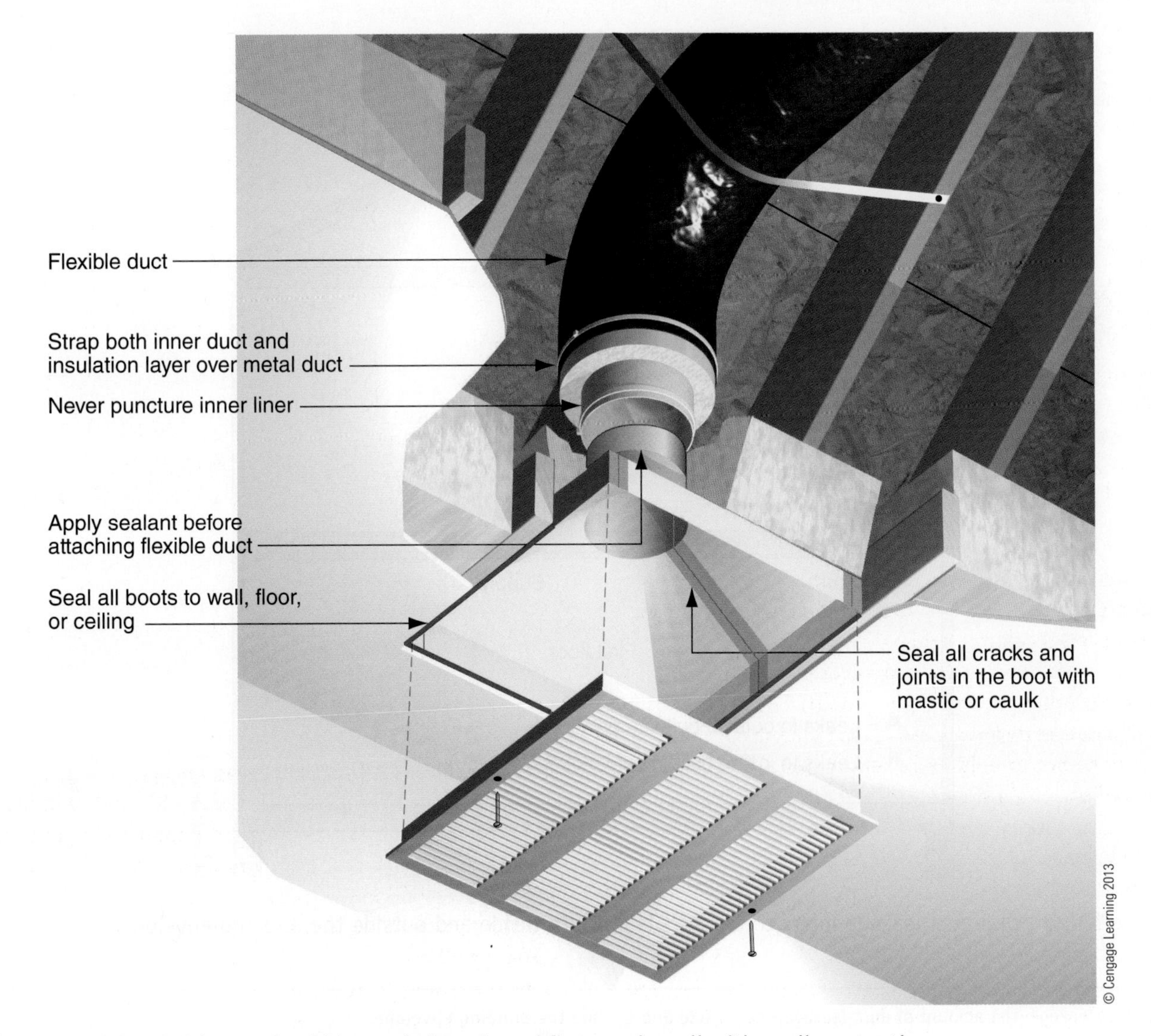

Figure 13.26 HVAC boots should be sealed to the subfloor or drywall with caulk or mastic.

with tape. The duct tester fan is installed to a large, centrally located return or to the air handler. Duct testing includes pressurizing or depressurizing the duct system to an industry-established pressure. Most of the United States pressurizes the ducts to 25 Pa with respect to the outside. Some areas of the country prefer to depressurize the duct system to −25 Pa because it tends to draw the tape tight to the registers.

Total Duct Leakage Total duct leakage refers to leakage throughout the whole duct system that is both inside and outside the building envelope (Figure 13.27). Duct leakage outside the building envelope is considered an energy penalty because the homeowner is losing air that they paid to heat and cool. Although duct leakage inside the building envelope is not an energy penalty, it may create comfort complaints and pressure imbalances. Increasingly, green building programs require homes meet thresholds for both total duct leakage and leakage to outside the building envelope.

Total duct leakage is calculated by opening a window or door to equalize the house and outdoor pressures. The duct tester either pressurizes or depressurizes the duct system to 25 Pa. The fan flow is calculated either by the manometer or from manufacturer charts. The results are reported in CFM25.

Leakage to the Outside Leakage to the outside refers only to the duct leakage that is outside the building envelope. To calculate only the leakage to the outside, all leakage to the interior must be eliminated. The duct system is prepared by turning off the air handler, sealing all registers with tape, and bringing the house pressure to 25 Pa with the blower door (Figure 13.28). The duct tester then pressurizes the duct system and creates a 0-Pa differential to the home. When the ducts are at 0 Pa with respect to the house, the ducts are also at 25 Pa with respect to the outside. The test can also be performed in reverse with the house and ducts depressurized. As covered in Chapter 2, air leakage requires an opening or penetration and pressure difference

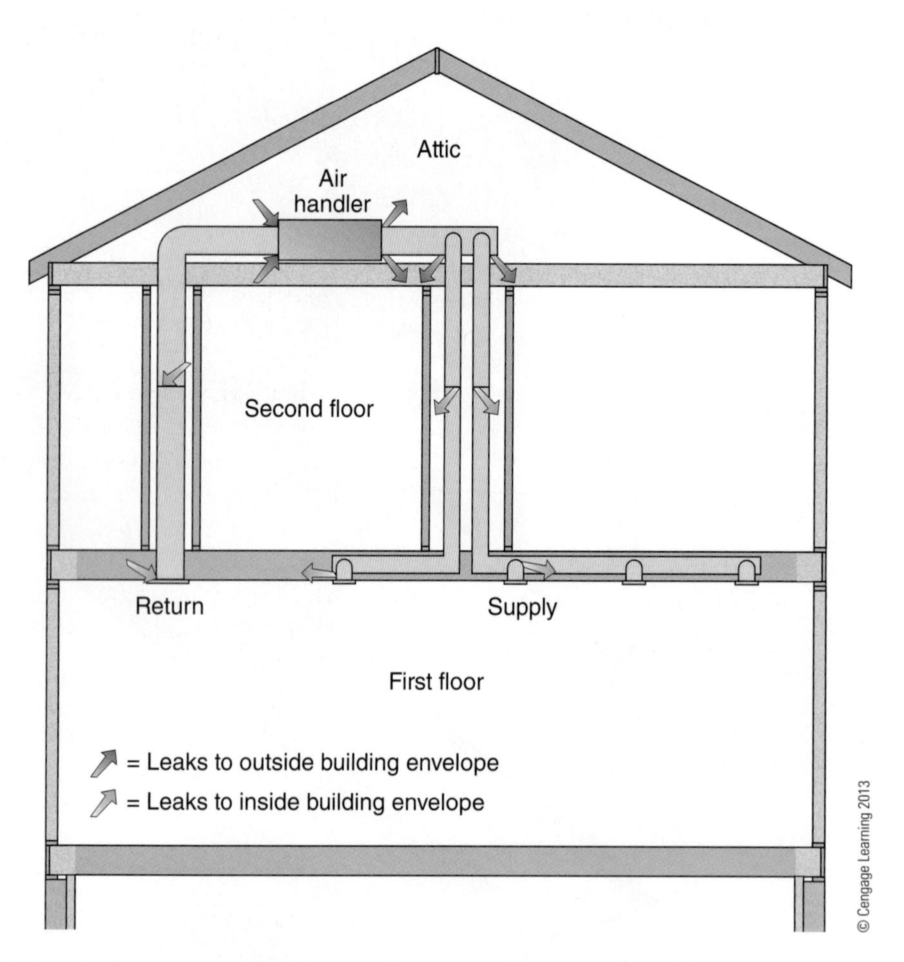

Figure 13.27 Total duct leakage refers to the duct leaks both inside and outside the building envelope.

---

total duct leakage the amount of duct leakage both inside and outside the building envelope.

leakage to the outside duct leakage that is located not within the building envelope.

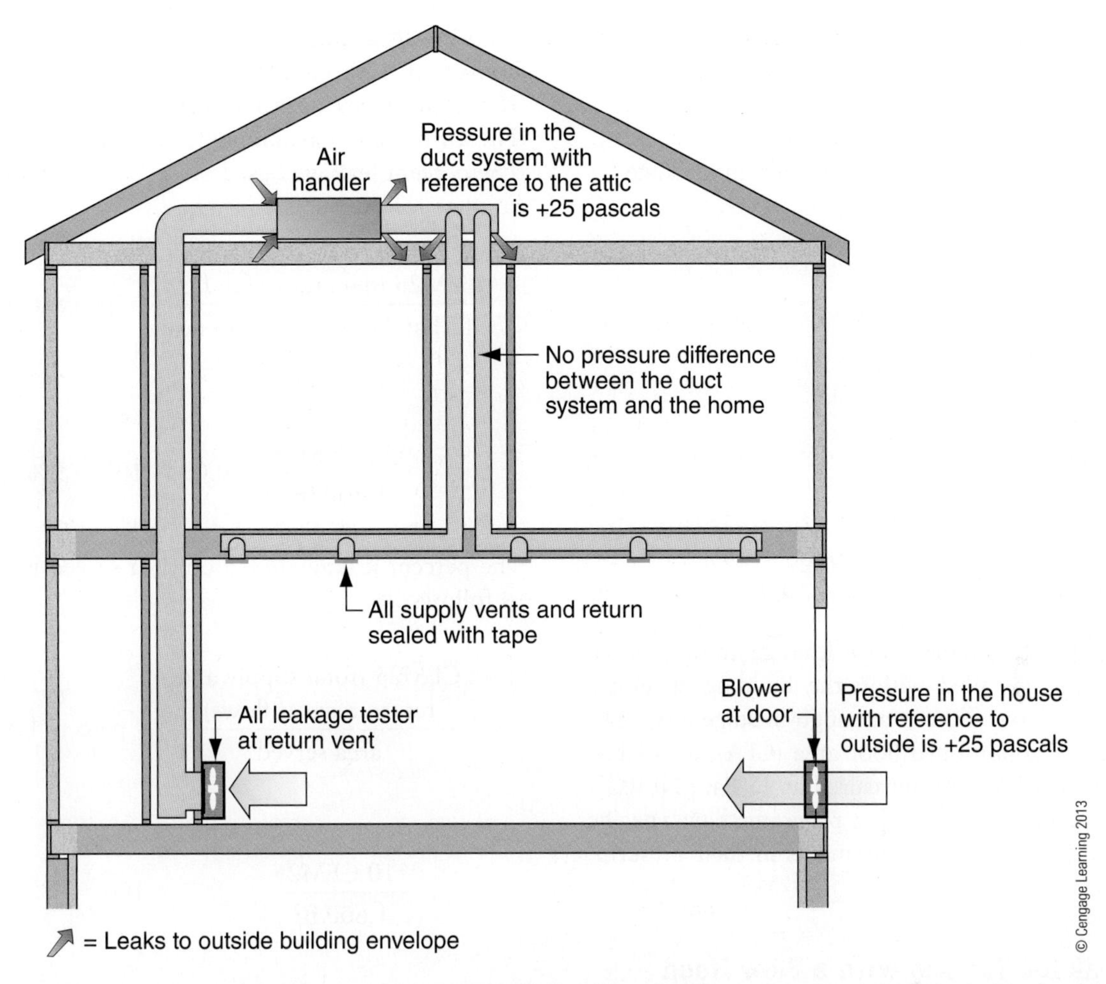

Figure 13.28 The duct leakage to the outside is calculated by using the blower door in conjunction with the duct blaster.

to occur. By making the ducts and the house the same pressure, no leakage occurs between the ducts and inside the building envelope. Because of this, the leakage to the outside is always less than or equal the total leakage.

Calculating Percentage Leak

Duct tester fan flows when used alone can only identify total duct leakage. When the fan flow is compared with either the air handler fan flow or the area served by the duct system, it identifies duct leakage to the outside. The percent leakage based on floor area served is calculated by dividing the duct tester fan flow by the area served:

$$\frac{\text{CFM25 (duct tester fan flow)}}{\text{area served}} = \text{percent leakage}$$

The duct leakage based on air handler fan flow is calculated by dividing the duct tester fan flow by the air handler fan flow:

$$\frac{\text{CFM25 (duct tester fan flow)}}{\text{air handler nominal fan flow}} = \text{percent leakage}$$

Step-by-Step Calculations Let's look at a 3,450 ft^2 home with two air handlers. The first floor is 1,400 ft^2, and the second floor is 1,050 ft^2. The first-floor duct system has 89 CFM25 leakage to the outside. The second-floor system has 171 CFM25 leakage to the outside. The first-floor air handler has a nominal air flow of 800 CFM, and the second floor has an air flow of 600 CFM. The percent leakage based on system fan flow is calculated as follows:

$$\frac{\text{CFM25 (duct tester fan flow)}}{\text{air handler nominal fan flow}} = \text{percent leakage}$$

First floor:

$$\frac{\text{89 CFM25}}{\text{800 CFM}} = 0.11 \times 100 = 10\%$$

Second floor:

$$\frac{\text{171 CFM25}}{\text{600 CFM}} = 0.285 \times 100 = 28.5\%$$

The percent leakage based on area served is calculated as follows:

$$\frac{\text{CFM25 (duct tester fan flow)}}{\text{area served}} = \text{percent leakage}$$

First floor:

$$\frac{89\ \text{CFM25}}{1{,}400\ \text{ft}^2} = 0.06 \times 100 = 6\%$$

Second floor:

$$\frac{171\ \text{CFM25}}{1{,}050\ \text{ft}^2} = 0.16 \times 100 = 16\%$$

The 2009 IECC requires duct leakage testing of all ductwork not installed within the building envelope. Maximum leakage allowed at rough-in stage is 8 CFM per 100 ft^2 of conditioned floor area (CFA), or 6 CFM per 100 ft^2 of CFA, both measured at 25 Pa. ENERGY STAR and most green building programs have specific maximum duct leakage requirements in their prescriptive paths.

Duct Leakage Testing with a Flow Hood

An alternate method of testing duct leakage to the outside can be done with a blower door and a flow hood. Using the blower door to depressurize the house to −25 Pa with respect to the outside, the flow hood is used to measure the amount of air moving through each supply and return grille in CFM. All the air flows are added together for each separate system to determine the total duct leakage, providing a total CFM25. Dividing the CFM25 by the total square footage of the floor area for the section served by each system provides the percentage of duct leakage to the outside for that system. Alternatively, the leakage as a percent of air handler fan flow can be calculated.

Duct leakage testing with a flow hood is performed as follows:

1. Walk the house and record all register locations.
2. Use the blower door to depressurize the house to −25 Pa with respect to outside.
3. Use the flow hood to measure the flow through each supply and return grille.
4. Add all flows to determine total duct leakage in CFM.
5. Calculate percentage of leakage by dividing the total CFM25 by the floor area (ft^2) of the zone that HVAC system serves.

Step-by-Step Calculations Let's look at a 1,600 ft^2 home with 1 air handler. The total leakage through the 18 supply vents and 3 returns vents is 110 CFM25. The air handler has a nominal air flow of 1,200 CFM. The percent leakage based on system fan flow is calculated as follows:

$$\frac{\text{CFM25 (duct tester fan flow)}}{\text{air handler nominal fan flow}} = \text{percent leakage}$$

$$\frac{110\ \text{CFM25}}{1{,}200\ \text{ft}^2} = 0.09 \times 100 = 9\%$$

The percent leakage based on area served is calculated as follows:

$$\frac{\text{CFM25 (total supply and return register flows)}}{\text{area served}} = \text{percent leakage}$$

$$\frac{110\ \text{CFM25}}{1{,}600\ \text{ft}^2} = 0.07 \times 100 = 7\%$$

Test and Balance

Duct systems are designed to provide specific amounts of air flow at supply and return registers. Finished installations often do not meet the design requirements due to constrictions, defective connections, and site changes that affect system performance. The total CFM flowing in and out of each register should be measured with a flow hood or pressure pan, comparing the total flow to the design criteria (Figure 13.29). Flows that vary more than 15% or 10 CFM from the design should be adjusted to meet the design through the use of dampers that decrease flow in some runs, causing the flow to increase in others. When all registers meet the designed flow rates, the system will perform at its maximum efficiency and minimize pressure differentials within and between rooms. The Air Conditioning Contractors of America (ACCA) provides guidance in Manual B on the testing and balancing of duct systems and hydronic systems.

Zoning

Forced-air systems that supply multiple floors or large areas with different heating and cooling requirements should use zone controls to provide different levels of heating and cooling to meet each areas needs (Figure 13.30).

Courtesy of TSI Incorporated

Figure 13.29 The amount of air flowing out of a supply register or into a return is measured with a flow hood.

Courtesy of EWC Controls–Englishtown, NJ

Figure 13.30 Zone damper control.

Zone controls use separate thermostats in each area to send information to a control panel that adjusts automatic dampers (Figure 13.31) that open and close trunk lines to direct air flow only to the zone that needs conditioning. Zone controls allow one central HVAC system to evenly condition multi-floor homes and enable homeowners to control the amount of HVAC supplied to different sections of their homes.

zone control a device that controls the amount of air flow to different areas of the home.

HVAC System Design

The Air Conditioning Contractors of America (ACCA) publishes standards for sizing heating and cooling systems (Manual J), selecting equipment (Manual S), designing duct systems (Manual D), and selecting registers (Manual T).

ACCA Manual J

Heating and cooling equipment sizing is calculated using the local climate, the size, shape, and orientation of the house; the amount and quality of the insulation; window size, location, and efficiency; air leakage rate; number of occupants; and internal gains from lighting and electronic equipment. Lighting and electronics can create significant amounts of heat, which adds to the cooling load while reducing the heating load. Loads are calculated using a system known as Manual J, created by ACCA. Currently in its eighth edition, Manual J provides formulas that can be used to calculate loads manually, but most professionals use one of several software versions available, such as EnergyGauge® and Wrightsoft's Right-J®. To create an accurate load calculation, data entered into the program must match the building specifications, including window and door National Fenestration Rating Council (NFRC) ratings, insulation quantities, building orientation, air leakage rate, local climate data, and other factors.

Manual J produces a report (Figure 13.32) that includes the amount of both heating and cooling Btu required. These data are used to select specific equipment that matches these requirements as closely as possible. Furnaces, boilers, and heat pump sizes are based on total Btu they produce. Forced-air furnaces and heat pumps are available in sizes ranging from about 38,000 to 115,000 Btu. Boiler sizes range from about 40,000 to 300,000 Btu.

ACCA Manual S

ACCA Manual S assists in the selection and sizing of heating and cooling equipment to meet Manual J loads on the basis of local climate and ambient

Air Conditioning Contractors Association (ACCA) a professional organization that publishes standards for heating and cooling systems.

ACCA Manual J a guide for sizing residential heating and cooling systems based on local climate and ambient conditions at the building site.

ACCA Manual S a guide in the selection and sizing of heating and cooling equipment to meet Manual J loads.

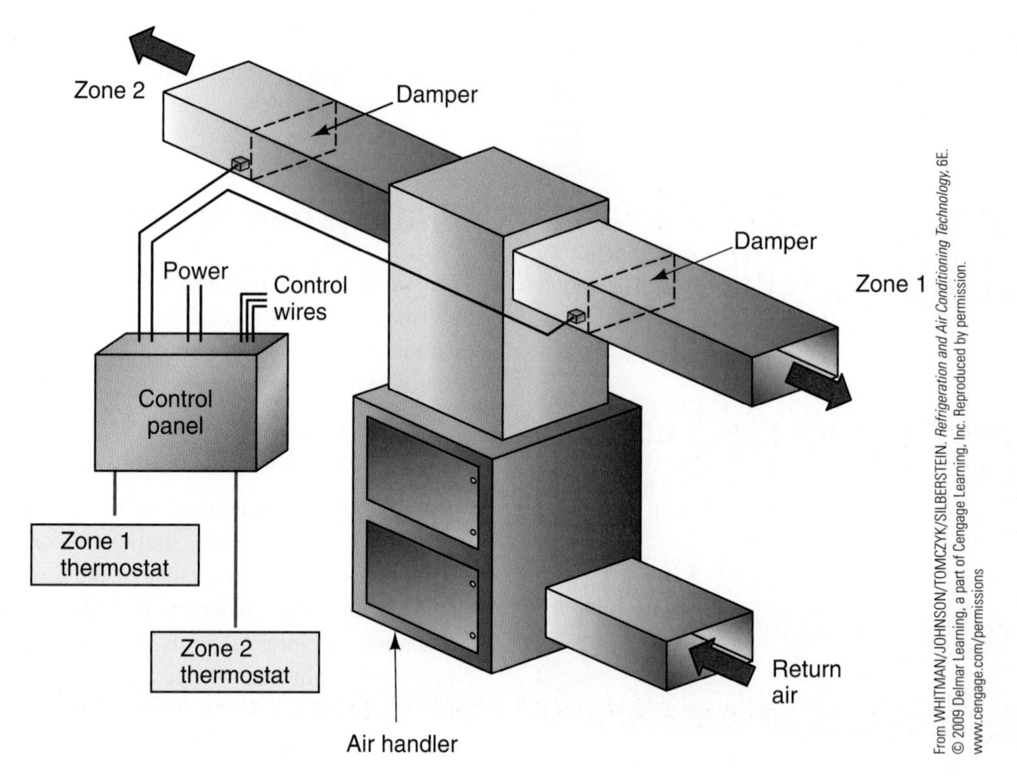

Figure 13.31 A common two-zone system layout.

conditions at the building site. Manual S covers selection strategies for all types of cooling and heating equipment.

ACCA Manual D

ACCA Manual D is a guide to the design of residential duct systems that helps ensure that the ducts will deliver the proper amount of heated or cooled air to each room. Duct designs are calculated on the basis of the total amount of CFM that is desired for delivery at the end of each duct run. Ducts create resistance against air flow depending on their length and fittings, such as elbows and boots. Additional resistance comes from filters and coils. Manual D helps determine the effective duct length, the total amount of pressure loss of the straight sections, and all fittings in each duct run. Individual fittings, such as elbows, can produce as much resistance as a single long run of straight duct, referred to as the equivalent length. Equivalent lengths can be added together in any run to determine the total effective duct length. For example, a metal elbow adds between 20' and 35' of equivalent length to a duct run, depending on the size of its radius.[1]

Manual J and D calculations are most often performed with automated software, which determines the total building loads and assists in selecting HVAC equipment (Figure 13.33).

ACCA Manual T

ACCA also publishes Manual T, which shows designers how to prevent drafts and stagnant air problems caused by improper sizing and incorrect equipment selection. It explains how to select, size, and locate the supply air registers, and return grilles. The manual also provides examples of how to use manufacturers' performance data to calculate pressure losses and control noise.

[1] ACCA Manual D, 1995, Appendix 3, P A3-20.

ACCA Manual D a guide to the design of residential duct systems that helps ensure that the ducts will deliver the proper amount of heated or cooled air to each room.

effective duct length the total amount of pressure loss of the straight sections and all fittings in each duct run.

equivalent length the comparable length of a single straight duct run when taking into account air flow resistance from duct compression, elbows, fittings, and other obstructions.

ACCA Manual T a guide for designers on how to select, size, and locate supply outlets and return inlets.

wrightsoft® Load Short Form
Entire House
Southface

Job:
Date: Feb 04, 2011
By: John Smith HVAC

241 Pine Steeet NE, Atlanta, GA 30308 Phone:404-872-3549 Fax: 404-872-5009 Web: www.southface.org

Project Information

For: ABC Builders
101 Sample Way, Atlanta, GA 30308

Design Information

	Htg	Clg
Outside db (°F)	24	92
Inside db (°F)	70	75
Design TD (°F)	46	17
Daily range	-	M
Inside humidity (%)	30	50
Moisture difference (gr/lb)	19	38

Infiltration	
Method	Simplified
Construction quality	Tight
Fireplaces	0

HEATING EQUIPMENT

Make		
Trade		
Model		
GAMA ID		
Efficiency	80AFUE	
Heating input	0	Btuh
Heating output	0	Btuh
Temperature rise	0	°F
Actual air flow	600	cfm
Air flow factor	0.060	cfm/Btuh
Static pressure	0.50	in H20
Space thermostat		

COOLING EQUIPMENT

Make		
Trade		
Cond		
Coil		
ARI ref no.		
Efficiency	0 SEER	
Sensible cooling	0	Btuh
Latent cooling	0	Btuh
Total cooling	0	Btuh
Actual air flow	600	cfm
Air flow factor	0.047	cfm/Btuh
Static pressure	0.50	in H20
Load sensible heat ratio	0.86	

ROOM NAME	Area (ft²)	Htg load (Btuh)	Clg load (Btuh)	HtgAVF (cfm)	ClgAVF (cfm)
Bedroom 1	100	***1243***	***1389***	74	66
Bedroom 2	110	***759***	***1234***	45	58
Bedroom 3	110	***733***	***1215***	44	57
Master Bed	154	***1500***	***1407***	89	66
Master Bath	48	***381***	***473***	23	22
Bath	70	***425***	***210***	25	10
Closet	36	***0***	***0***	0	0
Utility	70	***150***	***623***	9	29
Kitchen/Living	594	***4487***	***5763***	267	272
Hall	108	***0***	***0***	0	0
Crawl Space	1400	***400***	***400***	24	19
Entire House	2800	***10077***	***12715***	600	600
Other equip loads		497	338		
Equip. @ 0.97 RSM			12608		
Latent cooling			2164		
TOTALS	2800	***10574***	***14773***	600	600

Figure 13.32 There are numerous software programs that perform ACCA Manual J load calculations. This report shows the amount of heating and cooling required for each room of the home. The required amount of supply air is also presented.

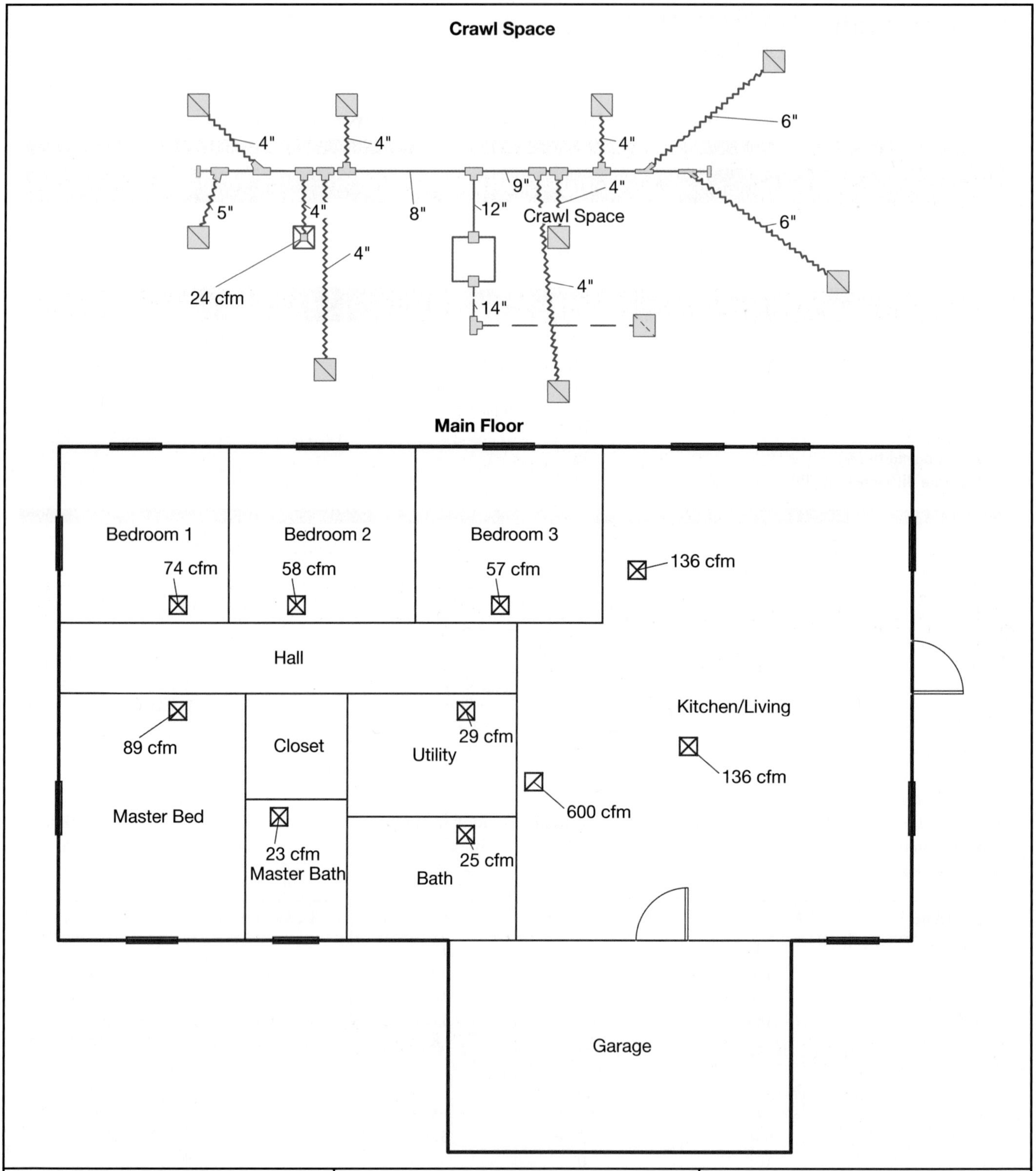

Crawl Space

Main Floor

Job #: **Performed by John Smith HVAC for:** ABC Builders 101 Sample Way Atlanta, GA 30308	**Southface** 241 Pine Street NE Atlanta, GA 30308 Phone: 404-872-3549 Fax: 404-872-5009 www.southface.org	Scale: 1 : 83 Page 1 Right-Suite® Universal 7.1.25 RSU02996 2011-Mar-16 20:05:16 ...\ECH Builder Training Sample.rup

Figure 13.33 This ACCA Manual D duct design corresponds to the same home as Figure 13.32. The software produces the duct layout along with duct sizes and register air flows.

(Continued)

Duct System Summary
Entire House
Southface

Job:
Date: Feb 04, 2011
By: John Smith HVAC

241 Pine Steeet NE, Atlanta, GA 30308 Phone:404-872-3549 Fax: 404-872-5009 Web: www.southface.org

Project Information

For: ABC Builders
101 Sample Way, Atlanta, GA 30308

	Heating		Cooling	
External static pressure	***0.50***	in H20	***0.50***	in H20
Pressure losses	0.33	in H20	0.33	in H20
Available static pressure	0.17	in H20	0.17	in H20
Supply / return available pressure	0.11/0.06	in H20	0.11/0.06	in H20
Lowest friction rate	0.040	in/100ft	0.040	in/100ft
Actual air flow	600	cfm	600	cfm
Total effective length (TEL)			422 ft	

Supply Branch Detail Table

Name		Design (Btuh)	Htg (cfm)	Clg (cfm)	Design FR	Diam (in)	H x W (in)	Duct Matl	Actual Ln (ft)	Ftg. Eqv Ln (ft)	Trunk
Bath	h	425	25	10	0.042	4.0	0 x 0	VIFx	16.0	250.0	st2
Bedroom 1	h	1243	74	66	0.045	4.0	0 x 0	VIFx	17.2	230.0	st3
Bedroom 2	c	1234	45	58	0.040	4.0	0 x 0	VIFx	12.0	265.0	st3
Bedroom 3	c	1215	44	57	0.045	4.0	0 x 0	VIFx	12.0	235.0	st2
Crawl Space	h	400	24	19	0.044	4.0	0 x 0	VIFx	14.0	240.0	st3
Kitchen/Living	c	2882	134	136	0.047	6.0	0 x 0	VIFx	21.6	215.0	st2
Kitchen/Living-A	c	2882	134	136	0.046	6.0	0 x 0	VIFx	18.8	225.0	st2
Master Bath	h	381	23	22	0.041	4.0	0 x 0	VIFx	19.0	250.0	st3
Master Bed	h	1500	89	66	0.047	5.0	0 x 0	VIFx	18.2	220.0	st3
Utility	c	623	9	29	0.044	4.0	0 x 0	VIFx	10.0	245.0	st2

Supply Trunk Detail Table

Name	Trunk Type	Htg (cfm)	Clg (cfm)	Design FR	Veloc (fpm)	Diam (in)	H x W (in)	Duct Material	Trunk
st3	Peak AVF	255	231	0.040	730	8.0	0 × 0	ShtMetl	st1
st2	Peak AVF	345	369	0.042	834	9.0	0 × 0	ShtMetl	st1
st1	Peak AVF	600	600	0.040	764	12.0	0 × 0	ShtMetl	

Return Branch Detail Table

Name	Grill Size (in)	Htg (cfm)	Clg (cfm)	TEL (ft)	Design FR	Veloc (fpm)	Diam (in)	H x W (in)	Stud/Joist Opening (in)	Duct Matl	Trunk
rb4	0 × 0	600	600	145.0	0.040	561	14.0	0 × 0		ShMt	

Figure 13.33 (Concluded)

Hydronic Heating Installation and Design Guide

Hydronic systems are designed and sized using the "Residential Hydronic Heating Installation and Design Guide," which is produced by the Air-Conditioning, Heating, and Refrigeration Institute (AHRI). This publication provides guidance for sizing heating equipment as well as choosing sizes and lengths of pipes, selecting radiator selection, and distinguishing manifold and pump details. Similar to Manual J, this guide covers the design, installation, and troubleshooting of both water- and steam-based hydronic heating systems and provides data to assist in designing hydronic systems with radiators and floor heating. Hydronic system designs are also detailed in the ACCA publication "Residential Hydronic Heating: Installation and Design."

Hydronic Distribution

Hydronic heating relies on hot water to be distributed throughout a house through metal or plastic pipe. In hydronic heating systems, the key decision involves the choice of radiant floors, wall-mounted radiators, or individual fan coils, which consist of a copper coil and a fan to distribute conditioned air through ducts or grilles. Although less common, hydronic radiant heat systems can also be installed in walls and ceilings. Hydronic systems should be designed with supply pipes that are as short as possible to reduce heat loss. Not unlike ducts, long pipe runs can lose heat, particularly if any are run in unconditioned spaces. Supply pipes that feed floor piping or radiators should be insulated between R-5 and R-11, depending on the pipe's length, location, and climate.

Hydronic systems take up much less space than forced-air ducts and do not create pressure imbalances that can cause problems with indoor air quality. Some systems use fan-coil distribution that can provide cooling as well as heat, but this is not a common residential application; in most cases, a separate system must be installed to provide air conditioning if required. The initial installation cost of hydronic systems is higher than that of forced-air systems.

Hydronic System Materials

Most modern hydronic systems use cross-linked polyethylene piping with copper or plastic manifolds to manage distribution to different areas of the house. Electric pumps and remote-controlled solenoid valves move water to rooms when heat is required. Older systems used copper pipe, which is still employed in some areas. The heat can be delivered through under-floor systems or by terminal units, commonly referred to as radiators and baseboard heaters.

Floor Installations

Radiant floor systems are typically installed in slabs or under wood-framed floors. Slabs should be insulated beneath to limit heat loss to the ground. Framed installations can use special subflooring sheets that have aluminum-lined channels for containing the pipes (Figure 13.34), or the tubing can be attached to the underside of the subfloor between joists. Metal plates can be used to help transfer heat from the pipes to the floor (Figure 13.35). Pipes can also be installed in thin, lightweight slabs placed on top of wood-framed floors. Insulation should be placed below pipes that are installed in or on the subfloor to limit heat loss. Due to the thermal mass of concrete slabs, these installations are slower to heat and cool than subfloor or underfloor applications. In climates where the outdoor temperature can experience rapid temperature swings, slab installations may not be able to respond quickly enough.

Finish Flooring over Radiant Heat Radiant floor heating can put some limitations on possible finish floor materials. Thick carpet decreases system efficiency, and solid wood flooring may warp or crack from the heat. Ceramic tile and stone floors over radiant heat are durable, transmit heat well to the living space, and can add additional thermal mass. Solid wood flooring should be quarter-sawn or an engineered product to avoid excessive movement from temperature changes. Review manufacturer's recommendations before using any flooring material over radiant heat floors.

Radiators

Most modern baseboard radiators use copper tubes with aluminum fins that dissipate the heat behind a metal cover (Figure 13.36). Warm air rises from the fins and is replaced by cooler air at floor level, creating a convective current that warms a room. Some units have adjustable vents that control the amount of air flowing past; opening and closing these vents can manage the amount of heat entering into the room through convection. Because these radiators do not have much mass, they warm and cool more quickly than floor radiant systems that change temperature slowly. Normally placed at windows and around the perimeter of a room, these radiators can also be placed in central locations in a very well-insulated and air sealed home. Doing so can potentially shorten pipe runs and allow for more flexibility in furniture location. Radiators

---

Air-Conditioning, Heating, and Refrigeration Institute (AHRI) a trade association representing heating, ventilation, air-conditioning, and commercial refrigeration manufacturers.

fan coils a simple device consisting of a heating or cooling coil and fan used to distribute heating and cooling into a space.

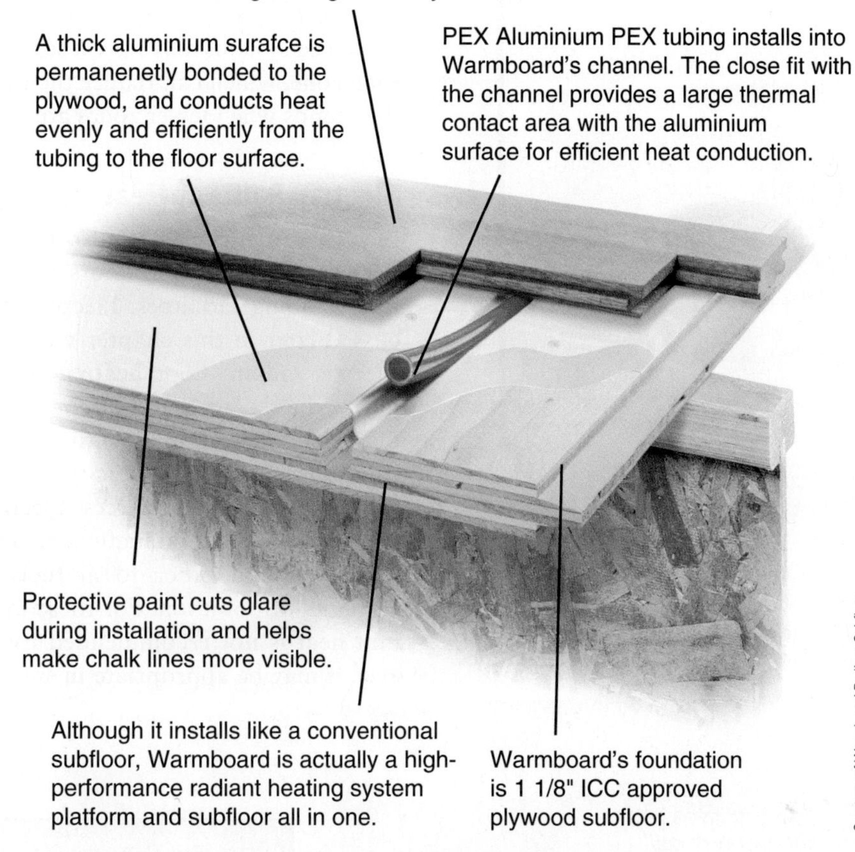

Figure 13.34 Radiant floor systems may be installed in special subfloor.

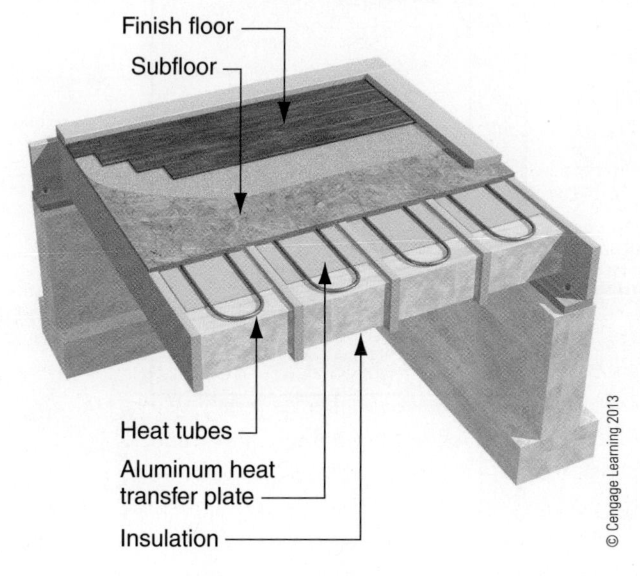

Figure 13.35 Radiant floor systems may be installed below the subfloor.

are available that can also serve as towel warmers for use in bathrooms. Radiators generally require water at 160°F or higher to heat effectively, whereas in-floor systems provide heat from water that is much cooler.

Older hot water and steam heat systems used cast iron radiators that heated and cooled more slowly than copper and aluminum models. New cast iron radiators are available for whole-house installations or for additions to older systems.

Fan Coils

Fan-coil distribution systems incorporate copper tubes and aluminum fins (similar to radiators) and a fan to distribute the conditioned air, either directly through grilles or via attached ductwork (Figure 13.37). Fan coils are not common in residential construction; however, they do provide an alternative method to offer both heating and cooling with a hydronic system.

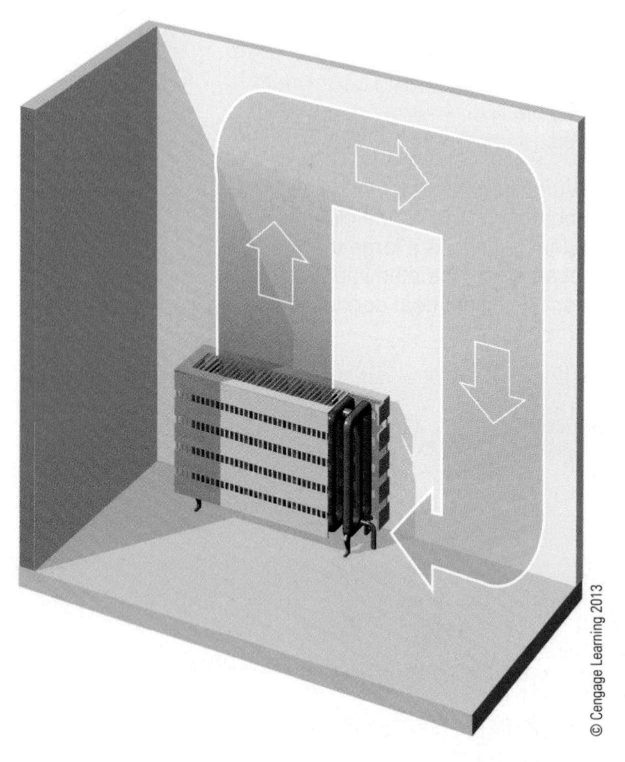

Figure 13.36 A traditional hydronic radiator showing how warm air rises from the fins and is replaced by cooler air at floor level.

Zoning

Hydronic heat can be zoned to heat different sections of house based on demand. Some systems have a single circulation pump with valves that open and close for each section according to thermostat demands. More advanced systems may have a separate smaller pump for each zone, which can reduce the total electrical used by the pumps when fewer zones are in use.

Electric Radiant Heat

Electric radiant heaters, also referred to as resistance heaters, can be used in floors and for wall-mounted or free-standing radiators. Electric radiators will be discussed later in this chapter with equipment selection. Electric radiant floor heating is typically installed in mats adhered to the subfloor below the finish flooring. Although electric heating converts 100% of the site energy to heat, electricity is generally more expensive than other fuel sources. Electrical generation and transmission is also inefficient and, when produced with coal and other fossil fuels, creates significant amounts of pollution. For these reasons, electric resistant heat is not recommended for most homes; however, it may be appropriate in superinsulated homes in

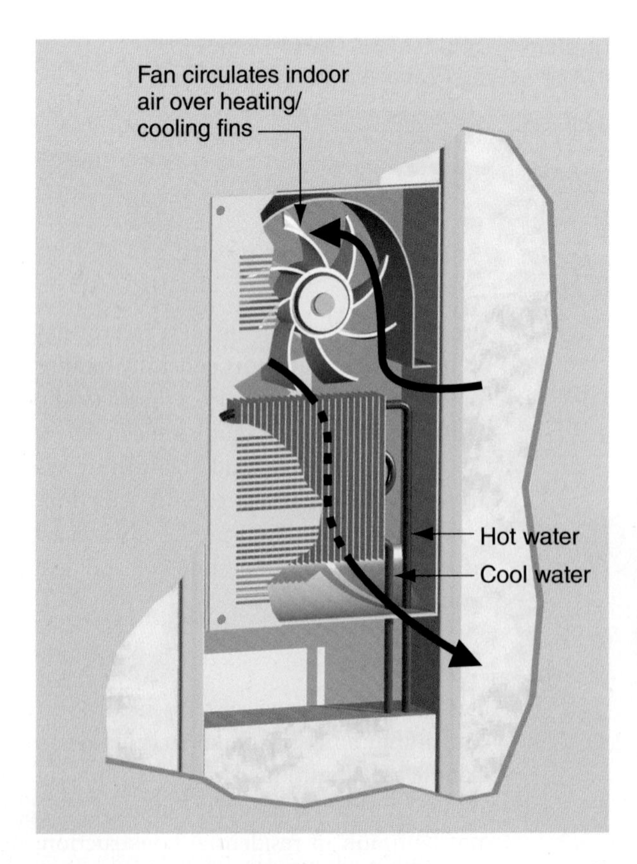

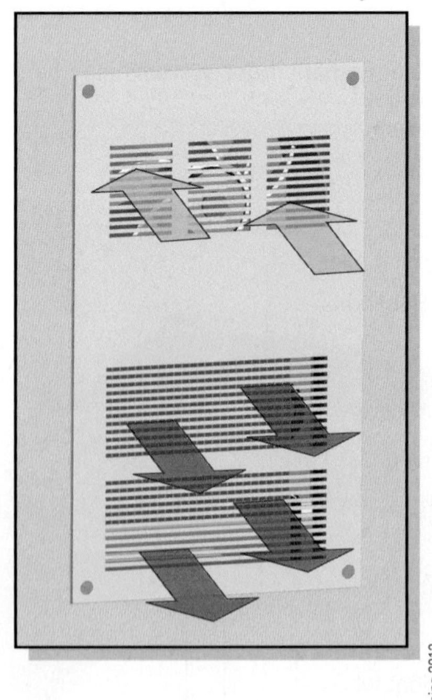

Figure 13.37 Hydronic fan coil systems are available in many different styles. Here is a unit that installed in the wall cavity with the surface flush with the drywall.

Reconsidering Radiant Floor Heating in High-Performance Homes

Alex Wilson

Radiant floor heating is a way of delivering heat through the floor—usually with hot-water tubing embedded in a concrete slab. It's a very popular heating system advanced by zealous proponents. If you want to pick a fight in the building industry, simply criticize such sacred cows as radiant-floor heating.

Don't get me wrong. Radiant-floor heating makes a lot of sense for the right applications. In fact, I think it's a great heating system, but for lousy houses. With new construction, if the house is designed and built to be highly energy-efficient it usually doesn't make sense.

Before explaining why radiant-floor heating is a poor choice in new construction, let me describe what I like about it. The heat is distributed over a large surface area, so it is delivered at a relatively low temperature. It's uniform, and it warms people directly, rather than having to heat the air. This means that radiant heat can provide comfort at a slightly lower air temperature than is required with forced-warm-air or baseboard hot water heat. You might be able to keep your thermostat lower—say 65 degrees—and be perfectly comfortable with radiant-floor heating, while 68 or even 70 degrees would be required with other systems.

Most people with radiant-floor heating absolutely love the warmth underfoot; you can walk around barefoot, even in the middle of winter. If we're used to drafty old houses, there's nothing nicer than a floor that's warm underfoot and gently radiates heat upward. Radiant heat also tends to have less of a drying effect than does forced-air heat. And because there aren't baseboard radiators, furniture can fit right up against the wall. So, what's wrong with radiant-floor heating?

I have two concerns, both of which apply only to very energy-efficient, superinsulated houses. First, in a highly insulated house, such a tiny amount of supplemental heat is needed that a radiant floor needs to be kept no more than a few degrees above the air temperature—or else overheating will occur. If a concrete-slab or tile floor surface is maintained at 72 or 75 degrees, it will likely feel cool underfoot—since it's at a lower temperature than your feet. So you may not get that delightful benefit of a warm floor surface. And, if you're delivering heat to the floor during the nighttime, and then have significant passive solar gain during the daytime, overheating is likely to occur. In short, radiant-floor heating just isn't a good fit with superinsulated houses.

My second issue with radiant-floor heating has to do with economics. Radiant-floor heating systems, with tubing embedded in a concrete slab, multiple pumps for different zones, and sophisticated controls, will easily cost $10,000. I'd rather see someone spend that $10,000 on better windows, more insulation, and so forth—then recoup some of that extra cost by spending less on the heating system. Homes built to the rigorous Passive House standard (see Katrin Klingenberg's "From Experience" earlier in this chapter) can be heated with, literally, a few incandescent light bulbs in each room. In a more typical superinsulated house, we can provide the desired comfort with one or two through-the-wall-vented gas space heaters or a few lengths of inexpensive electric baseboard heating element.

Again, these arguments apply to highly insulated houses—usually new construction—when you can pull out all the stops and far exceed typical insulation and air sealing standards. In existing houses or in new construction when standard energy details are being used, radiant-floor heating can make sense. In a house with a relatively large heating load, and especially in a drafty house, a radiant-floor heating system is a great option.

which the total heating load is very small, particularly if the electricity comes from a local or central renewable power source.

Non-Distributed HVAC Systems

As discussed earlier, individual heat and cooling units can condition individual rooms or whole homes without air or water distribution systems. These include several styles of heat pumps, fireplaces, stoves, and space heaters. Details of each of these will be discussed in the equipment selection section.

Selecting Heating and Cooling Equipment

The first question to answer when selecting the most appropriate HVAC system is whether to include heating, air conditioning, or both. Although air conditioning is common in much of the country today (even in cold climates that have very short cooling seasons) the need for central air conditioning may be avoided through careful building design and construction, owner behavior, and, in humid climates, the use of dehumidification. Heating is required in all but the hottest climates, but the size of the heating system can be minimized with careful attention to design, construction, and behavior. In the case of buildings built to Passive House standards, the heating system can sometimes even be eliminated, allowing a house to be heated only by the excess heat produced by the occupants and electric equipment.

Homes that require air conditioning must use some type of air distribution system, whereas those that require only heating have more options, including both air and radiant distribution. Understanding the pros and cons of each type of distribution system, equipment efficiency, available fuel sources, and the occupants' preferences will assist the project team in determining the most appropriate HVAC system.

Heating Fuel Types

The type of fuel and heating system best suited for a home depends on the following factors:

- The cost and availability of the fuel or energy source
- The type of appliance used to convert that fuel to heat, and how the heat is distributed through the house
- The cost to purchase, install, and maintain the heating appliance
- The efficiency of the heating appliance and heat delivery system
- The environmental impacts associated with the heating fuel

All of these factors affect the total cost of ownership of a given heating system. A cost per delivered Btu calculation allows for accurate comparisons between different types of heating equipment (Table 13.1).

Heating Fuel Availability

Heating fuel availability varies from region to region. Throughout every region of the United States except the south, the most common heating fuel type is natural gas. Heating fuel oil is primarily limited to the northeast, and electric heat is the most common in the south. Regardless of the region, options in rural areas tend to be limited to propane, wood, and electricity. Figure 13.38 shows the breakdown of heating sources according to a 2005 survey by the U.S. Energy Information Administration.

Heating Fuel Costs

Fuel costs vary regionally with local availability and demand. The heat content (Btu) in a given quantity of fuel also varies because, with the exception of electricity, all fuel types have slight variations in chemical composition. Table 13.1 presents average fuel prices and approximate energy content for the most common heating fuel types. To allow easy comparison, the cost per million Btu (mmBtu) is also calculated. Although the cost per mmBtu is a useful metric, the efficiency of delivering heating Btu must also be known to identify the most cost-effective heating system.

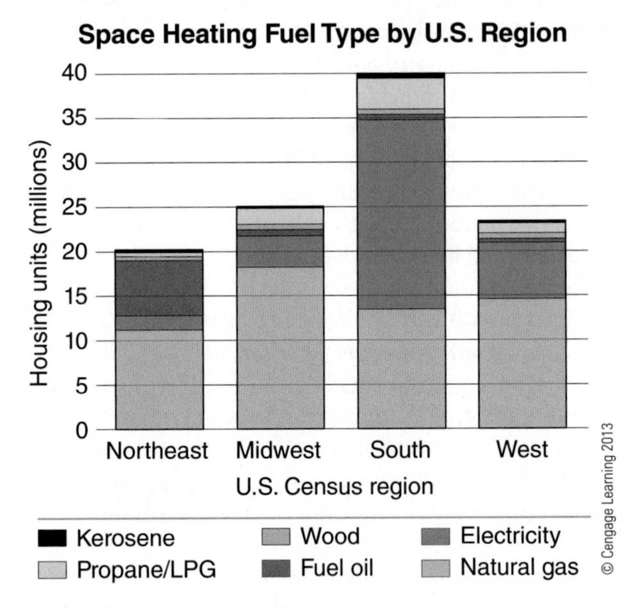

Figure 13.38 Space heating fuel sources. From the 2005 survey by the U.S. Energy Information Administration.

Table 13.1 Heating Fuel Types

Fuel Type	Fuel Unit	Fuel Price per Unit (dollars)	Fuel Heat Content per Unit (Btu)	Fuel Price per Million Btu (dollars)
Fuel oil (#2)	Gallon	$3.00	138,690	$21.63
Electricity	Kilowatt/hour	$0.116	3,412	$33.85
Natural gas	Therm	$1.23	100,000	$12.27
Propane	Gallon	$2.60	91,333	$28.47
Wood	Cord	$200.00	22,000,000	$9.09
Pellets	Ton	$250.00	16,500,000	$15.15
Kerosene	Gallon	$2.97	135,000	$21.96

Prices are approximate national averages for 2010. The fuel heat content is an approximate value with the exception of electricity. Fuel prices are based on national averages.

Heating Equipment

Depending on the type of distribution system selected for a house, a variety of heating equipment options are available (Table 13.2). Forced-air systems use electric or fuel-fire[…] heat pumps; […]ces; air-, ground-, and water-source Hydronic syst[…] and gas or electric water heaters. heat pumps, and […] heated with fuel-fired boilers, […] heaters.

Table 13.2 HVAC Distribution System Comparison

Distribution System Type	Pros	Cons	[…]ments
Forced air	Can provide heating, cooling, humidification, dehumidification, and filtration Least expensive way to distribute conditioning throughout house Responds quickly to temperature changes	Can cause pressure imbalances and bring pollutants into house when not properly designed and sealed Oversized air conditioning systems can be inefficient and not dehumidify properly Floor and low sidewall registers can be blocked by furniture Systems can be noisy and transfer noise between rooms through ducts	
Hydronic radiators	Responds quickly to temperature changes Cannot cause pressure imbalances Quiet	Only supplies heating Does not allow for air filtration Radiators can interfere with furniture placement Radiators covered by furniture will not heat effectively	
Hydronic floor	Heats with lower water temperature than radiators Warms people more than spaces Comfortable heat Cannot cause pressure imbalances Quiet	Slow to respond to temperature changes Only supplies heating Does not allow for air filtration	Hydronic cooling systems are rare and generally only effective in climates with limited cooling needs where the temperature of the water can be maintained below the dew point to avoid condensation
Hydronic fan coil	Can provide both heating and cooling Responds quickly to temperature changes	Does not allow for air filtration Fans create noise Wall-mounted units can interfere with furniture Uncommon in residential applications High installation costs	
Electric radiators	Responds quickly to temperature changes Cannot cause pressure imbalances Quiet	Only supplies heating Does not allow for air filtration Electric resistance generally more expensive method of heat than fossil fuels Radiators can interfere with furniture placement Radiators covered by furniture will not heat effectively	Can be very effective for superinsulated homes that require very little heating May be powered by on-site renewable electrical generation, but source power often comes from coal

(Continued)

Table 13.2 (Concluded) HVAC Distribution Sy[...]parison

Distribution System Type	Pros	Cons	Comments
Electric floor	Warms peop[...] than spaces Comforta[...] Cannot [...]essure imbalances Quiet	Slow to respond to temperature changes Electric resistance generally more expensive method of heat than fossil fuels Only supplies heating Does not allow for air filtration	May be powered by onsite renewable electrical generation, but source power often comes from coal
Unit heat pump/air conditioner	[...]le one-piece systems [...]ailable in very high-efficiency equipment Does not cause pressure imbalances Provides room-by-room temperature control Can provide heating and air conditioning without any efficiency loss through ductwork	Can be noisy Does not allow for location of registers for even HVAC distribution Oversized air conditioning systems can be inefficient and not dehumidify properly Provides limited air filtration	May be powered by onsite renewable electrical generation, but source power often comes from coal
[...]uctless mini-split	Available in very high-efficiency equipment Does not cause pressure imbalances Provides room-by-room temperature control Can provide heating and air conditioning without any efficiency loss through ductwork	Does not allow for location of registers for even HVAC distribution Oversized air conditioning systems can be inefficient and not dehumidify properly Provides limited air filtration	May be powered by on-site renewable electrical generation, but source power often comes from coal
Wood and pellet stoves	Quiet Cannot cause pressure imbalances	Only supplies heating Does not allow for air filtration Does not distribute conditioned air to rooms other than where located Requires monitoring and refilling wood or pellets Not thermostatically controlled	Fuel availability
Fuel-fired space heater	Quiet Cannot cause pressure imbalances	Only supplies heating Does not distribute conditioned air to rooms other than where located Does not allow for air filtration	Unvented fuel-fired space heaters should never be installed in a green home
Masonry fireplace	Quiet Cannot cause pressure imbalances	Only supplies heating Does not distribute conditioned air to rooms other than where located Does not allow for air filtration Generally inefficient Slow to respond to temperature changes Requires manual operation	
Fuel-fired fireplace	Quiet Cannot cause pressure imbalances	Only supplies heating Does not distribute conditioned air to rooms other than where located Does not allow for air filtration	

Locating Heating Equipment

Heating equipment operates most efficiently when it is installed in a central area of the house, reducing the length of duct and pipe runs. Additionally, fuel-fired equipment must be located where exhaust flues can be installed to meet manufacturer's requirements. Forced-air equipment performs best when located in conditioned space to minimize heat gain and loss and to minimize the opportunity for pollutants to be drawn into the duct system.

Heat Pump Equipment

Air-source heat pumps are available as split systems with a separate outdoor unit and air handler, or as package units (Figure 13.39a and Figure 13.39b). Air handlers are available with single-, dual-, and variable-speed fans. Dual- and variable-speed fans allow for more accurate temperature control, delivering an appropriate amount of air to condition the space. Outdoor units are available in single-speed and two-speed models that provide similar control over space conditioning. Ground- and water-source heat pumps and fuel-fired furnaces use an air handler typically located inside the house, available with the same fan options as air source heat pumps.

Air-Source Heat Pump Efficiency The heating seasonal performance factor (HSPF) is the most commonly used measure of the heating efficiency of air-source heat pumps. HSPF is a heat pump's estimated seasonal heating output in Btu divided by the amount of energy that it consumes in watt-hours. The key concept is that the HSPF is a *seasonal* measure, taking into account the fact that the heat pumps rarely operate for as long as is optimal during lower-load periods in the spring and fall. A heat pump with a high HSPF is more efficient than a heat pump with a low HSPF. The legal minimum HSPF is 7.7, with efficient units rated as high as 10 or more.

$$HSPF = \frac{\text{Average Annual Heating (Btu)}}{\text{Average Annual Power consumption (Watt} - \text{hours)}}$$

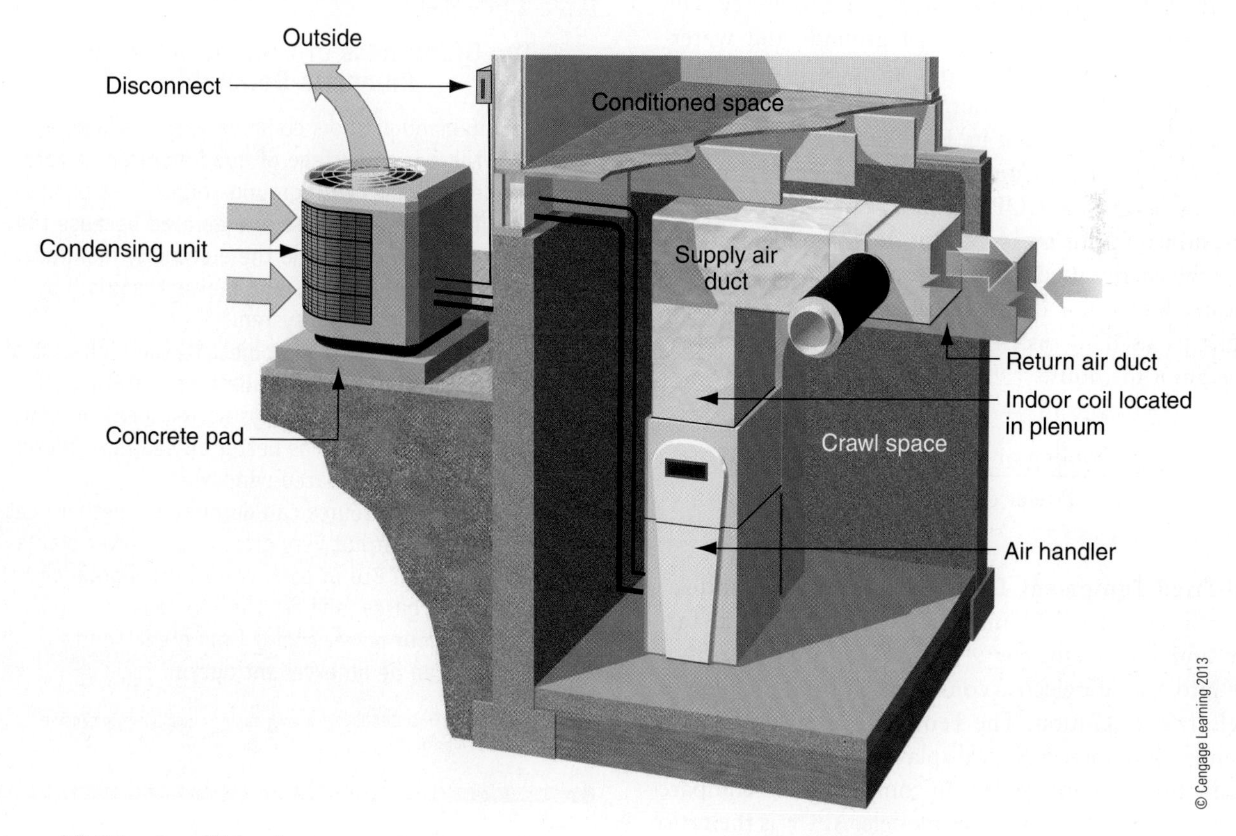

Figure 13.39a A split heat pump system.

heating seasonal performance factor (HSPF) a heat pump's estimated seasonal heating output in Btu divided by the amount of energy that it consumes in watt-hours.

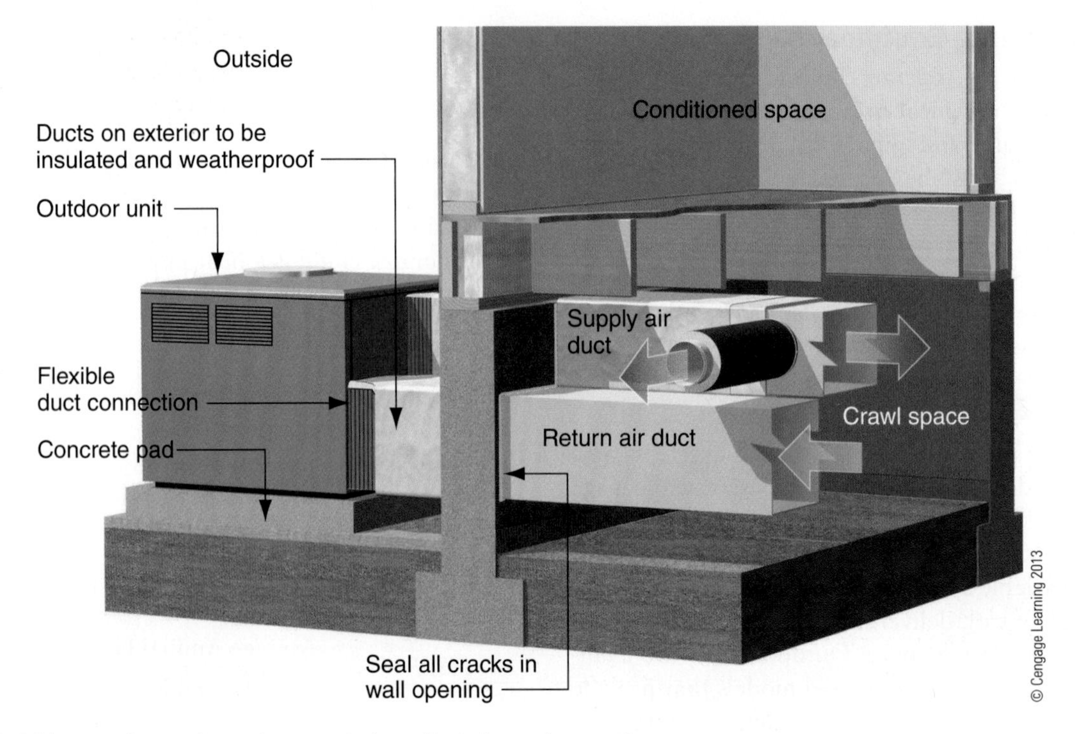

Figure 13.39b An air-to-air package unit installed through a wall.

Ground- and Water-Source Heat Pump Efficiency The measurement of the efficiency of ground- and water-source heat pumps is the coefficient of performance (COP), the ratio of energy output (heating or cooling) to the amount of energy put in. For example, a heat pump with a COP of 5 puts out 5 times more energy than it uses. The higher the COP, the more efficient the equipment; however, this measurement does not take into account the energy used to pump coolant through ground or water loops, nor does it include the energy for fans or pumps used in distributing the heating or cooling throughout the house.

$$COP = \frac{\text{Cooling or Heating Energy (Watts)}}{\text{Power consumption (Watts)}}$$

Fuel-Fired Equipment Efficiency Annual fuel utilization efficiency (AFUE) measures fuel-fired central furnace and boiler efficiency. The AFUE rating does not take into account electric consumption from the blower or electronic ignition. The Federal Trade Commission requires manufacturers to display the AFUE rating on all new furnaces and boilers so consumers can compare heating efficiencies of various models. AFUE is the ratio of heat output of the furnace or boiler compared with the total energy consumed by a furnace or boiler.

$$\text{AFUE} = \frac{\text{average annual heat produced (Btu)}}{\text{average annual fuel consumed (Btu)}}$$

The Green-ness of Ground-Source Heat Pumps Is Debatable

Heat pump manufacturers claim very high efficiencies, but there is a shortage of good monitoring data on completed homes with ground-source heat pumps. Some touted COP ratings are exaggerated because the calculations may not include the electricity required to operate all of the pumps and blowers required by a ground-source heat pump system.

Many energy consultants object to the high cost of ground-source heat pump installations, pointing out that the money is almost always better spent on envelope improvements, such as better air sealing, thicker insulation, and triple-glazed windows.

Your electricity source can determine whether heat pumps are green or not very green. Coal power plants waste 70% of the Btu in coal, contribute significantly to greenhouse gases, and are the top cause of mercury pollution. If your power comes from green sources, heat pumps can be an excellent option.

Source: Reprinted from GBA http://www.greenbuildingadvisor.com/green-basics/heat-pumps-basics

coefficient of performance (COP) the ratio of energy output (heating or cooling) to the amount of energy put in.

annual fuel utilization efficiency (AFUE) the ratio of heat output of the furnace or boiler compared with the total energy consumed by a furnace or boiler.

An AFUE of 78% means that 78% of the energy in the fuel becomes heat for the home, and the other 22% escapes up the chimney and elsewhere. High-efficiency equipment is available with AFUE ratings as high as 97%.

Condensing furnaces, those with AFUE ratings of 90% or above, use a secondary heat exchanger to condense the flue gas and capture additional heat, increasing their efficiency. The cooled flue gasses can be vented through polyvinyl chloride pipe instead of a metal flue,

Figure 13.40a High-efficiency condensing furnaces are easily distinguishable by their PVC combustion air pipes. Systems typically contain two PVC pipes (one for combustion air intake and the other for combustion exhaust gases), though they may have a double-walled single pipe (pipe inside a pipe). In this instance, the builder opted to not run the combustion air intake pipe to the exterior to reduce installation costs. Although the system will likely not backdraft, it does use air from the conditioned basement for combustion and may increase infiltration into the home. Even though the furnace is energy efficient the installation results in an energy penalty for the home.

Figure 13.40b Proper installation of a condensing furnace with both the combustion air intake and exhaust pipes run to the exterior.

using a blower motor that allows the gas to exit through a side wall instead of the roof (Figure 13.40a and Figure 13.40b). Most condensing furnaces have a sealed combustion chamber supplied with fresh air directly through a pipe to the outside. Sealed combustion furnaces eliminate the possibility of backdrafting carbon monoxide into the house and therefore can be safely installed inside the building envelope.

The AFUE does not include heat losses of the duct system or piping, which can be as much as 35% of the energy for output of the furnace when ducts are located in the attic. High-efficiency furnaces will not deliver the total desired efficiency when coupled with poorly designed and inefficient distribution systems. Like furnaces, boilers receive an AFUE rating, with federal regulations requiring a minimum of 80% efficiency. High-efficiency condensing sealed combustion boilers are available with AFUE ratings as high as 97%.

Combustion Safety

A fossil fuel furnace requires fuel, oxygen, and ignition for combustion to occur. Complete combustion is the process by which carbon from the fuel bonds with oxygen to form carbon dioxide (CO_2), water vapor, nitrogen, and air. Incomplete combustion occurs when there is an incorrect ratio of fuel to oxygen, causing carbon monoxide and aldehyde to be produced (Figure 13.42). Combus-

---

condensing furnace a combustion heating appliance that uses a secondary heat exchanger to condense vapor in the flue gas and capture additional heat, increasing their efficiency.

complete combustion the process by which carbon from fuel bonds with oxygen to form carbon dioxide (CO_2), water vapor, nitrogen, and air.

incomplete combustion occurs when ratio of fuel to oxygen is incorrect, causing carbon monoxide and aldehyde to be produced.

Combined Heat and Power

Combined heat and power (CHP) generation systems are commonly used in large commercial buildings but are now entering the residential market. CHP, also referred to as **cogeneration**, uses fuels like natural gas to produce heat and electricity simultaneously. The heat produced in the course of generating power is used for water or space heating. Efficiencies are in the range of 90% (similar to high-efficiency furnaces), which is more efficient than central power generation plants without additional transmission line losses. CHP units are normally installed in grid-tied electrical systems, returning power to the grid when not needed. The local utility purchases this electricity, thereby offsetting a home's power bill. CHP can also work in homes that are not connected to the power grid. CHP units are about the size of a furnace and are installed in conjunction with a central heating or water heating system in a home's mechanical room (Figure 13.41).

CHP is most appropriate in cold climates where there is more demand for heat, where electricity is more expensive than natural gas, and where utilities pay favorable rates for site-generated power. In warmer climates, the excess heat generated may be wasted, limiting the efficiency of the system.

Courtesy of ECR International

Figure 13.41 Combined heat and power (CHP) systems produce electricity and work as a furnace or a boiler for hydronic systems.

tion efficiency, system temperatures, and draft should be checked when furnaces are installed or serviced. The draft gauge measures the pressure of the flue gases in a furnace (Figure 13.43). The pressure over the fire in the furnace should be negative because the flue gases are moving away from the appliance. The typical draft pressures are in the range of −0.01 to −0.02 inches of water column. Flue pipe or stack temperatures are read with a stack thermometer (Figure 13.44). Carbon monoxide detectors are used to record the levels of CO in the flue pipe or surrounding areas (Figure 13.45).

Furnace combustion efficiency is determined by comparing the amount of useful heat produced with the total heat produced, including heat lost up the chimney. For example, a furnace operating at 80% efficiency is losing approximately 20% of the heat produced up the chimney. A combustion analyzer is a tool used on-site to measure furnace efficiency (Figure 13.46).

Carbon Monoxide Poisoning

Carbon monoxide (CO) is a colorless, odorless, poisonous gas that results from incomplete combustion of fuels (e.g., natural or liquefied petroleum gas, oil, wood, and coal). Sources of CO, such as furnaces, generators, gas space heaters, motor vehicles, and fireplaces, are common in homes or work environments. Most signs and symptoms of CO exposure are nonspecific and often mistaken for other causes, such as the flu and other viral illnesses. CO may be fatal if undetected or unsuspected. Common symptoms caused by carbon monoxide poisoning include:

---

combined heat and power (CHP) the use of a heat producing appliance or power station to simultaneously generate both electricity and useful heat, also known as *cogeneration*.

cogeneration the use of a heat producing appliance or a power station to simultaneously generate both electricity and useful heat, also known as *combined heat and power (CHP)*.

draft gauge measures the pressure of the flue gases in a furnace or water heater.

stack thermometer an instrument for measuring temperature within the flue pipe or stack of a combustion appliance.

carbon monoxide detector a device that records the levels of CO in a combustion appliance flue pipe or surrounding areas.

combustion analyzer a tool used on-site to measure furnace efficiency.

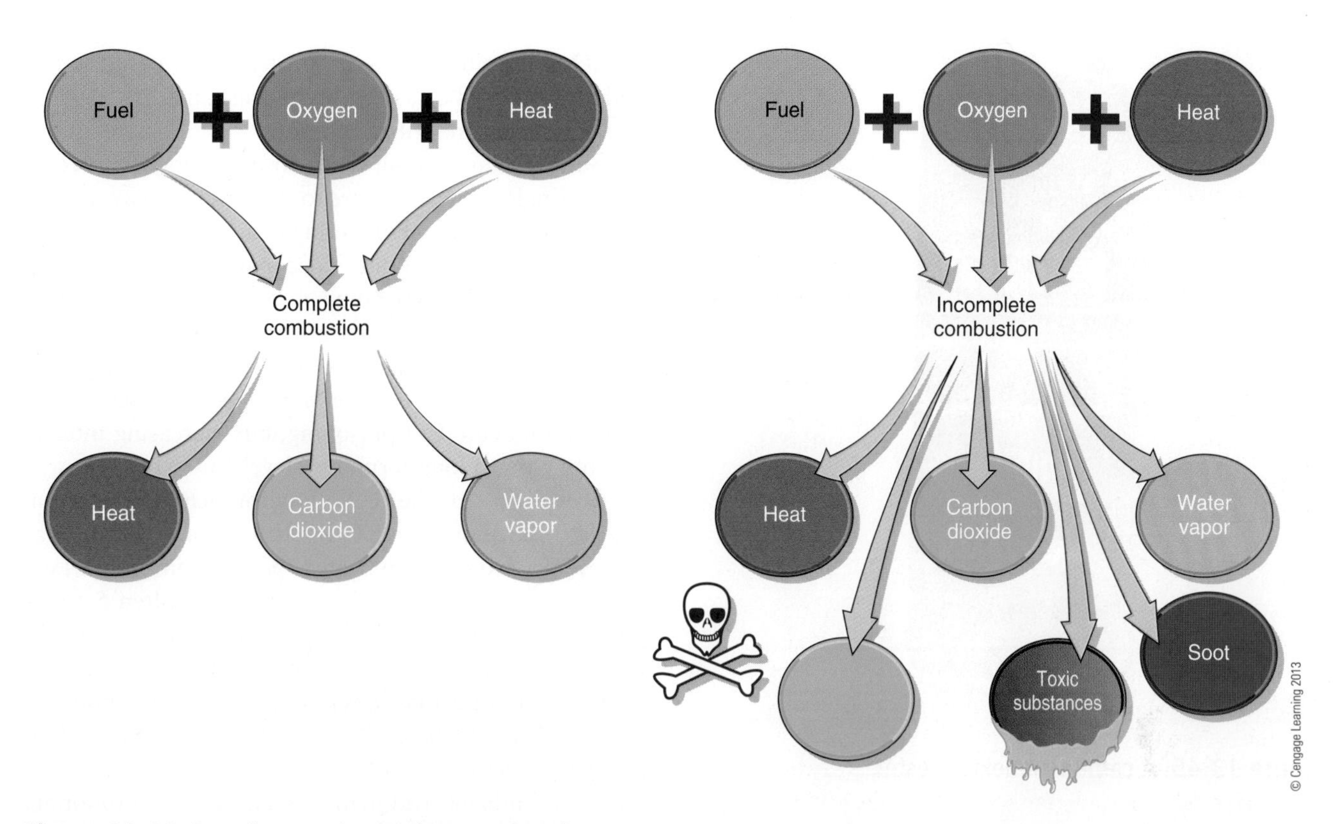

Figure 13.42 Complete vs. incomplete combustion.

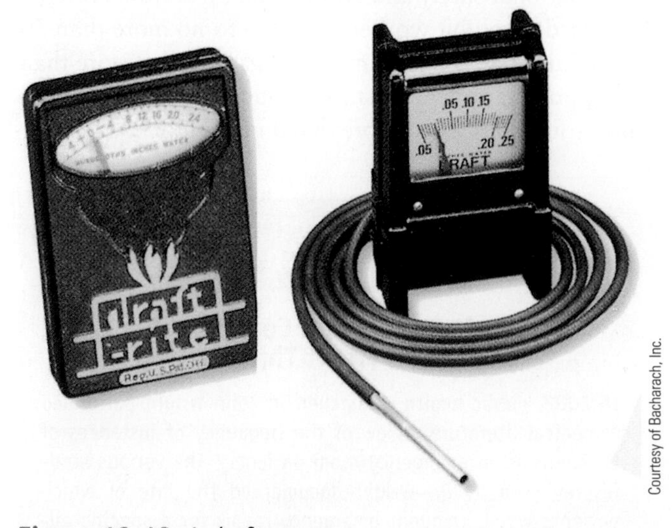

Figure 13.43 A draft gauge.

- Headache
- Dizziness
- Weakness
- Vomiting and diarrhea
- Loss of consciousness
- Seizure
- Confusion
- Chest pain (angina)
- Breathlessness

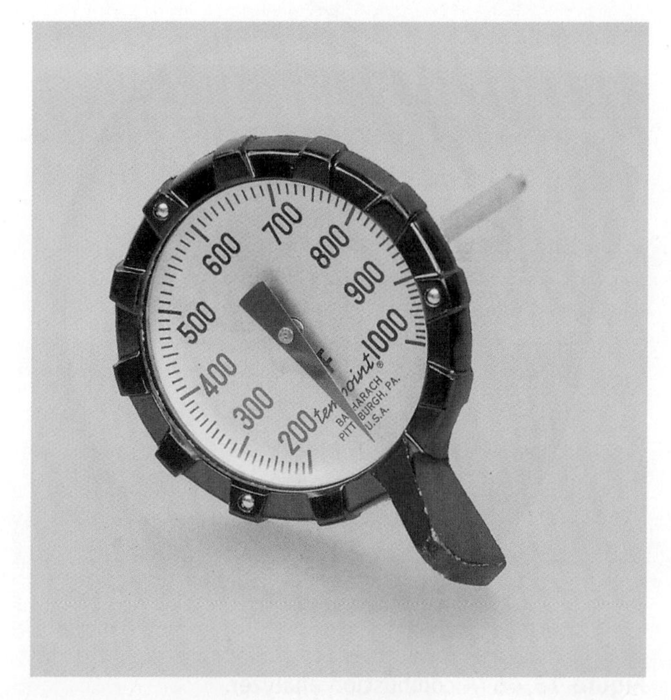

Figure 13.44 A stack thermometer.
From SILBERSTEIN. *Residential Construction Academy: HVAC*, 1E. © 2009 Delmar Learning, a part of Cengage Learning, Inc. Reproduced by permission. www.cengage.com/permissions

CO poisoning is a significant health threat in the United States and can result in long-term neurologic or cardiovascular complications if unrecognized. Although the precise incidence of CO poisoning is unknown, a 2005 report from the Centers for Disease Control and

Courtesy of Bacharach, Inc.

Figure 13.45 A carbon monoxide tester.

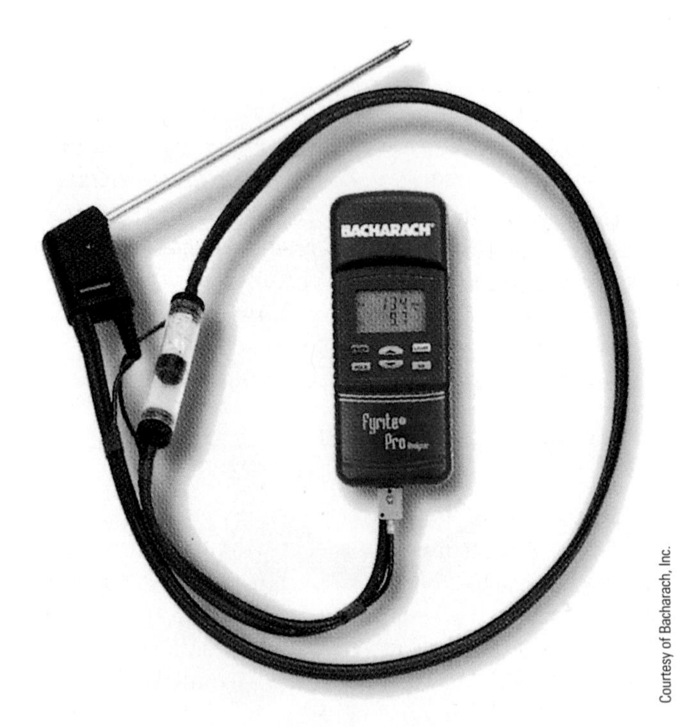

Courtesy of Bacharach, Inc.

Figure 13.46 A combustion analyzer.

Prevention estimated that 15,200 patients were treated annually in emergency departments during the years 2001 through 2003 for non-fire-related unintentional CO exposure, and approximately 480 people died yearly from CO poisoning during that period.[2] The incidence of CO poisoning increases during the winter months and after natural disasters when alternative heating and power sources are utilized.

Acute CO poisoning refers to poisoning that occurs after a single, large exposure to the gas, and may involve one or more people. For example, a backdrafting water heater will expose a home's occupants to high levels of CO. Chronic CO poisoning is used to indicate those cases in which individuals are exposed to the gas on more than one occasion—usually at comparatively low concentrations. To date, most medical research has focused on acute CO poisoning, but increasing interest is being paid to long-term, low-level CO exposure.

According to the U.S. Environmental Protection Agency (EPA), average CO levels in homes without gas stoves vary from 0.5 to 5 parts per million (ppm). Levels near properly adjusted gas stoves are often 5 to 15 ppm, and those near poorly adjusted stoves may be 30 ppm or higher. There is no universally accepted standard for CO concentrations in residential indoor air. Both the Building Performance Institute (BPI) and the Residential Energy Services Network (RESNET) reference the U.S. National Ambient Air Quality Standards for outdoor air, which are 9 ppm for 8 hours and 35 ppm for 1 hour. The Occupational Safety and Health Administration (OSHA) standards prohibit worker exposure to no more than 50 ppm during an 8-hour time period and to no more than 200 ppm at any time (instantaneous readings). The common sources of CO are presented in Table 13.3.

Did You Know?

CO Poisoning: More Common Than You Might Think

In 2002, public health researcher Dr. John Wright performed a medical literature review of the frequency of instances of CO exposure in emergency room patients.[3] The various studies reviewed by Dr. Wright documented the rate of which patients would frequent emergency rooms for a specific ailment (epileptic seizures, headaches, nausea, and flu) only to discover the cause was CO exposure. The prevalence of CO toxicity in the studies varied, but one study found that 23.6% of patients with flu-like symptoms actually were experiencing CO poisoning.

[2] CDC. Unintentional non-fire-related carbon monoxide exposures- United States, 2001–2003. MMWR 2005; 54:36–39.

[3] Wright J. Chronic and occult carbon monoxide poisoning: we don't know what we're missing. Emerg Med J. 2002 September; 19(5): 386–390.

acute CO poisoning refers to poisoning that occurs after a single, large exposure to the gas; may involve one or more people.

chronic CO poisoning is used to indicate those cases when individuals are exposed on more than one occasion to carbon monoxide—usually at comparatively low concentrations.

Table 13.3 Common Sources of Accidental CO Poisoning

Source	Cause	Prevention
Atmospherically vented water heater and furnace	Clogged burner, blocked vent, faulty pilot light, backdrafting	Regular maintenance and repairs, correct installation, isolation from the building envelope
Portable space heaters	All combustion products are vented into room	All combustion appliances must be vented to the outdoors
Fireplace	Poor drafting and incomplete exhaust	Regular maintenance, sealed combustion/direct vent models
Kitchen range/stove	Rust, clogged burner, dirt, improper installation, faulty device	Regular maintenance and repairs, correct installation, spot ventilation ducted to the outdoors
Attached garage	Running car engine in an attached garage, especially if door closed	Full isolation of garage from building envelope; removal of vehicles from garage to warm up

Source: Adapted from Wright, J., "Chronic and occult carbon monoxide poisoning: we don't know what we're missing." *Emergency Medicine Journal.* 2002;19:386–390.

Selecting Cooling Systems

The type of cooling system best suited for a home depends on the following factors:

- The cost to purchase, install, operate, and maintain the equipment
- The delivery system efficiency

Most new homes in the United States contain some form of air conditioning. The most common method is a ducted central air conditioner. Figure 13.47 shows the breakdown of cooling equipment type by U.S. Census region using 2005 government data.[4]

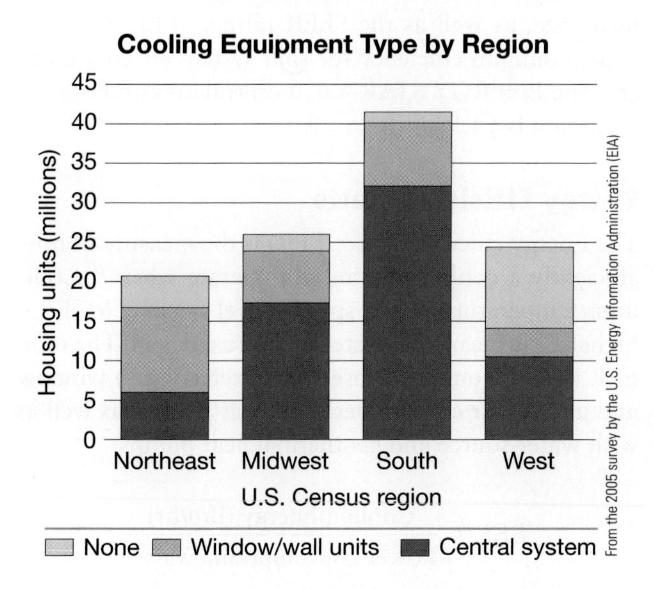

Figure 13.47 Cooling equipment type by region. From the 2005 survey by the U.S. Energy Information Administration.

[4] Energy Information Administration, Office of Energy Markets and End Use, Forms EIA-457 A, B, C of the 2005 Residential Energy Consumption Survey.

Because almost all cooling systems are fueled by electricity, and most homes require heating as well as air conditioning, the cooling system decision often follows the heating system design.

Cooling and Dehumidification

Air conditioners are designed to provide both latent and sensible cooling. Sensible heat is the energy associated with temperature change. Latent heat is the energy associated with removing moisture from the air. Latent heat is energy that can be released by a phase change, such as the condensation of water vapor. An air conditioner's ratio of sensible to latent capacity is referred to as sensible heat fraction (SHF) or sensible heat ratio (SHR). The SHF is the sensible capacity divided by the total capacity (sensible plus latent capacity).

$$\text{SHF} = \frac{\text{Sensible Capacity}}{\text{Total Capacity (Sensible + Latent)}}$$

The SHF may vary from one model to the next, and systems are optimized for particular climates. In a climate with a high latent load, such as Miami, Florida, or Savannah, Georgia, an air conditioner with a low SHF is desired. In Georgia, an SHF of 0.70 – 0.77 is common and means that 23% to 30% of an air conditioner's load is latent. Las Vegas, on the other hand, has relatively little latent load, so an SHF over 0.90 is common.

sensible heat the energy associated with temperature change.

latent heat the heat that produces a change of state without a change in temperature; the portion of the cooling load that results when moisture in the air changes from a vapor to a liquid (condensation).

sensible heat fraction (SHF) an air conditioner's ratio of sensible to latent capacity; also known as *sensible heat ratio (SHR)*.

sensible heat ratio (SHR) an air conditioner's ratio of sensible to latent capacity; also known as *sensible heat fraction (SHF)*.

SHF is determined by the combination of the coil, condenser, and the air flow through the coil. Manufacturers' specifications typically list the SHF, which can be used to select the most appropriate equipment for each climate.

Advanced HVAC controllers can regulate air conditioner operation based on temperature and relative humidity set points (see later in this chapter for more information on controllers).

Cooling Equipment

Depending on the type of distribution system selected for a house, a variety of options for cooling equipment are available. Forced-air systems, the most predominant form of air conditioning, use air-, ground-, or water-source heat pumps. Evaporative cooling is an alternative to traditional forced-air systems suitable for dry climates. Hydronic radiant cooling systems, although not common, can be used in dry climates; hydronic air conditioning with fan coil units, also not very common, can work in both humid and dry climates.

Forced-Air Cooling

Forced-air cooling systems are available as split systems or packaged units. Typically, when a distributed cooling system is installed, it is either a heat pump or air conditioner partnered with a central furnace and shared ductwork. The air handler should be centrally located to reduce duct run length, but also placed near the outdoor condensing unit to limit length of refrigerant line. Condensing units require adequate air flow to perform efficiently. This entails keeping all obstructions, including fences and shrubs, a minimum of 2' away from the sides of the unit to release heat effectively. Shading from direct sun with trees or trellises also helps increase efficiency. When a forced-air cooling system that does not also provide heat is installed in an unconditioned attic, the registers should be designed to be easily sealed during cold weather This strategy will prevent convective heat loss into the attic area from heated rooms below.

Hydronic Radiant Cooling

Radiant cooling systems rely on chilled water pipes to distribute cooling throughout a building, unlike a conventional system that uses chilled air and ductwork. Radiant cooling systems directly cool occupants through radiative heat transfer because the pipes, which are commonly run through ceilings, maintain the surface at temperatures of about 65°F. Through radiative heat transfer, people in the room will emit heat that is absorbed by the radiant cooling surface. The water temperature is carefully controlled to avoid condensation. To manage indoor humidity levels and air quality, a separate ventilation system to supply fresh air is needed. Due to the ease of controlling water flow, independent control of areas of the home is relatively simple. Radiant cooling systems are generally limited to dry climates with low cooling loads.

Air Conditioner Efficiency

Seasonal energy efficiency ratio (SEER) describes how efficiently air conditioning equipment works. A higher SEER means better efficiency and lower energy bills. SEER is calculated by dividing the amount of cooling supplied by the air conditioner or heat pump (Btu per hour) by the power (watts) used by the cooling equipment under a specific set of seasonal conditions.

$$SEER = \frac{\text{Seasonal Cooling Energy (Btu)}}{\text{Seasonal Power Consumption (Watt} - \text{hours)}}$$

SEER ratings are determined in a laboratory with precise indoor and outdoor conditions. This allows consumers to compare cooling equipment by a common standard. Although SEER ratings are useful, they do not provide a complete evaluation of cooling system performance in real-world conditions. The SEER rating reflects only the evaporator and condenser equipment; it does not take into account duct system efficiency and the length of refrigerant line sets. SEER ratings apply to coils and condensers that are matched according to AHRI standards (Figure 13.48).

The actual efficiency of an air conditioning system is dependent on air flow, charge, duct tightness, and equipment size, as well as the SEER rating. SEER 13 is the code minimum efficiency for split system air conditioners. The ENERGY STAR–rated central air conditioning minimum is 14.5 SEER.

Energy Efficiency Ratio

The Energy efficiency ratio (EER) is the measure of how efficiently a cooling system will operate when the outdoor temperature is at a specific level (usually 95°F). A higher EER means the system is more efficient. The term EER is most commonly used when referring to window and unitary air conditioners and heat pumps, as well as with water-source and geothermal heat pumps.

$$EER = \frac{\text{Cooling Energy (Btu/hr)}}{\text{Power consumption (Watts)}}$$

---

seasonal energy efficiency ratio (SEER) describes how efficiently air conditioning equipment works.

energy efficiency ratio (EER) the measure of how efficiently a cooling system will operate when the outdoor temperature is at a specific level (usually 95°F).

Certificate of Product Ratings

AHRI Certified Reference Number: 4237148 Date: 3/24/2011

Product: Split System: Air-Cooled Condensing Unit, Coil with Blower
Outdoor Unit Model Number: 6ACC3042E1
Manufacturer: Air Conditioning Co.
Indoor Unit Model Number: ACC061E1***
Manufacturer: Air Conditioning Co.
Trade/Brand name: Air Conditioning Co.

Manufacturer responsible for the rating of this system combination is Air Conditioning Co.

Rated as follows in accordance with AHRI Standard 210/240-2008 for Unitary Air-Conditioning and Air-Source Heat Pump Equipment and subject to verification of rating accuracy by AHRI-sponsored, independent, third party testing:

Cooling Capacity (Btuh):	42000
EER Rating (Cooling):	10.50
SEER Rating (Cooling):	13.00

* Ratings followed by an asterisk (*) indicate a voluntary rerate of previously published data, unless accompanied with a WAS, which indicates an involuntary rerate.

DISCLAIMER
AHRI does not endorse the product(s) listed on this Certificate and makes no representations, warranties or guarantees as to, and assumes no responsibility for, the product(s) listed on this Certificate. AHRI expressly disclaims all liability for damages of any kind arising out of the use or performance of the product(s), or the unauthorized alteration of data listed on this Certificate. Certified ratings are valid only for models and configurations listed in the directory at www.ahridirectory.org.

TERMS AND CONDITIONS
This Certificate and its contents are proprietary products of AHRI. This Certificate shall only be used for individual, personal and confidential reference purposes. The contents of this Certificate may not, in whole or in part, be reproduced; copied; disseminated; entered into a computer database; or otherwise utilized, in any form or manner or by any means, except for the user's individual, personal and confidential reference.

CERTIFICATE VERIFICATION
The information for the model cited on this certificate can be verified at www.ahridirectory.org, click on "Verify Certificate" link and enter the AHRI Certified Reference Number and the date on which the certificate was issued, which is listed above, and the Certificate No., which is listed below.

AHRI Air-Conditioning, Heating, and Refrigeration Institute

©2011 Air-Conditioning, Heating, and Refrigeration Institute

CERTIFICATE NO.: 129454481976080695

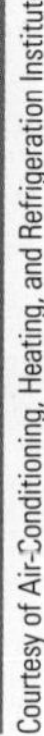

Figure 13.48 Sample Air-Conditioning, Heating, and Refrigeration Institute matching certificate showing the capacity and efficiency of two particular air conditioning coils.

Refrigerants

In the past, refrigerants used in air-source heat pumps and air conditioners were made of chemicals that attacked the earth's ozone layer. Over time, those chemicals have been replaced with newer, less damaging ones. R-22, a hydrochlorofluorocarbon (HCFC) refrigerant was phased out of newly manufactured equipment beginning in 2010, having been replaced with R-410A, a hydrofluorocarbon (HFC), non-ozone-depleting alternative.

Refrigerant Charge The refrigerant charge in a heat pump or air conditioner is the ratio of refrigerant required to the cooling capacity. This ratio is defined in the manufacturer's equipment specifications. Verifying that the charge is correct helps ensure that the equipment operates at maximum efficiency. Several studies, summarized in a 1999 report for Southern California Gas, determined that most systems are either over- or under-charged beyond manufacturer's recommendations, resulting in decreased efficiencies of between 10% and 20% and creating the potential for premature equipment failure.[5]

Thermostatic Expansion Valves Thermostatic expansion valves (TXVs) automatically adjust the refrigerant flow into the evaporator coil when a system is operating in the air conditioning mode. Systems with TXVs manage refrigerant flow more efficiently than non-TXV systems, and they also compensate for improper refrigerant charge better. The previously mentioned Southern California Gas study determined that TXV-equipped systems consistently performed better than non-TXV systems when both had improper refrigerant charges.

Minimum Equipment Efficiencies

Congress adopted the National Appliance Energy Conservation Act (NAECA) in 1992 to establish nationwide minimum efficiency levels for a variety of residential and commercial appliances that use energy and water. The U.S. Department of Energy (DOE) is responsible for implementing NAECA, including updating minimum required efficiency levels periodically. NAECA minimums generally supersede state and local regulations, including building/energy code provisions that are inconsistent.

The most recent updates to NAECA took effect on January 23, 2006 after being initially announced in 2001. Table 13.4 shows the NAECA and ENERGY STAR minimum standards.

[5] http://www.socalgas.com/calenergy/docs/hvac/references/proctornationalstudy.pdf

National Appliance Energy Conservation Act (NAECA) the federal law enacted in 1992 to establish nationwide minimum efficiency levels for a variety of residential and commercial appliances that use energy and water.

Proper Use of Ceiling Fans as Part of a Cooling Strategy

In theory, ceiling fans can reduce heating and costs in two ways. During the swing seasons (spring and fall), ceiling fans can be a substitute for an air conditioner. Ceiling fans cost much less to operate than air conditioners. In the summer, ceiling fans can allow occupants to set the thermostat higher and run the air conditioner less. Air movement is one of the primary factors of human comfort. Air blowing over your body increases the amount of evaporation, which is how we cool ourselves. In practice, however, most people do not use ceiling fans in such a way that reduces energy consumption.

Ceiling fans save energy if their use also results in reduced air conditioner use. In fact, since the fans themselves use energy, it may actually cost you more if air conditioner temperatures are not adjusted higher when using fans. The key with ceiling fans is they cool people and not the air. Fans should only operate when people are present in the room.

In Florida, using ceiling fans and raising a home's temperature 2°F will generate about a 14% net savings in annual cooling energy use (subtracting out the ceiling fan energy and accounting for internally released heat).[6] The same study found that most people do not adjust their thermostats when using ceiling fans, and may actually increase their energy consumption.

Motor and Pump Efficiency

In addition to the rated equipment efficiencies, HVAC systems use energy to run fans and pumps to deliver conditioned air and water through their distribution systems. Forced-air systems use either permanent split capacitor (PSC) or electronically commutated motors (ECM) to operate blower fans. ECMs run more efficiently than PSC motors and can provide savings of as much as 50% on fan energy.

Hydronic systems use pumps to circulate water, and, like forced-air motors, pump efficiency can vary. The European Union has established efficiency standards for water circulation pumps, some of which are available in the United States from such companies as Grundfos. Pump efficiencies range from "A" (the most efficient) to "G" the least efficient. An "A" pump uses 75% less energy than a "D", ultimately reducing the overall energy consumption of a hydronic system.

[6] James, P., Sonne, J., Vieira, R., Parker, D., Anello, M., "Are Energy Savings Due to Ceiling Fans Just Hot Air?" Presented at the 1996 ACEEE Summer Study on Energy Efficiency in Buildings.

Table 13.4 Minimum Equipment Efficiencies

Appliance Type	NAECA Minimum	ENERGY STAR Minimum
Central air-cooled air conditioner	13 SEER	14.5 SEER
Air conditioner, through-the-wall single package	10.6 SEER	Varies by size and configuration
Central air source heat pump	13 SEER/7.7 HSPF	14.5 SEER/8.2 HSPF
Electric furnace	N/A	Not eligible
Gas furnace	78 AFUE	90 AFUE
Oil furnace	78 AFUE	85 AFUE
Ground-source heat pump	Varies by size and configuration	Varies by size and configuration

Equipment Sizing

A properly sized heating and cooling system is designed to adequately condition a space according to the summer and winter loads placed on it. As discussed earlier in this chapter, loads are calculated using ACCA Manual J. Air conditioning equipment sizing is typically listed in tons, with one ton being equal to 12,000 Btu of cooling capacity per hour. Residential forced-air air conditioning systems are available in sizes ranging from 1.5 tons to 5 tons, increasing in half-ton increments (except 4.5-ton units are not normally available). HVAC contractors will frequently use a "rule of thumb" to determine AC sizing, often allowing about 500 to 600 ft^2 of conditioned space per ton of air conditioning. When a Manual J calculation is performed, particularly on a high-performance home, one ton of air conditioning can condition 1,000 to 1,200 ft^2 of space or more. Estimating equipment sizes without using the available tools often leads to incorrectly sized systems that do not perform effectively. The IRC requires equipment be sized per Manual J and S or an equivalent calculation.

Design Temperatures

Manual J specifies use of the 99% design temperature as developed by the American Society of Heating, Refrigeration, and Air-Conditioning Engineers (ASHRAE). This value is the outside summer temperature and coincident air moisture content that will be exceeded only 1% of the hours from June to September. In other words, this is the expected hottest temperature the home will experience. If the 99% design temperatures are unavailable, the 97.5% temperatures may be used. For example, if according to Manual J, the summer outdoor temperature in a specific climate is 93°F and the design temperature inside is 75°F, a system designed to these criteria will cool and dehumidify properly when set at 75°F when the outside temperature is 93°F. If the interior temperature in the Manual J calculation is set to 70°F, then the system may not properly dehumidify and may not run efficiently if set at 75°F. When systems are installed that exceed this differential during extreme heat, are usually too large to work properly during moderate weather.

Oversizing Cautions

Selecting HVAC equipment that is larger than needed costs more to install and, in the case of air conditioning, can reduce its efficiency and ability to remove humidity. Air conditioning works by lowering the air temperature and removing humidity. Properly designed systems run long enough to both cool and dehumidify. In humid climates, oversized systems cool the air too quickly, before it is fully dehumidified. This can make the indoor air feel too humid, requiring the thermostat to be set lower to dehumidify adequately. In extreme situations, the indoor temperature must be set below 70°F to achieve the required run times to remove enough moisture to be comfortable.

Air Conditioner Short Cycling

Oversized air conditioners are prone to short cycling, a condition in which they turn on and drop the indoor temperature very quickly, then turn off. Air conditioners are least efficient when starting and stopping, not unlike a car that gets better mileage on the highway than in stop-and-go traffic. Frequent starting and stopping of an air conditioner is less efficient than long run times. In periods of extreme heat, central air conditioning should run almost continuously, providing both adequate cooling and dehumidification. Short cycling may also shorten equipment life due to added stress.

Sizing of heating equipment, while important, is not as critical as cooling because there is no dehumidification required during heating seasons.

What Is a Ton of Air Conditioning?

Air conditioning systems are commonly sized in tons. One ton of air conditioning is equal to 12,000 Btu/hour, but why is it described in tons? It all comes from the ice-making industry. It takes 144 Btu of energy to transform 1 pound of 32°F water into ice. Multiply that by 2000 pounds, and you need 288,000 Btu to turn a ton of water into ice. Divide 288,000 by 24 hours in a day, and you get 12,000 Btu, which equals 1 ton of air conditioning.

Distribution Capacity

The total capacity a particular piece of equipment can produce must be matched with the distribution system to ensure that the heating and cooling it can produce is adequately delivered to the space requiring conditioning. If a furnace blower is not large enough to adequately deliver the hot or cold air throughout the duct system, the system will not perform as expected.

Thermostats

HVAC systems are controlled by thermostats, which measure the room temperature for the purpose of turning heating or cooling on and off to maintain the desired temperature. Standard thermostats are set manually and maintain their setting until changed. Clock-setback thermostats are programmable by day and hour to set the temperature up or down to reflect the expected use patterns. They can be programmed to turn the temperature down during the heating season or up during the cooling season when a house is unoccupied, returning it to the desired occupancy setting automatically before people return home in the evening. These thermostats can also be used to change settings in the entire house or in individual zones at night, or when areas are otherwise unused. Keep in mind that in extremely efficient homes, there is minimal heat gain and loss through the building envelope. Interior temperatures remain consistent, therefore setback thermostats do not provide as much benefit as in less-efficient homes where outdoor temperature swings can change indoor temperature more rapidly. Setback thermostats also do not work as effectively with radiant floor heating, which heats and cools much more slowly than forced air.

The 2009 IECC requires a minimum of one clock setback thermostat in each house. Heat pump thermostats are also required to have a built-in override to prevent the use of electric resistance heat when the outside temperature will allow the compressor to meet the required heating load.

More sophisticated thermostats can monitor humidity levels, controlling air conditioning and dehumidifiers to remove moisture when necessary. They can also manage fresh air ventilation by operating the furnace blower to bring in outside air as part of a whole house ventilation system (Figure 13.49).

Boilers can run more efficiently when an outdoor reset control is used. This equipment regulates the boiler fluid temperature based on outdoor temperature, the rate at which the boiler temperature is dropping, and the total heat load. These controls can reduce short cycling, allowing for more efficient operation.

Figure 13.49 On the left is a digital thermostat that can be programmed; on the right is an analog thermostat.
From SILBERSTEIN. *Residential Construction Academy: HVAC*, 1E. © 2009 Delmar Learning, a part of Cengage Learning, Inc. Reproduced by permission www.cengage.com/permissions

Advanced HVAC Controls

Advanced HVAC controls include systems that tell thermostats to set back automatically when alarms are set, and sensors that turn off HVAC systems when doors and windows are open. Web-enabled controls can allow owners to monitor and change temperature settings remotely. These systems can also control lighting, audio, and other equipment, adding to energy savings when properly managed.

Commissioning

Commissioning is the process of diagnosing and verifying building system performance. The process also involves proposing ways to improve the performance of systems in compliance with owner or occupants' requests. Commissioning is performed to keep the system in optimal condition through the life of the building from an environmental perspective and in terms of energy and facility usage. Commissioning can occur at any stage of a building's life.

Although commissioning is a standard practice in many commercial and industrial buildings it is not common in single-family residential projects. Commissioning in a home may include checking refrigerant charge, air handler air flows and static pressures, and ventilation system air flows.

Non-Distributed Heating and Cooling Systems

Heating and cooling can be provided without whole-house forced-air or hydronic distribution systems by using space heaters, fireplaces, stoves, and one- and

---

commissioning the process of diagnosing and verifying building system performance.

two-piece heat pumps and air conditioners. High-performance homes with excellent insulation and air sealing can often be heated and cooled adequately with space heaters and unit air conditioners. This strategy works particularly well with small, compact, open plan homes where a space heater can be centrally located. Distant, less-used rooms can be heated with individual small space heaters when needed, or by using ducts and fans to circulate conditioned air from rooms where equipment is located.

Direct Heating Equipment

Direct heating equipment includes both permanent and portable heaters (Figure 13.50), commonly referred to as space heaters, wall heaters, floor heaters, and room heaters. They are predominantly fired with natural gas or propane, although some are electrically operated.

All types of direct heating equipment convey heat without ducts, and the products of combustion must be vented outside. Heat is provided by radiation only or by convection with a fan that blows heated air into the room. Direct heating equipment comes in a wide variety of styles and models, including room heaters, gravity wall furnaces, fan wall furnaces, fireplaces, and floor furnaces.

Figure 13.50 Direct heaters may be portable or installed permanently.

Did You Know?

Unvented Direct Heaters and Fireplaces

Ventless gas space heaters have been around for many years. While they are not as common as they once were, ventless gas fireplaces have become more common (Figure 13.51). Any fuel-fired heater requires air to create combustion, and all can produce carbon monoxide. Gas-fired appliances release significant amounts of moisture when they operate. While these heaters are outlawed in some areas, they are still legal in many places. Regardless of their legality, ventless heaters and fireplaces should never be installed in a green home due to the risk to occupants from combustion gases mixing with the indoor air, as well as the humidity problems created as excess moisture is released. Most green building certification programs prohibit the installation of ventless heaters and fireplaces.

Figure 13.51 Ventless fireplaces pose serious health and safety risks because all combustion gases are released into the home.

Fuel-fired heaters should be treated as small furnaces with careful attention to combustion safety. Most efficient models have electronic ignition, but some may have a standing pilot that does not require electricity, which may be helpful where power outages are common.

Space Heater Efficiencies On April 16, 2010, the DOE issued a final rule for amended standards for residential direct heating equipment. The standards depend on unit type and input capacity, with the AFUE ranging from 57% for a small floor unit to 76% for a large, fan-assisted wall unit.

Electric heaters are 100% efficient but are not a very green solution unless powered by renewable energy. As mentioned earlier, the cost of electricity is high, and most of the source energy used at power plants is coal. Electric heaters can be free-standing, baseboard, radiant wall or ceiling panels, and even double as towel warmers in bathrooms. Electric heaters respond very quickly to calls for heat and may be an appropriate solution for rarely used rooms in very efficient homes.

Wood and Pellet Stoves

Similar in function to space heaters, stoves provide heat from a single, central location and use solid wood, wood pellets, or corn as fuel. Stoves are available with sealed combustion and intake air supplies to keep pollutants out of the indoor air. Although, wood stoves require regular attention, pellet and corn stoves are available with automatic feeders that can provide a steady supply of fuel for days at a time.

Fireplaces

Fireplaces can be fueled with natural gas, propane, or wood. Gas-fired units are available in both open and sealed combustion models with intake air supplies to keep pollutants out of the indoor air. Many are available with electronic ignition to avoid the need for a standing pilot light. Some have built-in fans to circulate heat through ducts if desired. Wood fireplaces are available in manufactured metal units and site-built masonry. Open combustion fireplaces, whether gas or wood, should be fitted with closable doors and outside intake air to limit backdrafting potential, although these efforts will not significantly increase their efficiency.

Masonry Heaters

Most residential masonry fireplaces are primarily decorative and do not efficiently heat homes, but certain designs can provide very effective heating in cold climates. Known as Russian or Finnish fireplaces, these fireplaces use a combination of a small firebox, large masonry mass, and long, twisted flues that run through the mass (Figure 13.52). A small, very hot fire heats the flues and transfers heat into the mass, which slowly releases the heat into the house throughout the day. Masonry heaters can be as much as 90% efficient.

Masonry heaters are available as either prefabricated manufactured units or custom site-built. If incorporated into a passive solar house, they can serve as some of the required thermal mass to store radiant heat. One disadvantage of masonry heaters is that they heat and cool very slowly and are not responsive to rapid temperature changes.

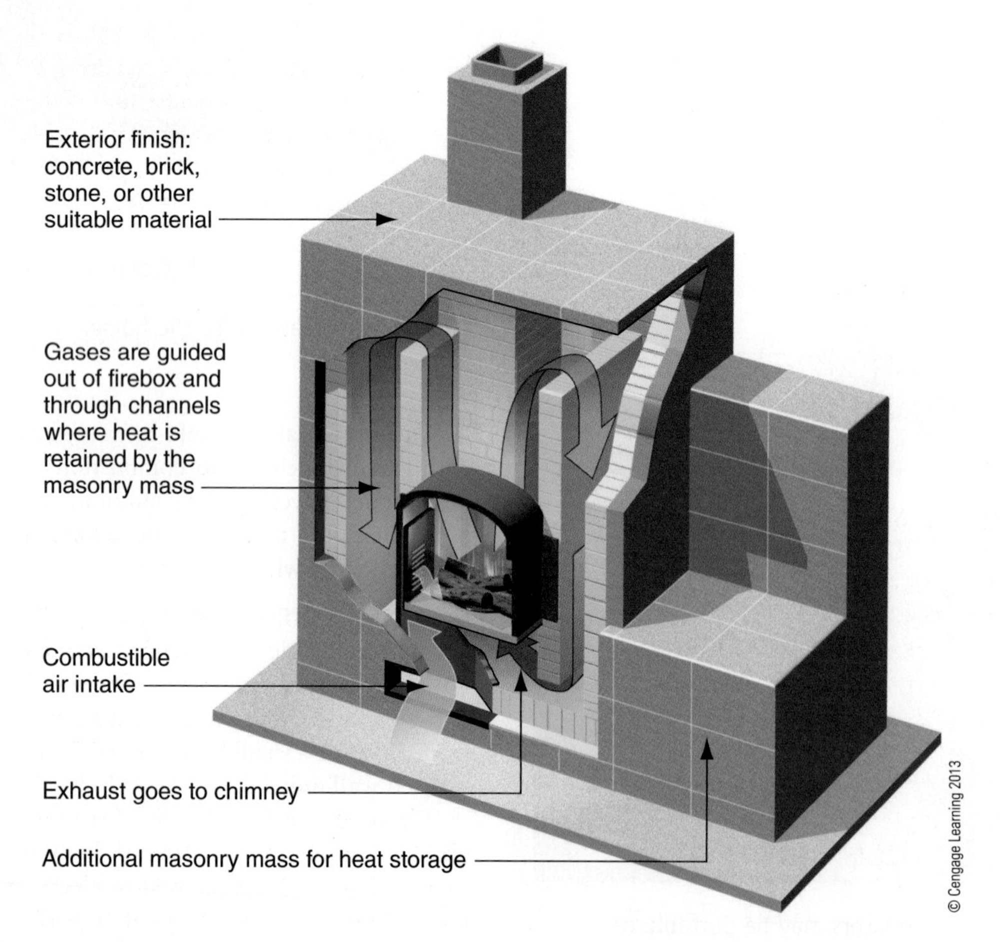

Figure 13.52 Masonry heater.

Fireplace Inserts

Fireplace inserts, installed into a new or existing masonry fireplace, work like sealed wood stoves. They offer the feel of a fireplace with the efficiency of a high-performance stove. When installed in an older fireplace they may require the installation of a flue liner for proper operation.

Wood Burning Restrictions

Some areas restrict the installation of wood-burning stoves and fireplaces in new construction because wood smoke is a major contributor to air pollution. Many new stoves and inserts are made with EPA-certified catalytic combustors that burn at lower temperatures. These new designs use fuel more efficiently while reducing the quantity of emissions and the amount of soot and creosote build-up on the inside of the flue.

Efficiency

The EPA rates the efficiency of a limited number of wood stoves and fireplaces. EPA-qualified units carry a hang-tag on the front of the fireplace to demonstrate that these models have met EPA qualifications to be considered cleaner burning. A white tag indicates units that have met more stringent requirements and are approximately 70% cleaner than older fireplace models. An orange tag indicates the unit has met requirements for the first phase of the voluntary program and is approximately 57% cleaner than older fireplace models.

Heat Pumps and Air Conditioners

Small homes and those that are superinsulated sometimes have such small heating and cooling loads that traditional split HVAC systems are only available in sizes that are too large to effectively heat and cool them. When selecting space-conditioning equipment for low-load homes, one-piece heat pumps and air conditioners or ductless mini-splits may be a good option for properly sized, cost-effective space conditioning.

One-Piece Heat Pumps and Air Conditioners One-piece heat pumps and air conditioners can be installed in windows or through walls to provide heating, cooling, or both to individual rooms. Powered by electricity, they are available in a range of efficiencies, and the lack of ductwork eliminates losses associated with duct leakage. One-piece units create the most noise of any system because they have both a fan and a compressor that run whenever heating or cooling is required.

Larger one-piece units, known as package terminal air conditioners (PTACs), are commonly used in hotels and motels to allow for easy control of individual room temperature. PTACs are often installed in metal sleeves that are built into exterior walls. One-piece units are generally available only in lower efficiencies than central and mini-split systems.

Mini-Split Heat Pumps Mini-split heat pumps consist of a condensing unit that is installed on the exterior of the home and connected to refrigerant lines that feed heated or cooled refrigerant to a fan-coil unit on the interior. A single condensing unit can handle one or as many as eight interior distribution units. The fan coils can be surface-mounted or recessed, and some are available with outlets for ducts that can provide conditioned air to multiple rooms.

Mini-splits provide conditioning to any room where a fan-coil unit is installed; individual temperature controls are included with each unit. Some systems are available with water heating options.

Efficiency Mini-split heat pumps must meet the same minimum efficiency requirements as split systems: 13 SEER and 7.7 HSPF. They are commonly available with efficiencies as high as 22 SEER and 10 HSPF, although they often obtain higher operating efficiencies than traditional ducted systems because there are no duct losses involved in distributing the space conditioning.

Ventilation

Homes need whole-house ventilation systems to ensure a steady supply of fresh air to dilute pollutants throughout the home. Spot ventilation is also needed to remove moisture and pollutants from bathrooms, kitchens, garages, laundry rooms, and storage areas.

Whole-House Ventilation Systems

Whole-house ventilation systems must provide enough fresh air to keep the house safe and comfortable but prevent over-ventilating to the point that energy is wasted in the process. In dry climates excessive ventilation can cause the indoor air to be too dry and during cold weather it may require additional heating to offset large amounts of heat loss. In hot and humid climates, excess outdoor air can result in high humidity and require additional cooling and dehumidification. Insufficient ventilation can result in odor and moisture problems, although there are few data to suggest that occupant health is compromised.

Fresh outside air may be provided through supply, exhaust, or balanced ventilation systems (Figure 13.53). Natural ventilation using operable windows is generally not considered an effective ventilation strategy because the air flow changes depending on wind speed and temperature differences between the interior and the exterior. Also, there is no way to effectively filter, heat, cool, or dehumidify air brought in through natural

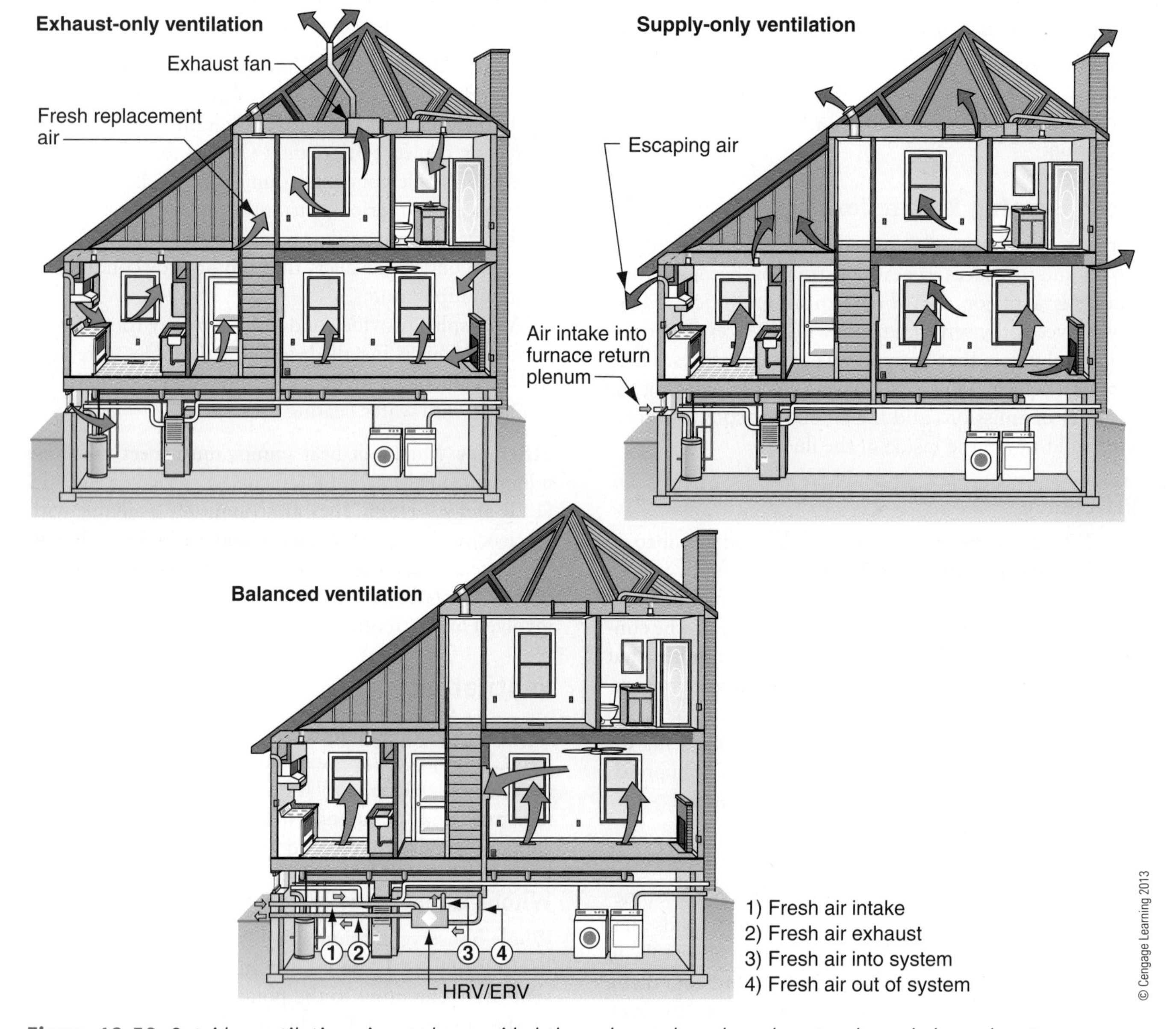

Figure 13.53 Outside ventilation air may be provided through supply-only, exhaust-only, or balanced systems.

ventilation. Regardless of these issues, opening windows to bring fresh air into a home and turning off HVAC systems is an excellent energy-saving strategy when outside temperature and humidity are moderate.

Supply-only ventilation provides outside air to the home without exhaust. Supply-only systems positively pressurize the home and are recommended for cooling-dominant climates. One common method of supply ventilation is the installation of a duct that brings outside air into the return plenum of the forced-air HVAC system. The supply air can be controlled by a barometric damper that opens whenever the blower runs or mechanical damper that is controlled by a timer, humidity sensor, and/or thermostat. The mechanical damper controls the amount of fresh air brought into the HVAC system and prevents the flow when the outside air has a high moisture content or is too cold or hot. To carefully control the amount of fresh air brought into the house, a fan cycling controller operates the blower fan motor. This mechanism turns on the motor and opens the intake damper whenever ventilation is needed, based on a set amount of minutes per hour that is determined by the desired amount of fresh air. When using

---

supply-only ventilation provides outside air to the home without exhaust.

the forced-air system for supply ventilation, ECM fan motors should be used for maximum efficiency. Outside air intakes should not be located on roofs where fumes from asphalt coatings could be drawn inside, and they should also avoid areas where vehicles may idle or other pollutants are present, such as exhaust vents and furnace and water heater flues.

Exhaust-only ventilation systems remove air from the home with equal parts of makeup air entering through an uncontrolled opening (i.e., envelope leaks). Exhaust ventilation is generally not recommended in humid climates to avoid the infiltration of moisture-laden air into building components from the exterior. Exhaust ventilation is most often accomplished by using a continuously operating exhaust fan in a bathroom, ducted to the exterior.

Balanced ventilation provides equal parts supply and exhaust air through the use of fans, or air-to-air heat exchangers, referred to as either heat recovery ventilators (HRVs) or energy recovery ventilators (ERVs). Generally used in cold climates, HRVs remove heat from stale air being exhausted, moving it to the incoming fresh air to reduce energy loss. In humid climates, ERVs remove moisture from fresh outside air, moving it to the stale air being exhausted. This transfer of heat and moisture helps reduce the energy required to condition the fresh air that may be colder, warmer, or more humid than desired. This equipment, however, does use energy to operate. In moderate climates, the net energy savings may be minimal; in some cases, the cost of using an HRV or ERV may be higher than if unconditioned fresh air is brought into the house. ERVs and HRVs require maintenance and, like any mechanical equipment, are subject to occasional breakdowns that may not be anticipated by homeowners, rendering them ineffective if not repaired.

Fan systems can be single-point, with one exhaust fan and one supply inlet or fan, or they may be incorporated into existing HVAC or separate duct systems. Although single-point systems will not effectively circulate fresh air throughout all the rooms in a house, ducted systems can provide even distribution of fresh air. Fan systems introduce unconditioned air into the house that may need to be heated, cooled, or dehumidified—increasing energy use. Table 13.5 shows the relative costs for different ventilation strategies and presents their appropriateness by climate.

ASHRAE Standard 62.2

ASHRAE developed a standard for residential ventilation, commonly referred to as 62.2. Originally issued in 2003, the 2010 version is used to calculate minimum whole-house and spot ventilation rates for homes. Most building codes use ASHRAE 62.2 as a requirement, and green building programs either require compliance with the standard, or provide points for meeting it. Mechanical ventilation is designed to ensure an adequate supply of fresh air and dilution of pollutants present in tightly sealed houses; however, it requires energy to operate. When outside air is cold, hot, or humid, it extracts an energy penalty, requiring additional heating, cooling, or dehumidification. While compliance with this standard is either required or recommended, in a home with minimal pollutants and low occupancy rates, the rate of ventilation may be greater than actually required in a home with minimal pollutants and low occupancy rates. Whole-house ventilation rates can be adjusted up or down to reach a level at which the indoor air is fresh and relative humidity is at appropriate levels, based on the occupants' comfort.

Calculating Whole-House Ventilation

ASHRAE 62.2 has established a ventilation formula based on the conditioned floor (CFA) area and the number of bedrooms. Ventilation in CFM is calculated as 1% of the CFA, plus 7.5 CFM times the number of bedrooms plus 1.

Table 13.5 Selecting Ventilation Equipment

Ventilation Strategy	Climate	Costs
Supply ventilation	Cooling dominated	Low initial costs and moderate to high operating costs
Exhaust ventilation	Heating dominated	Low initial costs and moderate operating costs
Balanced ventilation (HRV)	Heating dominated	High initial costs and low operating costs
Balanced ventilation (ERV)	Cooling dominated	High initial costs and low operating costs
Natural	Mild	No cost

exhaust-only ventilation systems that remove air from the home with equal parts make up air entering through uncontrolled opening (i.e., envelope leaks).

balanced ventilation a system that provides equal parts supply and exhaust air through the use of fans, or air-to-air heat exchangers, referred to as either heat recovery ventilators (HRV) or energy recovery ventilators (ERV).

For example, a 4-bedroom, 2,500 ft^2 house would require 62.5 CFM of ventilation, calculated as follows:

$$\text{Ventilation (CFM)} = (\text{Conditioned Floor Area [CFA]} \times 0.01) + 7.5\ (\text{number of bedrooms} + 1)$$

$$= (2{,}500 \times 0.01) + 7.5(4 + 1)$$

$$= 62.5\ \text{CFM}$$

Ventilation can be provided either continuously or intermittently. Continuous ventilation should provide the required CFM when running full-time. Intermittent ventilation is provided with a higher CFM for a portion of each hour.

For example, a 150 CFM fan running for 25 minutes of every hour provides the equivalent to a 62.5 CFM fan running continuously for one hour.

$$62.5\ \text{CFM} \times 60\ \text{minutes} = 3{,}750\ \text{CF/hour}$$

$$150\ \text{CFM} \times 25\ \text{minutes} = 3{,}750\ \text{CF/hour}$$

ASHRAE 62.2 limits the total ventilation in hot humid and very cold climates to a maximum of 7.5 CFM per 100 ft^2 of floor area to avoid over-ventilation in these extreme conditions.

Home Ventilating Institute

The Home Ventilating Institute (HVI) is an association of manufacturers that certifies ventilation equipment, provides information to consumers and professionals about home ventilation, and participates in the development of building codes. HVI provides recommendations for whole-house and spot ventilation that is broader than the minimum requirements of ASHRAE 62.2.[7]

Spot Ventilation

Spot ventilation exhausts rates are calculated according to ASHRAE Standard 62.2 or IRC requirements. The IRC requires minimum exhaust rates for kitchens of 100 CFM intermittently or 24 CFM continuously; for bathrooms, IRC rates are 50CFM intermittently or 20 CFM continuously. Most kitchen and bath ventilation in homes is intermittent rather than continuous, although bathroom vent fans that run continuously at a slow speed are often used as part of a whole-house ventilation system.

Kitchen Ventilation

Kitchen ventilation rates of between 100 and 300 CFM are usually adequate for most residential ranges; however, the use of larger, commercial style ranges and their manufacturer's recommendations for larger capacity range hoods often leads to the installation of kitchen spot ventilation systems with rates as high as 900 to 1,200 CFM. While this much ventilation may be appropriate when using 6 or 8 burners at the same time, it far exceeds typical everyday cooking needs and can lead to over-ventilation and depressurization. Forced-air HVAC systems typically move about 400 CFM for each ton of air conditioning, so a range hood that exhausts 1,200 CFM is removing the equivalent of 3 tons of conditioned air from the space when it is running. This not only requires additional heating or cooling, but it will frequently cause backdrafting of fireplaces and any open-combustion appliances, as well as infiltration of outside air through gaps in the building envelope.

Manufacturers of commercial style ranges often recommend ventilation rates in the range of 300 CFM per linear foot, leading to the prevalence of 1,200 CFM hoods for their large ranges. These estimates are based on commercial kitchen designs, which allow for and include makeup air. The needs of a residential kitchen are different than those of a commercial one, and makeup air is uncommon in homes. HVI recommends ventilation rates of 100 CFM per linear foot of range width when the range is mounted against the wall, and 150 CFM when the range is set in an island. For example, a 48" range would require a rate of 400 CFM by HVI standards, which is less than half of what most range manufacturers would recommend.

Kitchen Ventilation Makeup Air The 2009 IRC requires makeup air for range hoods exhausting in excess of 400 CFM at a rate approximately equal to the exhaust air rate. The makeup air system must be equipped with an automatically controlled damper or other closure mechanism that operates simultaneously with the exhaust system (Figure 13.54a and Figure 13.54b). Makeup air systems, while very common in commercial kitchens, are rare in homes. Little standard equipment is available, requiring makeup air to be provided through a custom-designed and installed system. Although the code requirement is a good start, it does not address heating or cooling the makeup air; if this air is unconditioned, it can add to the home's overall heating or cooling load. In addition, the location of the makeup air inlet is critical. The ASHRAE journal has published studies on commercial kitchen ventilation studies that show how improper location of makeup air inlets can lead to short circuiting of the air flow into the hood, reducing its effectiveness at removing pollutants.[8] Similar studies on residential applications are limited, but the same principles apply.

The best practice in a green home is to install the smallest range hood available and advise homeowners to use any hood at the lowest available setting whenever possible. If a house has outside intake air as part of the HVAC system (as previously described), it may provide

[7] http://www.hvi.org/resourcelibrary/HowMuchVent.html

[8] Richard T. Swierczyna, Paul A. Sobiski, "The Effect of Makeup Air on Kitchen Hoods," ASHRAE Journal, July 2003.

FROM EXPERIENCE

Whole-House Ventilation: How Much Is Enough?

Armin Rudd, Principal, Building Science Corporation It is important to size ventilation rates correctly, as ventilation rates that are too low may result in poor odor and moisture control as well as inadequate air for a healthy living environment. Ventilation rates that are higher than needed will tend to waste energy, cause homes to be too dry in dry climates or during the wintertime in cold climates, and add excess humidity in hot, humid climates, which has the effect of increasing cooling and dehumidification loads.

Contributed by Armin Rudd, Building Science Corp., www.buildingscience.com

Principal engineer at Building Science Corporation since 1999, Armin Rudd conducts analysis, design, inspection, research, and development of building mechanical systems and building envelopes. His activities focus on the overall production homebuilding industry and commercial building moisture investigations.

Best Practice Guidelines Based on Homebuilding Experience and Research

Over the past two decades, many hundreds of thousands of homes—both site-built and manufactured—have been constructed throughout the United States and Canada with varying amounts of whole-house mechanical ventilation. Research has shown that these homes have better perceived air quality and fewer moisture problems than homes that do not have whole-house ventilation.

Over the last decade, homebuilders in partnership with the U.S. DOE's Building America Program (http://www.buildingamerica.gov) have been on the leading edge of producing high-performance homes that use a systems engineering approach to obtain improved energy efficiency and comfort, better indoor air quality, higher durability, and lower risk. These homes include whole-building mechanical ventilation with whole-building distribution of ventilation air, even though there is no whole-house distribution requirement in the ASHRAE 62.2 Standard. This means, for example, that a house with a single local exhaust fan in the master bathroom gets the same whole-house ventilation performance credit as a fully ducted ventilation system. Experience with several hundred thousand such high-performance homes, having roughly 50% to 60% of the ventilation rate required by ASHRAE Standard 62.2 but with full whole-building distribution and mixing of ventilation air has proven successful. The lack of complaints by occupants indicates that the systems are working to provide acceptable indoor air quality. In our practice, we recommend this level of whole-house ventilation, along with more than enough available capacity to meet the ASHRAE 62.2 Standard upon occupant demand.

On the basis of three years of published research (ASHRAE Transactions), a simple approach has been proposed that could be applied to the ASHRAE Standard 62.2 minimum ventilation rate to obtain a minimum ventilation fan flow rate. Such a rate would account for the effect of ventilation system characteristics, such as balanced vs. unbalanced systems, single-point vs. multi-point systems related to ventilation air distribution, and whole-building mixing related to ventilation air distribution.

The next research area to improve ASHRAE Standard 62.2 should be to consider the effect that the source of ventilation air has on occupant exposure to contaminants. For example, whole-building ventilation systems that supply ventilation air drawn from known fresh air locations will have a different impact on occupant contaminant exposure than exhaust systems that rely on building depressurization to draw ventilation air from the unknown paths of least resistance. Such paths could be through walls from garages, through floors from crawl spaces or below-slab, among others. Likewise, supply ventilation systems that have duct leakage on the inlet side can have similar problems. Work in this area is expected to begin in the near future.

Contributed by Armin Rudd, Building Science Corp.
http://www.buildingscience.com

enough fresh air to balance a low-volume range hood. A barometric damper will automatically open when the house becomes depressurized, or a mechanically operated damper can be designed to open automatically whenever the range hood operates, providing makeup air through the home's duct system.

Bathroom Ventilation

Bathroom ventilation is an important component in controlling vapor in a home. Unvented and insufficiently vented shower and tub rooms build up excess moisture, which can cause mold and mildew to form and, in extreme cases, lead to structural damage in wood framing.

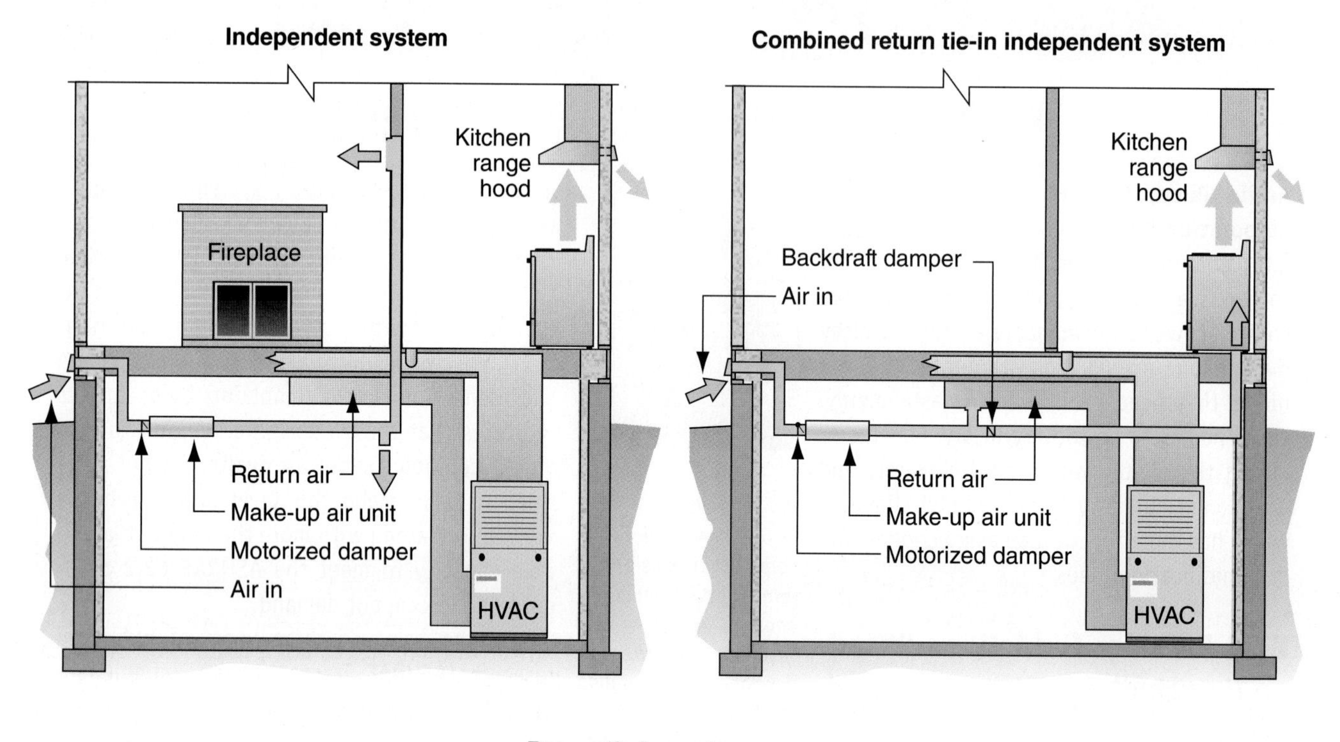

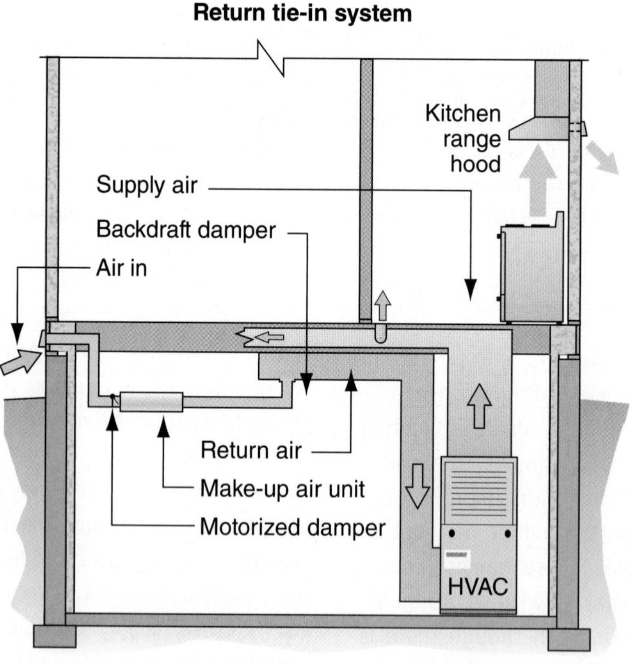

Figure 13.54a Three options for kitchen exhaust make-up air.

ASHRAE 62.2 defines a bathroom as any room containing a bathtub, shower, spa, or a similar source of moisture. These rooms are required to have mechanical ventilation or an alternative method, designed by a licensed design professional, that provides the minimum required exhaust rates. Vent fans are often installed in powder rooms and toilet compartments, but most jurisdictions allow for the fan to be eliminated when an operable window is present.

Bathroom ventilation fan sizing is an imprecise science. In addition to ASHRAE 62.2, HVI has guidelines that are calculated either by the room volume or by the number of fixtures. For bathrooms up to 100 ft^2, HVI recommends 1 CFM per square foot of floor area. For an 8' × 5' bathroom (40 ft^2), the recommended fan size is 40 CFM. For baths over 100 ft^2, 50 CFM is recommended for each toilet, shower, or tub, and 100 CFM

Courtesy of Dakota Supply Group

Figure 13.54b Ventilation fan for kitchen exhaust make-up air.

is recommended for each jetted tub. The recommended ventilation rate for a bathroom with one shower, one tub, and one toilet is 150 CFM. This could be accomplished with a single large fan or with two or more smaller fans. Locating a fan close to a shower compartment is recommended because this is where most of the vapor is created in a bathroom.

Fans are available in sizes ranging from about 50 CFM to over 200 CFM. These ratings are usually for the fan connected to a short length of duct. Long runs, elbows, and other restrictions in the exhaust duct reduce the actual amount of air that the fan is able to move to the exterior. ASHRAE 62.2, as well as several fan manufacturers, provide prescriptive duct design tables to help determine the maximum length of duct and the number of elbows allowed that will not restrict the total fan flow. Flexible duct, where allowed by local code, is more restrictive than smooth metal duct, therefore allowable lengths for flex duct are shorter than those for hard duct. In addition, each elbow and exterior wall cap adds the equivalent of 15 linear feet to the duct length, further reducing the allowed total length for each fan. Duct tables also help determine the minimum pipe diameter required for adequate flow. For example, ASHRAE only allows rigid ducts of 3" diameter with a maximum length of 5' for fans up to 50 CFM, effectively requiring 4" or larger ducts for all fans. For 50 CFM fans, 4" rigid ducts up to 105 linear feet are allowed; flex ducts for the same size fan are only allowed to be 70 linear feet. As larger ducts are installed, length limits are reduced and, in some cases, disappear. For example, a 50 CFM fan installed on a 5" hard duct system has no length limit. When fans cannot be located with short, direct routes to the exterior, larger ducts or higher capacity fans can be installed to provide adequate flow rates. Best practices call for rigid duct, the most direct path to the exterior, and as few elbows as possible. Fans can be tested to confirm that they meet the required flow rates. This can be done with a flow hood, a pressure pan, or similar tools.

Bath Fan Operation and Controls

Bathroom fans should be run for between 20 and 40 minutes following a shower or other event that produces large amounts of vapor. HVI recommends a minimum of 20 minutes. Fans can be operated with a manual switch; however, this often leads to fans being turned off prematurely (failing to adequately remove vapor), or allowed to run too long (wasting energy). Automatic controls include humidity sensors that are set to turn on when the relative humidity reaches a pre-set level and turn off when it drops below 50% or other setting. Simple turn-knob or push-button timers are an inexpensive way to ensure that the fan will run for long enough but not left on for too long. Fans can also be controlled with delay switches that control both the light and fan with a timer that keeps the fan running for a period of time after the light is turned off. Motion sensors can also be used to run the fan for a set length of time whenever someone enters the room.

HRVs and ERVs can incorporate bathroom ventilation with whole-house ventilation, usually running continuously at a low volume with return ducts in each bathroom.

Bath Ventilation Fans

The most efficient bath fans are ENERGY STAR–rated, which also limits the amount of noise they are allowed to make when running. Fans such as those available from Panasonic are available with efficient direct-current motors that automatically adjust their speed to provide a specified flow rate. This feature helps eliminate low flow problems associated with long and constricted duct runs. Broan offers SmartSense®, a whole-house ventilation system that incorporates bath and kitchen vent fans that communicate with each other through house wiring. The system automatically turns on fans when manual operation does not meet a pre-set minimum hourly ventilation rate. The SmartSense system includes an automatically operated damper that opens when any fan in the system operates, providing makeup air through the return air of a central duct system or a dedicated intake vent. Multiport fans, such as those made by Fantech, use a single fan mounted in a remote location with ducts run from the fan to each bathroom and a single exhaust to the exterior. Recirculating fans do not exhaust moisture or stale air and do not provide code-required ventilation.

Makeup Air Makeup air for bathroom fans is not as much of a problem as with kitchen ventilation due to the smaller fan size. Providing air flow by leaving the bathroom door open allows for much faster removal of moisture and shorter fan run times.

Garage Ventilation

Attached garages have the potential to allow carbon monoxide and other pollutants to enter the house through gaps in the building envelope. While a high-performance home can have an effective air seal between a garage and the house, minor gaps or doors left ajar eliminate effective separation of space. As a good practice, mechanical ventilation should be provided in any attached garages (Figure 13.55). The EPA recommends a rate of 50 CFM running continuously or 100 CFM intermittently. Intermittent fans can operate on motion sensors or by a switch tied to garage door operator.

Whole-House Fans

Although not generally considered part of a whole-house ventilation strategy, whole-house fans can be used for passive cooling and ventilation in moderate weather to make the house more comfortable. Traditional whole-house fans draw a high volume of air into an unconditioned attic, relying on soffit, gable, or ridge vents to exhaust it to the exterior. In practice, however, these fans can pressurize the attic and force hot air back down into the house through gaps in the ceiling, such as lights and HVAC registers. When a whole-house fan is installed, a thorough air seal of the ceiling is critical, and the roof vents must be large enough to exhaust all the air that the fan brings into the attic. In addition, most whole-house fans have automatic louvers that do not provide either an adequate air seal or insulation, requiring that a separate insulated and air-sealed cover be installed when the fan is not in use.

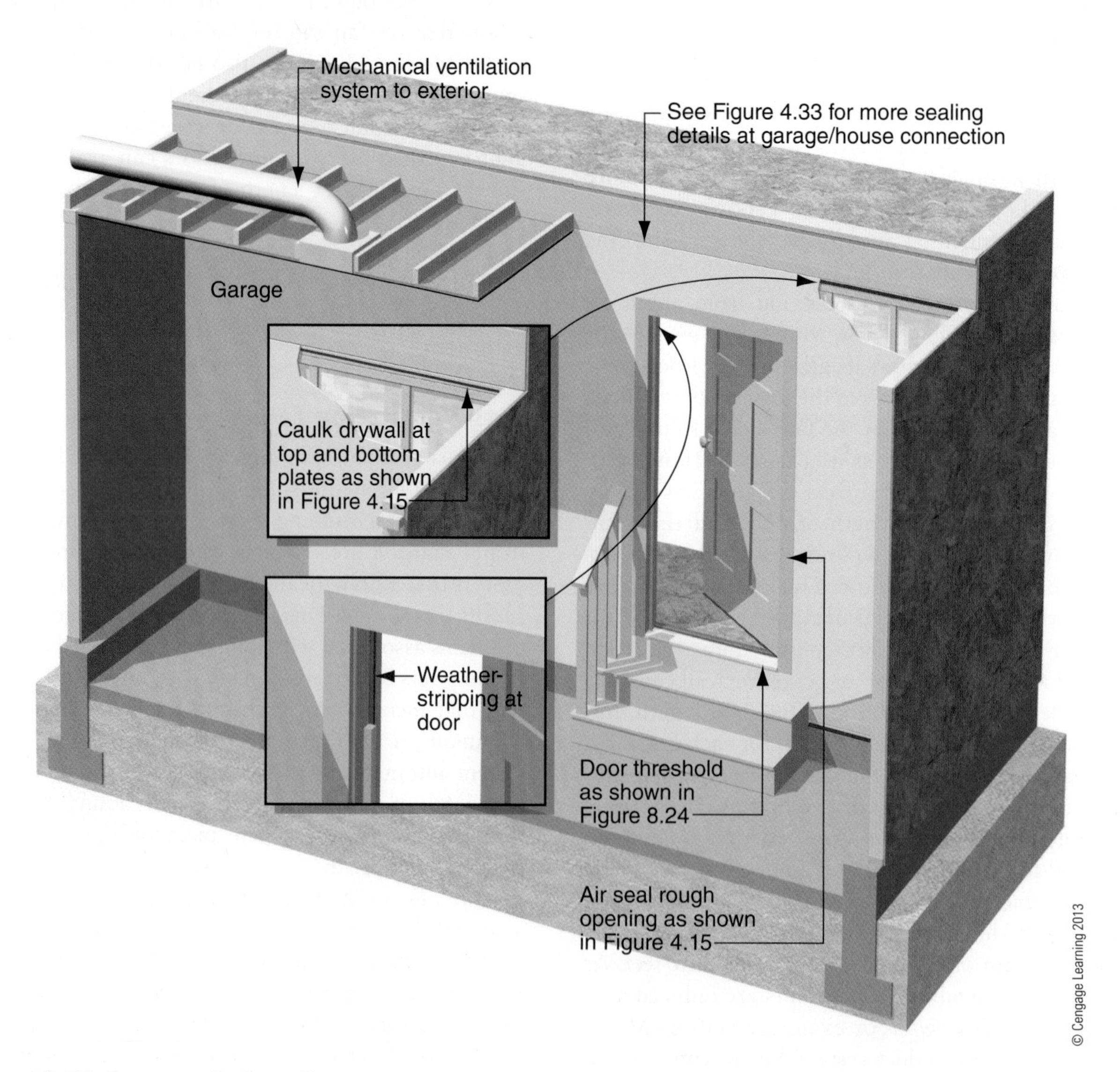

Figure 13.55 Garage wall air sealing.

Fans such as those manufactured by Tamarack Technologies have an automatic insulated cover that provides both the insulation and the air sealing that is required in a high-performance home. If a whole-house fan is installed in a home with an insulated roofline, it must have an insulated shaft or duct that directs the air to an exhaust louver, and a method to seal off the opening when not in use.

Natural Ventilation

Homes can take advantage of natural ventilation during moderate weather as a substitute for mechanical systems, although this is not considered a reliable method of consistent whole-house ventilation. Locating windows to provide cross-ventilation and designing thermal chimneys to take advantage of the natural stack effect are options available to designers. Natural ventilation is generally undesirable when humidity is extremely high or low, when temperatures are extreme, or when levels of pollen, smog, or other pollutants in outdoor air are high.

Keep in mind that most mechanical ventilation systems operate full-time or for a certain number of minutes every hour. A house designed with natural ventilation should include sensors that turn off mechanical vent systems when windows and doors are open for extended periods to avoid wasting energy.

Filters

Filters help improve indoor environmental quality by removing particulates and other pollutants from the indoor air. In homes with central duct systems, filters are near the air handler or in a central return grille. When there is no central duct system, filters can be part of separate whole-house ventilation system, such as an ERV, HRV, or a central dehumidifier. Homeowners will sometimes use individual room filters to purify indoor air, but the perceived need for these is typically an indication of other problems, such as excess air infiltration, duct leakage, and poor-quality whole-house filters.

Filter Types

Residential air filters that remove particulates from the air include fiberglass, pleated paper, and electrostatic precipitators (Figures 13.56a, b, and c). Filters collect particulates when air passes through them, requiring regular replacement as they become filled with dust and dirt. Fiberglass filters only trap the largest particles, allowing most to remain in the air. Pleated paper filters remove most particulates except the smallest, such as bacteria and viruses. All filters provide some level of resistance to air flow, which must be taken into account when designing duct systems and selecting air handlers so as not to create excess pressure on the fan motor and reduce air flow through the ducts. Fiberglass and thick pleated paper filters have the least resistance to air flow; thin pleated paper filters have the most. Many HVAC systems are designed for 1" fiberglass filters that provide adequate air flow; however, the filters are often replaced

Courtesy of Aerostar Filtration Group

Figure 13.56a Fiberglass filters.

Courtesy of Aerostar Filtration Group

Figure 13.56b Pleated filters.

Courtesy of Aerostar Filtration Group

Figure 13.56c A pleated HEPA filter.

with 1" thick pleated paper filters that do a better job of filtering particulates but also reduce air flow. Consequently, performance problems arise, including cooling coil freezing, premature blower failure, and system imbalances.

Electrostatic precipitators, also referred to as electronic air cleaners, use an electrically charged wire to charge particles in the air and attract them to a plate where they are collected (Figure 13.57). These permanent filters have high efficiency ratings but require regular maintenance and cleaning for effective operation.

Although less common, activated carbon filters and ultraviolet (UV) purification systems can remove volatile organic compounds (VOCs) and other contaminants. Activated carbon filters can remove VOCs but not particulates, which can damage them, so carbon filters must be combined with mechanical filters. They may be useful for people that have high chemical sensitivity, but a better strategy to improve indoor air quality is to avoid introducing VOCs in the first place. UV filters sterilize air as it passes through, killing pathogens, viruses, and mold. While not very common in homes, UV filtration used in combination with a high-efficiency mechanical filter may be effective for people with compromised immune systems.

Regardless of the type of filter installed, the performance of the duct system plays a large role in its effectiveness. A leaky duct system will allow contaminants to enter the air through the ducts both before and after the filter.

Figure 13.57 An electrostatic precipitator that is mounted by the air handler.

Filter Efficiency

ASHRAE developed a rating system for filter efficiency, referred to as the Minimum Efficiency Reporting Value (MERV). This scale ranges from 1 to 16; the higher the number, the more efficiently the filter removes particulates from the air (Table 13.6). High-efficiency particle arrestor (HEPA) filters meet EPA standards, removing at least 99.97% of airborne particles per 0.3 micrometers. Used most commonly in hospitals and manufacturing facilities, HEPA filters are also available for residential applications. HEPA filters typically have a MERV rating of 16, although not all MERV 16 filters meet HEPA requirements.

Filter Location

Filters should be designed for easy removal and located where they are accessible for maintenance and replacement. In homes with one or more central returns, filters can be installed behind hinged grilles. Another common location is at the air handler. Many air handlers have built-in 1" filter racks, many of which are difficult to operate and limit the thickness of filters. Best practices suggest installing a thick pleated paper filter in a separate cabinet that is installed between the return air plenum and the air handler. Due to the high pressures at the air handler, filter cabinets must be very well air sealed and have a tight-fitting cover to minimize air leaks.

Maintenance

HVAC systems must be maintained for maximum performance. Twice-annual inspections of equipment, occasional refrigerant charge check, regular filter replacement, and inspections for damage to ducts, pipes, and equipment will ensure that the entire system functions properly.

Remodeling Considerations

HVAC systems in existing homes provide an excellent opportunity to improve both efficiency and indoor environmental quality. Inefficient equipment, poorly designed and installed duct systems, combustion safety problems, and poor-quality filters waste energy and degrade air quality.

Table 13.6 MERV Parameters

MERV Value	Average Particle Size (0.3–1.0 Microns)	Average Particle Size (1.0–3.0 Microns)	Average Particle Size (3.0–10.0 Microns)
1	–	–	<20%
2	–	–	<20%
3	–	–	<20%
4	–	–	<20%
5	–	–	20%–35%
6	–	–	35%–50%
7	–	–	50%–70%
8	–	–	70%–85%
9	–	<50%	85%
10	–	50%–65%	85%
11	–	65%–80%	85%
12	–	80%	85%
13	<75%	90%	90%
14	75%–85%	90%	90%
15	85%–95%	90%	90%
16	95%	95%	90%

Source: Courtesy of American Society for Heating, Refrigerating, and Air-Conditioning Engineers, reproduced with permission in the format Textbook via Copyright Clearance Center: *ANSI/ASHRAE Standard 62.1-2007: Ventilation for Acceptable Indoor Air Quality*, copyright 1985.

Evaluating Existing Systems

Performing an energy and combustion safety audit on existing HVAC systems is the best way to determine the potential for improvements. Forced-air systems should have a duct leakage test performed, checking for leakage both within and outside the building envelope. Flow testing should also be performed to determine if there is adequate delivery of conditioned air. Duct systems should be repaired or replaced as appropriate for the project conditions.

Duct System Improvement

Older duct systems were often poorly designed, improperly installed, or suffered damage following installation, all of which led to inefficient performance. When evaluating an existing system, consider the overall distribution for efficiency and for supply and return balance, and inspect for restricted flow, poor sealing, disconnected joints, and major leaks, such as those that occur at panned joists. Old duct systems were frequently sealed with asbestos tape, which, if disturbed, must be removed or encapsulated in compliance with local and federal environmental regulations. Careless removal of asbestos tape can expose workers and occupants to fibers that, when inhaled, can lead to such health problems as mesothelioma and cancer.

Systems that were designed with single central returns can benefit from additional room returns, transfer grilles, or jumper ducts to balance supply and return pressures. If the basic distribution system design is sound, simply sealing leaks and repairing defects will increase performance. If the system is poorly designed, partial or complete replacement may be an appropriate solution, paying particular attention to panned joists and other areas that are difficult to seal effectively. In homes where duct testing shows significant leakage but ducts are concealed in walls and ceilings, a product such as Aeroseal® can be used to reduce duct leakage by blowing a sealant through the duct system. After sealing existing duct systems, total supply and return flows should be measured and balanced to avoid creating excessive positive or negative pressure in the house.

Relocation of the building envelope to incorporate ducts inside the building envelope can significantly increase efficiency; in humid climates, however, reducing the load on the system may lead to oversized air conditioning and the attendant problems of short cycling and poor dehumidification. When making significant changes to an existing HVAC system, a new load calculation should be performed, and appropriate changes made to match system size to the heating and cooling needs.

High-Velocity Duct Systems

High-velocity HVAC systems are primarily used in retrofit situations where space for standard sized ducts is limited or existing wall and ceiling finishes are to remain. The small diameter allows ductwork to be snaked through small spaces, such as stud or joist cavities, where standard ducts would not fit. These systems can also be installed in new construction where space is limited. High-velocity systems consist of an air handler, rigid trunk lines, 2" insulated flexible supply ducts, and sound-suppressing registers. The high-pressure air handlers, which are smaller than traditional units, include an indoor coil and are combined with traditional condensing units or hydronic heating systems.

Since the flow of air is so strong, each register has a zone of influence where streams of moving air can be felt up to about 5' away. This requires registers to be located so that the air flow will not blow directly on occupants to avoid discomfort issues. High-velocity systems are less common than traditional duct systems, and they are generally more expensive to install. They can be a good alternative, however, where duct space is limited and the ability to remove and replace existing interior finishes is limited.

Improving and Replacing Existing Equipment

Open combustion furnaces and boilers can either be replaced with more efficient, sealed combustion equipment or isolated from the air in the home with a combustion closet or other air-sealing method. When replacing combustion equipment that is not sealed combustion, the new equipment may not be able to be vented through an existing flue or chimney; a new vent system may be required, or the existing chimney may need to be upgraded to meet the requirements of the new equipment. Air conditioners and heat pumps, if retained, should have their refrigerant charge checked and adjusted to meet manufacturer's specifications.

The existing HVAC systems in many homes may not adequately condition the existing space, but they may be in good condition and efficient enough to consider reusing. When the building envelope and duct systems are inadequate, they put more demand on the HVAC system. Improving the building envelope and installing a more efficient duct system or repairing defects in the existing system can reduce the load enough to use the existing equipment. In some cases, additional space can be added without adding or replacing equipment. Strategies for load reduction can include moving ducts into conditioned space, duct sealing, insulation, air sealing, and window shading. Reducing the overall load and reusing existing equipment is a better long-term strategy than simply installing new, larger HVAC equipment without building envelope improvements. Even when the existing equipment is not as efficient as desired, it can still be replaced later with a smaller system than if the overall building load was not reduced during the initial renovations.

Summary

HVAC is one of the most critical components in creating a green home. Proper design, installation, and maintenance of heating, air conditioning, dehumidification, and ventilation systems are key components of a healthy and efficient building. By coordinating the HVAC design into the structure from the very beginning, rather than as an afterthought, you can create efficient duct and piping designs, choose properly sized equipment, and gain easy access for maintenance.

Review Questions

1. Use of which of the following is the best way to reduce a building's overall energy use for HVAC?
   a. High-efficiency equipment
   b. Geothermal heat pumps
   c. Duct sealing
   d. High-performance building envelope
2. Which of the following is the most common method for distributing heating and cooling throughout a home?
   a. Radiant cooling
   b. Forced air
   c. Solar
   d. Fan coil units
3. Which of the following is not an example of a non-distributed HVAC system?
   a. Window air conditioner
   b. Fireplace
   c. Forced air heating
   d. Ductless mini-split
4. Which of the following is not a primary energy source for heat pumps?
   a. Water
   b. Air
   c. Ground
   d. Natural gas
5. Which of the following is the best option to consider when the indoor air in a house is too dry?
   a. Install a humidifier
   b. Upgrade HVAC system
   c. Air seal the building envelope
   d. Install a dehumidifier
6. Which of the following is the best strategy for improving indoor air quality?
   a. Run vent fans continuously
   b. Avoid introducing pollutants into the house
   c. Use high-quality filters
   d. Install high-efficiency air conditioning

7. The best strategy for a balanced duct system includes which of the following?
   a. Single central return grille
   b. Flex duct
   c. Rigid duct
   d. Individual room returns
8. What is the best location for HVAC ducts?
   a. Unconditioned basement
   b. Outside the building envelope
   c. Vented attic
   d. Inside the building envelope
9. Which of the following is the safest heating system in regard to indoor air quality?
   a. 80% AFUE gas furnace
   b. Wood pellet furnace
   c. 7 HSPF heat pump
   d. 80% AFUE oil boiler
10. Which of the following whole-house ventilation strategies is the best way to meet ASHRAE 62.2 requirements in a hot, humid climate?
   a. Exhaust only
   b. Supply only
   c. Balanced
   d. Natural

Critical Thinking Questions

1. How does the location of the heating equipment and distribution system affect the overall efficiency of the HVAC?
2. What are the pros and cons of different types of air vs. water distribution systems in homes?
3. How do the design and construction of the home, including insulation, air sealing, and window placement, affect the size and location of HVAC systems?

Key Terms

ACCA Manual D, 350
ACCA Manual J, 349
ACCA Manual S, 349
ACCA Manual T, 350
acute CO poisoning, 366
air conditioner, 326
air conditioning (AC), 331
Air Conditioning Contractors Association (ACCA), 349
Air-Conditioning, Heating, and Refrigeration Institute (AHRI), 354
air-source heat pump, 330
annual fuel utilization efficiency (AFUE), 362
balanced ventilation, 377
boiler, 326
carbon monoxide detector, 364
chronic CO poisoning, 366
coefficient of performance (COP), 362
cogeneration, 364
combined heat and power (CHP), 364
combustion analyzer, 364
commissioning, 372
complete combustion, 363
compressor, 329
condenser, 329
condensing furnace, 363
condensing unit, 332
direct evaporative cooler, 332
draft gauge, 364
ductless mini-splits, 329
effective duct length, 350
electric furnace, 326
electronic air filters, 337
energy efficiency ratio (EER), 368
equivalent length, 350
evaporative cooling, 331
evaporator, 329
exhaust-only ventilation, 377
fan coils, 354
furnace, 326
ground-source heat pump (GSHP), 330
heating seasonal performance factor (HSPF), 361
heat pump, 326
hydronic, 335
incomplete combustion, 363
indirect evaporative cooler, 333
jumper ducts, 339
latent heat, 367
leakage to the outside, 346
mechanical air filters, 337
National Appliance Energy Conservation Act (NAECA), 370
pleated air filters, 337
plenums, 339
radial duct system, 340
radiant heating, 334
refrigerant, 329
seasonal energy efficiency ratio (SEER), 368
sensible heat, 367
sensible heat fraction (SHF), 367
sensible heat ratio (SHR), 367
spider duct systems, 339
spot ventilation, 336
stack thermometer, 364
static pressure, 339
supply-only ventilation, 376
total duct leakage, 346
transfer grilles, 339
trunk-and-branch duct systems, 339
water-source heat pump (WSHP), 330
zone control, 349

Additional Resources

Building Science: *Ventilation Guide* by Armin Rudd. Building Science Press, 2006. http://www.buildingscience.com

American Society of Heating, Refrigerating and Air-Conditioning Engineers, Inc.: *ANSI/ASHRAE Standard 62.2-2010 —Ventilation and Acceptable Indoor Air Quality in Low-Rise Residential Buildings.* American Society of Heating, Refrigerating, and Air-Conditioning Engineeers, Inc., 2010. http://www.ashrae.org

Home Ventilating Institute (HVI): http://www.hvi.org

Air Conditioning Contractors of America (ACCA): *Manual J Residential Load Calculation,* 8th edition, by Hank Rutkowski, 2001. http://www.acca.org

Manual D Residential Duct Systems, by Hank Rutkowski, 1995. http://www.acca.org

Manual S Residential Heating and Cooling Equipment Selection, by Hank Rutkowski, 1995. http://www.acca.org

Manual T Air Distribution Basics for Residential and Small Commercial Buildings by Hank Rutkowski, 2009. http://www.acca.org

Building Science Tech (Producer), 2009. Building Analyst Field Training Video: How to Perform a Home Energy Audit [DVD]. Available from http://www.buildingsciencetech.com/

Electrical

This chapter covers lighting, appliances, wiring, controls, and other electrical equipment used in homes. Lighting and appliances compose a significant portion of residential energy use, which can be reduced through careful design and management. Strategies including fixture and lamp selection, control systems, and daylighting will be reviewed. Outdoor light pollution, wiring layout, and electromagnetic fields are also discussed. Ventilation fans, including exhaust fans, ceiling fans, and whole-house fans are discussed in Chapter 13.

LEARNING OBJECTIVES

Upon completion of this chapter, the student should be able to:

- Explain how to reduce electrical consumption in homes
- Describe how efficient lighting systems work
- Explain how to specify efficient appliances

Green Building Principles

 Energy Efficiency

 Resource Efficiency

 Indoor Environmental Quality

 Homeowner Education and Maintenance

Electricity Use in Homes

As discussed in Chapter 1, residential buildings consume approximately 21% of the energy produced in the United States. Although buildings are more efficient than they used to be, they are also larger and have more electrically powered equipment than in the past, such as appliances and electronics. Much of this equipment is operated by remote control and has digital displays that use electricity even when not in use. This power use is referred to as phantom loads and can amount to as much as 5% of the total electrical use in a home. This equipment, combined with more light fixtures and larger homes, significantly offsets much of the savings from reduced heating and cooling loads in more efficient homes. New and existing appliances and electronics, many of them quite inexpensive, add to the overall demand on our electrical systems.

Phantom Loads

Phantom loads, also referred to as vampires, are the small amounts of electricity that many appliances and electronics use even when they appear to be turned off. Phantom loads include electric clocks, displays on audio and video equipment and appliances, chargers, and transformers for cordless phones and similar equipment.

Most modern electronics must be in a full-time standby state for their remote controls to operate. This standby state uses a small amount of electricity, which over time and the number of devices, amounts to a significant amount of power. The U.S. Environmental Protection Agency (EPA) has estimated that one half of the power used by the average television during its lifetime is used when it is off and in the standby mode.

---

phantom loads also referred to as vampires, are the small amounts of electricity that many appliances and electronics use even when they appear to be turned off.

FROM EXPERIENCE

From Sell-a-Lot to Save-a-Watt!

Robert S. Mason, Jr., P.E., Vice President, Energy Efficiency, GoodCents Why do electric utilities want to encourage their customers to use less of the product they produce? The simple answer is . . . at one time they didn't. But, after decades of growth and declining cost, the electric utility industry entered the 1970s, a period where the business engine began to sputter. Impacted by increasing inflation, an Arab Oil Embargo followed by rising fossil-fuel prices, Federal Clean Air legislation, and the passage of The National Energy Conservation Policy Act of 1978, electric utilities entered a new business paradigm of increasing cost and business risks that has matured and expanded for 40 years.

Today's business world is an issues-balancing act for electric utilities, regulators, legislators, and consumers; balancing volatile fuel costs, high construction costs, and public concerns against legislative mandates, environmental impacts, and growth in demand for electric energy. The overall cost to produce and deliver electric energy at certain times is frequently higher than the cost to ensure it is consumed efficiently. As touted by Jim Rogers, CEO of Duke Energy, energy efficiency is the "Fifth Fuel," a reference to energy efficiency being on par with energy production. Furthering the growth of energy efficiency, Duke Energy developed the "Save-a-Watt" value proposition, a process that equates the dollar value of conserving (not selling) a kilowatt-hour of electricity to the financial value of generating one. Promotion of energy efficiency measures coupled with increasing efficiency requirements for appliances, equipment, controls, and building standards and practices has become a major focus for energy utilities.

Retail electric and natural gas utilities now provide a portfolio of cost-effective programs that enable customers to secure information, analysis, guidance, and incentives to help in reducing energy bills and make their homes and lifestyles more energy efficient. Popular among utility customers is the ability to have trained energy technicians come to their homes and/or businesses to conduct on-site examination, testing, and analysis to determine where they stand relative to energy efficiency and the steps they should take toward improvement. Especially desirable to customers are programs that offer financial rebates to customers who implement significant energy savings measures. Every customer should look at what their utility can do to help them be more efficient. It's the best way to keep utility costs down and ensure that we meet the needs of a growing world.

This increased demand becomes critical in periods of peak load, such as hot summer afternoons when people arrive home, turn on the air conditioning, and run appliances and electronics. Utilities are required to meet these peak loads, if only for a short period of time for a few days a year. This requires construction of additional power plants that only run during these peak load periods, which increases the use of coal and gas and adds to air pollution and greenhouse gases. Reducing our peak load demands provides long-term benefits by limiting the need for additional power plants.

Reducing Electricity Use as a Green Home Strategy

Strategies to reduce overall electricity use include efficient lighting, equipment, and appliances, and a combination of control systems and owner education to minimize power use. In addition to individual actions to reduce electricity use, utilities are developing their own strategies to reduce demand. Utility controlled load control switches can turn off air conditioning, refrigerators, and other high-demand appliances for short periods of time, managing peak load at critical times to avoid power shortages and reducing the need for additional power generation (Figure 14.1).

Smart Grids, Meters, and Appliances

Electric utilities nationwide are developing smart grid technology that is designed to communicate with smart meters and individual appliances to help manage electrical demand and reduce peak loads when necessary (Figure 14.2). Smart meters use information they receive from the smart grid to control electrical appliances and

---

smart meter **a broad term defining electrical meters that include two-way communication and other advanced capabilities.**

© Cengage Learning 2013

Figure 14.1 The load control switch is the gray box on the wall to the left of the condenser. The switch is able to communicate with the utility, which then can cycle the air conditioner on and off during periods of high usage.

equipment in the house, turning them off and on to manage system-wide loads. Combined with variable, time-of-day pricing, smart grids are designed to flatten out the peak demand periods and reduce the need for additional power generation for those peaks. Homeowners will be able to program their appliances to run only when prices and demand are lower, saving money and peak load demand. For example, when you turn on your electric oven when rates are high, it could tell your refrigerator to stop the defrost cycle or raise its temperature temporarily to save energy. Smart appliances can also respond to commands from utility central offices to reduce demand when necessary. Such appliances as televisions and audio equipment could be programmed to turn off automatically at certain times. Smart grid (and smart appliance) technology is advancing steadily and will likely become more common as utilities expand their construction.

Green Power

Many utilities have green power programs in which they purchase a certain amount of their energy or its equivalent from renewable sources, such as solar and wind. Consumers can specify that a portion or all of their electricity be supplied by these renewable sources, usually for a premium. The Green-e Energy certification program has developed definitions for renewable electricity and renewable energy certificates (RECs), which are tradable energy commodities that identify that 1 megawatt-hour of electricity was generated with renewable resources. Administered by the nonprofit Center for Resource Solutions, Green-e verifies renewable energy in competitive markets and sold in local utility green programs.

Household Electrical Systems

Within individual homes, electrical systems can be separated into *infrastructure*—the wiring, receptacles, switches, timers and other controls—and *equipment*—the lighting, fans, appliances (e.g., stoves, refrigerators, and microwaves), and electronics (e.g., televisions, audio components, computers, and chargers). The design and installation of infrastructure affect the amount of material used, the impact on the building envelope, and the occupant's ability to easily control the equipment. Equipment selection and installation affect a building's total electrical demand, the cooling load, and the impact on the building envelope.

Wiring

Home electrical systems consist of both high-voltage and low-voltage wiring. High-voltage systems include lighting, equipment, and some electronics. Low-voltage systems include audio, video, cable, satellite, computer, telephone, intercom, alarm, and related equipment.

High-Voltage Wiring

A key to designing high-voltage wiring systems for green homes is to reduce the amount of wire required to adequately service the installed equipment. The placement

---

smart grid a nickname for an ever-increasing number of utility applications that enhance and automate the monitoring and control of electrical distribution.

renewable energy certificates (RECs) tradable energy commodities that identify that 1 megawatt-hour of electricity was generated with renewable resources.

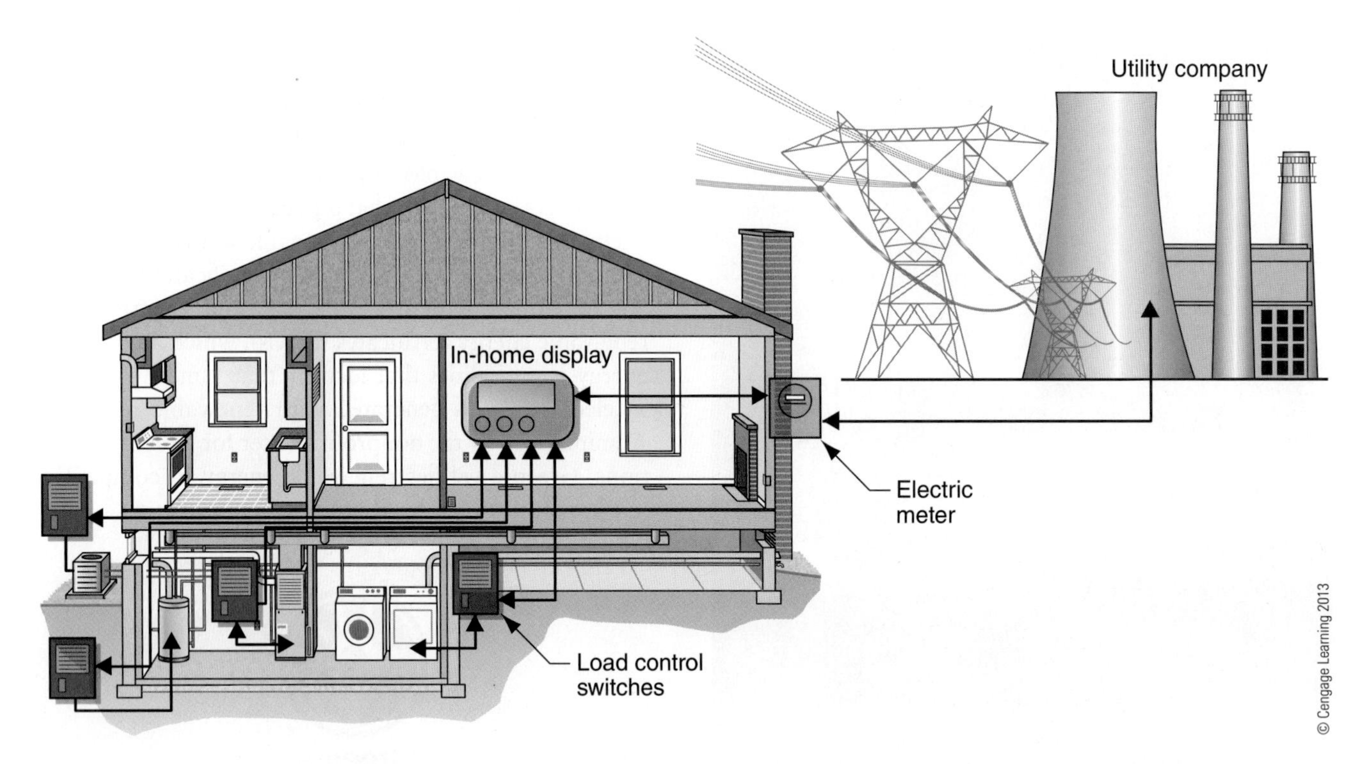

Figure 14.2 The smart grid allows the utility to communicate with the home's electrical meter and any in-home displays. The load control switches are installed to control major energy consumers, such as central air conditioners, electric water heaters, and refrigerators.

of the main electrical service panel can have an impact on the thermal envelope and influence the amount of wire required. When a panel is located in a garage or exterior wall, it displaces insulation and requires careful air sealing to the air barrier. Additionally, when a panel is located on one side of a house, individual circuit wires must run longer distances than would be required if the panel were located at a central location (Figure 14.3). Standard residential wiring has plastic coatings on the wire as well as a plastic sheathing over the entire cable. Like most plastics, these cable sheathings contain such chemicals as polyethylene, polyvinyl chloride, fluoropolymers, and lead that can leach out over time and produce toxic gases when burned. A greener alternative is metal-shielded cable which still has a plastic covering on individual wires. Wires should be run in the most direct path to equipment to limit the amount of material required, and wires should be cut accurately to reduce waste.

Wiring should not be installed in exterior walls, ceilings, and floors whenever possible to avoid interfering with insulation. Where wiring must be installed in an insulated area, the thermal insulation should be installed to Grade 1 specifications for proper performance (see Chapter 4). Electrical boxes for switches and receptacles should also avoid exterior walls wherever possible to avoid displacing insulation and penetrating the air barrier. Insulation must be carefully installed around boxes to avoid compression and voids, and boxes must be caulked to the air barrier, and holes in the box should be sealed. Special airtight boxes can be used for high-performance air sealing (Figure 14.4). Holes drilled for wires that are routed between exterior walls and interior partition walls should be air sealed to prevent air infiltration.

Boxes that penetrate the WRB must be carefully sealed to avoid moisture penetration (see Chapter 4 for more information). Coordinating and assigning responsibility for air and moisture sealing between the electrical contractor and other trades can help ensure that the thermal envelope and weather-resistive barrier are complete and correct.

Low-Voltage Wiring

Many low-voltage systems use central panels from which wires are run to equipment. The same issues associated with high-voltage wiring apply to low-voltage systems: centrally locate panels to reduce wiring runs and avoid penetrating the thermal envelope. Because the technology for low-voltage systems changes frequently, installing empty conduit or other chases can make future upgrades simpler.

Figure 14.3 Electrical panel in an exterior wall displaces insulation, creating a thermal bridge and contributing to air infiltration.

Electromagnetic Fields

Electromagnetic fields (EMF), also referred to as *electrical and magnetic fields*, are invisible forces created by the transmission of electricity through wires. EMFs are created by outdoor power lines, interior wiring, lighting, and motors. For many years, there have been concerns that EMF exposure can lead to health problems ranging from minor discomfort to increased incidence of cancer. Over 30 years of research by such groups as the World Health Organization has not shown any definitive relationship between household EMFs and human illness; however, research is continuing and further results may alter these opinions.

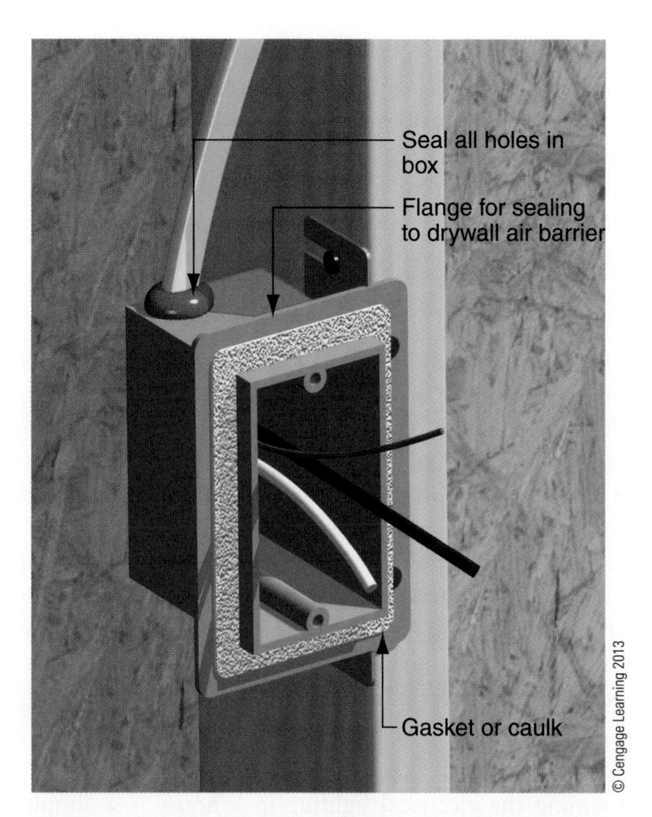

Figure 14.4 The amount of leakage from electrical boxes can be significant, considering the total number of boxes in a home.

Controls

Control systems can be as simple as wall switches and timers and as complicated as computer-controlled whole-house management systems. Most advanced control systems require power to operate, creating additional electrical load. When selecting and installing control systems, the design team must understand how much load a system requires to operate and avoid systems that have very high loads.

Hard-Wired Switches

Standard light switches control lights and receptacles by opening and closing the electrical circuit. Traditional wiring plans include switches for lights or a receptacle in each room. Multiway switching, an interconnected set of switches that control lighting from multiple locations (e.g., the top and bottom of a stairwell), can help save energy by enabling occupants to easily turn lights off from numerous locations instead of walking across a room or up and down stairs to access a switch. Multiway

---

electromagnetic fields (EMF) invisible forces created by the transmission of electricity through wires; also called *electrical* and *magnetic fields*.

multiway switching an interconnected set of switches that control lighting from multiple locations.

switching with two locations is referred to as three-way, switching at three locations is referred to as four-way, and so on. With proper wiring, an unlimited number of switches can control a single light. Switches that control receptacles can be used to turn off audio, video, and computer equipment, some of which may draw power even when switched off to operate in standby mode.

Power Strips

Televisions, computers and peripherals, and audio equipment can be plugged into switched power strips, allowing groups of equipment to be turned off from a single location. Smart power strips have a primary outlet for a television or computer and multiple switched outlets for auxiliary equipment, such as printers, wireless routers, and DVD players. Special circuitry detects when the primary device is turned off and automatically turns off the auxiliary receptacles (Figure 14.5). Newer power strips also include motion sensors for automatic controlling of select outlets.

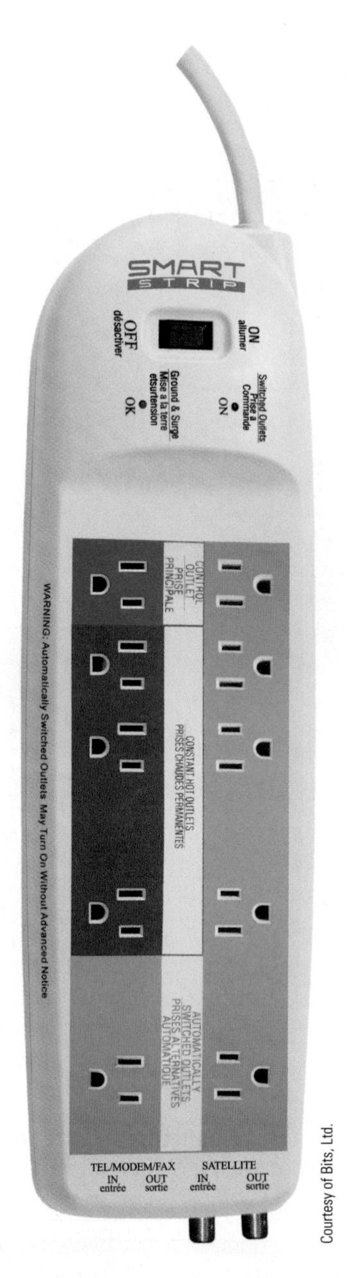

Figure 14.5 Intelligent power strips detect how much electricity your appliances use to prevent idle use or "vampire loads." One common step up is for the power strip to detect when the primary appliance, such as a television or computer, is turned off. At that time, the peripherals, like printers or DVD players, are also shut off to prevent idle use.

Dimmers

Dimming the electrical lighting in a house is a simple way to reduce energy use. Dimming lamps by 25% reduces energy use by approximately 20%; setting dimmers at 50% of full light output saves about 40% of the energy required. Dimmers must be paired with compatible, dimmable lights for proper operation.

Occupancy and Vacancy Sensors

Occupancy sensors turn circuits on and off automatically when they sense heat or movement in a room and can help reduce energy usage by turning lights and fans on or off as needed, depending on whether people are in a room. Occupancy sensors can switch on when people enter a room and turn off a specified amount of time after the room is vacant. Vacancy sensors have manual-on controls, requiring operation of a switch to turn them on, and they also turn off automatically after the room is vacated. Vacancy sensors work well in locations where there is enough ambient light that an individual does not need to turn on a light if they are only planning to be in the room for a few minutes. Occupancy sensors can be set to run bathroom vent fans for a specified period of time after the room is vacated to remove moisture (see Chapter 13), creating an effective way to obtain adequate ventilation without wasting power by leaving fans on for too long. Such sensors can also be used to turn off ceiling fans left running in unoccupied rooms (see Chapter 13). Closet lights can be controlled by switches installed in doorjambs, turning lights on when the door is opened and off when it is closed; if the door is left open, however, the light would remain turned on.

---

occupancy sensors a device that turns circuits on or off automatically when they sense that occupants have entered or left a room.
vacancy sensors lighting controls that require operating a switch to turn them on that turn off automatically after a room is vacated.

Timers

Timers are useful for bathroom vent and ceiling fans to minimize energy use when they are not needed. As discussed in Chapter 13, bathroom vents should be run long enough to remove moisture from the room, but they are often left on longer than is necessary, wasting energy in the process. Using a timer instead of a toggle switch provides assurance that the fan does not run too long. Timers are also an effective alternative to occupancy sensors for ceiling fans. Long-run timers that can be set to up to eight hours allow for a fan to run overnight in a bedroom, turning it off in the morning after people leave.

Outdoor Lighting Controls

Outdoor lighting that is left on during daylight hours does not serve any purpose and wastes electricity. Controlling outdoor lighting with timers, motion sensors, and photocells are effective ways to reduce energy use. Timers can be useful for security and decorative outdoor lighting; however, they should be programmed to adjust with the changing seasons and often need to be reprogrammed after power outages. **Photocell** switches are used on exterior lighting to turn them off during daylight hours when they are not needed. Some are combined with motion sensors that turn lights on when a person or vehicle approaches the control, allowing the light to come on only when it is dark. Some systems have a setting at which the lights are allowed to burn at 50% power for decorative purposes until detected motion turns them up to full power.

Automatic Control Systems

Systems that manage lighting, heating and air conditioning, alarms, and other household equipment can range from simple arrangements of a few programmable switches to sophisticated computer-controlled whole-house systems, some even operated by web-enabled remote control. These automatic control systems enable a variety of lighting scenes, manage heating and cooling systems according to occupancy, monitor overall energy use, and manage the home's systems remotely.

Control systems can reduce the amount of wire required by using radio frequency communication instead of high-voltage wiring to operate switches. Room lighting can be operated by radio frequency–controlled multiway switches with direct wiring from only one switch to the fixture, reducing the need for wiring to each individual switch. These systems can be used as part of a strategy to reduce the amount of wire required in a house. Receptacles can be switched using remote controllers rather than hard wiring (Figure 14.6), providing flexibility as uses change or when hard-wired switching is not installed. Individual remote control switches require small amounts of energy to operate (in the range of one-quarter to one-half watt per switch), creating phantom loads. These systems also must be programmed when set up and to make changes in how they operate, which often requires the help of specialized consultants.

Photograph Courtesy of Lutron® Electronics Co., Inc.

Figure 14.6 Lighting fixtures can be plugged into a remote-controlled switch. This saves the time, money, and resources of running wiring.

Whole-house systems may use a central processor or dedicated computer running full-time to operate them, adding further to the home's electrical load. Central processor loads range from approximately 7 to 14 watts at the low end to as high as 200 to 300 watts for a full-time computer-based system. Some of these systems have touch screens that use in the range of 10 to 14 watts each to operate, as well as online interfaces for operation while away from home. It is important to understand how much energy these systems use compared with how much energy they can save. Even the best system can use more energy than it saves if not operated properly.

Energy Monitors

Energy monitors that provide instant reports of energy use in a house are effective feedback devices that can help change behavior and reduce power use. The Energy Detective™ (TED) and Lucid Design Group's Building Dashboard® are two monitors that connect to the electrical panel and show total energy use in a house (Figure 14.7a and Figure 14.7b). Some models have Internet-based interfaces to monitor energy use from anywhere. Individual equipment energy use can be tracked using such monitors as the P3 International's Kill A Watt™ or Watts Up plug-load meters (Figure 14.8). Most whole-house control systems have energy monitors as available options.

---

photocell switches are used on exterior lighting to turn them off during daylight hours when they are not needed.

Figure 14.7a The Energy Detective™ (TED) is a home energy monitor that displays electricity usage in real time.

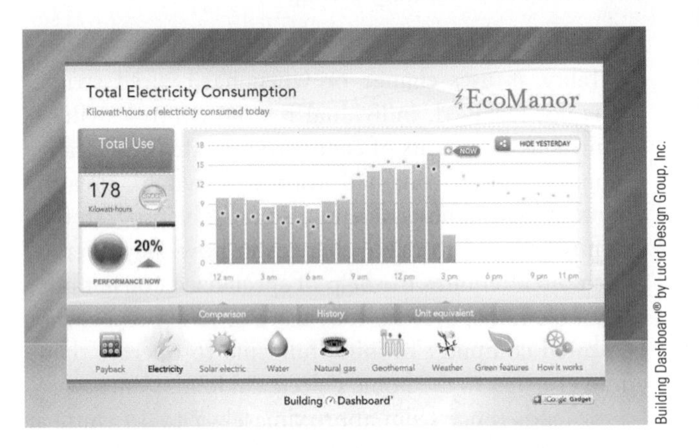

Figure 14.7b The Building Dashboard® by Lucid Design Group, Inc. serves as the educational centerpiece of EcoManor, the first LEED-certified residence in the Southeast, where real-time performance and financial payback of solar electric, water-recycling, and geothermal systems are displayed via a touch-screen display in the kitchen and on the web.

Figure 14.8 Using P3 International's Kill A Watt™, it was discovered that a cell phone still drew 1 watt of electricity even when fully charged.

Did You Know?

Energy Feedback Devices Can Change Behavior

Southern California Edison provided a small group of customers with a device called the Energy Orb, a small plastic ball that glowed green when the energy grid was underused (and prices were lower) and pulsed red when the grid was heavily used (and rates were higher). These signals gave homeowners immediate, visual feedback on energy use and prompted them to use less energy at critical times. The average customer with an Energy Orb reduced their peak-period energy use by 40% after it was installed.

Electrical Systems and Green Building

Building codes and green building programs address efficiency in residential electrical and lighting systems in very limited ways, with the exception of California Title 24. ENERGY STAR, LEED for Homes, the National Green Building Standard (NGBS), and most local green programs offer points for efficient lighting and equipment, but few programs require that they be included to achieve certification.

ENERGY STAR

ENERGY STAR is a label created by the U.S. Environmental Protection Agency to identify products that have demonstrated energy efficiency above and beyond minimum standards. In addition to homes, ENERGY STAR labels are given to qualifying products used in homes and commercial buildings, such as appliances, electronics, certain lighting, ceiling fans, windows, water heaters, and heating, ventilating, and air conditioning (HVAC) equipment.

The ENERGY STAR label provides guidance to the consumer on energy efficiency when purchasing electrical equipment. In addition to the ENERGY STAR label, most appliances and equipment subject to federal minimum efficiency standards have Energy Guide labels that show their performance relative to other products in the same category, as well as estimated annual energy costs.

Selecting ENERGY STAR equipment and comparing the energy use estimates in the Energy Guide label are useful in choosing energy-efficient products for a green home. ENERGY STAR does not label certain appliances, including clothes dryers, ranges, ovens, and microwave ovens, because the energy use between models does not vary significantly. Minimum requirements for ENERGY STAR labeling are updated periodically to remain current with minimum efficiency standards and product advances.

Advanced Lighting Package

Similar to the ENERGY STAR home rating system, the EPA offers the Advanced Lighting Package (ALP) designation for homes that meet the program requirements. ALP is a construction option that can be offered by homebuilders to upgrade light fixtures and ceiling fans commonly used in the home with ENERGY STAR–qualified models.

The ALP designation applies to lighting packages that consist of a minimum of 60% ENERGY STAR–qualified hard-wired fixtures and 100% ENERGY STAR–qualified ceiling fans. ENERGY STAR–qualified recessed downlights, ceiling fan light kits, and ventilation fans with lighting can be counted toward the fixture requirement.

Electrical Systems and Green Building Programs

ENERGY STAR for homes and green building certification programs either include requirements for efficient electrical equipment or provide points in their programs for including them. Among and within programs, requirements vary for the performance and prescriptive paths to certification.

LEED for Homes has no minimum requirements for the performance path. For certification through the prescriptive path, a home must include at least four ENERGY STAR fixtures or lamps in high-use rooms, such as the kitchen, living room, dining room, family room, or hallways. Points are available for installing three additional interior lights and automatic controls on exterior lights. Additional points are awarded for meeting ALP certification or installing 80% ENERGY STAR fixtures and 100% ENERGY STAR fans throughout the home.

The NGBS has no minimum requirements in either the performance or prescriptive paths. Points are awarded for installing ENERGY STAR fixtures, lamps, and appliances, and for installing occupancy sensors.

ENERGY STAR for homes has no minimum requirements for the performance path. The prescriptive path requires all appliances and fans to be ENERGY STAR rated. In addition, the project must achieve ALP certification, or 80% of all fixtures in RESNET-defined qualifying fixture locations must be ENERGY STAR certified.

2009 International Energy Conservation Code

The only electrical efficiency requirement in the 2009 International Energy Conservation Code (IECC) is for 50% of all permanent lighting fixtures in homes to have high-efficacy lamps, such as compact fluorescent lights, tubular fluorescent bulbs, or otherwise meet equivalent efficacies.

California Title 24

California's building energy code, Title 24, requires specific energy efficiency measures in all residential new construction and renovation projects requiring a building permit. The code requires that 50% of the connected load in the kitchen to be "high-efficacy" luminaires, and other rooms and the exterior should have "high-efficacy" luminaires or controls, such as dimmers and occupancy sensors. Traditionally, California has one of the most stringent energy codes in the nation and is influential in national policy.

Lighting

To reduce our consumption of electricity, we need to light our homes only when daylight is not adequate. Designing a lighting system that produces the appropriate amount of light for all activities using the least amount of energy is a key component of a green home. Lighting can be broken down into three main categories:

- **Ambient:** general interior lighting for daily activities and as needed outdoors for security and safety
- **Task:** lighting directed to a particular work area, such as kitchen counters and desks, and not throughout a room
- **Accent:** lighting on walls, artwork, and architectural features on the interior and exterior to enhance the visual appeal of an area; the lighting itself may also serve a decorative function

Design for Efficiency

Reducing the amount of artificial lighting required through the use of daylighting (see Chapter 8) is an important component of energy efficiency; however, artificial lighting is necessary much of the time. Maximizing energy efficiency requires more than selecting the most efficient fixtures and lamps. Installing the correct amount of lighting for ambient, task, and accent lighting (i.e., not too much, not too little) is just as important; automatic controls can help reduce energy use, and occupants must also be willing to conserve.

Lamp Types

Lamps, commonly referred to as light bulbs, are the replaceable part of a fixture that produces light from electricity. Different lamp types produce a variety of lighting color and quality at varying levels of efficiency. Lighting designs should strive for the best balance of light quality and efficiency.

Incandescent

Incandescent lamps (Figure 14.9) create light when electricity passes through a tungsten filament encased in glass, causing it to glow and create light. Having changed little since they were originally invented in the nineteenth century, incandescent lamps are very inefficient because they convert only about 3% of energy used into light, with the balance producing heat. This extra heat increases cooling loads whenever space heating is not required. Installing large quantities of incandescent lamps can require larger air conditioning systems just to offset the heat they create. Federal regulations effectively call for replacement of common incandescent lamps with significantly more efficient alternatives by 2014. If manufacturers do not develop more efficient incandescent lamps, they may disappear from the market entirely. Most incandescent bulbs have a screw-in base for installation in a wide range of fixtures.

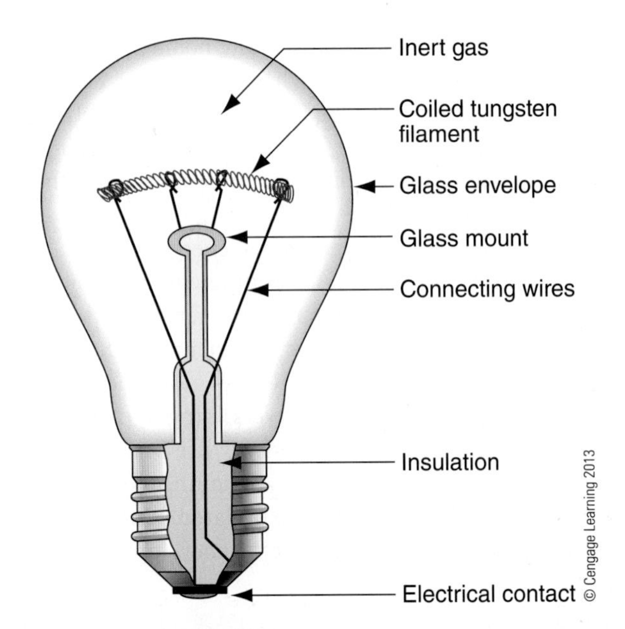

Figure 14.9 Incandescent lamp.

Halogen

Halogen lamps (Figure 14.10) are a variation on incandescent, using a filament inside a compact glass envelope filled with gas. Halogen lamps are somewhat more efficient than incandescents. They have a longer service life, and they provide a more attractive light. They are often used for accent lighting. Most halogen bulbs have a pin base, which limits their use to fixtures that are designed for the specific bulb type.

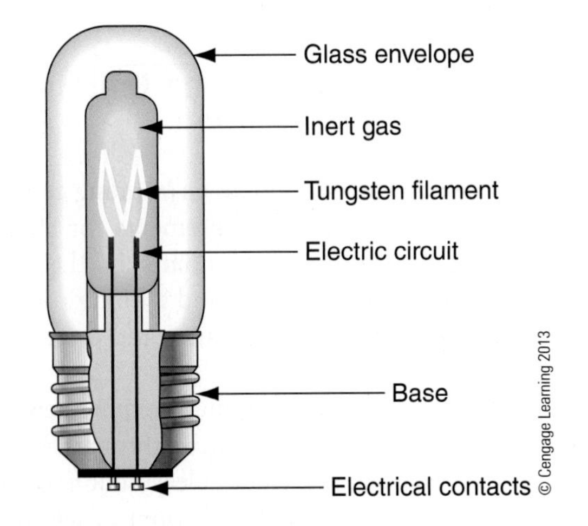

Figure 14.10 Halogen lamp.

---

ambient lighting general interior lighting for daily activities and as needed outdoors for security and safety.

task lighting lighting directed to a particular work area, such as kitchen counters and desks, and not throughout a room.

accent lighting lighting on walls, artwork, and architectural features on the interior and exterior to enhance the visual appeal of an area.

lamp commonly referred to as light bulbs, are the replaceable part of a fixture that produces light from electricity.

incandescent lamps that create light when electricity passes through a tungsten filament encased in glass, causing it to glow and create light.

halogen a variation on incandescent lamps that uses a filament inside a compact glass envelope filled with gas.

Fluorescent

Fluorescent lamps create light when electricity passes through gas inside a phosphor-coated glass tube, causing it to glow (Figure 14.11). They are available in tubular and compact designs and are as much as five times more efficient than incandescent lamps, creating much less heat (Figure 14.12a and Figure 14.12b). Compact fluorescent lamps (CFLs) are common replacements for incandescent lamps in residential interiors and tubular fluorescent lamps are frequently used in utility areas, closets, and some kitchens. All fluorescent lamps require a ballast, a device that increases the frequency of line power delivered to a fixture, to control the starting and operating voltages of the lamp. Fluorescent lamps installed in garages or other cold exterior locations should be designed for outdoor use. Standard lamps may not operate well in cold conditions. Fluorescent light is generally considered to be less desirable than incandescent; however, the quality of fluorescent lamps continues to improve while incandescents are being regulated out of the market. Table 14.1 compares the cost of a 60-watt incandescent lamp with a 15-watt CFL.

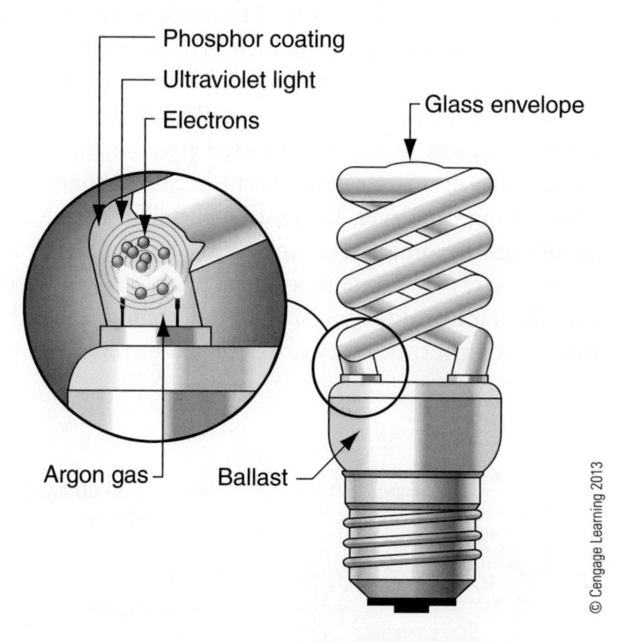

Figure 14.11 Fluorescent lamp.

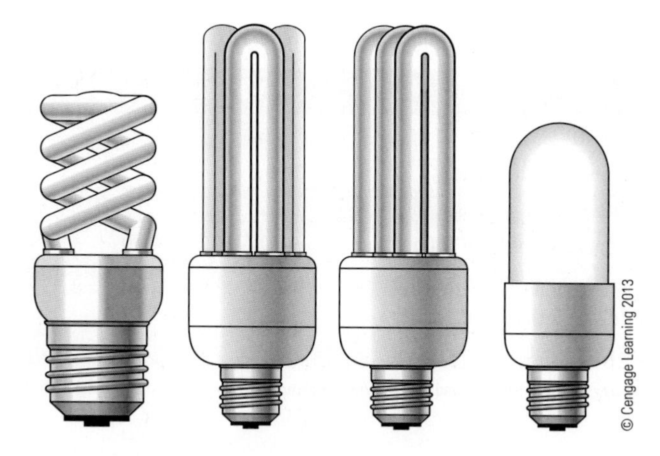

Figure 14.12a Types of CFLs.

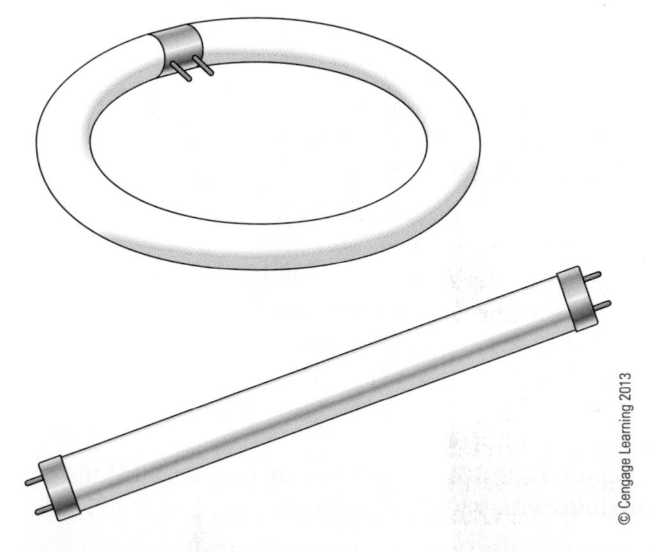

Figure 14.12b Types of fluorescent tube lighting.

Although fluorescent lamps have a long life, lamp life can be shortened by extremely short on–-off cycles. CFLs should be designed for the fixture in which they are installed; some are not designed to be installed base-end up, as in a recessed light fixture. Improper applications can reduce lamp life. Fluorescent lamps are not normally dimmable, although some models are designed to work with standard dimmers. Dimmable CFLs that work with standard dimmers are readily available. Tubular fluorescent lamps can be dimmed when paired with dimmable ballasts and matching dimmers. Fluorescent lamps may not operate properly with some occupancy sensors.

Tubular lamps and nonballasted CFLs frequently have pin bases (Figure 14.13), which limits their installation to fixtures for which they are designed. Some codes and standards require the use of pin bases when utilizing CFLs. CFLs with integrated ballasts have screw-in bases that are designed to replace incandescent lamps in standard fixtures. ENERGY STAR allows manufacturers to use their label on CFLs that meet minimum requirements for efficacy, color temperature, lamp life, and other criteria.

---

fluorescent a type of lamp that creates light when electricity passes through gas inside a phosphorous-coated glass tube, causing it to glow.

compact fluorescent lamps (CFLs) a common lighting alternative to incandescent lamps that give off light when a mixture of three phosphors are exposed to ultraviolet light from mercury atoms.

ballast a device that regulates the frequency of electricity delivered to a fixture to control the starting and operating voltages of the lamp.

Table 14.1 Cost Comparison of a 60-Watt Incandescent Light Bulb vs. a 15-Watt ENERGY STAR–Qualified CFL

Variable	60-Watt Incandescent	ENERGY STAR–Qualified 15-Watt CFL
Initial cost (a)	$0.50	$3.00
Light output (lumens)	800	800
Life of bulb (hours)	1,000	10,000
Replacement light bulbs (b)	9 @ $0.50 = $4.50	–
Lifetime electricity cost (c)	10,000 hours × 60 watts × $0.10/kWh = $60.00	10,000 hours × 15 watts × $0.10/kWh = $15.00
Total lifetime cost (a + b + c)	$65.00	$18.00
Savings	–	**$47.00**

Source: ENERGY STAR, Canada, http://www.oee.nrcan.gc.ca/energystar/english/pdf/basic-facts-residential-e.pdf

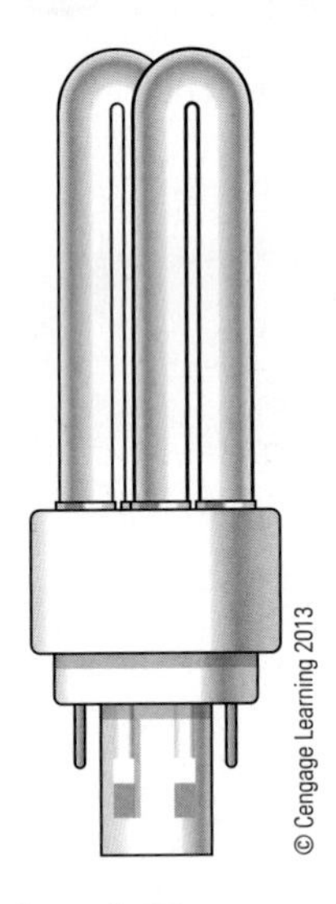

Figure 14.13 Pin-based CFLs are only installed in fixtures with integrated ballasts.

All fluorescent lamps contain small amounts of mercury, a known toxin. Fluorescent lamps can release this mercury into the atmosphere if broken or discarded improperly. Electricity generated from coal also produces mercury, and a single fluorescent lamp typically offsets more mercury through energy savings than is contained in the lamp itself.

Recycling Fluorescent Lamps

All fluorescent lamps, including tubular and CFLs as well as high-density discharge and magnetic induction lamps, contain mercury and should be recycled to avoid contaminating landfills. Lamps that contain mercury should not be discarded with standard residential and commercial waste. The EPA provides guidelines for proper disposal of these lamps. Many retailers, including IKEA and Home Depot, now accept lamps for recycling.

Light-Emitting Diodes

Light-emitting diodes (LEDs) are semiconductors that glow when electrical current passes through them (Figure 14.14). Long a standard in electronic displays like clocks and stereos, they are an alternative to incandescent and fluorescent lamps. LED lamp efficiency is improving but remains slightly less than that of fluorescent lamps with as good or better light quality and longer life; the lamp cost, however, is higher. As LED technology continues to improve, more options will be available at lower costs. LED lamps are available with pin bases to replace halogen lamps, with screw-in bases to replace incandescents, and fully integrated into recessed lights and other fixtures (Figure 14.15). Typical LED lamps provide a directed light, making them potentially more effective for task rather than ambient lighting; however, lamps that distribute light more evenly are becoming available. LED lamps placed in a tubular array to simulate a linear fluorescent lamp can provide a longer life and warmer light temperature than the fluorescent tubes they are designed to replace.

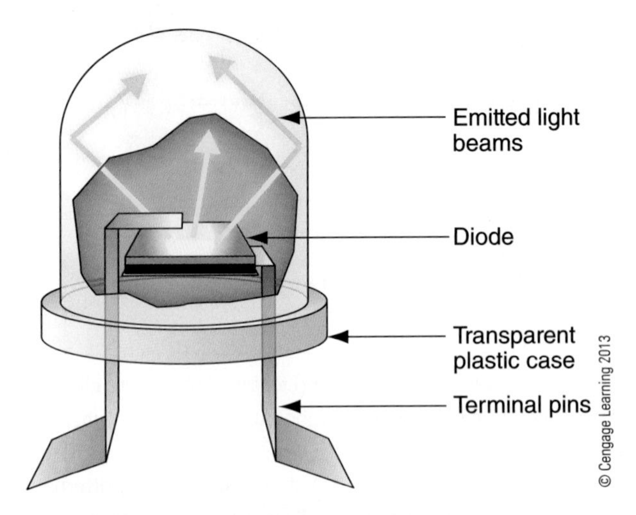

Figure 14.14 Light-emitting diodes (LEDs).

---

light-emitting diodes (LEDs) semiconductors that glow when electrical current passes through them.

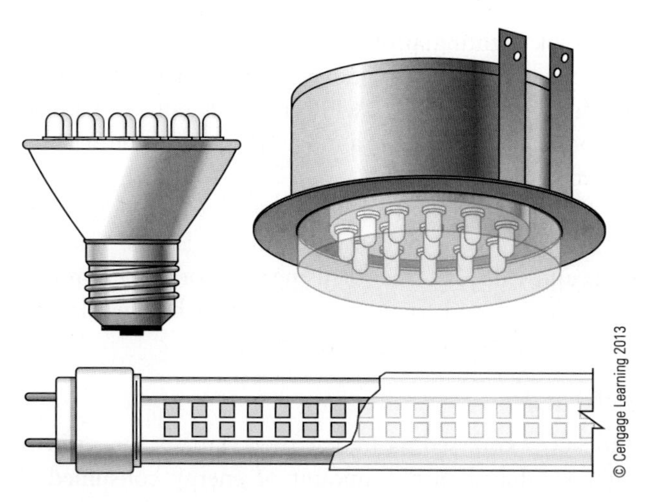

Figure 14.15 LED lamps are now available in a variety of styles.

Magnetic Induction

Magnetic induction lamps, also known as internal inductor lamps, (Figure 14.16) are a variation on fluorescent technology and use an electromagnet to cause gas in the lamp to glow. Traditionally only available for commercial fixtures, these lamps are now available as screw-in replacements for incandescent lamps, similar to CFLs. They are somewhat more efficient and have a longer life than CFLs. Like CFLs, they contain mercury and must be disposed of properly.

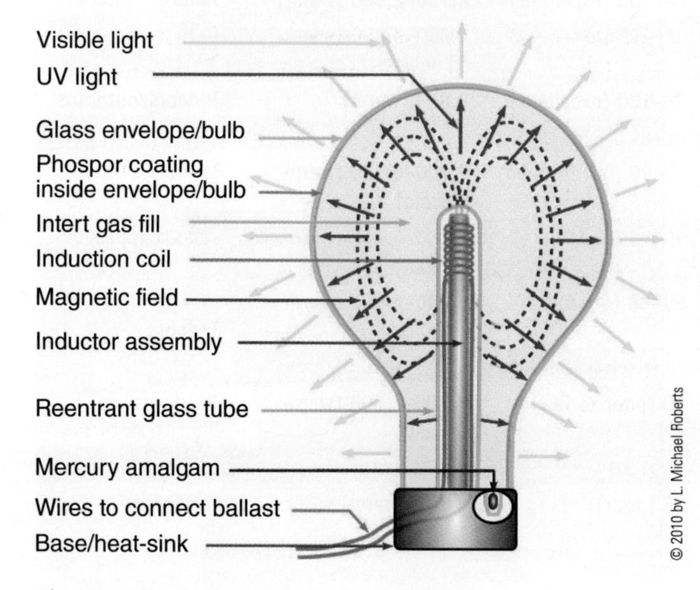

Figure 14.16 Magnetic induction lamp.

magnetic induction lamps a variation on fluorescent technology and use an electromagnet to cause gas in the lamp to glow.

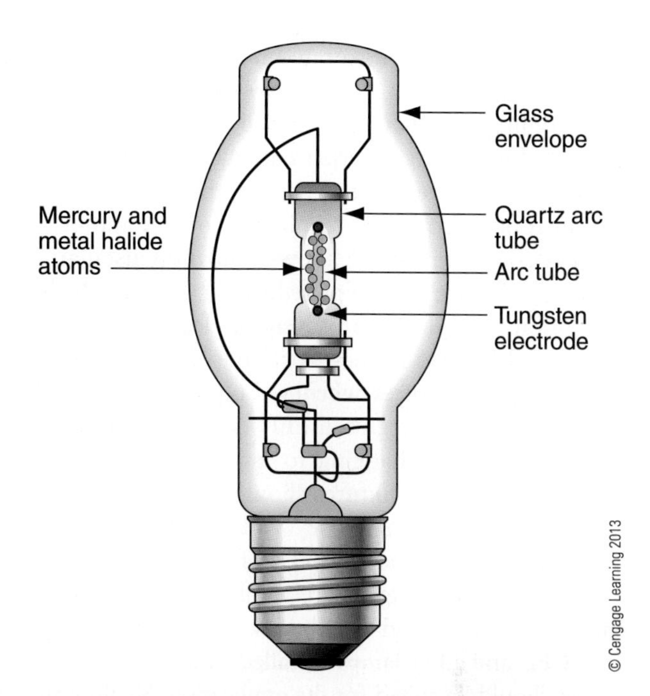

Figure 14.17 High-intensity discharge (HID) lamp.

High-Intensity Discharge

High-intensity discharge (HID) lamps, a variation on incandescent lamps that are made with mercury vapor, metal halide, and high- and low-pressure sodium gases, are options for residential outdoor applications (Figure 14.17). HID lamps are very efficient; however, they are not commonly used inside homes. They are limited mostly to street and parking lot lights and the interiors of industrial spaces, such as gymnasiums, and arenas. HID lamps have a slow warm-up time, making them inappropriate for lights that turn on and off regularly. HID lamps may be a good choice for exterior lighting that must remain on throughout the night.

Selecting Fixtures

Light fixtures, also referred to as luminaires, are available in a wide variety of designs for every application. Typical fixtures used in homes include recessed ceiling lights, surface-mounted ceiling and wall lights, track lighting, and free-standing floor and table lamps. Recessed fixtures can provide ambient, accent, and task lighting; however, the area to which they provide light is limited by the lamp location. They are available with screw-in bases for incandescent, CFL, or LED lamps; with pin bases for halogen lamps; with ballasts and pin bases for CFLs; and

high-intensity discharge (HID) a variation on incandescent lamps that are made with mercury vapor, metal halide, and high- and low-pressure sodium gases; used most commonly for residential outdoor applications.

luminaires a light fixture; the complete lighting unit, including lamp, reflector, ballast, socket, wiring, diffuser, and housing.

with LED lamps fully integrated into the fixture. When installing screw-base CFL and LED lamps in recessed lights, confirm that they are designed for a base-up application so they do not experience premature failure.

Recessed fixtures that penetrate the thermal envelope must be rated IC for insulation contact and be an airtight design. This design allows insulation installation against the fixture and helps reduce air leakage. Proper installation is required to prevent air leakage and heat loss around fixtures in these locations. Limit the quantity of recessed fixtures whenever possible, and install pin-based CFL or integrated LED types to keep homeowners from later changing to less-efficient incandescent lamps. Recessed fixtures should not be installed in insulated rooflines where they will displace insulation and reduce the efficiency of the roof assembly.

Surface-mounted lights are available with incandescent, halogen, CFL, and LED lamps. As with recessed fixtures, CFL and LED lamps installed in enclosed surface fixtures should be suited for the application. Shallow surface lights for undercabinet kitchen task lighting are available with LED, halogen, and fluorescent lamps. Ambient lighting can be created with linear fluorescent lamps above cabinets or in coves around the perimeter of a room.

Track lighting provides decorative accent lighting that can be adjusted to fit changing room arrangements. Traditionally designed with incandescent or halogen lamps, LED lamps offer a more efficient option without sacrificing light quality.

Freestanding and table lamps are available with incandescent, CFL, LED, and halogen lamps. Selecting CFL and LED fixtures (or using CFL or LED lamps in fixtures designed for incandescent and halogen lamps) are good strategies for saving energy.

Efficacy

Efficacy, expressed as lumens per watt, is the ratio of light produced to the amount of energy consumed. A lumen is a measurement of the power of light as perceived by the human eye. A 100-watt incandescent lamp produces about 1,750 lumens. Dividing the lumens by the watts determines that the lamp has an efficacy of 17.5 (1,750/100 = 17.5). A comparable CFL may produce 1,400 lumens while using only 23 watts, with an efficacy of about 65 (1,400/23 = 65.2). The higher the efficacy, the more light is produced for a given amount of energy. Table 14.2 lists efficacy ranges for standard lamp types.

Table 14.2 Efficacy of Different Lighting Technologies

Lighting Type	Efficacy (Lumens/Watt)	Lifetime (Hours)	Color Rendition Index (CRI)	Color Temperature (K)	Indoors/Outdoors
Incandescent					
Standard "A" bulb	10–17	750–2,500	98–100 (excellent)	2,700–2,800 (warm)	Indoors/outdoors
Tungsten halogen	12–22	2,000–4,000	98–100 (excellent)	2,900–3,200 (warm to neutral)	Indoors/outdoors
Reflector	12–19	2,000–3,000	98–100 (excellent)	2,800 (warm)	Indoors/outdoors
Fluorescent					
Straight tube	30–110	7,000–24,000	50–90 (fair to good)	2,700–6,500 (warm to cold)	Indoors/outdoors
CFL	50–70	10,000	65–88 (good)	2,700–6,500 (warm to cold)	Indoors/outdoors
Circline	40–50	12,000	65–88 (good)	2,700–6,500 (warm to cold)	Indoors
High-Intensity Discharge					
Mercury vapor	25–60	16,000–24,000	50 (poor to fair)	3,200–7,000 (warm to cold)	Outdoors
Metal halide	70–115	5,000–20,000	70 (fair)	3,700 (cold)	Indoors/outdoors
High-pressure sodium	50–140	16,000–24,000	25 (poor)	2,100 (warm)	Outdoors
Light-Emitting Diodes					
Cool white LEDs	60–92	35,000–50,000	70–90 (fair to good)	5,000 (cold)	Indoors/outdoors
Warm white LEDs	27–54	35,000–50,000	70–90 (fair to good)	3,300 (neutral)	Indoors/outdoors

Source: U.S. Department of Energy, http://www.energysavers.gov/

efficacy the ratio of light produced to the amount of energy consumed; expressed as lumens per watt.

Illumination

Illumination is the quantity of light reaching a task or work surface, which is referred to as *illuminance*. Regardless of the type or efficacy of lighting, the purpose, almost always, is to produce adequate illumination for a task. The illuminance at a task is measured in foot-candles, the amount of light produced by one lumen over a 1 ft^2. Illumination requirements range from as low as 5 foot-candles where no work is being performed other than circulation, to about 50 for typical kitchen work, and as high as 500 for very detailed tasks for long periods of time. Age, visual acuity, and personal preference will affect the decision of how much light to provide. An appropriate amount of light must be provided for all work areas without over-lighting to avoid wasting energy.

Determining how much illumination a particular lighting design will provide in a space involves careful calculations that consider fixture location and quantity, lamp type, fixture design, reflectivity of surfaces, and other factors. Lighting design software, such as Visual from Acuity Brands, can assist in determining the appropriate amount of light required for a home (Figure 14.18).

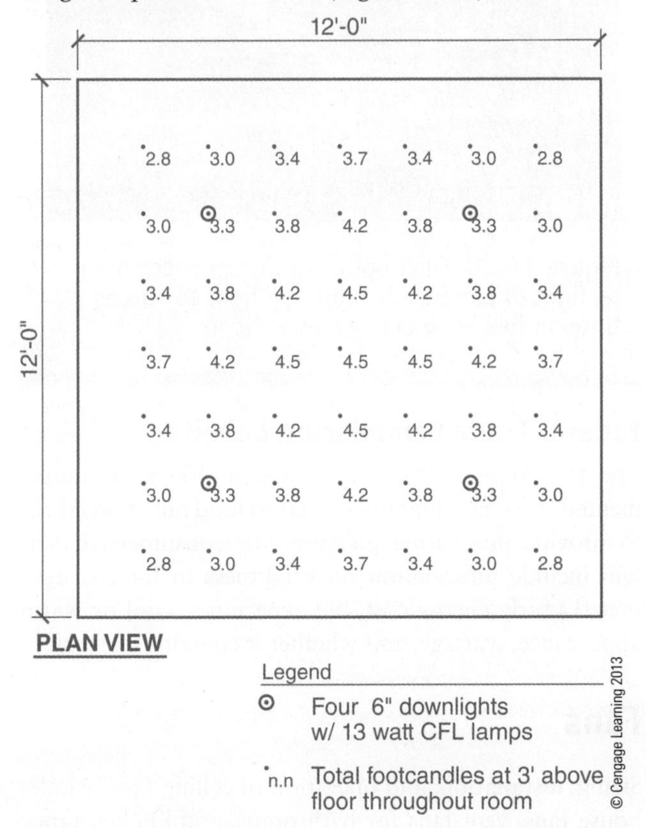

Figure 14.18 This report from Visual software shows the amount of illumination throughout a given room produced by four ceiling-mounted lamps. Lighting designers can use modeling programs to prevent over-lighting a room.

Color Temperature

Color temperature is an indicator of the color of a light source and compares the color of a lamp to natural sunlight measured in Kelvin (K) (Figure 14.19). The Kelvin scale is a scientific measurement used to determine the color temperature of different light types. Lower temperatures are considered warm light, and higher temperatures cool. Incandescent and halogen lamps fall in the range of 2700 to 3200K, fluorescent and magnetic induction lamps range between 2700 and 6500K, and LED lamps range from as low as 2800 to as high as 8000K. Cooler light is generally better for detailed tasks

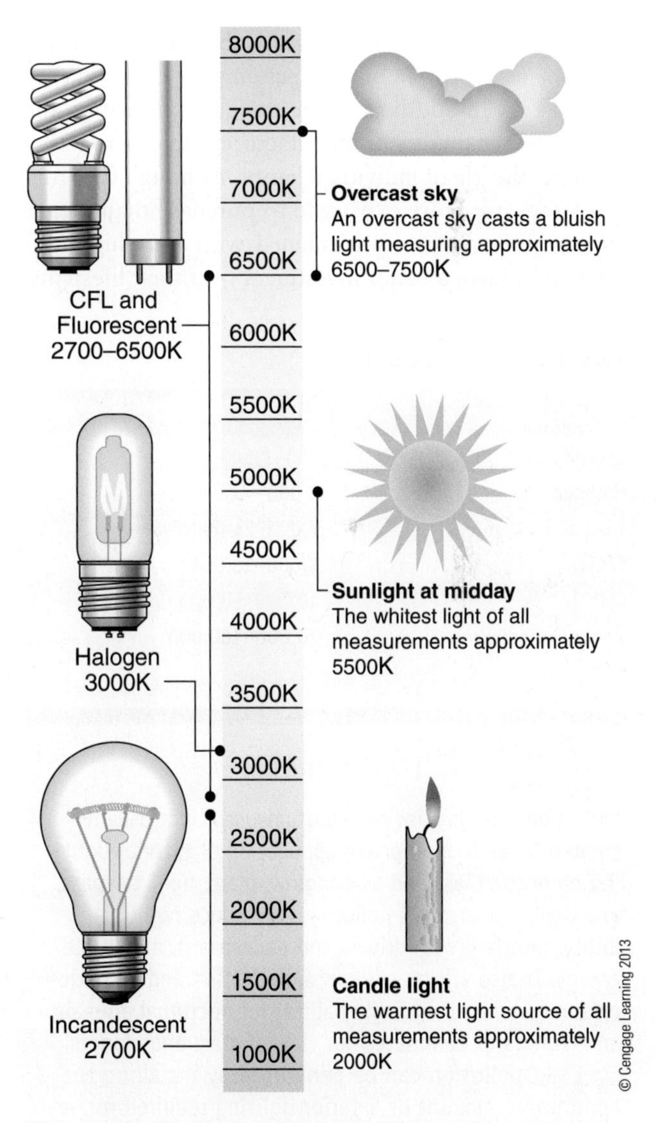

Figure 14.19 The color temperature scale rates the warmth of the light from a given lamp.

---

illumination is the quantity of light reaching a task or work surface, which is referred to as *illuminance*.

foot-candles the amount of light produced by one lumen over a 1 ft^2 area.

color temperature an indicator of the color of a light source and compares the color of a lamp to natural sunlight; measured in degrees kelvin.

Kelvin a temperature scale used to determine the color temperature of different light types.

because it offers greater contrast, whereas warmer temperatures are preferred for general lighting because it produces more flattering skin tones.

Color Rendering Index After efficacy, the most important factor in selecting lamps is the Color Rendering Index (CRI), which quantifies a lamp's ability to render colors the same as a reference lamp type. Based on a scale of 0 to 100, a CRI rating of 100 is based on a 100-watt incandescent bulb. A CRI over 80 is considered suitable for residential lighting. Fluorescent tube lamps range from 50 to 90, CFLs from 65 to 88, and HID lighting from 25 to 50.

Lamp Life

Lamp types have estimated lifespans that range from 750 to as long as 100,000 hours, depending on the lamp type (Table 14.3). In addition to the design of the lamp, the type of use, the quality of manufacture, and other factors will affect the life of individual lamps. Although CFL and LED lamps are more expensive to purchase than incandescents, their longevity combined with their high efficacy makes them a better investment over their lifespans.

Table 14.3 Lamp Lifespans

Lamp Type	Lifespan (Hours)
Incandescent	750–2,500
Halogen	2,000–4,000
Tube fluorescent	7,000–24,000
CFL	7,000–10,000
LED	40,000–50,000
Magnetic induction	70,000–100,000

Light Pollution

Light pollution is misdirected or misused light usually created by an inappropriate application of exterior lighting products. Classified as sky glow, glare, light trespass, and light clutter, light pollution can reduce night visibility, interfere with drivers and pedestrians, and waste energy. It also affects wildlife reproduction and migration patterns, reduces available habitat for nocturnal animals, and reduces or eliminates our view of stars and planets.

Light pollution can be minimized by installing the appropriate amount of exterior lighting required for security and safety without over-lighting an area. Lighting design should minimize light pollution by directing light to avoid sky glow, glare and light trespass.

Fiber Optic Lighting Systems

Fiber optic lighting systems use very fine, flexible glass or plastic fibers to transmit light (Figure 14.20). Fiber optic lighting can provide natural light to interior spaces in buildings by concentrating sunlight through lenses, sending it through fiber optic cables to areas where it is needed, and distributing it within a room through fixtures that have the appearance of traditional lights. Although lighting systems using this technology are becoming available in the marketplace, they are designed for commercial buildings and are very costly. As the technology improves and prices come down, fiber optic lighting may become an appropriate solution for residential projects.

Figure 14.20 Fiber optic lighting provides natural light or electrically produced light to spaces through flexible glass or plastic fibers.

Federal Trade Commission Labeling

The U.S. Federal Trade Commission (FTC) has implemented labeling for lamps, similar to food nutrition labels, to provide purchasing guidance for consumers. Labels will include information on brightness in lumens, estimated yearly energy cost, life expectancy, cool or warm appearance, wattage, and whether it contains mercury.

Fans

Sizing, installation, and operation of ceiling fans, whole-house fans, vent fans for bathrooms, and kitchen range hoods are covered in Chapter 13; however, they all use electricity to operate and affect the overall home performance, so fans should also be considered when designing electrical systems.

Color Rendering Index (CRI) quantifies a lamp's ability to render colors the same as a reference lamp type.

fiber optic lighting lighting systems that use very fine, flexible glass or plastic fibers to transmit light

Vent Fans

ENERGY STAR certifies fans that meet their minimum specifications on efficiency and noise level. Fan efficiency, or more accurately efficacy, is measured in cubic feet per minute of air flow per watt of energy (CFM/W). Bathroom vent fans up to 89 CFM must have an efficacy of at least 1.4 CFM/W and larger vent fans and range hoods must have an efficacy of at least 2.8 CFM/W. Noise level is rated in sones, a unit of perceived loudness. Small vent fans and range hoods must not exceed 2 sones, and larger vent fans must not exceed 3 sones. Many green building programs require vent fans to not exceed 1 sone. ENERGY STAR vent fans with lights must use CFL or LED lamps for certification and have a minimum 1-year warranty.

Ceiling Fans

Ceiling fans can provide comfort through convective cooling with higher indoor air temperatures in warm weather, offering energy savings if managed properly. Fans should not be left running when rooms are unoccupied, and their use should be limited to those times when air movement allows for lower HVAC settings because their motors create heat. Similar to ventilation fans, ENERGY STAR certifies ceiling fans that meet their minimum specifications of CFM/W. The requirements for ENERGY STAR fans are listed by fan speed in Table 14.4.

Appliances

Appliances are categorized as large and small. Large appliances include ovens, refrigerators, washers, dryers, dishwashers, and other similar equipment that is either permanently installed in place or rarely moved. Small appliances include items such as toaster ovens, microwave ovens, blenders, hair dryers, and coffee makers. ENERGY STAR appliances should be selected for highest efficiency whenever possible.

Table 14.4 ENERGY STAR Air Flow Efficiency Requirements for Ceiling Fans

Fan Speed	Minimum Air Flow (CFM)	Efficiency Requirement (CFM/W)
Low	1,250	155
Medium	3,000	100
High	5,000	75

Source: U.S. Environmental Protection Agency, ENERGY STAR program.

sone a unit of perceived loudness.

Large Appliances

Refrigerators use the most power of any household appliance, followed by washing machines, clothes dryers, freezers, and electric ranges. The FTC requires many common appliances to have EnergyGuide labels (Figure 14.21) showing how their energy consumption compares with other similar products. Use of these labels, in conjunction with ENERGY STAR labels, can help the consumer to select the most efficient appliances for a home. Ranges, ovens, and clothes dryers are not required to have EnergyGuide labels because there is no significant difference in efficiency between models.

Refrigerators

With refrigerators, not only the efficiency but also the configuration affects overall energy use. Top-freezer models are more efficient than side-by-side models. Models with icemakers and through-the-door ice and water dispensers are less efficient than models without these features. To compare performance, consult the EnergyGuide label that can be found inside each refrigerator and available online for all models. When considering units of comparable efficiency, a smaller unit will use less energy. Refrigerators that are full run more efficiently than empty ones because cooled and frozen food holds temperature better than air, so selecting a right-sized unit that will remain full most of the time is the most efficient choice. Automatic defrost functions on refrigerators use more energy than regular manual defrosting, but the efficiency and utility of manual defrost units decrease if they are not defrosted regularly. Avoid locating refrigerators next to ovens, and provide as much ventilation space around the coils for most efficient operation.

Recycle old, inefficient refrigerators instead of keeping them running in the basement or garage for the occasional overflow requirements. When more capacity is needed, a single larger refrigerator is more efficient than two separate units.

Clothes Washers

Clothes washers use significant amounts of energy and water, and their ability to wring water out of clothes in the spin cycle affects how much energy is required

EnergyGuide a bright yellow label, created by the Federal Trade Commission, that is required by law to appear on many new household appliances to show the relative energy consumption as compared with similar products.

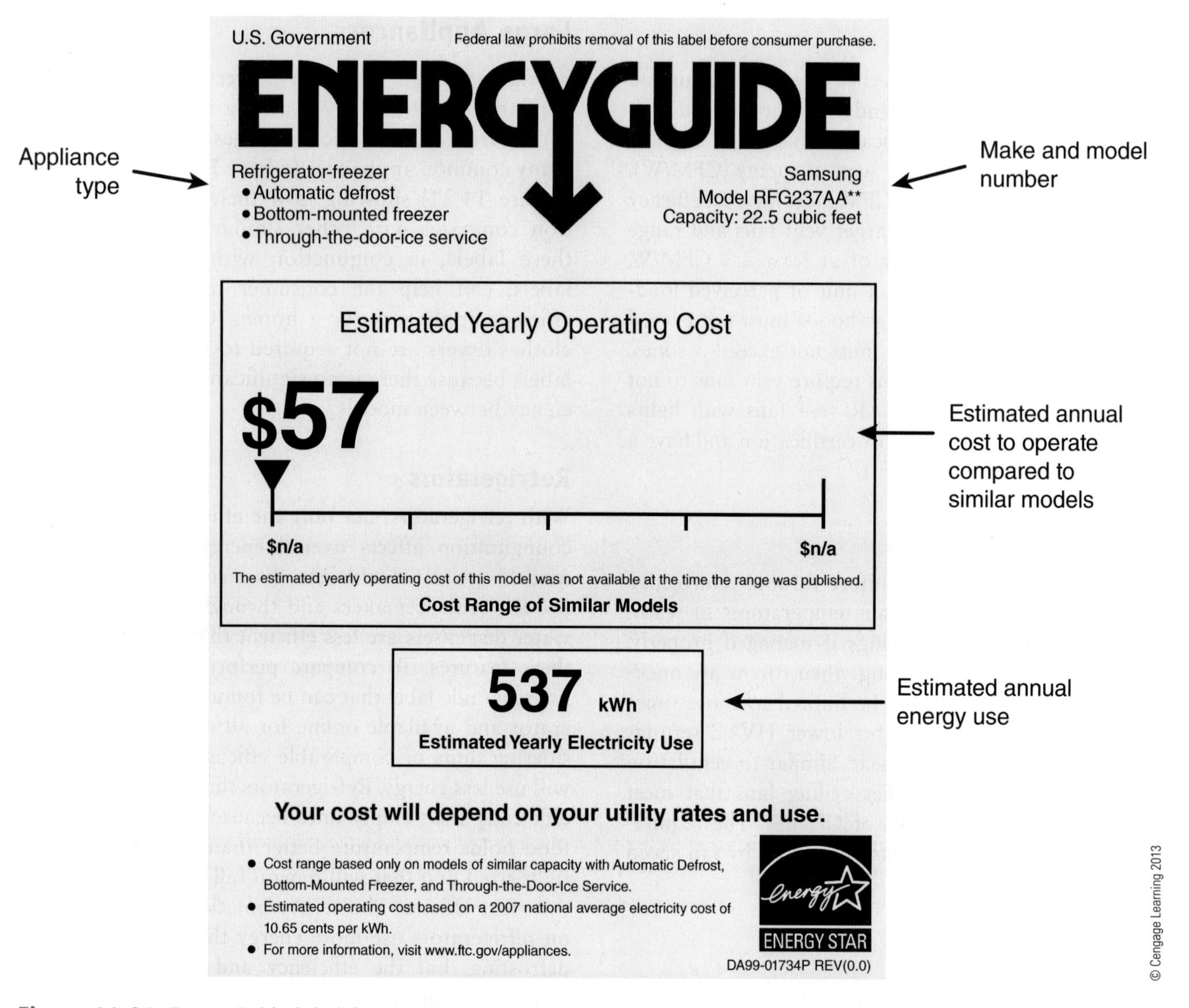

Figure 14.21 EnergyGuide label for an ENERGY STAR–certified refrigerator.

for drying. Washer efficiency is rated with a modified energy factor (MEF). The MEF considers the energy required to operate the washer, heat the water, and dry the clothes depending on how much water is removed in the spin cycle. Higher MEFs designate more efficient washers. The federal minimum MEF standard is 1.28. Washers must have an MEF of at least 1.8 to qualify for the ENERGY STAR label. Washers are also rated by water factor (WF) that identifies the number of gallons of water needed per cycle per cubic foot of laundry. The lower the WF, the more efficiently a washer uses water.

In general, front-loading washers are more efficient than top-loaders because they use less water for each wash cycle. New high-efficiency top-loading units may be a cost-effective alternative. Select an ENERGY STAR–rated washer and compare efficiencies between units by using the EnergyGuide label. How the washer is used also has an effect on its efficiency. Washing full rather than partial loads, setting manual water levels to match the size of the load, and using the lowest water temperature setting are all strategies the consumer can use to save energy when doing laundry.

Dishwashers

Dishwashers use energy to run pumps and, in some cases, auxiliary water heaters, and they use water to wash and rinse. ENERGY STAR dishwashers must exceed the minimum federal efficiency standards for both energy

Modified Energy Factor (MEF) the efficiency rating for clothes washers that considers the energy required to operate the washer, heat the water, and dry the clothes based on how much water is removed in the spin cycle.

and water use. Review specifications among similar units for energy and water efficiency ratings (Table 14.5). Models with more cycle choices offer an opportunity for more efficient operation, and soil sensors that automatically adjust water levels provide for additional water savings. Running full loads and using the no-heat dry setting allows for most efficient operation.

Clothes Dryers

Traditional clothes dryers work by convection: heated air removes the moisture from the clothes. Because there is no significant difference in efficiency between competing models, there are no ENERGY STAR ratings or EnergyGuide labels for clothes dryers. Selecting a model with a moisture sensor function that turns off the dryer when clothes are dry avoids over- or under-drying loads. Vented dryers remove air from the house as they exhaust heat and moisture to the outside, requiring replacement makeup air to be provided to the house or laundry area. Proper venting is necessary for most efficient operation. Follow manufacturer's instructions for type of pipe, distance, number of elbows, and cap type to avoid creating too much pressure on the vent system. Dryers work most efficiently when the clothes are as dry as possible, usually determined by the efficiency of the washer spin cycle. The most common dryers run on electricity, but natural gas and propane models are available.

Ventless condensing dryers are an alternative that are useful in multifamily projects and locations where it is not possible to vent to the exterior. They are not common in the United States, and they are slower and slightly less efficient than convection dryers. In addition, some models use large quantities of water for the condensing function. Combination washer/dryer units are available with condensing technology.

Clotheslines as an Energy Efficiency Strategy Clotheslines, while not as common as they once were, are an easy way to save energy. Providing a clothesline that is in close proximity to the laundry may encourage homeowners to use their dryer less when the weather cooperates. Some homeowners' associations prohibit the use of clotheslines for aesthetic reasons, and some areas have passed legislation overriding this ban in the name of efficiency.

Cooking Appliances

Cooking appliances include ranges, wall ovens, cooktops, and specialty equipment, such as indoor barbecue grills. Most wall-mounted ovens are fueled by electricity, including some with convection functions that circulate hot air around the food for faster cooking. Ranges are available in gas, electric, or a combination with gas burners and an electric oven. Cooktops are available with gas or electric burners as well as electric induction, which, rather than heating a burner, creates a magnetic field that causes iron and steel cookware to heat to cook the food within. Induction cooktops do not get hot to the touch and provide accurate and fast control of the heat level.

ENERGY STAR does not rate cooking appliances, nor are EnergyGuide labels available for them. Selecting cooking appliances is primarily a matter of personal preference. Table 14.6 shows the costs of operating different appliances to cook a casserole.

Maintenance

As with any equipment, proper maintenance extends the life of an appliance and keeps it running efficiently. Following manufacturers' maintenance schedules helps keep equipment in top shape. Typical maintenance includes cleaning coils and door gaskets on refrigerators, cleaning clothes dryer filters regularly, and checking the dryer vent for excess lint buildup where it exits the house.

Small Appliances

Small appliances, such as microwave ovens, toaster ovens, and crockpots, are more efficient than ovens and cooktops for cooking small portions. They also can reduce heat buildup in warm weather, lowering cooling bills. When selecting small appliances, look for models without electric displays, remote controls, or other always-on features to reduce phantom loads. Table 14.7 presents the energy consumption associated with operating different home appliances.

Table 14.5 ENERGY STAR Dishwasher Specifications

Product Type	Federal Standard	Federal Standard	ENERGY STAR Criteria (July 1, 2011)
Standard	EF ≥ 0.46	≤355 kWh/yr ≤6.5 gallons/cycle	≤307 kWh/yr ≤5.0 gallons/cycle
Compact	EF ≥ 0.62	≤260 kWh/yr ≤4.5 gallons/cycle	≤222 kWh/yr ≤3.5 gallons/cycle

Source: U.S. Environmental Protection Agency, ENERGY STAR program.

Table 14.6 Costs of Cooking a Casserole

Appliance	Temperature	Time	Energy	Cost*
Electric oven	350	1 hr	2.0 kWh	$0.16
Electric convection oven	325	45 min	1.39 kWh	$0.11
Gas oven	350	1 hr	0.112 therm	$0.07
Electric frying pan	420	1 hr	0.9 kWh	$0.07
Toaster oven	425	50 min	0.95 kWh	$0.08
Electric crockpot	200	7 hr	0.7 kWh	$0.06
Microwave	"High"	15 min	0.36 kWh	$0.03

*Assumes the cost of gas is $0.60/therm, and the cost of electricity is $0.08/kWh.

Source: Courtesy of ACEEE (www.aceee.org). Reprinted with permission.

Table 14.7 Energy Consumption of Typical Home Appliances

Appliance	Nameplate Wattages (Watts)
Aquarium	50–1,210
Clock radio	10
Clothes washer (does not include hot water)	350–500
Clothes dryer	1,800–5,000
Coffee maker	900–1,200
Dehumidifier	785
Dishwasher (does not include hot water)	1,200–2,400 (using the drying feature greatly increases energy consumption)
Electric blanket (single/double)	60/100
Fan (ceiling)	65–175
Fan (furnace)	750
Fan (whole-house)	240–750
Fan (window)	55–250
Hair dryer	1,200–1,875
Heater (portable)	750–1,500
Microwave oven	750–1,100
Personal computer (processor awake/sleep)	120/30 or less
Personal computer (monitor awake/sleep)	150/30 or less
Personal computer (laptop)	50
Radio (stereo)	70–400
Refrigerator (frostfree 16 cubic feet)	725
Television (color, 19")	65–110
Television (color, flatscreen)	120
Toaster oven	1,225
Vacuum cleaner	1,000–1,440
VCR/DVD	17-21/20-25
Water bed (with heater, no cover)	120–380
Water heater (40-gallon electric)	4,500–5,500
Water pump (deep well)	250–1,100

Source: U.S. Department of Energy, National Renewable Energy Laboratory http://www.energysavers.gov

Electronics

The expanded use of electronic equipment in homes is a primary cause of increased energy consumption. Typical electronics include televisions, digital video recorders (DVRs), cable and satellite control boxes, audio receivers, DVD players, computers, printers, scanners, and cordless phones. Certain equipment, such as DVRs and cable boxes, must be left on around the clock to operate properly, although they can be disconnected when not in use for several days over vacation periods to save energy. Televisions, audio, computers, printers, and similar equipment should have their power disconnected at the plug whenever they are not in use to reduce phantom loads. This can be accomplished by using switched or smart power strips or by installing a kill switch that turns off the power to the receptacles into which they are plugged (Figure 14.22). Computers, monitors, and printers normally have a standby mode that uses less power

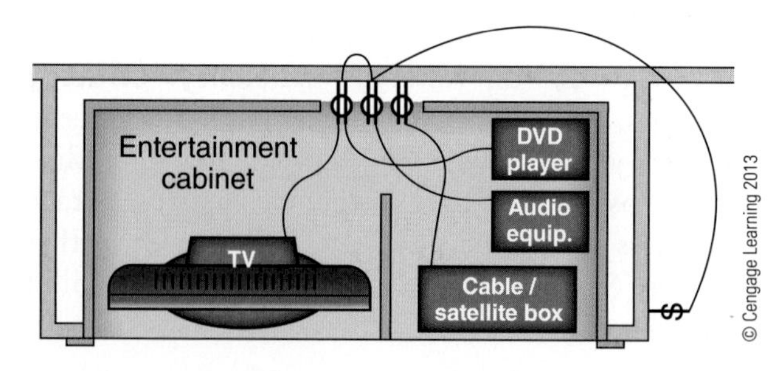

Figure 14.22 Switched receptacles turn off the television, audio components, DVD player, and other equipment with plug loads used only for remote controls. Cable and satellite receivers that must remain on to retain programming are plugged into unswitched outlets.

kill switch a control that turns off the power to a set of receptacles that are plugged into it.

than when the unit is in operation but still wastes energy. Electronics also create heat when they are on, even in standby mode, so disconnecting power to them also helps keep a home cooler and reduces the need for air conditioning in warm climates. Any equipment that must remain turned on should be plugged into non-switched receptacles.

Limiting the number of electronics in a home, using switched receptacles or other control devices to disconnect power when in standby mode, and selecting the most efficient models are good strategies to reduce overall energy use.

Specialty Equipment

In addition to lighting, appliances, and electronics, many homes have special electrically operated equipment that can add to the electrical load or, in some cases, reduce it. Pool and spa pumps should be selected for highest efficiency and configured to run for maximum efficiency. Combined heat and power systems can be an effective strategy in cold climates to take advantage of generating power while heating a home.

Pool and Spa Pumps

Pool and spa pumps can have a big effect on energy use in a home. Pump efficiency varies significantly, and selecting the most efficient pump that is sized properly rather than oversized will lead to more efficient operation. Larger diameter and shorter pipes and a larger filter all reduce resistance throughout the system; using 45-degree sweeps instead of 90-degree elbows also allows for pumps to operate as much as 40% more efficiently, according to the U.S. Department of Energy. Limit pump running times to the minimum required to keep the water clean; in some cases, as little as three hours per day is adequate. A study at Florida Atlantic University determined that employing these efficiencies and replacing pumps with smaller, more efficient models helped to reduce energy use by as much as 75%. Chapter 16 covers pool equipment in more depth.

Occupant Education

Educating owners or renters of a house or apartment is a key component of reducing energy use. Understanding how their behavior affects the amount of energy being used is the first step in making effective changes in use patterns. Provide them with energy monitors and feedback devices so they can track their energy use. Instruct them on how to use control systems in their home, and remind them to turn off lights, fans, and equipment when not in use. Leave them with extra lamps for their fixtures, and give them model numbers and suppliers for replacements. When occupants have the information they need to efficiently operate their home, it becomes easier for them to make the effort to live more efficiently.

Remodeling Considerations

In existing homes undergoing renovation, electrical systems (and lighting in particular) offer opportunities to improve the efficiency. In rooms that may have too much lighting for the tasks that take place in them, consider installing smaller lamps, changing fixtures, or installing dimmers to reduce energy use. Changing incandescent lamps to CFL or LED replacements is a simple way to save energy. Select ENERGY STAR–rated fixtures where replacements or new fixtures are installed. Inspect all fixtures that penetrate the building envelope, particularly recessed lights in insulated ceilings. Fixtures that are not designed as airtight and IC should have covers installed over them to provide an air seal and protect from the heat created by insulation (Figure 14.23). Look for phantom loads that can be reduced by using power strips, switched receptacles, or replacing equipment with more efficient models.

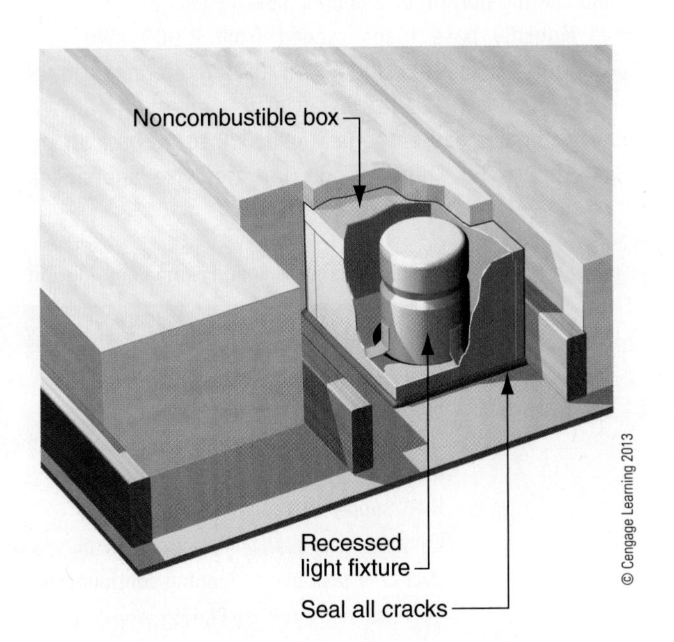

Figure 14.23 Non-airtight recessed can lights should be enclosed in a non-combustible air impermeable box. The box is then sealed to the ceiling.

FROM EXPERIENCE

Beyond Building Science

David Wasserman, Mechanical Engineer, Servidyne Imagine the surprise when you receive the first summer electric bill in your new super-efficient green-certified home and it is for $1,200. Granted, the house is 7,500 ft², but you spent extra money on high-performance windows, walls of insulating concrete forms, a roof sealed with spray polyurethane foam, high-efficiency heat pumps, solar water heating, ENERGY STAR appliances, and perhaps even tankless water heaters. And you paid extra to have the house certified under a national program, like ENERGY STAR or LEED for Homes, or a regional program like EarthCraft House, and the house passed with a low HERS Rating.

David Wasserman is a mechanical engineer with degrees from Cornell University and the University of Tennessee. He spent 20 years at Oak Ridge National Laboratory as a research engineer in energy efficiency as well as in pollution prevention. More recently, he has worked at Southface and Servidyne in Atlanta, Georgia, doing residential and commercial building energy assessments and modeling.

This is not an abstract scenario but a very real occurrence, particularly with larger green-certified homes. As a senior engineer at Southface in Atlanta, I had the opportunity to inspect a half-dozen large certified homes with owner complaints of high utility bills. In every case, I found a home with a tight, energy-efficient shell and ductwork, an efficient HVAC system, and ENERGY STAR appliances. In every case, I also found a very high base load—the non-heating and cooling portion of a home's monthly usage.

Monthly base loads ranged from 2,000 kWh in a 4,000 ft² home to 7,000 kWh in the aforementioned 7,500 ft² home. Where was all of this energy going? All of the homes had an estimated 1,000 to 1,200 kWh per month use for "normal" interior uses like appliances, lighting, and electronics, and, in some cases, wine coolers and elevators. Two of the homes had electric water heaters. In various combinations, the rest of the base load could be accounted for by the peripherals described in Table 14.8.

The first four of these wasteful energy practices are owner controlled and can be addressed by education. The rest, however, can be addressed by the builder or trades at installation. Ground-source heat pump water loop pumps should cycle with the compressors, pool pumps should be sized properly, energy-efficient variable speed pumps should be chosen, and timers should be installed to run the pumps only long enough to filter all water once daily. Hot water circulation loops are probably the biggest energy hogs of all, particularly if they run continuously and have little or no pipe insulation. Timers need to be installed and used on circulation pumps, and pipes should be heavily insulated; better yet, a strategy should be developed to get hot water to faucets quickly without a circulation loop.

Creating an energy-efficient home today takes more than building science. It takes an understanding of all the major energy uses in the home and a strategy for limiting energy consumption without adversely affecting the occupant's comfort or lifestyle. This approach needs to include building commissioning and an owner's manual for energy-efficient operation of the home.

Table 14.8 Base Load Energy Consumption

Energy Use	Monthly kWh
Ceiling fans running continuously	20 each
Refrigerators in garages (summer)	100 each
Dehumidifiers in basements running continuously because humidistat set too low	235
HVAC supply fans running continuously	115/ton cooling
Ground-source heat pump water loop pumps always on	150/ton cooling
Oversized pool pumps running continuously	1,200 to 1,500
Domestic hot water circulation loops	1,500 to 4,000

Source: Field measurements performed by David Wasserman, mechanical engineer, Servidyne.

Creating multiway switching for lights and switched receptacles can be done by installing new wiring or with remote radio frequency units. Standard switches can be replaced with timers or occupancy sensors to control lights and fans. When installing new devices and fixtures in the building envelope, make certain that the air barrier is maintained and insulation is properly installed around each item.

In addition to efficiency improvements, existing electrical wiring and equipment should be evaluated for compatibility with any changes in insulation being considered. Old **knob and tube wiring**, composed of individual wires running in pairs connected to framing with ceramic insulators (Figure 14.24), should be inspected before insulation is added to walls and ceilings and replaced if necessary to avoid overheating. Older cables often had cloth shielding and paper or cloth insulation, which degrades over time and should be inspected and replaced if deteriorated. Finally, review the occupants' behavior and offer recommendations for simple changes they can make that will save them energy.

Figure 14.24 Knob and tube wiring can pose a significant fire hazard and should be replaced prior to insulation.

knob and tube wiring an old system of electrical wiring composed of individual wires running in pairs and connected to framing with ceramic insulators.

Summary

The selection of lighting and appliances has a major effect on energy efficiency. Installing efficient equipment, reducing phantom loads, and educating occupants on managing usage can improve efficiency significantly. All electrically operated devices should be evaluated for the impact they will have on home energy use. Installing the appropriate amount of lighting for the tasks in each room of a house saves energy by avoiding over-lighting areas. Using timers, motion and photo sensors, and other control systems reduces energy use by turning off devices when they are not needed.[1]

Review Questions

1. Which of the following are not appropriate for a green electrical system strategy?
   a. Smart Meter
   b. Renewable energy certificates
   c. Incandescent recessed lights
   d. Dimmers
2. Which of the following controls are most effective for interior light controls?
   a. Vacancy sensor
   b. Photocell
   c. Timer
   d. Remote control

[1] Thompson, C. *Clive Thompson Thinks: Desktop Orb Could Reform Energy Hogs*. http://www.wired.com/techbiz/people/magazine/15-08/st_thompson; personal correspondence with inventor, Mark Martinez (September 28–October 1, 2010).

3. With whole-house control systems, which of the following is of most concern in a green home?
   a. Ability to dim lights efficiently
   b. Energy used to operate controls
   c. Interconnection with alarm system
   d. Additional wiring required
4. Which of the following is not a benefit of a whole-house energy monitor?
   a. It can encourage less energy use
   b. It shows how much energy is being used throughout the house
   c. It provides instant feedback on amount of energy being used
   d. It controls how much power is being used
5. Which type of lighting is used for kitchen counters?
   a. Ambient
   b. Accent
   c. Task
   d. HID
6. Which of the following lamps are the least efficient?
   a. CFL
   b. Tubular fluorescent
   c. Halogen
   d. LED
7. Which of the following criteria determines the efficiency of a lamp?
   a. Lumens/watt
   b. CRI
   c. Color temperature
   d. Lamp life
8. What color temperature looks most like the mid-day sun?
   a. 7000K
   b. 5000K
   c. 1000K
   d. 2500K
9. Which of the following does not typically create a phantom load?
   a. Electric clock radio turned off
   b. Television with remote control turned off
   c. Computer in standby mode
   d. Ceiling fan without remote control turned off
10. Which of the following is the most important factor in conserving energy?
    a. Number of lights in a house
    b. ENERGY STAR appliances
    c. Percentage of CFL bulbs
    d. Occupant behavior

Critical Thinking Questions

1. What would you consider the highest priority when planning electrical systems for a green home?
2. Discuss the pros and cons of different lamp types.
3. Consider strategies for reducing phantom loads.

Key Terms

accent lighting, 398
ambient lighting, 398
ballast, 399
Color Rendering Index (CRI), 404
color temperature, 403
compact fluorescent lamps (CFLs), 399
efficacy, 402
electrical and magnetic fields, 00
electromagnetic fields (EMF), 393
EnergyGuide, 405
fiber optic lighting, 404
fluorescent, 399
foot-candles, 403
halogen, 398
high-intensity discharge (HID), 401
illumination, 403
incandescent, 398
Kelvin, 403
kill switch, 408
knob and tube wiring, 411
lamp, 398
light-emitting diodes (LEDs), 400
luminaires, 401
magnetic induction lamps, 401
Modified Energy Factor (MEF), 406
multiway switching, 393
occupancy sensors, 394
photocell, 395
phantom loads, 389
renewable energy certificates (RECs), 391
smart grid, 391
smart meter, 390
sones, 405
task lighting, 398
vacancy sensors, 394

Additional Resources

ENERGY STAR: http://www.energystar.gov

Federal Trade Commission: http://www.ftc.gov

Green By Design Lighting Guide:
http://greenbydesign.com/2008/12/30/lighting/

California Energy Commission Consumer Energy Center:
http://www.consumerenergycenter.org/

15

Plumbing

Proper plumbing design helps reduce water consumption and the energy used to heat water. This chapter describes how to specify and install efficient plumbing systems, including supply and drain piping, fixtures, fittings, and water heaters. In addition, we will discuss advanced green building techniques, such as rainwater and gray water harvesting. The implications of remodeling on plumbing systems are also discussed.

LEARNING OBJECTIVES

Upon completion of this chapter, the student should be able to:

- Explain the importance of water conservation
- Identify methods of conserving water
- Describe the core concepts of efficient hot water delivery systems
- Identify different options for on-site treatment of gray and black wastewater
- Explain the core concepts of rainwater harvesting
- Specify appropriate water heating equipment for a project
- Explain how to specify water-efficient fixtures and fittings
- Describe how water filtration and softening systems operate
- Explain how to make water-efficient improvements to existing homes

Green Building Principles

Energy Efficiency

Indoor Environmental Quality

Resource Efficiency

Water Efficiency

Homeowner Education and Maintenance

Worldwide Water Resources

Over 97% of all the water on earth is saltwater, unsuitable for human consumption (Figure 15.1). Of the remaining 3%, approximately two thirds is fresh water ice, leaving only 1% of the total as fresh liquid water. Water is a finite resource. We reuse the same fresh water over and over again, returning it to the earth via evaporation, waterways, and infiltration to underground aquifers. The continually increasing worldwide population, combined with increased per capita water use, reduces available fresh water supplies in many areas.

Water Use in the United States

According to a 2006 United Nations' study of worldwide water use, the United States consumes an average of approximately 124 gallons per day (GPD) per person, the highest of any country. By comparison, Spain uses 70 GPD, France 65 GPD, and Niger only 6.5 GPD. Americans may not care to reduce their consumption to the level of a country like Niger, however, lowering it to a level more comparable to that of other developed nations is not unreasonable.

Water Conservation

Water that we do not use through conservation is the least expensive source of additional water. Conservation also saves energy that municipal water supplies use

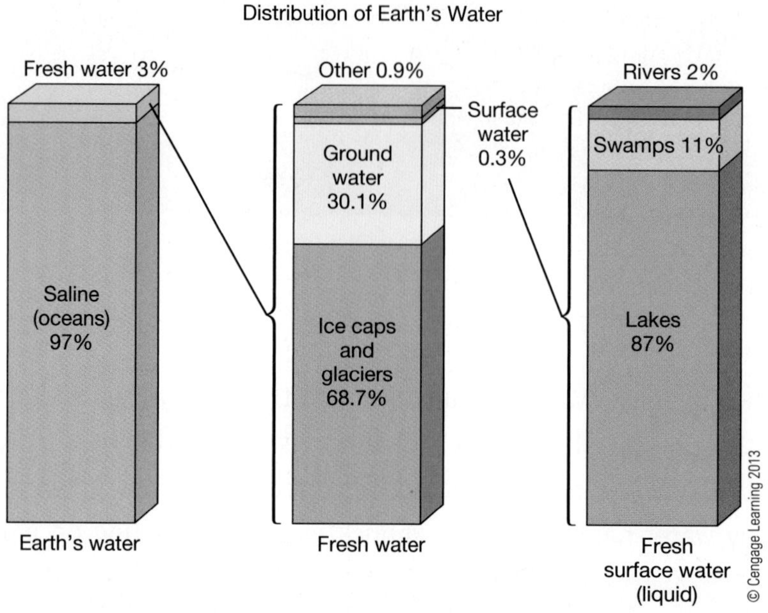

Figure 15.1 World water resources.

for water treatment and distribution. As populations increase, conservation can reduce the need for additional water treatment, storage, and distribution piping. Water that is not needed for daily consumption remains in rivers and streams to maintain critical habitats for wildlife and fish and freshwater flows to larger waterways.

The Water/Energy Connection

According to the U.S. Environmental Protection Agency (EPA), an estimated 3% of the nation's consumed energy —approximately 56 billion kWh—is used to treat and transport drinking and wastewater. If 1% of American households replaced old plumbing fixtures with more efficient models, the total savings would be about 100 million kWh of electricity per year.

Not only is power used to treat and transport water, but water is used to generate power. The U.S. Geological Survey (USGS) calculated that 137,000 million gallons of fresh water was used each day for power generation. According to the National Energy Technology Laboratory of the U.S. Department of Energy, fossil fuel plants use about 39% of all water consumed in the United States, second only to the agricultural industry.

Considering the energy used to manage our water supply and water used in electricity generation, the two are obviously very closely connected. Water conservation helps save energy, and energy efficiency helps conserve water.

Where Water Is Wasted

Water is wasted through excessive irrigation, leaks, homeowner behavior, inefficient plumbing fixtures, toilet tank overfilling, and slow or ineffective hot water distribution to fixtures. Irrigation systems in relation to landscaping are covered in detail in Chapter 11. According to the EPA, the typical American household loses an average of 10,000 gallons of water each year through water leaks, dripping faucets, and running toilets. Many leaks go undetected for years in crawl spaces, walls, and underground, causing problems with building structure and indoor environmental quality. Leaks at sink faucets and toilet valves are visible but often ignored, wasting additional water. Testing for leaks can be as simple as turning off all plumbing and observing the water meter. If the meter is moving with all fixtures off, then water is leaking somewhere on the property. If the source of the leak is not apparent, a plumbing professional should be consulted to locate and repair any problems. High water pressure, typically in excess of 80 pounds per square inch (psi) can cause premature failure of faucet seals and toilet valves as well as excessive water consumption. Where municipal water supplies provide high street pressure, a **pressure-reducing valve** should be installed where the water service enters the house to maintain water pressure inside the home at a consistent level.

---

pressure-reducing valve a device that maintains water pressure inside the home at a consistent level.

Homeowner behavior can have a significant effect on water use. Simple actions, such as not running water while brushing teeth and shaving and taking shorter showers, can save significant amounts of water over the course of a year. Washing dishes in a full dishwasher generally uses less water than hand washing. Choosing water saver settings on clothes and dishwashers further reduce water use.

To help mitigate the effects of waste, water demand should be reduced as much as possible through efficient fixtures, fittings, appliances, and water-efficient behavior. Selecting high-efficiency plumbing faucets, showerheads, and toilets are addressed later in this chapter.

Fresh Water Supply

Most homes obtain their water from a municipal supply that is tested regularly for pathogens and certain toxic chemicals according the Safe Drinking Water Act, enacted by Congress in 1974, providing a consistent supply of healthy, potable water. In some areas, reclaimed non-potable water may be available for irrigation and other applications. Where a public water supply is not available, potable water is typically obtained from a well drilled on the property by way of pumps and storage tanks. Well water should be tested regularly for contamination to ensure that it remains healthy and potable. Alternative sources of water include rainwater and gray water.

Rainwater harvesting is the collection, storage, and use of precipitation from roofs and other surfaces. This practice has been used for centuries as an alternative to well and municipal water supplies. A source of free, relatively clean water, rainwater harvesting can reduce the amount of energy required to treat and transport water and can help reduce flooding and erosion.

Rainwater harvesting is generally used for irrigation, in toilets, and for other non-potable purposes because water quality regulations in most areas preclude its use for drinking. Where allowed, appropriately filtered and sterilized rainwater can provide potable water supplies. In some areas, including the U.S. Virgin Islands and Santa Fe, New Mexico, rainwater harvesting is required for all new buildings. Some states restrict or prohibit rainwater harvesting because they consider it a public resource that does not belong solely to the individual on whose property it lands.

Gray water is the waste from lavatories (bathroom sinks), showers, tubs, and laundries, which can be treated on- or off-site and reused for filling toilets and certain types of irrigation.

---

rainwater harvesting the collection, storage, and use of precipitation from roofs and other surfaces.

Fresh Water Distribution

Fresh water distribution pipes are available in copper and a selection of plastics, including chlorinated polyvinyl chloride (CPVC), cross-linked polyethylene (PEX), and polypropylene (PP). Polyvinyl chloride (PVC) is not used for supply water because it does not tolerate high heat. Until recently, copper was the most common (and one of the least expensive) pipe materials, but higher copper prices have led to rising installation costs and more job site thefts of installed pipe, making the plastic alternatives less expensive to install and less likely to be stolen.

Copper pipe is available in either rigid or flexible forms, both of which require soldered joints to make connections. Copper is very durable, has a high recycled content, and is fully recyclable; however, it does have higher embodied energy than its plastic counterparts and is susceptible to deterioration if exposed to highly acidic (low pH) water. Copper should not be installed if the water has a pH of 6.5 or less. Because most public water supplies are treated to have a pH between 7.2 and 8.0, most problems with copper pipe arise on properties with local well water. Where the water cannot be treated to raise the pH above 6.5, plastic pipe should be used instead of copper.

Chlorinated polyvinyl chloride (CPVC) is chemically similar to PVC but contains added chlorine to increase its rigidity at high temperatures, making it suitable for hot water piping. A rigid pipe, CPVC requires solvent-based glue at fittings (Figure 15.2); these glues are highly

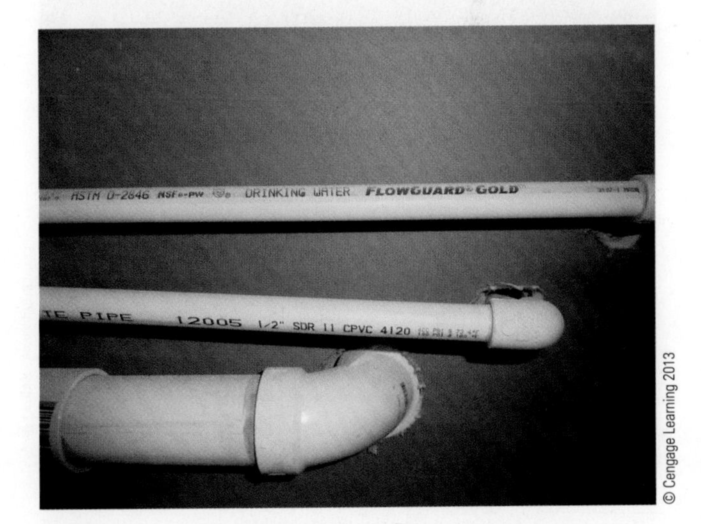

Figure 15.2 The top two chlorinated polyvinyl chloride (CPVC) pipes are supplying drinking water. The bottom PVC pipe is the waste water drain.

---

chlorinated polyvinyl chloride (CPVC) piping material that is chemically similar to PVC but contains added chlorine to increase rigidity at high temperatures.

toxic, so personal protection should be used during installation. It is inexpensive and therefore not likely to be stolen for scrap value. While theoretically recyclable, facilities that will accept surplus materials for recycling are uncommon. Like all vinyl products, CPVC releases dioxins when incinerated. Both copper and CPVC are common in trunk-and-branch piping systems.

Cross-linked polyethylene (PEX) is made by cross-linking, or knitting together long chains of ethylene in a process that allows the material to be both strong and flexible at high temperatures. No chlorine or vinyl is used in its manufacture. PEX is attached mechanically using either crimp rings or push-in connectors (Figure 15.3); it cannot be attached with solvents like other plastics. Since PEX is flexible and available in long lengths, it is the most commonly used pipe in manifold systems. Its primary benefit is that it can be installed in a single piece from a manifold to a fixture with no intermediate connections. Many manufacturers provide the pipe and manifolds as a single system. Although not recyclable, PEX can be burned without releasing harmful gases because it is composed of carbon and hydrogen.

Polypropylene (PP) is sometimes referred to as PP-R, with the "R" standing for "random" because it is made with a combination of both long and short molecules. Its composition provides a combination of strength and flexibility and is less toxic than CPVC because there is no vinyl or chlorine used in its manufacture (Figure 15.4).

Figure 15.3 PEX pipe and fittings.

---

cross-linked polyethylene (PEX) specialized type of polyethylene plastic that is strengthened by chemical bonds formed in addition to the usual bonds in the polymerization process.

PEX *see* cross-linked polyethylene.

polypropylene (PP) a plastic plumbing material made from a combination of both long and short molecules that provides both strength and flexibility.

Figure 15.4 PP-R pipe.

PP-R has been used in Europe for over 35 years and is available in the United States but is not commonly used in residential construction here, primarily due to its high cost. PP-R is a rigid pipe, joined at connections via heat fusion that bonds the connections at a chemical level as they cool. It has low embodied energy and is chemically inert. It is available in several forms for both potable and non-potable water systems, including gray water and rainwater collection and distribution.

There is no single best solution for water supply piping for a green home. Copper has a high recycled content and is fully recyclable, but it has much higher embodied energy than plastic pipe. Plastic pipes are not recyclable and use petrochemicals in their manufacturing. The long lengths and limited connectors required with PEX piping provide a smooth, uninterrupted flow of hot water, providing faster hot water delivery. Rigid pipe with elbows and tees slows down water velocity, which may lead to greater heat loss. Of the plastics, PEX and PP-R are less toxic than CPVC, but the impact of product selection on a project's sustainability is modest because the overall quantity of pipe in a house is a small percentage of the total materials.

Hot Water Delivery

Reducing the amount of water wasted while waiting for hot water to arrive at fixtures is one of the most critical components in residential water efficiency. As the average size of new homes has increased in recent years, the number of hot water fixtures per home has also increased. Although lower flow rates have made fixtures more efficient, this gain may be negated by waste in other areas. According to GreenPlumbers USA, the average distance from a water heater to the farthest fixture in single-family homes increased from approximately

30' to 80' between 1970 and 2010. These longer distances waste water by having occupants wait for hot water to arrive and by allowing more time for water in pipes to cool between hot water uses, which requires more energy to reheat.

The simplest way to avoid wasting water in hot water lines is to locate all fixtures close to the water heater. Smaller homes can be designed with a single plumbing core, a design that places all plumbing close together to reduce pipe lengths, with short pipe runs to the fixtures (see Figure 3.22 in Chapter 3). Larger homes can incorporate multiple cores with separate heaters or pumps to bring hot water to fixtures rapidly. Home design plays a major role in water efficiency. Designing central plumbing cores in close proximity to water heaters allows for efficient, short runs of supply and drain piping, which saves materials as well as water. When bathrooms and kitchens are spread far apart, more pipe is required and more water may be wasted.

Water Supply Piping Systems

The most common residential water piping system is referred to as trunk-and-branch, consisting of 3/4" or 1" hot and cold water pipes (or trunks) that run to each plumbing location, from which smaller 1/2" pipes (or branches) run to individual fixtures (Figure 15.5). An alternative design is a manifold system, also referred to as a home run system, which consists of 1/2" or 3/8" pipes that run directly from the water source to individual fixtures (Figure 15.6). When manifold systems are installed with short runs, the smaller pipe allows the hot water to reach fixtures faster than would be possible with typical trunk-and-branch designs. Manifold systems have the added benefit of providing individual valves for each fixture, allowing for easy cutoff of water supply when needed for maintenance. Larger homes may incorporate separate manifolds near each plumbing location, with heaters or pumps at each manifold to reduce water waste.

Selecting the appropriate water piping system depends on the distance between the water heater and the fixtures using the hot water, as well as the household usage patterns. When all the fixtures are close to the hot water source, a trunk-and-branch system can be very efficient, particularly when several people may be using hot water in rapid succession, as when an entire family showers each morning. A 3/4" insulated trunk line will remain hot longer than smaller manifold home run pipes, allowing for multiple uses in succession to take advantage of the hot water remaining in the trunk. The first time hot water is used for the day (or the first time

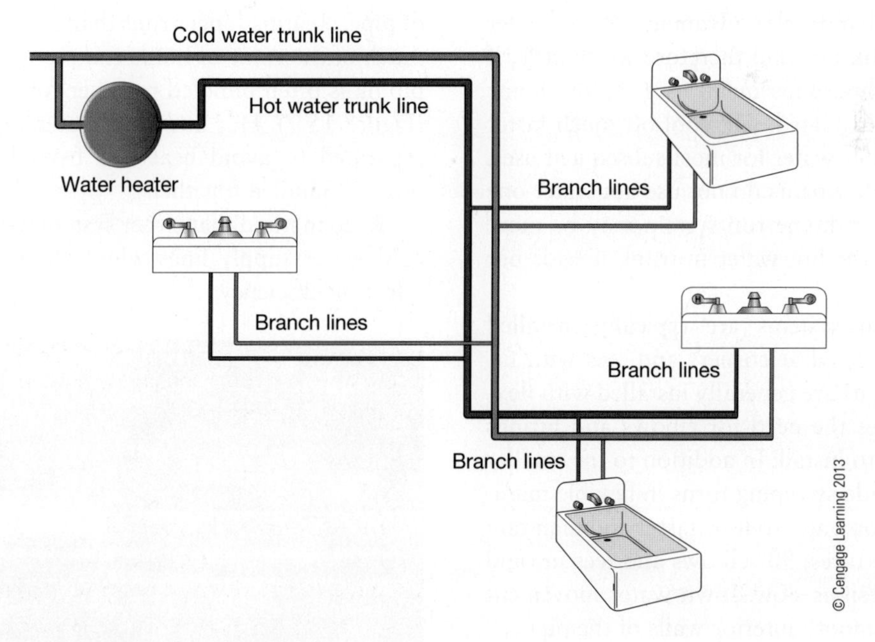

Figure 15.5 Trunk-and-branch plumbing layout.

plumbing core a design that places all plumbing close together to reduce pipe lengths and enable short pipe runs to the fixtures.
trunk-and-branch a plumbing design consisting of 3/4"or 1" hot and cold water pipes or trunks that run to each plumbing location and from which smaller 1/2" pipes, or branches, run to individual fixtures.
manifold system a plumbing design consisting of 1/2" or 3/8" pipes that run directly from the water source to individual fixtures; also referred to as a *home run system*.
home run system see *manifold system*.

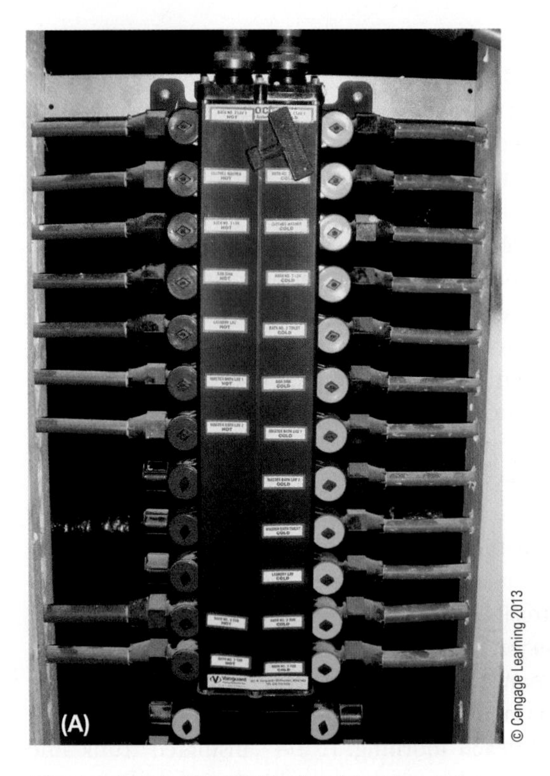

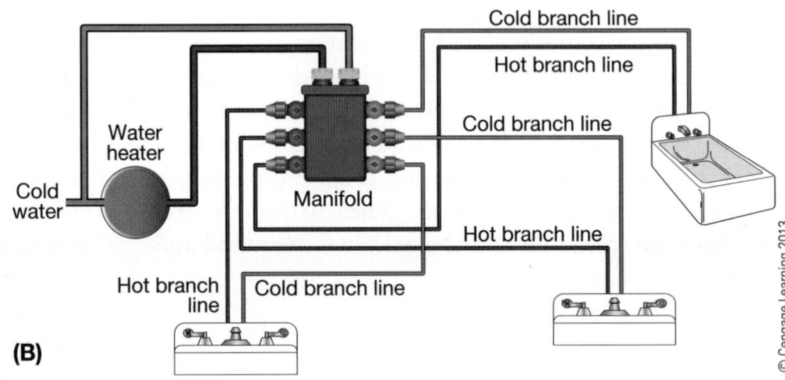

Figure 15.6 Manifold plumbing layout.

it is used after the trunk line has cooled) is known as the cold start. Cold starts require filling up the trunk with hot water, wasting some cold water in the process. Hot starts, on the other hand, take advantage of the water remaining in the trunk line and therefore waste little, if any, cold water for these subsequent uses. A 3/8" home run line, even when insulated, will cool off much faster and must refill with hot water for most subsequent uses. In homes where the occupants do not use hot water one person after another, a home run system may be more appropriate because the hot water in trunk lines is not used before it cools.

Trunk-and-branch systems are typically installed with rigid pipe connected at corners and tees with fittings; manifold systems are generally installed with flexible pipe that reduces the need for elbows and fittings that take more time to install. In addition to the smaller pipe diameter, the wide sweeping turns in flexible manifold systems allow hot water to flow faster and maintain higher pressure at fixtures; 90° elbows and tees in rigid trunk-and-branch designs slow down water movement by interrupting the smooth interior walls of the pipe.

cold start the first time hot water is used for the day or after the trunk line has cooled.

hot start the temperature in the hot water supply line when a fixture is turned on; occurs when hot water remains in the supply line and is available for use.

Manifold systems work best when pipes are run in the most direct route between the manifold and fixtures. Poorly planned installations can use excessive amounts of pipe, creating longer runs than necessary and negating much of the value of home run piping. Manifold system piping is often bundled together for ease of installation (Figure 15.7). Hot and cold water pipes must be kept separated to avoid heat loss from hot to cold water pipes if bundled together.

Keep in mind that either system works effectively for cold water supply lines where the pipe length has no effect on efficiency.

Figure 15.7 When bundled together, the hot water lines lose heat to the cold lines.

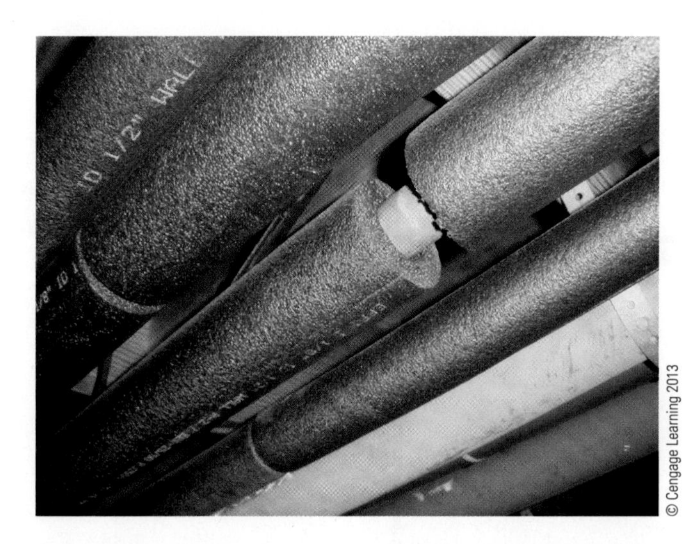

Figure 15.8 For greatest performance, insulation needs to provide complete coverage with no gaps. This image clearly shows a gap where two pieces of insulation are not in contact.

Pipe Insulation

Installing insulation on hot water lines helps keep the water warm between uses, saving energy; and reduces the amount of water wasted in trunk-and-branch and manifold. Insulation also helps eliminate condensation on cold water pipes and prevents them from freezing in areas such as unconditioned crawl spaces. Pipe insulation must be installed tightly around all pipes and fittings and sealed completely for highest performance (Figure 15.8). Gaps and holes will reduce efficiency and may allow pipes to freeze when installed in unconditioned spaces in cold climates.

Slab Installation

Homes built with slab-on-grade foundations often require water pipes to be installed in the slab to supply centrally located kitchens and bathrooms. Slab pipes should be insulated to keep water hot and avoid condensation. Consider installing pipes in large sleeves to allow for future replacement if required. Without the ability to replace a pipe, a single failure would require significant amounts of demolition and reconstruction for an otherwise minor plumbing repair.

Hot Water Circulation Systems

A common solution to long waits for hot water with trunk-and-branch systems is to install a **hot water circulation loop**, a continuous pipe from the water heater that runs close to every fixture and back to the bottom of the heater (Figure 15.9). A loop like this can provide practically instant hot water at every fixture, saving water in the process. However, this water savings comes with a significant energy penalty when hot water is circulated but not used. This can occur with continuous pumps, timer-operated pumps, temperature pumps, and thermosiphon loops, which are described as follows:

- **Continuous pumps** operate 24 hours a day, keeping a supply of hot water in trunk lines at all times. This method requires the water heater to operate more frequently than otherwise necessary to reheat water that is continuously cooling in the pipes. See David Wasserman's "From Experience" in Chapter 14 for more information.
- **Timer-operated pumps** are set to run for certain periods of time when occupants will require hot water. Although preferable to continuous pumps, timed pumps waste energy if hot water is not used during the periods that they operate; when hot water is needed when they are not operating, and more waste is generated as cold water is cleared out of the pipes.
- **Temperature pumps** have sensors near fixtures that tell the pump to circulate hot water when the temperature drops below a set level. These systems are more efficient than continuous pumps, but they also waste significant amounts of energy reheating water in the pipes.
- **Thermosiphon loops** employ a loop without a pump, using the natural convection of rising hot water to circulate to the top of the loop and allowing cooler water to flow down and back to the water heater (Figure 15.10). Like continuous pumps, thermosiphon systems waste energy by constantly reheating water.

A more efficient solution to the circulation systems described above is a **demand pump** that moves hot water to the fixtures only when needed, eliminating

---

hot water circulation loop a continuous pipe from the water heater that runs close to every fixture and back to the bottom of the heater.

continuous pumps pumps that circulate hot water throughout the trunk lines 24 hours a day.

timer-operated pumps pumps that can be set to run for certain periods of time when occupants will require hot water.

temperature pumps a recirculating pump that has sensors near fixtures that tell it when to circulate hot water based on when the temperature drops below a set level.

thermosiphon loops a passive heat exchange system that operates without a mechanical pump, using the natural convection of rising hot water to circulate to the top of the loop, thereby allowing cooler water to flow down and back to the water heater.

demand pump moves hot water to plumbing fixtures only when needed, eliminating standby losses from the constant reheating of water in pipes.

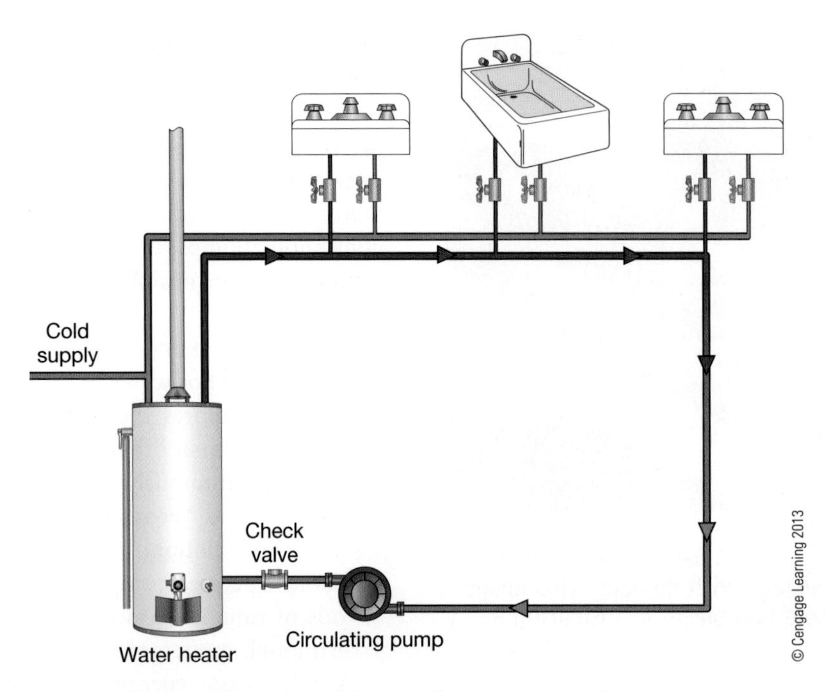

Figure 15.9 Traditional hot water circulation loop with dedicated return pipe.

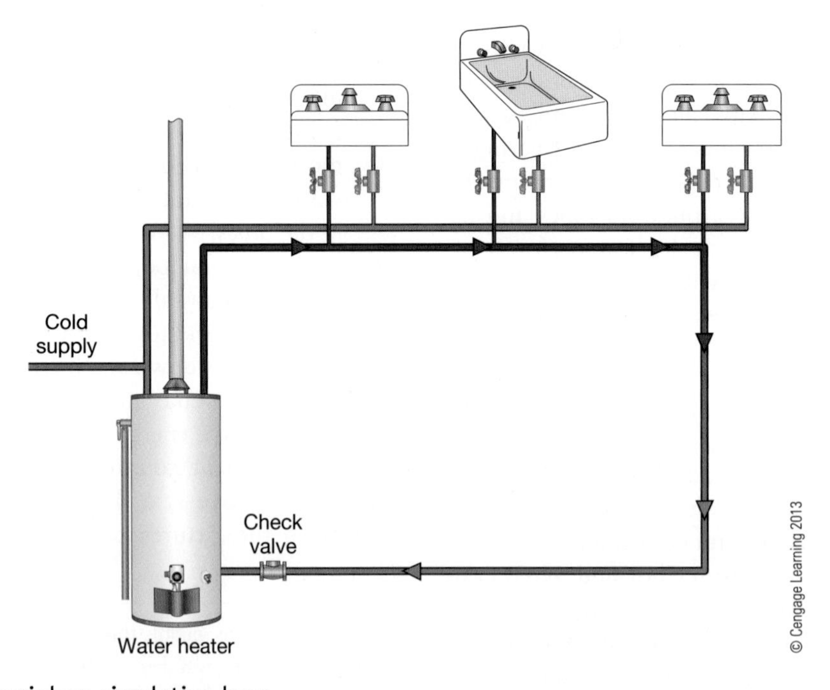

Figure 15.10 Thermosiphon circulation loop.

standby losses from the constant reheating of water in pipes (Figure 15.11). Demand pumps operate by pressing a button or activating a motion sensor. The pump then draws hot water through the pipes, pushing the cold water back through the existing cold water lines or a dedicated return line (Figure 15.12). Because the pump moves water through the lines faster than it would flow out of high-efficiency faucets or showerheads, the overall wait time is much less than if water were allowed to run from fixtures until hot water arrives. A temperature sensor in the pump turns it off as soon as hot water arrives. Demand pumps can be installed in a variety of locations, including at the last fixture at the end of a trunk-and-branch system, using the existing cold water line for the return (Figure 15.13); at the water heater with a dedicated return line (Figure 15.14); or at another convenient location. On multiple manifold systems, a demand pump can be installed at each manifold with remote control operators placed near each fixture.

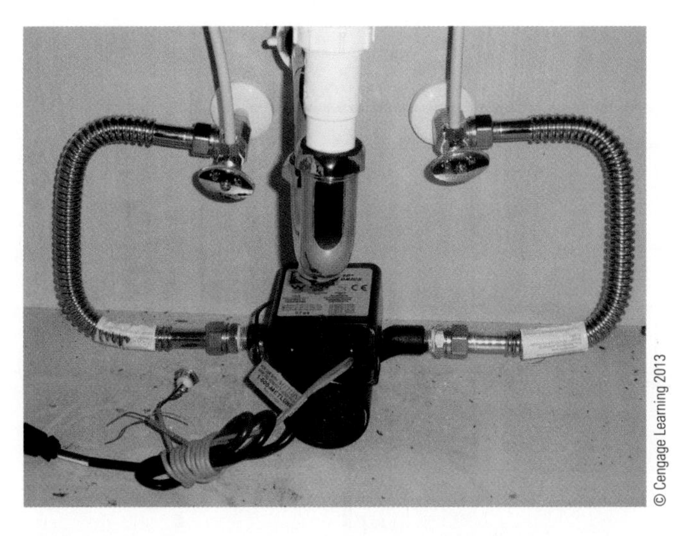

Figure 15.11 A circulation pump installed under a sink.

Demand pumps do require a certain amount of behavior change in homeowners. They must press a button or trigger a motion sensor and then wait briefly for the pump to deliver hot water to their fixtures, often leaving enough time to brush their teeth or use the toilet. When they understand the amount of energy and water saved by using a demand pump, as well as the convenience of not waiting for hot water, most homeowners should be able to make the necessary adjustments.

Pumps and Water Heaters

Water heaters will be addressed in detail later in this chapter, but it is useful to consider how circulation systems work with different heater types. Water heaters break down into two types: tank and tankless

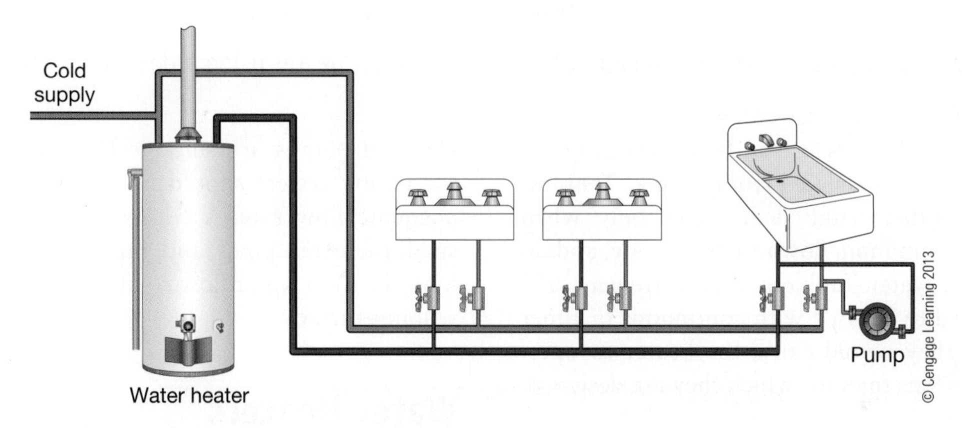

Figure 15.12 Circulating system using a cold water line return.

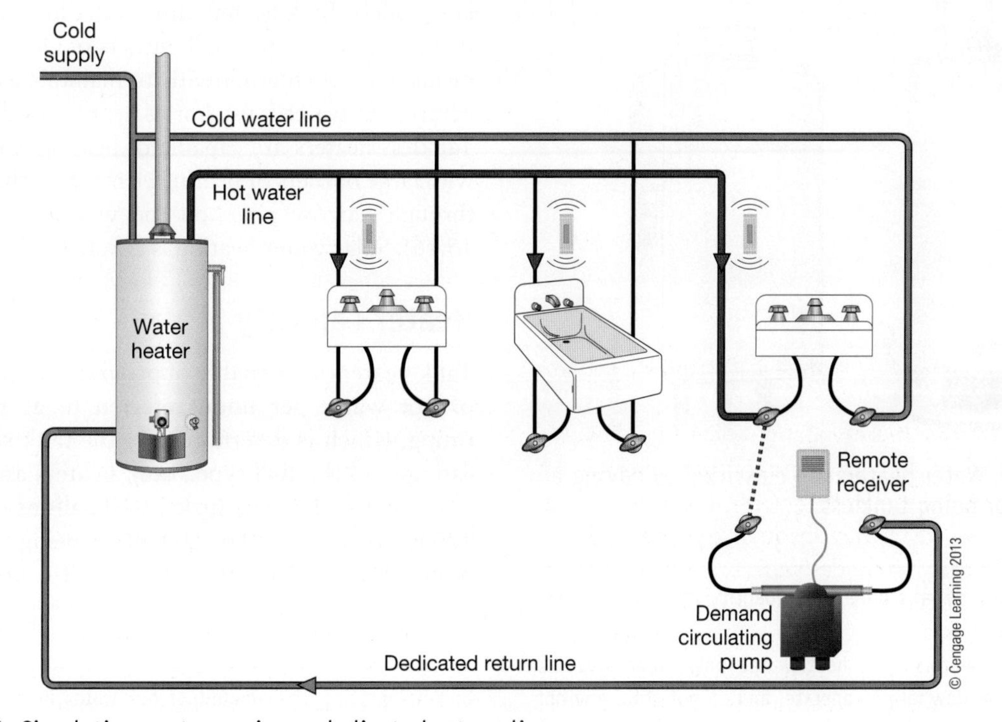

Figure 15.13 Circulating system using a dedicated return line.

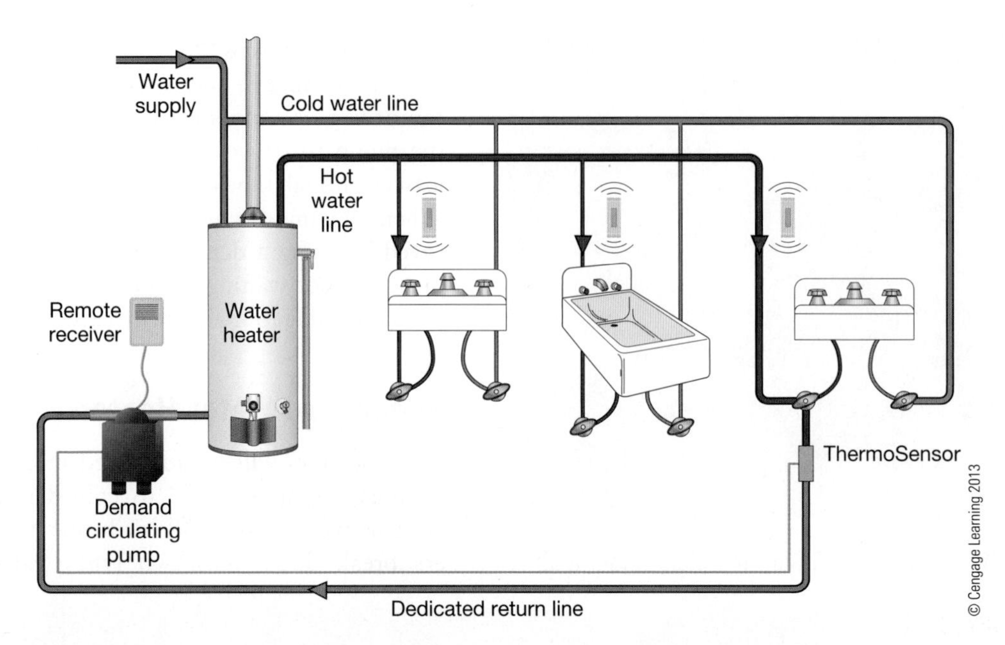

Figure 15.14 Circulating system with the pump installed at the water heater using a dedicated return line.

(Figure 15.15). Tank heaters heat and store a large quantity of hot water in an insulated storage tank. Tankless heaters, on the other hand, heat water only when needed, require a minimum flow rate to operate, and do not store any hot water. Tankless heaters are generally not recommended to be used with continuous or timer pumps because they would cause the heater to operate for longer periods than for which they are designed.

Figure 15.15 Water heaters are classified as having a storage tank or being tankless.

Demand pumps are suitable for tankless heaters, but the piping system should be designed to allow for an adequate flow rate. A constricted system design with small-diameter pipes and numerous sharp turns can limit the total water flow so much that a tankless heater will not turn on.

Water Heaters

Residential water heaters are available in tank and tankless models. Tank heaters store water in an insulated tank that is kept hot with small burners or electric heating elements that run intermittently to maintain a constant temperature as water is used or as heat is lost during storage. Tankless heaters are capable of heating water as needed with large burners or heating elements as the water passes through narrow pipes on the way to fixtures (Figure 15.16). Solar water heating is covered in Chapter 16.

Heater Capacity

Tank heaters are capable of providing a specific amount of hot water per hour, referred to as the first-hour rating, which is determined by the tank size, efficiency rating, and the fuel type. Tank heaters are available in sizes ranging from as little as 10 gallons to as much as 120 gallons, with 40 or 50 gallons being the most common sizes in residential projects. If a household uses

---

tank heaters heaters that store a large quantity of hot water in an insulated storage tank.

tankless heaters heaters that heat water only when needed, require a minimum flow rate to operate, and do not store any hot water.

first-hour rating the amount of hot water in gallons the heater can supply per hour (starting with a tank full of hot water).

FROM EXPERIENCE

How Does Your Green Program Deal with Hot Water?

Gary Klein, Managing Partner, Affiliated International Management, LLC

Hot water is a system. It includes the water heater, the piping, the choice of fixtures and appliances, the water that runs down the drain, and the behaviors of the people using the hot water who, in turn, are influenced by the components of the system. The components must work together, as a system, to provide the desired hot water services.

Gary has been intimately involved in energy efficiency and renewable energy since 1973. One fifth of his career was spent in the Kingdom of Lesotho, the rest in the United States. He has a passion for hot water: getting into it, getting out of it, and efficiently delivering it to meet customer's needs. After recently completing 19 years with the California Energy Commission, Gary formed a new firm, Affiliated International Management LLC, to provide consulting services on sustainability through an international team of affiliates. Gary received his bachelor's degree from Cornell University in 1975 with an independent major in technology and society and an emphasis on energy conservation and renewable energy.

The goal is to create a water-saving, energy-conserving, and time-efficient hot water system. One often-ignored key to accomplishing this is to reduce the volume of water in the piping between the source of hot water and the hot water outlets. Sometimes the source of hot water is a water heater; sometimes it is a trunk line that has been primed with hot water. In addition to minimizing the volume in the branches and twigs, the volume in the trunk line should be kept as small as is practical. The result of applying these principles is to think about installing more than one water heater per building and to create zones in the hot water distribution system.

Several green programs and codes have begun to address the implementation of these principles, including LEED for Homes, National Green Building Standard (NGBS), WaterSense for Homes, International Association of Plumbing and Mechanical Officials' (IAPMO's) Green Plumbing and Mechanical Supplement, 2012 International Energy Conservation Code (IECC), and the International Green Construction Code (IgCC), Version 2. In addition to these national efforts, several state and local programs include recommendations to improve the performance of hot water distribution systems, including the City of Austin, Texas, and Build-It-Green in California. Each of these programs looks at the hot water distribution system a bit differently, with most offering optional credits or points for efficient hot water systems rather than mandating certain system requirements.

The NGBS gives the user three choices of system types: central core, home run manifold, and structured plumbing—each of which will waste very different amounts of water, energy, and time. The home run manifold option is allowed to waste more than three times the water of the central core option and six times that of the structured plumbing option, which gives the builder the most flexibility regarding the location of the hot water outlets. These large variations in allowable performance are odd, given that this comes from a standard in which the options should be very similar.

LEED for Homes does a very good job of limiting the volume of water in the piping by addressing the trunks, branches, and twigs. The effect of the rules is to force the installation of multiple water heaters and tighter plumbing cores, both of which are the right things to do. For some builders, however, LEED's rules are a bit too good, probably because the layout of the hot water distribution system is not established until late in the design process, at which time it becomes too difficult (expensive) to add a water heater or relocate the hot water outlets. There is some concern that the limitations on the trunk length are too restrictive, particularly for single-story homes. There is also some confusion about whether or not you can zone the demand-controlled pumping alternative (you can).

The IAPMO and IECC codes focus on limiting volume by either limiting the length of uninsulated piping (IECC) or by specifying the allowable volume (IAPMO and IgCC). The allowable amounts of water waste are relatively small; no more than 32 ounces (0.25 gallons) can be in the piping for central core and home run manifold systems, and no more than 16 ounces (0.125 gallons) is allowed when the source of hot water is a primed trunk line.

WaterSense takes a very different approach. This program says that no more than 0.6 gallons of cold water can come out of any hot water outlet before hot water can arrive. This includes not only the water in the faucet, showerhead, and piping but also the water that runs through a typical tankless water heater while it ramps up to temperature. To meet this measured volume when the building is finished, no more than 0.25 gallons can be present in the piping.

(Continued)

All of these programs either require or provide points for insulation on the hot water piping. Perhaps the biggest benefit of having pipe insulation is to increase the cool-down time of the hot water in the piping. Insulation on 1/2" pipe doubles this time from 10 to 20 minutes; on 3/4" pipe, this time triples from 15 to 45 minutes. This conservation strategy makes hot water more quickly available during "morning rush hour" and "evening plateau" periods of hot water use. The IAPMO and the IgCC have innovative requirements for pipe insulation: the wall thickness of the pipe insulation shall be equal to the nominal pipe diameter for all piping greater than 1/4" up to and including 2". For piping larger than 2", the minimum wall thickness shall be 2".

Virtually all of these programs prohibit the use of any control strategy for recirculation other than demand controlled, which is far and away the most energy-efficient way to prime trunk lines with hot water.

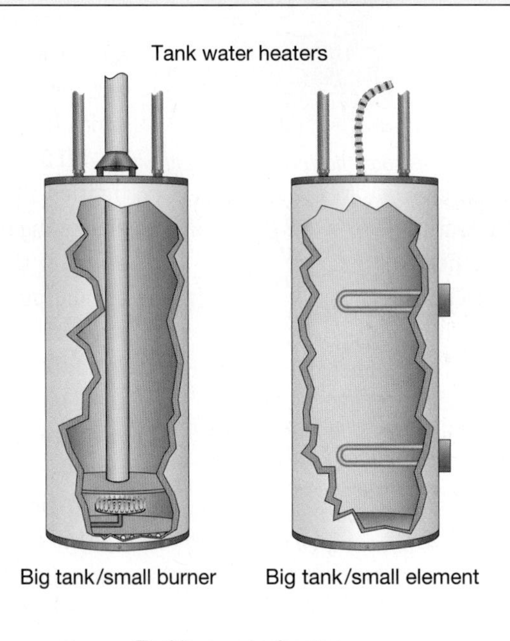

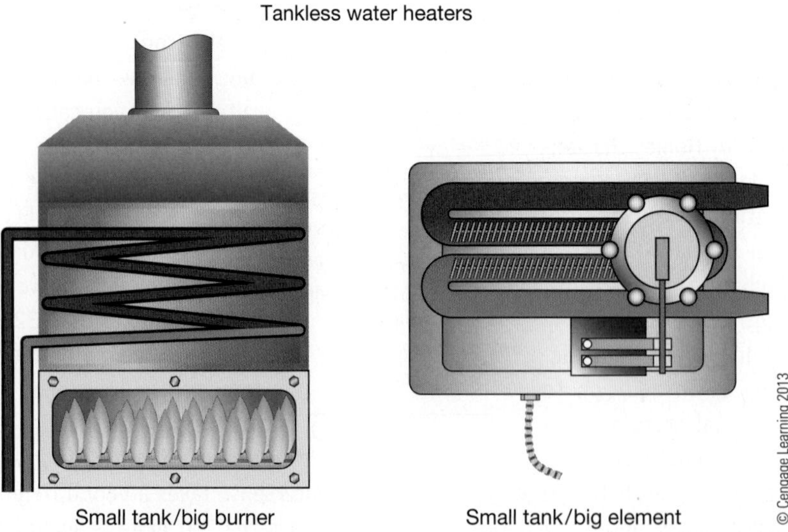

Figure 15.16 Tank water heaters contain heating elements or burners that are relatively small compared with those of tankless models, which must heat the water to the desired temperature nearly instantaneously.

more hot water in an hour than their heater is capable of delivering, the last users will not have enough hot water. The Energy Guide label on water heaters lists the first hour rating, providing consumers with information to help them select an appropriate size tank (Figure 15.17).

Tankless heaters provide continuous hot water at a specific number of gallons per minute (GPM) based on the amount of **temperature rise**, which is the difference

temperature rise the difference between the temperature of water entering the house and that of the hot water leaving the water heater.

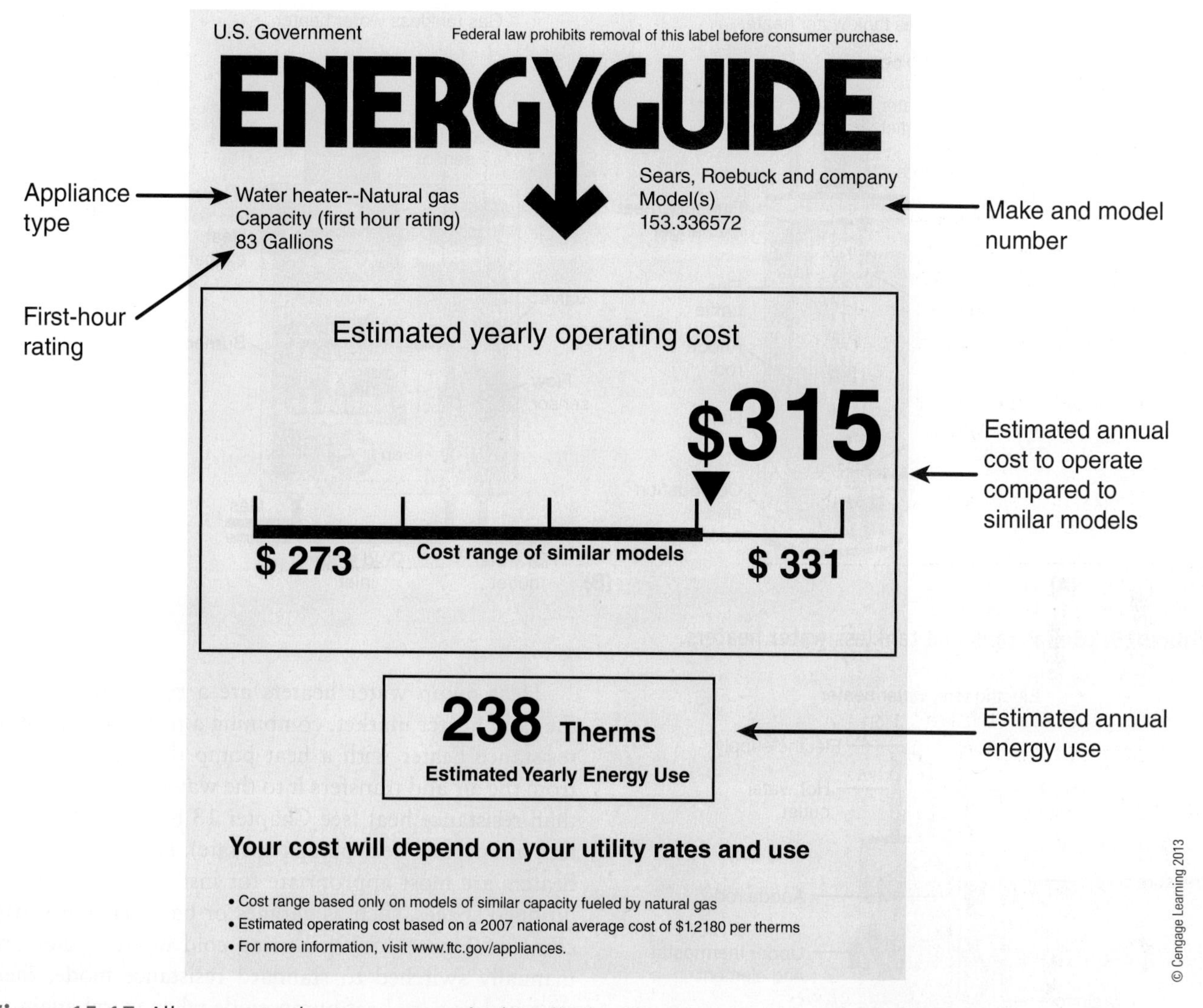

Figure 15.17 All new water heaters are required to display the EnergyGuide at time of sale.

between the water entering the house and the hot water leaving the heater. The larger the temperature rise, the lower the GPM for the heater. Manufacturers' literature lists maximum GPM ratings at different temperature rises. Although tankless heaters do not lose heat from stored water, their ability to provide continuous and endless hot water may encourage overuse of water. A study by the Partnership for Advanced Technology in Housing (PATH) determined that energy savings of 10% to 20% is possible by using tankless heaters on the basis of eliminating standby tank losses; however, usage patterns have a significant effect on overall energy consumption.[1] In a home that is unoccupied much of the day with only one or two residents, a tankless heater may provide noticeable savings, a home with more occupants and regular hot water use will not experience the same level of savings because the standby losses would be minimal. Tankless heaters require a minimum amount of water flow to operate. When the hot water tap is opened just slightly, as may occur when shaving or washing dishes, there may not be enough flow to make the heater operate.

Fuel

Most water heaters use gas or electric fuel to operate. Gas can be either natural gas or propane, depending on local availability. Gas tank heaters have a burner at the bottom, operated by a thermostat that turns on and off as required to maintain the desired tank temperature. Gas-fueled tankless heaters have burners that heat water as it flows through small pipes in the heater cabinet (Figure 15.18). Electric tank heaters have resistance heating elements that turn on and off to maintain the tank temperature; their tankless counterparts have heating elements wrapped around the pipes to heat the water as it flows through them (Figure 15.19).

[1] NAHB Research Center. http://www.toolbase.org/Building-Systems/Plumbing/tankless-water-heaters

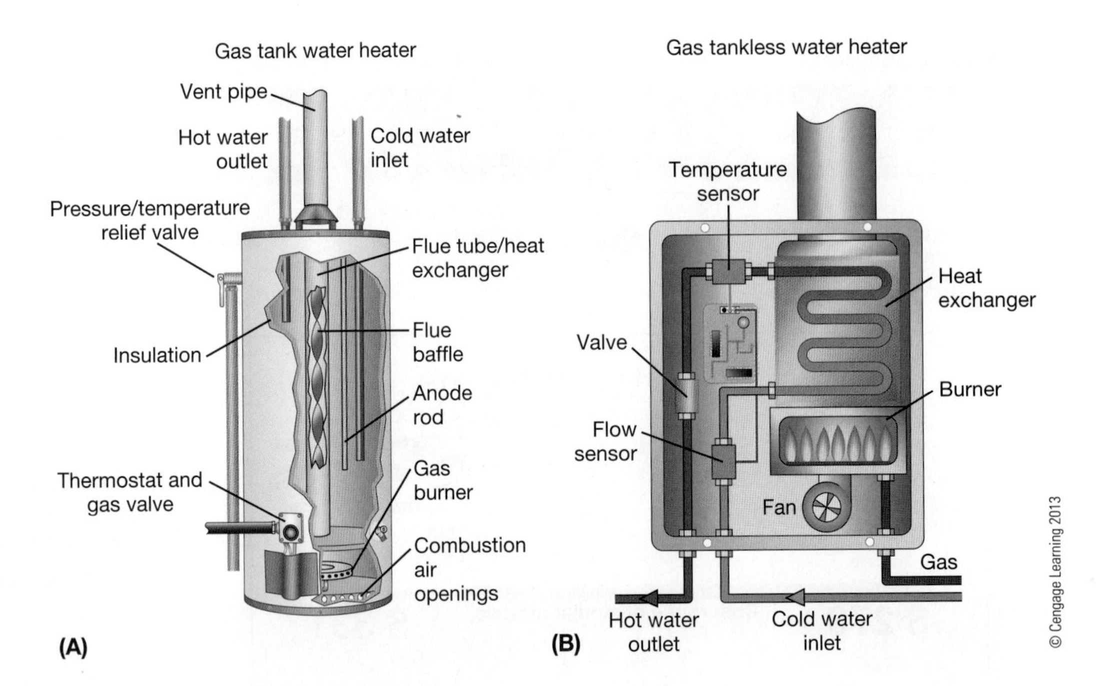

Figure 15.18 Gas tank and tankless water heaters.

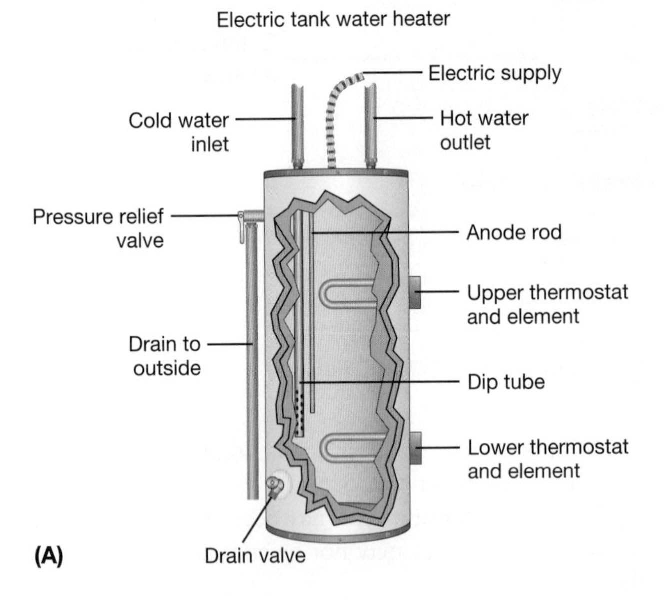

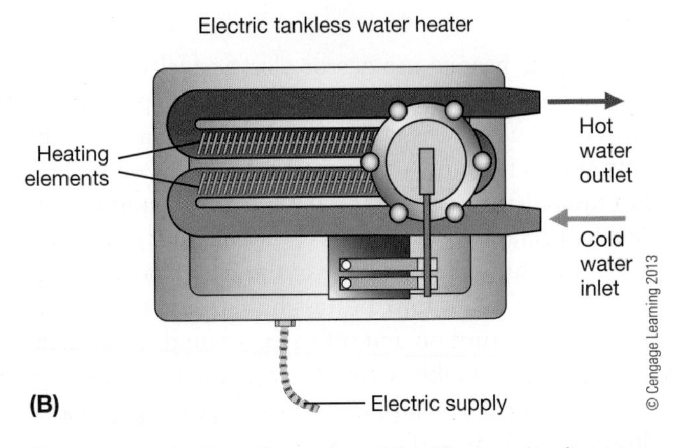

Figure 15.19 Electric tank and tankless water heaters.

Heat pump water heaters are a recent addition to the tank heater market, combining a traditional electric resistance heater with a heat pump that removes heat from the air and transfers it to the water more efficiently than resistance heat (see Chapter 13 for more information about how heat pumps operate). Heat pump water heaters are most appropriate for installation in unconditioned spaces, such as garages or basements in warm climates (Figure 15.20). During cold weather, they are manually switched to standard resistance mode, then switched back to heat pump mode when warm again.

Water heating can also be provided by ground-source heat pumps (see Chapter 13) through a desuperheater, a device that recovers excess heat from the cooling and heating process and diverts it to heat water in a storage tank very efficiently (Figure 15.21). In the cooling season, the desuperheater enables the ground-source heat pump to redirect the home's heat to a water heater storage tank instead of to the ground.

Heater Efficiency

Water heater efficiency is measured with the energy factor (EF), which takes into account the heating efficiency and standby losses. The higher a heater's EF, the more efficiently it operates. Gas water heaters have EFs

---

desuperheater a device that recovers excess heat from the cooling and heating process, diverting it to heat water in a storage tank very efficiently.

energy factor (EF) the energy efficiency rating for a water heater; based on the amount of hot water produced per unit of fuel consumed over a typical day.

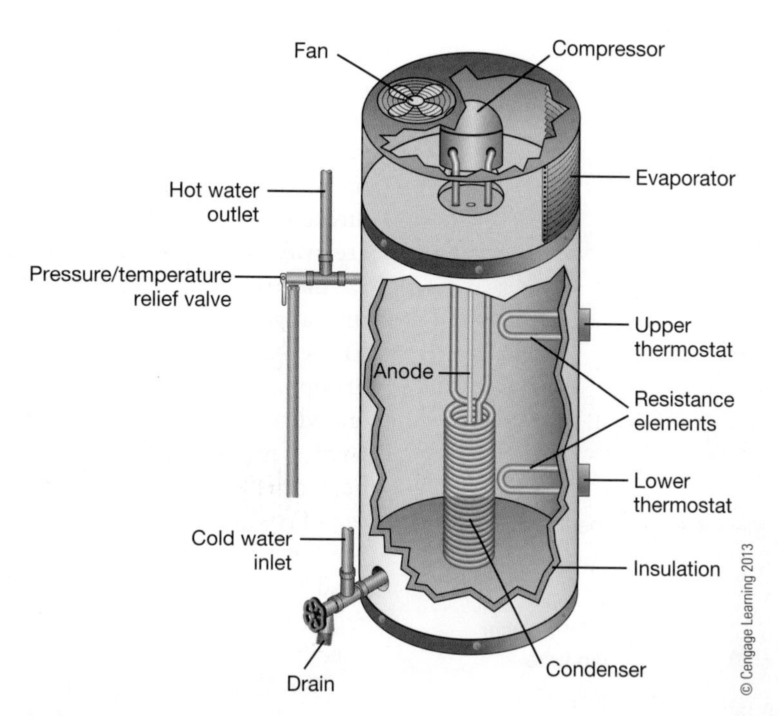

Figure 15.20 Heat pump water heater.

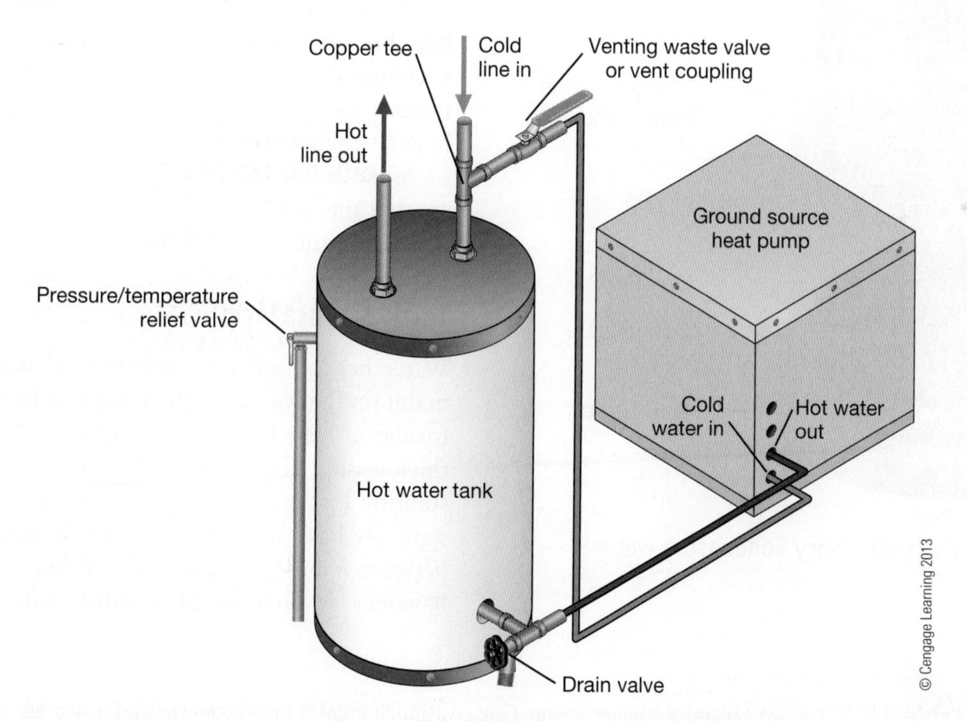

Figure 15.21 Typical desuperheater installation.

that range from 0.59 for open-combustion units with pilot lights to as high as 0.96 for sealed-combustion units. As discussed in Chapter 13, the use of open-combustion water heaters in well-sealed homes poses the risk of backdrafting and carbon monoxide poisoning. When using open-combustion heaters, they should be installed outside the building envelope. **Condensing heaters**, also known as sealed-combustion heaters, capture excess heat that is otherwise wasted and exhausted through the flue, causing water from burned fuel to form as condensation (Figure 15.22). Condensing heaters have burners that are isolated from the ambient air in the house, reducing the risk of backdrafting

---

condensing heaters high-efficiency water heaters that remove large amounts of heat from the flue gases, resulting in the condensing of the flue gases.

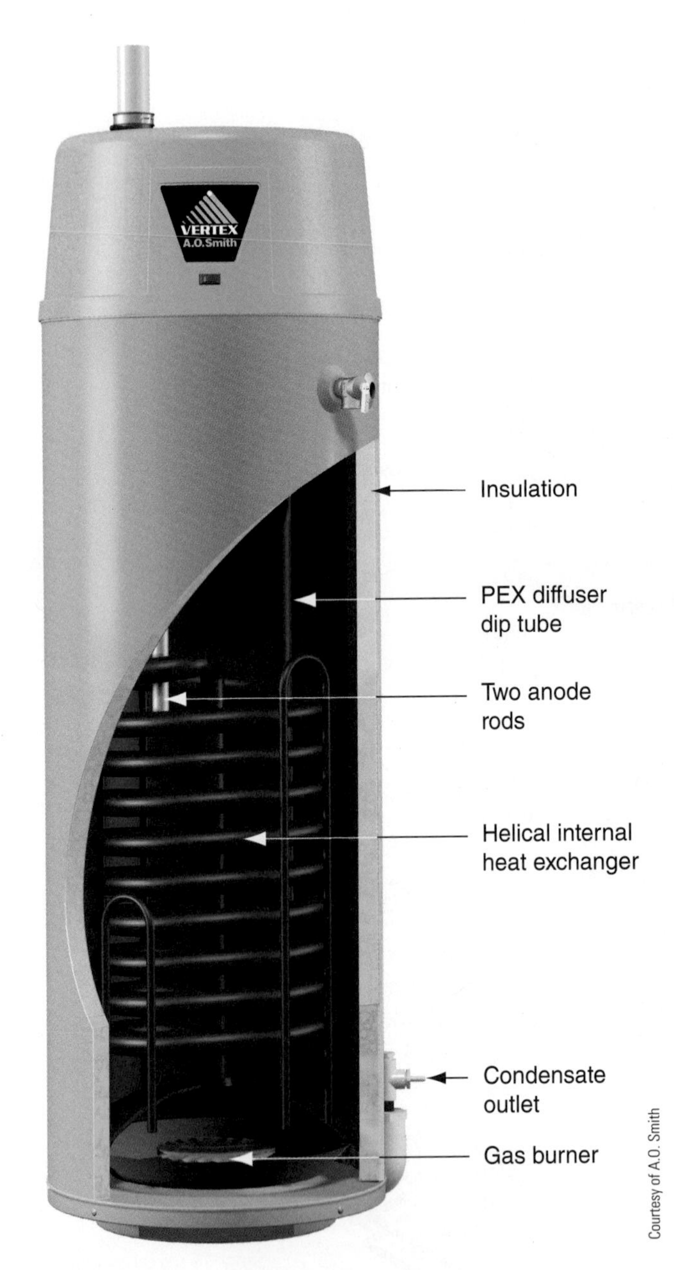

Figure 15.22 High-efficiency condensing water heater.

in tight homes. Both gas tank and tankless heaters are available as condensing units. Most gas tankless heaters are of the condensing type, with many using PVC vents instead of metal ones. Tank heaters are available in open-combustion types with traditional metal vents or direct vents; condensing or sealed-combustion heaters are available in power-vented models.

Electric resistance heaters typically have EFs in the range of 0.9 to 0.97. Heat pump heaters have an EF of as high as 2.0 when used in heat pump mode, dropping back to the EF of electric resistance heaters when operated in standard mode. Electric heaters have a higher EF, but the total fuel cost for electric heaters is generally higher than gas units, except in regions where electricity is very inexpensive and gas is expensive. The EF is a laboratory calculation that is used to compare different heaters and should not be regarded as a reflection of the amount of energy that a specific heater will use.

The U.S. Department of Energy (DOE) establishes national efficiency standards for water heaters that are based on fuel type, tank type, and tank size (Table 15.1).

In homes with central boilers used for space heating, hot water may be heated indirectly with heat exchange coils that run from the boiler to the water tank (Figure 15.23), in which case the tank usually does not have its own heat source.

Since 2008, ENERGY STAR has labeled water heaters that meet certain minimum efficiency and hot water delivery requirements (Table 15.2).

Tank Insulation

Water heater tanks are constructed with insulation to maintain the water temperature and to reduce heat loss. Insulation can be either fiberglass or foam, in varying thicknesses. The more insulation used, the lower the standby losses in tank heaters. Standard water heaters typically have integrated insulation with an R-value of between 8 and 16, while very efficient heaters may have R-values as high as 24. Heaters with less than R-16

Table 15.1 Current Federal Minimum Energy Conservation Standards for Residential Water Heaters

Water Heater Type	EF as of January 20, 2004
Storage, gas-fired	0.67 – (0.0019 × rated storage volume in gallons)
Storage, oil-fired	0.59 – (0.0019 × rated storage volume in gallons)
Storage, electric	0.97 – (0.00132 × rated storage volume in gallons)
Tabletop	0.93 – (0.00132 × rated storage volume in gallons)
Instantaneous, gas-fired	0.62 – (0.0019 × rated storage volume in gallons)
Instantaneous, electric	0.93 – (0.00132 × rated storage volume in gallons)

Source: Energy Conservation Standards for Residential Water Heaters, Direct Heating Equipment, and Pool Heaters: Final Rule, Federal Register, 75 FR 20112, April 16, 2010.

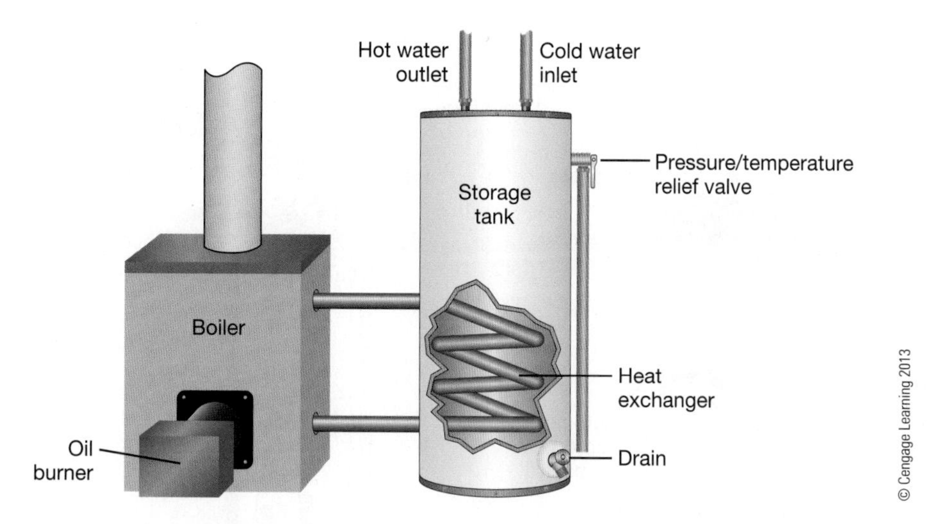

Figure 15.23 Indirect water heating.

Table 15.2 ENERGY STAR Water Heaters

ENERGY STAR Criteria	≥6 Years on Sealed System EF	Solar Fraction	First-Hour Rating	Gallons per Minute	Warranty
High-efficiency gas storage					
Gas storage	≥0.67	NA	≥67 gallons per hour	NA	≥6 years on sealed system
Gas condensing	≥0.8	NA	≥67 gallons per hour	NA	≥8 years on sealed system
Heat pump	≥2.0	NA	≥50 gallons per hour	NA	≥6 years on sealed system
Whole-home gas tankless	≥0.82	NA	NA	2.5 over a 77°F rise	≥10 years on heat exchanger; 5 years on parts
Solar	NA	≥0.5	NA		≥10 years on solar collector; 6 years on storage tank; 2 years on control; 1 year on piping, parts

Source: U.S. EPA ENERGY STAR program, http://www.energystar.gov

insulation can have an insulation blanket installed to further improve efficiency (Figure 15.24). In addition, installing insulation on the first 3' to 4' of hot water pipe exiting the heater can help to save energy.

Heat Traps

Used only on tank heaters, heat traps are special one-way valves or loops of pipe that keep hot water from flowing naturally out of the top of the heater via convection, saving energy in the process (Figure 15.25). Many heaters include heat traps as part of the tank. When not integrated into the tank, these traps can be installed separately.

heat traps special one-way valves or loops of pipe that keep hot water from flowing naturally out of the top of the heater via convection, saving energy in the process.

Heater Temperature

In the past, water heaters were often set at 140°F, usually to provide hot enough water for a dishwasher to work properly. Water heated to 140°F wastes energy and creates a scalding risk. Most dishwashers have integrated heating elements that pre-heat water, allowing us to keep our water heaters set safely at 120°F.

Water Heater Maintenance

Water heaters do not require significant maintenance. Both tank and tankless heaters can benefit from regular cleaning to remove debris from tanks or pipes. Where water is very hard, lime may form and reduce heater efficiency, leading to premature failure of tankless heaters. Follow manufacturers' instructions for maintenance for maximum performance.

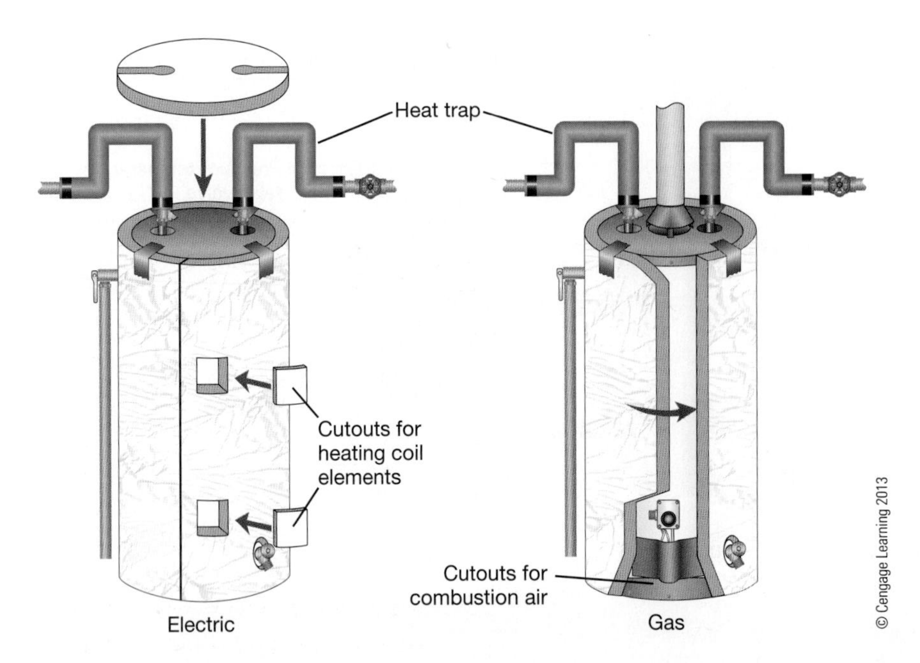

Figure 15.24 Water heaters contain some insulation built into the tank, but more insulation may be added on the exterior.

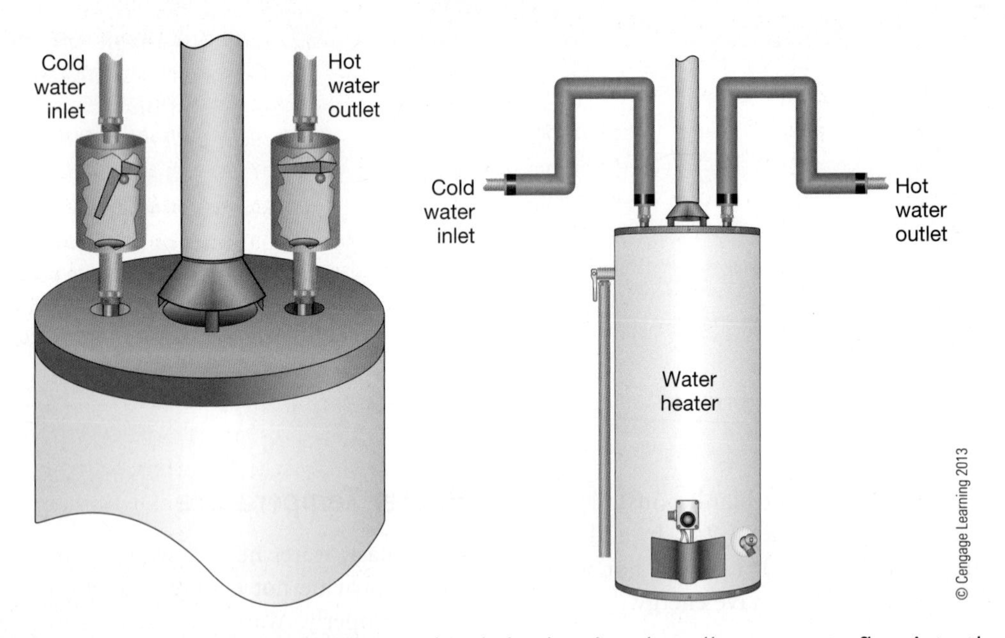

Figure 15.25 Heat traps may be one-way valves or bends in the pipe that allow water to flow into the water heater tank while preventing unwanted hot water flow out of the tank.

Instant Hot Water Dispensers

Instant hot water dispensers are a luxury item installed in many kitchens. Employing a tank that keeps a small supply of water hot at all times, they operate like a small electric water heater to provide hot water for tea, coffee, and similar needs. Manufacturers claim that they use the equivalent of a 40-watt incandescent lamp to operate. Such dispensers can save water if there is a long wait for water at a faucet, but water can usually be heated more efficiently when needed on the stove or in a microwave oven.

Wastewater Removal

In addition to inlets for fresh water supply, all plumbing fixtures have outlets for drain water that must be collected and delivered to a central location for treatment.

Courtesy of Scotts Valley Water District

Figure 15.26 Recycled water may be used for toilets and irrigation, but additional warnings may be required.

That central location can be either a municipal treatment facility or an on-site treatment system where there is no central sewer treatment facility. Municipal sewer treatment plants collect raw sewage from residential, commercial, and industrial buildings at a central location where it is filtered, sanitized, and released back into local waterways. Waste treatment facilities in some areas provide recycled or reused water to local residents for non-potable uses, such as irrigation. Warnings and appropriate signage must be present to prevent health risks when recycled water is used because it is not treated to potable standards (Figure 15.26). Municipal recycled water systems are increasingly common in the arid portions of the United States.

Black Water and Gray Water

Wastewater is classified as black or gray. Black water comes from toilets and kitchens and contains food and human waste that must be fully treated and sanitized before being released into waterways or reused. Most homes combine black and gray water in the same drain pipes, but separate gray water treatment and reuse systems are becoming more common. Gray water drains, usually colored purple to distinguish from those receiving black water, are separated from black water drains throughout the house and brought to a central location for connection to a treatment system. Some communities require that gray and black water be separated, and they provide treated gray water supply separate from potable water for use in homes.

---

black water waste that comes from toilets and kitchens containing human or food waste, which must be fully treated and sanitized before being released into waterways or reused.

Did You Know?

The first water reclamation system in the United States was built in St. Petersburg, Florida, and it remains one of the largest in the world. The city's system provides more than 37 million gallons per day to over 10,600 customers, primarily for lawn irrigation.[2] The city also has 316 fire hydrants that utilize reclaimed water. Another well-established reclamation system is in Tucson, Arizona.[3] The system delivers reclaimed water to about 900 sites, including 18 golf courses, 39 parks, 52 schools, and more than 700 single-family homes.

On-Site Treatment

Septic systems are underground treatment systems for household sewage in rural areas and other locations where municipal sewage services are unavailable. The most common treatment systems are anaerobic septic systems, where solids are removed in the absence of oxygen with no direct treatment of effluent, the wastewater produced. Alternatives include more complicated systems, such as aerobic systems and constructed wetlands. Aerobic septic systems mechanically inject air into a waste collection tank, encouraging decomposition to provide a higher-quality effluent. Constructed wetlands simulate natural wastewater treatment, allowing effluent to flow through water beds filled with plants that break down contaminants (Figure 15.27).

The standard on-site anaerobic treatment method is a septic system consisting of a holding tank and a drain field consisting of perforated pipe buried in an absorptive soil (Figure 15.28). The holding tank provides a place for

---

septic system an underground treatment system for human sewage.

anaerobic septic system a wastewater treatment system in which solids are removed in the absence of oxygen with no direct treatment of effluent.

effluent wastewater flow before or after treatment.

aerobic septic system a wastewater treatment system that mechanically injects air into a waste collection tank to encourage decomposition, which provides a higher-quality effluent.

constructed wetlands a means of treating wastewater that simulate natural wastewater treatment, allowing effluent to flow through water beds filled with plants that break down contaminants.

drain field the final component of a septic system in which effluent flows through perforated pipe into the soil, where it is filtered through the ground as it moves downward towards the local aquifer.

holding tank the component of septic systems that allows solid waste to settle while letting liquid effluent flow out to the soil.

---

[2] City of St. Petersburg, Florida. http://www.stpete.org/water/reclaimed_water/index.asp

[3] City of Tuscon, Arizona. http://cms3.tucsonaz.gov/water/reclaimed

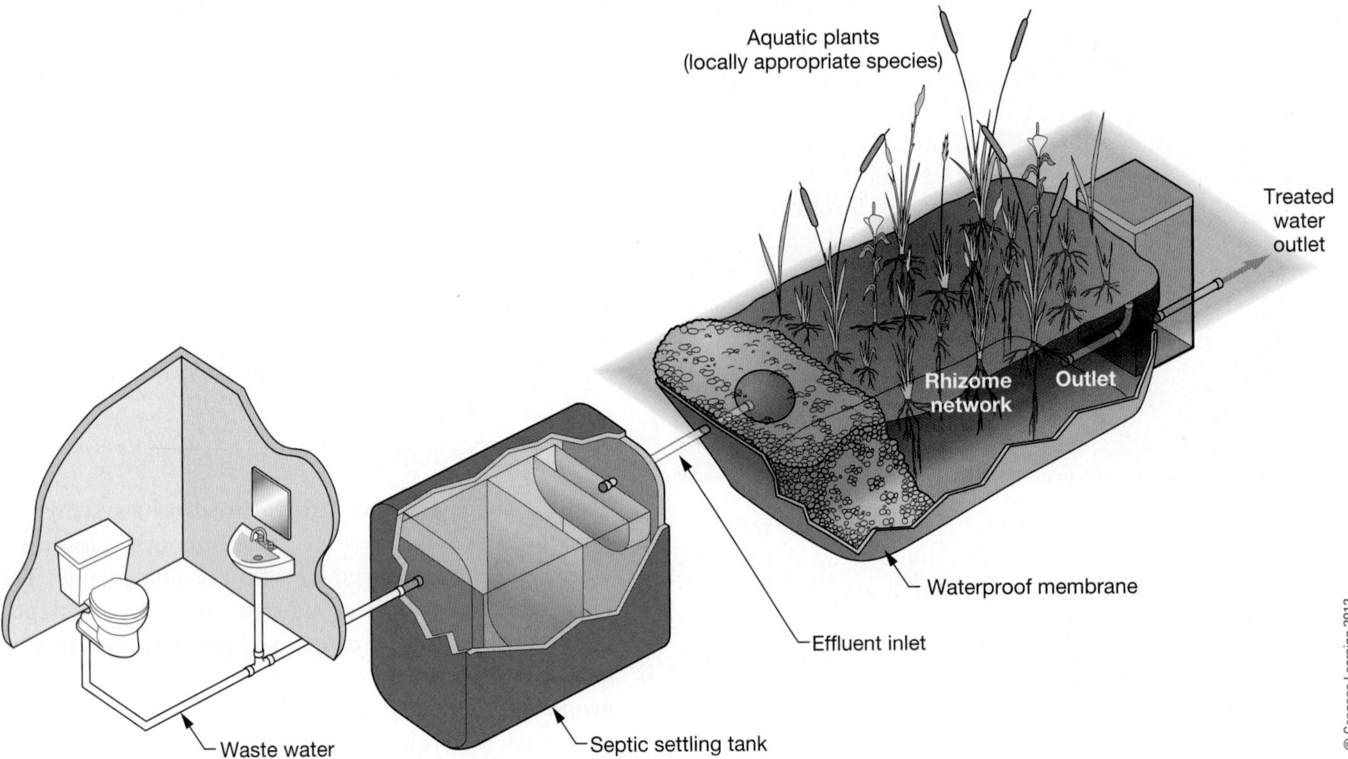

Figure 15.27 Constructed wetlands mimic natural ecosystems to treat wastewater on-site and provide a wildlife habitat.

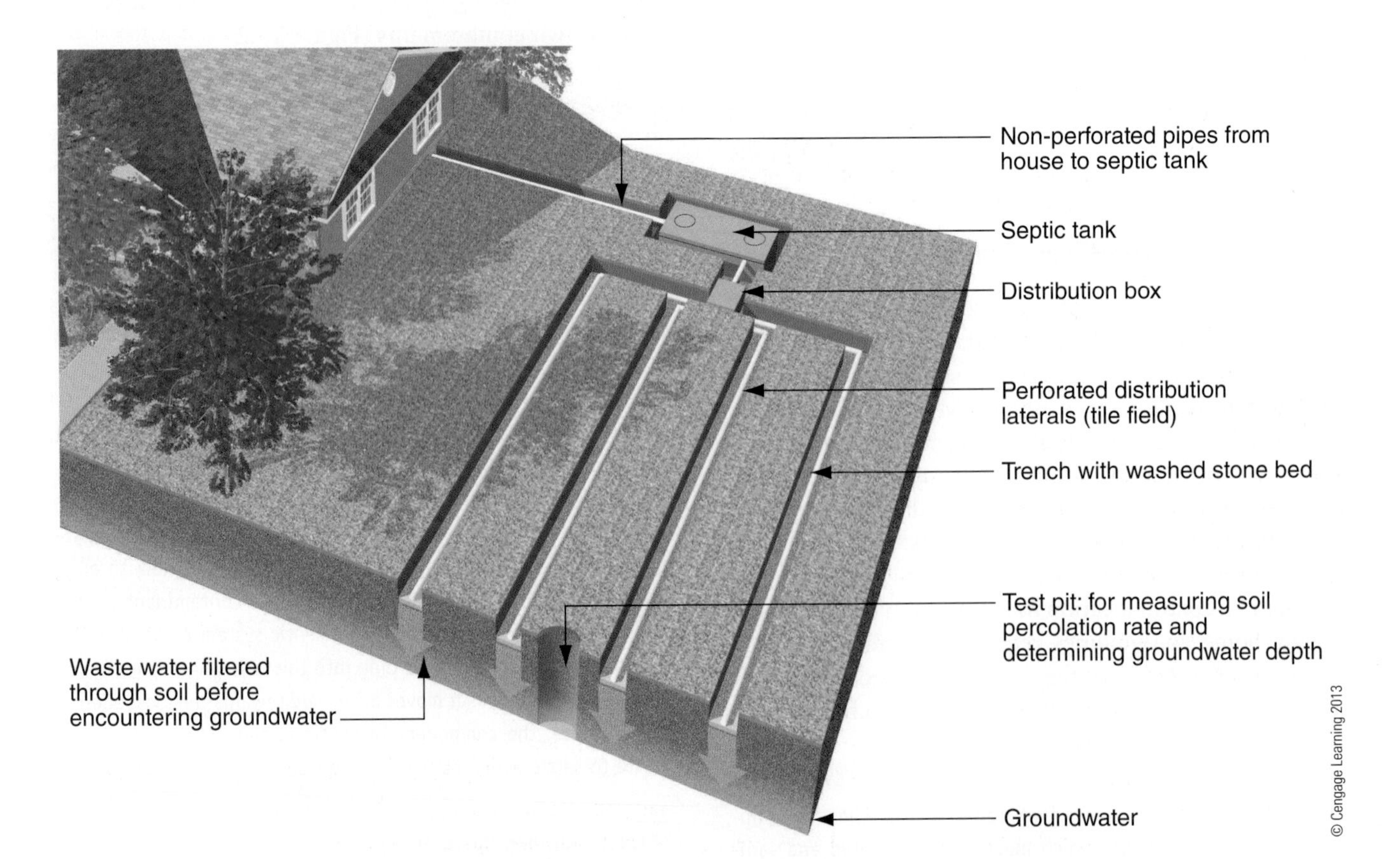

Figure 15.28 A conventional septic system.

solid waste to settle while letting liquid effluent migrate to the drain field where it flows through perforated pipe into the soil. The wastewater is filtered through the ground as it moves downward toward the local aquifer. Septic systems can only be installed where there is an adequate amount of soil that is porous enough to accept the effluent produced. The porosity of the soil is determined by a percolation ("perc") test. Percolation tests determine the speed at which soil will absorb wastewater. The rate of percolation and the size of the building determine the size of the drain field required. The more porous the soil and the fewer number of bedrooms in a house, the smaller the field can be. Septic tanks require maintenance, including periodic removal of collected solid waste. Cleaning frequency depends on the size of the tank and the number of people using the system. High-use systems with small tanks may need to be cleaned yearly, large tanks with light usage may last many years between cleanings. Improperly maintained septic systems can lead to overflows, excessive amounts of pollutants entering soil and groundwater, and in some cases, complete system failure.

Aerobic systems can be installed when the amount and porosity of the soil does not allow for a traditional septic drain field. Compared with a septic system, aerobic systems offer increased rates of decomposition due to injected air that reduces the size of the drain field required. Aerobic systems can be used to retrofit a failed septic system or where conditions preclude the installation of a septic system, such as when there is not enough available land or the groundwater level is too high. Aerobic treatment is more complicated than anaerobic systems, requiring sewage pretreatment to reduce the amount of solids, followed by an aeration process, final treatment, and disinfection. Aerobic systems also consume electricity to operate the pump used to inject air into the tank. Typically costing two to three times more than anaerobic systems, aerobic systems require less excavation and less land area, and they can reduce groundwater pollution. Their versatility can allow for construction on a site that would otherwise not be buildable due to unsuitability for a septic system. Aerobic systems require more maintenance than standard septic systems and can fail if the power is disconnected or if harmful chemicals are introduced into the system.

Constructed wetlands use the natural tendency of the roots of certain plants to provide an aerobic environment that breaks down contaminants in sewage effluent. Similar to aerobic systems, they can be a suitable alternative where site conditions preclude the use of a septic system. Constructed wetlands are typically custom-designed for a specific site. They can either discharge the treated water to the surface or retain it all underground. Similar to septic and aerobic systems, they use holding tanks to remove solids but do not exclusively use drain fields to release effluent below ground. In addition, they employ pumps, pipes, and lined ponds filled with plants that thrive in water, and, if required by local authorities, a drainage field. Water treated in constructed wetlands can be of very high quality and may be suitable for release into waterways. Because wetlands rely on constant flow of effluent, they are not suitable for seasonal residences. In cold climates, ponds must be large enough to prevent freezing, and steep slopes may require significant amounts of cut and fill to slow water flow. When properly constructed, wetlands are low maintenance and long-lasting solutions to on-site sewage treatment.

Wastewater Piping

Most wastewater piping installed in buildings is made of PVC, acrylonitrile butadiene styrene (ABS) (an alternative plastic pipe) or cast iron (Table 15.3). Exterior pipe can be made of PVC, ABS, or vitrified clay pipe, a dried and fired clay product. Both PVC and ABS are made from petroleum products and produce toxic chemicals in the manufacturing process, although ABS poses less of a health risk to workers because a safer solvent is used for site assembly of pipe and fittings. Cast iron has a high recycled content, but the manufacturing process uses coke, a coal by-product that produces toxic and carcinogenic emissions in its manufacture. From both an embodied energy and a toxicity standpoint, plastic pipe is considered preferable to cast iron. One of the primary benefits of cast iron drainpipe is its sound-dampening qualities. Plastic drainpipe must be thoroughly insulated to provide the same level of sound reduction as cast iron.

Drain System Design

Just as centrally locating bathrooms reduces the amount of water supply pipe required, it also saves on the amount of drainpipe required. Where gray water is reclaimed,

---

percolation test the speed at which soil will absorb wastewater, also known as a *perc test*.

septic drain field a drain field used to remove contaminants and impurities from the liquid that emerges from the septic tank.

acrylonitrile butadiene styrene (ABS) rigid piping material used for drain lines.

vitrified clay pipe plumbing pipe made from clay that has been subjected to vitrification, a process that fuses the clay particles to a very hard, inert, glass-like state.

coke a solid byproduct from the combustion of coal.

Table 15.3 Comparing Various Pipe Materials

Pipe Material	Weight of 4" Pipe (lb/ft)	Embodied Energy of Mat'l (Btu/lb)	Embodied Energy of 4" Pipe (Btu/ft)	Cost of 4" Pipe ($/ft)
PVC (4" Sch. 40)	2.0	34,000	68,000	1.20
ABS (4" Sch. 40)	1.6	47,700	75,843	1.20
Cast iron (4" no hub)	7.4	14,891	110,200	3.00
Vitrified clay (4")	8.9	2,706	24,080	1.50

Source: Adapted from Building Green, http://www.buildinggreen.com/auth/article.cfm/1994/1/1/Should-We-Phase-Out-PVC/

separate drain systems for black and gray water must be installed. Gray water drain lines are typically required to be purple in color to distinguish them from black water lines. In addition to drainpipe, waste systems must also have vents that prevent suction from forming in drains, allowing them to operate properly and quietly.

Vent pipes normally run through the roof of a house to allow air to enter the system. However, the use of **air admittance valves**, special one-way valves that allow air to enter the vent system as needed, can reduce the amount of pipe required and limit the total number of roof penetrations (Figure 15.29).

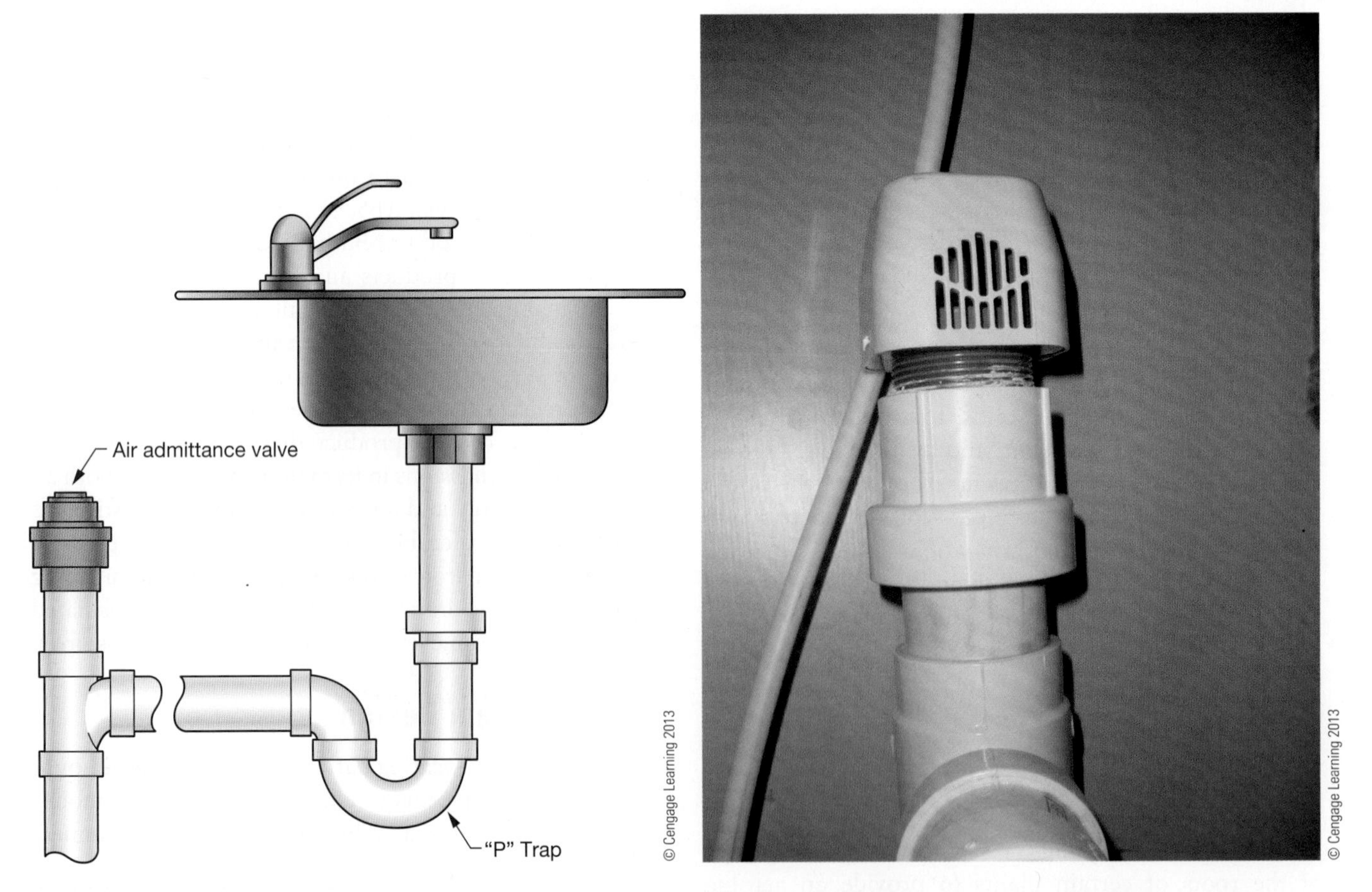

Figure 15.29 Although air admittance valves are common in locations where conventional roof venting is not feasible, they are beneficial in many other applications.

air admittance valve a one-way valves that allows air to enter the plumbing vent system as needed.

Drain-Water Heat Recovery

Whenever we use hot water, most of the heat in it is lost as it goes down the drain. Drain-water heat recovery (DHR) systems can be used to recapture a significant portion of that heat through the use of a gravity-film heat exchanger (GHX), which is a special metal drain pipe wrapped in coils through which fresh water flows, allowing the heat to transfer from the drain-water to the supply water (Figure 15.30). DHR systems cannot be installed in single-story slab-on-grade homes, but they work in two-story homes or one-story homes with basements. They are simple devices that work effectively with showers. Energy savings between 16% and 30% have been reported from several studies completed between 1997 and 2006. The typical DHR system does not store water but rather heats incoming water when a shower or sink is running, making it less effective for use with bathtubs. Storage tank systems that capture heat from all hot water draining from fixtures are available, but more expensive and less common.

Gray Water Reclamation

The Water Conservation Alliance of Southern Arizona estimates that the average household can save between 30,000 and 50,000 gallons of fresh water annually through the use of a whole-house gray water system. Residential gray water systems are available in several configurations, ranging from a tank that collects water from a single sink to supply an adjacent toilet (Figure 15.31),

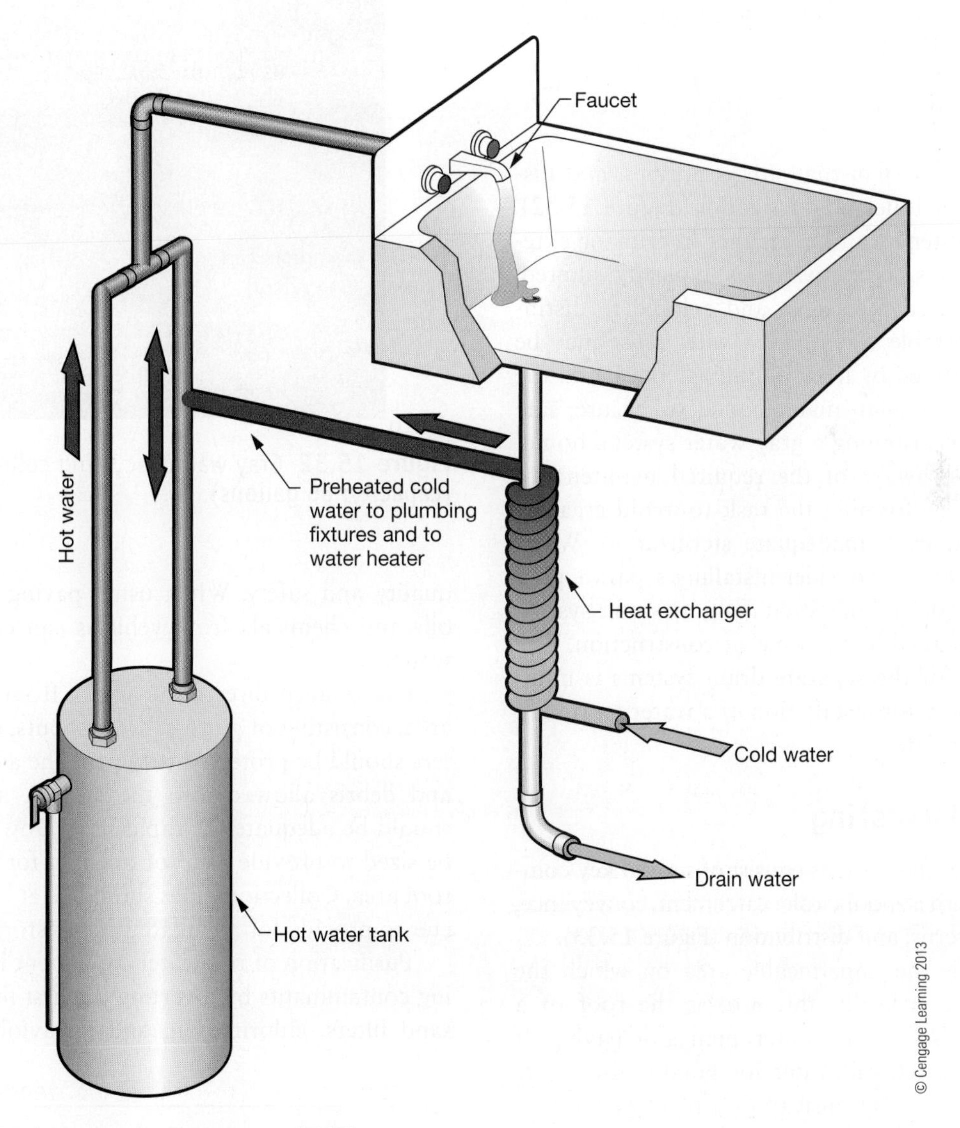

Figure 15.30 Drain-water heat recovery (DHR).

drain-water heat recovery (DHR) the use of a capturing heat for reuse from wastewater.
gravity-film heat exchanger (GHX) a heat transfer device used with drain-water heat recovery systems.

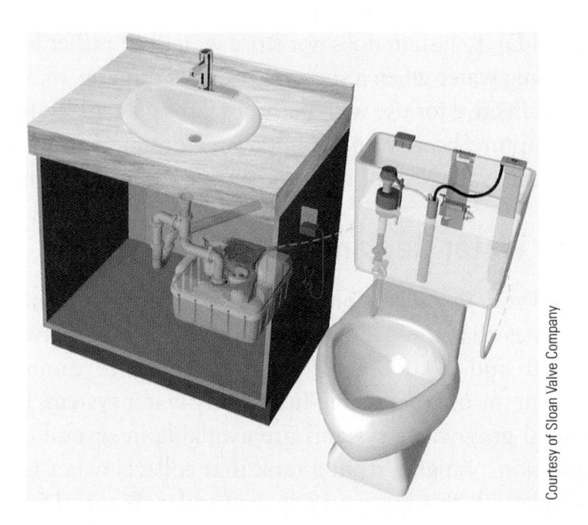

Courtesy of Sloan Valve Company

Figure 15.31 This gray water system reuses wastewater from the sink to flush the toilet. The simple system includes a small filter and pump.

© Cengage Learning 2013

Figure 15.32 Gray water recycling collection system (capacity: 66 gallons).

to a whole-house system that stores, purifies, and distributes water for toilets and irrigation (Figure 15.32). Whole-house systems can have either interior or exterior storage tanks. Gray water is typically filtered, chemically sterilized for safety, and colored to distinguish it from potable water. Gray water reuse may be limited or prohibited by local plumbing codes, and the systems require regular maintenance to ensure safe operation. When installing a gray water system, homeowners must be aware of the required maintenance and capable of performing the task to avoid creating a health hazard from inadequate sterilization. When building a new house, consider installing separate gray and black water drain lines, even if a gray water system is not being installed at the time of construction. The additional cost for the separate drain systems is minimal and will allow for installation of a water treatment system at a later date.

Rainwater Harvesting

Rainwater harvesting systems consist of several key components, each with a specific role: catchment, conveyance, purification, cisterns, and distribution (Figure 15.33).

Catchment is the impermeable area on which the rainwater lands. Typically, this area is the roof of a house but can also be rock outcroppings or paving. If the intent is to catch rainwater for potable use, roofs should be made of metal instead of asphalt or plastics, and lead flashing should be avoided to ensure water quality and safety. When using paving for catchment, oils and chemicals from vehicles can contaminate the water.

Conveyance directs the water from the catchment area, consisting of gutters, downspouts, and pipes. Gutters should be protected to reduce the amount of leaves and debris allowed into the system, and their slope should be adequate for rapid flow. Downspouts should be sized to provide 1 in^2 of opening for each 100 ft^2 of roof area. Collection pipes should be at least 4" in diameter to provide for adequate flow to storage.

Purification of rainwater catchment involves removing contaminants by diverting the first flush or by using sand filters, chlorination, and ultraviolet sterilization.

catchment the impermeable area on which rainwater lands.

conveyance a system that directs rainwater from the catchment area, consisting of gutters, downspouts, and pipes.

purification the process of removing contaminants from collected rainwater by diverting the first flush or by using sand filters, chlorination, and ultraviolet sterilization.

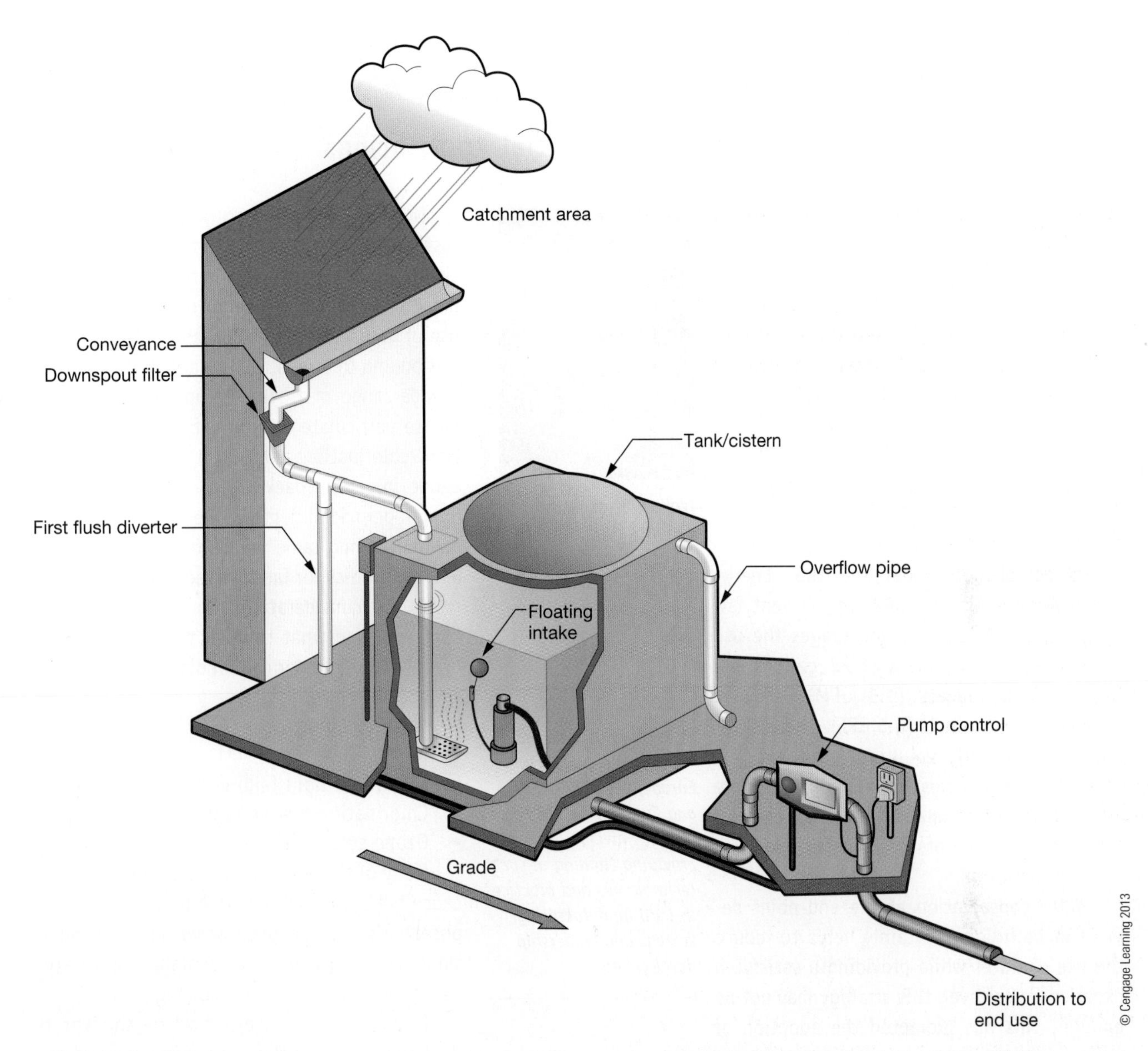

Figure 15.33 Typical rainwater harvesting system. Courtesy of the Georgia Rainwater Harvesting Guidelines, 2009; adapted with permission.

The **first flush** is the initial flow of water coming off a roof that often includes bird droppings, leaves, and other debris that is diverted away from the storage containers for cleanliness.

Cisterns are used for storing the collected rainwater. They can be made of metal, plastic, fiberglass, wood, or concrete. Flexible fabric bladders, like the Rainwater Pillow, can be installed in crawl spaces or under decks where underground or above-ground tanks are not appropriate. Simple systems may use rain barrels placed at individual downspouts (Figure 15.34). Tanks may be installed above or below ground (Figure 15.35). Most underground tanks are made of plastic or fiberglass. Plastic tanks installed above ground should be made of an ultraviolet-resistant material to avoid premature deterioration. If water is for potable use, tanks must be approved by the U.S. Food and Drug Administration. Cisterns can be sized according to the **supply approach**, in which a tank large enough to provide the total monthly need for a house is installed, or the **deficit approach**, which

first flush the initial flow of water coming off a roof that often includes bird droppings, leaves, and other debris.
cisterns used for storing collected rainwater.

supply approach a method of sizing cisterns in which a tank large enough to provide the total monthly need for a house is installed.
deficit approach a method of sizing cisterns, which involves calculating the total need during periods of little or no rainfall.

FROM EXPERIENCE

Gray Water in Residential Applications

William L. Strang, Senior Vice President of Operations, TOTO

Water availability is a significant challenge for our world and will continue to be in the next several decades. In the United States, a strong move to adopt water conservation is driving a range of technical solutions in residential, institutional and commercial installations. One example is the U.S. EPA WaterSense program, which provides consumers with water-conserving bathroom products that reduce consumption below the requirements established under the U.S. Energy Policy Act (EPAct) of 1992, the current law of the land. WaterSense encourages the use of high-efficiency toilets (1.28 gallons per flush [GPF]), faucets (1.5 GPM), showers (2.0 GPM), and urinals (0.5 GPF). These standards are currently voluntary on a national level, but many states and municipalities are adopting these as mandatory requirements, including the city of Los Angeles and the state of California.

Courtesy of William Strang, Toto

William Strang's experience spans many industries, including automotive engine manufacturing, military submarine components, pharmaceuticals, petrochemicals, and agricultural equipment manufacturing. He has studied and worked in Europe, the Middle East, and Canada. He has also taught university classes, providing training in such technologies and practices as lean manufacturing, kaizen, and leadership development.

Water conservation at the end-point devices in bathrooms certainly helps to reduce the use of water while providing a satisfying experience; however, this strategy may not be enough. This has prompted the adoption of gray water systems to recycle household water for other uses. Gray water presents a unique array of challenges, which vary with the source of the water:

- Bathroom sinks (the kitchen sink is not recommended due to the heavy bio loads)
- Bathtub and shower drains
- Clothes washers

Each one of these effluent water streams bring with it a range of by-products that require some type of filtering, cleaning, and treatment before the water can be rerouted for reuse. The greatest opportunity to collect water used in the United States is from the shower. Studies indicate that U.S. residents spend an average of 7.6 minutes to 10 minutes in the shower each day. At a flow rate of 2.5 GPM, a showerhead produces 19 to 25 gallons of water in a single shower. For a family of four, that can amount to 76 to 100 gallons per day.

Now for the most important decision in the system's design: Shall we use the gray water inside the home (flushing toilets) or outside the home (irrigating the lawn or garden)? Internal use of gray water requires a higher level of filtration and treatment. Gray water systems have a wide range of treatment options, including coarse particulate filtration, fine carbon filtration, chlorination, ultraviolet light treatment, ozone injection, backflow preventers, and hydrogen peroxide injectors. The system may also have a holding tank, pumps for head pressure, and electronics for fault protection.

Other considerations in system design center around what to do if the system must operate under abnormal conditions, such as:

- Loss of power
- Filter clogging
- Pump failure
- Ultraviolet light failure
- Chlorination system failure
- Ozone system failure

Many of these faults or system failures can cause the system to not work properly and expose the system to poor water quality conditions. This may expose residents to unsanitary conditions, which is not recommended.

Of significant concern is where the water will be used in the home. If the gray water is used to flush toilets, several very important issues must be addressed. When routing water into the toilet tank, chlorine levels in the water should not exceed 5 parts per million because excessive chlorine will destroy the internal parts of the toilet tank. Water should not contain particulates, lint, hair, or other contaminant that may cause clogging of internal fill valves and flush valve mechanisms. Oils or lighter-than-water compounds from the shower also present challenges and should not be in the water supply to the toilet tank. Most importantly, treated gray water should not be held in a tank for longer than 72 hours. This limit prevents the incidence of stagnation since oils can cause a film to build up on the water surface, reducing oxygen exchange. This buildup allows the water to become septic, an anaerobic condition that can promote bacteria growth in the water.

(Continued)

Homeowner Challenges

Many of the turnkey gray water systems on the market today require routine maintenance. This may include adding chlorine (via tablets or injector); cleaning coarse strainers, filters, or lenses on ultraviolet light sources; and servicing the pumps that provide head pressure to route the treated gray water to points of use in the home.

Unfortunately, many homeowners are terrible at home maintenance. Some do not realize what is needed, and others are not motivated. In fact, many homeowners do not realize that their refrigerator icemaker has a filter that must be routinely maintained to ensure quality of the water that they drink. To expect homeowners to service and maintain a gray water system located somewhere in the basement will be a hurdle that needs to be addressed. Because routine maintenance is critical to the success of any mechanical system, it also presents an opportunity for system designers to find clever low-cost solutions that will eliminate or minimize homeowner obligations.

Certainly many of these challenges are lessened if gray water systems are used to provide water to subsurface irrigation lines, often with just a simple lint or hair filter to keep the lines free of obstructions. With subsurface irrigation, the requirements for treatment are dramatically lessened; in some cases, a simple gravity feed can achieve movement of the water from inside the home to the garden or lawn outside. Using graywater for external purposes tends to lessen not only the cost of the system but also the challenges to the homeowner's maintenance skills.

Figure 15.34 Rainwater catchment may be as simple as a 55-gallon rain barrel under a gutter downspout.

involves calculating the total need during periods of little or no rainfall. Using the supply approach, you may calculate that a family of three requires 6500 gallons per month, suggesting a minimum tank size of 6500 gallons. Using the deficit approach, you may determine that, during a typical three-month dry spell with no significant rainfall, the same family of three will require 19,500 gallons (6,500 gallons per month × 3). This suggests a tank size of at least 19,500 gallons.

Distribution is the piping that delivers harvested and filtered water for use. Using the same materials and system designs as fresh water supply, distribution systems move collected water from cisterns to its final use. Harvested water can be used in irrigation systems, in toilets, or as potable water where allowed by local health departments. Unless tanks are above ground and high enough to create sufficient water pressure, a pressurized

Figure 15.35 Large cisterns may be buried below ground, or installed above ground and left as an architectural element.

distribution the piping that delivers harvested and filtered rainwater for use.

distribution tank or an automatic pump that pressurizes supply lines is required. Usually a cistern will have a backup feed from a municipal supply or well to refill it when rainwater supplies run low.

Air Conditioner Condensate Collection

Air conditioning systems produce condensate, the fresh water that condenses and is collected when they operate. In hot, humid climates, the amount of condensate can provide additional water at no additional cost. Condensate can be collected in rain barrels, small containers, or rainwater cisterns. Like gray water, condensate is suitable for reuse as irrigation, but it should not be used for toilet tanks because its low pH may create corrosion problems.

Fixtures and Fittings

Plumbing fixtures (e.g., toilets, sinks, and tubs), and plumbing fittings (the metal valves and showerheads that supply water to the fixtures) are available in a variety of materials and efficiencies. Federal regulations set maximum flow rates for toilets, urinals, showerheads, faucets, and faucet aerators in order to save our water resources (Table 15.4). Many products are available that exceed these federal minimum standards to further reduce water use.

Fittings

Plumbing fittings include valves, faucets, and showerheads that are used in kitchens and bathrooms. Fittings are available in a wide range of finishes and efficiencies, many significantly exceeding minimum federal standards.

The EPAct of 1992 and updated in 2005 requires that all lavatory faucets, the fittings used in bathroom sinks, deliver a maximum of 2.2 GPM. The requirements also apply to kitchen faucets and replacement aerators, which are devices that mix water coming out of a faucet with air to reduce water flow. Bathtub faucets are not required to meet any maximum flow rates because tubs are generally filled with no water wasted while the faucets are running. WaterSense certification is available for lavatory faucets that deliver a maximum of 1.5 GPM. WaterSense does not certify kitchen faucets. High-performance faucets and aerators that produce as little as 0.5 GPM are available and can be an effective component in household water savings.

Faucet Controls

Automatic or foot controls for faucets can be effective in reducing water usage where manual controls may be difficult to operate. This often occurs in kitchens when hands are full or contaminated with food, as well as for persons with disabilities. Both kitchen and bath faucets are available with motion sensors that turn them on for a specified length of time at a predetermined temperature when

Table 15.4 National Efficiency Standards and Specifications for Residential Water-Using Fixtures and Appliances

Residential Fixtures and Appliances	Current Federal Standards (EPAct 1992, EPAct 2005 or Backlog NAECA Updates)	WaterSense or ENERGY STAR
Toilets	1.6 GPF	WaterSense: 1.28 GPF with at least 350 gram waste removal
Bathroom faucets	2.2 GPM at 60 psi	WaterSense: 1.5 GPM at 60 psi (no less than 0.8 GPM at 20 psi
Showerheads	2.5 GPM at 80 psi	2.0 GPM
Clothes washers	Modified energy factor ≥1.26 ft^3/kWh/cycle; water factor ≤9.5 gal/cycle/ft^3	ENERGY STAR Modified energy factor ≥2.0 ft^3/kWh/cycle; water factor ≤6.0 gal/cycle/ft^3

Source: EPA WaterSense, National Efficiency Standards and Specifications for Residential and Commercial Water-Using Fixtures and Appliances, available at http://www.epa.gov/WaterSense/docs/matrix508.pdf

condensate the moisture removed from the air by an air conditioning or dehumidification system.

plumbing fixtures an item for the distribution and use of water in homes, including toilets, sinks, and tubs.

plumbing fittings used in pipe and plumbing systems to connect pipe or tubing sections, to adapt to different sizes or shapes, and to regulate fluid flow.

lavatory faucets the fittings used in bathroom sinks.

aerators a device installed on faucets to increase spray velocity, reduce splash, and save both water and energy.

activated. Other control options include touch sensors that turn faucets on and off by tapping any portion of them with the back of a hand or an elbow or other body part. While these controls can reduce water use, they often use batteries that must be replaced or recharged periodically.

Foot pedals can be used to operate faucets. Installed between the water supply and the faucet, usually in the toe space of a cabinet, they offer hands-free operation. They can save both water and energy, although many use a small amount of electricity to operate.

Exterior Faucets

Hose bibbs, or exterior faucets, are prone to freezing in cold climates, causing water damage if not immediately repaired. Where freezing can occur, freeze-resistant hose bibbs or interior cutoffs should be installed. These faucets are also susceptible to water intrusion when not properly integrated and flashed into the weather-resistive barrier (see Chapter 9).

Showerhead Water Efficiency

Federal standards require showerheads to deliver a maximum of 2.5 GPM; however, there is no limit on the number of showerheads allowed in a single stall. Multihead shower spas that can deliver 20 GPM or more are not uncommon.

Early low-flow showerheads used flow restrictors and other technologies that often produced less-than-satisfactory showers, leading many people to remove the flow restrictors or retain older, inefficient showerheads. Newer technologies provide high-quality showers that meet the 2.5 GPM maximum flow requirement.

WaterSense requires 2 GPM maximum flow rate, with a total allowable flow for any individual shower compartment of no more than 2.5 GPM. For shower compartments that exceed 15 ft^2, one additional showerhead that does not exceed 2.5 GPM is allowed for each additional 15 ft^2 or smaller area. Several manufacturers have developed high-performance showerheads that deliver between 1.5 and 2 GPM that provide equivalent experience to higher-flow showerheads.

All showerheads are required to be controlled by pressure balance or thermostatic valves, controls that maintain a safe shower temperature to avoid scalding due to changes in water pressure or temperature. Most pressure balance valves have a single on/off setting, providing a consistent volume of water with temperature control integrated into the valve. Thermostatic valves allow for control of temperature and work with volume control valves that control water flow separately, providing a broader range of options than simpler pressure balance valves. Individual showerheads must be matched with specific valves for them to operate properly. Using mismatched showerheads and valves can make the anti-scald function ineffective, leading to scald injuries.

The U.S. DOE is considering rules similar to WaterSense to reduce the total water flow in shower compartments and limit the use of multi-head spa showers. As of this writing, these rules have not been finalized.

hose bibb a faucet with hose threads on the spout found outside or near clothes washers and wash basins.

pressure balance valve controls that maintain a safe shower temperature to avoid scalding due to changes in water pressure or temperature.

thermostatic valve a valve that maintains a safe shower temperature to prevent scalding by adjusting the mix of hot and cold water; also known as a *thermostatic compensating valve*.

Fixtures

Plumbing fixtures are made of cast iron, porcelain, fiberglass, metal, or plastic; with the exception of toilets, fixture selection has no effect on the amount of water consumed. Selection of non-toilet fixtures is primarily an aesthetic decision, however toilet selection can have a significant effect on total household water consumption.

Sinks

Sinks made of metal may be available with recycled content and may be able to be recycled when removed. Solid china and porcelain sinks can be ground into aggregate in lieu of disposing of them at the end of their lives. Fiberglass, plastic, and cast iron sinks are not easily recycled. Regardless of the material used for sinks, they last many years and represent only a small percentage of the overall materials in a house, which reduces the impact of their selection on the overall sustainability of a project.

Tubs

Tubs are available in porcelain-coated cast iron or steel, as well as solid acrylic and fiberglass. Few tubs are recyclable or made of recycled materials. Tubs that serve as showers require waterproof walls above the tub lip to protect walls from deterioration. Acrylic and fiberglass tubs may have attached or separate walls that provide an inexpensive waterproof surface. The more traditional solution is to install ceramic tile over a water-resistant base that directs shower water away from walls and into the tub. As noted in Chapter 4 (Insulation and Air Sealing), a complete thermal barrier, including insulation and air sealing, is critical behind tubs located on exterior walls to avoid thermal bypasses in this area.

volume-control valves valves that control the flow water to fixtures.

Showers

Showers can be constructed entirely of ceramic tile with a waterproof membrane at the base and walls. They can also be constructed with acrylic or cast bases, combined with acrylic, cast, or ceramic tile walls. Fiberglass shower systems are available in one-piece or multi-piece designs.

Shower Curtains and Doors

Tubs with shower fittings and showers must be designed to limit the quantity of water allowed to flow out onto bathroom floors to prevent damage to finishes and structure. Most tubs use shower curtains or sliding glass doors. Showers typically use sliding or swinging glass doors. Showers can also be designed to eliminate the need for a door by locating the showerhead far enough away from the opening to avoid water leaking out. Regardless of the materials used to control water flow, they must be operated and maintained properly for durability of the structure.

Toilets

Historically, toilets have used large volumes of water to remove human waste in a sanitary manner for health and safety. Before the 1950s, toilets often used as many as 7 GPF. Improvements in design and technology reduced this rate to about 5 GPF in the 1960s and to about 3.5 GPF in the 1970s. The EPAct of 1992 required that by 1994, all residential toilets use no more than 1.6 GPF. Many early high-efficiency toilets did not perform well, often requiring multiple flushes to evacuate solid waste. Over the next several years, various improvements ultimately solved the flushing problems through the use of improved mechanisms and wider drains within the toilets.

High-efficiency toilets (HETs) are defined by the EPA as those that use an average of 20% less water than the industry standard 1.6 GPF; a maximum of 1.28 GPF is allowable to receive this designation. Using a HET for a family of four is estimated to save as much as 8760 gallons of water per year. The HET criteria can be met by a standard single-flush toilet that uses less than 1.28 GPF, or by a dual-flush toilet (Figure 15.36) that provides one setting for solid waste (typically between 1.3 and 1.6 GPF) and a lower setting for liquid waste (typically 0.8

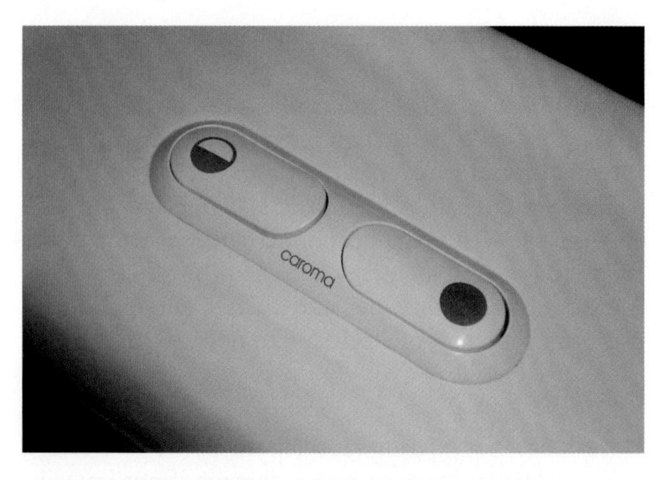

Figure 15.36 Controls on a dual-flush toilet. The user is able to select a one-half or full-flush depending on which is appropriate.

to 1.1 GPF). The HET rating is calculated by assuming that there will be two liquid flushes for every solid flush, on average.

Toilet Efficiency Ratings In 2003, water efficiency and plumbing fixture specialists in the United States and Canada cooperatively developed a new toilet testing protocol. Referred to as the maximum performance (MaP) test, this protocol determines that each toilet brand and model not only uses the specified amount of water but also effectively removes a minimum of 250 grams of solid waste from the bowl. The solid waste used for testing is a soybean paste encased in plastic. The idea behind the MaP test is to provide objective advice to consumers so that they can purchase HETs that do their job properly every time. The MaP test is also used to classify toilets that meet WaterSense criteria. Similar to the HET criteria, WaterSense requires that a HET be able to remove 350 grams of waste without plastic casing and also utilize a particular type of flush valve; performance and total tank capacity criteria are also specified. Continued improvement in toilet technology has led to units that remove as many as 1000 grams of waste with as little as 1 GPF. Begun in 2003, MaP testing is continued regularly on new toilets as they are introduced into the market.

Toilet Types Toilets are available with gravity flush mechanisms, which use the weight of the water in the

---

high-efficiency toilets (HETs) are defined by the EPA as those that use an average of 20% less water than the industry standard of 1.6 GPF.

dual-flush a high-efficiency toilet that gives users the choice of flushing at full capacity or with less water.

maximum performance (MaP) test determines how well toilets perform bulk removal by using a realistic test media; each toilet model is graded according to this performance.

gravity flush toilets that use the weight of the water in the tank to evacuate the bowl.

tank to evacuate the bowl, or with pressure-assist flush mechanisms, which use the supply water to build up pressure in a small container inside the tank and provide a forcible downward flush into the bowl (Figure 15.37). Some toilets (typically, very low-profile models) are available with a power-assisted flush that uses a small electric motor to provide effective flushing. Most of these designs are available in both single- and dual-flush models.

Toilet Fill Valves Fill valves, the internal parts of a toilet that measure the amount of water filling the tank and bowl, are designed to deliver just the right amount of water each time the toilet is flushed without waste. Improperly adjusted fill valves and replacement valves that are not designed to work with a particular toilet may waste water by allowing the tank or bowl to overfill, letting water run down the drain unnecessarily (Figure 15.38).

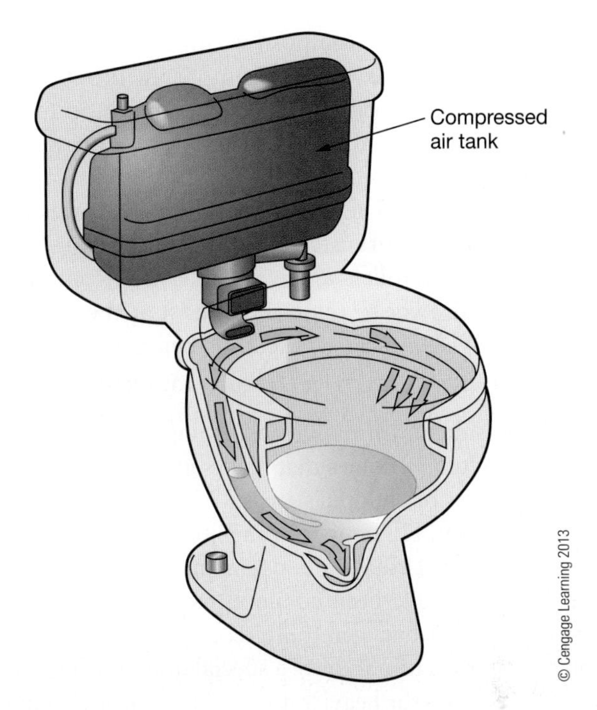

Figure 15.37 Pressure-assist flush toilet.

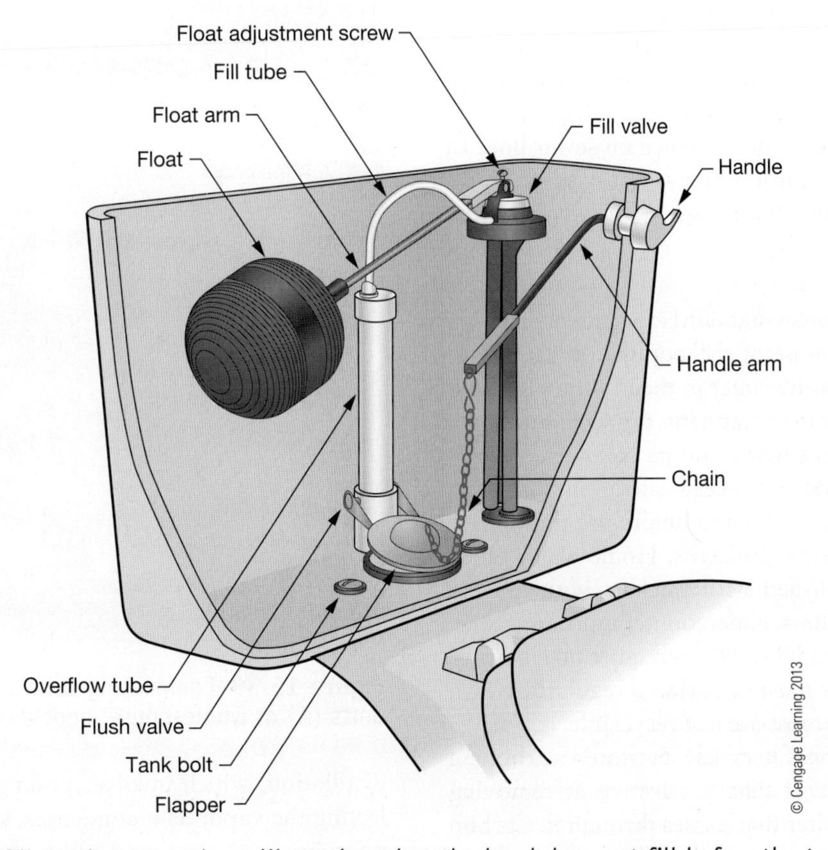

Figure 15.38 Toilet filler valves must be calibrated so that the bowl does not fill before the tank. Furthermore, water flowing into the bowl after it is full simply flows out the drain and is wasted.

---

pressure-assist flush a high-efficiency toilet that uses air pressure generated by water line pressure stored in a small tank to produce a more forceful flush.

power-assisted flush a high-efficiency toilet that uses a small electric pump to provide effective flushing.

fill valve the internal parts of a toilet that measure the amount of water filling the tank and bowl.

Composting Toilets Composting toilets convert human waste to sanitary usable compost without using any water in the process. While not common in many residences, they are an effective way to dramatically reduce water use while providing a consistent source of garden compost. They may also be an effective solution when the cost of installing a drain line is prohibitive. Composting toilets are available as self-contained units with a drawer to remove compost at the bottom (Figure 15.39(A)), or as whole-house units that collect compost from several toilets (Figure 15.39(B)).

Urinals Residential urinals are becoming more common, primarily in luxury homes that have separate bathrooms in master suites or children's areas. Urinals can save even more water than HETs by using as little as 10 ounces of water per flush. Urinals are also available in waterless models, which have been common fixtures in commercial construction but were only recently introduced for homes. Waterless urinals have a special drain filled with a liquid that allows the heavier urine to flow through it and down the drain while keeping the trap sealed from sewer gas odors (Figure 15.40). Waterless urinals use no water except for cleaning purposes; however, they do require periodic replacement of the trap liquid as well as the entire trap mechanism after it clogs with salts. Waterless urinals may lead to more maintenance on sewer lines to remove salt buildup because water is not used regularly to wash waste down the drain lines.

Water Treatment

The EPA sets and enforces standards for public drinking water, ensuring that most of the population receives a constant supply of healthy water in their homes. Regardless of the quality of water treatment, drinking water can be expected to contain small amounts of some contaminants as long as they do not exceed safe levels according to the EPA. Water may be filtered before use to improve taste or because of health concerns. Home water filtration may be accomplished with pitchers with built-in filters or with faucet filters, undercounter units, or whole-house systems (Figure 15.41). Filters require no energy to operate; however, they must be replaced regularly, which creates waste because most are not recyclable.

The most common filters use activated carbon, a block of porous mineral that is effective at removing contaminants from water that passes through it. Carbon filters can improve taste, and some will remove lead and other contaminants. Other filter media include fabrics, ceramic screening, and fiber. Water can be filtered by distillation, which involves boiling water and then collecting the vapor as it condenses, killing disease-causing microbes and removing most chemical contaminants. Distilled water often tastes flat because many natural minerals that improve taste have been removed. Distillation uses significant amounts of energy to purify water.

Figure 15.39 Composting toilets may be standalone units (A) or whole-house models (B).

---

composting toilets toilets that convert human waste to sanitary usable compost without using any water in the process.

activated carbon a form of carbon specially formulated for filtration.

distillation the process of purifying a liquid by boiling it and condensing its vapors.

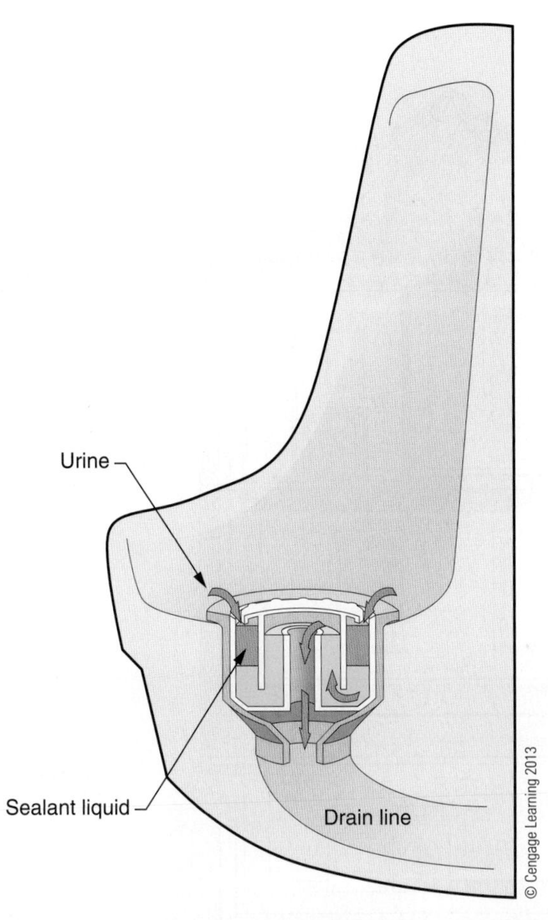

Figure 15.40 Waterless urinals allow liquid to drain away while preventing gas and odors from escaping into the bathroom.

Reverse osmosis (RO) filters force water through semi-permeable membranes under pressure, leaving contaminants behind in the process. RO filters can waste as much as three times as much water as they filter and also use energy in the process.

Filtering local water supply may be desired for personal preference or for health reasons by people with compromised immune systems. Generally, filtering does not play a significant role in a green home, and it does consume resources where it is employed. Filter media must be replaced, energy is used in processing, and water is lost through RO, depending on the type of system selected.

Water Softeners

The EPA establishes standards for drinking water that are based on health considerations and such aesthetics as taste, odor, color, or corrosivity. There are no national standards for water hardness. Hard water, defined as water having a high mineral content, can make washing difficult because soaps do not lather as easily as in soft water. Hard water can also create mineral deposits in faucets, showerheads, and water heaters, degrading their performance. Water hardness is measured in grains of minerals per gallon of water (GPG) (Table 15.5). One grain equals 64.8 milligrams. Water containing over 7 GPG is typically defined as hard, but there is no universally accepted classification of water hardness. Water softeners are appliances that remove minerals from water to reduce hardness, typically installed to treat the entire water supply for a home.

Water softeners consume energy to operate, use salts (e.g., sodium chloride or potassium chloride), and create a certain amount of wastewater in the softening process. They can be operated based on demand and a hardness sensor, or by the less-efficient timer clock method that treats water regardless of household demand. WaterSense has developed criteria for efficient whole-house water softener certification, certifying only units that operate on demand. Certified units must meet minimum efficiency requirements in the amount of salt they use in processing and how much wastewater they create during each softening cycle.

Washer Valves and Drain Pans

Clothes washers installed in or above living space can leak, causing significant damage to finishes in the process. Hoses and washers can leak or break, dumping large quantities of water in the process. Washer valves are typically installed in a wall-mounted box that includes a location for the drain hose to be installed. Safety valves can detect leaks when the washer is not in use (Figure 15.42), and single-throw control valves promote turning off the water supply when the washer is not in use (Figure 15.43); these strategies help reduce the risk of leaks in hoses and the subsequent damage they may cause. Drain pans below washers can be piped to the exterior or to an internal drain (Figure 15.44) to remove water from a failed washer before it causes damage inside a home.

---

reverse osmosis (RO) filters force water through semi-permeable membranes under pressure, leaving contaminants behind in the process.

hard water water that has a high mineral content.

water softeners appliances that remove minerals from water to reduce hardness; typically installed to treat the entire water supply for a home.

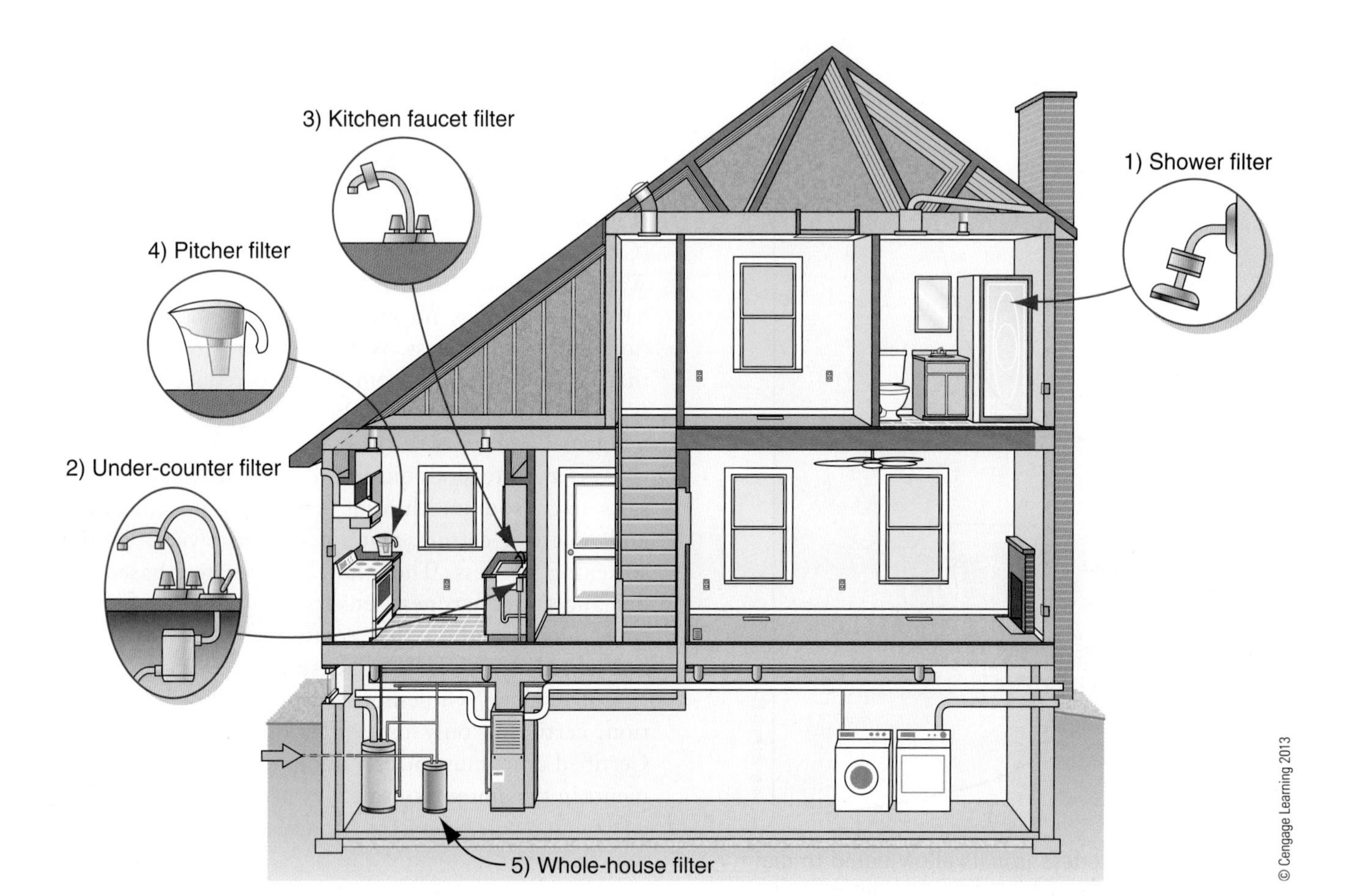

Figure 15.41 Locations for water filtration.

Table 15.5 Typical Classifications of Water Hardness

Classification	Grains/Gallon
Soft	0–1
Slightly hard	1–3.5
Moderately hard	3.5–7.0
Hard	7.0–10.5
Very hard	>10.5

Kitchen Food Waste Management

Garbage disposals are electrically operated grinders that allow food waste to be removed through the sewer lines, a process that requires both water and energy. Disposals increase the amount of solid waste, sometimes requiring additional filtering at water treatment

garbage disposal electrically operated grinders that allow food waste to be removed through the sewer lines, a process that requires both water and energy.

Figure 15.42 Automatic clothes washer shutoff valves, like the Watts IntelliFlows™, are electronic control devices that sense the machine's current flow. When the washing machine is turned on, the controls detect the current flow to the washer and open both hot and cold water inlet valves to allow water to flow to the washing machine. When the clothes washer completes the full cycle, the device senses the lack of current and closes the water inlet valves. These valves remain closed until the machine is used again.

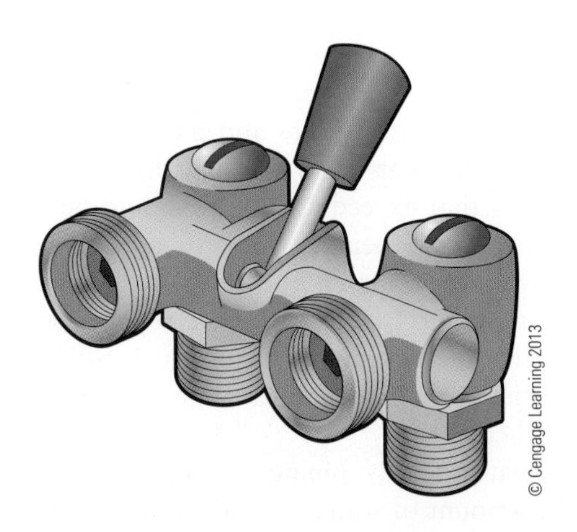

Figure 15.43 Single-lever washing machine valve.

Figure 15.44 Washing machine drain pans prevent damage from leaky or overflowing appliances.

facilities. Some regions have banned them to avoid overtaxing their treatment plants; other areas with limited landfill space may encourage their use. Most food waste can be composted and turned into nutrient-rich soil while avoiding the impact of disposal in solid waste or water treatment facilities. If a homeowner does not compost their waste and their local water treatment facility uses their solid waste to generate methane for energy production, then a garbage disposal can be a sustainable decision. If the local facility does not recover this solid waste, a better choice is to place food waste in the garbage to reduce the amount of energy and water used.

Figure 15.45 Electronic composters use fans to draw air into the machine, providing oxygen to the cultures and accelerating the decomposition process. A filter removes any lingering odors.

Kitchen Waste Composting

Most kitchen scraps, except for meats and certain other products, can be turned into compost, reducing the amount of solid waste produced in a home. Composting can be done outdoors in open or closed bins, or in small quantities with an electrically operated composter (Figure 15.45). The indoor units do require power to operate, but they may be a good option for a homeowner who would otherwise be unlikely to manage a large compost pile and would discard their food waste with their trash.

Remodeling Considerations

Many existing homes waste water unnecessarily. Inefficient fixtures, long hot water pipe runs, and leaks are key contributors to wasted water. The U.S. Geological Survey reports that one in every 318 homes has a water leak, which can account for as much as 20% of the total water consumption. A faucet that drips as few as 30 drops per minute can waste 54 gallons of water per month. Remodeling projects can provide excellent opportunities for water savings through improving hot water distribution systems and installing high-efficiency fixtures and fittings.

Water Audits

The best place to start looking for water savings in an existing building is by performing a water audit to identify leaks and inefficiencies in the plumbing system. Water audits begin by turning off all plumbing and inspecting the water meter. If the meter shows water use, then a leak is present that needs to be identified and repaired. Leaks can occur underground between the meter and the house; in supply lines where small failures in joints occur; at connections to toilets, sinks, and icemakers; and in irrigation systems, hose bibbs, water heaters, pools, spas, and other fixtures. In addition to leaks, minor drips in faucets should be identified and repaired. Toilets should be inspected for leaks by placing dye tablets in the tank. If any dye appears in the bowl before flushing, this indicates that a slow leak is wasting water, and the fittings in the tank should be repaired or replaced. Toilets can also be inspected to confirm that the tank and bowl complete their filling cycles without overflowing water down the drain.

After any leaks have been identified and repaired, fixture efficiency provides the best opportunity for savings. Toilets typically have the total GPF listed on the bowl or tank. Toilets that use more than 1.6 GPF should be replaced with new high-performance models. Single-flush 1.6 GPF toilets can be retrofitted with dual-flush conversion kits to save water; however, these should not be used on older toilets that use 3.5 or more GPF because they may not adequately clear the bowl. Many municipal water systems provide rebates and other incentives for replacing inefficient fixtures that help minimize the costs to homeowners. Showerheads and faucets may have their GPM flow rate listed on them. If not, simply capture and measure (in gallons) all of the water that flows out of it for one full minute. Fittings that exceed 2.5 GPM should be replaced with new models, being careful to confirm that the new showerheads match the pressure balance valve to avoid scalding. Faucets can have replacement aerators installed that allow as little as 0.5 GPM to flow through them, saving significant amounts of water at a minimal cost.

Hot water supply piping can be tested to determine the amount of water wasted as the user waits for hot water to arrive. Run showers and sink water into a bucket until hot water arrives to measure the total amount of water wasted waiting for hot water. If more than a few cups, consider relocating the hot water piping or water heater, adding pipe insulation, or installing an on-demand pump at individual fixtures to reduce wait time. Water heaters should be the most efficient models available and within the project budget; however, replacing properly operating water heaters without first improving the existing hot water delivery system and replacing fixtures with low-flow models may not be a prudent investment.

Additional Concerns

Older homes may contain materials that pose risks for building failures or occupant health and well-being. Homes may contain lead or copper pipes with high lead content solder that should be removed. Lead can cause health problems, particularly in young children and pregnant women.

Polybutylene is a soft plastic pipe that was used in many homes and for water service lines in the 1980s. This material is no longer recommended because of a chemical breakdown that caused premature failure. If polybutylene pipe is identified in a home during renovation, it is a good practice to replace it completely to avoid the possibility of future leaks.

Summary

Plumbing systems and codes were developed to protect health by providing potable water and removing waste without polluting the local water supply, with little regard for the amount of water used. Potable water is used to remove waste and irrigate land, an expensive and wasteful practice. Instead, look for ways to limit the use of potable water wherever we can to ensure that we will always have enough available.

Review Questions

1. Which of the following is the least expensive way to increase available water supply?
   a. Rainwater reclamation
   b. Gray water reclamation
   c. Conservation
   d. Water filtration
2. Which of the following is not one of the ways that household water is typically wasted?
   a. Cold water delivery to fixtures
   b. Toilet tank overfilling
   c. Leaks
   d. Irrigation
3. Which hot water pump is the most water- and energy-efficient type?
   a. Timer-operated
   b. Thermosiphon
   c. Demand-operated
   d. Continuous
4. Where local regulations allow, reclaimed gray water can be used to supply which of the following?
   a. Showers
   b. Toilets
   c. Kitchen sinks
   d. Dishwashers
5. Which of the following water heaters is the most efficient when located in unconditioned space in a warm climate?
   a. Gas condensing tankless
   b. Electric heat pump
   c. Electric tankless
   d. Sealed-combustion gas storage
6. Which of the following is the maximum flow rate allowed in a shower up to 15 ft^2 for WaterSense certification?
   a. 2.5 GPM per compartment
   b. 1.5 GPM per head
   c. 3.5 GPM per head
   d. 1.75 GPM per head
7. A toilet with which of the following flush rates would not meet the criteria for high efficiency?
   a. 1.28 GPF
   b. 1.1 GPF
   c. 1.6 GPF
   d. Dual-flush 1.6 GPF and 1.1 GPF
8. Which of the following is not a problem with hard water?
   a. Soap does not lather easily
   b. Water heater performance diminishes
   c. Showerhead clogging
   d. Low mineral content in water
9. Installation of which of the following is not a strategy to reduce household water use?
   a. WaterSense toilets
   b. Reverse osmosis water filtration system
   c. Demand hot water pumps
   d. Foot pedal faucet controls
10. Use of which of the following is not a strategy for energy savings in a home?
    a. Demand hot water pumps
    b. Waste water heat recovery
    c. Pipe insulation
    d. Water filtration

Critical Thinking Questions

1. How does the placement of bathrooms, laundry, and kitchen in relationship to the water heater affect water and energy use in a home?
2. Compare and contrast the different type of water heaters. What are the pros and cons of each?
3. Discuss the pros and cons of rainwater harvesting vs. gray water reclamation.

Key Terms

acrylonitrile butadiene systems (ABS), 435
activated carbon, 446
aerators, 442
aerobic septic system, 433
air admittance valve, 436
anaerobic septic system, 433
black water, 433
catchment, 438
chlorinated polyvinyl chloride (CPVC), 417
cisterns, 439
cold start, 420
coke, 435
composting toilets, 446
condensate, 442
condensing heaters, 429
constructed wetlands, 433
continuous pumps, 421
conveyance, 438
cross-linked polyethylene (PEX), 418
deficit approach, 439
demand pump, 421
desuperheater, 428
distribution, 441
distillation, 446
drain field, 433
drain-water heat recovery (DHR), 437
dual-flush, 444
effluent, 433
energy factor (EF), 428
fill valve, 445
first flush, 439
first-hour rating, 424
garbage disposal, 448
gravity-film heat exchanger (GHX), 437
gravity flush, 444
hard water, 447
heat traps, 431
high-efficiency toilets (HETs), 444
holding tank, 433
home run system, 419
hose bibb, 443
hot start, 420
hot water circulation loop, 421
lavatory faucets, 442
manifold system, 419
maximum performance (MaP) test, 444
percolation test, 435
PEX, 418
plumbing core, 419
plumbing fittings, 442
plumbing fixtures, 442
polypropylene (PP), 418
power-assisted flush, 445
pressure-assist flush, 445
pressure balance valve, 443
pressure-reducing valve, 416
purification, 438
rainwater harvesting, 417
reverse osmosis (RO), 447
septic drain field, 435
septic system, 433
supply approach, 439
tank heaters, 424
tankless heaters, 424
temperature pumps, 421
temperature rise, 426
thermosiphon loops, 421
thermostatic valve, 443
timer-operated pumps, 421
trunk-and-branch, 419
vitrified clay pipe, 435
volume control valves, 443
water softeners, 447

Additional Resources

Green Plumbers USA: http://www.Greenplumbersusa.org
Water Conservation Alliance of Southern Arizona: http://watercasa.org/
California Urban Water Conservation Council: http://www.cuwcc.org/
EPA WaterSense: http://www.epa.gov/WaterSense/
MaP Test Data: http://www.map-testing.com/

16

Renewable Energy

This chapter covers small-scale active and passive renewable energy systems that are installed on residential buildings. Active renewable energy comes in the form of solar photovoltaic cells, solar thermal systems, wind, water power, fuel cells, and biofuels. Passive solar is reviewed here and builds upon the content in previous chapters (in particular, Chapters 3 and 8). Renewable energy systems should only be considered after high-performance construction techniques have been incorporated into a project to minimize overall power demand.

LEARNING OBJECTIVES

Upon completion of this chapter, the student should be able to:

- Identify appropriate renewable energy strategies for a particular project
- Explain how a photovoltaic cell works
- Describe the different types of photovoltaic systems and their components
- Describe the different types of wind turbines
- Identify strategies to passively heat and cool a home
- Identify appropriate renewal energy systems to include in a project
- Explain how a solar water heating system works
- Describe the different types of solar water heating systems

Green Building Principles

Energy Efficiency

Indoor Environmental Quality

Community Impact

Renewable Energy Sources

The sun, wind, and moving water all contain energy that, if harnessed, can provide consistent sources of power without the need to burn fossil fuels for heat or electricity. We can use the sun to heat water and air and to create electricity. Wind and water can be used to create electricity. Other renewable energy sources include biofuels, such as wood and biodiesel. Fuel cells, which generate electricity from natural gas or hydrogen, are generally considered a renewable energy source. Geothermal or ground-source heat pumps, addressed in Chapter 13, can be considered a renewable energy source, particularly on a very large scale when they are used by utilities to generate heat and electricity. A small-scale renewable energy system may be located within a single home or serve a cluster of homes. Typically electric utility companies operate large-scale renewable energy systems.

Solar Power

Solar power can heat water for plumbing or air for space heating, and it can generate electricity. Systems that heat water and air are described as solar thermal, and those that create electricity are called photovoltaic (PV).

---

solar thermal a system that converts sunlight into heated air or water.

photovoltaic (PV) a device that converts the energy of sunlight directly into electricity.

Passive solar heating involves using the sun to heat an entire building through proper orientation and shading of windows and including thermal mass to contain and balance the heat gain.

Solar thermal systems for domestic hot water and PV for electrical generation are the most common systems used in residential construction. Government and utility financial incentives are frequently available to offset portions of the installation costs, improving the return on investment (ROI).

Solar power can also used for a portion of a home's space heating and for heating pools and spas to extend their seasons of use while reducing fossil fuel use.

Passive Solar Design

Passive solar design incorporates appropriate building orientation, careful location and shading of windows, and the integration of thermal mass to collect and distribute heat. It usually requires active participation of occupants to open and close windows and operate insulating shades and blinds to optimize the use of solar energy. In hot and mixed humid climates, passive solar is less effective for space conditioning because electrically powered dehumidification and cooling are often required for much of the year for occupant comfort, indoor air quality, and building durability.

Requirements for Solar Power

For a project to take advantage of solar power, the total amount of solar exposure on the site and in the region as a whole must be evaluated. The Department of Energy's National Renewable Energy Laboratory (NREL) has developed maps showing the amount of solar exposure throughout the United States, which can be used to determine the amount of energy you can expect to generate in any part of the country. Additionally, such obstructions as mature trees and adjacent buildings can affect the amount of solar potential on a specific lot. Finally, local zoning and historic restrictions may limit the installation of solar equipment on a project.

Wind Power

Wind power is used to turn windmills or turbines, which generate electricity whenever they are rotating fast enough. The general rule for effective wind power is having an average wind speed of at least 10 miles per hour (MPH). NREL has produced wind speed maps of the entire United States that can be used to determine the potential for wind power in a region; however, wind speed on specific lots can vary significantly with geography, trees, adjacent buildings, and the microclimate. Zoning restrictions may limit the installation of wind turbines in many areas.

Hydropower

Flowing water from a stream or river can be used to create electricity, referred to as hydropower. This alternative energy source is most commonly seen in utility scale projects, but local micro-hydro systems, generally defined as producing 100 kW or less, can be employed where there is adequate flow and vertical drop in a

Did You Know?

Origins of Renewable Energy in the United States

The solar power industry in the United States dates back to the time of the Arab oil embargo in the 1970s. Manufacturers of solar panels and complete systems appeared in response to energy shortages and government incentives for installing renewable energy systems. Contractors and homeowners created custom, site-built water heaters, solar greenhouses, and complete passive solar homes. While many of them did not meet energy-saving expectations, they did provide useful information about what did and did not work. When oil began to flow freely again, concerns regarding energy savings diminished, government incentives disappeared, and much of the solar energy industry faded from view. Even though interest in solar power waned, theories that developed out of these early experiments continued to develop through the late twentieth and early twenty-first centuries, evolving into what we now refer to as *building science*, one of the roots of green building. Another result of the failure of many solar systems was the creation of the Solar Rating and Certification Council to certify component and system performance. As interest in green building, energy efficiency, and a renewed interest in solar power developed at the end of the twentieth century, more manufacturers entered the renewable energy market, and incentives began to reenergize the industry.

The U.S. wind power industry has its origins in rural areas in the early twentieth century that wanted to take advantage of the recent development of electrical power but were not served by utility power lines due to their remote locations. Several manufacturers offered simple and effective wind turbines that operated alone or in combination with fuel-fired generators that provided power where otherwise unavailable. The Rural Electrification Act of 1936 funded the installation of power lines to remote areas but required building owners to remove their wind turbines and generators before they were connected to the electrical grid. Inexpensive and reliable grid power and subsidized power line installation led to a quick demise of decentralized wind power systems.

stream or river and local zoning and environmental laws allow its installation. Hydropower has the unique benefit of producing energy whenever water flows; solar power is generated only during the day, and wind-driven systems are subject to variations in wind speed. Micro-hydro systems that run 24 hours per day can produce the same amount of power in a single day as solar and wind systems with much larger capacities because they generate electricity for shorter periods of time.

Biofuels

Biofuels are non-fossil fuels, such as wood, ethanol, and biodiesel, and are generally considered renewable; however, biofuels can have significant environmental impact if not produced sustainably. Wood is available as split logs or in pellet form. Ethanol is a fuel similar to alcohol that is distilled from corn, sugar cane, or other plants. Biodiesel is typically made from virgin or recycled food oils, or animal fats remaining after rendering.

Corn ethanol is common in automobile fuel, and its producers benefit from significant government incentives; however, the amount of energy extracted from the corn does not significantly exceed the amount of energy required to produce the ethanol in the first place. The production of corn ethanol not only uses significant quantities of water and fertilizer but also uses corn, which drives up the price of the commodity. Alternative sources of ethanol include noncultivated plants, such as switchgrass, pine trees, straw, corn waste, and other sources, all of which are still evolving technologies.

Wood-fired furnaces, water heaters, boilers, and furnaces are available sources of renewable energy but are not common in residential projects. See Chapter 13 for more information on wood burning equipment.

Little, if any ethanol or biodiesel–fueled equipment is currently available for home energy or heat production. Most current uses are for vehicle fuels. More options may become available if these fuels become more common in the future.

Fuel Cells

A **fuel cell** is a device that uses fossil fuels, such as natural gas or propane, to produce electricity without combustion by extracting hydrogen, which is used to create power similar to how a battery works (Figure 16.1). Extracting hydrogen from fuels does create by-products, such as carbon dioxide and, in some cases, small quantities of nitric oxide and sulfur oxide. Fuel cells are approximately twice as efficient as standard electric power plants and produce less pollution. Once the hydrogen is extracted from the fuel, the only by-products of a fuel cell are heat and water. Distributed fuel cells also reduce the transmission losses associated with central power plants by producing power where it is consumed. Fuel cells are becoming available for residential applications and can benefit from subsidies and incentives to promote their use.

Courtesy of ClearEdge Power

Figure 16.1 The ENE Farm, which is currently only available in Japan, uses natural gas to provide electricity. The otherwise wasted heat is captured for space or water heating.

Pros and Cons of Renewable Energy

Renewable energy provides an opportunity to reduce dependence on fossil fuels and can help insulate homeowners from increasing energy costs. However, they must be

---

biofuels non-fossil fuels, such as wood, ethanol, and biodiesel, which are generally considered renewable.

fuel cell a device that uses fossil fuels, such as natural gas or propane, to produce electricity without combustion by extracting hydrogen, which is used to create power similar to how a battery works.

considered in the context of the project as a whole, taking into account overall building efficiency, occupant behavior, installation cost, regular maintenance, repair, and eventual replacements. Products that generate electricity require interconnection with the power grid or batteries for storage. Locally generated renewable electricity reduces the amount of fossil fuel burned in power plants as well as the losses due to transmission; however, these installations can be more costly than energy-efficient improvements that can save more energy than they can generate. Finding the balance between building performance and renewable energy is the key to a successful project. Table 16.1 presents the benefits and drawbacks to the different types of renewable energy.

For all renewable energy sources, any unused power must be sold back to the grid, stored on-site with batteries, or left unused. Some projects also offer opportunities to utilize multiple systems. For example, a photovoltaic array can be used to generate the majority of the power on sunny days and a wind turbine can be the primary producer at night.

Selecting Renewable Energy Systems

When choosing renewable energy systems for a home, an understanding of the project requirements and the available site resources is necessary to make the best decision. Project requirements include:

- The type of energy needed to operate the home's mechanical systems: hot water, space heating, or electricity
- The availability of utility connections on-site
- The funds available to pay for renewable energy systems
- The occupants' requirements for comfort and management of renewable systems

Projects that want or need to be energy independent must provide a place to store energy, such as batteries, when the sun, wind, and water are not available. Energy independent projects are referred to as **off-grid**, meaning

Table 16.1 Renewable Energy Sources

Renewable Energy Source	Pros	Cons
Solar photovoltaic	• Free fuel source • Fairly reliable fuel • Nonpolluting site energy generation • Most effective when sunny or clear and cold, which matches demand for air conditioning or heating when it is needed • Incentives may offset a portion of the installation costs	• Not available 24 hours a day • High first cost • Poor weather can decrease production • Roof size and lot constraints
Wind	• Free fuel source • Fairly reliable in proper locations • Nonpolluting site energy generation	• Local climactic conditions can affect wind capacity • Inconsistent energy source • Not available 24 hours a day • High first cost • Requires large land area free of obstructions and tall towers for best performance • Local zoning may limit installations
Micro-Hydro	• Free fuel source • Fairly reliable in proper locations • Nonpolluting site energy generation	• Site specific • Local regulations may restrict installations
Biofuels	• Relatively affordable fuel source • Typically produced locally	• Fuels are location specific • Little residential equipment available that uses biofuels
Fuel Cells	• Nonpolluting site energy generation • Affordable fuel source	• Equipment is expensive • Limited systems available on the market • Not completely renewable • Requires fossil fuel to operate

off-grid homes that they are not connected to electric and natural gas utilities.

that they are not connected to electric and natural gas utilities. Grid-tied systems, which are those that are connected to utility lines, usually sell all the energy generated back to the power company and use grid power for daily operation.

Available site energy is determined by investigating the solar, wind, and water resources. Solar and wind resource maps (such as those available from NREL) can be used to determine the site's suitability to renewable energy systems, and the site should be surveyed to determine where the sun and wind are unobstructed. If a site has flowing water, measuring the flow and vertical drop will help determine the suitability for micro-hydro. Finally, if wood is desired as a fuel, local sustainable sources for wood or pellets must be identified.

Net- and Near-Zero Energy Homes

A house that produces as much energy as it needs on a monthly or annual basis is referred to as a net-zero energy home. Most net-zero energy homes are grid-tied and sell power to the utilities when available and use grid power when necessary, averaging zero or negative net energy use. Homes that produce most, but not quite all, of their energy are called near-zero energy homes. Some net- and near-zero energy homes may be off-grid, using batteries, wood, propane, or other fuels when required to meet power needs.

Cost of Renewable Energy Systems

Most available renewable energy systems are expensive to install, require some level of maintenance, and eventually must be replaced, which are all costly endeavors. When a home is carefully designed, built, and managed to minimize total energy use, costs are minimized. It is always less expensive to reduce power requirements through energy efficiency and behavior change than to add renewable systems to a project.

As we discussed in Chapter 1, you can calculate the payback period or the ROI of energy efficiency investments. You can do the same for renewable energy systems by comparing the cost of efficiency and renewable systems to make the best decision for your project. In addition to the cost of fuels or electricity displaced with renewable energy systems, other factors must be considered to determine the ROI of a particular system, such as the cost to install power and gas lines, future fuel increases, and fuel availability.

Financial Incentives

Incentives for renewable power include tax credits, tax deductions, cash grants, lower utility rates, accelerated depreciation, low-interest financing, and premium rates for energy purchased by utilities, among others. They can be offered by federal, state, and local governments, utilities, and, in the case of financing, private entities. Most incentives require that installers or products meet specific requirements in order to obtain them. The most current list of incentives is available on the Database for State Incentive for Renewable Energy (DSIRE). The organization's website (http://www.dsireusa.org) is updated regularly.

Equipment Durability

As with all mechanical and electronic equipment, the more complicated the design, the more frequently it requires repair, replacement, and maintenance. Systems with moving parts, such as pumps, valves, and bearings, will require more frequent maintenance and repair. Homeowners' insurance policies should include replacement cost of renewable energy systems that may be damaged by wind, hail, falling trees, and other hazards.

Integration with the Building

Any renewable energy systems that are connected to the structure must be designed and installed to minimize any air and water intrusion. Proper wall and roof flashing and air sealing at all connections is critical to maintaining the building envelope. Systems that are attached to the house should be engineered to meet local wind conditions. Ground-mounted solar panels and wind turbines require foundations that provide support for all weather conditions. Roof-mounted systems may require additional structural bracing for proper support.

Certifications

Third-party certifications are available for many renewable energy products and the professionals that install them. Purchasing products that are evaluated and rated by a qualified third party provides some assurance that they will perform as expected. Hiring professionals who have obtained voluntary certifications will not guarantee a flawless installation; however, it likely indicates that they are serious professionals committed to high-quality work.

---

grid-tied refers to homes that are connected to the central electrical grid.

net-zero energy a building that produces as much energy as it needs on a monthly or annual basis.

Product Certifications

Solar thermal products are tested and certified by the Solar Rating and Certification Corporation (SRCC), which provides information for comparing system performance. Founded in 1980, the SRCC is a non-profit corporation that provides certification, rating, and labeling for solar thermal collectors and complete solar water heating systems. Equipment that is rated and certified by the SRCC must have their certification label, which lists the performance rating for that specific product. They also have a product directory on their website, listing both specifications and the certified performance rating. Many incentive programs require SRCC certification to assure that systems perform as expected. There is currently no single standard efficiency rating for, or organization that certifies, photovoltaic panels or systems.

Courtesy of the Solar Rating and Certification Corporation

Individual Certifications

Individuals involved in sales and installation of solar and wind systems can obtain voluntary certification in their field through the North American Board of Certified Energy Practitioners (NABCEP). NABCEP certifications are available for solar thermal and photovoltaic installers, photovoltaic technical sales professionals, and small wind installers. In addition to voluntary certifications, most local building departments require licensed plumbers, electricians, and heating, ventilating, and air conditioning (HVAC) contractors to make the appropriate connections to house plumbing, heating, and electrical systems.

Solar Rating and Certification Corporation (SRCC) a non-profit organization whose primary purpose is the development and implementation of certification programs and national rating standards for solar energy equipment.

North American Board of Certified Energy Practitioners (NABCEP) a national certification organization for professional installers in the field of renewable energy.

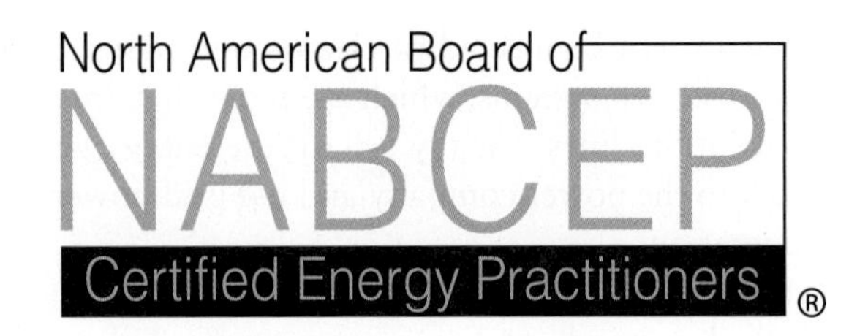

Courtesy of The North American Board of Certified Energy Practitioners (NABCEP)

Designing a Home for Renewable Energy

The most important part of planning for renewable energy is to reduce the total building energy loads, thereby minimizing the amount of renewable energy required. This is accomplished by limiting the building size, providing appropriate solar orientation, constructing a high-performance building envelope, reducing plug loads, installing an efficient HVAC system, and installing a high-performance hot water distribution system. In addition to creating a high-performance building, the site must provide adequate exposure to the sun, wind, or water to generate the energy required.

Most renewable energy systems are visible on the exterior of a home and may be subject to zoning and historic district restrictions. Solar panels, wind turbines, and micro-hydro installations may be restricted or prohibited in many areas, or require special permission to install. In addition to legal restrictions, physical restraints can limit the use of renewables.

Determining a Site's Solar Potential

The solar potential, the amount of sunlight a particular site can capture, is a combination of the latitude, local weather conditions, and the amount of unobstructed southern sky. Solar resource maps available from NREL show the amount of energy that can be expected in each region of the United States based on location, cloud cover, and other climate factors (Figure 16.2).

In the northern hemisphere, solar collectors are most efficient when facing solar south, rather than magnetic south. Adjustments to compass readings based on latitude are required to determine an accurate location of solar south. Mature trees and adjacent buildings may create too much shade for solar panels to work effectively. Using a tool like the Solar Pathfinder™ can

solar potential the amount of sunlight a particular site can capture based on latitude, local weather conditions, and the amount of unobstructed southern sky.

United States Photovoltaic Solor Resource: Flat Plate Tilted at Latitude

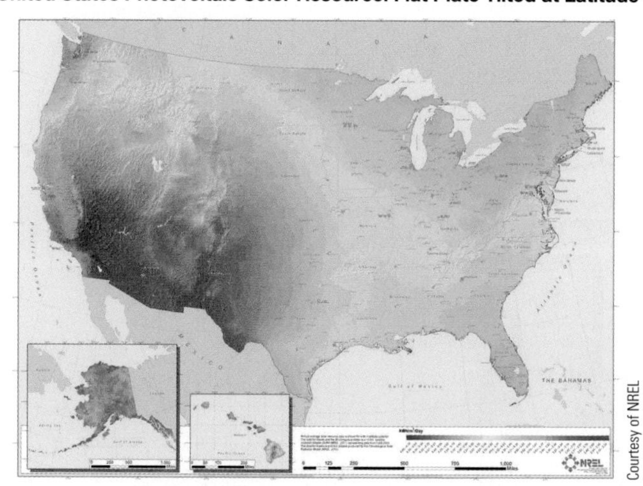

Courtesy of NREL

Figure 16.2 U.S. solar potential map from the National Renewable Energy Lab (NREL).

United States 50 Meter Wind Power Resource

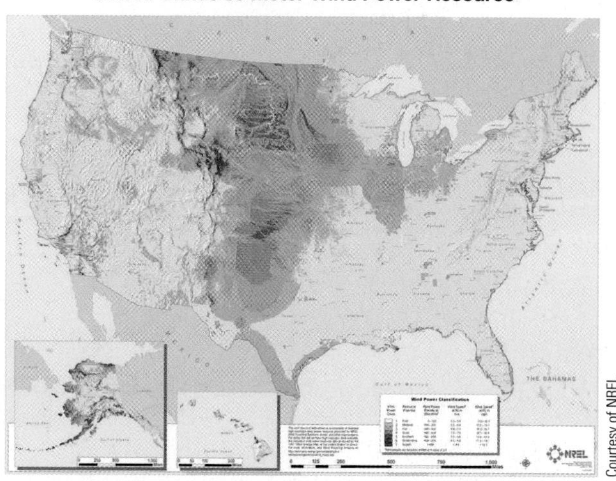

Courtesy of NREL

Figure 16.3 U.S. wind resource map from the National Renewable Energy Lab (NREL).

assist in determining the most appropriate place for solar collectors by evaluating how obstructions affect a specific location on a site.

Determining a Site's Wind Potential

Wind potential is determined through national average wind speed maps as well as by measurements of wind speed at a specific site. NREL wind power resource maps (Figure 16.3) classify every region in the United States by wind power class, ranging from 1 (poor) to 7 (superb), based on the average annual wind speed at heights of 33 feet and 164 feet above the ground. Before an investment in wind power is made, a local site wind survey should be conducted that covers several seasons to determine average wind speed. As with solar potential, obstructions near a turbine can reduce wind power. A general rule, a wind turbine should be installed at least 30 vertical feet higher than any buildings, trees, or other obstructions within a 300-foot horizontal radius.

Determining a Site's Hydro Potential

Hydropower requires the correct combination of stream flow, expressed in gallons per minute (GPM), and head (or vertical drop), expressed in feet. High stream flows with low head are found in shallow sloping, fast-moving streams. High-head, low-flow conditions are found in steeply sloped small streams. Productive low-head streams can have as little as 2 to 10 feet of head as long as they have adequate and consistent GPM. High-head streams can have thousands of feet of head with low flow and still produce adequate energy. Seasonal streams that run low or dry up periodically may not be appropriate for hydropower, and local legal restrictions may limit the ability to install hydropower.

Solar Thermal Systems

Solar thermal refers to using the power of the sun to heat water. Most commonly used for domestic hot water heating, solar thermal systems have the widest array of options available and, in some cases, can also provide space and pool heating. Solar photovoltaic, wind, and hydro systems have fewer variations and only provide electrical power. Matching a specific system with the local climate, site orientation, and building ensures the highest performance.

Types of Solar Thermal Systems

Solar thermal systems are classified as either indirect or direct and as active or passive. Direct systems circulate potable water through collectors and into a storage tank from where it is piped to faucets (Figure 16.4). Indirect systems use a heat transfer solution, typically an ethylene glycol/water mixture, which circulates between collectors and a heat exchanger in the storage tank (Figure 16.5). Indirect systems, which are more common, provide more reliable freeze protection than most direct systems.

Active systems use pumps to circulate the liquid between collectors and storage tanks. Passive systems use a combination of building water pressure and the tendency of hot water to rise by convection to operate.

Indirect Systems

Indirect systems use a food-grade propylene glycol antifreeze solution to transfer heat from collectors to a heat exchanger in a storage tank, allowing them to operate throughout the winter without freezing. A temperature sensor operates pumps that move the solution from

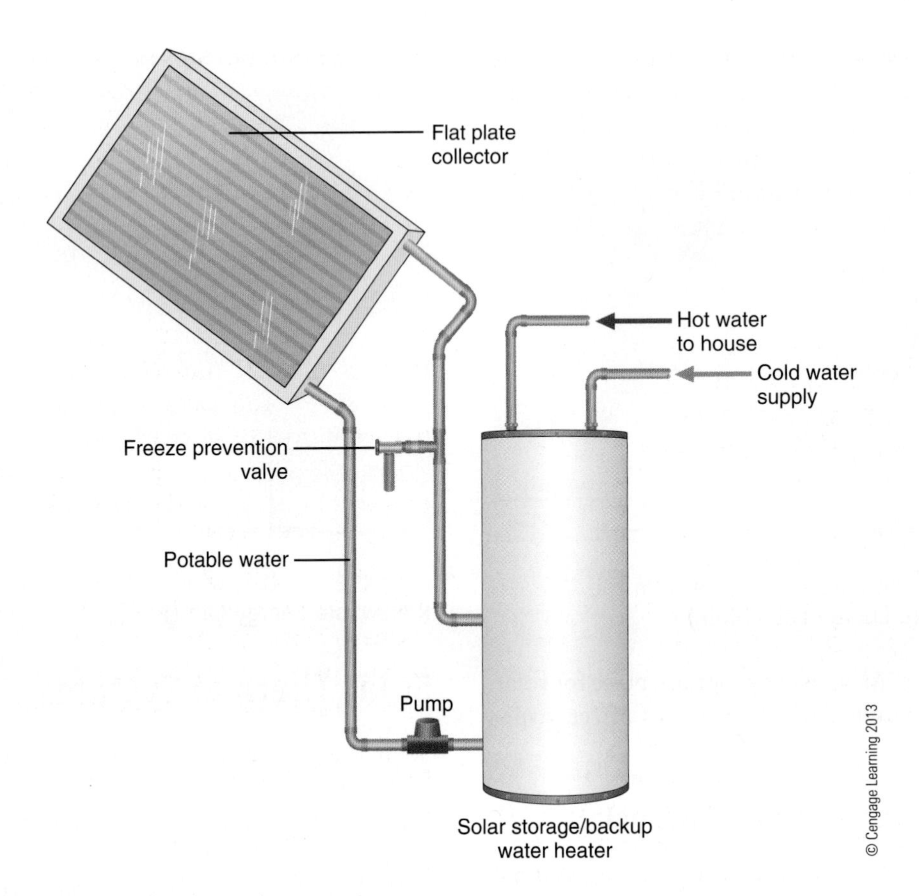

Figure 16.4 Active, direct solar thermal system.

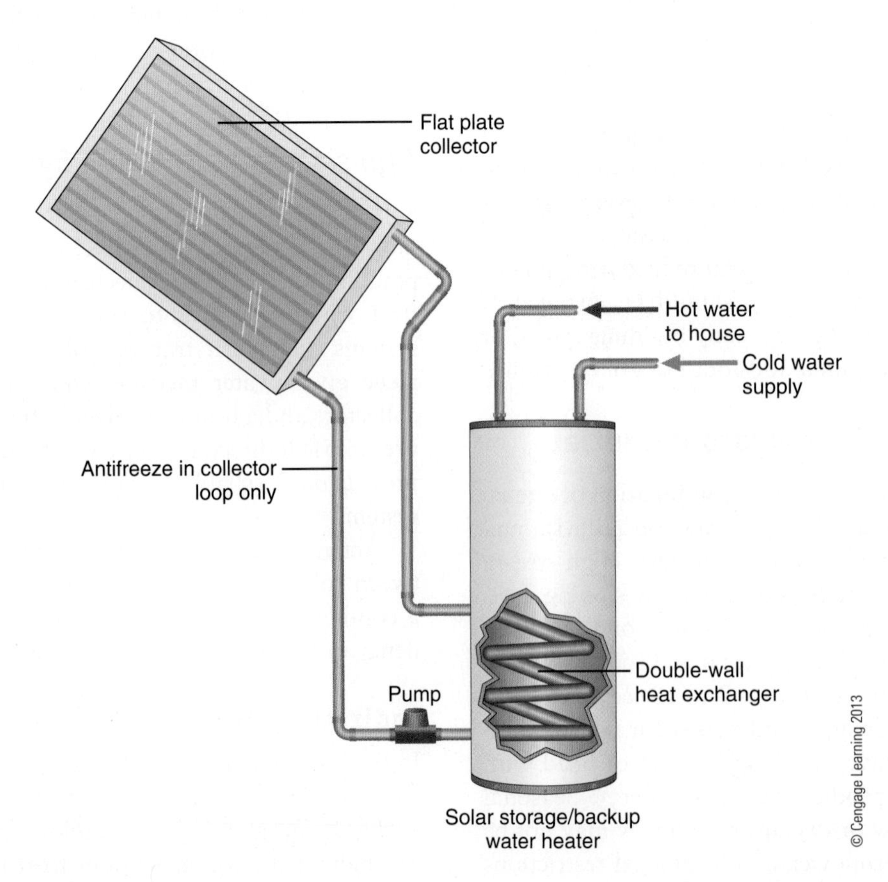

Figure 16.5 Active, indirect solar thermal system.

collectors to heat water in the storage tank when the collector temperature is higher. Antifreeze systems must be protected from overheating or extreme pressure, and the pH level of the coolant must be monitored. Coolant should be periodically replaced if it falls below the manufacturer's recommended pH levels.

Direct Systems

Direct systems heat potable water directly, which is stored in a tank before it is used. Water pressure forces cold water into collectors when faucets are turned on, pushing heated water through the storage tank into the hot water pipes.

Batch Heaters The simplest direct systems are batch heaters, which use a large black tank installed in a box with a sealed glass cover (Figure 16.6). The sun heats the water in the tank, which is then transferred to an indoor storage tank with backup heat using house water pressure. Batch heaters are simple and inexpensive but are mostly limited to custom and site-built projects. There are few manufactured batch heaters that are commonly used in the United States. Like all direct systems, batch heaters are prone to freezing. They also can lose heat quickly at night and in cold weather, requiring manual operation of insulated covers to keep stored water from cooling.

Roof-Mounted Storage Tanks Some direct systems include a rooftop-mounted storage tank integrated with collectors (Figure 16.7). While simple, these systems may require structural reinforcement to support the extra weight load on the roof and are not very common in the United States.

Active Systems

Active direct systems use pumps and temperature sensors to circulate water from the collectors to the storage tank as needed. A controller constantly monitors the water temperature in the storage tank, compares it with the collector temperature, moves hot water from the collectors to the storage tank, and moves colder water back to the collectors to be heated. The sensors can also prevent heated water from being moved into the storage tank if no additional heat is needed, limiting overheating of the storage tank. Direct systems that are installed in cold climates employ freeze protection through a drain-back system, which opens valves to allow the water in the collectors to drain down to a storage tank when cloudy and cold so that water will not freeze inside the collectors or pipes.

Passive Systems

Passive solar thermal systems (not to be confused with passive solar house design), often referred to as thermosiphon systems, use the natural tendency of hot

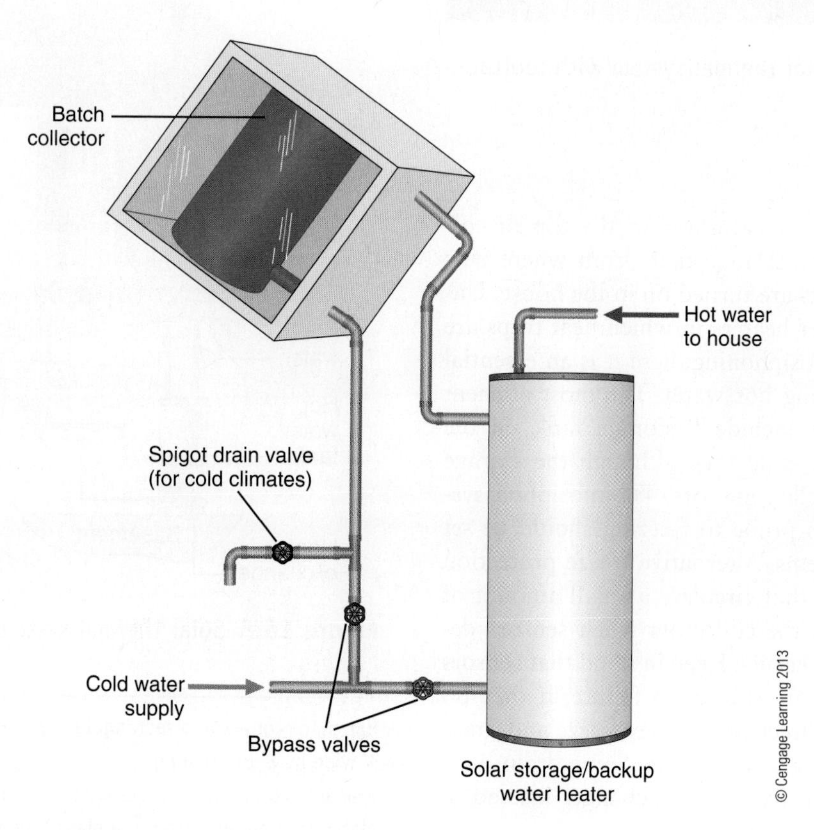

Figure 16.6 Passive, direct batch solar thermal system.

Figure 16.7 Direct solar thermal system with rooftop-mounted storage tank.

water to rise. Heated water moves to the top of collectors and then to the storage tank from where it is circulated when faucets are turned on in the house. Unlike conventional water heaters in which heat traps are used to prevent thermosiphoning, here it is an essential component of delivering hot water. The most efficient thermosiphon systems include a storage tank on the roof at the top of the collectors, although the storage tank can also be installed indoors. Thermosiphon systems installed in areas prone to freezing should be set up as drain-back systems. Alternative freeze protection systems include those that circulate a small amount of heated water through the collectors when sensors detect the potential for freezing. Keep in mind that sensors and automatic valves are subject to failure; if they do not work correctly, a direct system may freeze and cause damage to the system and the house. Some drain-back systems are indirect, using a heat exchanger instead of heating the potable water directly.

Solar Thermal Components

Solar thermal systems consist of *collectors* that absorb energy from the sun and transfer it into a liquid used for *storage*, a *heat transfer* mechanism to move the heat from the storage liquid into the water to be used, and various pumps and sensors that manage the movement of liquid throughout the system (Figure 16.8).

The most important components in a solar thermal system are the collectors. They are mounted directly to the roof with integrated flashing or installed on brackets. In cases where the best solar exposure is not on the roof of a building,they can be ground-mounted. When selecting a solar thermal system, consider the life of the system, maintenance requirements, complexity, reliability, total installed cost, and efficiency. Cold climate systems are most often indirect, providing reliable freeze protection while a direct system can work well where there is no risk of freezing. SRCC provides ratings for complete solar thermal systems that can be used to compare the efficiency between different systems.

Solar Thermal Collectors

The most common solar thermal collectors are flat plate, which are glazed insulated boxes that absorb solar energy, and evacuated tube, consisting of rows of glass

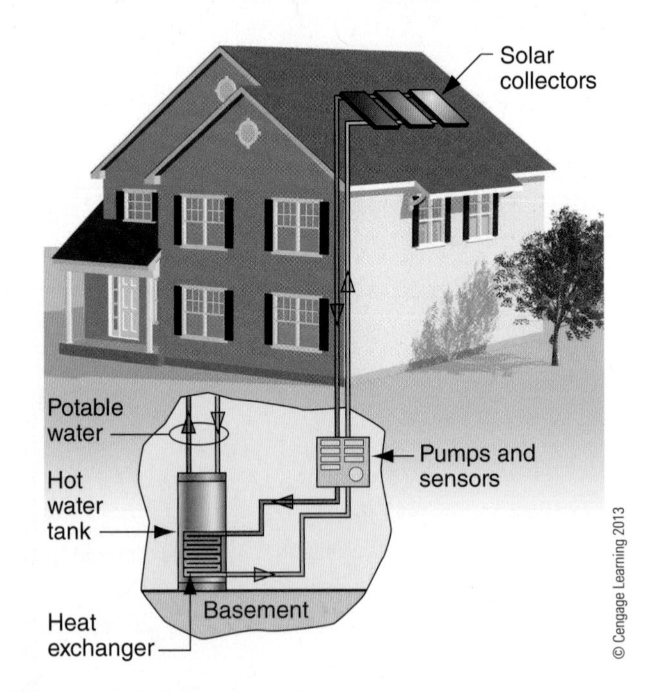

Figure 16.8 Solar thermal system components.

flat plate collector a rectangular solar thermal collector, typically 4' wide by 8' or 10' long.

evacuated tube collector a solar thermal collector that uses absorber plates that are enclosed in a glass tube with a vacuum inside.

tubes containing a second glass tube with a heat absorber inside. A third, less common type, parabolic concentrating, uses U-shaped troughs to concentrate sunlight on a tube that is placed on the focal line of the trough. Unglazed and plastic collectors are often used for pool heating in moderate climates.

Flat Plate Collectors Flat plate collectors function like small greenhouses, allowing heat from the sun to penetrate a clear glass cover and trapping it behind the glazing where it heats up a dark heat-absorbing plate. This plate transfers the heat to pipes that circulate water or antifreeze (Figure 16.9). The most common pipe in flat plate collectors is copper, which efficiently conducts heat from the absorber plate to the flowing liquid. High-performance collectors have double-glazing to retain more heat. Early collectors had black, painted heat absorbers; however, the development of selective coatings has improved efficiency significantly. Selective coatings, thin metallic layers applied to the absorber plate, are designed to absorb the maximum amount of the sun's heat to make collectors more efficient.

Evacuated Tube Collectors Evacuated tube collectors have a heat absorber bonded to a heat pipe inside glass tube. Sunlight striking the absorber material causes liquid in the heat pipe to boil, transferring energy to a condenser bulb at the top of each tube (Figure 16.10). The vacuum between the inner and outer glass tubes keep heat from escaping, making them very efficient in both cold and cloudy weather.

Parabolic Concentrating Collectors Parabolic concentrating collectors are the most efficient type of solar thermal collector; however, they are not common in residential applications. Limited primarily to large industrial installations, they generate steam used to operate turbines for electricity generation. Parabolic collectors can have tracking systems that follow the sun to further increase efficiency.

Collector Efficiency Ratings The SRCC rates collectors on basis of the total megajoules or million Btu (MBtu) of energy created per panel per day on clear, partly cloudy, and fully cloudy days according to the climate zone in which they are installed. These ratings can be used to compare the effectiveness of different collectors and determine the most appropriate product for a particular installation. SRCC collector ratings are not directly comparable to the complete system ratings they provide.

Collector Location Collectors should face as close to true south (as opposed to magnetic south) as possible with an unobstructed view of the sun between 9 a.m. and 3 p.m. every day of the year. The collector angle is usually equal to the latitude of the installation; however, some evidence suggests that placing collectors at a steeper angle captures more winter sun, providing better overall energy production. Due south is the preferred location, but collectors that are oriented as much as 45° toward the west or east will still provide up to 75% of the energy that they would gain if facing due south, providing some flexibility for installation on roofs that do not face directly south.

Collector Installation Collectors are installed on roof-mounted racks (Figure 16.11) or integrated directly into the roofing material, such as those available from VELUX, the skylight manufacturer (Figure 16.12). Rack-mounted systems can be adjusted for maximum solar exposure, although integrated mounting is effective on roofs that

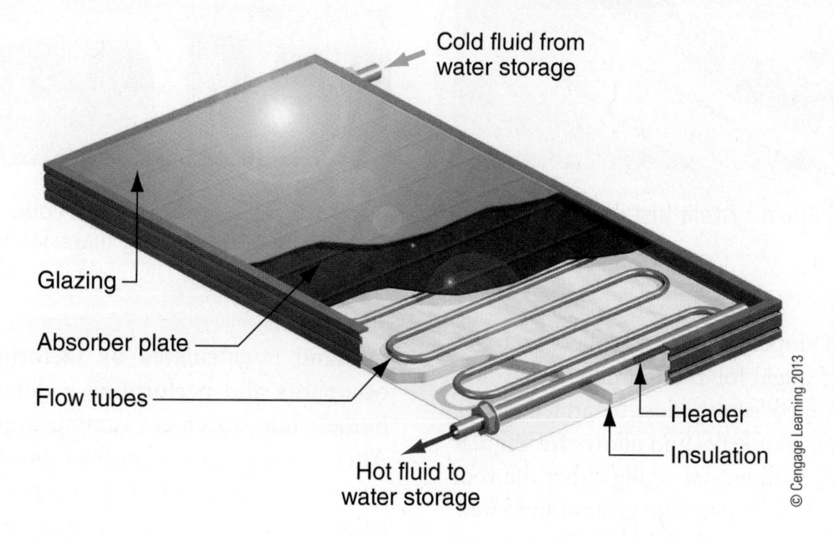

Figure 16.9 Flat-plate collector.

parabolic concentrating a type of solar thermal collector that uses U-shaped troughs to concentrate sunlight on a tube that is placed on the focal line of the trough.

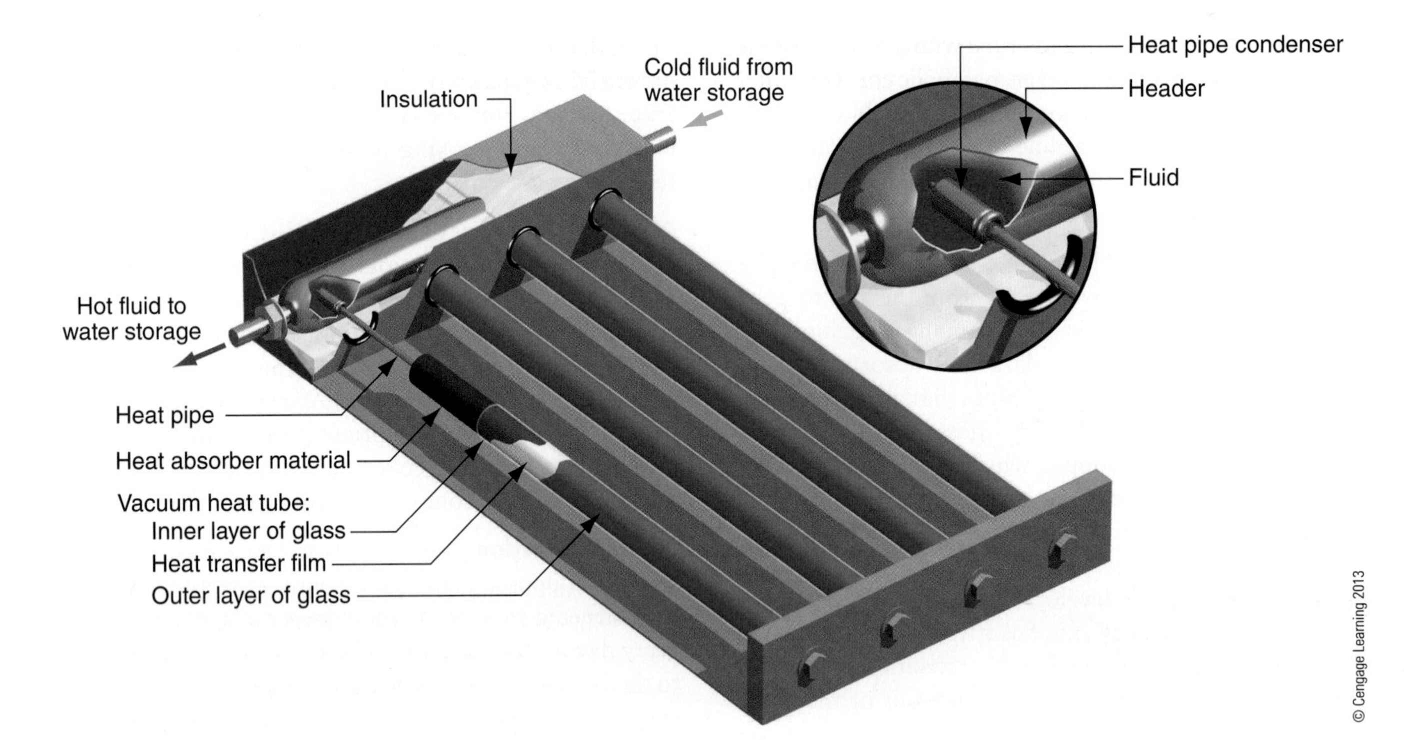

Figure 16.10 Evacuated tube collector.

Figure 16.11 Solar thermal system installed with roof-mounted racks.

Figure 16.12 Solar thermal collectors may be installed directly into the roofing material without the need for mounting racks.

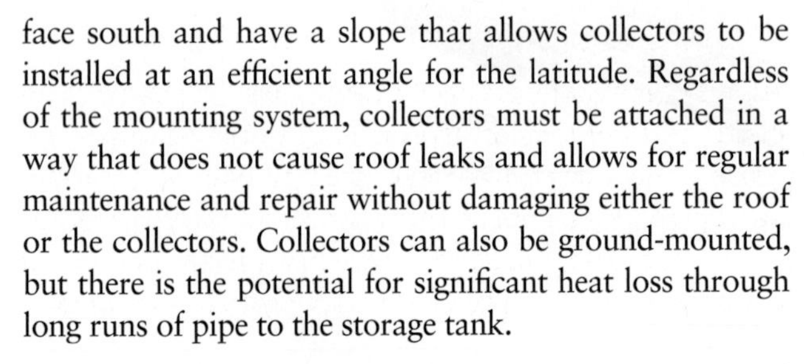

face south and have a slope that allows collectors to be installed at an efficient angle for the latitude. Regardless of the mounting system, collectors must be attached in a way that does not cause roof leaks and allows for regular maintenance and repair without damaging either the roof or the collectors. Collectors can also be ground-mounted, but there is the potential for significant heat loss through long runs of pipe to the storage tank.

Collector Size The first step in sizing a solar thermal system is to calculate the home's hot water demand. Demand is calculated by factoring in the number of occupants and performing a detailed inventory of the home's hot water–consuming appliances and fixtures. Alternatively, many installers simply use a national average of hot water consumption per occupant, such as 20 gallons of hot water a day. Be aware that as the number of occupants increases, the amount consumed per person actually decreases since there is some economy of scale. Most systems are not designed to provide 100% of the hot water demand because this would generally result in systems that are greatly oversized for part of the year. In general, there should be one gallon of hot water storage capacity for each gallon consumed daily.

The size of the collector array is calculated based on collector efficiency, orientation, installation angle, and the local climate. The amount of hot water produced ranges from 3/4 to 2 gallons per square foot of collector area per day. Most collectors come in standard sizes, and multiple units can be connected as needed to create an adequate hot water supply. Manufacturers provide sizing instructions and guidance for their products since sizing rules can vary between products.

Solar Thermal Storage

All solar thermal systems use storage tanks, often using a traditional tank water heater to store the heated water and to provide backup heat when solar energy is not adequate for the demand. Most systems use a single storage tank with electric backup heat; however, condensing natural gas heaters can also be used. Atmospherically vented water heaters are not recommended for solar thermal storage since heat in the water will escape through the open flue, requiring more backup heat than sealed units. Gas-fired tankless heaters are not commonly used with solar systems because the solar-heated water may be too warm to create enough of a temperature differential to operate a tankless heater; however they can be used with a heat exchanger to provide indirect backup heat. Electric tankless heaters that are designed to work with higher temperature water are able to provide the required range of additional heat and can be a good option as a backup for solar thermal systems. Most complete solar thermal systems are sold with collectors, coolant or water lines, pumps and controllers, and a single electric storage tank. Solar thermal systems are most efficient when there is a large temperature differential between the collectors and the tank. This suggests that a more efficient way to store and use solar hot water is to employ two tanks, using the solar heated tank as a pre-heater for an electric or fuel-fired tank. When using a single tank, the backup heat will run more frequently overnight and during periods of low solar energy, lowering the efficiency of the solar portion of the system.

Solar Thermal Pumps and Controllers

Most systems come with matched controllers and pumps as well as check valves, gauges, thermostats, and other required equipment. Look for high-efficiency pumps, such as those available for hydronic heating systems (see Chapter 13), to reduce energy used in moving water or antifreeze through the system. Some systems include photovoltaic panels that provide power for pumps that generally only need to run when the sun is shining, further reducing energy consumption.

Solar Thermal Maintenance

Antifreeze-based systems should have their pH checked annually, and new coolant should be added or replaced when necessary to avoid corrosion and for highest performance. As with any water heater, the tank should be drained periodically to remove any sediment, and the manufacturer's recommended maintenance schedule should be followed.

Solar Thermal Ready Homes

Preparing a house for a future solar array is not complicated and can simplify future installation. A south facing roof should have a minimum of 200 square feet of unshaded area with no mechanical penetrations (Figure 16.13). There should be an area 4' × 2' × 7' in the utility room, adjacent to the conventional water heater, for the future solar hot water tank and a solid wall space of 3' × 2' for the controls and pumps. Two 1" fully insulated copper lines should be run from the attic beneath the roof area where the system will be installed to a location for the future storage tank. If the structure permits, install two 3" or 4" straight pipe chases (instead of copper lines) from the attic to the utility room to allow for installing hot water lines in the future.[1]

Solar Space Heating

Solar thermal systems can be used to provide heat for hydronic or forced-air heating systems, and individual rooms can be heated with air collectors. As previously mentioned, water-based systems are considered active, requiring pumps and controls to move heat from collectors to conditioned space. Collectors that heat air are referred to as active because they normally use fans to circulate air, although passive models can work by convection. Although nearly impossible to completely heat a home with solar energy in most climates, solar space heating can often provide 25% to 75% of the annual space heating load. Solar space heating systems are sized to meet the heating load based on Air Conditioning Contractors of America Manual J load calculation.

Active Solar Thermal Space Heating

Active solar thermal space heating uses the same principles and equipment as domestic water heating with more collector and storage area to meet the larger energy requirements. Heated water is stored in large tanks that can be used for space heating when it is hot enough, or used to pre-heat water that runs through a

[1] The Canadian Solar Industries Association (canSIA) and the U.S. National Renewable Energy Laboratory have additional resources on "Solar Ready Homes."

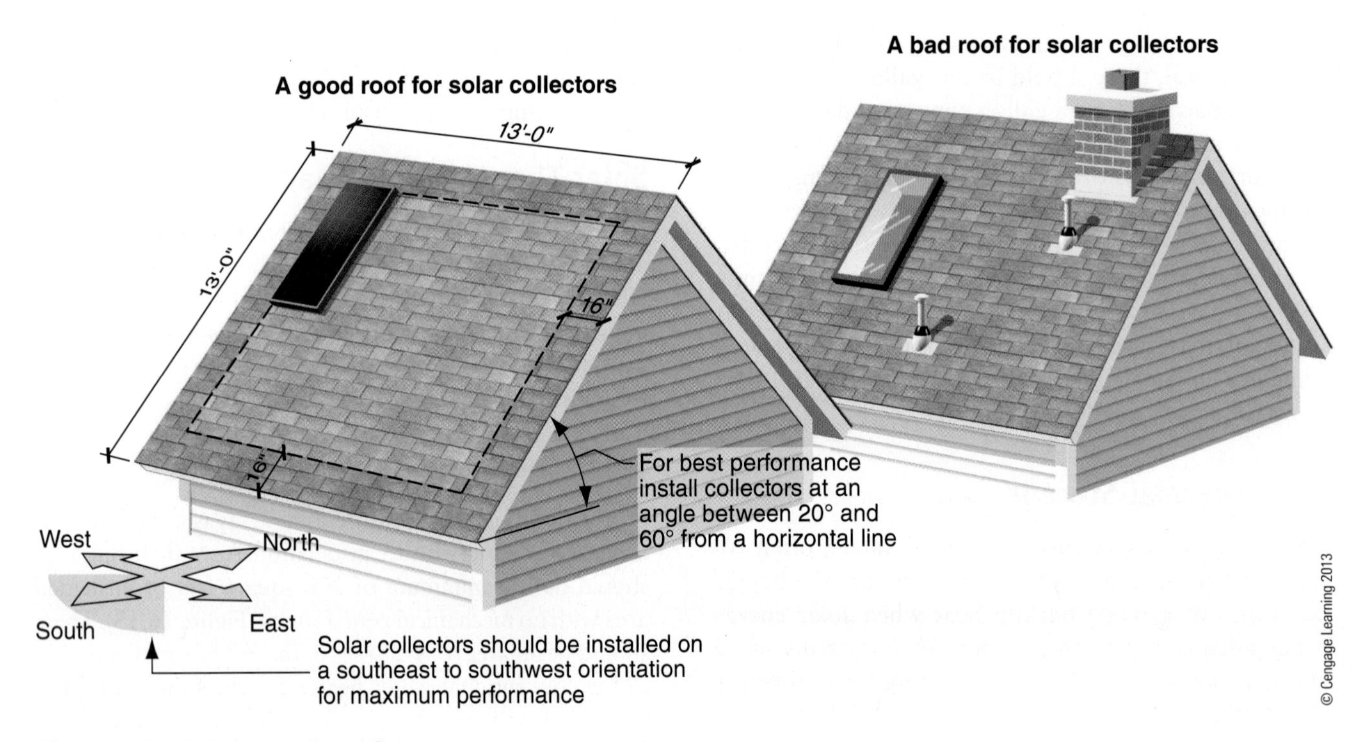

Figure 16.13 Solar-ready roofs.

boiler, reducing energy use. Depending on the climate, building size and efficiency, and how much energy is to be offset, space heating systems may require as few as 4 and as many as 20 collectors combined with 250 to 5,000 gallons of water storage capacity. Solar heated water can be circulated to in-floor radiant pipes, surface-mounted radiators, coils in central air handlers, or fan-coil units.

Solar Air Space Heating

Solar air collectors are mounted vertically on south-facing walls or on a roof slope. When the sun is shining, outside air passes into the collector and rises as it is heated, flowing into the house by convection or with a fan. The heated air can also be connected to a forced- air system for circulation through the house. Air collectors, such as those available from Cansolair and SolarSheat, are self-contained units that can be installed on any south-facing exterior wall (Figure 16.14). Solar air heating can provide supplementary heat in cold, sunny climates, but it is not typically used as a primary heating source.

Site-Generated Electricity

When generating electricity with the sun, wind, water, or fuel cells, it must be either stored for future use on-site, used when generated, delivered to the power grid, or a divided among combination of these. Renewable energy systems in homes work in either grid-tied or off-grid arrangements, regardless of the method used for creating power. Grid-tied systems are connected to utility electric service lines. Off-grid systems have no connection to service lines. All systems employ inverters to convert power from direct current (DC) to alternating current (AC), and most off-grid systems have batteries to store energy for use when needed.

Solar Dehumidification

As discussed in Chapter 13, maintaining low humidity can allow homes to be comfortable with higher summer temperatures; the use of dehumidification instead of, or in addition to, air conditioning can be an energy efficiency strategy in humid climates. While most dehumidifiers are powered with electricity, solar-powered models are a more efficient alternative. **Desiccants**, materials such as calcium oxide and silica that absorb water, can be used to remove humidity from indoor air that is passed through it with fans. As the desiccant becomes saturated with water, the air becomes dryer and cooler. Solar energy can be used to dry out the desiccant, allowing it to remove more moisture from the air while using little or no energy. When photovoltaic power is used for operating the fans, desiccant dehumidifiers can be operated entirely with solar power.

---

desiccant materials such as calcium oxide and silica that absorb moisture that are commonly used to remove humidity from indoor air that is passed through it with fans.

Figure 16.14 Solar space heaters are available as wall- and roof-mount models. Both of these units have their own independent thermostatically controlled impeller fan. For this residence, the roof-mount is facing southeast and turns on first thing in the morning; the wall-mount faces southwest and turns off last thing in the evening. For approximately 4 hours during the day, both units are running at the same time.

Grid-Tied

Grid-tied systems do not normally function during power failures to prevent islanding, the introduction of power into a dead grid, which can pose an electrocution hazard to workers. Disconnect switches are required to use power from grid-tied systems when the central power is out of service. Grid-tied systems are classified as either feed-in-tariff or net-metered. In feed-in-tariff (FIT) systems, utilities purchase renewable energy at variable rates, which are usually higher than the sales price. This is tracked by using two meters: one to measure electricity going to the grid, and another for electricity coming from the grid. Net-metered systems buy and sell electricity at the same rate, using a single meter, which runs either forward or reverse, depending on the direction of power flow. Most net-metered systems do not provide homeowners with credit for any electricity they generate beyond what they use. FIT systems, however, provide 100% credit for power put into the grid, allowing homeowners to receive a check from their utility when their production exceeds their use. Regulations for grid-tied systems are established by the local municipality or state. Not all utilities allow FIT or net-metering.

feed-in-tariff (FIT) system utilities purchase renewable energy at variable rates, which are usually higher than the rates at which the energy is sold.

net-metered systems homes that produce some of their electrical use on-site and use a single meter, which runs in either forward or reverse depending on the direction of power flow, to sell or buy power from the utility.

Off-Grid

Off-grid systems use a charge controller between the power source and storage batteries to make sure that they are charged properly according the manufacturer's specifications. Power leaving the batteries must go through an inverter to change the current from DC to AC, which then runs to the electrical panel for distribution through standard wiring to household electrical equipment. Off-grid systems other than fuel cells often use fuel-fired backup generators to provide power when the batteries are drained.

Inverters

Inverters convert site-generated DC power to AC for use in the home and to direct back to the power grid for FIT and net-metered systems. The inverter first converts the DC current to AC, then increases the voltage to the required 120 volts for use in the house. Inverters can also include controllers that manage power from a backup generator, a charger for batteries, and circuit breakers for safely managing both grid and site-generated power.

Batteries

Off-grid systems typically charge batteries with the site-generated DC current using a charge controller, a device that manages the amount of current going to the batteries to keep them properly charged and prevents damage from overcharging. Power stored in batteries must run through an inverter to change it to 120V AC power for use in the house.

Sizing Renewable Systems

Sizing, or selecting the amount of electricity a specific system will generate, begins with calculating the home's actual or estimated energy consumption. This is accomplished through the following steps:

1. Prepare an inventory of all electrical lights and appliances, including electric space and water heating, cooling, ventilation fans, and other equipment.
2. Estimate the amount of daily usage for each item.
3. Multiply the total light or appliance wattage by the estimated daily use.

This provides the daily energy requirement in watt-hours. Importantly, some lights and appliances have seasonally varied usage rates, which should be factored into the calculation. For example, air conditioners are only used

inverter a device that converts site-generated DC power to AC current for use in the home.

during the cooling season, and lights tend to be used more in the winter. The U.S. Department of Energy and other organizations provide average energy consumption, annual hourly use, and estimated annual energy consumption for lights and appliances. In Chapter 14, Table 14.7 shows typical wattages for home appliances. This total watt-hours quantity provides a guideline for the capacity of a photovoltaic system, allowing up to 100% of the estimated load to be from solar power.

Solar Photovoltaic Systems

Solar photovoltaic systems use the sun's energy to generate electricity for use in a house, replacing all or a portion of power required from electric generation plants. Photovoltaic systems are used on individual homes as well as in industrial scale installations to generate electricity for distribution through the power grid.

Solar Photovoltaic Components

Photovoltaic systems consist of panels or modules that collect solar energy and turn it into DC current, an inverter that converts DC into AC current to run lighting, appliances, and other equipment, and switching equipment to direct power back to the grid or to batteries for storage (Figure 16.15).

Photovoltaic Panels

Photovoltaic panels fall into two primary types: wafered silicon and thin film. When sunlight strikes these materials, electrons are knocked loose from their atoms, causing them to flow through the material to create an electric current (Figure 16.16). Silicon panels are made of a collection of individual silicon semiconductor cells, each in the range of 150 mm wide and 350 microns thick, wired together and protected by a layer of glass in a frame. A group of panels is installed and wired together into an array to generate a specific amount of energy. Thin film can be made from fine layers of silicon or from other semiconductors (such as cadmium telluride) that are produced in thicknesses in the range of 1 micron, requiring less material, reducing costs, and allowing them to be flexible. Thin film can be applied to roofing, glass, or similar smooth surfaces, in both flat and curved configurations. Thin films are less expensive to manufacture than silicon panels; however, they are not yet as efficient. Thin film can be used in clear glazing for windows and skylights that provide natural light while generating electricity.

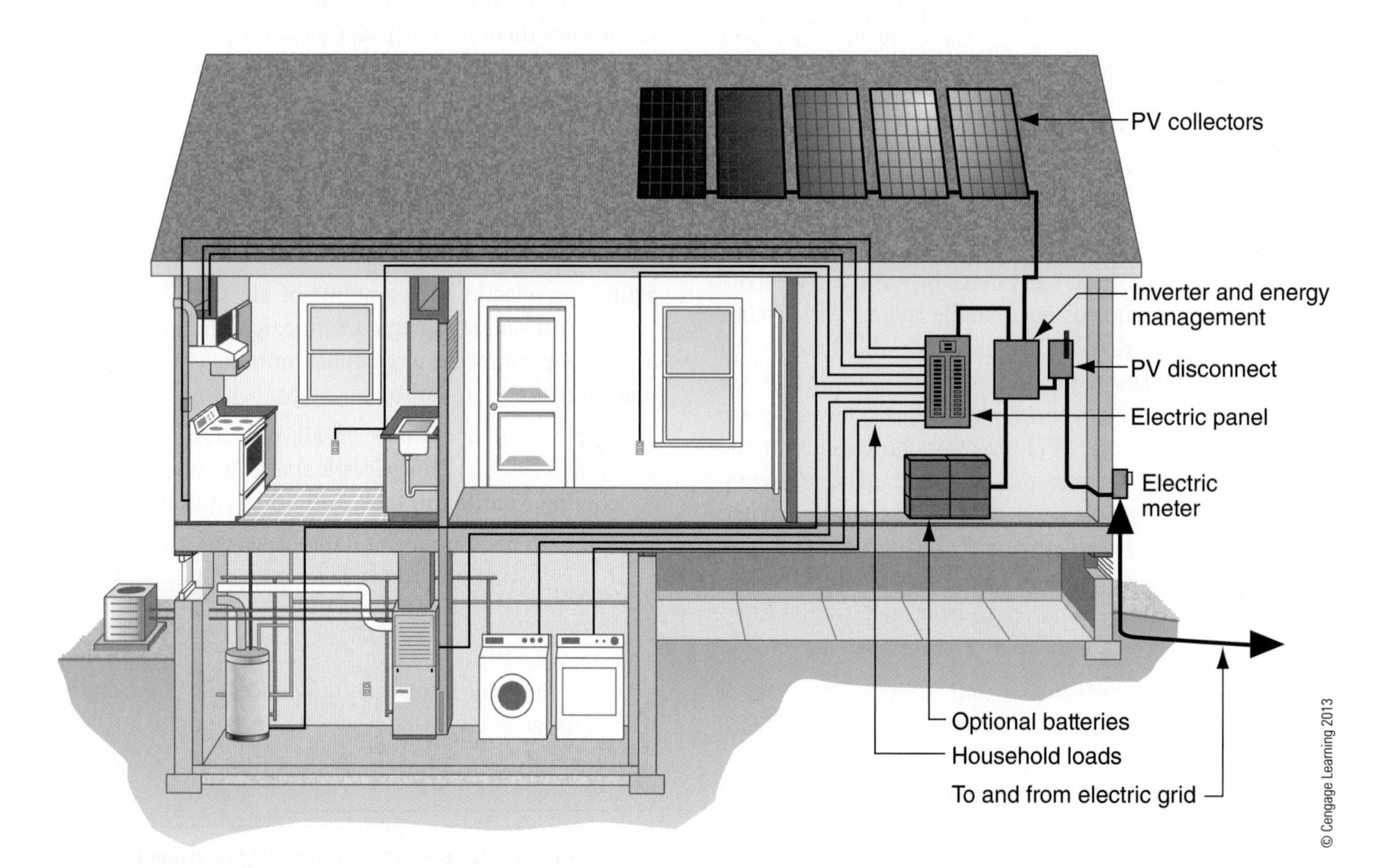

Figure 16.15 Solar photovoltaic system.

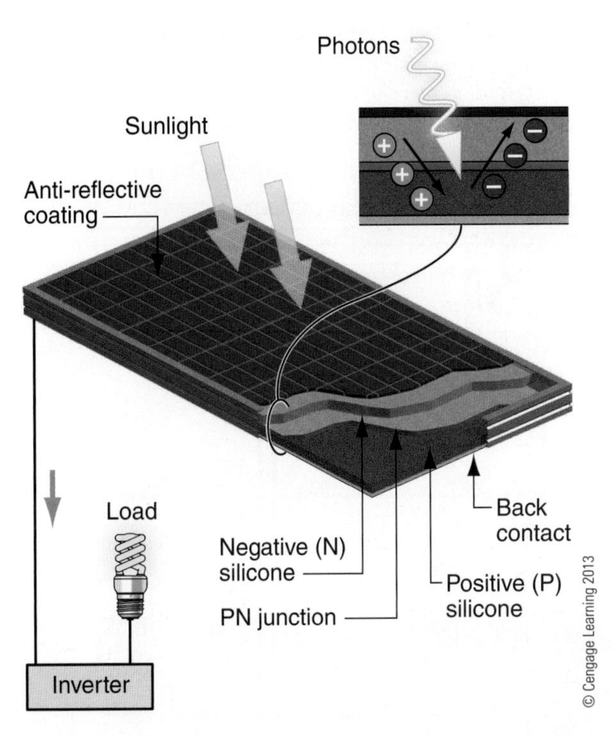

Figure 16.16 Solar panels are composed of small solar cells, which produce electricity. The panels, in turn, make a solar array.

New designs for solar panels are being tested with solar inks, dyes, and plastics. Advanced technologies include concentrators that focus the sun's energy onto smaller areas of very high-efficiency photovoltaic material. Photovoltaic systems continue to evolve and are expected to increase in efficiency and decrease in price over time.

Collector Efficiency Ratings Manufacturers of wafered silicon panels claim efficiencies in the range of 13% to 18%; and thin film products claim efficiencies of between 5% and 10%. Experimental thin films have produced efficiencies of almost 20%, and both types continue to increase efficiency as technology improves.

Collector Location As with solar thermal collectors, photovoltaic arrays should face due south. The general recommendation is for fixed panels to be set at an angle equal to the latitude for the best average solar exposure. Tracking arrays that automatically follow the angle of the sun increase effectiveness, but they increase costs and can be subject to mechanical breakdown. Adjustable arrays can be manually moved to a more vertical position during the winter and relocated in the summer. Integrated thin film arrays are installed as part of roofing shingles, slates, tiles, or metal panels. In regions with regular snowfall, arrays may need to be cleaned off regularly to produce optimum power, so consider locating them in an easily and safely accessible location.

Collector Installation The most common collector installation is on a roof using a racking system, which allows a flow of air below the panels to keep them cool and extend their life. Arrays can also be ground-mounted if the house roof does not face south or if too many obstructions are present. Any connections to the roof must be designed for easy maintenance of both the panels and the roof and be installed with waterproof connections to avoid any penetration of rain into the structure. Excessive heat can reduce the efficiency of photovoltaic panels, so an array that is attached to the roof should be designed with an air space on the underside to allow for air to flow to help keep them cool in hot weather. Thin film materials can be field or factory applied to roofing materials, which can be wired together into an array. Shingle products with photovoltaic modules are now available from various manufacturers, including Dow. Arrays can serve as awnings, providing shade and protection from the weather (Figure 16.17). Skylights with photovoltaic panels integrated into glazing are becoming available, providing both daylight and power generation. When considering solar glazing products, locate and shade windows to avoid overheating the building interior in warm climates, a situation that may require additional mechanical cooling that could offset some or all of the power generated.

Figure 16.17 Solar photovoltaic arrays can be integrated into awnings and canopies to provide window shading.

Collector Size Using solar potential maps from NREL, the total energy available (measured in kWh/m²/day) can be determined for a specific location. For example, if the total kWh/m²/day is 6 and wafered silicon panels are 15% efficient, then the total expected output will be 0.9 kWh/m²/day. If there is a total of 10 m² of panels, then the system can be expected to generate on average 9 kWh of energy per day. The solar potential, the panel efficiency, and the amount of energy required from the photovoltaic system must be calculated to determine the total area of panels required for an installation. Keep in mind that most panel ratings are based on ideal sun conditions that are uncommon and unreliable. A common practice is to de-rate the performance of arrays 20% to 40% to allow for local conditions.

Solar Photovoltaic Maintenance

Photovoltaic panel maintenance is typically limited to removing snow and cleaning pollen and dirt that may accumulate. Inverters, charge controllers, batteries, and the panels themselves may be subject to occasional breakdown or failure; however, required regular maintenance is minimal.

Solar Photovoltaic–Ready Homes

Sometimes photovoltaic areas are not in the project budget, but the homeowner is interested in a future installation. Preparing a house for a future solar array is not complicated and can simplify the eventual installation if done properly (Figure 16.18). Where there is a south- or near-south–facing roof, it should be designed to have a minimum of 200 ft² of unshaded area with no mechanical penetrations, such as plumbing vents, wires, or other obstructions. Sloped roofs should be angled as close to the latitude of the project as possible. Flat roofs provide a good opportunity to install panels either flat or angled towards the south. Conduits or chases that are accessible immediately below the roof and in a mechanical area will allow for easy installation of connecting wires. Providing space for installation of an inverter and other equipment, such as a FIT meter, should be considered as well.

The Environmental Impact of Solar Photovoltaic Manufacturing

All products and processed materials carry an environmental impact. This is particularly true for renewable energy systems and should be considered when selecting on-site generation systems.

The production of silicon used in photovoltaic systems requires the use of toxic and dangerous chemicals and produces such by-products as silicon tetrachloride that can be hazardous to the environment and individuals.

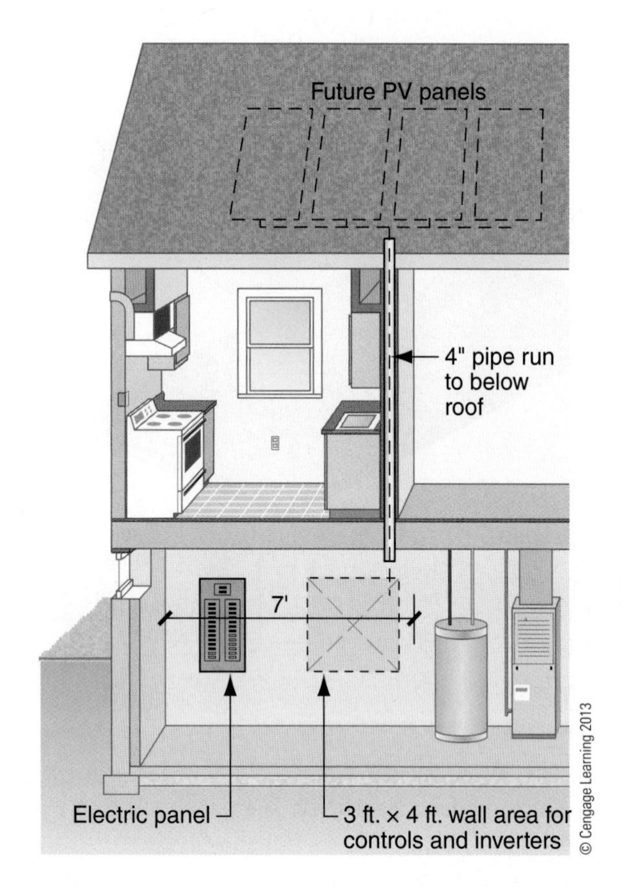

Figure 16.18 A home ready for solar photovoltaic arrays has space to accommodate future equipment near the electrical panel and the ability to easily run electrical wiring to future panels.

Many solar photovoltaic panels are manufactured overseas where costs are lower. In his book *The Ecology of Commerce,* Paul Hawken points out that China has not only lower labor costs but also more lenient environmental controls. These conditions allow them to produce silicon photovoltaic panels at a lower cost than in the United States and Germany where the silicon tetrachloride must be processed and recycled rather than simply discarded. This recycling is costly and requires significant amounts of energy. Hawken estimates that western companies' costs of producing silicon are between two and four times higher than Chinese companies due to the cost of pollution controls. It is worth considering the overall environmental impact of such products and their place of origin in addition to the energy savings they provide.

Wind Power Systems

The most appropriate site for wind power is one that has a combination of consistent, high-velocity winds, few obstructions, available space for wind turbines, and no legal restrictions against installing them. Power increases with the cube of the wind, giving a site with an average wind speed of 18 MPH wind eight times more power than one with an average speed of 9 MPH.

While the most common wind turbine installations are large, utility-scale operations, individual small scale turbines can, under the right conditions, generate between 300 and 2,000 kWh, depending on local wind speed, turbine size, and tower height.

Wind Power Components

Wind power systems consist of a tower, the turbine blades, a generator, and the same combination of inverters, batteries, and charge conditioners as all site-generated power systems (Figure 16.19). Towers can be self-supporting or reinforced with guy wires, consisting of a single smooth post or an open web structure. Self-supporting towers require larger foundations but do not require guy wires spreading out around the base. When wind turns a turbine's blades, it moves an internal generator, producing electricity.

Turbines are available for rooftop installation but they are generally not considered to produce enough consistent energy to be viable competitors to traditional renewable energy systems.

Wind Turbines

Turbines rotate on either a horizontal or vertical axis, with the majority being horizontal. Most turbines have three blades, an integrated generator, and a tail to direct the blades into the wind. The larger the diameter of the blades, the more energy a turbine can capture and turn into electricity.

Wind Turbine Effectiveness Wind turbine performance is difficult to evaluate and compare, with ratings often overstated by those selling and installing them.

Courtesy of Southwest Windpower

Figure 16.19 A relatively small backyard wind turbine.

A 2008 report for the Massachusetts Technology Collaborative investigated a group of small wind turbine installations and determined that the actual energy production was less than one third of initial estimates, ranging from as little as 2% to as much as 59% of original projections.

Wind Turbine Maintenance As with any mechanical devices, wind turbines are subject to mechanical failure and require maintenance. Since they are typically mounted on tall towers, wind turbines are much less accessible than other renewable energy systems, increasing the costs of maintenance and repair. These costs should be taken into account when choosing between different methods of power generation.

Micro-Hydro

A site with a consistent volume of flowing water and no legal or environmental restrictions may be a good candidate for a micro-hydro power system (Figure 16.20a and Figure 16.20b). High-head systems produce a low flow of high-pressure water, whereas low-head systems produce a high flow of low-pressure water. Since the installation of a micro-hydro system restricts water flow, it will have a downstream effect. The overall flow of water should not be reduced so much that it has a negative effect on aquatic life and the natural stream path.

Hydropower Components

Hydropower systems consist of the water source, a flume or a dam to capture and direct water, pipes, intake screens to keep fish and debris out of the system, the turbine, a generator, and various valves, pressure release devices, and gauges. Water is directed to the turbine through a flume or a pipe running through a dam, employing valves to turn off water for installation, maintenance, and replacement operations. High-head systems may use a penstock conduit, which is a pipe flowing a long distance downhill, directing high-pressure water to a small turbine. Low-head systems may use a flume to direct a large volume of low-pressure water to a large turbine. Costs and maintenance considerations include regularly cleaning filters, repairing and replacing mechanical equipment, and removing silt from streams.

Fuel Cell Systems

Although residential fuel cell systems are not yet readily available, prototypes are being tested and should become commercially available in the near future

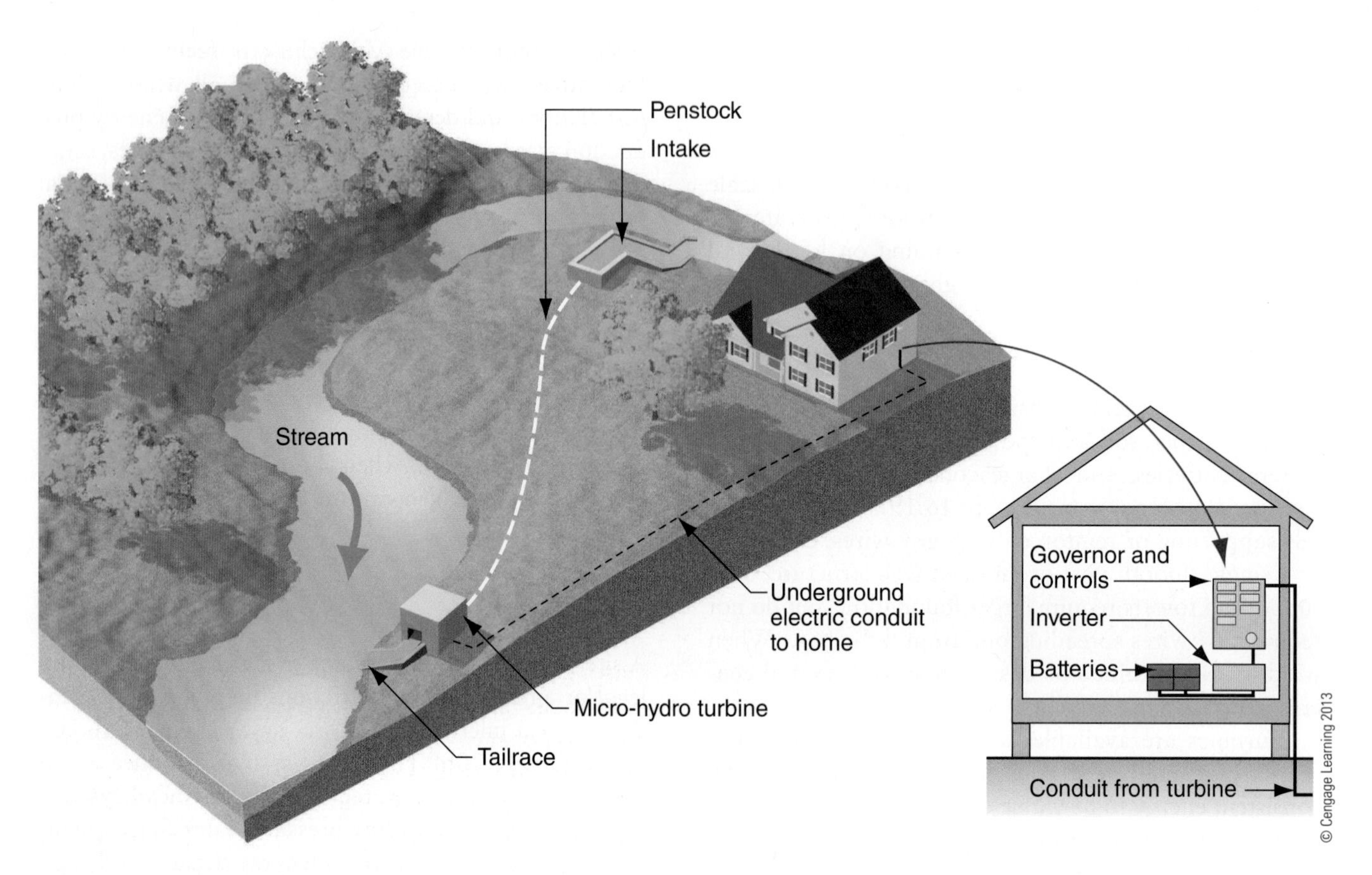

Figure 16.20a Micro-hydro systems divert water from a stream or river and direct it down hill to a turbine that generates electricity.

Figure 16.20b A close up of the Gentleman's Hydro System.

(Figure 16.21). Residential fuel cells are expected to be able to produce in the range of 5 to 7 kW from a unit roughly the size of a chest freezer, typically located on the exterior of a home. Fuel cells can function similarly to combined heat and power systems, producing both electricity as well as heat that can be used for space or water heating. Commercial and large-scale residential fuel cells are available from such companies as ClearEdge and Bloom Energy.

Although fuel cells are expected to be available at a similar cost per kW as solar photovoltaic panels, they still require fuel to operate. As a result, operating costs are ongoing, unlike most other renewable energy systems.

Passive Solar House Design

Passive solar design was discussed in detail in Chapter 3. While passive solar takes advantage of the sun's energy, entire passive solar homes are complex systems that exceed the scope of this text. When passive solar design principles were being explored in the 1970s, one common design concept was a **sunspace**, a passive solar addition to a new or existing house to provide additional winter heat through solar energy. Like all passive solar designs, sunspaces require very active management by occupants, such as opening and closing window shades and insulated

---

sunspace a passive solar addition to a new or existing house to provide additional winter heat through solar energy.

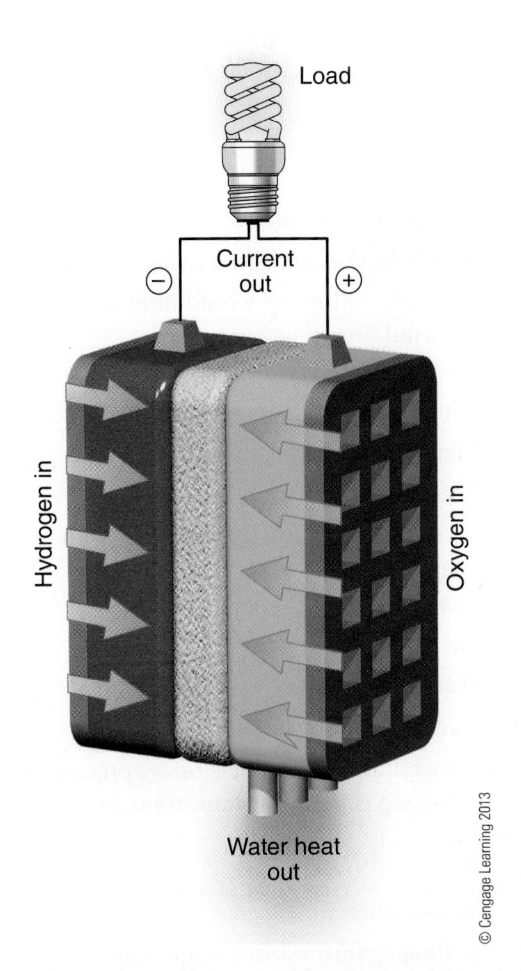

Figure 16.21 A fuel cell converts the chemical energy from a fuel, typically hydrogen, into electricity through a chemical reaction with oxygen or another oxidizing agent.

covers and operating vents to take advantage of the heat they generate and to prevent overheating. Many homeowners with passive solar sunspaces ended up installing HVAC systems for convenience, which in turn, negated their value for passive space heating and ultimately defeated their purpose.

When a homeowner is willing and able to manage a truly passive solar home or sunspace appropriately, they can provide significant energy savings. Keep in mind, however, that they have not proven to be very successful with mainstream homeowners and are typically reserved for custom projects.

Renewable Energy in Remodeling Projects

Installing renewable energy projects is not difficult for existing homes, provided the site meets a system's requirements and space is available on the interior for installation of the required pipes and wires necessary for operation. Regardless of the solar or wind potential of a site, investing in building envelope improvements is critical in existing buildings (just as with new homes) to first reduce the overall energy load as much as possible before installing renewables.

Summary

Renewable energy systems can be an effective addition to any green building project, provided that they are considered in the context of the entire building. The total energy use of the building—based on the size, climate, orientation, envelope efficiency, and owner behavior—must be considered, and all available methods to reduce energy use should be incorporated into the project before renewable energy options are taken. When renewables are to be included, the entire building design, from the roof orientation to accessibility for required pipes, wires, ducts, and other necessary connections, must be taken into consideration from the earliest stages of the design for maximum building performance.

Review Questions

1. Which of the following is the most important factor when deciding to use renewable energy in a project?
   a. Wind speed
   b. Solar potential
   c. Load reduction
   d. Cost of photovoltaic panels

2. Which of the following terms describes a house that uses renewables to produce as much energy as it consumes?
   a. Passive house
   b. Passive solar
   c. Near-zero energy
   d. Net-zero energy

3. Which of the following is not a component of an indirect solar thermal system?
   a. Collectors
   b. Storage tanks
   c. Inverter
   d. Heat exchanger
4. Which of the following compass directions for collector location will provide the most efficient solar thermal heating?
   a. Northeast
   b. Southeast
   c. East
   d. West
5. Which of the following is the most common type of renewable energy system used in residential construction in the United States?
   a. Fuel cells
   b. Wind
   c. Micro-hydro
   d. Solar thermal
6. Which of the following types of grid connection provides the most potential financial benefit for a homeowner?
   a. Net metering
   b. Feed-in-tariff
   c. Batteries
   d. Micro-hydro
7. Which of the following is the most efficient method of converting solar energy into electricity in homes?
   a. Thin film
   b. Wafered silicon
   c. Flat-plate collectors
   d. Parabolic concentrating collectors
8. Which of the following would limit a site's potential for wind power generation?
   a. Wind power class rating of 5
   b. Ring of 50' tall trees around site
   c. Net metering availability
   d. Feed-in-tariff availability
9. Which of the following is the most common use for solar thermal power?
   a. Space heating
   b. Generating electricity
   c. Water heating
   d. Radiant floor heating
10. When installing solar panels on a home, which of the following is the most important issue to consider?
   a. Panel size
   b. Wind uplift
   c. Sealing against moisture intrusion
   d. Access for maintenance

Critical Thinking Questions

1. How does the size, orientation, interior layout, and other features in a house affect the need for renewable energy systems?
2. What are the pros and cons of different systems to generate on-site electricity?
3. How does the home's geographic location affect the selection of renewable energy systems? Provide specific examples.

Key Terms

biofuels, 455
desiccant, 466
evacuated tube collector, 462
feed-in-tariff (FIT) systems, 467
flat plate collector, 462
fuel cell, 455
grid-tied, 457
inverter, 467
net-metered systems, 467
net-zero energy, 457
North American Board of Certified Energy Practitioners (NABCEP), 458
off-grid, 456
parabolic concentrating, 463
photovoltaic (PV), 453
solar potential, 458
Solar Rating and Certification Corporation (SRCC), 458
solar thermal, 453
sunspace, 472

Additional Resources

Database of State Incentives for Renewables & Efficiency (DSIRE): http://DSIREusa.org

Find Solar: http://www.findsolar.com/

North American Board of Certified Energy Practitioners (NABCEP): http://www.nabcep.org/

Solar Energy Industries Association (SEIA): http://www.seia.org/

Solar PathFinder: http://www.solarpathfinder.com/

Solar Rating and Certification Corporation (SRCC): http://www.solar-rating.org/

Epilogue

Putting It All Together

As you make your way through the green building maze, consider how you can contribute to making all of your projects as green as possible from the start. What can you do differently from what you are used to, or differently from the way others work, to create a project that incorporates all the tenets of green building from the very beginning of the design process? Whatever your role in the project—designer, developer, owner, contractor, or consultant—you will contribute to its sustainability by bringing your knowledge, experience, and commitment to green building to the team.

Your Path to Green Building

There are many routes to becoming experienced in green building, and many of them can run parallel. You might start by obtaining a professional designation or accreditation, such as those offered by the GBCI, NARI, NAHB, BPI, RESNET, and other organizations. Undergraduate and graduate degree programs in architecture, design, engineering, and building construction provide opportunities to study sustainable design. Regardless of the training you receive, you will learn the most from real-world experience. Work with or for an experienced professional as part of a team on a project seeking LEED for Homes, NGBS, Passive House, or other certifications. You may choose to build, renovate, or develop your own project, bringing in experienced professionals to assist you, learning from them throughout the project.

Understanding the Challenges

As we have stressed throughout this book, green building is a complex process, often with several correct answers to a single problem. Knowing which answer is best requires a solid understanding of the green building principles and properly applying them to the specific requirements of each particular project. The climate, building design, owner priorities, budget, and many other factors all play roles in making the most appropriate decisions regarding materials and methods.

Making the Right Decisions

The breadth and scope of green building and the wide range of options can prove daunting for even experienced professionals. New products claiming to be "green" compete for our business almost daily. As we discussed in Chapter 1, dozens of competing green product certification programs can provide guidance as well as increase confusion.

Industry experts do not always agree, and individual opinions often evolve over time. Continuing research into building materials and methods can lead us to embrace new technologies, but sometimes we also discover new challenges that may lead us back to older techniques that we believe provide, on balance, a better combination of performance, occupant health, durability, or sustainable material use.

Overcoming Objections from the Mainstream

While much of the building industry is embracing green building, many professionals remain resistant to change, sometimes avoiding adoption of even established high-performance practices. Improved energy codes, market demand, and education will help convince people to make necessary changes. Seek out open-minded professionals who are willing to be part of a team, listen to others, and be open to changing their methods when appropriate. Team members who are unwilling to consider appropriate alternatives to their standard practices are not the kind of professionals you want working on your green project. Find people who can work collaboratively with the rest of the team, helping to ensure a high-performance project.

Plan for Personal Success

Strive for a balance between achieving overall sustainability and managing change effectively. Trying to make too many changes simultaneously may lead to problems for the entire project team throughout the process and with the completed building. When you are fortunate to have enough experienced team members and a forward-thinking owner who push for the highest performance, take advantage of this and set high goals. When your team is less experienced and more conservative, you should be comfortable with more modest changes that ensure better performance without sacrificing overall project quality. Understand your limitations, build on previous successes, and learn from challenges and failures—both yours and others.

FROM EXPERIENCE

Two Journeys into Green Building

Abe Kruger, CEO and Founder, Kruger Sustainability Group; Author From an early age, I was drawn to construction and the environment. Throughout high school, a typical weekend would include hiking, volunteering with a local Habitat for Humanity affiliate, or building sets in the theater. I enrolled in Oberlin College, fully intending to change the world, though not yet understanding how. Environmental Studies was the logical first step. My courses ranged from international renewable energy policy to the geology of northern Ohio. In addition to my degree in Environmental Studies, I discovered a love of history. I was fascinated with the interplay of natural and built environments. My history courses explored how natural elements, such as soil fertility and proximity to water, influenced the location of cities and how, in response, cities changed their surroundings.

Abe Kruger, CEO and founder, Kruger Sustainability Group; author.

One of the reasons I was drawn to Oberlin is their Adam Joseph Lewis Center (AJLC) for Environmental Studies, which was designed by William McDonough. The AJLC was designed to be a cutting-edge teaching tool and remains one of the most ecological buildings in the country. David W. Orr, professor and director of Oberlin's Environmental Studies Program, coordinated efforts to design, fund, and build the AJLC. The overarching goal was to physically demonstrate that buildings can be designed "so well and so carefully that they do not cast a long ecological shadow over the future that our students will inherit . . . buildings can be designed to give more than they take."[1]

While at Oberlin, I learned the global need for green building and studied many of the strategies and technologies that go into such structures. But it was at Southface Energy Institute in Atlanta that I really learned about the construction process and received hands-on experience. Initially, I coordinated all services for existing homes and home energy modeling (both HERS ratings and heating and air conditioning load calculations). This provided me with valuable experience in evaluating homes. Over time, I found my niche in builder, contractor, and homeowner education. I was the lead HERS Trainer and helped administer the ENERGY STAR program.

After leaving Southface, I worked for one of the largest waterproofing contractors in the southeast. The company was committed to keeping structures dry through high-performance moisture management systems. Saving the environment was not the primary goal of the company even though its services contributed to green buildings. One of the owners described green building simply as "the right way to build" and said that their goal was to create long-lasting, healthy buildings. They also installed conditioned crawl spaces and spray polyurethane foam insulation. I managed their green building certification and inspection services. During my time with the company, I inspected over a hundred new and existing homes. I witnessed first-hand the range of home designs to which green building principles can be applied, as well as the spectrum of homeowners they attract.

In 2009, I started my own consulting company to provide green building training, consulting, and curriculum development for colleges, utilities, corporations, and nonprofits. We are committed to strengthening and expanding the green building community through education and outreach. We partner with organizations to empower builders, contractors, and homeowners to green their homes and work places.

My journey has taken me down many surprising paths. One week, I may be conducting a green building training for builders, and the next consulting with an electric utility on its demand-side management or energy efficiency programs. The green building industry is constantly evolving and is filled with energetic, talented, and passionate individuals who are working hard to move our industry ahead. Their energy can be felt at numerous conferences and trade shows around the country.

The beauty of green building is that you do not need to be a tree hugger to see the value. Green building is a sound economic investment for homeowners; it can directly improve occupant health and well-being, and yes, it is good for the environment. Green building is not about politics or income. Green building is simply the right way to build.

[1] David Orr, http://www.oberlin.edu/ajlc/ajlcHome.html

FROM EXPERIENCE

Carl Seville, Seville Consulting; Author When I was about 8 years old, I decided I wanted to be an architect when I grew up. After several years of technical and architectural drawing classes in high school, I enrolled in the Rhode Island School of Design, intending to become an Architect. I left school with a degree in Industrial Design, having taken a year off to work as a carpenter and leaving the architecture profession in my rear-view mirror. I returned to carpentry after college, eventually founding and co-owning SawHorse, Inc., one of the largest design-build residential remodeling firms in the Atlanta area. Interestingly, for a time, one of my duties was management of our team of up to six architects and designers, although I never did any design work myself. During my career as a contractor, I frequently came upon high-performance and green building concepts through classes I attended and projects I worked on, but I was challenged to integrate them effectively into our then 25-plus person design-build firm. While we were always tuning up our operations—something my business partner described as upgrading from a biplane to a 747 while in mid-flight—I never felt confident enough to make the transition to green building myself.

Carl Seville, Seville Consulting; author.

In 2000, I learned about the EarthCraft House program, recently developed by Southface Energy Institute and the Greater Atlanta Homebuilders Association. While EarthCraft House was originally designed exclusively for new construction, I saw an opportunity to use it as a blueprint for green renovations. After much persuasion, Southface was convinced to create a renovation program. As part of my introduction, I attended a full-day EarthCraft House training class, along with about half of our staff. The things we learned in that training were best summed up by one of my project managers, who, at the end of the day declared that he wanted to go home and tear his house apart and start over, a sentiment that most of us agreed with. After several months of meetings, EarthCraft Renovation was put into a pilot phase, and my company, with assistance from the Southface team, completed most of the pilot projects that received certification. Over the next few years, I was invited to give presentations about my experiences in green renovation and, in short order, found out that I had, somewhat unwittingly, become one of very few national experts on the subject.

Throughout my then 20 years as a remodeler, I periodically sought out opportunities for a career change, with no success. Once I had learned about green building and remodeling, I had finally found my passion. I spent several years working to grow green remodeling in my own business and the rest of the industry; however, I believe that the marketplace was not yet ready for mainstream green remodeling, and I no longer had the energy to both run a business and promote sustainable building and remodeling. I chose to leave my company in 2005 to become a full-time consultant and trainer for the then-emerging green building industry. Since then, building certifications have become an increasingly larger portion of my practice, although I still provide consulting and training and serve as an advocate for all forms of residential green building and remodeling.

My journey into green building and remodeling has brought me into a wonderful and ever-expanding network of like-minded professionals from around the country, many of whom have become good, lifelong friends. We all share a passion for sustainable building; we regularly fuss with each other about where the industry is going; and we argue about what is coming next. It has been an incredible journey, and I eagerly anticipate where it will take me in the future.

Look to learn something from every project you do. Each time you take on a new technique and evaluate its success (or failure), you have valuable information that you can use on future projects. Once you have effectively implemented a new process, it can become your standard, with appropriate reductions in cost and effort to use the next time. Adding new techniques when you are ready will provide continuous, effective advances in performance in your projects.

Plan for Project Scale Success

The success of a project is directly dependent on making the right decisions as early in the process as possible. As we discussed in Chapter 3, proper site orientation, climate considerations, appropriate building size, proper accommodations for mechanical systems, and other integrated design principles can help reduce both construction costs and lifetime operating expenses. Conversely, waiting until plans are almost complete to consider incorporating green principles can lead to increased costs and reduced performance.

Understand the Project Focus

Every project has priorities that should be used as a guide throughout the process. While none of the core principles of green building should be ignored, they can be prioritized and weighted differently for each project. For a homeowner whose priority is physical health, their project could focus on a combination of indoor environmental quality (IEQ) and construction in a walkable neighborhood. A different homeowner may be focused on energy efficiency and sustainable material use. A third may focus on water efficiency. At first glance, these projects may appear to have dramatically different requirements, but there are significant interconnections that make them more similar than may be expected.

Interconnections in Green Building

When the project focus is on IEQ, effective air and duct sealing is critical to reduce infiltration of pollutants; combustion gases must be kept separate from the indoor air, and low-emitting materials should be used. High-performance air and duct sealing leads to improved energy efficiency, as does the selection of sealed combustion space and water heating equipment. Where IEQ is a priority, energy efficiency comes along as part of the package.

If resource efficiency is a priority, advanced framing or other alternative structural systems are often employed. These often lead directly to improved insulation, thereby improving energy efficiency.

Where durability is the focus, careful water and vapor management helps maintain structural integrity and reduce or eliminate mold and mildew. Healthy humidity levels will also be maintained, directly leading to improved IEQ.

There are many scenarios where stressing one facet of green building will also improve others. Taking advantage of this in setting project priorities helps improve the overall sustainability of a project.

The Future of Green Building

Green building is continually evolving. Stay current with any certification programs that are relevant to your role because most are updated on a regular basis. These updates are a combination of clarifications of issues that users raise during implementation, routine improvements to strengthen the program, and efforts to stay ahead of increasingly stringent energy codes. As new energy codes are adopted, some green certification programs find themselves barely meeting legal minimums, requiring that they change their specifications to maintain a position of higher performance.

Government and utility incentives are available in most areas to help create demand for green building services. Incentives raise awareness in consumers; even when they do not apply to a specific project, they can help secure clients for businesses that feature them in their marketing efforts. Like green certifications, incentives change frequently. Staying current with available incentives helps you to be an effective resource for your clients. The online Database of State Incentives for Renewables and Efficiency (http://www.dsireusa.org) provides comprehensive up-to-date listings of local, state, and national incentives.

Going Further

What should we expect for the future of green building? As minimum codes improve, consumer demand increases, and incentives evolve, we will see most (if not all) mainstream professionals move toward at least a minimal level of green building. Companies and individuals who have already embraced sustainable practices will continue to expand their expertise, producing higher-performing buildings. The most progressive practitioners are moving towards net-zero energy buildings. As basic and advanced green building techniques become mainstream for individual companies and the industry as a whole, there will always be new challenges.

Energy Guarantees

Some experienced green builders offer energy and comfort guarantees to their customers, providing them with assurance that their utility bills will not exceed a specified amount and that temperatures will be consistent throughout the home. Builders may develop their own programs or work with national programs like Comfort Home and Environments for Living. Such guarantees can provide a marketing advantage to contractors offering them.

Reducing Liability

As we have learned, high-performance buildings require less maintenance, are more durable, and provide healthier indoor environments. These factors help reduce liability for the professionals involved in the design and construction of these projects. Reduced liability leads to increased profitability through lower insurance rates, less litigation, and fewer warranty calls, leaving more time and money to concentrate on running a profitable business.

FROM EXPERIENCE

Green Building: The Human Element

Michael Anschel, CEO, Verified Green, Inc.

The study of green building is no doubt one of the broadest and most complex fields: covering ecological principles; measuring embodied energy and environmental impacts of raw material extraction, conversion, manufacturing, and disposal; examining watershed, land use, and construction site impact; exploring the composition of products and their impact on human health; understanding the fundamentals of physics, moisture, and building science; computing the performance values of building materials to select the proper mechanical systems; and understanding the value of energy transference in all its forms. At the end of the day, however, one component of green building trumps all the formulas and measurements. No monetary value can be placed on this element: human life.

Michael Anschel is the owner and principal of Otogawa-Anschel Design Build, a nationally recognized and award-winning design-and-build firm and a committed leader to the green building movement in Minnesota. Among other professional affiliations, Michael is vice-chair of the board of directors of Minnesota GreenStar and CEO of Verified Green, Inc., which consults with builders, remodelers, architects, and state and city officials on green building.

A home may seem to be fairly green if it carries a HERS score of 14, if 98% of storm water is retained, and if the foundation concrete mix contains 60% flyash. Maybe careful selection of plumbing fixtures will reduce water consumption to 20 gallons per person per day. Maybe 90% of the floors are hard surfaces, and the walls are painted with zero-carcinogen paints. If, however, the slate used on the floors came from a quarry in India where the life expectancy is 30 years because health and safety standards are nonexistent, subsistence wages are barely paid, and silicosis is rampant, there is a problem. Selecting materials on the basis of price alone without considering the human impact, even if it is displaced, is unacceptable. Comprehending displaced impact can be hard, like comprehending garbage in a landfill, so try focusing on something closer to home that you can control. Plumbers who use polyvinyl chloride pipe expose themselves to a variety of neurotoxins, endocrine disruptors, and carcinogens through the handling, modification, and binding of the materials. A respirator and gloves would dramatically reduce their exposure risk, and you can require their use on your job sites. Likewise, carpenters expose themselves to unhealthy levels of formaldehyde. Drywall installers and tapers expose themselves to high levels of fine silica dust and should be required to use proper respirators. These are just three examples of how you can impact the social equity of the people who create that green home.

There are many decision-making tools that you can use to select the right product or determine the best on-site practices for your team. Asking questions about the origins of a product, and taking an ethical stand on which products your company will and will not install is a good first step. Publish a list of materials and best practices for your team to reference. Exceed the minimum requirements set forth by such agencies as the Occupational Safety and Health Administration, National Institute for Occupational Safety and Health, and Environmental Protection Agency. Look for ways to improve the quality of life for everyone in the supply and installation chain. To get you started, here are the four questions[1] that we use in our company to make decisions:

1. Is it the truth?
2. Is it fair to all concerned?
3. Will it build goodwill and better friendships?
4. Is it beneficial to all concerned?

You can be sure that as the green building programs evolve and become more science-based that they will also become more socially responsible. Of course, we should not need a program to tell us to respect our fellow human beings, but it would not be the first either.

[1] Herbert J. Taylor, Rotary International's "The Four-Way Test," 1932.

Moving from Sustainability to Regenerative Design

Much of the discussion among the most advanced green building practitioners involves advancing beyond sustainability to a truly regenerative paradigm. As we recognize that the earth's resources have limits, we should consider how to not only reduce the resources we use for construction and operation of buildings but also create buildings that have a net-zero impact on the environment. With the right planning, these buildings can actually add benefit to the environment over their lifetime.

Is Durability Always Desirable?

One major tenet of green building is durability. By making our buildings durable, we extend their lives, reduce the amount of raw materials used for repair and replacement, and reduce the landfill space required to dispose of materials being replaced. Indefinite durability in individual materials is not always desirable, however, particularly when they are not recyclable. Plastics in particular are a long-term environmental problem because they do not degrade and are not easily or regularly recycled, and some of the chemicals used in their manufacture are known to cause physical problems in humans and animals. Chemically treated wood does not deteriorate readily, is not recyclable, and can leach toxic chemicals after disposal. Avoiding use of materials such as these is, in the long term, more sustainable than using materials that will never deteriorate.

Non-treated wood is not in itself particularly durable, but when installed and finished properly as part of a water-managed assembly, it can last many lifetimes and be reused, recycled, or allowed to deteriorate into earth when no longer needed. Masonry and concrete are extremely durable as well as reusable and recyclable. Metals are also durable and most are easily and regularly recycled.

An argument can be made that we should not use any materials in our buildings that cannot be easily recycled or reused or that will not naturally decompose. Buildings constructed in such a way that they can be deconstructed and most or all of their components reused or allowed to decompose will ultimately have the least impact on the environment.

Living Building Challenge

The Living Building Challenge (LBC) is one of the most advanced sustainable building programs operating today. A certification developed by the International Living Building Institute, the LBC is a certification program that promotes the development of buildings that use net-zero energy and net-zero water, employ fully integrated systems, and embrace beauty. The key tenets of the LBC are Site, Water, Energy, Health, Materials, Equity, and Beauty. A central component of the LBC is that buildings can only be certified after a full year of occupancy, ensuring that they meet the program criteria as they are operated.

Green Building: Keeping the Core Concepts in Mind

As we discussed at the beginning of this book, the core concepts of green building should be used as guiding principles in all your projects. Understanding what each of these concepts mean, how they interact with each other, and giving each of them the appropriate emphasis is critical to achieving the most sustainable buildings possible.

Green Building Principles

- Energy Efficiency
- Resource Efficiency
- Durability
- Water Efficiency
- Indoor Environmental Quality
- Reduced Community Impact
- Homeowner Education and Maintenance
- Sustainable Site Development

Additional Resources

Database of State Incentives for Renewables & Efficiency: http://dsireusa.org/

International Living Building Institute: http://ilbi.org/

Appendix

R-Value Table

Material	R/Inch	R/Thickness
Insulation Materials		
Fiberglass Batt	3.14–4.30	
Fiberglass Blown (attic)	2.20–4.30	
Fiberglass Blown (wall)	3.70–4.30	
Mineral Wool Batt	3.14–4.00	
Mineral Wool Blown (attic)	3.10–4.00	
Mineral Wool Blown (wall)	3.10–4.00	
Cellulose Blown (attic)	3.13	
Cellulose Blown (wall)	3.70	
Vermiculite	2.13	
Autoclaved Aerated Concrete	1.05	
Urea Terpolymer Foam	4.48	
Rigid Fiberglass (> 4lb/ft^3)	4.00	
Expanded Polystyrene (beadboard)	4.00	
Extruded Polystyrene	5.00	
Polyurethane (foamed-in-place)	6.25	
Polyisocyanurate (foil-faced)	7.20	
Construction Materials		
Concrete Block 4"		0.80
Concrete Block 8"		1.11
Concrete Block 12"		1.28
Brick 4" (common)		0.80
Brick 4" (face)		0.44
Poured Concrete	0.08	
Soft Wood Lumber	1.25	
2" nominal (1 1/2")		1.88
2 x 4 (3 1/2")		4.38
2 x 6 (5 1/2")		6.88
Cedar Logs and Lumber	1.33	
Sheathing Materials		
Plywood	1.25	
1/4"		0.31
3/8"		0.47
1/2"		0.63
5/8"		0.77
3/4"		0.94
Fiberboard	2.64	
1/2"		1.32
25/32"		2.06
Fiberglass (3/4")		3.00
(1")		4.00

(Continued)

Material	R/Inch	R/Thickness
(1 1/2")		6.00
Extruded Polystyrene (3/4")		3.75
(1")		5.00
(1 1/2")		7.50
Foil-faced Polyisocyanurate (3/4")		5.40
(1")		7.20
(1 1/2")		10.80
Siding Materials		
Hardboard (1/2")		0.34
Plywood (5/8")		0.77
(3/4")		0.93
Wood Bevel Lapped		0.80
Aluminum, Steel, Vinyl (hollow backed)		0.61
(w/ 1/2" insulating board)		1.80
Brick 4"		0.44
Interior Finish Materials		
Gypsum Board (drywall 1/2")		0.45
(5/8")		0.56
Paneling (3/8")		0.47
Flooring Materials		
Plywood	1.25	
(3/4")		0.93
Particle Board (underlayment)	1.31	
(5/8")		0.82
Hardwood Flooring	0.91	
(3/4")		0.68
Tile, Linoleum		0.05
Carpet (fibrous pad)		2.08
(rubber pad)		1.23
Roofing Materials		
Asphalt Shingles		0.44
Wood Shingles		0.97
Windows		
Single Glass		0.91
w/storm		2.00
Double insulating glass (3/16") air space		1.61
(1/4" air space)		1.69
(1/2" air space)		2.04
(3/4" air space)		2.38
(1/2" w/low-E 0.20)		3.13
(w/suspended film)		2.77
(w/2 suspended films)		3.85
(w/suspended film and low-E)		4.05
Triple insulating glass (1/4" air spaces)		2.56
(1/2" air spaces)		3.23
Addition for tight fitting drapes or shades, or closed blinds		0.29

Material	R/Inch	R/Thickness
Doors		
Wood Hollow Core Flush (1 3/4")		2.17
Solid Core Flush (1 3/4")		3.03
Solid Core Flush (2 1/4")		3.70
Panel Door w/7/16" Panels (1 3/4")		1.85
Storm Door (wood 50% glass)		1.25
(metal)		1.00
Metal Insulating (2" w/urethane)		15.00
Air Films		
Interior Ceiling		0.61
Interior Wall		0.68
Exterior		0.17
Air Spaces		
1/2" to 4" approximately		1.00

Courtesy of ColoradoENERGY.org & R.L. Martin & Associates, Inc.

Glossary

A

ACCA Manual D a guide to the design of residential duct systems that helps ensure that the ducts will deliver the proper amount of heated or cooled air to each room.

ACCA Manual J a guide for sizing residential heating and cooling systems based on local climate and ambient conditions at the building site.

ACCA Manual S a guide in the selection and sizing of heating and cooling equipment to meet Manual J loads.

ACCA Manual T a guide for designers on how to select, size, and locate supply outlets and return inlets.

accent lighting lighting on walls, artwork, and architectural features on the interior and exterior to enhance the visual appeal of an area.

acrylonitrile butadiene styrene (ABS) rigid piping material used for drain lines.

activated carbon a form of carbon specially formulated for filtration.

acute CO poisoning refers to poisoning that occurs after a single, large exposure to the gas; may involve one or more people.

advanced framing a methodology of construction designed to conserve construction materials by using alternative framing methods; see also *optimum value engineering (OVE)*.

aerators a device installed on faucets to increase spray velocity, reduce splash, and save both water and energy.

aerobic septic system a wastewater treatment system that mechanically injects air into a waste collection tank to encourage decomposition, which provides a higher-quality effluent.

air admittance valve a one-way valves that allows air to enter the plumbing vent system as needed.

air barrier a protective, air-resistant material that controls air leakage into and out of the building envelope.

air changes per hour at 50 pascals (ACH_{50}) the number of times that the total volume of a home is exchanged with outside air when the home is depressurized or pressurized to 50 pascals.

air changes per hour natural ($ACH_{Natural}$) the number of times the total volume of a home is exchanged with outside air under natural conditions.

air conditioner a home appliance, system, or mechanism that dehumidifies and extracts heat from an area.

air conditioning (AC) the process of cooling indoor air by transferring indoor heat to a refrigerant, which then moves it to the exterior.

Air Conditioning Contractors Association (ACCA) a professional organization that publishes standards for heating and cooling systems.

Air-Conditioning, Heating, and Refrigeration Institute (AHRI) a trade association representing heating, ventilation, air-conditioning, and commercial refrigeration manufacturers.

air leakage (AL) a measurement of the total amount of air leakage, equivalent to the total cubic feet of air passing through 1 ft^2 of window area per minute (cfm/ft^2).

air sealing the process of tightening the building envelope by reducing air leakage into and out of a home.

air-source heat pump a heating and cooling system that consists of a compressor and two coils made of copper tubing, one located inside and one outside, and surrounded by aluminum fins to aid heat transfer.

airtight drywall approach (ADA) an air barrier system that connects the interior finish of drywall to the homes framing to form a continuous air barrier.

albedo the ratio of electromagnetic energy that an object or surface reflects.

algal bloom the rapid, excessive growth of the algae population in an aquatic system within a short time period.

ambient lighting general interior lighting for daily activities and as needed outdoors for security and safety.

anaerobic septic system a wastewater treatment system in which solids are removed in the absence of oxygen with no direct treatment of effluent.

annual fuel utilization efficiency (AFUE) the ratio of heat output of the furnace or boiler compared with the total energy consumed by a furnace or boiler.

applied insulation is thermal or sound insulation material that is placed between or on top of structural members after they are installed.

argon an inert gas commonly added to the air space between glass panes to lower the U-factor.

asbestos a naturally occurring fibrous material once commonly used for fireproofing; extremely harmful when inhaled.

attic knee wall a vertical wall separating conditioned interior space from an unconditioned attic area.

autoclaved aerated concrete (AAC) a lightweight, precast building material that provides structure, insulation, fire, and mold resistance.

awning window an operable window with a sash hinged at the top that swings outward.

B

balanced ventilation a system that provides equal parts supply and exhaust air through the use of fans, or air-to-air heat exchangers, referred to as either heat recovery ventilators (HRV) or energy recovery ventilators (ERV).

ballast a device that regulates the frequency of electricity delivered to a fixture to control the starting and operating voltages of the lamp.

balloon framing a system of wood-frame construction, first used in the 19th century, in which the studs are continuous from the foundation sill to the top wall plate.

batt insulation a thermal or sound insulation material, generally of fiberglass or cotton, which comes in varying widths and thickness (R-value) to conform to standard framing of walls and joists.

best management practices (BMPs) strategies for keeping soil and other pollutants out of streams and lakes; BMPs are designed to protect water quality and prevent new pollution.

bio-based insulation insulation products that contain materials from renewable resources as substitutes for petroleum and other nonrenewable products.

biofuels non-fossil fuels, such as wood, ethanol, and biodiesel, which are generally considered renewable.

black water waste that comes from toilets and kitchens containing human or food waste, which must be fully treated and sanitized before being released into waterways or reused.

blower door a diagnostic tool designed to measure the air tightness of buildings and to identify air leakage locations.

blown insulation a material composed of loose insulating fibers such as fiberglass, foam, or cellulose that is pumped or injected into walls, roofs, and other areas.

boiler piece of heating equipment designed to heat water (using electricity, gas, or oil as a heat source) for the purpose of providing heat to conditioned space or potable water.

British thermal units (Btu) the amount of heat required to raise a pound of water 1°F.

brominated flame retardants a group of chemicals that inhibit the spread of fire and consist of organic compounds containing bromine.

brownfield development an abandoned or underused industrial or commercial facility available for re-use.

building envelope the separation between the interior and the exterior environments of a building, consisting of an air barrier and thermal barrier that are continuous and in contact.

Building Performance Institute (BPI) a nonprofit that provides nationally recognized training, certification, accreditation, and quality-assurance programs for contractors and business.

building science the study of the interaction of building systems and components, occupants, and the surrounding environment; focuses on the flows of heat, air, and moisture.

C

capillary break an air space or material that prevents the movement of moisture between two surfaces by capillary action.

carbon monoxide (CO) a colorless, odorless, poisonous gas that results from incomplete combustion of fuels (e.g., natural or liquefied petroleum gas, oil, wood, and coal).

carbon monoxide detector a device that records the levels of CO in a combustion appliance flue pipe or surrounding areas.

carbon sink an environmental reservoir that absorbs and stores carbon, thereby removing it from the atmosphere.

Carpet and Rug Institute (CRI) a nonprofit industry association that created second-party certification programs, Green Label and Green Label Plus, which provide guidelines for VOC content and off-gassing of carpet and padding materials.

casement window a side hinged window that swings open to interior or exterior.

catchment the impermeable area on which rainwater lands.

cathedral ceiling an insulated vaulted ceiling with roof above; sometimes referred to as a "roof-ceiling" combination and commonly found in living rooms and attics with insulated rooflines.

cavity insulation insulation placed between wall studs or joists.

cellulose insulation made from recycled newspaper and an added fire retardant.

ceramic tile a thin surfacing unit composed of various clays fired to hardness.

charrette a design meeting consisting of all project stakeholders.

chlorinated polyvinyl chloride (CPVC) piping material that is chemically similar to PVC but contains added chlorine to increase rigidity at high temperatures.

chromated copper arsenate (CCA) a chemical wood preservative containing chromium, copper, and arsenic.

chronic CO poisoning is used to indicate those cases when individuals are exposed on more than one occasion to carbon monoxide—usually at comparatively low concentrations.

cisterns used for storing collected rainwater.

closed-cell foam a type of spray polyurethane foam installed at a rate of approximately 2 lb/ft^3 and sometimes referred to as "2-pound foam"; see also *spray polyurethane foam (SPF)*.

coal fly ash the very fine portion of ash residue that results from the combustion of coal that may be used as a Portland cement substitute.

cob a combination of natural clay and straw that is mixed together and built in a free-form style into structural walls.

coefficient of performance (COP) the ratio of energy output (heating or cooling) to the amount of energy put in.

cogeneration the use of a heat producing appliance or a power station to simultaneously generate both electricity and useful heat, also known as *combined heat and power (CHP)*.

coke a solid byproduct from the combustion of coal.

cold start the first time hot water is used for the day or after the trunk line has cooled

Color Rendering Index (CRI) quantifies a lamp's ability to render colors the same as a reference lamp type.

color temperature an indicator of the color of a light source and compares the color of a lamp to natural sunlight; measured in degrees kelvin.

column an upright pillar, typically cylindrical and made of stone, fiberglass, or wood, supporting an arch or other structure, or acting simply as an architectural feature.

combined heat and power (CHP) the use of a heat producing appliance or power station to simultaneously generate both electricity and useful heat, also known as *cogeneration*.

combustion analyzer a tool used on-site to measure furnace efficiency.

commissioning the process of diagnosing and verifying building system performance.

compact fluorescent lamps (CFLs) a common lighting alternative to incandescent lamps that give off light when a mixture of three phosphors are exposed to ultraviolet light from mercury atoms.

complete combustion the process by which carbon from fuel bonds with oxygen to form carbon dioxide (CO_2), water vapor, nitrogen, and air.

composition shingles made of asphalt and glass fiber; the most popular steep-slope roofing material.

composting toilets toilets that convert human waste to sanitary usable compost without using any water in the process.

compressor a mechanical pump that uses pressure to change a refrigerant from liquid to gas.

concrete a masonry product composed of cement or pozzolans, sand, and gravel or other coarse aggregate.

concrete masonry unit (CMU) a large rectangular block of concrete used in construction.

condensate the moisture removed from the air by an air conditioning or dehumidification system.

condensation resistance (CR) a measurement of a particular window unit's resistance to condensation forming on the interior.

condenser the component in a refrigeration system that transfers heat from the system by condensing refrigerant.

condensing furnace a combustion heating appliance that uses a secondary heat exchanger to condense vapor in the flue gas and capture additional heat, increasing their efficiency.

condensing heaters high-efficiency water heaters that remove large amounts of heat from the flue gases, resulting in the condensing of the flue gases.

condensing unit the portion of a refrigeration system where the compression and condensation of refrigerant is accomplished.

conditioned crawl space is a foundation without wall vents that encloses an intentionally heated or cooled space; insulation is located at the exterior walls.

conduction the transfer of heat from one substance to another by direct contact.

constructed wetlands a means of treating wastewater that simulate natural wastewater treatment, allowing effluent to flow through water beds filled with plants that break down contaminants.

continuous insulation insulation that is not interrupted by structural members, typically placed on the outside surface of wood framing or concrete walls.

continuous pumps pumps that circulate hot water throughout the trunk lines 24 hours a day.

convection the transfer of heat through a fluid (liquid or gas).

convective loop the continuous circulation of air (or another liquid) in an enclosed space as it is heated and cooled.

conveyance a system that directs rainwater from the catchment area, consisting of gutters, downspouts, and pipes.

cooling degree-day (CDD) a measure of how warm a location is over a period of time relative to a base temperature, most commonly specified as 65°F.

Cool Roof Rating Council (CRRC) an independent, nonprofit organization that maintains a third-party rating system for radiative properties of roof-surfacing materials.

cornice the entire finished assembly where the walls of a structure meet the roof; sometimes called *eaves* or *soffits.*

cotton insulation a thermal or sound insulation material, generally from clothing manufacturing waste, that comes in varying widths and thickness (R-value) to conform to standard framing of walls and joists.

cricket a small, false roof built behind a chimney or other roof obstacle for the purpose of shedding water; also called a saddle.

cross-linked polyethylene (PEX) specialized type of polyethylene plastic that is strengthened by chemical bonds formed in addition to the usual bonds in the polymerization process.

cultured stone a cast masonry unit made of cement and various additives that simulates the look of natural stone.

D

damp-proofing a treatment used on concrete, masonry, or stone surfaces to repel water and reduce the absorption of water in the absence of hydrostatic pressure.

daylighting the use of natural light to supplement or replace artificial lighting.

deck a raised, roofless structure adjoining a house.

decoupling membrane a flexible plastic sheet that is placed between ceramic tile and the subfloor to provide strength and crack resistance.

deficit approach a method of sizing cisterns, which involves calculating the total need during periods of little or no rainfall.

demand pump moves hot water to plumbing fixtures only when needed, eliminating standby losses from the constant reheating of water in pipes.

Department of Energy (DOE) the federal department responsible for maintaining the national energy policy of the United States.

desiccant materials such as calcium oxide and silica that absorb moisture that are commonly used to remove humidity from indoor air that is passed through it with fans.

desuperheater a device that recovers excess heat from the cooling and heating process, diverting it to heat water in a storage tank very efficiently.

direct evaporative cooler a device that reduces the temperature of air by passing it through water-soaked pads.

distillation the process of purifying a liquid by boiling it and condensing its vapors.

distribution the piping that delivers harvested and filtered rainwater for use.

dormer a structure that projects out from a sloping roof to form another roofed area to provide a surface for the installation of windows.

double-hung window a window that has two vertically operating sashes.

downspout a vertical member used to carry water from the gutter downward to the ground; also called a conductor or leader.

draft gauge measures the pressure of the flue gases in a furnace or water heater.

drainage mat a material that creates a gap between the soil and the foundation walls; this space relieves hydrostatic pressures and provides water with a path of least resistance to drain away from the home.

drain field the final component of a septic system in which effluent flows through perforated pipe into the soil, where it is filtered through the ground as it moves downward towards the local aquifer.

drain-water heat recovery (DHR) the use of a capturing heat for reuse from wastewater.

drip cap a horizontal molding or flashing installed over the frame for a door or window to direct water away from the frame.

drip edge a metal strip that extends beyond the other parts of the roof and is used to direct rainwater away from the structure.

drip groove a cutout on the underside of the projection intended to prevent water from traveling beyond it and back to the face of the wall.

drought-tolerant species a tree or plant that is able to grow and thrive in arid conditions.

dry-bulb temperature the temperature of air indicated on an ordinary thermometer; does not account for the effects of humidity.

drywall a construction material used for finished wall and ceiling surfaces; made of kiln-dried gypsum pressed between fiberglass mat or paper facing; also referred to as *gypsum board.*

dual-flush a high-efficiency toilet that gives users the choice of flushing at full capacity or with less water.

duct leakage test a diagnostic test designed to measure the air tightness of heating and air conditioning duct systems and to identify air leakage locations.

ductless mini-splits compact, wall-mounted air-conditioner or heat pump connected to a separate outdoor condensing unit via refrigerant lines.

dynamic glazing (DG) products either glass that changes properties electronically via an electric current, or glazing with blinds between glass layers that control light and heat.

E

eave baffles materials that prevent attic insulation wind washing by directing soffit air flow over attic insulation; also known as *positive-ventilation chutes.*

edge development a site with 25% or less of the property boundary adjacent to existing development.

effective duct length the total amount of pressure loss of the straight sections and all fittings in each duct run.

effective leakage area (ELA) the area of a special nozzle-shaped hole (similar to the inlet of a blower door fan) that would leak the same amount of air as the building does at a pressure of 4 pascals.

efficacy the ratio of light produced to the amount of energy consumed; expressed as lumens per watt.

effluent wastewater flow before or after treatment.

electric furnace a heating system that uses electric resistance to convert electricity into heat.

electromagnetic fields (EMF) invisible forces created by the transmission of electricity through wires; also called *electrical and magnetic fields.*

electronic air filters use electricity to attract smaller molecules, such as smoke, mold, and pet odors, to metal fins.

embodied energy the energy required to manufacture or harvest, package, and ship a material to a job site; may refer to an individual material or an entire home.

energy a measurable quantity of heat, work, or light.

energy efficiency ratio (EER) the measure of how efficiently a cooling system will operate when the outdoor temperature is at a specific level (usually 95°F).

energy-efficient mortgage (EEM) uses the energy savings from a new energy-efficient home to increase the home buying power of consumers, capitalizing the energy savings in the appraisal.

energy factor (EF) the energy efficiency rating for a water heater; based on the amount of hot water produced per unit of fuel consumed over a typical day.

EnergyGuide a bright yellow label, created by the Federal Trade Commission, that is required by law to appear on many new household appliances to show the relative energy consumption as compared with similar products.

energy improvement mortgage (EIM) finances the energy upgrades of an existing home in the mortgage loan by using monthly energy savings.

ENERGY STAR a joint program of the U.S. Environmental Protection Agency and the U.S. Department of Energy that sets standards for energy-efficient products and buildings.

energy truss a roofing truss designed to span an area and provide adequate space for full-depth attic insulation across the full area; see also *raised-heel truss.*

engineered flooring a product that is made up of multiple layers of wood and glued together as one board.

engineered lumber is wood that is manufactured by bonding together wood strands, veneers, lumber or fiber to produce a stronger and more uniform composite; also known as *manufactured wood product.*

environmentally preferable products (EPP) products that have a reduced effect on human health and the environment when compared with traditional products or services that serve the same purpose.

environmental product declarations (EPDs) the quantified environmental data for a product with pre-set categories of parameters based on the ISO 14040 series of standards, but not excluding additional environmental information.

equivalent length the comparable length of a single straight duct run when taking into account air flow resistance from duct compression, elbows, fittings, and other obstructions.

erosion the removal of solids (e.g., sediment, soil, rock, and other particles) by wind, water, or ice in the natural environment.

ethylene propylene diamine monomer (EPDM) a single-ply membrane consisting of synthetic rubber; commonly used for flat roofs.

evacuated tube collector a solar thermal collector that uses absorber plates that are enclosed in a glass tube with a vacuum inside.

evaporative cooling a means of temperature reduction that operates on the principle that water absorbs latent heat from the surrounding air when it evaporates.

evaporator the system component responsible for performing the actual cooling or refrigerating of the occupied space.

exfiltration air flow outward through a wall, building envelope, window, or other material.

exhaust-only ventilation systems that remove air from the home with equal parts make up air entering through uncontrolled opening (i.e., envelope leaks).

expanded polystyrene (EPS) a foam insulation board made of expanded polystyrene beads.

extensive vegetated roofs a type of green roof that uses a thin layer of a special growing medium (usually placed over a drainage mat) and requires special low-growing, short-rooted plants, such assedum.

exterior insulated finish system (EIFS) a synthetic stucco finish applied over foam insulation.

extruded polystyrene (XPS) a closed cell insulation foam board.

F

faced batts batt insulation that contain a foil or kraft paper vapor retarder covering.

fan coils a simple device consisting of a heating or cooling coil and fan used to distribute heating and cooling into a space.

fascia the vertical trim between the soffit and roof.

feed-in-tariff (FIT) system utilities purchase renewable energy at variable rates, which are usually higher than the rates at which the energy is sold.

fenestration describes all the products that fill openings in a building envelope, including windows, doors, and skylights that allow air, light, people, or vehicles to enter.

fiberboard is an engineered lumber product used primarily as an insulating board and for decorative purposes, but may also be used as wall sheathing.

fiberglass insulation a blanket, loose fill, or rigid board insulation composed of glass fibers bound together with a binder.

fiber optic lighting lighting systems that use very fine, flexible glass or plastic fibers to transmit light.

fill valve the internal parts of a toilet that measure the amount of water filling the tank and bowl.

first flush the initial flow of water coming off a roof that often includes bird droppings, leaves, and other debris.

first-hour rating the amount of hot water in gallons the heater can supply per hour (starting with a tank full of hot water).

first-party certification when a single company develops its own rules, analyzes its performance, and reports on its compliance.

flash and batt (FAB) a hybrid insulation system that combines a 1" to 2" layer of closed-cell SPF that is covered with fiberglass batts to fill the framing cavity.

flat plate collector a rectangular solar thermal collector, typically 4' wide by 8' or 10' long.

fluorescent a type of lamp that creates light when electricity passes through gas inside a phosphorous-coated glass tube, causing it to glow.

foot-candles the amount of light produced by one lumen over a 1 ft^2 area.

footing the widened support, usually concrete, at the base of foundation walls, columns, piers, and chimneys that distributes the weight of these elements over a larger area and prevents uneven settling.

Forest Stewardship Council (FSC) is an independent, non-governmental, not-for-profit organization established to promote the responsible management of the world's forests.

foundation drainage the process of directing groundwater away from the foundation and the home.

foundation walls walls constructed partially below ground that support the weight of the building above and enclose the basement or crawl space.

French balcony a set of doors with an exterior rail that can open the interior space to the outside.

frost-protected shallow foundations provides protection against frost damage without the need for excavating below the frost line.

fuel cell a device that uses fossil fuels, such as natural gas or propane, to produce electricity without combustion by extracting hydrogen, which is used to create power similar to how a battery works.

furnace an appliance fired by gas, oil, or wood in which air is heated and circulated throughout a building in a duct system.

G

gable roof a type of roof that slopes in two directions.

garbage disposal electrically operated grinders that allow food waste to be removed through the sewer lines, a process that requires both water and energy.

glazing a transparent part of a wall or door assembly that is usually made of glass or plastic.

grade beam a perimeter load-bearing support of a structure that spans between piers without relying on the ground below for support.

grain a unit of measurement for moisture content; 1 pound contains 7,000 grains.

gravity-film heat exchanger (GHX) a heat transfer device used with drain-water heat recovery systems.

gravity flush toilets that use the weight of the water in the tank to evacuate the bowl.

grayfield development previously developed and underutilized real estate assets or land.

gray water nonpotable water reclaimed from sinks, baths, and washing machines that may be used to flush toilets and for irrigation.

green building an environmentally sustainable building, designed, constructed, and operated to minimize the total environmental impacts.

greenfield development a previously undeveloped land, in a city or rural area that is currently used for agriculture, landscape design, or wilderness area.

greenhouse effect the buildup of heat in an interior space caused by energy input through a transparent membrane such as glass; also refers to process by which planets maintain temperature through the presence of an atmosphere containing gas that absorbs and emits infrared radiation.

greenhouse gas (GHG) any of the atmospheric gases, such as carbon dioxide (CO_2), sulfur oxides (SO_x), and nitrous oxides (NO_x), that contribute to the greenhouse effect.

green roof a roof that is partially or completely covered with vegetation and a growing medium, planted over a waterproofing membrane; also known as a *vegetated* or *living roof.*

grid-tied refers to homes that are connected to the central electrical grid.

grille the bars that divide the sash frame into smaller lites or panes of glass.

ground granulated blast furnace slag a by-product of iron and steel making that is used to make durable concrete in combination with ordinary Portland cement or other pozzolanic materials.

ground-source heat pump (GSHP) a central heating and cooling system that pumps heat to and from the ground.

gutter a wood, metal, or plastic trough used at the roof edge to carry off rainwater and water from melting snow.

H

halogen a variation on incandescent lamps that uses a filament inside a compact glass envelope filled with gas.

hardscape the nonvegetated elements of a landscape, including paving, walkways, roads, retaining walls, street amenities, fountains, and pools.

hard water water that has a high mineral content.

head flashing the flashing over a projection, protrusion, or window opening.

heating degree-day (HDD) a measure of how cold a location is over a period of time relative to a base temperature, most commonly specified as 65°F.

heating seasonal performance factor (HSPF) a heat pump's estimated seasonal heating output in Btu divided by the amount of energy that it consumes in watt-hours.

heat island effect the phenomenon of urban areas being warmer than more rural areas primarily due to the increased use of paving and building materials that effectively retain heat.

heat pump a heating and cooling unit that draws heat from an outdoor source and transports it to an indoor space for heating purposes or, inversely, for cooling purposes.

heat traps special one-way valves or loops of pipe that keep hot water from flowing naturally out of the top of the heater via convection, saving energy in the process.

HERS rater a nationally accredited individual who performs HERS Ratings and evaluates the energy efficiency of homes; sometimes referred to as Home Energy Rater.

high-efficiency toilets (HETs) are defined by the EPA as those that use an average of 20% less water than the industry standard of 1.6 GPF.

high-intensity discharge (HID) a variation on incandescent lamps that are made with mercury vapor, metal halide, and high- and low-pressure sodium gases; used most commonly for residential outdoor applications.

hip roof a three or four-sided roof having sloping ends and sides.

holding tank the component of septic systems that allows solid waste to settle while letting liquid effluent flow out to the soil.

Home Energy Rating System (HERS) a nationally recognized measurement of a home's energy efficiency.

Home Performance with ENERGY STAR (HPwES) a joint program of the U.S. Environmental Protection Agency and the U.S. Department of Energy that sets standards evaluating and improving the energy efficiency of existing homes.

home run system see *manifold system.*

hose bibb a faucet with hose threads on the spout found outside or near clothes washers and wash basins.

hot roof an unvented attic containing insulation on the underside or directly above the roof decking; also known as a *cathedral attic, conditioned attic,* or *insulated roofline.*

hot start the temperature in the hot water supply line when a fixture is turned on; occurs when hot water remains in the supply line and is available for use.

hot water circulation loop a continuous pipe from the water heater that runs close to every fixture and back to the bottom of the heater.

housewrap a synthetic water-resistive barrier designed to shed bulk moisture and allow vapor to pass through; see also *water-resistive barriers.*

hydronic a space-conditioning system that circulates heated or cooled water through wall- or baseboard-mounted radiators, in-floor tubing, or a combination.

hydrostatic pressure the force exerted on a foundation by groundwater.

hygroscopic materials that readily attract and retain moisture.

I

ice dam ice that forms at the eave of sloped roof, causing water build-up behind it to back up under roofing materials.

I-joist a structural building component consisting of a wide vertical web of OSB with engineered wood flanges at the top and bottom. I-joists can be used for floor, roof, and wall framing.

illumination is the quantity of light reaching a task or work surface, which is referred to as *illuminance.*

impermeable a material or assembly that does not allow air or moisture to flow through.

incandescent lamps that create light when electricity passes through a tungsten filament encased in glass, causing it to glow and create light.

incomplete combustion occurs when ratio of fuel to oxygen is incorrect, causing carbon monoxide and aldehyde to be produced.

indirect evaporative cooler similar to direct evaporative cooling but uses some type of heat exchanger to prevent the cooled moist air from coming in direct contact with the conditioned environment.

infill development the insertion of additional housing units into an already approved subdivision or neighborhood.

infiltration the uncontrolled process by which air or water flows through the building envelope into the home.

insulated concrete form (ICF) insulating foam or mineralized wood forms that are left in place after the concrete is poured for a foundation or wall.

insulated glass a window unit made up of at least two panes separated by a sealed space that is filled with air or other gases.

insulation a material that reduces or prevents the transmission of heat.

insulation board a rigid insulation product available in varying widths and thickness (R-value).

insulation contact (IC) units recessed light fixtures that dissipate heat into the room, permitting insulation to be in contact with the fixture without causing a fire hazard or gaps in the thermal envelope.

integrated design a collaborative method for designing buildings that emphasizes the development of a holistic design.

integrated pest management (IPM) a strategy to first limit the use of pesticides, and only when necessary, use the least hazardous products sparingly.

intensive vegetated roofs a type of green roof containing deep layers of soil that can support shrubs and small trees.

internal drains are openings in the surface of a low-slope roof that lead to downspouts placed inside the building structure to remove water from the roof.

intumescent paint a fire-retardant paint.

invasive species non-native plants that tend to spread aggressively.

inverter a device that converts site-generated DC power to AC current for use in the home.

Ipe a South American hardwood that is very heavy and naturally resistant to insects and rot.

J

jalousie a window with operable parallel glass, acrylic, or wooden louvers set in a frame.

joist horizontal framing members used in a spaced pattern that provide support for the floor or ceiling system.

jumper ducts small sections of duct installed in ceilings that allow air to flow between rooms.

K

Kelvin a temperature scale used to determine the color temperature of different light types.

kick-out flashing a flashing piece installed at the bottom of a roof slope that is adjacent to a wall, preventing roof rain water from getting behind the wall cladding material and WRB.

kill switch a control that turns off the power to a set of receptacles that are plugged into it.

knob and tube wiring an old system of electrical wiring composed of individual wires running in pairs and connected to framing with ceramic insulators.

krypton an inert gas commonly added to the air space between glass panes to lower the U-factor.

L

lamp commonly referred to as light bulbs, are the replaceable part of a fixture that produces light from electricity.

landscape all outdoor features of the home, including natural and built elements.

latent heat the heat that produces a change of state without a change in temperature; the portion of the cooling load that results when moisture in the air changes from a vapor to a liquid (condensation).

lavatory faucets the fittings used in bathroom sinks.

Leadership in Energy and Environmental Design (LEED) a system to categorize and certify the level of environmentally sustainable construction in sustainable buildings.

leakage to the outside duct leakage that is located not within the building envelope.

LEED Accredited Professional (AP) Homes an individual who has passed a national test and displayed the knowledge necessary to participate in the LEED design and certification process.

life cycle assessment (LCA) process of evaluating a product or building's full environmental cost, from harvesting raw materials to final disposal.

light-emitting diodes (LEDs) semiconductors that glow when electrical current passes through them.

linoleum durable flooring materials made of linseed oil and cork and wood particles.

low-E coating a microscopic layer of metal applied to the glass surface that acts as a radiant barrier, reducing the amount of infrared energy that penetrates through the metallic surface.

low emissivity (low-E) a surface that radiates, or emits, low levels of radiant energy.

low-slope roof a roof angle or pitch that is 30° (2:12) or less.

luminaires a light fixture; the complete lighting unit, including lamp, reflector, ballast, socket, wiring, diffuser, and housing.

M

magnetic induction lamps a variation on fluorescent technology and use an electromagnet to cause gas in the lamp to glow.

manifold system a plumbing design consisting of 1/2" or 3/8" pipes that run directly from the water source to individual fixtures; also referred to as a *home run system.*

manufactured homes buildings that are fully finished in the factory and delivered on a permanent steel chassis.

Material Safety Data Sheets (MSDS) are documentation available for most products that fit the Occupational Safety and Health Administration definition of hazardous; they identify potentially dangerous content, exposure limits, safe handling instructions, clean-up protocol, and other factors to consider in selecting and using products.

maximum performance (MaP) test determines how well toilets perform bulk removal by using a realistic test media; each toilet model is graded according to this performance.

mechanical air filters filters that use synthetic fibers, fiberglass, or charcoal to remove particulates; most common type of filter used in homes.

millwork refers to the wood and composite products used to create doors, trim, cabinets, and similar finishes in homes.

mineral wool insulation a manufactured wool-like material consisting of fine inorganic fibers made from slag and used as loose fill or formed into blanket, batt, block, board, or slab shapes for thermal and acoustical insulation; also known as *rock wool* or *slag wool.*

Modified Energy Factor (MEF) the efficiency rating for clothes washers that considers the energy required to operate the washer, heat the water, and dry the clothes based on how much water is removed in the spin cycle.

modular construction the practice of factory fabrication of complete sections of a house including floors, walls, ceilings, mechanical systems, and finishes, which are delivered by truck to the job site where they are placed on a foundation and finished in place.

multiple-point testing a blower door procedure involving testing the building over a range of pressures (typically 60 pascals to 15 pascals) and analyzing the data using a blower door test analysis computer program.

multiway switching an interconnected set of switches that control lighting from multiple locations.

muntins the actual bars that comprise a grille and divide the sash frame into smaller lites of glass.

N

National Appliance Energy Conservation Act (NAECA) the federal law enacted in 1992 to establish nationwide minimum efficiency levels for a variety of residential and commercial appliances that use energy and water.

National Association of Home Builders (NAHB) national trade association representing home builders.

National Fenestration Research Council (NFRC) a nonprofit organization that administers a uniform, independent rating and labeling system for the energy performance of windows, doors, skylights, and attachment products.

National Oceanic and Atmospheric Administration (NOAA) the U.S. federal agency that conducts research and gathers data about the global oceans, atmosphere, space, and sun.

naturally decay-resistant wood wood species, such as redwood, western cedar, cypress, black locust, Pacific yew, and Ipe, that are not prone to moisture damage.

net-metered systems homes that produce some of their electrical use on-site and use a single meter, which runs in either forward or reverse depending on the direction of power flow, to sell or buy power from the utility.

net-zero energy a building that produces as much energy as it needs on a monthly or annual basis.

new urbanism an urban design strategy that promotes walkable neighborhoods that contain a range of housing and job types; highly influenced by traditional neighborhood development and transit-oriented development.

non-reservoir walls wall assemblies that cannot absorb water without risking structural damage.

North American Board of Certified Energy Practitioners (NABCEP) a national certification organization for professional installers in the field of renewable energy.

O

occupancy sensors a device that turns circuits on or off automatically when they sense that occupants have entered or left a room.

off-gassing the process by which many chemicals volatilize, or let off molecules in a gas form into the air; see also *volatile organic compounds.*

off-grid homes that they are not connected to electric and natural gas utilities.

old-growth a forest or woodland area having a mature or overly mature ecosystem that is more or less uninfluenced by human activity.

open-cell foam a type of spray polyurethane foam installed at a rate of approximately 0.5 lb/ft^3 and sometimes referred to as "half-pound foam"; see also *spray polyurethane foam.*

open-web floor truss an engineered assembly of dimensional lumber with metal connector plates that replaces a larger single structural member with one using fewer materials while providing equivalent strength.

operable window a window with movable sashes.

optimum value engineering (OVE) a methodology of construction designed to conserve construction materials by using alternative framing methods; see also *advanced framing.*

oriented-strand board (OSB) an engineered lumber product that is often used as a substitute for plywood in the exterior wall and roof sheathing.

overhang an architectural feature, such as a soffit, awning, or porch, that protects glazing from the sun and walls and doors from rain.

P

panelized framing consists of wall, floor, ceiling, and roof panels constructed in a controlled environment and delivered to the site ready for installation.

parabolic concentrating a type of solar thermal collector that uses U-shaped troughs to concentrate sunlight on a tube that is placed on the focal line of the trough.

parapet a low wall at the edge of a roof, terrace, balcony, or other structure.

Pascal (Pa) a unit of pressure equal to one newton per square meter.

passive solar design the practice of designing a home to utilize the sun's energy for heating and cooling.

patio an on-grade outdoor space for dining or recreation that adjoins a home and is often paved.

percolation test the speed at which soil will absorb wastewater, also known as a *perc test.*

permanent wood foundations (PWF) foundation systems consisting of pressure-treated wood walls.

permeability the measure of air or moisture flow through a material or assembly.

permeable a material or assembly that allows air or moisture to flow through.

perm rating the rate of water vapor passage through a material under fixed conditions.

pervious asphalt a method of asphalt paving that allows water to enter the ground below.

pervious concrete a method of concrete paving that allows water to enter the ground below.

pervious paving paving materials that allow water to enter the ground below.

PEX *see* cross-linked polyethylene.

phantom loads also referred to as vampires, are the small amounts of electricity that many appliances and electronics use even when they appear to be turned off.

phenol formaldehyde is a potentially harmful chemical binder commonly used in fiberglass insulation and engineered wood products.

photocell switches are used on exterior lighting to turn them off during daylight hours when they are not needed.

photovoltaic (PV) a device that converts the energy of sunlight directly into electricity.

pier foundation is a grid system of girders (beams), piers, and footings used in construction to elevate the superstructure above the ground plane or grade; the piers serve as columns for the superstructure.

plaster a mixture of lime or gypsum with sand and water that hardens into a smooth solid; used to cover walls and ceilings.

plate top or bottom horizontal member of a wall frame.

platform framing a method of wood frame construction in which walls are erected on a previously constructed floor deck or platform.

pleated air filters high-efficiency paper mechanical air filters that contain more fiber per square inch than disposable fiberglass filters.

plenums rectangular boxes attached to the furnace that receive heated or cooled air which is then distributed to the trunks and individual ducts.

plumbing core a design that places all plumbing close together to reduce pipe lengths and enable short pipe runs to the fixtures.

plumbing fittings used in pipe and plumbing systems to connect pipe or tubing sections, to adapt to different sizes or shapes, and to regulate fluid flow.

plumbing fixtures an item for the distribution and use of water in homes, including toilets, sinks, and tubs.

plywood a piece of wood made of three or more layers of veneer joined with glue, and usually laid with the grain of adjoining plies at right angles.

polypropylene (PP) a plastic plumbing material made from a combinationof both long and short moleculesthat provides both strength and flexibility.

porch an unconditioned open structure with a roof attached to the exterior of a building that often forms a covered entrance or outdoor living space.

porous pavers pavers with openings between and within the pavers that are filled with vegetation or gravel, allowing water to drain through to the ground.

Portland cement the most common form of cement consisting of certain minerals that form the binder in concrete and plasters; see also *pozzolan*.

post-consumer recycled content the portion of a product that is reclaimed after consumer use.

post-industrial recycled content the portion of a product that contains manufacturing waste material that has been reclaimed; also called *pre-consumer recycled content.*

power-assisted flush a high-efficiency toilet that uses a small electric pump to provide effective flushing.

pozzolan a material which, when combined with calcium hydroxide, exhibits cementitious properties; examples include Portland cement, coal fly ash, and ground granulated blast furnace slag.

precast concrete a construction technique in which concrete components are cast in a factory or on the site before being lifted into their final position on a structure.

prefabricated foundations are foundation walls that are manufactured in a factory and assembled on site.

pre-plumb the process of installing plumbing distribution systems during construction to meet future technology need.

pressure-assist flush a high-efficiency toilet that uses air pressure generated by water line pressure stored in a small tank to produce a more forceful flush.

pressure balance valve controls that maintain a safe shower temperature to avoid scalding due to changes in water pressure or temperature.

pressure-reducing valve a device that maintains water pressure inside the home at a consistent level.

pre-wire the process of installing electric wiring during construction to meet future technology demands.

psychrometric chart a chart or graph showing the relationship between a particular sample of air's dew point temperature, dry-bulb temperature, wet-bulb temperature, humidity ratio, and relative humidity.

purification the process of removing contaminants from collected rainwater by diverting the first flush or by using sand filters, chlorination, and ultraviolet sterilization.

R

radial duct system a distribution system where the branch ducts that deliver conditioned air to individual supply outlets are connected directly to a small supply plenum.

radiant barrier a material that inhibits heat transfer by thermal radiation; commonly found in attics.

radiant heating a heating system in which the heating source (electric resistance or hot water) is installed under the finish flooring or individual radiators.

radiation heat energy that is transferred through air.

radon a naturally occurring radioactive gas that is present in the ground at varying concentrations across the country.

radon ventilation systems that prevent the entry of radon and other soil gases into the home by ventilation to the outside.

rain garden a depressed vegetated area that collects surface runoff from impervious surfaces and allows infiltration into the groundwater supply or return to the atmosphere through evaporation.

rainscreen a method of constructing walls in which the cladding is separated from the water-resistive barrier by an air space that allows pressure equalization to prevent rain from being forced inside.

rainwater harvesting the collection, storage, and use of precipitation from roofs and other surfaces.

raised-heel truss a roofing truss designed to span an area and provide adequate space for full-depth attic insulation across the full area; see also *energy truss.*

rammed earth an ancient form of construction in which soil and additives, such as straw, lime or cement, are placed within forms in multiple 6- to 8-inch thick layers and compacted to create a structural mass wall.

recycled content the amount of pre- and post-consumer recovered material introduced as a feedstock in a material production process, usually expressed as a percentage.

refrigerant a chemical that transfers heat as it changes from a liquid to a gas and back to a liquid.

relative humidity (RH) the ratio of the amount of water in the air at a given temperature to the maximum amount it could hold at that temperature; expressed as a percentage.

renewable energy electricity generated from resources that are unlimited, rapidly replenished, or naturally renewable (e.g., wind, water, sun, geothermal [ground heat], wave, and refuse) and not from the combustion of fossil fuels.

renewable energy certificates (RECs) tradable energy commodities that identify that 1 megawatt-hour of electricity was generated with renewable resources.

Residential Energy Services Network (RESNET) a nonprofit member-based organization that strives to ensure the success of the building energy performance certification industry, set the standards of quality, and increase the opportunity for ownership of high-performance buildings.

retaining wall a structure that holds back soil or rock from a building, structure, or area.

reverse osmosis (RO) filters force water through semi-permeable membranes under pressure, leaving contaminants behind in the process.

rock wool another name for *mineral wool.*

roof decking the wooden or metal surface to which roofing materials are applied.

roof slopes the angle of a roof, described by the rise (vertical height) over the run (horizontal length).

R-value quantitative measure of resistance to heat flow or conductivity, the reciprocal of U-value.

S

sash a structure that holds the panes of a window in the window frame.

scupper an opening in the side of a building, such as a parapet, that allows water to flow outside.

seasonal energy efficiency ratio (SEER) describes how efficiently air conditioning equipment works.

second-growth a forest or woodland area that has regrown after the removal of all or a large part of the previous stand by cutting, fire, wind, or other force; typically, a long enough period will have passed so that the effects of the disturbance are no longer evident.

second-party certification when an industry or trade association fashions its own code of conduct and implements reporting mechanisms.

semi-permeable a material or assembly that allows some air or moisture to flow through.

sensible heat the energy associated with temperature change.

sensible heat fraction (SHF) an air conditioner's ratio of sensible to latent capacity; also known as *sensible heat ratio (SHR).*

sensible heat ratio (SHR) an air conditioner's ratio of sensible to latent capacity; also known as *sensible heat fraction (SHF).*

septic drain field a drain field used to remove contaminants and impurities from the liquid that emerges from the septic tank.

septic system an underground treatment system for human sewage.

sheathing boards or sheet material that are fastened to joists, rafters, and studs and on which the finish material is applied.

shed roof a type of roof that slopes in one direction only.

sill first horizontal wood member resting on the foundation supporting the framework of a building; also, the lowest horizontal member in a window or door frame.

simple payback the amount of time it takes to recover the initial investment for energy-efficiency improvements through energy savings, dividing initial installed cost by the annual energy cost savings.

single-ply membrane roofing material that comes in sheets, which are attached to the roof deck and seamed together with mechanical fasteners, heat, or chemical solvents.

single-point blower door testing blower door testing that utilizes a single measurement of fan flow needed to create a 50-Pa change in building pressure.

slab a single door panel, excluding the jamb, hinges, threshold, and door hardware.

slag wool another name for *mineral wool.*

slope disturbance a process of destabilizing a slope through grading and site development.

smart grid a nickname for an ever-increasing number of utility applications that enhance and automate the monitoring and control of electrical distribution.

smart meter a broad term defining electrical meters that include two-way communication and other advanced capabilities.

soffit a horizontal surface that projects out from an exterior wall.

softscape the vegetated elements of a landscape, including plants and soil.

soil gas includes air, water vapor, radon, methane, and other soil pollutants that can enter a building though gaps in the foundation or crawlspace floor.

solar gain the heat provided by solar radiation.

solar heat gain coefficient (SHGC) the fraction of solar radiation admitted through glazing.

solar orientation the cardinal direction in which the home and its glazing faces.

solar potential the amount of sunlight a particular site can capture based on latitude, local weather conditions, and the amount of unobstructed southern sky.

Solar Rating and Certification Corporation (SRCC) a non-profit organization whose primary purpose is the development and implementation of certification programs and national rating standards for solar energy equipment.

solar reflectance is a decimal number less than one that represents the fraction of light reflected off the roof; see also *albedo.*

Solar Reflectance Index (SRI) measure of a material's ability to reject solar heat, thereby staying cool; typical values range on a scale from 0 to 100, from low to high ability to reject heat.

solar thermal a system that converts sunlight into heated air or water.

solid surface a manufactured product typically used for countertops that emulates stone, created by combining natural minerals with resin and additives.

sone a unit of perceived loudness.

spectrally selective glazing a coated or tinted glass with optical properties that are transparent to some wavelengths of energy and reflective to others.

spider duct systems individual runs of ductwork directly from the furnace or air handler to individual rooms.

spot ventilation the mechanical process of removing moisture, odors, and pollutants directly at the source.

spray polyurethane foam (SPF) insulation a spray-applied insulating foam plastic that is installed as a liquid and then expands many times its original size; see also *open-cell foam* and *closed-cell foam.*

stack effect the draft established in a building from air infiltrating low and exfiltrating high.

stack thermometer an instrument for measuring temperature within the flue pipe or stack of a combustion appliance.

standing seam a roof assembled from metal panels with vertical seams that are snapped or crimped together to form a seal.

static pressure the pressure inside the duct system and an indicator of the amount of resistance to air flow within the system; usually measured in inches of water column (IWC) or Pascals (Pa).

steep-slope roof a roof whose angle is more than 30° (2:12).

step flashing individual metal pieces installed behind the water-resistive barrier (WRB) and interlaced with the roof shingles.

storage reservoir walls wall assemblies that can absorb water without risking structural damage.

stormwater the flow of water that results from precipitation following rainfall or as a result of snowmelt.

straw bale construction using baled straw from wheat, oats, barley, rye, rice, and other agricultural waste products in walls covered by stucco or earthen plaster.

structural insulated panels (SIP) are composite building materials made from solid foam insulation sandwiched between two sheets of oriented-strand board to create construction panels for floors, walls, and roofs.

structurally integrated insulation an insulation system that is part of a building structure as opposed to applied to a structural component.

sunspace a passive solar addition to a new or existing house to provide additional winter heat through solar energy.

sun-tempered design a passive solar design strategy where the majority of glazing is oriented on the north-south axis.

supply approach a method of sizing cisterns in which a tank large enough to provide the total monthly need for a house is installed.

supply-only ventilation provides outside air to the home without exhaust.

surface area of building envelope (SFBE) the total area (in ft^2) of the building envelope or building shell.

sustainable a pattern of resource use that aims to meet human needs while preserving the environment so that these needs can be met, not only in the present but also for future generations.

sustainable forest management is forestry management practices that maintain and enhance the long-term health of forest ecosystems while providing ecologic, economic, social, and cultural opportunities for the benefit of present and future generations.

Sustainable Forestry Initiative (SFI) a nonprofit organization responsible for maintaining, overseeing, and improving a sustainable forestry certification program.

T

tank heaters heaters that store a large quantity of hot water in an insulated storage tank.

tankless heaters heaters that heat water only when needed, require a minimum flow rate to operate, and do not store any hot water.

task lighting lighting directed to a particular work area, such as kitchen counters and desks, and not throughout a room.

temperature pumps a recirculating pump that has sensors near fixtures that tell it when to circulate hot water based on when the temperature drops below a set level.

temperature rise the difference between the temperature of water entering the house and that of the hot water leaving the water heater.

thermal barrier a boundary to heat flow (i.e., insulation).

thermal break a material of low thermal conductivity placed in an assembly to reduce or prevent the flow of heat between conductive materials.

thermal bridge a thermally conductive material that penetrates or bypasses an insulation system, such as a metal fastener or stud.

thermal bypass the movement of heat around or through insulation, frequently due to missing air barriers or gaps between the air barriers and the insulation.

thermal emittance a decimal number less than one that represents the fraction of heat that is re-radiated from a material to its surroundings.

Thermal Enclosure System Rater Checklist an inspection of building details for thermal bypasses; for a home to qualify for the ENERGY STAR label, a this checklist must be completed by a certified HERS Rater.

thermally treated lumber produced by exposing wood to very high temperatures and steam, transforming it into a product that is not affected by insects and decay.

thermosiphon loops a passive heat exchange system that operates without a mechanical pump, using the natural convection of rising hot water to circulate to the top of the loop, thereby allowing cooler water to flow down and back to the water heater.

thermostatic valve a valve that maintains a safe shower temperature to prevent scalding by adjusting the mix of hot and cold water; also known as a *thermostatic compensating valve.*

third-party certification a review and confirmation by a nonaffiliated, outside organization that a product meets certain standards.

timer-operated pumps pumps that can be set to run for certain periods of time when occupants will require hot water.

torch-down modified bitumen a rolled roof material with a heat-activated adhesive.

total duct leakage the amount of duct leakage both inside and outside the building envelope.

tracer gas a nontoxic gas used to measure envelope air leakage.

traditional neighborhood development (TND) the design of a complete neighborhood or town using traditional town planning principles that emphasize walkability, public space, and mixed-use development; see also *new urbanism.*

transfer grilles vents placed in walls to allow air to flow between rooms.

transit oriented development (TOD) the design of communities to be within walking distance of public transit, mixing residential, retail, office, open space, and public uses in a way that makes it convenient to travel on foot or by public transportation instead of by car.

tree plantation an actively managed crop of trees that, unlike a forest, contains one or two tree species and provides little wildlife habitat.

trellis a structure used to provide shading or support for climbing plants; usually made from interwoven pieces of wood, bamboo, or metal.

true divided light (TDL) window or doors in which multiple individual panes of glass or lites are assembled in the sash using muntins.

trunk-and-branch a plumbing design consisting of 3/4" or 1" hot and cold water pipes or trunks that run to each plumbing location and from which smaller 1/2" pipes, or branches, run to individual fixtures.

trunk-and-branch duct systems large ducts (trunks) installed through the center of the house with smaller ducts (branches) that run off the trunks to supply individual rooms.

truss an engineered lumber product consisting of wood or wood and metal members used to support roofs or floors that reduces the amount of lumber required to support a specific load.

tubular daylighting devices (TDD) a cylindrical skylight with a reflective tube to provide daylight to interior rooms.

U

unfaced batts cotton or fiberglass batt insulation that does not contain a vapor retarder covering.

unvented roof an attic assembly that does not contain ventilation.

urea formaldehyde a potentially toxic chemical commonly used as a binder or adhesive in engineered building products.

U.S. Green Building Council (USGBC) a nonprofit that promotes green building and developed the LEED rating systems; see also *LEED.*

U-value thermal transmittance or thermal conductance of a material; the reciprocal of R-value.

V

vacancy sensors lighting controls that require operating a switch to turn them on that turn off automatically after a room is vacated.

vapor barrier a Class I vapor retarder (0.1 perm or less).

vapor diffusion retarder (VDR) a material that reduces the rate at which water vapor can move through a material.

vitrified clay pipe plumbing pipe made from clay that has been subjected to vitrification, a process that fuses the clay particles to a very hard, inert, glass-like state.

volatile organic compounds (VOC) are chemical compounds that have a high vapor pressure and low water solubility; many VOCs are man-made chemicals that are used and produced in the manufacture of paints, pharmaceuticals, refrigerants, and building materials; VOCs are common indoor air pollutants and ground water contaminants.

volume-control valves valves that control the flow water to fixtures.

W

walk-off mat an abrasive floor mat installed over a shallow receptacle that catches debris as it is scraped off shoes.

wallcovering a covering on a wall, such as vinyl or paper wallpaper.

waterproofing a treatment used on concrete, masonry, or stone surfaces that prevents the passage of water under hydrostatic pressure.

water-resistive barrier (WRB) the material behind the cladding that forms a secondary drainage plane for liquid water, commonly referred to as the weather-resistive barrier, weather-resistant barrier, and water-resistant barrier.

WaterSense the US Environmental Protection Agency–administered program that provides guidelines for installation of efficient irrigation systems and certification of irrigation professionals.

water softeners appliances that remove minerals from water to reduce hardness; typically installed to treat the entire water supply for a home.

water-source heat pump (WSHP) a heating and cooling unit that exchanges heat between the ground or water and the home interior; also referred to as geothermal heat pumps.

weep screed a metal or plastic flashing at the base of walls that allows moisture to drain away from behind masonry and stucco.

wet-bulb temperature the temperature recorded by a thermometer whose bulb has been covered with a wetted wick and whirled on a sling psychrometer; used to determine relative humidity, dew point, and enthalpy.

X

xeriscaping a landscaping technique that uses drought-tolerant plants to minimize the need for water, fertilizers, and maintenance.

Z

zone control a device that controls the amount of air flow to different areas of the home.

Glosario

A

ACCA Manual D guía para el diseño de sistemas de conductos residenciales que ayuda a asegurarse de que los conductos proporcionen la cantidad adecuada de aire calefaccionado o refrigerado a cada habitación.

ACCA Manual J guía para dimensionar los sistemas de calefacción y enfriamiento residenciales según las condiciones climáticas y ambientales en el lugar de la obra.

ACCA Manual S guía para la selección y la dimensión de los equipos de calefacción y enfriamiento, a fin de cumplir las cargas indicadas en el Manual J.

ACCA Manual T guía para diseñadores sobre la forma de seleccionar, dimensionar y ubicar salidas de alimentación y entradas de retorno.

iluminación de acento iluminación en paredes, obras de arte y elementos arquitectónicos del interior o exterior para realzar el atractivo visual de un área en particular.

acrilonitrilo butadieno estireno (ABS) material rígido para tuberías que se utiliza en líneas de drenaje.

carbón activado forma de carbón especialmente formulada para la filtración.

intoxicación por monóxido de carbono se refiere a la intoxicación que tiene lugar después de una única exposición importante al gas; puede afectar a una o más personas.

armado avanzado metodología de construcción destinada a conservar los materiales de construcción utilizando métodos alternativos de armazón; véase también *ingeniería de valor óptimo (OVE)*.

aireador dispositivo instalado en los grifos para aumentar la velocidad de rocío, reducir las salpicaduras y ahorrar agua y energía.

sistema séptico aeróbico sistema de tratamiento de aguas residuales que inyecta aire de forma mecánica en un tanque de recolección de agua para fomentar la descomposición, lo que suministra un efluente de calidad superior.

válvula de entrada de aire válvula unidireccional que permite el ingreso de aire al sistema de ventilación de la plomería, según sea necesario.

barrera de aire material protector y resistente al aire que controla las fugas de aire que entran y salen de la envolvente del edificio.

cambios de aire por hora a 50 pascales (ACH_{50}) la cantidad de veces que el volumen total de un hogar se intercambia con el aire exterior-cuando el hogar está descomprimido o comprimido a 50 pascales.

cambios de aire por hora natural ($ACH_{Natural}$) la cantidad de veces que el volumen total de un hogar se intercambia con aire exterior en condiciones naturales.

aire acondicionado electrodoméstico, sistema o mecanismo que deshumidifica y extrae calor de una zona particular.

aire acondicionado (AC) el proceso de enfriamiento del aire interior mediante la transferencia de calor bajo techo a un refrigerante, que luego se desplaza hacia el exterior.

Air Conditioning Contractors Association—Asociación de Contratistas de Aire Acondicionado (ACCA) organismo profesional que publica normas para sistemas de calefacción y enfriamiento.

Air-Conditioning, Heating, and Refrigeration Institute — Instituto de Aire acondicionado, Calefacción y Refrigeración (AHRI) asociación gremial que representa a los fabricantes de calefacción, ventilación, aire acondicionado, y refrigeración comercial.

fuga de aire (AL) medición de la cantidad total de fuga de aire, equivalente al total de pies cúbicos de aire que pasan a través de 1 pie^2 de superficie de ventana por minuto (pies cúbicos por minuto/pie^2).

sello de aire el proceso de cerrar herméticamente la envolvente del edificio reduciendo la fuga de aire tanto hacia dentro como hacia fuera de un hogar.

bomba de calor por compresión sistema de calefacción y enfriamiento que consiste de un compresor y dos bobinas de tubo de cobre, una ubicada adentro y otra afuera, con aletas de aluminio alrededor para ayudar a la transferencia de calor.

método de paneles de yeso herméticos (ADA) sistema de barrera de aire que conecta la terminación interior de los paneles de yeso (tablarroca) a la estructura de las casas para formar una barrera de aire continua.

albedo coeficiente de energía electromagnética que refleja un objeto o superficie.

floración algácea crecimiento rápido y excesivo de la población de algas de un sistema acuático dentro de un período breve.

iluminación ambiental iluminación general para actividades cotidianas en el interior y, según sea necesario, en el exterior por motivos de seguridad.

sistema séptico anaeróbico sistema de tratamiento de aguas residuales en el cual se eliminan los sólidos en ausencia de oxígeno sin tratamiento directo de los efluentes.

eficiencia anual de consumo de combustible (AFUE) coeficiente de producción de calor del calentador o la caldera comparado con la energía total consumida por el calentador o la caldera.

aislamiento aplicado material de aislamiento térmico o sonoro que se coloca entre los elementos estructurales o encima de esllos después de haberlos instalados.

argón gas inerte que suele agregarse al espacio aéreo entre paneles de vidrio para reducir el factor U.

asbesto material fibroso natural que solía usarse como tratamiento contra el fuego; extremadamente peligroso cuando se inhala.

pared de ático pared vertical que separa el espacio interior acondicionado de la superficie del ático sin acondicionar.

concreto aireado autoclave (AAC) material liviano premoldeado que ofrece estructura, aislamiento y resistencia al fuego y al moho.

toldo cubierta secundaria adosada a la pared exterior de un edificio para proteger una ventana del sol y la lluvia.

B

ventana basculante ventana accionable que tiene en la parte superior un bastidor de hoja con bisagra que se abre hacia afuera.

ventilación balanceada sistema que proporciona suministro y escape en partes iguales de aire mediante el uso de ventiladores o intercambiadores de calor aire-aire, conocido como ventiladores de recuperación de calor (HRV) o ventiladores de recuperación de energía (ERV).

balasto dispositivo que regula la frecuencia de la electricidad que se suministra a un accesorio de iluminación para controlar las tensiones inicial y de funcionamiento de la lámpara.

armazón de globo sistema de construcción con estructura de madera, que comenzó a utilizarse en el siglo XIX, en el cual se emplean montantes continuos desde los cimientos hasta el larguero superior.

placas de aislamiento material de aislamiento térmico o sonoro, generalmente de fibra de vidrio o algodón, que viene en distintos anchos y espesores (valor R) para adaptarse a las estructuras estándar de paredes y vigas.

mejores prácticas de administración (BMP) estrategias para mantener la suciedad y otros contaminantes alejados de arroyos y lagos; las BMP fueron concebidas con el objeto de proteger la calidad del agua y prevenir una nueva contaminación.

bioaislamiento productos para aislamiento que contienen materiales fabricados a partir de recursos renovables como sustitutos de productos a base de petróleo y otros productos no renovables.

biocombustibles combustibles no fósiles, como la madera, el etanol y el biodiesel que, por lo general, se consideran renovables.

aguas negras desechos que provienen de baños y cocinas, que contienen desechos humanos o alimentarios, y deben ser tratadas y desinfectadas antes de liberarlas a vías fluviales o de volverlas a utilizar.

soplador en puerta herramienta de diagnóstico diseñada para medir la calidad hermética de los edificios y para identificar los lugares de fuga de aire.

aislamiento soplado material compuesto de fibras de aislamiento sueltas, como fibra de vidrio, espuma o celulosa, que se bombea o inyecta hacia dentro de las paredes, los techos y otros sitios.

caldera equipo de calefacción diseñado para calentar agua (mediante electricidad, gas o aceite como fuente de calor), con el fin de suministrar calor a un espacio acondicionado o al agua potable.

unidad térmica británica (Btu) cantidad de calor necesaria para elevar una libra de agua 1°F.

retardadores de llama bromados grupo de productos químicos que inhiben la propagación del fuego y constan de compuestos orgánicos que contienen bromo.

desarrollo de zona industrial abandonada (brownfield) instalación industrial o comercial abandonada o subutilizada que está disponible para un nuevo uso.

envolvente del edificio separación entre los ambientes interno y externo de un edificio, que consiste de una barrera de aire o barrera térmica con características continuas y en contacto.

Building Performance Institute—Instituto de Desempeño de Edificios (BPI) organismo sin fines de lucro que ofrece programas reconocidos de capacitación, certificación, acreditación y aseguramiento de la calidad para contratistas y empresas.

ciencia de la construcción estudio de la interacción de los sistemas y componentes de los edificios, los ocupantes y el entorno que los rodea; se concentra en la circulación de calor, aire y humedad.

C

interrupción capilar espacio o material que impide el movimiento de la humedad entre dos superficies por intervención de la capilaridad.

monóxido de carbono (CO) gas incoloro, inodoro y tóxico que es produto de la combustión incompleta de los combustibles (p. ej., gas licuado de petróleo o gas natural, petróleo, madera y carbón).

detector de monóxido de carbono dispositivo que registra los niveles de CO en el tiro de un aparato de combustión o en las zonas adyacentes.

sumidero de carbono depósito ambiental que absorbe y almacena carbono, retirándolo de la atmósfera.

Carpet and Rug Institute—Instituto de la Alfombra (CRI) asociación industrial sin fines de lucro, creadora de programas de certificación de terceros, Etiqueta verde y Etiqueta verde plus, que brindan lineamientos concernientes al contenido orgánico volátil y los gases que se desprenden de las alfombras y los materiales de relleno.

ventana abatible ventana con bisagras laterales que se abre hacia adentro o hacia afuera.

captación zona impermeable en la cual cae el agua de lluvia.

cielorraso estilo catedral cielorraso abovedado y aislado con techo por encima; a veces se le conoce como combinación "techo-cielorraso" y suele encontrarse en salas y áticos con techos aislados.

aislamiento de cavidad aislamiento que se coloca entre los montantes o las vigas de la pared.

aislamiento de celulosa fabricado con papel periódico reciclado y con agregado de retardador de fuego.

baldosa cerámica unidad de revestimiento superficial delgada compuesta de varias arcillas cocidas para endurecerlas.

charrette asamblea de diseño en la que participan todas las partes interesadas de un proyecto.

cloruro de polivinilo clorado (CPVC) material para tuberías que es similar químicamente al PVC, pero contiene cloro para aumentar la rigidez a altas temperaturas.

arseniato de cobre cromatado (CCA) conservante químico de la madera que contiene cromo, cobre y arsénico.

intoxicación por CO crónica se utiliza para indicar los casos de personas expuestas en más de una ocasión al monóxido de carbono, por lo general en concentraciones relativamente bajas.

cisternas se utilizan para almacenar agua de lluvia que se ha recolectado.

espuma de celda cerrada tipo de espuma poliuretánica en rociador, que se instala a una velocidad aproximada de 2 libras/pie^3 y a veces se conoce como "espuma de 2 libras"; véase también *espuma de poliuretano rociada.*

cenizas volantes de carbón la porción muy fina o el residuo ceniciento de la combustión de carbón que puede utilizarse como sustituto del cemento Portland.

ladrillo crudo combinación de arcilla natural y paja que se mezcla y con la que se construyen muros estructurales en estilo libre.

coeficiente de rendimiento (COP) coeficiente de la energía producida (de calefacción o enfriamiento) con respecto a la cantidad de energía consumida.

cogeneración uso de un artefacto para producir calor o de una central eléctrica para generar simultáneamente electricidad y calor aprovechable, algo que también se conoce como *centrales combinadas para la producción de electricidad y calor (CHP).*

coque subproducto sólido de la combustión del carbón.

arranque en frío primera vez que se utiliza agua caliente en el día o después de que se ha enfriado la línea principal.

Índice de rendimiento de color (CRI) cuantifica la capacidad de una lámpara de reproducir colores al igual que un tipo de lámpara de referencia.

temperatura de color indicador del color de una fuente de luz que compara el color de una lámpara con la luz natural del sol; se mide en grados kelvin.

columna pilar vertical, por lo general, cilíndrico y hecho de piedra, fibra de vidrio o madera, que sostiene un arco u otra estructura, o simplemente actúa como elemento arquitectónico.

central combinada para la producción de electricidad y calor (CHP) uso de un artefacto para producir calor o de una central eléctrica para generar simultáneamente electricidad y calor aprovechable, algo que también se conoce como *cogeneración.*

analizador de combustión herramienta utilizada in situ para medir la eficiencia de una caldera.

puesta a punto proceso de diagnosticar y verificar el rendimiento del sistema edilicio.

lámparas fluorescentes compactas (CFL) alternativa de iluminación común a las lámparas incandescentes que emiten luz cuando una mezcla de tres fósforos se expone a la luz ultravioleta de átomos de mercurio.

combustión completa proceso por el cual el carbono del combustible se une con el oxígeno para formar dióxido de carbono (CO_2), vapor de agua, nitrógeno y aire.

tejas sintéticas de asfalto y fibra de vidrio; es el material más utilizado para techos con vertiente pronunciada.

sanitario compostero sanitario que convierte los desechos humanos en compuesto sanitario utilizable y sin usar agua para efectuar el proceso.

compresor bomba mecánica que utiliza presión para cambiar un refrigerante de líquido a gas.

concreto producto de mampostería compuesto de cemento o puzolana, arena y grava u otro elemento grueso.

unidad de mampostería de concreto (CMU) bloque grande y rectangular de concreto utilizado en la construcción.

condensación humedad extraída del aire mediante un sistema de deshumidificación o acondicionador de aire.

resistencia a la condensación (CR) medida de la resistencia de una ventana en particular a la formación de condensación en el interior.

condensador del componente en un sistema de refrigeración que transfiere calor desde el sistema por la condensación del refrigerante.

caldera de condensación aparato de calefacción por combustión que emplea un intercambiador de calor secundario para condensar vapor en el gas de combustión y capturar calor adicional, aumentando su eficiencia.

calentadores por condensación calentadores de agua de alta eficiencia que extraen grandes cantidades de calor de los gases de combustión, lo que provoca la condensación de estos gases.

unidad condensadora la porción de un sistema de refrigeración donde se realiza la compresión y condensación del refrigerante.

espacio subyacente acondicionado cimiento sin ventilación en las paredes que encierra un espacio calefaccionado o refrigerado intencionalmente; el aislamiento se coloca en las paredes exteriores.

conducción transferencia de calor de una sustancia a otra por contacto directo.

humedales construidos medio de tratamiento de aguas residuales que simula un tratamiento natural de estas aguas, al permitir que los efluentes circulen a través de cauces de agua llenos de plantas que descomponen los contaminantes.

aislamiento continuo aislamiento que no ha sido interrumpido por elementos estructurales, por lo general se coloca en la superficie exterior del armazón de madera de las paredes de concreto.

bombas continuas bombas que circulan agua caliente a través de líneas troncales 24 horas al día.

convección transferencia de calor a través de un fluido (líquido o gas).

bucle convectivo circulación continua de aire (o de otro líquido) en un espacio encerrado a medida que se calefacciona o refrigera.

conducción de agua de lluvia sistema que dirige el agua de lluvia desde la zona de captación, compuesto por canaletas, tubos de bajada de aguas y tubería.

grados día de enfriamiento (CDD) medida de lo caliente que está un lugar en un período de tiempo relativo a una temperatura base, por lo general especificada como 65°F.

Cool Roof Rating Council—Consejo de Calificación de Techo Frío (CRRC) organismo independiente, sin fines de lucro, que mantiene un sistema de calificación de terceros de las propiedades radiactivas de los materiales de recubrimiento superficial del techo.

cornisa ensamble entero terminado donde las paredes de una estructura se unen con el techo; a veces se denomina *alero* o *sofito*.

aislamiento de algodón material de aislamiento térmico o sonoro, por lo común, confeccionado con residuos de fabricación textil, que viene en anchos y espesores (valor R) variados para adaptarse a las estructuras estándar de paredes y vigas.

tejadillo falso pequeño techo falso construido detrás de una chimenea u otro obstáculo para el techo, con el propósito de repeler el agua; también denominado chaflán.

polietileno reticulado (PEX) tipo especial de plástico polietilénico que se fortifica con uniones químicas formadas además de las uniones habituales en el proceso de polimerización.

piedra cultivada unidad de mampostería moldeada hecha de cemento y varios aditivos que simula el aspecto de la piedra natural.

D

resistencia a la humedad tratamiento utilizado en superficies de concreto, mampostería o piedra para repeler el agua y reducir la absorción de agua en ausencia de presión hidrostática.

iluminación natural uso de la luz natural para complementar o reemplazar la iluminación artificial.

cubierta de una estructura elevada, techo contiguo a una casa.

membrana de desacoplamiento lámina de plástico flexible que se coloca entre las baldosas cerámicas y la capa base del suelo para proporcionar fuerza y resistencia al agrietamiento.

método del déficit método para medir cisternas, por el cual se calcula la necesidad total durante períodos de precipitaciones escasas o nulas.

bomba por demanda traslada el agua caliente a los artefactos de plomería cuando se necesita, con lo que elimina las pérdidas del recalentamiento constante del agua en la tubería.

Department of Energy—Departamento de Energía (DOE) departamento federal responsable de la política energética nacional de los Estados Unidos.

desecante materiales como el óxido de calcio y el sílice que absorben la humedad y que suelen usarse para extraer la humedad del aire interior que se hace circular a través de ellos con ventiladores.

desupercalentador dispositivo que recupera el exceso de calor del proceso de refrigeración y calefacción, y lo desvía para calentar agua en un tanque de almacenamiento con gran eficiencia.

enfriador por evaporación directa dispositivo que reduce la temperatura del aire haciéndolo circular a través de almohadillas empapadas en agua.

destilación proceso de purificación de un líquido hirviéndolo y condensando sus vapores.

distribución tubería que entrega agua de lluvia recolectada y filtrada para su utilización.

buhardilla estructura que se proyecta hacia fuera de un techo con vertiente para formar otra zona techada que brinda una superficie para la instalación de ventanas.

ventana de guillotina doble ventana con dos hojas de funcionamiento vertical.

bajada de aguas elemento vertical utilizado para transportar agua desde la canaleta hasta el suelo; también se denomina tubo conductor o guía.

medidor de presión de gas mide la presión de los gases de combustión en una caldera o calentador de agua.

estera de drenaje material que crea un espacio entre el suelo y las paredes de los cimientos; este espacio libera presiones hidrostáticas y le otorga al agua un camino de menor resistencia para drenar fuera de la vivienda.

campo de drenaje componente final de un sistema séptico en el cual los efluentes circulan a través de una tubería perforada en el suelo, donde se filtran a medida que avanzan hacia el acuífero local.

recuperación de calor de agua de drenaje (DHR) reutilización del calor que se captura de las aguas residuales.

goterón moldura o tapajuntas horizontal que se instala sobre el marco para que una puerta o ventana expulsen el agua fuera de éste.

borde de goteo banda de metal que se extiende más allá de las otras partes del techo y que se utiliza para expulsar el agua de lluvia fuera de la estructura.

acanaladura de goteo corte en la parte inferior de la proyección con el objeto de evitar que el agua se traslade más allá de esta y regrese a la cara de la pared.

especies tolerantes a la sequía árbol o planta que puede crecer y prosperar en condiciones áridas.

temperatura de bola seca temperatura del aire indicada en un termómetro común; no tiene en cuenta los efectos de la humedad.

mampostería seca (drywall) material de construcción utilizado para paredes terminadas y superficies de cielorrasos; hecho de yeso secado en horno presionado entre placas de fibra de vidrio o revestimiento de papel; también se le conoce como *cartón-yeso o tablarroca.*

descarga dual sanitario de alta eficiencia que permite que los usuarios elijan si desean hacer una descarga de agua total o utilizar menos agua.

prueba de fugas de conductos prueba de diagnóstico diseñada para medir la calidad hermética de los sistemas de conductos de calefacción y aire acondicionado e identificar los lugares con fuga de aire.

minisplit sin conductos bomba de aire acondicionado o calefacción compacta, montada en la pared, conectada a una unidad de condensación exterior mediante líneas refrigerantes.

productos de vidriado dinámico (DG) vidrio que cambia de propiedades electrónicamente mediante una corriente eléctrica o también es un vidrio con persianas entre capas de vidrio que controlan la luz y el calor.

E

aislante del alero material que evita que el aislamiento del ático se lave con el viento, dirigiendo al flujo de aire del sofito (plafón) por encima del aislamiento del ático; también se conoce como *canal de ventilación positiva.*

construcción de orilla predio con un 25% o menos de los límites de las propiedades adyacentes a una construcción existente.

longitud efectiva de conducto cantidad total de pérdida de presión de las secciones rectas y todos los accesorios de cada tramo de conducto.

superficie efectiva de fuga (ELA) superficie de un orificio especial con forma de boquilla (similar a la entrada de un ventilador de soplador en puerta) que perdería la misma cantidad de aire que la que pierde un edificio a una presión de 4 pascales.

eficiencia coeficiente de luz producida sobre la cantidad de energía consumida; se expresa en lúmenes por vatio.

efluente flujo de agua residual antes o después del tratamiento.

caldera eléctrica sistema de calefacción que utiliza resistencia eléctrica para convertir electricidad en calor.

espectro electromagnético (EMF) fuerzas invisibles creadas por la transmisión de electricidad a través de cables; también se denomina *campos eléctricos y magnéticos.*

filtros de aire electrónicos utilizan electricidad para atraer moléculas pequeñas, como humo, moho y olores de mascotas, a aletas metálicas.

energía incorporada la energía necesaria para fabricar o recolectar, empacar y enviar material a una obra en construcción; puede referirse a un solo material o a la vivienda entera.

energía cantidad mensurable de calor, trabajo o luz.

coeficiente de eficiencia energética (EER) medición de la eficiencia de un sistema de refrigeración cuando la temperatura exterior está a un nivel específico (por lo general 95°F).

hipoteca de eficiencia energética (EEM) utiliza el ahorro energético de una vivienda nueva con eficiencia de energía para aumentar la capacidad de adquisición de viviendas de los consumidores, capitalizando el ahorro energético en la valoración.

factor energético (EF) calificación de la eficiencia energética de un calentador de agua basándose en la cantidad de agua caliente producida por unidad de combustible consumida en un día normal.

EnergyGuide etiqueta de color amarillo brillante, creada por la Comisión Federal de Comercio, que, por ley, debe aparecer en muchos electrodomésticos nuevos para mostrar el consumo relativo de energía comparado con productos similares.

hipoteca de mejoramiento energético (EIM) financia las mejoras energéticas de una vivienda existente e hipotecada utilizando los ahorros mensuales de energía.

ENERGY STAR programa conjunto de la Agencia de Protección Ambiental y el Departamento de Energía de los Estados Unidos que establece estándares para productos y edificios eficientes desde el punto de vista energético.

armazón de eficiencia energética armazón de techo diseñado para abarcar una superficie y brindar espacio adecuado para un aislamiento del ático en toda su profundidad y área; véase también *armazón de talón elevado.*

piso multilaminado maquinado producto compuesto por varias capas de madera adheridas como si fuera una sola tabla.

madera maquinada es madera que se elabora uniendo fibras de madera, enchapado, madera o fibra para producir un compuesto más robusto y uniforme; también se conoce con el nombre de *producto laminado de madera.*

productos preferibles para el medio ambiente (EPP) productos que tienen un efecto reducido sobre la salud humana y el medio ambiente comparados con productos o servicios tradicionales que sirven el mismo propósito.

declaraciones de producto medioambiental (EPD) datos medioambientales cuantificados para un producto con categorías predefinidas de parámetros basados en la serie de normas ISO 14040, pero sin excluir información medioambiental adicional.

longitud equivalente longitud comparable de un solo tramo de conducto recto cuando se tiene en cuenta la resistencia del flujo de aire desde la compresión del conducto, los codos, los accesorios y otras obstrucciones.

erosión la extracción de sólidos (p. ej., sedimentos, suelo, roca y otras partículas) por parte del viento, el agua o el hielo en el entorno natural.

monómero etileno propileno dieno (EPDM) una mebrana unicapa compuesta de caucho sintético, de uso común en techos planos.

colector de tubos vaciados colector térmico solar que utiliza placas absorbentes encerradas en un tubo de vidrio con vacío interior.

refrigeración por evaporación medio de reducción de la temperatura que funciona bajo el principio de que el agua absorbe calor latente del aire que la rodea cuando se evapora.

evaporador (vaporizador) componente de un sistema responsable de la refrigeración o el enfriamiento real del espacio ocupado.

exfiltración circulación de aire hacia fuera a través de una pared, envolvente de edificio, ventana u otro material.

ventilación sólo escape sistemas que extraen el aire de una vivienda con partes iguales de compensación de aire que ingresa a través de aberturas sin controlar (p. ej., fugas de la envolvente).

poliestireno expandido (EPS) placa de espuma para aislamiento elaborada con cuentas de poliestireno expandido.

techos verdes extensivos tipo de techo verde en el que se utiliza una capa delgada de algún medio de cultivo (es habitual colocarla sobre una estera de drenaje), que requiere plantas especiales de raíces cortas y poco crecimiento, como las del género sedum.

sistema de terminación con aislamiento exterior (EIFS) estuco sintético que se aplica sobre la espuma para aislamiento.

poliestireno extruido (XPS) placa de espuma para aislamiento de celda cerrada.

F

placas de aislamiento con retardador de vapor aislamiento en placas (bloques) que contiene un revestimiento retardador de vapor laminado metálico o de papel Kraft.

bobinas de ventilador dispositivo simple que consiste de una bobina para calefacción o refrigeración que se utiliza para distribuir la calefacción o la refrigeración en un espacio.

fascia también se denomina imposta; panel vertical entre el sofito y el techo.

sistema de tarifas energéticas reguladas (FIT) las compañías de servicios públicos adquieren energía renovable a tarifas variables, que suelen ser mayores que las tarifas a las que se vende la energía.

fenestración describe a todos los productos que cubren aberturas en una envolvente de edificio, como ventanas, puertas y tragaluces, que permiten el ingreso de aire, luz, personas o vehículos.

cartón madera producto de madera maquinada utilizado principalmente como placa para aislamiento y con fines decorativos, pero también puede usarse como revestimiento de paredes.

aislamiento con fibra de vidrio aislamiento mediante mantas, relleno suelto, o placas rígidas compuesto de fibras de vidrio unidas con un aglutinante.

iluminación por fibra óptica sistemas de iluminación que emplean fibras de vidrio o plástico flexibles y muy delgadas para transmitir la luz.

válvula de llenado parte interna que mide la cantidad de agua que llena el tanque y la taza del inodoro.

calificación de primera hora cantidad de agua caliente en galones que un calentador puede suministrar por hora (comenzando con un tanque lleno de agua caliente).

certificación interna cuando una compañía establece sus propias reglas, analiza su desempeño e informa su cumplimiento.

primera descarga el flujo inicial de agua que sale de un techo que suele incluir excremento de aves, hojas y otros tipos de suciedad.

tapajuntas y placa (FAB) sistema híbrido de aislamiento que combina una capa de 1 a 2 pulgadas de SPF de celda cerrada, cubierto con placas de fibra de vidrio para rellenar la cavidad del armazón.

colector de placa plana colector térmico solar rectangular, por lo general de 8 a 10 pulgadas de largo.

fluorescente tipo de lámpara que genera luz cuando la electricidad pasa a través de un gas que está dentro de un tubo de vidrio recubierto con fósforo y hace que brille.

bujía-pie cantidad de luz producida por un lumen sobre una superficie de 1 pie^2.

apoyo del cimiento apoyo ensanchado, generalmente de concreto, en la base de las paredes de cimiento, columnas, pilares y chimeneas, que distribuye el peso de estos elementos en una superficie de mayor tamaño e impide un asentamiento desigual.

Forest Stewardship Council—Consejo de Manejo Forestal (FSC) organismo no gubernamental, independiente y sin fines de lucro fundado para promover el manejo responsable de los bosques del mundo.

drenaje de los cimientos proceso de dirigir el agua subterránea fuera de los cimientos y la vivienda.

paredes de cimiento paredes construidas parcialmente bajo tierra que soportan el peso del edificio sobre ellas y encierran el sótano o el espacio subyacente al piso.

balcón Francés puertas con baranda externa que permiten abrir el espacio interno al exterior.

cimientos poco profundos protegidos de la escarcha protección contra los daños provocados por la escarcha sin necesidad de excavar por debajo de la línea de escarcha.

pila de combustible dispositivo que utiliza combustibles fósiles, como gas natural o propano, para producir electricidad sin combustión, mediante la extracción de hidrógeno, que se utiliza para generar energía de una forma similar a la de las baterías.

caldera dispositivo que funciona con gas, aceite o madera en el cual se calienta aire y se hace circular por un edificio en un sistema de conductos.

G

techo a dos aguas tipo de techo con vertiente en dos direcciones.

eliminación de residuos trituradoras eléctricas que permiten eliminar los restos de alimentos a través de la cloaca, proceso que requiere agua y electricidad.

vidriado parte transparente de una pared o puerta que suele estar confeccionada en vidrio o plástico.

viga de cimiento soporte de cargas perimetral de una estructura que se extiende entre los pilares sin apoyarse en el suelo que se encuentra debajo.

grano unidad de medida de contenido de humedad; 1 libra contiene 7,000 granos.

termocambiador con película de gravedad (GHX) dispositivo de transferencia térmica utilizado con sistemas de recuperación de calor de agua de drenaje.

descarga por gravedad sanitarios (inodoros) que utilizan el peso del agua del tanque para evacuar la taza.

aguas grises agua no potable recuperada de fregaderos, tinas y máquinas de lavar que podría utilizarse para las descargas de los sanitarios o para irrigación.

edificio ecológico edificio ambientalmente sostenible, diseñado, construido y operado para reducir al mínimo el impacto ambiental total.

desarrollo nuevo (greenfield) tierra que no ha tenido ninguna construcción, en una zona urbana o rural, y que actualmente se utiliza para agricultura, paisaje o parque natural.

efecto invernadero acumulación de calor en el espacio interior provocado por el ingreso de energía a través de una membrana transparente como el vidrio; se refiere al proceso por el cual los planetas mantienen la temperatura mediante la presencia de una atmósfera que contiene gas que absorbe y emite radiación infrarroja.

gas invernadero (GHG) cualquiera de los gases atmosféricos, como el dióxido de carbono (CO_2), los óxidos de azufre (SO_x) y los óxidos nitrosos (NO_x), que contribuyen al efecto invernadero.

techo ecológico techo parcial o totalmente cubierto con vegetación y un medio de cultivo, plantado sobre una membrana impermeable; también se conoce como *techo verde* o *techo vivo.*

desarrollo gris (greyfield) bienes inmuebles o tierra que han sido subutilizados.

interconexión a la red se refiere a viviendas conectadas a la red eléctrica central.

grilla barras que dividen un marco guillotina en paños de vidrio más pequeños.

escoria de alto horno granulada y molida subproducto de la fabricación del hierro y el acero que se utiliza para hacer concreto duradero en combinación con cemento Portland común un otros materiales puzolánicos.

bomba de calor geotérmica (GSHP) sistema de calefacción y refrigeración central que bombea calor desde el suelo y hacia este.

canaleta canal de madera, metal o plástico utilizado en la orilla del techo para vaciar el agua de lluvia y la nieve derretida.

H

halógeno variación de lámpara incandescente que utiliza un filamento dentro de una estructura de vidrio compacta rellena con gas.

paisajes duros elementos no vegetados del paisaje, que incluyen pavimento, aceras, caminos, muros de contención, construcciones, fuentes y piscinas.

agua dura agua con alto contenido mineral.

tapajuntas de dintel tapajuntas que se coloca sobre una proyección, protuberancia o abertura para ventana.

grados día de calefacción (HDD) medición de lo frío que está un lugar en un período determinado y relativo a una temperatura base, por lo general especificada como 65°F.

factor de rendimiento estacional de la calefacción (HSPF) la producción de calor estacional estimada de una bomba de calor en Btu dividida entre la cantidad de energía que consume en horasvatio.

efecto isla de calor el fenómeno de que las zonas urbanas son más calurosas que las zonas rurales, especialmente por el aumento en la utilización de pavimento y materiales de construcción que retienen calor con eficiencia.

bomba de calor unidad de calefacción y refrigeración que extrae calor de una fuente exterior y lo transporta a un espacio interior para calefaccionarlo o, al contrario, para enfriarlo.

trampas de calor válvulas o bucles de tubería especiales unidireccionales que impiden la circulación natural del agua caliente fuera de la parte superior del calentador por convección, con lo que ahorran energía.

calificador HERS persona acreditada a escala nacional que lleva a cabo calificaciones HERS para evaluar la eficiencia energética de las viviendas; a veces se conoce como calificador de energía doméstica.

sanitarios de alta eficiencia (HET) la EPA los define como aquellos que utilizan un 20% menos de agua, en promedio, que el estándar de la industria que es de 1,6 GPF.

descarga de alta intensidad (HID) variación de lámparas incandescentes fabricadas con vapor de mercurio, haluro metálico y gases sódicos de presión alta y baja, de uso común en aplicaciones residenciales exteriores.

tejado a tres o cuatro aguas techo con tres o cuatro vertientes y lados.

tanque colector componente de los sistemas sépticos que permite que los residuos sólidos se asienten, pero deja fluir los efluentes líquidos hacia el suelo.

Sistema de calificación de la energía residencial (HERS) medición con reconocimiento nacional de la eficiencia energética de una vivienda.

Rendimiento doméstico con ENERGY STAR (HPwES) programa conjunto de la Agencia de Protección Ambiental y el Departamento de Energía de los Estados Unidos que establece estándares para evaluar y mejorar la eficiencia energética de las viviendas existentes.

sistema doméstico de colectores véase *sistema de colectores.*

grifo de manguera llave con mangueras en el pico de salida, que se encuentra en el exterior de lavaderos o fregaderos, o cerca de ellos.

ático acondicionado ático sin ventilar que contiene aislamiento por debajo o directamente por encima el substrato del techo; también se conoce como *ático estilo catedral* o *ático aislado.*

arranque en caliente temperatura de la línea de suministro de agua caliente cuando se abre un grifo; se da cuando el agua caliente permanece en la línea y está disponible para su uso.

bucle de circulación de agua caliente tubería continua desde el calentador de agua distribuida cerca de todos los artefactos y regresa al fondo del calentador.

barrera envolvente barrera sintética resistente al agua diseñada para repeler la humedad y permitir el paso del vapor; véase también *barrera resistente al agua.*

hidrónico sistema de acondicionamiento de espacios que hace circular agua calefaccionada o refrigerada a través de radiadores montados en la pared o en una placa base, de tubería que recorre el piso o de una combinación de ambos métodos.

presión hidrostática fuerza que ejerce el agua subterránea sobre los cimientos.

higroscópico material que atrae y retiene fácilmente la humedad.

I

reborde de hielo hielo que se forma en el alero de un tejado con vertiente, que hace que la acumulación de agua detrás de este sea retenida debajo de los materiales del techo.

viga en I componente estructural de la construcción que consiste de un armazón ancho de paneles de fibra orientada verticalmente con alas de madera maquinada en las partes superior e inferior. Las vigas en I pueden usarse para el suelo, el techo o el armazón de la pared.

iluminación es la cantidad de luz que llega a una superficie de trabajo o tarea, suele denominarse también *iluminancia.*

impermeable material o conjunto de elementos que no permite el paso del aire o la humedad.

incandescente lámparas que generan luz cuando la electricidad pasa por un filamento de tungsteno encerrado en vidrio, haciéndolo brillar.

combustión incompleta tiene lugar cuando la proporción combustible-oxígeno es incorrecta, lo que ocasiona la producción de monóxido de carbono y aldehído.

enfriador por evaporación indirecta enfriador similar al de evaporación directa, pero que utiliza algún tipo de intercambiador de calor para evitar que el aire húmedo enfriado entre en contacto directo con el ambiente acondicionado.

unidades de relleno inserción de unidades habitacionales adicionales en un barrio o subdivisión ya aprobados.

infiltración proceso no controlado por el cual el aire o el agua circulan a través de la envolvente de un edificio hacia el interior del hogar.

encofrado de concreto aislado (ICF) encofrado de madera mineralizada o espuma aislante que se dejan colocados después de verter concreto para cimientos o paredes.

vidrio aislado ventana compuesta por un mínimo de dos paños separados por un espacio sellado, lleno de aire u otros gases.

aislamiento material que reduce o impide la transmisión de calor.

unidades para contacto con aislamiento (IC) aparatos de iluminación empotrados que disipan calor en el ambiente, lo que permite que el aislamiento esté en contacto con ellos sin que haya riesgo de incendio ni interrupciones en la envolvente térmica.

placa de aislamiento producto de aislamiento rígido, disponible en varios anchos y espesores (valor R).

diseño integrado método colaborativo para diseñar edificios que destaca el empleo de un diseño holístico.

control de plagas integrado (IPM) estrategia para primero limitar el uso de pesticidas y únicamente en casos de necesidad, usar aquellos productos que representen el menor riesgo y usarlos con moderación.

techos verdes intensivos tipo de techos verdes con capas profundas de suelo que pueden soportar arbustos y pequeños árboles.

desagües internos aberturas en la superficie de un techo de vertiente baja, que conducen a bajadas de agua dentro de la estructura del edificio para eliminar el agua del techo.

pintura intumescente pintura con retardador de incendio.

especies invasivas plantas no autóctonas que tienden a diseminarse de forma agresiva.

inversor dispositivo que convierte energía de corriente continua generada en planta a energía de corriente alterna para uso doméstico.

lapacho madera dura sudamericana que es muy pesada y se caracteriza por su resistencia natural a los insectos y la putrefacción.

J

persiana ventana con tablillas paralelas operables de vidrio, acrílico o madera colocadas en un marco.

viga de soporte miembro horizontal de la estructura utilizado en un patrón espaciado que da soporte al sistema de piso o techo.

conductos de puente pequeñas secciones de conductos instaladas en los cielorrasos que permiten que el aire circule entre las habitaciones.

K

Kelvin escala de temperatura utilizada para determinar la temperatura del color de diferentes tipos de luz.

lámina de escurrimiento preventiva lámina de escurrimiento instalada en la parte inferior de una vertiente de techo, que se encuentra junto a una pared y evita que el agua de lluvia del techo llegue detrás del material de revestimiento de la pared y la barrera resistente al agua (WRB).

interruptor de corte control que interrumpe el suministro eléctrico a un conjunto de receptáculos.

cableado de perilla y tubo antiguo sistema de cableado eléctrico compuesto de cables individuales distribuidos en pares y que se conectan a la estructura mediante aisladores de cerámica.

lámparas de gas inerte que suele agregarse al espacio aéreo entre paneles de vidrio para reducir el factor U.

L

Criptón comúnmente conocidas como bombillas luminosas, son las partes reemplazables de un artefacto que produce luz mediante electricidad.

paisaje todas las características exteriores de la vivienda, que incluyen los elementos naturales y construidos.

calor latente el calor que produce un cambio de estado sin un cambio en la temperatura; la parte de la carga de enfriamiento que se produce cuando la humedad del aire cambia de vapor a líquido (condensación).

grifos del lavabo accesorios utilizados en los lavabos.

Leadership in Energy and Environmental Design—Liderazgo en Energía y Diseño Ambiental (LEED) sistema para categorizar y certificar el nivel de construcción ambientalmente sostenible en edificios sostenibles.

fuga hacia el exterior fuga de un conducto que no se encuentra ubicada dentro de la envolvente del edificio.

Profesional con acreditación LEED Homes (LEED AP Homes) persona que ha aprobado un examen nacional y exhibido los conocimientos necesarios para participar en el proceso de diseño y certificación del LEED.

evaluación del ciclo de vida (LCA) proceso de evaluación del costo ambiental total de un producto o un edificio, desde la recolección de las materias primas hasta su eliminación final.

diodos emisores de luz (LED) semiconductores que brillan cuando la corriente eléctrica pasa a través de ellos.

linóleo materiales duraderos para pisos hechos de aceite de linaza, corcho y partículas de madera.

revestimiento de bajo E una capa microscópica de metal aplicada a la superficie de vidrio que actúa como barrera radiante reduciendo así la cantidad de energía infrarroja que penetra a través de la superficie metálica.

baja emisividad superficie que irradia, o emite, niveles bajos de energía radiante.

techo de vertiente baja ángulo o inclinación del techo que es de 30° (2:12) o menos.

luminaria artefacto de iluminación; la unidad de iluminación completa, que incluye la lámpara, el reflector, el balasto, el zócalo, el cableado, el difusor y la carcasa.

M

lámpara de inducción magnética variante de la tecnología fluorescente que utiliza un electroimán para hacer brillar el gas de la lámpara.

sistema de múltiple diseño de tuberías que consiste de tubos de ½ o 3/8 de pulgadas y está distribuido directamente desde la fuente de agua hasta los accesorios particulares; también se conoce como *sistema doméstico.*

casas prefabricadas edificios que son totalmente terminados en la fábrica y se entregan sobre un chasis de acero permanente.

Material Safety Data Sheets—Folletos informativos de seguridad del material (MSDS) documentación disponible para la mayoría de los productos que satisfacen la definición de peligroso de la Occupational Safety and Health Administration (OSHA); identifican los contenidos potencialmente peligrosos, los límites de exposición, las instrucciones de manipulación segura, el protocolo de limpieza y otros factores a considerar en la selección y el uso de los productos.

prueba de rendimiento máximo (MaP) determina la forma en que los sanitarios realizan la eliminación de los excrementos mediante el uso de un medio de comprobación realista; cada modelo de sanitario es clasificado de acuerdo con este rendimiento.

filtros de aire mecánicos filtros que utilizan fibras sintéticas, fibra de vidrio o carbón vegetal para filtrar las partículas infinitesimales; el tipo más común de filtros que se utiliza en las construcciones residenciales.

carpintería mecánica se refiere a la madera y los productos compuestos utilizados para crear puertas, molduras, armarios y elementos similares en las construcciones residenciales.

aislamiento de lana mineral material sintético similar a la lana, compuesto de finas fibras inorgánicas hechas de escoria y que se utiliza como relleno suelto o con forma de manta, placa, bloque, tabla o losa para aislamiento térmico y acústico, también conocido como *lana de roca* o *lana de escoria.*

Factor de Energía Modificado (MEF) índice de eficiencia para las lavadoras de ropa que toma en cuenta la energía requerida para operar la lavadora, calentar el agua y secar la ropa sobre la base de la cantidad de agua que se elimina en el ciclo de centrifugado.

construcción modular práctica de emplear la fabricación de secciones completas de una casa como pisos, paredes, cielorrasos, sistemas mecánicos y acabados, que son entregadas por camión en la obra, donde son colocadas sobre una base y terminadas en su lugar.

comprobación de puntos múltiples procedimiento de soplador en puerta que comprende la puesta a prueba de un edificio en una gama de presiones (generalmente entre 60 y 15 pascales) y el análisis de los datos mediante un software especial.

conmutación multivía conjunto interconectado de interruptores que controlan la iluminación desde múltiples ubicaciones.

montantes barras que componen una rejilla y dividen el bastidor en paños de vidrio más pequeños.

N

National Appliance Energy ConservationAct—Ley Nacional para la Conservación de Energía con Electrodomésticos (NAECA) ley federal promulgada en 1992 para establecer en toda la nación niveles mínimos de eficiencia energética en una variedad de artefactos de uso doméstico y comercial que utilizan energía y agua.

National Association of Home Builders—Asociación Nacional de Constructores de Viviendas (NAHB) asociación gremial de ámbito nacional que representa a los constructores de viviendas.

National Fenestration Research Council—Consejo Nacional de Investigación de Ventanaje (NFRC) organización sin fines de lucro que administra un sistema uniforme e independiente de clasificación y etiquetado del rendimiento energético de ventanas, puertas, tragaluces y productos de fijación.

National Oceanic and Atmospheric Administration—Administración Nacional Oceánica y Atmosférica (NOAA) agencia federal de los EE.UU. que realiza investigaciones y recopila datos sobre los océanos, la atmósfera, el espacio y el sol.

madera naturalmente resistente a la putrefacción especies de madera, tales como la secoya, el cedro occidental, el ciprés, el algarrobo negro, el tejo del Pacífico, y el iapacho, que no son proclives al daño ocasionado por la humedad.

sistemas medidos por la red casas que producen parte de su consumo eléctrico in situ y utilizan un medidor único, que funciona ya sea hacia adelante o hacia atrás según sea la dirección de la corriente de energía, para vender o comprar suministro eléctrico a la red pública.

energía neta cero edificio que produce energía en la medida que la necesita, ya sea mensual o anualmente.

nuevo urbanismo estrategia de planificación urbana que promueve barrios peatonales con una amplia gama de tipos de construcciones residenciales y laborales; sumamente influenciado por la planificación de barrios tradicionales y transporte público.

paredes no de depósito módulos de pared que no pueden absorber el agua sin riesgo de sufrir daños estructurales.

North American Board of Certified Energy Practitioners—Consejo de Profesionales de Energía Certificada de los EE.UU. (NABCEP) organización nacional de certificación para instaladores profesionales en el campo de las energías renovables.

O

sensores de ocupación dispositivo que activa y desactiva automáticamente los circuitos cuando detectan que los ocupantes han entrado o salido de una habitación.

liberación de gases el proceso por el cual muchos productos químicos se volatilizan, o dejan escapar hacia el aire moléculas en forma de gas; véase también *compuestos orgánicos volátiles.*

fuera de la red casas que no están conectadas a los servicios públicos de electricidad y gas natural.

de edad madura bosque o zona forestal que cuenta con un ecosistema maduro o demasiado maduro que está más o menos intocado por la actividad humana.

espuma de celdas abiertas tipo de espuma de poliuretano en rocío instalada a una velocidad de aproximadamente 0,5 lb/pie^3 y a veces denominada "espuma de media de libra"; véase también *espuma de poliuretano rociada.*

entramado de piso de red abierta módulo diseñado de madera dimensionada con placas metálicas conectoras que reemplaza un elemento estructural individual mayor con otro que utiliza menos materiales, a la vez que proporciona una resistencia equivalente.

ventana operable ventana con bastidores móviles.

ingeniería de valor óptimo (OVE) metodología de construcción destinada a conservar los materiales de construcción utilizando métodos alternativos de armazón; véase también *armado avanzado.*

placa de hebras orientadas (OSB) diseño de productos de madera que se utiliza a menudo como sustituto del plywood (madera terciada) en la pared exterior y la lámina (base) del techo.

voladizo elemento arquitectónico, tal como un sofito, toldo o porche, que protege las superficies vidriadas del sol y las paredes y las puertas de la lluvia.

P

estructura panelizada consiste de pared, piso, cielorraso y paneles de techo construidos en un ambiente controlado y entregados al pie de la obra listos para ser instalados.

concentración parabólica tipo de colector térmico solar que utiliza bandejas en forma de U para concentrar la luz solar en un tubo que se coloca en la línea focal de la bandeja.

pared parapeto pared baja en la orilla de un techo, terraza, balcón u otra estructura.

Pascal (Pa) unidad de presión que equivale a un newton por metro cuadrado.

diseño solar pasivo práctica de diseñar una casa de modo que utilice la energía solar para la calefacción y el enfriamiento.

patio espacio al aire libre y al nivel del suelo para comer o recrearse, que está junto a una casa y a menudo está embaldosado.

prueba de percolación la velocidad a la que el suelo absorbe las aguas residuales; también conocido como *prueba de perc.*

cimientos de madera permanentes (PWF) sistemas de cimientos compuestos de paredes de madera tratadas con presión.

permeabilidad medida del flujo de aireo humedad a través de un material o módulo.

permeable material o módulo que permite que el aire o la humedad lo traspase.

tasa de permeabilidad tasa del pasaje del vapor de agua a través de un material en condiciones fijas.

asfalto permeable método de pavimentación con asfalto que permite que el agua ingrese al suelo que está debajo.

concreto (hormigón) permeable método de pavimentación con concreto (hormigón) que permite que el agua ingrese al suelo que está debajo.

pavimento permeable materiales de pavimentación que permiten que el agua ingrese al suelo que está debajo.

PEX *véase* polietileno reticulado.

cargas fantasma también conocidas como vampiros, son las pequeñas cantidades de electricidad que utilizan muchos electrodomésticos y artefactos electrónicos, incluso cuando aparentan estar apagados.

fenol-formaldehído adhesivo químico potencialmente dañino utilizado habitualmente en el aislamiento con fibra de vidrio y los productos hechos de madera.

célula fotoeléctrica se utilizan como interruptores en la iluminación exterior para apagarla durante el día cuando no resulta necesaria.

fotovoltaico (PV) dispositivo que convierte la energía de la luz solar directamente en electricidad.

cimentación con pilares sistema reticular compuesto de vigas maestras, pilares y cimientos utilizados en la construcción para elevar la superestructura por encima del terreno plano o en pendiente; los pilares sirven como columnas para la superestructura.

yeso mezcla de cal o yeso con arena y agua que se endurece formando un sólido suave; se utiliza para revestir paredes y cielos.

placa miembro horizontal superior o inferior de la estructura de una pared.

estructura sobre plataforma método de construcción de estructuras de madera en el cual las paredes son levantadas sobre una plataforma previamente construida.

filtros de aire plisados filtros mecánicos de aire de alta-eficiencia hechos de papel, que contienen más fibra por pulgada cuadrada que los filtros de fibra de vidrio desechables.

cámaras de mezcla cajas rectangulares fijadas al horno que reciben aire caliente o frío que luego se distribuye hacia las troncales y los conductos individuales.

fontanería centralizada diseño que coloca de cerca todas las instalaciones sanitarias para reducir la longitud de los tubos y permitir recorridos cortos de los tubos hacia los artefactos.

acoples de tuberías utilizados en sistemas de tuberías e instalaciones sanitarias para conectar los tubos o las secciones de tubos, para adaptarlos a diferentes tamaños o formas y para regular el flujo de los líquidos.

accesorios de instalación sanitaria elemento para la distribución y uso del agua en las construcciones residenciales, que incluye los inodoros (sanitarios), los lavabos y las bañeras.

plywood (madera terciada) trozo de madera hecho de tres o más capas de chapas de madera unidas con goma, y por lo general dispuesto con el grano de las chapas adyacentes en ángulo recto.

polipropileno (PP) material plástico para fontanería hecho de una combinación de moléculas tanto largas como cortas, que proporciona fuerza y flexibilidad.

porche estructura abierta no condicionada con un techo unido al exterior de un edificio que a menudo forma una entrada techada o un espacio habitable al aire libre.

adoquines porosos adoquines con aberturas entre y dentro de ellos que son llenadas de vegetación o grava, lo que permite que el agua fluya a través ellas hacia el suelo.

cemento Portland la forma más común del cemento, consistente de ciertos minerales que forman el aglutinante del hormigón (concreto) y el yeso; véase también *puzolana.*

contenido reciclado postconsumidor la parte de un producto que se recupera luego de ser utilizado por los consumidores.

contenido reciclado postindustrial la parte de un producto que contiene desechos del material de fabricación que ha sido recuperado; también llamado *contenido reciclado preconsumidor.*

limpieza con corriente de agua y bomba eléctrica sanitario (inodoro) de alta eficiencia que utiliza una bomba eléctrica pequeña para que la limpieza con corriente de agua sea eficaz.

puzolana material que, cuando se combina con el hidróxido de calcio, exhibe propiedades similares a las del cemento: algunos ejemplos incluyen el cemento Portland, las cenizas volantes de carbón y la escoria granulada de alto horno.

prefabricados de hormigón técnica de construcción en la que los componentes del hormigón se vacían en fábrica o en el sitio antes de ser izados hasta su posición final en una estructura.

cimientos prefabricados son paredes de cimiento que son fabricadas en una planta y ensambladas a pie de obra.

preinstalación sanitaria el proceso de instalación de sistemas de distribución de tuberías durante la construcción para satisfacer necesidades tecnológicas futuras.

limpieza con corriente de agua y presión sanitario de alta eficiencia que utiliza la presión del aire generada por la presión de la línea de agua almacenada en un tanque pequeño para inducir la limpieza con una corriente de agua más brusca.

válvula de balanceo de presión control que mantiene una temperatura segura de la ducha para evitar quemaduras debidas a los cambios de presión o de temperatura del agua.

válvula reductora de presión dispositivo que mantiene la presión del agua dentro de la casa a un nivel constante.

precableado proceso de instalación del cableado eléctrico durante la construcción para satisfacer demandas tecnológicas futuras.

cuadro psicrométrico cuadro o gráfico que exhibe la relación entre una muestra particular de la temperatura del punto de rocío del aire, la temperatura de bola seca, la temperatura de bola húmeda, el porcentaje de humedad y la humedad relativa.

purificación proceso de eliminación de contaminantes del agua de lluvia recogida ya sea desviando la primera recogida, o utilizando filtros de arena, cloración y esterilización ultravioleta.

R

sistema de conductos radiales sistema de distribución en el que los conductos del ramal que proporcionan aire acondicionado a las salidas de suministro individuales son conectados directamente a una pequeña cámara de distribución.

barrera de radiación térmica material que inhibe la transferencia de calor por radiación térmica; habitualmente encontrado en los áticos.

calefacción radiante sistema de calefacción en el que la fuente de calor (resistencia eléctrica o agua caliente) se instala debajo del piso terminado o en radiadores individuales.

radiación energía térmica que es transferida a través del aire.

radón gas radiactivo producido naturalmente que está presente en el suelo en diferentes concentraciones en todo el país.

ventilación anti-radón sistemas que impiden el ingreso a la casa de radón y otros gases del suelo mediante la ventilación hacia el exterior.

jardín pluvial zona deprimida de vegetación que recoge el escurrimiento superficial de las áreas impermeables y permite la infiltración en el suministro de agua subterránea o el retorno a la atmósfera a través de la evaporación.

barrera pluvial método de construcción de paredes en el cual el revestimiento se separa de la barrera resistente al agua mediante un espacio de aire que permite la igualación de la presión para evitar que la lluvia sea forzada a su interior.

recolección de agua de lluvia la recolección, almacenamiento y uso de la precipitación de agua en los techos y otras superficies.

armazón de talón elevado (raised-heel) armazón de techo diseñado para abarcar cierta área y proporcionar un espacio adecuado para el aislamiento del ático en toda su profundidad y toda su área; véase también *armazón de eficiencia energética.*

tierra apisonada antigua forma de construcción en la que la tierra y los aditivos, tales como paja, cal o cemento, se colocan dentro de moldes en múltiples capas de seis a ocho pulgadas de espesor y se compactan para crear un muro de masa estructural.

contenido reciclado la cantidad de material de pre y postconsumo recuperado e introducido como materia prima en un proceso de producción de materiales, expresada generalmente en porcentaje.

refrigerante producto químico que transfiere calor a medida que cambia de líquido a gas y nuevamente a líquido.

humedad relativa (HR) proporción entre la cantidad de agua presente en el aire a una temperatura determinada y la cantidad máxima que podría sostener a esa temperatura, expresada en porcentaje.

energía renovable electricidad generada a partir de recursos que son ilimitados, rápidamente reabastecidos, o renovables de forma natural (por ejemplo, el viento, el agua, el sol, los elementos geotérmicos [calor de la tierra], las olas y la basura) y no de la combustión de combustibles fósiles.

certificados de energía renovables (REC) productos comercializables de energía que identifican que un megavatio-hora de energía eléctrica fue generado con recursos renovables.

Residential Energy Services Network—Red de Servicios Energéticos a Residencias (RESNET) organización sin fines de lucro conformada por afiliados que se esfuerza por asegurar el éxito de la industria de la certificación del rendimiento energético en la construcción, establecer las normas de calidad pertinentes e incrementar las oportunidades de propiedad de edificios de alto rendimiento.

pared de retención estructura que mantiene la tierra o la roca separada de un edificio, estructura o área.

ósmosis inversa (OI) los filtros fuerzan el agua a través de membranas semipermeables bajo presión, dejando contaminantes durante el proceso.

lana de roca otro nombre para la *lana mineral.*

substrato del techo la superficie de madera o metal a la cual se aplican los materiales para el techado.

vertiente del techo ángulo de un techo, descrito por la elevación (altura vertical) respecto a la luz (longitud horizontal).

valor R medida cuantitativa de la resistencia al flujo de calor o la conductividad, la recíproca del valor U.

S

bastidor estructura que sostiene los paños de una ventana en el marco.

imbornal abertura en el costado de un edificio, tal como un pretil, que permite que el agua fluya hacia afuera.

índice de eficiencia energética estacional (SEER) describe la eficiencia operativa de los equipos de aire acondicionado.

segundo crecimiento un bosque o zona forestal que ha vuelto a crecer después de la extracción de toda o gran parte de su composición anterior debido a la tala, el fuego, el viento, o cualquier otra fuerza; por lo general, significa que ha transcurrido un tiempo prolongado de manera que los efectos de la perturbación ya no resultan evidentes.

certificación gremial cuando una asociación gremial o industrial establece su propio código de conducta e implementa mecanismos de rendición de cuentas.

semi permeable material o módulo que permite que cierta cantidad de aire o humedad pase a través de él.

calor sensible energía asociada con los cambios de temperatura.

fracción de calor sensible (SHF) relación en un aparato de aire acondicionado entre la capacidad sensible y la capacidad latente, también conocida como *relación de calor sensible (SHR).*

relación de calor sensible (SHR) relación en un aparato de aire acondicionado entre la capacidad sensible y la capacidad latente, también conocida como *fracción de calor sensible (SHF).*

campo de drenaje séptico terreno de drenaje utilizado para eliminar los contaminantes y las impurezas del líquido que emerge del tanque séptico.

sistema séptico un sistema de tratamiento subterráneo de las aguas cloacales originadas por el hombre.

revestimiento tablas o material laminado que se fija a las viguetas, maderados y salientes y a los cuales se aplica el material de acabado.

techo de un agua tipo de techo que se inclina en una sola dirección.

durmiente primer miembro de madera horizontal que descansa sobre el cimiento y soporta la estructura de un edificio; también, el elemento horizontal más bajo en el marco de una ventana o puerta (alféizar).

recuperación de la inversión sencilla la cantidad de tiempo que se tarda en recuperar la inversión inicial para mejorar la eficiencia energética a través de ahorros de energía, dividiendo el costo inicial de instalación entre los ahorros anuales de costos de energía.

membrana de una sola hoja material para techos que viene en láminas, que son colocadas sobre el substrato del techo y unidas entre sí mediante elementos de fijación mecánica, calor o solventes químicos.

prueba de soplador en puerta de punto único prueba de soplador en puerta que utiliza una sola medición del flujo de ventilación necesaria para crear un cambio de 50 Pa en la presión del edificio.

plancha panel de puerta individual, con exclusión del marco, las bisagras, el umbral y la tornillería de la misma.

lana de escoria (lana mineral) otro nombre para la *lana mineral.*

perturbación de una pendiente proceso de desestabilización de una pendiente a través de nivelación y obras.

red inteligente una forma de designación para un número cada vez mayor de aplicaciones de utilidad para mejorar y automatizar la supervisión y el control de la distribución eléctrica.

medidor inteligente un término amplio que define los medidores eléctricos que incluyen comunicación bidireccional y otras capacidades avanzadas.

sofito (plafón) superficie horizontal que se proyecta hacia fuera de una pared exterior.

paisajes suaves elementos de un paisaje de vegetación, entre ellos las plantas y el suelo.

gas del suelo incluye el aire, el vapor de agua, el radón, el metano y otros contaminantes del suelo que pueden ingresar a un edificio a través de los cimientos o el espacio subyacente al piso.

ganancia solar calor generado por la radiación solar.

coeficiente de ganancia de calor solar (SHGC) fracción de la radiación solar admitida a través de las superficies vidriadas.

orientación solar dirección cardinal hacia la que apunta la casa y sus superficies vidriadas.

potencial solar cantidad de luz solar que puede capturar un sitio en particular, basada en la latitud, las condiciones climáticas locales y la cantidad de cielo del sur no obstruido.

Solar Rating and Certification Corporation—Corporación para la Calificación y Certificación de la Energía Solar (SRCC) organizaciónsin fines de lucro cuyo objetivo principal es el desarrollo e implementación de programas de certificación y normas nacionales de clasificación para los equipos de energía solar.

reflectancia solar es un número decimal menor que uno, que representa la fracción de luz reflejada por el techo; véase también *albedo.*

Índice de Reflectancia Solar (SRI) medida de la capacidad de un material para rechazar el calor solar, y por lo tanto mantenerse frío; los valores típicos están en una escala de 0 a 100, desde baja a alta capacidad de rechazo del calor.

solar térmico sistema que convierte la luz solar en aire o agua calefaccionados.

superficie sólida producto manufacturado, normalmente utilizado para las superficies que imitan la piedra, creado mediante la combinación de minerales naturales con resinas y aditivos.

sone unidad de intensidad de ruido.

superficie vidriada selectiva espectralmente vidrio revestido o tonalizado con propiedades ópticas que son transparentes a ciertas longitudes de onda y reflectantes a las demás.

sistemas de conductos en araña tramos individuales de conductos que van directamente desde la caldera o el controlador del aire hacia las habitaciones individuales.

ventilación puntual proceso mecánico de eliminación de la humedad, los olores y los contaminantes directamente en la fuente.

aislamiento por espuma de poliuretano rociada (SPF) plástico aislador en forma de espuma aplicado por rocío que se instala como un líquido y luego se expande múltiples veces su tamaño original; véase también *espuma de celdas abiertas* y *espuma de celdas cerradas.*

efecto de chimenea corriente de aire establecida en un edificio debido al ingreso del aire a bajo nivel y su salida a alto nivel.

termómetro de chimenea instrumento para medir la temperatura del tubo de chimenea o el tubo de escape de un artefacto de combustión.

junta alzada techo armado a partir de paneles metálicos con costuras verticales que son calzadas a presión o dobladas hacia adentro para formar un sello.

presión estática presión dentro del sistema de conductos y un indicador de la cantidad de resistencia al flujo de aire dentro del sistema; por lo general se mide en pulgadas de columna de agua o pascales (Pa).

techo de vertiente pronunciada techo cuyo ángulo es de más de 30° (2:12).

botaguas escalonado piezas individuales de metal instaladas detrás de la barrera resistente al agua (WRB) y entrelazadas con las tejas del techo.

paredes del depósito de almacenamiento de pared ensamblados que pueden absorber el agua sin riesgo de daños estructurales.

aguas pluviales el flujo de agua que resulta de la precipitación que sigue a una lluvia o a consecuencia del derretimiento de la nieve.

fardo de paja construcción realizada utilizando fardos de paja de trigo, avena, cebada, centeno, arroz y otros productos agrícolas de desecho en las paredes, y cubiertos de estuco o yeso de barro.

paneles estructurales aislados (SIP) materiales compuestos de construcción que consisten de un aislador de espuma sólida colocado entre dos láminas de placa de hebras orientadas para crear paneles de construcción para pisos, paredes y techos.

aislamiento integrado estructuralmente sistema de aislamiento que forma parte de la estructura de un edificio, en lugar de ser aplicado a un componente estructural.

espacio solar adición solar pasiva a una casa nueva o ya existente para proporcionar calor adicional en el invierno a través de la energía solar.

diseño solar atemperado estrategia de diseño solar pasivo, donde la mayoría de las superficies vidriadas se orientan en el eje norte-sur.

abastecimiento estratégico método de dimensionar cisternas en el que se instala un tanque lo suficientemente grande como para abastecer la totalidad de las necesidades mensuales de una casa.

ventilación solo de suministro proporciona aire exterior a la casa sin escape.

área superficial de la envolvente del edificio (SFBE) la superficie total (en pies^2) de la envolvente o caparazón de un edificio.

sostenible patrón de uso de los recursos que tiene como objetivo satisfacer las necesidades humanas a la vez que se preserva el medio ambiente de modo que estas necesidades puedan ser satisfechas no solo en el presente sino también para las generaciones futuras.

gestión forestal sostenible prácticas de gestión forestal que mantienen y mejoran la salud a largo plazo de los ecosistemas forestales a la vez que proporcionan oportunidades ecológicas, económicas, sociales y culturales para beneficio de las generaciones presentes y futuras.

Sustainable Forestry Initiative—Iniciativa Forestal Sostenible (SFI) organización sin fines de lucro responsable del mantenimiento, la supervisión y la mejora de un programa de certificación forestal sostenible.

T

calefactores de tanque calefactores que almacenan una gran cantidad de agua caliente en un tanque de almacenamiento aislado.

calefactores sin tanque calefactores que calientan el agua solo cuando se necesita, requieren un caudal mínimo para operar, y no almacenan agua caliente.

iluminación focalizada iluminación dirigida a un área de trabajo específica, tal como las superficies de las cocinas y los escritorios, y no a toda una habitación.

bomba de temperatura bomba de recirculación que tiene sensores cerca de los artefactos, que le indican que debe hacer circular agua caliente cuando detectan que la temperatura cae por debajo de un nivel establecido.

aumento de temperatura la diferencia entre la temperatura del agua que ingresa a la casa y la del agua caliente que sale del calentador de agua.

barrera térmica límite al flujo de calor (p. ej., aislamiento).

interrupción térmica material de baja conductividad térmica colocado en un módulo, para reducir o prevenir el flujo de calor entre los materiales conductores.

puente térmico material térmicamente conductor que penetra o elude un sistema de aislamiento, tal como un sujetador o un montante metálico.

derivación térmica desplazamiento del calor en torno o a través del aislamiento, frecuentemente debido a barreras de aire faltantes o brechas entre las barreras de aire y el aislamiento.

emisión térmica número decimal menor que uno, que representa la fracción de calor que es vuelta a irradiar desde un material hacia su entorno.

lista de verificación del calificador del compartimiento térmico inspección de los detalles de la construcción respecto de la derivación térmica: para que una casa califique para la etiqueta ENERGY STAR, esta lista de verificación debe ser completada por un calificador HERS.

madera tratada térmicamente producida exponiendo la madera a temperaturas muy altas y vapor, transformándola en un producto que no es afectado por los insectos y no se pudre.

bucle de termosifón sistema de intercambio pasivo de calor que opera sin una bomba mecánica, utilizando la convección natural de la elevación del agua caliente para circular hacia la parte superior del bucle, lo que permite que el agua más fría fluya hacia abajo y de regreso al calentador de agua.

válvula termostática válvula que mantiene una temperatura segura de la ducha para evitar quemaduras, mediante el ajuste de la mezcla de agua caliente y fría; también se conoce como *válvula de compensación termostática.*

certificación externa revisión y confirmación por parte de una organización externa no afiliada, de que un producto satisface ciertas normas.

bombas temporizadas bombas que se pueden configurar para funcionar durante ciertos períodos de tiempo en que los ocupantes requerirán agua caliente.

alquitrán modificado con soplete material para entechar con alquitrán modificado que se aplica con calor.

total de fugas del conducto cantidad de fugas en los conductos dentro y fuera de la envolvente del edificio.

gas trazador gas no tóxico utilizado para medir las fugas de aire en la envolvente.

urbanización tradicional de vecindarios (TND) planificación de un barrio o una ciudad completa utilizando los principios tradicionales de planificación urbana que hacen hincapié en la transitabilidad, el espacio público y el desarrollo de uso mixto; véase también *nuevo urbanismo.*

rejillas de transferencia aberturas colocadas en las paredes para permitir que el aire fluya entre las habitaciones.

urbanización orientada al transporte público (TOD) planificación de comunidades que estén a poca distancia del transporte público, combinando la utilización de viviendas, comercios, oficinas, espacios abiertos y espacios públicos de una manera tal que haga cómodo transitar a pie o usando transporte público en lugar de recurrir al automóvil.

plantación de árboles una plantación de árboles gestionada activamente que, a diferencia de los de un bosque, contienen solo una o dos especies y proporcionan pocas condiciones para la vida silvestre.

espaldera estructura que se utiliza para proporcionar sombra o apoyo a las plantas trepadoras; generalmente construida de piezas entrelazadas de madera, bambú o metal.

luz dividida genuina (TDL) ventana o puerta en la que varias secciones o paños individuales de vidrio son armados en el bastidor utilizando parteluces.

troncal y ramales diseño de fontanería que consiste de tuberías o troncales de agua caliente y fría de ¾ o 1 pulgada que van hacia cada lugar de instalación sanitaria y desde donde tuberías más pequeñas, de ½ pulgada, o ramales, se dirigen hacia los artefactos individuales.

sistemas de conductos de troncal y ramales conductos de gran tamaño (troncales) instalados a través del centro de la casa con conductos más pequeños (ramales) que parten de las troncales para abastecer a las habitaciones individuales.

armazón diseño de productos de madera que consiste de elementos de madera o de madera y metal, utilizado para soportar los techos o pisos y que reduce la cantidad de madera requerida para soportar una carga específica.

dispositivos tubulares de iluminación con luz natural (TDD) tragaluz cilíndrico con un tubo reflectante para proporcionar luz natural a las habitaciones interiores.

U

placas de aislamiento sin retardador de vapor aislamiento en placas (bloques) que no contiene un revestimiento retardador de vapor de algodón o fibra de vidrio.

techo no ventilado ático que no tiene ventilación.

urea-formaldehído producto químico potencialmente tóxico utilizado como aglutinante o adhesivo en los productos diseñados para la construcción.

U.S. Green Building Council—Consejo para las construcciones ecológicas de los EE.UU. (USGBC) organización sin fines de lucro que promueve la construcción ecológica y creó los sistemas de clasificación LEED; véase también *LEED.*

Valor U transmitancia térmica o conductividad térmica de un material; la recíproca del valor R.

V

sensores de vacancia controles de iluminación que requieren la operación de un interruptor para activarlos, y que se apagan automáticamente después que una habitación queda vacía.

barrera de vapor retardador de vapor Clase I (0,1 perm o menos).

retardador de la difusión del vapor (VDR) material que reduce la velocidad a la cual el vapor de agua puede desplazarse a través de un material.

tubería de arcilla vitrificada tubería para instalaciones sanitarias hecha de arcilla que ha sido sometida a vitrificación, un proceso que funde las partículas de arcilla hasta un estado muy duro, inerte, similar al del vidrio.

compuestos orgánicos volátiles (VOC) compuestos químicos que tienen una alta presión de vapor y baja solubilidad en agua; muchos COV son sustancias químicas sintéticas que son utilizadas y producidas en la fabricación de pinturas, productos farmacéuticos, refrigerantes y materiales de construcción; los COV son contaminantes comunes del aire en interiores y del agua subterránea.

válvulas de control de volumen válvulas que controlan el flujo de agua a los artefactos.

W

felpudo de umbral felpudo abrasivo instalado sobre un receptáculo poco profundo que atrapa los desechos a medida que son raspados de los zapatos.

empapelado revestimiento de pared, tal como vinilo o papel de empapelar.

impermeabilización tratamiento utilizado en las superficies de hormigón, mampostería o piedra, que impide el paso del agua bajo presión hidrostática.

barrera resistente al agua (WRB) material ubicado detrás del revestimiento y que forma un plano secundario de drenaje del agua en estado líquido, comúnmente conocido como barrera resistente al clima, barrera resistente a la intemperie y barrera resistente al agua.

WaterSense programa administrado por la Agencia de Protección Ambiental de los EE.UU. que proporciona pautas para la instalación de sistemas de riego eficientes y certificación para los profesionales del riego.

ablandadores de agua aparatos que eliminan los minerales del agua para reducir la dureza; generalmente instalados para tratar todo el suministro de agua de una casa.

bomba de calor con fuente de agua (WSHP) unidad de calefacción y enfriamiento que intercambia calor entre el suelo o el agua y el interior de la casa; también conocida como bomba de calor geotérmica.

lámina de drenaje lámina de escurrimiento metálica o plástica en la base de las paredes que permite que la humedad se escurra de detrás de la mampostería y el estuco.

temperatura de bola húmeda temperatura registrada por un termómetro cuyo bola ha sido cubierta con una mecha humedecida y retorcida sobre un psicrómetro de eslinga; se utiliza para determinar la humedad relativa, el punto de rocío y la entalpía.

X

paisaje seco técnica de paisajismo que utiliza plantas tolerantes a la sequía para reducir al mínimo la necesidad de agua, fertilizantes y mantenimiento.

Z

control zonal dispositivo que controla la cantidad de flujo de aire a diferentes áreas de la casa.

Index

D

E

J

K

L